z	0.00	0.01	0.02	0.03	0.04	0.05	0.06	0.07	0.08	0.09
0.0	0.5000	0.5040	0.5080	0.5120	0.5160	0.5199	0.5239	0.5279	0.5319	0.5359
0.1	0.5398	0.5438	0.5478	0.5517	0.5557	0.5596	0.5636	0.5675	0.5714	0.5753
0.2	0.5793	0.5832	0.5871	0.5910	0.5948	0.5987	0.6026	0.6064	0.6103	0.6141
0.3	0.6179	0.6217	0.6255	0.6293	0.6331	0.6368	0.6406	0.6443	0.6480	0.6517
0.4	0.6554	0.6591	0.6628	0.6664	0.6700	0.6736	0.6772	0.6808	0.6844	0.6879
0.5	0.6915	0.6950	0.6985	0.7019	0.7054	0.7088	0.7123	0.7157	0.7190	0.7224
0.6	0.7257	0.7291	0.7324	0.7357	0.7389	0.7422	0.7454	0.7486	0.7517	0.7549
0.7	0.7580	0.7611	0.7642	0.7673	0.7704	0.7734	0.7764	0.7794	0.7823	0.7852
0.8	0.7881	0.7910	0.7939	0.7967	0.7995	0.8023	0.8051	0.8078	0.8106	0.8133
0.9	0.8159	0.8186	0.8212	0.8238	0.8264	0.8289	0.8315	0.8340	0.8365	0.8389
1.0	0.8413	0.8438	0.8461	0.8485	0.8508	0.8531	0.8554	0.8577	0.8599	0.8621
1.1	0.8643	0.8665	0.8686	0.8708	0.8729	0.8749	0.8770	0.8790	0.8810	0.8830
1.2	0.8849	0.8869	0.8888	0.8907	0.8925	0.8944	0.8962	0.8980	0.8997	0.9015
1.3	0.9032	0.9049	0.9066	0.9082	0.9099	0.9115	0.9131	0.9147	0.9162	0.9177
1.4	0.9192	0.9207	0.9222	0.9236	0.9251	0.9265	0.9278	0.9292	0.9306	0.9319
1.5	0.9332	0.9345	0.9357	0.9370	0.9382	0.9394	0.9406	0.9418	0.9429	0.9441
1.6	0.9452	0.9463	0.9474	0.9484	0.9495	0.9505	0.9515	0.9525	0.9535	0.9545
1.7	0.9554	0.9564	0.9573	0.9582	0.9591	0.9599	0.9608	0.9616	0.9625	0.9633
1.8	0.9641	0.9649	0.9656	0.9664	0.9671	0.9678	0.9686	0.9693	0.9699	0.9706
1.9	0.9713	0.9719	0.9726	0.9732	0.9738	0.9744	0.9750	0.9756	0.9761	0.9767
2.0	0.9772	0.9778	0.9783	0.9788	0.9793	0.9798	0.9803	0.9808	0.9812	0.9817
2.1	0.9821	0.9826	0.9830	0.9834	0.9838	0.9842	0.9846	0.9850	0.9854	0.9857
2.2	0.9861	0.9864	0.9868	0.9871	0.9875	0.9878	0.9881	0.9884	0.9887	0.9890
2.3	0.9893	0.9896	0.9898	0.9901	0.9904	0.9906	0.9909	0.9911	0.9913	0.9916
2.4	0.9918	0.9920	0.9922	0.9925	0.9927	0.9929	0.9931	0.9932	0.9934	0.9936
2.5	0.9938	0.9940	0.9941	0.9943	0.9945	0.9946	0.9948	0.9949	0.9951	0.9952
2.6	0.9953	0.9955	0.9956	0.9957	0.9959	0.9960	0.9961	0.9962	0.9963	0.9964
2.7	0.9965	0.9966	0.9967	0.9968	0.9969	0.9970	0.9971	0.9972	0.9973	0.9974
2.8	0.9974	0.9975	0.9976	0.9977	0.9977	0.9978	0.9979	0.9979	0.9980	0.9981
2.9	0.9981	0.9982	0.9982	0.9983	0.9984	0.9984	0.9985	0.9985	0.9986	0.9986
3.0	0.9987	0.9987	0.9987	0.9988	0.9988	0.9989	0.9989	0.9989	0.9990	0.9990
3.1	0.9990	0.9991	0.9991	0.9991	0.9992	0.9992	0.9992	0.9992	0.9993	0.9993
3.2	0.9993	0.9993	0.9994	0.9994	0.9994	0.9994	0.9994	0.9995	0.9995	0.9995
3.3	0.9995	0.9995	0.9995	0.9996	0.9996	0.9996	0.9996	0.9996	0.9996	0.9997
3.4	0.9997	0.9997	0.9997	0.9997	0.9997	0.9997	0.9997	0.9997	0.9997	0.9998

Probability and Statistics for Engineering and the Sciences

SECOND EDITION

Jay L. Devore
California Polytechnic State University
San Luis Obispo

Brooks/Cole Publishing Company
Monterey, California

To Carol, Allie, and Teri

Brooks/Cole Publishing Company
A Division of Wadsworth, Inc.

Printed in the United States of America

10 9 8 7 6 5 4 3 2 1

Library of Congress Cataloging-in-Publication Data

Devore, Jay L.
  Probability and statistics for engineering and the sciences.

  Includes bibliographies and index.
    1. Probabilities. 2. Mathematical statistics.
I. Title.
QA273.D46 1987   519.5   86-20731
ISBN 0-534-06828-6

Sponsoring Editor: John Kimmel
Editorial Assistant: Maria Alsadi
Production Editor: Michael G. Oates
Manuscript Editor: Sandy Spiker
Permissions Editor: Carline Haga
Interior and Cover Design: Victoria A. Vandeventer
Art Coordinator: Lisa Torri
Interior Illustration: Lori Heckelman
Typesetting: Allservice Phototypesetting Company
Printing and Binding: R.R. Donnelley & Sons Company

Preface

The use of probability models and statistical methods for analyzing data has become common practice in virtually all scientific disciplines. This book attempts to provide a comprehensive introduction to those models and methods most likely to be encountered and used by students in their careers in engineering and the natural sciences. Although the examples and exercises have been designed with scientists and engineers in mind, most of the methods covered are basic to statistical analyses in many other disciplines, so that students of business and the social sciences will also profit from reading the book.

Students in a statistics course designed to serve other majors may be initially skeptical of the value and relevance of the subject matter, but my experience is that students *can* be turned on to statistics by the use of good examples and exercises which blend their everyday experiences with their scientific interests. Consequently, I have worked hard to find examples of real, rather than artificial, data—data that someone thought was worth collecting and analyzing. Many of the methods presented, especially in the later chapters on statistical inference, are illustrated by analyzing data taken from a published source, and many of the exercises also involve working with such data. Sometimes the reader may be unfamiliar with the context of a particular problem (as indeed I often was), but I have found that students are more attracted by real problems with a somewhat strange context than by patently artificial problems in a familiar setting.

The exposition is relatively modest in terms of mathematical development. Substantial use of the calculus is made only in Chapter 4 and parts of Chapters 5 and 6. In particular, with the exception of an occasional remark or aside, calculus appears in the inference part of the book only in the second section of Chapter 6. Matrix algebra is not used at all. Thus almost all the exposition should be accessible to those whose mathematical background includes one semester or two quarters of differential and integral calculus.

v

Although the book's mathematical level should give most science and engineering students little difficulty, working toward an understanding of the concepts and gaining an appreciation for the logical development of the methodology may sometimes require substantial effort. To help students gain such an understanding and appreciation, I have provided numerous exercises ranging in difficulty from many that involve routine application of text material to some that ask the reader to extend concepts discussed in the text to somewhat new situations. There are many more exercises than most instructors would want to assign during any particular course, but I recommend that students be required to work a substantial number of them; in a problem-solving discipline, active involvement of this sort is the surest way to identify and close the gaps in understanding that inevitably arise.

The second edition incorporates the following major changes, many of which were suggested by reviewers and users of the first edition.

- Chapters 7–9 have been reorganized so that one-sample confidence intervals now appear in Chapter 7 (prior to hypothesis testing), one-sample test procedures are presented in Chapter 8, and all parametric two-sample methods are gathered in Chapter 9.
- A new section on probability plots has been added at the end of Chapter 4, there is now a separate section on P-values in Chapter 8, and material on box plots has been included in Chapter 1.
- The section on hypothesis testing using the binomial distribution has been eliminated from Chapter 3, where many had found its placement somewhat awkward. The introduction to hypothesis testing at the outset of Chapter 8 is somewhat less terse than what appeared in the first edition.
- Supplementary exercises now appear at the end of all 15 chapters. The total exercise count is about 15% higher than in the first edition.
- To keep the length under control, some of the more advanced methodology that was included in the first edition, including such specialized topics as ridge regression and loglinear models for contingency table data, has been deleted. Matrix algebra has been completely removed from the sections on multiple regression.

There is enough material in the book for a full-year (30-week) course, so in courses of shorter duration a selection of topics will be necessary. Because goals, backgrounds, and abilities of students—and instructors' tastes—vary widely, I hesitate to make specific recommendations regarding coverage of topics. Experience in teaching a two-quarter sequence at Cal Poly (three lectures per week and no quiz or discussion sections) may provide helpful guidelines.

First-quarter coverage includes Chapter 1, most of Chapters 2 and 3, the first three sections of Chapter 4 (with little time spent on continuous families of distributions other than the normal), the last two sections of Chapter 5 (joint distributions and expected values, presented in the first two sections, are de-emphasized), Section 1 of Chapter 6, and the first one or two sections of Chapters 7 and 8. In the second quarter we cover the remainder of Chapters 7 and 8, Chapter 9, selected material from Chapters 10 and 11 (typically Section 1 and parts of Sections 2 and 3 of Chapter 10, and brief mention of multifactor analyses), Chapter 12, selected portions of Chapter 13, and material from the first three sections of Chapter 14. There always seems to be too little time in lectures to discuss all the topics that we statisticians think ought to be discussed. I hope that my presentation of material is readable enough so that students can be asked to read on their own selected portions of the text which have not or will not be covered in lecture.

Acknowledgments

I gratefully acknowledge the numerous suggestions and constructive criticisms provided by the following reviewers: C. E. Antle, Pennsylvania State University; William Astle, Colorado School of Mines; Thomas J. Boardman, Colorado State University; Lyle Broemeling, Oklahoma State University; Maurice C. Bryson, Los Alamos National Laboratory; Kenneth Constantine, University of New Hampshire; Louis J. Cote, Purdue University; Jonathan Cryer, University of Iowa; Arthur Dayton, Kansas State University; Angela Dean, Ohio State University; Janice Dubien, Western Michigan University; Bennett Eisenberg, Lehigh Community College; William Griffith, University of Kentucky; Fred Hoppe, University of Michigan; Thomas R. Jefferson, University of Pittsburgh; Bruce Johnson, University of Victoria; Bryan Johnson, University of Victoria; Bryan Karney, University of Calgary; F. Klebauer, University of Michigan; Melvin Lang, Walla Walla College; William Lesso, University of Texas, Austin; Frederick Morgan, Clemson University; Dennis Pearl, Ohio State University; Wolfgang Pelz, University of Akron; John Philpot, University of Tennessee; Thomas Reiland, North Carolina State University; Larry Ringer, Texas A&M University; Terry Sandifer, Western New England College; Paul Shaman, University of Pennsylvania; Martyn Smith, Michigan Technical University; Richard Van Nostrand, Eastman Kodak Company; and Donald Weber, Miami University of Ohio.

Professor S. C. Wirasinghe of the University of Calgary provided some new problem settings for the probability chapters, and Professor Roxy Peck did likewise for later chapters. Professor John Groves gave invaluable help by checking much of the new material.

Typing chores were admirably handled by Lynda Alamo, Carol Devore, Fran Fairbrother, Pat Fleischauer, and Meri Kay Gurnee. I very much appreciate the editorial and production services rendered by John Kimmel, Michael Oates, and the staff at Brooks/Cole. Finally, words cannot adequately express my gratitude toward my family—Carol, Allie, and Teri—for their ongoing support of my writing efforts.

Jay L. Devore

Contents

ix

Contents

Probability and Statistics
for Engineering and the Sciences

SECOND EDITION

CHAPTER 1

Introduction and Descriptive Statistics

1.1　An Overview of Probability and Statistics

In our own work, through conversations with others, and through contact with media of various sorts (books, television, newspapers, and the like), we are continually being confronted with collections of facts or **data.** Statistics is the branch of scientific inquiry that provides methods for organizing and summarizing data, and for using information in the data to draw various conclusions.

Frequently we wish to acquire information or draw some conclusion about an entire **population** consisting of all individuals or objects of a particular type. The population of interest might consist of all radial tires manufactured by a particular company during the previous calendar year, or it might be the collection of all individuals who had been inoculated with a particular flu vaccine, or it might consist of all U.S. colleges and universities. In this last case, we might be interested specifically in the number of students enrolled at each school. If so, rather than think of the schools as the population members, we may speak of the **numerical population** in which each population member is an enrollment figure such as 1536 or 21,311. In the inoculation example, we might wish to focus on whether or not individuals had subsequently exhibited a certain condition (a rash, dizziness, and so on). We might then visualize the population as consisting of Y's (for yes, the condition was present) and N's (no, the condition was not present). This is an example of a **dichotomous** (two-valued) population. In general, we will define the population to reflect our particular interests at the time of the investigation.

The data at our disposal frequently consists of a portion or subset of the population; any such subset is called a **sample.** If the population is all U. S. colleges and universities, one sample would be {Stanford University, University of Washington, Oberlin College, California Institute of Technology, Iowa State

University}. If the population comprises all college and university enrollment figures, a sample might consist of {13,043, 35,234, 2756, 21,831}.

The objectives of organization and summarization of data have been pursued for hundreds of years. The part of statistics that deals with methods for performing these operations is called **descriptive statistics.** Descriptive methods can be used either when we have a list of all population members (a **census**), or when the data consists of a sample.

When the data is a sample and the objective is to go beyond the sample to draw conclusions about the population based on sample information, methods from **inferential statistics** are used. With a few isolated exceptions, the development of inferential statistics has occurred only since the early 1900s, making it of much more recent vintage than descriptive statistics. Yet much of the interest and activity in statistics today, particularly as it relates to scientific activity and experimentation, concerns inferences rather than just description. The psychologist who tries her behavior modification technique on a sample of obese individuals would like to infer something about what the technique would do if applied to all such individuals. The engineer who accumulates data on a sample of computer systems will ultimately wish to draw conclusions about all such systems. The medical team that develops a new vaccine for a disease threatening a particular population is interested in what would happen if the vaccine were administered to all people in the population. The marketing expert may test a product in a few "representative" areas; from the resulting information he will draw conclusions about what would happen if the product were made available to all potential purchasers.

The main focus of this book is on presenting and illustrating methods of inferential statistics that are useful in scientific work. The three important types of inferential procedures—point estimation, hypothesis testing, and estimation by confidence intervals—are introduced in Chapters 6–8, and then used in more complicated settings in Chapters 9–15. The remainder of this chapter presents methods from descriptive statistics that are most used in the development of inference.

Chapters 2–5 present material from the discipline of probability. This material ultimately forms a bridge between the descriptive and inferential techniques and leads to a better understanding of how inferential procedures are developed and used, how statistical conclusions can be translated into everyday language and interpreted, and when and where pitfalls can occur in applying the methods. Probability and statistics both deal with questions involving populations and samples, but do so in an "inverse manner" to one another.

In a probability problem, properties of the population under study are assumed known (in a numerical population, for example, some specified distribution of the population values may be assumed), and questions regarding a sample taken from the population are posed and answered. In a statistics problem, characteristics of a sample are available to the experimenter, and this information enables the experimenter to draw conclusions about the population. The relationship between the two disciplines can be summarized by saying that probability reasons from the population to the sample (deductive reasoning), whereas statistics reasons from the sample to the population (inductive reasoning). This is illustrated in Figure 1.1.

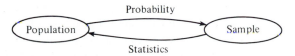

Figure 1.1 The relationship between probability and inferential statistics

Before we can understand what a particular sample can tell us about the population, we should first understand the uncertainty associated with taking a sample from a given population. This is why we study probability before statistics. The following two examples pursue the distinction between the two disciplines and the types of questions asked within each.

Example 1.1 A new type of gasoline pump nozzle has been developed in order to minimize the emission of pollutants into the atmosphere during pumping. One potential design defect is that "topping off" the tanks may cause gasoline to be sucked back into the underground storage tank. (This phenomenon was described in an article that appeared in a May 1981 issue of the *Los Angeles Times*.) Suppose that 10,000 nozzles manufactured by a particular company are currently in use at Southern California gas stations. In probability, we might assume that 10% of them have the defect described above (an assumption about the population of nozzles) and ask, "How likely is it that a sample of 25 nozzles will include at least 5 that are defective?" or, "How many defective nozzles can we expect in a sample of 25?" On the other hand, in statistics we might find that there were 5 defectives in a sample of 25 nozzles (sample information) and then ask, "Does this strongly indicate that at least 10% of all nozzles currently in use are defective?" The last question concerns the population as a whole. ∎

Example 1.2 The following measurements (reported in *Science 167*, pp. 277–279) of the ratio of the mass of the earth to that of the moon were obtained during several different spacecraft flights: 81.3001 (*Mariner 2*), 81.3015 (*Mariner 4*), 81.3006 (*Mariner 5*), 81.3011 (*Mariner 6*), 81.2997 (*Mariner 7*), 81.3005 (*Pioneer 6*), and 81.3021 (*Pioneer 7*). These numbers differ from one another (and presumably from the true ratio) because of measurement error. In probability, we might assume that the distribution of all possible measurements is bell shaped and centered at the true value of 81.3035, and then ask, "How likely is it that all seven measurements actually made result in observed ratios that are less than the true value?" In statistics, having been given the above seven measurements, we might ask, "Does this sample information conclusively demonstrate that the ratio is something other than 81.3035?" or, "How confident can we be that the true ratio is in the interval (81.2998, 81.3018)?" ∎

In Example 1.1, the population is a well defined concrete or existing one—all nozzles presently in use. In Example 1.2, however, while the sample consists of seven observations, the population does not actually exist; instead, the population here is one consisting of all possible measurements that might be made under similar experimental conditions. Such a population is referred to as a **conceptual** or **hypothetical population.** There are a number of problems in which we fit questions into the framework of inferential statistics by conceptualizing a population. As another example, imagine that a sample of five catalytic converters with a new design is experimentally manufactured and tested. Then the inferences made will refer to the conceptual population consisting of all converters of this type that could be manufactured.

Many newcomers to the study of statistics are unaware of the broad potential application of probability and statistical methods. To remedy this, a number of statisticians contributed short, nontechnical essays to a book of readings edited by Judith Tanur, entitled *Statistics: A Guide to the Unknown* (Wadsworth & Brooks/Cole Advanced Books and Software, 1978). These essays are about areas of application, rather than particular methods of analysis, and serve as an excellent supplement to the problem-oriented textbook.

Exercises / Section 1.1 (1–4)

1. List a sample of size four from each of the following populations:
   a. All daily newspapers published in the United States.
   b. All U. S. corporations.
   c. All companies that manufacture hand-held calculators.
   d. All radio stations.

2. For each of the following hypothetical populations, list a plausible sample of size four.
   a. All distances that might result when you throw a football.
   b. Page lengths of books you might read during the next year.

   c. All possible earthquake strength measurements (Richter scale) that might be recorded in California during the next year.
   d. All pH measurements made on soil samples from a particular region.

3. List three different examples of concrete populations and three different examples of hypothetical populations.

4. For one each of your concrete and your conceptual examples (Exercise 3), give an example of a probability question and an example of an inferential statistics question.

1.2   Pictorial and Tabular Methods in Descriptive Statistics

Descriptive statistics can be divided into two general subject areas. In this section we will discuss the first of these areas—representing a data set using visual techniques. In Sections 1.3 and 1.4, we will develop some numerical summary measures for data sets. Many visual techniques may already be familiar to you: frequency tables, tally sheets, histograms, pie charts, bar graphs, scatter diagrams, and the like. Here we focus on a selected few of these techniques that are most useful and relevant to probability and inferential statistics.

Notation

Some general notation will make it easier to apply our methods and formulas to a wide variety of practical problems. The number of observations in a single data set will often be denoted by n, so that $n = 4$ for the sample of universities {Stanford, Iowa State, Wyoming, Rochester} and also for the sample of pH measurements {6.3, 6.2, 5.9, 6.5}. If two data sets are simultaneously under consideration, either m and n or n_1 and n_2 may be used to denote the numbers of observations. Thus if {29.7, 31.6, 30.9} and {28.7, 29.5, 29.4, 30.3} are thermal efficiency measurements for two different types of diesel engines, then $m = 3$ and $n = 4$.

Given a data set consisting of n observations, the observations themselves will be represented by subscripting a selected letter. We will frequently represent the observations by $x_1, x_2, x_3, \ldots, x_n$ (though any other letter could be used in place of x). The subscript bears no relation to the magnitude of a particular observation, so that x_1 will not in general be the smallest observation in the set, nor will x_n typically be the largest. In many applications x_1 will be the first observation gathered by the experimenter, x_2 the second, and so on. The ith observation in the data set will be denoted by x_i.

Stem and Leaf Displays

Suppose we have a data set $x_1, x_2, \ldots, x_n$ for which each x_i consists of at least two digits. A quick way to obtain an informative visual representation of the data set is to construct a stem and leaf display. To do this, split each x_i into two parts: a stem, consisting of one or more of the leading digits, and a leaf, which consists of the remaining digits. Thus if the data set consists of exam scores between 0 and 100, then we would split the score 83 into the stem 8 and the leaf 3. If the data set consists of automobile gas mileages, each recorded to a tenth of a mile per gallon, lying between 7.1 and 47.8, then a reasonable choice of stems would be 0, 1, 2, 3, and 4; 32.6 would then have stem 3 and leaf 2.6, and 7.1 would have stem 0 and leaf 7.1. In general, the stems should be chosen so that there are relatively few stems compared with the number of observations— between 5 and 20 stems is usually desirable. For the gas mileages, choosing stems 7, 8, ..., 47 and each leaf as the digit to the right of the decimal point (so 32.6 has stem 32 and leaf .6) results in too many stems unless the data set is huge.

Once the set of stems has been established, the stem values are listed out along the left-hand margin of the page, and beside each stem all leaves corresponding to data values are listed out in the order in which they are encountered as we proceed through the set.

Example **1.3**   The following data on motor octane ratings for various gasoline blends is taken from an article in *Technometrics* (vol. 19, p. 425), a journal devoted to applications of statistics in the physical sciences and engineering:

88.5, 87.7, 83.4, 86.7, 87.5, 91.5, 88.6, 100.3, 95.6, 93.3, 94.7, 91.1, 91.0, 94.2, 87.8, 89.9, 88.3, 87.6, 84.3, 86.7, 88.2, 90.8, 88.3, 98.8, 94.2, 92.7, 93.2, 91.0, 90.3,

93.4, 88.5, 90.1, 89.2, 88.3, 85.3, 87.9, 88.6, 90.9, 89.0, 96.1, 93.3, 91.8, 92.3, 90.4, 90.1, 93.0, 88.7, 89.9, 89.8, 89.6, 87.4, 88.4, 88.9, 91.2, 89.3, 94.4, 92.7, 91.8, 91.6, 90.4, 91.1, 92.6, 89.8, 90.6, 91.1, 90.4, 89.3, 89.7, 90.3, 91.6, 90.5, 93.7, 92.7, 92.2, 92.2, 91.2, 91.0, 92.2, 90.0, 90.7

Since the smallest observation is 83.4 and the largest is 100.3, we choose as stem values the numbers 83, 84, . . . , 100. The resulting stem and leaf display is given in Figure 1.2, which shows immediately that most of the octane ratings lie between 86 and 95, and that the middle value (which divides the data set into two sets of equal size) is someplace between 90 and 92. It also demonstrates that the octane ratings are distributed in a roughly symmetric fashion about the middle value. The display thus allows us to extract some of the salient features of the data set.

```
 83 ‖ .4
 84 ‖ .3
 85 ‖ .3
 86 ‖ .7, .7
 87 ‖ .7, .5, .8, .6, .9, .4
 88 ‖ .5, .6, .3, .2, .3, .5, .3, .6, .7, .4, .9
 89 ‖ .9, .2, .0, .9, .8, .6, .3, .8, .3, .7
 90 ‖ .8, .3, .1, .9, .4, .1, .4, .6, .4, .3, .5, .0, .7
 91 ‖ .5, .1, .0, .0, .8, .2, .8, .6, .1, .1, .6, .2, .0
 92 ‖ .7, .3, .7, .6, .7, .2, .2, .2
 93 ‖ .3, .2, .4, .3, .0, .7
 94 ‖ .7, .2, .2, .4
 95 ‖ .6
 96 ‖ .1
 97 ‖
 98 ‖ .8
 99 ‖
100 ‖ .3
```

Figure 1.2 Stem and leaf display for octane data   ■

Example 1.4   As a second example of a stem and leaf display, Figure 1.3 presents in this format a random sample of yardages of golf courses that have been designated by *Golf Magazine* as among the most challenging in the United States. Among the sample of 40 yardages, the shortest course is 6433 yards long while the longest is 7280 yards. The yardages appear to be distributed in a roughly uniform fashion over the range of values in the sample. Notice that a stem choice here of either a single digit (6 or 7) or three digits (643, . . . , 728) would yield an uninformative display, the first because of too few stems and the latter because of too many.

The book *Exploratory Data Analysis* listed in the chapter bibliography provides many interesting examples, variations, and extensions of stem and leaf displays, as well as a wealth of information on other new descriptive techniques.

```
64 ‖ 35, 64, 33, 70
65 ‖ 26, 27, 06, 83
66 ‖ 05, 94, 14
67 ‖ 90, 70, 00, 98, 70, 45, 13
68 ‖ 90, 70, 73, 50
69 ‖ 00, 27, 36, 04
70 ‖ 51, 05, 11, 40, 50, 22
71 ‖ 31, 69, 68, 05, 13, 65
72 ‖ 80, 09
```

Figure 1.3 Stem and leaf display for golf course yardage ∎

Frequency Distributions for Quantitative Data

A frequency distribution provides an even more compact summary of a data set than does a stem and leaf display. Rather than retaining the entire data set in the display, a frequency distribution essentially provides a count of only the number of observations associated with each stem choice. The first step in constructing a frequency distribution is to divide the relevant measurement axis into a collection of disjoint (nonoverlapping) intervals such that each observation in the data set is contained in one of these intervals. If, for example, the data set consists of heights of professional basketball players and all heights are between 70 in. and 86 in., one choice of intervals would be 70 in.–under 71 in., 71 in.–under 72 in., . . . , 85 in.–86 in. Each of the resulting intervals is called a **class interval,** or simply a class.

Once the class intervals have been chosen, they are listed along the left-hand margin, and then beside each interval a tally mark is recorded for every number falling in the interval as it is encountered in the data set. Finally, the number of tally marks beside each interval is recorded in a column marked "frequency," and the numbers in that column, along with the class intervals, make up the frequency distribution.

Example 1.5

The following data consists of lifetimes (in thousands of cycles) of 101 rectangular strips of aluminum subjected to repeated alternating stress at 21,000 psi, 18 cycles per second. The information is taken from an article in the *Journal of the American Statistical Association* (1958, p. 159).

1293, 1567, 1222, 1199, 1782, 1055, 797, 1420, 1016, 2100, 930, 1505, 1310, 1115, 1881, 1238, 370, 1522, 1419, 1262, 1763, 1020, 1102, 1730, 1893, 1594, 716, 1475, 1540, 1203, 988, 2268, 1485, 1792, 1330, 886, 1602, 2023, 2440, 1102, 990, 1390, 1502, 1270, 1140, 1910, 1000, 1450, 1608, 2130, 1290, 706, 1313, 1578, 1478, 1258, 1010, 1018, 1820, 1530, 1420, 2215, 1269, 746, 1134, 1522, 1235, 960, 1768, 1315, 844, 1452, 1940, 1252, 1781, 1108, 785, 1200, 1416, 886, 1750, 1923, 1355, 1085, 858, 1674, 1200, 1890, 1513, 1945, 1120, 1604, 1750, 1481, 855, 1868, 1560, 1300, 1630, 1895, 1642

Since the smallest observation is 370 and the largest is 2440, a reasonable choice of class intervals (in that it produces neither too few nor too many intervals) is obtained by starting the first interval at 350 and letting each interval have length 200. The resulting frequency distribution is given in Table 1.1. The frequency distribution indicates a small number of relatively small and relatively large lifetimes, with the distribution of values in the middle intervals being relatively flat.

Table 1.1   A frequency distribution for the lifetimes of Example 1.5

Class interval	Tally	Frequency = f_i	Relative freq. = f_i/n
1. 350–under 550	I	1	.010
2. 550–under 750	III	3	.030
3. 750–under 950	N III	8	.079
4. 950–under 1150	N N N II	17	.168
5. 1150–under 1350	N N N IIII	19	.188
6. 1350–under 1550	N N N IIII	19	.188
7. 1550–under 1750	N N I	11	.109
8. 1750–under 1950	N N N II	17	.168
9. 1950–under 2150	III	3	.030
10. 2150–under 2350	II	2	.020
11. 2350–under 2550	I	1	.010
		$n = 101$	1.000

There are no hard and fast rules for selecting class intervals for a frequency distribution. A distribution constructed from either very few or a great many intervals will not typically be very informative; most distributions are based on between five and 20 intervals. It is not even necessary that all intervals have the same length; often there are very few observations on either end of the distribution, so that the end intervals are chosen to be longer than the middle intervals. Thus in Example 1.5 we could have combined the two lowest intervals, 350–550 and 550–750, into one larger interval, and done the same for the two highest intervals.

We can obtain the **relative frequency distribution** from the frequency distribution by dividing each number in the frequency column by the number of observations n in the data set. The relative frequency distribution thus gives the *proportion* of the data set falling into each of the class intervals. In Example 1.5 each frequency is divided by $n = 101$, and the resulting relative frequencies lie between 0 and 1 and (unless there is round-off error) add to 1. That is, if f_i denotes the frequency for the ith interval, then the ith relative frequency is f_i/n and $\Sigma f_i/n = 1$. If the data set constitutes a sample from a population and the sample size is large, the relative frequency f_i/n will tend to be close to the proportion of the entire population falling in the ith interval. More will be said about this subsequently.

Histograms

A pictorial representation of a frequency distribution can be obtained by constructing a histogram. First draw a horizontal line to represent the measurement axis, and then mark the boundaries of adjacent class intervals on the axis. Now above each class interval draw a rectangle whose area is proportional to the frequency of that interval. Thus if interval i contains twice as many observations as interval j, the rectangle above interval i would have twice the area of the rectangle above interval j. Figure 1.4 pictures the histogram corresponding to the frequency distribution of Table 1.2. Often relative frequencies (or frequencies) are included either just above each rectangle or inside each rectangle.

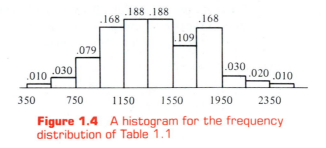

Figure 1.4   A histogram for the frequency distribution of Table 1.1

A histogram is easiest to construct and gain information from if the class intervals are of equal length, for then we need only construct rectangles whose heights, rather than areas, are proportional to the frequencies. It is much easier to look at two rectangles with the same base length and judge that one is twice as high as the other than it is to compare areas of two rectangles that have different bases and heights.

A vertical axis to the left of the histogram can be used instead of entering each relative frequency on the histogram. One possibility is to mark relative frequencies on this axis. Another frequently used method is to mark the axis so that the area of each rectangle is its relative frequency. For this to be the case, the height of a rectangle should equal (relative frequency)/(length of base). This results in the sum of the areas of all rectangles being 1. In Figure 1.5, the histogram of Figure 1.4 has been redrawn. Each base length is 200, so the height of the highest rectangle is $.188/200 = .00094$, and so on.

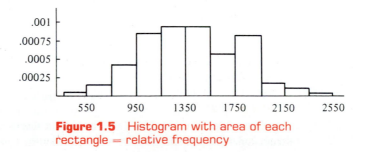

Figure 1.5   Histogram with area of each rectangle = relative frequency

Qualitative Data

Both a frequency distribution and a histogram can be constructed when the data set is qualitative (categorical) in nature. In some cases there will be a natural ordering of classes—for example, freshmen, sophomores, juniors, seniors, graduate students—while in other cases the order will be arbitrary—for example, Catholic, Jewish, Protestant, and the like. With such categorical data the intervals above which rectangles are constructed should have equal length.

Example 1.6   Each member of a sample of 120 individuals owning motorcycles was asked for the name of the manufacturer of his or her bike. The frequency distribution and histogram for the resulting data are given in Table 1.2.

Table 1.2   Frequency distribution and histogram for motorcycle data

Frequency distribution			Histogram
Manufacturer	*Frequency*	*Relative frequency*	
1. Honda	41	.34	
2. Yamaha	27	.23	
3. Kawasaki	20	.17	
4. Suzuki	18	.15	
5. Harley-Davidson	3	.03	
6. Other	11	.09	
	120	1.01	

Multivariate Data

The techniques presented so far have been exclusively for situations in which each observation in a data set has been either a single number or a single category. Often, however, the data is multivariate in nature. That is, if we obtain a sample of individuals or objects, and on each one we make two or more measurements, then each "observation" would consist of several measurements on one individual or object. The sample is bivariate if each observation consists of two measurements, so that the data set can be represented as $(x_1, y_1), \ldots, (x_n, y_n)$. For example, x might refer to engine size and y to miles per gallon, or x might refer to brand of calculator owned and y to academic major. In Chapters 11–14 we shall analyze multivariate data sets of this sort, so we will postpone a detailed discussion until that time.

Exercises / Section 1.2 (5–14)

5. The following data set refers to the tonnage (in thousands of tons) for a sample of large oil tankers. These numbers were taken from the 1976 *International Petroleum Encyclopedia*.

229, 232, 239, 232, 259, 361, 220, 260, 231, 229, 249, 254, 257, 214, 237, 253, 274, 230, 223, 253, 195, 269, 231, 268, 189, 290, 218, 313, 220, 270, 277, 374, 222, 290, 231, 258, 227, 269, 220, 224

a. Construct a stem and leaf display in which the stems are 18, 19, 20, ...

b. Use the display of (a) to construct a frequency distribution consisting of eight class intervals of equal length, with the first being 175–under 200.

c. Draw the histogram corresponding to the distribution of (b).

6. The following observations are values of the cost of living index for various cities of the world (New York City = 100), as given in the *1978 World Almanac* (p. 602), published by the Newspaper Enterprise Association.

129, 136, 96, 88, 112, 91, 80, 94, 80, 99, 79, 115, 77, 142, 90, 72, 121, 77, 131, 88, 98, 67, 75, 92, 67, 132, 119, 81, 101, 127, 93, 129, 68, 133, 84, 77, 112, 82, 117, 76, 132, 99, 138, 84, 101, 124, 84, 123, 84, 83, 97, 95, 84, 75, 58, 89, 103, 79, 89, 82, 73, 100, 115, 68, 135, 111, 87, 114, 91, 70, 79, 81, 95, 87, 87, 84, 100, 93, 87, 85, 97, 102, 78, 101, 98, 89, 97, 130, 109, 99, 96, 68, 118, 79, 93, 109

Construct:

a. A stem and leaf display for this data set. Does the distribution of cost of living index appear to be bell shaped or uniform over the set of recorded values?

b. A frequency and relative frequency distribution for the data.

c. A histogram for the distribution of (b).

7. The accompanying data set consists of observations on shear strengths (lb) of ultrasonic spot welds made on a certain type of alclad sheet. Construct a relative frequency distribution and histogram based on 10 class intervals with the first interval having lower limit 4000 and upper limit 4200 (the histogram will agree with the one that appeared in the paper "Comparison of Properties of Joints Prepared by Ultrasonic Welding and Other Means," *J. Aircraft*, 1983, pp. 552–556).*

5434, 4948, 4521, 4570, 4990, 5702, 5241, 5112, 5015, 4659, 4806, 4637, 5670, 4381, 4820, 5043, 4886, 4599, 5288, 5299, 4848, 5378, 5260, 5055, 5828, 5218, 4859, 4780, 5027, 5008, 4609, 4772,

5133, 5095, 4618, 4848, 5089, 5518, 5333, 5164, 5342, 5069, 4755, 4925, 5001, 4803, 4951, 5679, 5256, 5207, 5621, 4918, 5138, 4786, 4500, 5461, 5049, 4974, 4592, 4173, 5296, 4965, 5170, 4740, 5173, 4568, 5653, 5078, 4900, 4698, 5248, 5245, 4723, 5275, 5419, 5205, 4452, 5227, 5555, 5388, 5498, 4681, 5076, 4774, 4931, 4493, 5309, 5582, 4308, 4823, 4417, 5364, 5640, 5069, 5188, 5764, 5273, 5042, 5189, 4986

8. The following data set constitutes measurements made on the steam rate (lb/hr) of a distillation tower ("A Self-Descaling Distillation Tower," *Chem. Eng. Prog.*, 1968, pp. 79–84).

1170, 1350, 1640, 1800, 1800, 1260, 1440, 1730, 1710, 1350, 1440, 1710, 1530, 1800, 1530, 1170, 1440, 1350, 1260, 1530, 1350, 1440, 1170, 1350, 1170, 1620, 1800, 1170, 1440, 1800, 1260, 1170, 1260, 1710, 1710, 1530, 1350, 1530, 1440, 1530, 1170, 1350, 1620, 1495, 1440, 1260, 1540, 1170, 1170, 1440

Construct:

a. A stem and leaf plot of the data using the leading two digits as stems (so that 1160 has stem 11 and leaf 60).

b. A relative frequency distribution and histogram for the data using class intervals of length 100 with the first interval beginning at 1100.

9. Each of the numbers in the accompanying data set is the length of service in years of a U. S. Supreme Court Justice who has terminated his court service.

5, 1, 20, 8, 6, 9, 1, 13, 15, 4, 31, 34, 30, 16, 18, 33, 22, 20, 2, 32, 14, 32, 28, 4, 28, 15, 19, 27, 5, 23, 6, 8, 23, 18, 28, 14, 34, 8, 10, 21, 9, 14, 34, 6, 7, 20, 11, 5, 21, 20, 15, 10, 2, 16, 13, 26, 29, 19, 3, 4, 5, 26, 5, 10, 10, 26, 22, 5, 8, 15, 16, 7, 16, 11, 15, 6, 34, 19, 23, 36, 9, 5, 1, 12, 6, 13, 7, 18, 7, 16, 16, 5, 3, 4

a. Construct a frequency and relative frequency distribution for this data set using class intervals 0–under 5, 5–under 10, . . . , 35–under 40.

b. Draw a histogram corresponding to the distributions of (a) so that area = relative frequency.

c. Regarding the above data set as a sample from the conceptual population of all possible choices for Supreme Court Justices, does it

---

*Throughout this book, we will consistently cite articles from areas other than statistics to show the wide practical applications statistics has. In these cases because the data drawn from the articles is often not pertinent to the main thrust of the articles' authors, we have chosen not to cite these authors by name.

appear as though a randomly selected Justice can be expected to serve 10 years or more? Justify your answer.

10. An article in *Environmental Concentration and Toxicology* ("Trace Metals in Sea Scallops," vol. 19, pp. 326–1334) reported the amount of cadmium in sea scallops observed at a number of different stations in North Atlantic waters. The observed values follow:

5.1, 14.4, 14.7, 10.8, 6.5, 5.7, 7.7, 14.1, 9.5, 3.7, 8.9, 7.9, 7.9, 4.5, 10.1, 5.0, 9.6, 5.5, 5.1, 11.4, 8.0, 12.1, 7.5, 8.5, 13.1, 6.4, 18.0, 27.0, 18.9, 10.8, 13.1, 8.4, 16.9, 2.7, 9.6, 4.5, 12.4, 5.5, 12.7, 17.1

a. Construct a frequency and relative frequency distribution for the data set using 0.0–under 4.0 as the first class interval.

b. Draw a histogram corresponding to the distributions of (a) so that area = relative frequency.

11. If in Example 1.5 the relative frequencies are computed to only two decimal places, do they add up to 1?

12. In a study of warp breakage during the weaving of fabric (*Technometrics*, 1982, p. 63), 100 pieces of yarn were tested. The number of cycles of strain to breakage was recorded for each yarn sample. The resulting data is given below.

86, 146, 251, 653, 98, 249, 400, 292, 131, 169, 175, 176, 76, 264, 15, 364, 195, 262, 88, 264, 157, 220, 42, 321, 180, 198, 38, 20, 61, 121, 282, 224, 149, 180, 325, 250, 196, 90, 229, 166, 38, 337, 65, 151, 341, 40, 40, 135, 597, 246, 211, 180, 93, 315, 353, 571, 124, 279, 81, 186, 497, 182, 423, 185, 229, 400, 338, 290, 398, 71, 246, 185, 188, 568, 55, 55, 61, 244, 20, 284, 393, 396, 203, 829, 239, 236, 286, 194, 277, 143, 198, 264, 105, 203, 124, 137, 135, 350, 193, 188

a. Using class intervals 0–<100, 100–<200, and so on, construct a relative frequency distribution for breaking strength.

b. If weaving specifications require a breaking strength of at least 100 cycles, what proportion of the yarn samples would be considered satisfactory?

13. Every score in the following batch of exam scores is in the 60's, 70's, 80's, or 90's. A stem and leaf display with only the four stems 6, 7, 8, and 9 would not give a very detailed description of the distribution of scores. In such situations it is desirable to use repeated stems. Here we could repeat the stem 6 twice, using 6*l* for scores in the low 60's (leaves 0, 1, 2, 3, and 4) and 6*h* for scores in the high 60's (leaves 5, 6, 7, 8, and 9). Similarly, the other stems can be repeated twice (7*l*, 7*h*, 8*l*, 8*h*, 9*l*, and 9*h*) to obtain a display consisting of eight rows. Construct such a display for the given scores. What feature of the data is highlighted by this display?

74, 89, 80, 93, 64, 67, 72, 70, 66, 85, 89, 81, 81, 71, 74, 82, 85, 63, 72, 81, 81, 95, 84, 81, 80, 70, 69, 66, 60, 83, 85, 98, 84, 68, 90, 82, 69, 72, 87, 88

14. For quantitative data, the **cumulative frequency** and cumulative relative frequency for a particular class interval are the sum of frequencies and relative frequencies, respectively, for that interval and all intervals lying below it. If, for example, there are four intervals with frequencies 9, 16, 13, and 12, then the cumulative frequencies are 9, 25, 38, and 50 and the cumulative relative frequencies are .18, .50, .76, and 1.00. Compute the cumulative frequencies and cumulative relative frequencies for the data of Exercise 7.

1.3   Measures of Location

Having briefly studied tabular and pictorial methods for organizing and summarizing data, in this section and the next we will focus on numerical summary measures for a given data set. That is, from the data we try to extract several summarizing numbers, numbers that might serve to characterize the data set and convey some of its salient features. Our primary concern will be with numerical data, though some comments regarding categorical data appear at the end of the section.

Suppose, then, that our data set is of the form $x_1, x_2, \ldots, x_n$, where each x_i is a number. What features of such a set of numbers are of most interest and deserve emphasis? One important characteristic of a set of numbers is its location, and in particular its center. This section presents methods for describing the location of a data set, while in Section 1.4 we will turn to methods for measuring the variability in a set of numbers.

The Mean

For a given set of numbers $x_1, x_2, \ldots, x_n$, the most familiar and useful measure of the center is the mean, or arithmetic average of the set. Because we will almost always think of the x_i's as constituting a sample, we will often refer to the arithmetic average as the **sample mean** and denote it by $\bar{x}$.

Definition

> The **sample mean** $\bar{x}$ of a set of numbers $x_1, x_2, \ldots, x_n$ is given by
>
> $$\bar{x} = \frac{x_1 + x_2 + \cdots + x_n}{n} = \frac{\sum_{i=1}^{n} x_i}{n}$$

The value of $\bar{x}$ is in a sense more precise than the accuracy associated with any single observation. For this reason, we will customarily report the value of $\bar{x}$ using one digit of decimal accuracy beyond what is used in the individual x_i's.

Example **1.7**

The amount of light reflectance by leaves has been used for various purposes, including evaluation of turf color, estimation of nitrogen status, and measurement of biomass. The paper "Leaf Reflectance-Nitrogen-Chlorophyll Relations in Buffelgrass" (*Photogrammetric Engineering and Remote Sensing,* 1985, pp. 463–466) gave the following observations, obtained using spectrophotogrammetry, on leaf reflectance under specified experimental conditions:

15.2, 16.8, 12.6, 13.2, 12.8, 13.8, 16.3, 13.0, 12.7, 15.8, 19.2, 12.7, 15.6, 13.5, 12.9

The sum of these 15 x_i's is $\Sigma x_i = 15.2 + 16.8 + \cdots + 12.9 = 216.1$, so the value of the sample mean is

$$\bar{x} = \frac{\sum_{i=1}^{15} x_i}{15} = \frac{216.1}{15} = 14.41 \qquad \blacksquare$$

A physical interpretation of $\bar{x}$ will demonstrate how it measures the location (center) of a sample. Think of drawing and scaling a horizontal measurement axis, and then represent each sample observation by a one-pound weight placed at the corresponding point on the axis. The only point at which a fulcrum can be placed to balance the system of weights is the point corresponding to the value of $\bar{x}$ (see Figure 1.6).

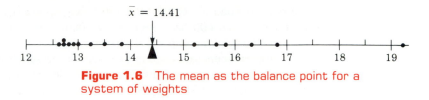

Figure 1.6   The mean as the balance point for a system of weights

Just as $\overline{x}$ represents the average value of the observations in a sample, the average of all values in the population can be calculated. This average is called the **population mean** and is denoted by the Greek letter μ. When there are N values in the population (a finite population), then $\mu =$ (sum of the N population values)$/N$.

Example 1.8

Exercise 9 listed the length of service of each U. S. Supreme Court Justice whose service terminated before 1978. Regarding these $N = 94$ numbers as constituting the population of interest, the population mean is

$$\mu = \frac{5 + 1 + 20 + \cdots + 3 + 4}{94} = \frac{1406}{94} = 15.0$$

If a sample of size $n = 5$ is selected and the result is $x_1 = 8$, $x_2 = 23$, $x_3 = 16$, $x_4 = 7$, and $x_5 = 18$, then the sample mean is $\overline{x} = 14.4$. If we did not know μ, we could estimate it as 14.4, the value of the sample mean.   ■

In Chapters 3 and 4 we will discuss models for infinite populations, so we will postpone until then a general definition of μ. Just as $\overline{x}$ is an interesting and important measure of sample location, μ is an interesting and important (often the most important) characteristic of a population. In the chapters on statistical inference, we will present methods based on the sample mean for drawing conclusions about a population mean. For example, we might use the sample mean $\overline{x} = 14.41$ computed in Example 1.7 as a point estimate (a single number which is our "best" guess) of $\mu =$ the true average reflectance for all leaves under the specified conditions.

The sample mean does possess one property that renders it a somewhat unsatisfactory measure of location for some data sets. The computed value of $\overline{x}$ can be greatly influenced by the presence of just one observation that lies very far to one side or the other of the other values.

Example 1.9

Suppose that we randomly select five recordings of classical music from the *Schwann Record Catalog* (which lists all current recordings of classical and nonclassical music), and determine the listening time for each. If the data values are (rounded to the nearest minute) $x_1 = 37$, $x_2 = 46$, $x_3 = 40$, $x_4 = 57$, and $x_5 = 50$, then $\overline{x} = 46.0$ minutes. However, if the fifth recording selected is not a Tchaikovsky symphony but instead a Wagner opera, so that $x_5 = 200$ (and seems much longer), then $\overline{x} = 76$. Since most of the data values are considerably smaller than 76, many would feel that $\overline{x}$ here is not a reliable measure of location. Of course, this effect would be even more pronounced if x_5 were, for example, 400 (so $\bar{x} = 116$) or 1000. ∎

A sample of incomes often produces a few such outlying values (those lucky few who earn astronomical incomes), and the use of average income as a measure of location will often be misleading. Such examples suggest that we look for a measure that is less sensitive to outlying values than $\bar{x}$, and we shall momentarily propose one. However, while $\bar{x}$ does have this potential defect, it is still the most widely used measure, largely because there are many populations for which an extreme outlier in the sample would be highly unlikely. When sampling from such a population (a normal or bell shaped population being the most important example), the sample mean will be stable and quite representative of the sample.

The Median

The word "median" is synonymous with "middle," and the sample median is indeed the middle value when the observations are ordered from smallest to largest in magnitude. When the observations are denoted by $x_1, \ldots, x_n$, we will use the symbol $\tilde{x}$ to represent the sample median.

Definition

> Given the sample $x_1, x_2, \ldots, x_n$, rearrange the observations in increasing order (from most negative to most positive). Then the **sample median** is given by
>
> $$\tilde{x} = \begin{cases} \text{single middle value in the ordered list if } n \text{ is odd;} \\ \text{the average of the two middle values in the ordered list if } n \text{ is even} \end{cases}$$

Example 1.10

The following data refers to active repair times (hours) for an airborne communication receiver.

> 1.1, 4.0, .5, 5.4, 2.0, .5, .8, 9.0, 5.0, 3.3, .3, .7, 2.2, 22.0, 4.0, 2.7, 1.0, 3.0, 1.0, 1.5, 24.5, 1.5, 3.3, 2.5, 1.0, .8, 1.5, .6, 10.3, 1.3, .7, 2.0, 4.7, 3.0, .8, 1.0, 8.8, .6, 4.5, 7.0, .7, 5.4, 1.5, .2, 7.5, .5

Because $n = 46$ is even, $\tilde{x}$ will be the average of the two middle values in the ordered list. Ordering $x_1, \ldots, x_{46}$ gives

> .2, .3, .5, .5, .5, .6, .6, .7, .7, .7, .8, .8, .8, 1.0, 1.0, 1.0, 1.0, 1.1, 1.3, 1.5, 1.5, 1.5, 1.5, 2.0, 2.0, 2.2, 2.5, 2.7, 3.0, 3.0, 3.3, 3.3, 4.0, 4.0, 4.5, 4.7, 5.0, 5.4, 5.4, 7.0, 7.5, 8.8, 9.0, 10.3, 22.0, 24.5

The two middle values are 1.5 and 2.0, so that $\tilde{x}$ is 1.75. Notice that if we had not listed all replications of data values, $\tilde{x}$ would have been much larger, since these replications occur for values of small magnitude but not values of large magnitude. ∎

The data set in Example 1.10 illustrates an important property of $\tilde{x}$ in contrast to $\bar{x}$; the sample median is very insensitive to a number of extremely small or extremely large data values. If, for example, we increased the two largest x_i's from 22.0 and 24.5 to 122.0 and 424.5, $\tilde{x}$ would be unaffected. Thus in the treatment of outlying data values, $\bar{x}$ and $\tilde{x}$ are at opposite ends of a spectrum: $\bar{x}$ is sensitive to even one such value, while $\tilde{x}$ is insensitive to a large number of outlying values.

Because the large values in the sample of Example 1.10 affect $\bar{x}$ more than $\tilde{x}$, $\tilde{x} < \bar{x}$ for that data. While $\bar{x}$ and $\tilde{x}$ both provide a measure for the center of a data set, they will not in general be equal, since they focus on different aspects of the sample.

Analogous to $\tilde{x}$ as the middle value in the sample, there is a middle value in the population, the **population median,** denoted by $\tilde{\mu}$. As with $\bar{x}$ and μ, we can think of using the sample median $\tilde{x}$ to make an inference about $\tilde{\mu}$. In Example 1.10 we might use $\tilde{x} = 1.75$ as an estimate of the median number of hours necessary to repair all such communication receivers when they fail. A median is often used to describe income or salary data (because it is not greatly influenced by a few large salaries), so if the median salary for a sample of engineers were $\tilde{x} = \$26{,}416$, we might use this as a basis for concluding that the median salary for all engineers exceeds $\$25{,}000$.

The population mean μ and median $\tilde{\mu}$ will not generally be equal to one another. If the population distribution is either positively or negatively skewed, as pictured in Figure 1.7, then $\mu \neq \tilde{\mu}$. When this is the case, in making inferences we must first decide which of the two population characteristics is of most interest and then proceed accordingly.

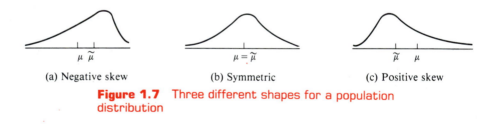

(a) Negative skew   (b) Symmetric   (c) Positive skew

Figure 1.7   Three different shapes for a population distribution

Other Measures of Location: Quartiles, Percentiles, and Trimmed Means

The median (population or sample) divides the data set into two parts of equal size. To obtain finer measures of location, we could divide the data into more than two such parts. Roughly speaking, quartiles divide the data set into four equal parts, with the observations above the third quartile constituting the upper quarter of the data set, the second quartile being identical to the median, and the first quartile separating the lower quarter from the upper three-quarters. Similarly, a data set (sample or population) can be divided into 100 equal parts using percentiles; the ninety-ninth percentile separates off the highest 1% from the bottom 99%, and so on. Unless the number of observations is a multiple of 100, care must be exercised in obtaining percentiles. We will use percentiles in Chapter 4 in connection with certain models for infinite populations, so we shall postpone discussion until that point.

As emphasized above, the sample mean and sample median are influenced by outlying values in a very different manner: the mean greatly and the median not at all. Since extreme behavior of either type might not always be desirable, we briefly consider alternative measures that are neither as sensitive as $\bar{x}$ nor as insensitive as $\tilde{x}$. To motivate these alternatives, note that $\bar{x}$ and $\tilde{x}$ are at opposite extremes of the same "family" of measures. After ordering the data, $\tilde{x}$ is computed by throwing away as many values on either end as one can without eliminating everything (leaving just one or two middle values) and averaging what is left, while to compute $\bar{x}$ one throws away nothing before averaging. To paraphrase, the mean involves trimming 0% from either end of the sample, while for the median the maximum possible amount is trimmed from either end. A **trimmed mean** is a compromise between $\bar{x}$ and $\tilde{x}$. A 10% trimmed mean, for example, would be computed by eliminating the smallest 10% and the largest 10% of the sample and then averaging what is left over.

Example 1.11   Consider the following 20 observations, ordered from smallest to largest, each one representing the lifetime in hours of a certain type of incandescent lamp.

612, 623, 666, 744, 883, 898, 964, 970, 983, 1003, 1016, 1022, 1029, 1058, 1085, 1088, 1122, 1135, 1197, 1201

The average of all 20 observations is $\bar{x} = 965.0$, while $\tilde{x} = 1009.5$. The 10% trimmed mean is obtained by deleting the smallest two observations (612 and 623) and the largest two (1197 and 1201), and then averaging the remaining 16 to obtain $\bar{x}_{tr(10)} = 979.1$. The effect of trimming here is to produce a "central value" that is somewhat above the mean ($\bar{x}$ is pulled down by a few small lifetimes) and yet considerably below the median. Similarly, the 20% trimmed mean averages the middle 12 values to obtain $\bar{x}_{tr(20)} = 999.9$, even closer to the median. (See Figure 1.8.) ■

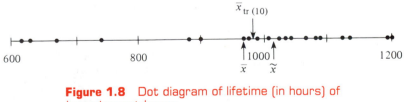

Figure 1.8   Dot diagram of lifetime (in hours) of incandescent lamps   ■

Generally speaking, using a trimmed mean with a moderate trimming proportion (10 or 20%) will yield a measure that is neither as sensitive to outliers as the mean (since any small number of outliers will be deleted before averaging) nor as insensitive as the median. For this reason trimmed means have merited increasing attention from statisticians both for descriptive and inferential purposes. More will be said about trimmed means when point estimation is discussed in Chapter 6. As a final point, if the trimming proportion is denoted by α and $n\alpha$ is not an integer, then it is not obvious how the $100\alpha\%$ trimmed mean should be computed. For example, if $\alpha = .10$ (10%) and $n = 22$, then $n\alpha = (22)(.10) = 2.2$, and we cannot trim 2.2 observations from each end of the ordered sample. In this case, the 10% trimmed mean would be obtained by first trimming two observations from each end and calculating $\bar{x}_{tr}$, then trimming three and calculating $\bar{x}_{tr}$, and finally interpolating between the two values to obtain $\bar{x}_{tr(10)}$.

Categorical Data and Sample Proportions

When the data is categorical, a frequency distribution or relative frequency distribution provides an effective tabular summary of the data. The natural numerical summary quantities in this situation are the individual frequencies and the relative frequencies. For example, if a survey of individuals who own stereo receivers is undertaken to study brand preference, then each individual in the sample would identify the brand of receiver that he or she owned, from which we could count the number owning Sony, Marantz, Pioneer, and so on. Consider sampling a dichotomous population—one which consists of only two categories (such as voted or did not vote in the last election, does or does not own a stereo receiver, and so on). If we let x denote the number in the sample falling in category one, then the number in category two is $n - x$. The relative frequency or *sample proportion* in category one is x/n and the sample proportion in category two is $1 - x/n$. To see that x/n *can be regarded as a sample mean,* identify a category one response by a 1 and a category two response by a 0. A sample of size $n = 10$ might then yield the responses 1, 1, 0, 1, 1, 1, 0, 0, 1, 1. The sample mean for this numerical sample is (since number of 1's $= x = 7$)

$$\frac{x_1 + \cdots + x_n}{n} = \frac{1 + 1 + 0 + \cdots + 1 + 1}{10} = \frac{7}{10} = \frac{x}{n} = \frac{\text{sample}}{\text{proportion}}$$

This result can be generalized and summarized as follows: *If in a categorical data situation we focus attention on a particular category and code the sample results so that a 1 is recorded for an individual in the category and a 0 for an individual not in the category, then the sample proportion of individuals in the category is the sample mean of the sequence of 1's and 0's.* Thus a sample mean can be used to summarize the results of a categorical sample. These remarks also apply to situations in which categories are defined by grouping values in a numerical sample or population (for example, we might be interested in knowing whether or not individuals have owned their present automobile for at least five years, rather than studying the exact length of ownership).

Analogous to the sample proportion x/n of individuals falling in a particular category, let p represent the proportion of individuals in the entire population falling in the category. As with x/n, p is a quantity between 0 and 1. While x/n is a sample characteristic, p is a characteristic of the population; the relationship between the two parallels the relationship between $\tilde{x}$ and $\tilde{\mu}$ and between $\bar{x}$ and μ. In particular, we shall subsequently use x/n to make inferences about p. If, for example, a sample of 100 car owners reveals that 22 owned their car at least five years, then we might use $22/100 = .22$ as a point estimate of the proportion of all owners who have owned their car at least five years. We will study the properties of x/n as an estimator of p and see how x/n can be used to answer other inferential questions. With k categories ($k > 2$) we can use the k sample proportions to answer questions about the population proportions $p_1, \ldots, p_k$.

Exercises / Section 1.3 [15–24]

15. The nine measurements that follow are temperature determinations at various locations beneath the surface of Lake Ontario, as reported in *Limnology and Oceanography* (vol. 22, pp. 158–159):

4.45, 3.91, 3.86, 3.93, 3.94, 3.90, 3.80, 3.73, 3.69

Compute:
a. The sample mean of these data values.
b. The sample median of these data values.

16. Bacterial mutants of the bacterium *E. Coli* that are resistant to the virus T_1 are called *Ton^r* variants. A recent paper in *Genetics* reported the following results on the number of *Ton^r* bacteria found in 10 different bulk cultures when a particular experimental culture method was used:

14, 15, 13, 21, 15, 14, 26, 16, 20, 13

Compute:
a. The sample mean number of bacteria per culture.
b. The sample median number of bacteria per culture.
c. The 10% trimmed mean for the data set.

17. The 20-year average annual costs for $50,000 5-year-renewable term insurance policies (men, age 35) for ten companies are

214, 352, 379, 344, 337, 347, 338, 300, 300, 361

Compute:
a. The sample mean cost.

b. The sample median.
c. The 10% trimmed mean.

18. The paper "Penicillin in the Treatment of Meningitis" (*J. Amer. Medical Assn.*, 1984, pp. 1870–1873) reported the body temperatures (in degrees Fahrenheit) of patients hospitalized with meningitis. Ten of the observations were as follows.

104.0, 104.8, 101.6, 108.0, 103.8, 100.8, 104.2, 100.2, 102.4, 101.4

a. Compute the sample mean.
b. Do you think the 10% trimmed mean would differ much from the sample mean computed in (a)? Why? Answer without actually computing the trimmed mean.

19. Although blood pressure is a continuous variable, its value is often reported to the nearest 5 mmHg (100, 105, 110, and so on). Suppose that the actual blood pressure values for nine randomly selected individuals are 118.6, 127.4, 138.4, 130.0, 113.7, 122.0, 108.3, 131.5, and 133.2.

a. What is the median of the reported blood pressure values?
b. Suppose that the blood pressure of the second individual is 127.6 rather than 127.4 (a small change in a single value). How does this affect the median of the reported values? What does this say about the sensitivity of the median to rounding or grouping in the data?

20. The propagation of fatigue cracks in various aircraft parts has been the subject of extensive study in recent years. The accompanying data consists of propagation lives (flight hours/10^4) to reach a given crack size in fastener holes intended for use in military aircraft ("Statistical Crack Propagation in Fastener Holes under Spectrum Loading," *J. Aircraft*, 1983, pp. 1028–1032).

.736, .863, .865, .913, .915, .937, .983, 1.007, 1.011, 1.064, 1.109, 1.132, 1.140, 1.153, 1.253, 1.394

a. Compute the values of the sample mean and median.

b. By how much could the largest sample observation be decreased without affecting the value of the median?

21. Compute the sample median, 25% trimmed mean, 10% trimmed mean, and sample mean for the data given in Exercise 10.

22. In an attempt to study the effect of choice of postage stamps on response rate in a mail survey, W. E. Hensley (*Public Opinion Quarterly*, vol. 38, pp. 280–283), reported the following data on number of mailings n_i and number of returns x_i both when inside and outside stamps were dissimilar ($i = 1$) and similar ($i = 2$): $n_1 = 354$, $x_1 = 217$, $n_2 = 176$, $x_2 = 89$.

a. Compute the sample proportion of returns both for dissimilar stamps and similar stamps.

b. The sample proportions of (a) can be viewed as estimates of true return proportions p_1 and p_2 for hypothetical populations. If there is actually no difference in response rate due to the types of stamps, then $p_1 = p_2$, and we have a sample of size $n = 354 + 176 = 530$ from a single hypothetical population, with $x = 306$ returns. Compute the sample proportion for this pooled (combined) sample.

23. In Exercise 17, obtain the sample proportion of insurance companies that charge at least $350 for such policies. What is the sample proportion of companies that charge between $300 and $350 inclusive?

24. a. If a constant c is added to each x_i in a sample, yielding $y_i = x_i + c$, how do the sample mean and median of the y_i's relate to the mean and median of the x_i's?

b. If each x_i is multiplied by a constant c, yielding $y_i = cx_i$, answer the question of part (a).

1.4 Measures of Variability

No single measure of location can give a complete summarization of a data set. Consider the x data set 20, 100, 0, 60, 70 and the y data set 60, 20, 80, 60, 30. Since $\bar{x} = \bar{y} = 50$ and $\tilde{x} = \tilde{y} = 60$, the two standard measures of location by themselves do not distinguish between the two sets. Yet the plots of these two data sets in Figure 1.9 show that the observations in the y data set cluster more closely about their center than do the observations from the x data set. That is, there is more variability or dispersion in the x's than in the y's.

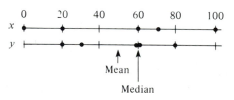

Figure 1.9 Data sets with the same center and differing variability

One simple measure of variability is the **sample range,** defined as the difference between the largest observation and the smallest observation—that is, sample range = $\max(x_i) - \min(x_i)$. We might then say that a small (large) range indicates little (great) variability. A defect of the sample range, though, is that it depends only on the two most extreme observations and disregards the positions of the middle observations. For example, the two samples 0, 5, 5, 5, 10 and 0, 1, 5, 9, 10 both have the same sample range, yet in the first sample there is variability only in the two extreme values, while the second sample has more variability in its middle values. We would like a measure that somehow depends on all observations rather than just a few.

Deviations from the Mean

The quantity $x_i - \bar{x}$ is called the deviation of the ith observation from the mean, or just the *i***th deviation.** A positive deviation indicates an observation to the right of $\bar{x}$ on the measurement axis, while a negative deviation indicates an observation to the left of $\bar{x}$. If all the deviations $x_1 - \bar{x}, \ldots, x_n - \bar{x}$ are small in absolute magnitude, then all x_i's are close to $\bar{x}$ and thus to one another, suggesting a relatively small amount of variability in the sample. On the other hand, if some of the $(x_i - \bar{x})$'s are large in absolute magnitude, then some of the x_i's lie far from $\bar{x}$, suggesting large variation. A simple way of combining the n deviations into a single quantity is to average them (sum them and divide by n). However, since some deviations will be negative and others positive, adding results in cancellation.

Proposition

$$\sum_{i=1}^{n} (i\text{th deviation}) = \sum_{i=1}^{n} (x_i - \bar{x}) = 0$$

so the average deviation from the mean is always 0.

Proof. The verification uses several standard rules of summation and the fact that $\bar{x}$ is a constant in the summation:

$$\Sigma (x_i - \bar{x}) = \Sigma x_i - \Sigma \bar{x} = \Sigma x_i - n\bar{x} = \Sigma x_i - n\left(\frac{1}{n} \Sigma x_i\right) = 0 \qquad \blacksquare$$

To obtain an informative measure of variability, we need to change the deviations to nonnegative quantities before combining. One possibility is to average the absolute deviations $|x_1 - \bar{x}|, \ldots, |x_n - \bar{x}|$. Because this leads to a multitude of theoretical difficulties, consider instead the squared deviations $(x_1 - \bar{x})^2, \ldots, (x_n - \bar{x})^2$. We might now use the average of the squared deviations. There is, however, a technical reason (to be discussed shortly) for dividing the sum of squared deviations by $n - 1$ rather than n.

The Sample Variance and Standard Deviation

Definition

> The **sample variance** of the set $x_1, \ldots, x_n$ of numerical observations, denoted by s^2, is given by
>
> $$s^2 = \frac{\sum_{i=1}^{n}(x_i - \overline{x})^2}{n - 1}$$
>
> The **sample standard deviation,** denoted by s, is the positive square root of the sample variance.

The divisor $n - 1$ in s^2 is smaller than n, so s^2 is somewhat larger than the average squared deviation. Whatever the units in which the x_i's are expressed, s^2 is given in squared units (e.g., in.^{2} when the x_i's are heights in inches, sec^2 when observations are stopping times expressed in seconds, and so on). The sample standard deviation has the desirable property of measuring dispersion in units identical to those in which the x_i's themselves are given.

Example **1.12**   Strength is an important characteristic of materials used in prefabricated housing. Each of $n = 11$ prefabricated plate elements was subjected to a severe stress test and the maximum width (mm) of the resulting cracks was recorded. The given data (Table 1.3) appeared in the paper "Prefabricated Ferrocement Ribbed Elements for Low-Cost Housing" (*J. Ferrocement,* 1984, pp. 347–364).

Table 1.3

x_i	$x_i - \overline{x}$	$(x_i - \overline{x})^2$
.684	−.9841	.9685
2.540	.8719	.7602
.924	−.7441	.5537
3.130	1.4619	2.1372
1.038	−.6301	.3970
.598	−1.0701	1.1451
.483	−1.1851	1.4045
3.520	1.8519	3.4295
1.285	−.3831	.1468
2.650	.9819	.9641
1.497	−.1711	.0293
$\Sigma x_i = 18.349$	$\Sigma(x_i - \overline{x}) = -.0001$	$\Sigma(x_i - \overline{x})^2 = 11.9359$

$$\overline{x} = \frac{18.349}{11} = 1.6681$$

Effects of rounding account for the sum of deviations not being exactly zero. The numerator of s^2 is 11.9359, so $s^2 = 11.9359/(11 - 1) = 11.9359/10 = 1.19359$ and $s = \sqrt{1.19359} = 1.0925$ mm.   ■

Admittedly s^2 and s do not lend themselves to the intuitive understanding that measures of location do. Because variability is a less familiar characteristic than is location, the usefulness of measures of variability is less obvious. In particular, we are not in a position to say that $s = 1.9025$ indicates a large or a small amount of variability. All you should believe at this point is that if "eye-balling" two different samples suggests that the first clearly has less variability than the second, then s^2 and s for the first sample should be smaller than the corresponding quantities for the second sample.

Motivation for s^2

In order to explain why s^2 rather than the average squared deviation is used to measure variability, note first that whereas s^2 measures sample variability, there is a measure of variability in the population called the population variance. We shall use σ^2 (the square of the lowercase Greek letter sigma) to denote the population variance and σ to denote the population standard deviation (the square root of σ^2). When the population is finite and consists of N values, $\sigma^2 = \Sigma \, (i\text{th population value} - \mu)^2/N$, which is the average of all squared deviations from the population mean (for the population, the divisor is N and not $N - 1$). More general definitions of σ^2 appear in Chapters 3 and 4.

Just as $\bar{x}$ will be used to make inferences about the population mean μ, we should define the sample variance so that it can be used to make inferences about σ^2. Now note that σ^2 involves squared deviations about the population mean μ. If we actually knew the value of μ, then we could define the sample variance as the average squared deviation of the sample x_i's about μ. However, the value of μ is almost never known, so the sum of squared deviations about $\bar{x}$ must be used. But *the x_i's tend to be closer to their average $\bar{x}$ than to the population average μ, so to compensate for this the divisor $n - 1$ is used rather than n.* Said another way, if we used a divisor n in the sample variance, then the resulting quantity would tend to underestimate σ^2 (produce estimated values that are too small on the average), while dividing by the slightly smaller $n - 1$ corrects this underestimating.

It is customary to refer to s^2 as being based on $n - 1$ **"degrees of freedom."** This terminology results from the fact that while s^2 is based on the n quantities $x_1 - \bar{x}, x_2 - \bar{x}, \ldots, x_n - \bar{x}$, these sum to 0, so specifying the values of any $n - 1$ of the quantities determines the remaining one. For example, if $n = 4$ and $x_1 - \bar{x} = 8$, $x_2 - \bar{x} = -6$, and $x_4 - \bar{x} = -4$, then automatically we have $x_3 - \bar{x} = 2$, so only three of the four $x_i - \bar{x}$'s are freely determined (3 degrees of freedom).

The Computation of s^2

With n observations the computation of s^2 from the definition involves n subtractions and n squaring operations. If $\bar{x}$ is not an integer, the subtractions to obtain the deviations can be tedious, and the deviations themselves may be unpleasant to square. There is an equivalent expression for the numerator of s^2 that yields a more efficient method for calculating s^2 when computations are done by hand or hand-held calculator.

Proposition

$$s^2 = \frac{\sum\limits_{i=1}^{n}(x_i - \overline{x})^2}{n - 1} = \frac{\sum\limits_{i=1}^{n}x_i^2 - \left(\sum\limits_{i=1}^{n}x_i\right)^2 / n}{n - 1}$$

Proof.   Because $\overline{x} = \Sigma x_i / n$, $n\overline{x}^2 = (\Sigma x_i)^2 / n$. Then

$$\sum_{i=1}^{n}(x_i - \overline{x})^2 = \sum_{i=1}^{n}(x_i^2 - 2\overline{x} \cdot x_i + \overline{x}^2) = \sum_{i=1}^{n}x_i^2 - 2\overline{x}\sum_{i=1}^{n}x_i + \sum_{i=1}^{n}(\overline{x})^2$$

$$= \sum_{i=1}^{n}x_i^2 - 2\overline{x} \cdot n\overline{x} + n(\overline{x})^2 = \sum_{i=1}^{n}x_i^2 - n(\overline{x})^2$$

$$= \sum_{i=1}^{n}x_i^2 - \frac{\left(\sum\limits_{i=1}^{n}x_i\right)^2}{n}$$

∎

To use this method for computing s^2, square each x_i (before subtraction), then add the squares, and subtract $(\Sigma x_i)^2 / n$ from Σx_i^2. While this involves squaring $n + 1$ numbers (Σx_i in addition to each x_i), only one subtraction is necessary. We shall refer to this formula for computing s^2 as the "shortcut method for s^2." The shortcut for s involves computing s^2 using the shortcut and then taking the square root.

Example 1.13   Table 1.4 displays the $n = 15$ reflectance observations first introduced in Example 1.7.

Table 1.4

Observation	x_i	x_i^2
1	15.2	231.04
2	16.8	282.24
3	12.6	158.76
4	13.2	174.24
5	12.8	163.84
6	13.8	190.44
7	16.3	265.69
8	13.0	169.00
9	12.7	161.29
10	15.8	249.64
11	19.2	368.64
12	12.7	161.29
13	15.6	243.36
14	13.5	182.25
15	12.9	166.41
	$\Sigma x_i = 216.1$	$\Sigma x_i^2 = 3168.13$

Thus the shortcut formula gives

$$s^2 = \frac{\Sigma x_i^2 - \dfrac{(\Sigma x_i)^2}{n}}{n-1} = \frac{3168.13 - \dfrac{(216.1)^2}{15}}{15-1} = \frac{54.85}{14} = 3.92$$

and $s = \sqrt{3.92} = 1.98$. ∎

The shortcut method can yield values of s^2 and s that differ from the values computed using the definitions. These differences are due to effects of rounding and will not be important in most problems. In order to minimize the effects of rounding when using the shortcut formula, particularly when there is little variability in the data, intermediate calculations should be done using several more significant digits than are to be retained in the final answer. Because the numerator of s^2 is the sum of nonnegative quantities (squared deviations), s^2 is guaranteed to be nonnegative. Yet if the shortcut method is used, particularly with data having little variability, a slight numerical error can result in a negative numerator (Σx_i^2 smaller than $(\Sigma x_i)^2/n$). If your value of s^2 is negative you have made a computational error.

There are several other properties of s^2 that can sometimes be used to increase computational efficiency. These are summarized in the following proposition.

Proposition

> Let $x_1, x_2, \ldots, x_n$ be a sample and c be any nonzero constant.
>
> **1.** If $y_1 = x_1 + c, y_2 = x_2 + c, \ldots, y_n = x_n + c$, then $s_y^2 = s_x^2$, and
> **2.** If $y_1 = cx_1, \ldots, y_n = cx_n$, then $s_y^2 = c^2 s_x^2$, $s_y = |c| s_x$,
>
> where s_x^2 is the sample variance of the x's and s_y^2 is the sample variance of the y's.

In words, (1) says that if a constant c is added to (or subtracted from) each data value, the variance is unchanged. This is intuitive, since adding or subtracting c shifts the location of the data set but leaves distances between data values unchanged. According to (2), multiplication of each x_i by c results in s^2 being multiplied by a factor of c^2. These properties can be proved by noting in (1) that $\bar{y} = \bar{x} + c$ and in (2) that $\bar{y} = c\bar{x}$.

Example 1.14

Recall the spacecraft data (Example 1.2) $x_1 = 81.3001$, $x_2 = 81.3015$, $x_3 = 81.3006$, $x_4 = 81.3011$, $x_5 = 81.2997$, $x_6 = 81.3005$, and $x_7 = 81.3021$. If we subtract the smallest value 81.2997 from all observations, the resulting values are .0004, .0018, .0009, .0014, .0000, .0008, and .0024; the variance of this set is the same as that of the original data and is easier to compute. Now if we multiply each value by 10,000, the variance of the resulting set is $(10,000)^2$ times the original variance. The new data set is 4, 18, 9, 14, 0, 8, and 24 with mean 11.0 and sum of squares 1257, so the variance is $[1257 - (77)^2/7]/6 =$

68.33. The variance of the original data set is therefore $68.33/(10,000)^2 \doteq$ .0000006833. ■

Boxplots

Stem and leaf displays and histograms convey rather general impressions about a data set, whereas a single summary such as the mean or standard deviation focuses on just one aspect of the data. In recent years, a pictorial summary called a *boxplot* has been used successfully to describe several of a data set's most prominent features. These features include (a) center, (b) spread, (c) the extent and nature of any departure from symmetry, and (d) identification of "outliers," observations that lie unusually far from the main body of the data. Because even a single outlier can drastically affect the value of some numerical summaries (such as $\bar{x}$ and s), a boxplot is based on measures that are "resistant" to the presence of a few outliers—the median and a measure of spread called the *fourth spread*.

Definition

> After ordering the n observations in a data set from smallest to largest, the **lower fourth** and **upper fourth** are given by
>
> $$\begin{aligned} \text{lower} \\ \text{fourth} \end{aligned} = \begin{cases} \text{median of the smallest } n/2 \text{ observations} & n \text{ even} \\ \text{median of the smallest } (n+1)/2 \text{ observations} & n \text{ odd} \end{cases}$$
>
> $$\begin{aligned} \text{upper} \\ \text{fourth} \end{aligned} = \begin{cases} \text{median of the largest } n/2 \text{ observations} & n \text{ even} \\ \text{median of the largest } (n+1)/2 \text{ observations} & n \text{ odd} \end{cases}$$
>
> That is, the lower (upper) fourth is the median of the smallest (largest) half of the data, where the median $\tilde{x}$ is included in both halves if n is odd. A measure of spread that is resistant to outliers is the **fourth spread** f_s, given by
>
> $$f_s = \text{upper fourth} - \text{lower fourth}$$

Roughly speaking, the fourth spread is unaffected by the positions of those observations in the smallest 25% or the largest 25% of the data.

A boxplot can now be constructed via the following sequence of steps:

1. Draw and mark a horizontal measurement axis.
2. Construct a rectangle whose left edge lies above the lower fourth and whose right edge lies above the upper fourth.
3. Draw a vertical line segment inside the box above the median.
4. Extend lines from each end of the box out to the furthest observations that are still within $1.5f_s$ of the corresponding edges.
5. Draw an open circle to identify each observation that falls between $1.5f_s$ and $3f_s$ from the edge to which it is closest; these are called **mild outliers.**
6. Draw a solid circle to identify each observation that falls more than $3f_s$ from the closest edge; these are called **extreme outliers.**

Example **1.15**   The accompanying data on oxygen concentration for a sample of silicon wafers was reported in the paper "Determination of Conversion Factor for Infrared Measurement of Oxygen in Silicon" (*Solid State Science and Technology,* 1985, pp. 1707–1713). The sample size is $n = 22$, so the smallest (largest) half of the data consists of the 11 smallest (largest) observations:

smallest: 2.68 3.06 4.31 4.71 5.71 5.99 6.06 7.04 7.17 7.46 7.50
largest: 8.27 8.42 8.73 8.84 9.14 9.19 9.21 9.39 11.28 15.19 21.06

The median, lower, and upper fourths are

$$\tilde{x} = \frac{7.50 + 8.27}{2} = 7.885, \quad \begin{matrix} \text{lower} \\ \text{fourth} \end{matrix} = 5.99, \quad \begin{matrix} \text{upper} \\ \text{fourth} \end{matrix} = 9.19$$

Thus $f_s = 9.19 - 5.99 = 3.20$, $1.5f_s = 4.80$, and $3f_s = 9.60$. Adding 4.80 and 9.60 to the upper fourth gives 13.99 and 18.79, respectively, so a line segment extends from the box's right edge out to 11.28, and 15.19 is a mild outlier and 21.06 an extreme outlier. Because there are no observations more than 4.80 below the lower fourth, the sample contains no outliers on the lower end. The boxplot appears in Figure 1.10. The median is near the middle of the relatively narrow box, giving the impression that the middle 50% of the data is reasonably symmetrically distributed and not too spread out. The upper tail appears much longer than the lower tail because of the two outliers.

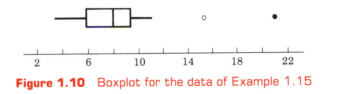

Figure 1.10   Boxplot for the data of Example 1.15   ■

Exercises / Section 1.4 (25–36)

25. The amount of leaf protein (mg/g fresh weight) in soybean plants of a particular variety was determined for a sample of six plants, yielding the following data (*Science,* vol. 199, p. 974):

11.7, 16.1, 14.0, 6.1, 5.1, 4.9

Compute:
a. The sample range.
b. The sample variance s^2 from the definition (that is, by first computing deviations, then squaring them, and so on).
c. The sample standard deviation.
d. s^2 using the shortcut method.

26. A sample of eight resistors of a certain type resulted in the sample resistances (ohms) $x_1 = 40$, $x_2 = 43$, $x_3 = 39$, $x_4 = 35$, $x_5 = 37$, $x_6 = 43$, $x_7 = 46$, $x_8 = 37$.
a. Compute s^2 and s directly from the definitions.
b. Compute s^2 and s using the shortcut formula.
c. Subtract 35 from each x_i and then compute s^2.
d. If the resistances were 400, 430, 390, 350, 370, 430, 460, and 370, how would you use the results of (a), (b), or (c) to compute s^2 and s?

27. The following observations are weight gains (mg/cm³) of Ti-Cr alloy samples due to oxidation when exposed to CO_2 for one hour at 1000°C: 6.4, 5.9, 6.1, 5.8, 6.6, 6.0.

a. Using $\bar{x}$ computed to two decimal places, compute s^2 directly from the definition, and then compute s.

b. Repeat (a) with $\bar{x}$ rounded to three decimal places.

c. Compute s^2 using the shortcut formula.

28. Compute s^2 and s for the bacteria data from Exercise 16.

29. Compute s^2 and s for the temperature data from Exercise 18 by first subtracting 100 from each data value. Is there anything special about the choice of 100, or would other choices work just as well?

30. The accompanying data on bearing load-life (million revs.) for bearings of a certain type when subjected to a 9.56 kN load appeared in the paper "The Load-Life Relationship for M50 Bearings with Silicon Nitride Ceramic Balls" (*Lubric. Engr.*, 1984, pp. 153–159): 14.5, 25.6, 52.4, 66.3, 69.3, 69.8, 76.2.

a. Calculate the values of the sample mean and median. What do their values relative to one another tell you about the sample?

b. Compute the values of the sample variance and standard deviation.

31. Refer back to the fatigue crack propagation data given in Exercise 20.

a. Calculate the values of the sample variance and sample standard deviation.

b. Calculate the value of the fourth spread.

c. Construct a boxplot. Does the data appear to be symmetrically distributed along the measurement axis? Are there any outliers?

32. The paper "A Thin-Film Oxygen Uptake Test for the Evaluation of Automotive Crankcase Lubricants" (*Lubric. Engr.*, 1984, pp. 75–83) reported the following data on oxidation induction time (min) for various commercial oils:

87, 103, 130, 160, 180, 195, 132, 145, 211, 105, 145, 153, 152, 138, 87, 99, 93, 119, 129

a. Calculate the sample variance and standard deviation.

b. If the observations were reexpressed in hours, what would be the resulting values of the sample variance and sample standard deviation?

Answer without actually performing the reexpression.

33. Refer back to the oxidation induction time data given in Exercise 32. Summarize the data by constructing a boxplot and comment on its most prominent features.

34. Cardiac output and maximal oxygen uptake typically decrease with age in sedentary individuals, but these decreases are at least partially arrested in middle-aged individuals who engage in a substantial amount of physical exercise. To understand better the effects of exercise and aging on various circulatory functions, the paper "Cardiac Output in Male Middle-Aged Runners" (*J. Sports Medicine*, 1982, pp. 17–22) presented data from a study of 21 male middle-aged runners. A stem and leaf display of oxygen uptake values (mL/kg · min) while pedaling at 100 watts on a bicycle ergometer is given.

12 ‖	81
13 ‖	
14 ‖	95
15 ‖	97, 83
16 ‖	
17 ‖	90
18 ‖	34, 27
19 ‖	94, 82
20 ‖	99, 93, 98, 62, 88
21 ‖	15
22 ‖	16, 24 Stems: ones
23 ‖	16, 56 Leaves: hundredths
HI:	35.78, 36.73

Construct a boxplot and discuss its characteristics.

35. In Exercise 15, after subtracting 3.70 from each data value, multiply each resulting number by 100 and then compute s^2 for the resulting set of numbers. How can you now obtain s^2 for the original data set?

36. The standard deviation and variance are both measures of variability that depend on the units of measurement, with s expressed in the original units of measurement. The **coefficient of variation**, defined by $c.v. = s/\bar{x}$, is a dimensionless quantity that measures the amount of variability relative to the value of the mean. Compute the value of $c.v.$ for the leaf protein data of Exercise 25 and the weight gain data of Exercise 27. Which data set has more relative variation?

Supplementary Exercises / Chapter 1 (37–45)

37. The **mode** of a numerical data set is the value that occurs most frequently in the set.
  a. What is the mode for the Supreme Court service length data from Exercise 9?
  b. Repeat (a) for the cost of living data from Exercise 6.
  c. For a categorical sample, how would you define the modal category?

38. The task completion time in a learning experiment was recorded both on the first trial (x) and the tenth trial (y) for six individuals:

  individual i: 1   2   3   4   5   6
  x_i: 23.7 36.9 25.5 30.2 28.0 34.8
  y_i: 15.0 25.4 21.0 22.3 25.2 28.8

  a. Compute the decrease $d_i = x_i - y_i$ for each individual, and then compute the sample average decrease.
  b. Compute the sample average time on the first trial, the sample average time on the tenth trial, and then the difference in the two sample averages.
  c. Show in general using rules of summation that the average of the differences $d_i = x_i - y_i$ equals the difference $\bar{x} - \bar{y}$ of the averages.

39. Fifteen air samples from a certain region were obtained and for each one the carbon monoxide concentration was determined. The results were (ppm)

  9.3, 10.7, 8.5, 9.6, 12.2, 15.6, 9.2, 10.5, 9.0, 13.2, 11.0, 8.8, 13.7, 12.1, 9.8

  Using the interpolation method suggested in Section 1.3, compute the 10% trimmed mean.

40. a. For what value of c is the quantity $\Sigma (x_i - c)^2$ minimized? *Hint:* Take the derivative with respect to c, set equal to 0, and solve.
  b. Using the result of (a), which of the two quantities $\Sigma (x_i - \bar{x})^2$ and $\Sigma (x_i - \mu)^2$ will be smaller than the other (assuming that $\bar{x} \neq \mu$)?

41. a. Let a and b be constants and let $y_i = ax_i + b$ for $i = 1, 2, \ldots, n$. What is the relationship between $\bar{x}$ and $\bar{y}$ and between s_x^2 and s_y^2?
  b. A sample of temperatures for initiating a certain chemical reaction yielded a sample average (°C) of 87.3 and a sample standard devi-

ation of 1.04. What are the sample average and standard deviation measured in °F? *Hint:* $F = \frac{9}{5}C + 32$.

42. Reconsider the data presented in Example 1.10. Construct:
  a. A stem and leaf display in which the two largest values are displayed separately in a row labeled HI.
  b. A histogram based on six class intervals with zero as the lower limit of the first interval and interval lengths of 2, 2, 2, 4, 10, and 10, respectively.

43. Consider a sample $x_1, x_2, \ldots, x_n$, and suppose that the values of $\bar{x}$, s^2, and s have been calculated.
  a. Let $y_i = x_i - \bar{x}$ for $i = 1, \ldots, n$. How do the values of s^2 and s for the y_i's compare to the corresponding values for the x_i's? Explain.
  b. Let $z_i = (x_i - \bar{x})/s$ for $i = 1, \ldots, n$. What are the values of the sample variance and sample standard deviation for the z_i's?

44. Let $\bar{x}_n$ and s_n^2 denote the sample mean and variance for the sample $x_1, \ldots, x_n$, and let $\bar{x}_{n+1}$ and s_{n+1}^2 denote these quantities when an additional observation x_{n+1} is added to the sample.
  a. Show how $\bar{x}_{n+1}$ can be computed from $\bar{x}_n$ and x_{n+1}.
  b. Show that

  $$ns_{n+1}^2 = (n-1)s_n^2 + \frac{n}{n+1}(x_{n+1} - \bar{x}_n)^2$$

  so that s_{n+1}^2 can be computed from x_{n+1}, $\bar{x}_n$, and s_n^2.
  c. Use the results of (a) and (b) to compute $\bar{x}$ and s^2 recursively (first $\bar{x}_2$ and s_2^2, then $\bar{x}_3$ and s_3^2, and so on) for the data of Exercise 25. *Note:* These relationships are used by some computer packages to increase computational efficiency and accuracy.

45. Consider numerical observations $x_1, \ldots, x_n$. It is frequently of interest to know whether the x_i's are (at least approximately) symmetrically distributed about some value. If n is at least moderately large, the extent of symmetry can be assessed from a stem and leaf display or histogram. However, if n is not very large, such pictures are not particularly informative. Consider the following alternative. Let y_1 denote the smallest x_i, y_2 the second smallest x_i, and so on. Then plot the following pairs as points on a two-dimensional coordinate system: $(y_n - \tilde{x}, \tilde{x} - y_1)$, $(y_{n-1} - \tilde{x}, \tilde{x} - y_2)$, $(y_{n-2} - \tilde{x}, \tilde{x} - y_3)$, ... There are $n/2$ points when n is even and $(n-1)/2$ when n is odd.

a. What does this plot look like when there is perfect symmetry in the data? What does it look like when observations stretch out more above the median than below it (a long upper tail)?

b. The accompanying data on rainfall (acre-feet) from 26 seeded clouds is taken from the paper "A Bayesian Analysis of a Multiplicative Treatment Effect in Weather Modification" (*Technometrics,* 1975, pp. 161–166). Construct the plot and comment on the extent of symmetry or nature of departure from symmetry.

4.1, 7.7, 17.5, 31.4, 32.7, 40.6, 92.4, 115.3, 118.3, 119.0, 129.6, 198.6, 200.7, 242.5, 255.0, 274.7, 274.7, 302.8, 334.1, 430.0, 489.1, 703.4, 978.0, 1656.0, 1697.8, 2745.6

Bibliography

Chambers, John, Cleveland, William, Kleiner, Beat, and Tukey, Paul, *Graphical Methods for Data Analysis,* Wadsworth, Belmont, CA (1983). A highly recommended presentation of both older and more recent graphical and pictorial methodology in statistics.

Devore, Jay, and Peck, Roxy, *Statistics: The Exploration and Analysis of Data,* West, St. Paul, MN (1986). The first few chapters give a very nonmathematical survey of methods for describing and summarizing data.

Freedman, David, Pisani, Robert, and Purves, Roger, *Statistics,* Norton, New York (1978). An excellent very nonmathematical survey of basic statistical reasoning and methodology.

Hoaglin, David, Mosteller, Frederick, and Tukey, John, *Understanding Robust and Exploratory Data Analysis,* Wiley, New York (1983). Discusses why, as well as how, exploratory methods should be employed; it is good on details of stem and leaf displays and boxplots.

Hoaglin, David, and Velleman, Paul, *Applications, Basics, and Computing of Exploratory Data Analysis,* Duxbury, Boston (1980). A good discussion of some basic exploratory methods; it's easier to read than the book by Tukey, though less comprehensive.

Moore, David, *Statistics: Concepts and Controversies* (2nd ed.), W. H. Freeman, San Francisco (1985). An extremely readable and entertaining paperback that contains an intuitive discussion of problems connected with sampling and designed experiments.

Tanur, Judith (ed.), *Statistics: A Guide to the Unknown* (2nd ed.), Wadsworth & Brooks/Cole Advanced Books and Software, Monterey, CA (1978). Contains many short nontechnical articles describing various applications of statistics.

Tukey, John, *Exploratory Data Analysis,* Addison-Wesley, Reading, MA (1977). Introduces and illustrates a broad range of newly developed descriptive and "data-snooping" methods.

Probability

Introduction

The term **probability** refers to the study of randomness and uncertainty. In any situation in which one of a number of possible outcomes may occur, the theory of probability provides methods for quantifying the chances or likelihoods associated with the various outcomes. The language of probability is constantly used in an informal manner in both written and spoken contexts. Examples include such statements as "It is quite likely that the Dow-Jones average will increase by the end of the year," "There is a 50-50 chance that the incumbent will seek reelection," "There will probably be at least one section of that course offered next year," "The odds favor a quick settlement of the strike," and "It is expected that at least 20,000 concert tickets will be sold." In this chapter we introduce some elementary probability concepts, indicate how probabilities can be interpreted, and show how the rules of probability can be applied to compute the probabilities of many interesting events. The methodology of probability will then permit us to express in precise language such informal statements as those given above.

The study of probability as a branch of mathematics goes back over 300 years, where it had its genesis in connection with questions involving games of chance. Many books are devoted exclusively to probability, but our objective here is to cover only that part of the subject that has the most direct bearing on problems of statistical inference.

2.1 Sample Spaces and Events

An **experiment** is any action or process that generates observations. Although the word "experiment" generally suggests a planned or carefully controlled laboratory-testing situation, we use it here in a much wider sense. Thus, experiments that may be of interest include tossing a coin once or several times, selecting a card or cards from a deck, weighing a loaf of bread, ascertaining the commuting time from home to work on a particular morning, obtaining blood types from a group of individuals, or measuring the compressive strengths of different steel beams.

The Sample Space of an Experiment

Definition

> The **sample space** of an experiment, denoted by $\mathcal{S}$, is the set of all possible outcomes of that experiment.

Example **2.1**

The simplest experiment to which probability applies is one with two possible outcomes. One such experiment consists of examining a single fuse to see whether or not it is defective. The sample space for this experiment can be abbreviated as $\mathcal{S} = \{N, D\}$, where N represents not defective, D represents defective, and the braces are used to enclose the elements of a set. Another such experiment would involve tossing a thumbtack and noting whether it landed point up or point down, with sample space $\mathcal{S} = \{U, D\}$, and yet another would consist of observing the sex of the next child born at the local hospital, with $\mathcal{S} = \{B, G\}$. ■

Example **2.2**

If we examine three fuses in sequence and note the result of each examination, then an outcome for the entire experiment is any sequence of N's and D's of length 3, so

$$\mathcal{S} = \{NNN, NND, NDN, NDD, DNN, DND, DDN, DDD\}$$

If we had tossed a thumbtack three times, the sample space would be obtained by replacing N by U in $\mathcal{S}$ above, with a similar notational change yielding the sample space for the experiment in which the sexes of three newborn children are observed. ■

Example **2.3**

If a six-sided die is tossed once and the outcome is the number on the upturned face, then $\mathcal{S} = \{1, 2, 3, 4, 5, 6\}$. If both a red die and a green die are tossed together, then a particular outcome specifies both the number on the red die and the number on the green die. An outcome consists of a *pair* of numbers, such as (1, 1) or (3, 2), and the sample space consists of the 36 outcomes listed in the accompanying table.

		\multicolumn{6}{c}{Green die}					
		1	2	3	4	5	6
	1	(1, 1)	(1, 2)	(1, 3)	(1, 4)	(1, 5)	(1, 6)
	2	(2, 1)	(2, 2)	(2, 3)	(2, 4)	(2, 5)	(2, 6)
	3	(3, 1)	(3, 2)	(3, 3)	(3, 4)	(3, 5)	(3, 6)
Red die	4	(4, 1)	(4, 2)	(4, 3)	(4, 4)	(4, 5)	(4, 6)
	5	(5, 1)	(5, 2)	(5, 3)	(5, 4)	(5, 5)	(5, 6)
	6	(6, 1)	(6, 2)	(6, 3)	(6, 4)	(6, 5)	(6, 6)

Note that the outcome (2, 1) is different from (1, 2) and that both are included in the sample space; a similar comment applies to any other pair for which the first toss and the second toss yield different numbers. ■

Example 2.4 If a new type-D flashlight battery has a voltage that is outside certain limits, that battery is characterized as a failure (F); if the battery has a voltage within the prescribed limits, it is a success (S). Suppose that an experiment consists of testing each battery as it comes off an assembly line until we first observe a success. Although it may not be very likely, a possible outcome of this experiment is that the first 10 (or 100 or 1000 or ...) are F's and the next one is an S. That is, for any positive integer n, we may have to examine n batteries before seeing the first S. The sample space is $\mathcal{S} = \{S, FS, FFS, FFFS, \ldots\}$, which contains an infinite number of possible outcomes. The same abbreviated form of the sample space is appropriate for an experiment in which, starting at a specified time, the sex of each newborn infant is recorded until the birth of a female is observed. ■

Events

In our study of probability, we will be interested not only in the individual outcomes of $\mathcal{S}$ but also in any collection of outcomes from $\mathcal{S}$.

Definition

> An **event** is any collection (subset) of outcomes contained in the sample space $\mathcal{S}$. An event is said to be **simple** if it consists of exactly one outcome and **compound** if it consists of more than one outcome.

When an experiment is performed, a particular event A is said to occur if the resulting experimental outcome is contained in A. In general, exactly one simple event will occur, but many compound events will occur simultaneously.

Example 2.5 Consider an experiment in which each of three automobiles taking a particular freeway exit turns left (L) or right (R) at the end of the exit ramp. The eight possible outcomes that comprise the sample space are LLL, RLL, LRL, LLR, LRR, RLR, RRL, and RRR. Thus, there are eight simple events, among which are $E_1 = \{LLL\}$ and $E_5 = \{LRR\}$. Some compound events include

$A = \{RLL, LRL, LLR\} =$ the event that exactly one of the three cars turns right,

$B = \{LLL, RLL, LRL, LLR\} =$ the event that at most one of the cars turns right, and

$C = \{LLL, RRR\} =$ the event that all three cars turn in the same direction.

Suppose that when the experiment is performed, the outcome is LLL. Then the simple event E_1 has occurred, and so also have the events B and C (but not A). ■

Example **2.6**
(Example 2.3
continued)

When two dice are tossed, there are 36 possible outcomes, so there are 36 simple events: $E_1 = \{(1, 1)\}$, $E_2 = \{(1, 2)\}$, ..., $E_{36} = \{(6, 6)\}$. Examples of compound events are

$A = \{(1, 1), (2, 2), (3, 3), (4, 4), (5, 5), (6, 6)\} =$ the event that both tosses show the same face,

$B = \{(1, 3), (2, 2), (3, 1)\} =$ the event that the sum of the numbers resulting from the two tosses is four, and

$C = \{(6, 6), (6, 5), (5, 6)\} =$ the event that the sum of the two numbers is at least 11. ∎

Example **2.7**
(Example 2.4
continued)

The sample space for the battery examination experiment contains an infinite number of outcomes, so there are an infinite number of simple events. Compound events include

$A = \{S, FS, FFS\} =$ the event that at most three batteries are examined, and

$E = \{FS, FFFS, FFFFFS, \ldots\} =$ the event that an even number of batteries are examined. ∎

Some Relations from Set Theory

An event is nothing but a set, so that relationships and results from elementary set theory can be used to study events. The following concepts from set theory will be used to construct new events from given events:

1. The **union** of two events A and B, denoted by $A \cup B$ and read "*A or B*," is the event consisting of all outcomes that are *either in A or in B or in both events* (so that the union includes outcomes for which both A and B occur as well as outcomes for which exactly one occurs).
2. The **intersection** of two events A and B, denoted by $A \cap B$ and read "*A and B*," is the event consisting of all outcomes that are in *both A and B*.
3. The **complement** of an event A, denoted by A', is the set of all outcomes in $\mathscr{S}$ that are not contained in A.

Example **2.8**
(Example 2.3
continued)

For the experiment consisting of a single die toss, let

$A = \{1, 2, 3\} = \{$outcome is $\leq 3\}$
$B = \{1, 2, 5, 6\}$
$C = \{1, 3, 5\} = \{$outcome is odd$\}$

Then

$A \cup B = \{1, 2, 3, 5, 6\}$, $A \cup C = \{1, 2, 3, 5\}$, $A \cap B = \{1, 2\}$
$A \cap C = \{1, 3\}$, $A' = \{4, 5, 6\}$, and $C' = \{2, 4, 6\}$ ∎

Example 2.9
(Example 2.4
continued)

In the battery experiment, define A, B, and C by

$$A = \{S, FS, FFS\}$$
$$B = \{S, FFS, FFFFS\}$$

and

$$C = \{FS, FFFS, FFFFFS, \ldots\}$$

Then

$$A \cup B = \{S, FS, FFS, FFFFS\}$$
$$A \cap B = \{S, FFS\}$$
$$A' = \{FFFS, FFFFS, FFFFFS, \ldots\}$$

and

$$C' = \{S, FFFS, FFFFS, \ldots\} = \{\text{an odd number of batteries are examined}\} \quad \blacksquare$$

Sometimes A and B have no outcomes in common, so that the intersection of A and B contains no outcomes.

Definition

> When A and B have no outcomes in common, they are said to be **mutually exclusive** or **disjoint** events.

Example 2.10

A small city has three automobile dealerships: a GM dealer selling Chevrolets, Pontiacs, and Buicks; a Ford dealer selling Fords and Mercurys; and a Chrysler dealer selling Plymouths and Chryslers. If an experiment consists of observing the brand of the next car sold, then the events $A = \{\text{Chevrolet, Pontiac, Buick}\}$ and $B = \{\text{Ford, Mercury}\}$ are mutually exclusive, since the next car sold cannot be both a GM product and a Ford product. $\quad \blacksquare$

The operations of union and intersection can be extended to more than two events. For any three events A, B, and C, the event $A \cup B \cup C$ is the set of outcomes contained in at least one of the three events, while $A \cap B \cap C$ is the set of outcomes contained in all three events. Given events $A_1, A_2, A_3, \ldots$, these events are said to be mutually exclusive (or pairwise disjoint) if no two events have any outcomes in common.

A pictorial representation of events and manipulations with events is obtained by using Venn diagrams. To construct a Venn diagram, draw a rectangle whose interior will represent the sample space $\mathcal{S}$. Then any event A is represented as the interior of a closed curve (often a circle) contained in $\mathcal{S}$. Examples of Venn diagrams appear in Figure 2.1.

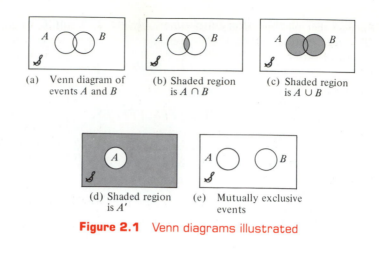

(a) Venn diagram of
     events A and B

(b) Shaded region
     is $A \cap B$

(c) Shaded region
     is $A \cup B$

(d) Shaded region
     is A'

(e) Mutually exclusive
     events

Figure 2.1   Venn diagrams illustrated

Exercises / Section 2.1 (1–10)

1. A family that owns two cars is selected, and for both the older car and the newer car we note whether the car was manufactured in America, Europe, or Asia.

a. What are the possible outcomes of this experiment?

b. What outcomes are contained in the event that one car is American and the other is foreign?

c. What outcomes are contained in the event that at least one of the two cars is foreign? What is the complement of this event? Is either of these two events a simple event?

2. Each of a sample of four life insurance companies is classified as a mutual (M) or nonmutual (N) company.

a. What are the 16 outcomes in $\mathscr{S}$?

b. What outcomes are in the event that exactly three of the companies are of type M?

c. What outcomes are in the event that all four companies are of the same type?

d. What outcomes are in the event that at most one of the four companies is of type M?

e. What is the union of the events in (c) and (d), and what is the intersection of these two events?

f. What are the union and intersection of the two events in (b) and (c)?

3. A family consisting of three persons—A, B, and C—belongs to a medical clinic which always has

a doctor at each of stations 1, 2, and 3. During a certain week each member of the family visits the clinic once and is assigned at random to a station. The experiment consists of recording the station number for each member. One outcome is $(1, 2, 1)$ for A to station 1, B to station 2, and C to station 1.

a. List the 27 outcomes in the sample space.

b. List all outcomes in the event that all three members go to the same station.

c. List all outcomes in the event that all members go to different stations.

d. List all outcomes in the event that no one goes to station 2.

4. A small town has both an Arco gas station and a Conoco gas station. Each station sells its own brand of motor oil and also Pennzoil and Quaker State. An experiment consists of recording both the station and brand of oil chosen by someone who has stopped to purchase a single can of oil.

a. List in abbreviated fashion the sample space of this experiment.

b. Is the event $P = \{$the chosen brand is Pennzoil$\}$ a simple event? Why or why not?

c. Is the event $A = \{$the chosen brand is Arco$\}$ simple? Why or why not?

d. List the outcomes in the events $A \cup P$, A', and P'.

e. Are the events A and P mutually exclusive? Why or why not?

f. Suppose that the type of gasoline purchased (R = regular, U = unleaded, E = premium) is also recorded. What is the new sample space?

5. To demonstrate my skill at clairvoyance, I shuffle a deck consisting of three cards marked 1, 2, and 3 and place the deck face down on a table. I then guess at the number on the first card and turn it over. If my guess is correct, the experiment terminates. If my first guess is incorrect, I guess again and then turn over the second card. At this point I know what the third card is, so at most the first two guesses and cards need to be reported. An outcome is described by a sequence of either two or four numbers in which a guess and card alternate.
a. List all outcomes in the event that the experiment terminates after just one guess.
b. List all outcomes in the event that the two guesses are both incorrect (I will only guess a number that is possible).

6. Jack Twain has just visited a friend at location A below. He wishes to get to location I. He will travel only along indicated straight-line paths and will only go either north (up) or east (to the right). At each intersection at which there is a choice, he will toss a coin to decide in which of the two possible directions he should go. An outcome for Jack's experiment consists of specifying the corners visited on his trip.
a. List all outcomes in the sample space.
b. List all outcomes for which three coin tosses are necessary.

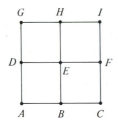

7. An academic department has just completed voting by secret ballot for a department head. The ballot box contains four slips with votes for candidate A and three slips with votes for candidate B. Suppose that these slips are removed from the box one by one.
a. List all possible outcomes.
b. Suppose that a running tally is kept as slips are removed. For what outcomes does A remain ahead of B throughout the tally?

8. An engineering construction firm is currently working on power plants at three different sites. Let A_i denote the event that the plant at site i is completed by the contract date. Use the operations of union, intersection, and complementation to describe each of the following events in terms of A_1, A_2, and A_3, draw a Venn diagram, and shade the region corresponding to each one.
a. At least one plant is completed by the contract date.
b. All plants are completed by the contract date.
c. Only the plant at site 1 is completed by the contract date.
d. Exactly one plant is completed by the contract date.
e. Either the plant at site 1 or both of the other two plants are completed by the contract date.

9. Use Venn diagrams to verify the following two relationships for any events A and B (these are called De Morgan's laws):
a. $(A \cup B)' = A' \cap B'$
b. $(A \cap B)' = A' \cup B'$

10. a. In Example 2.10, identify three events that are mutually exclusive.
b. Suppose that there is no outcome common to all three of the events A, B, and C. Are these three events necessarily mutually exclusive? If your answer is yes, explain why, while if your answer is no, give a counterexample using the experiment of Example 2.10.

2.2   Axioms, Interpretations, and Properties of Probability

Given an experiment and a sample space $\mathcal{S}$, *the objective of probability is to assign to each event A a number $P(A)$, called the probability of the event A, which will give a precise measure of the chance that A will occur.* To ensure that the probability assignments will be consistent with our intuitive notions of probability, all assignments should satisfy the following axioms (basic properties) of probability.

Axiom 1

For any event A, $P(A) \geq 0$.

Axiom 2

$P(\mathcal{S}) = 1$.

Axiom 3

a. If $A_1, A_2, \ldots, A_n$ is a finite collection of mutually exclusive events, then

$$P(A_1 \cup A_2 \cup \ldots \cup A_n) = \sum_{i=1}^{n} P(A_i)$$

b. If $A_1, A_2, A_3, \ldots$ is an infinite collection of mutually exclusive events, then

$$P(A_1 \cup A_2 \cup A_3 \cup \ldots) = \sum_{i=1}^{\infty} P(A_i)$$

Axiom 1 reflects the intuitive notion that the chance of A occurring should be at least 0, so that negative probabilities are not allowed. The sample space is by definition an event that must occur when the experiment is performed ($\mathcal{S}$ contains all possible outcomes), so Axiom 2 says that the maximum possible probability of one is assigned to $\mathcal{S}$. The third axiom formalizes the idea that if we wish the probability that at least one of a number of events will occur and no two of the events can occur simultaneously, then the chance of at least one occurring is the sum of the chances of the individual events.

Example 2.11   In the experiment in which a single coin is tossed, the sample space is $\mathcal{S} = \{H, T\}$. The axioms specify $P(\mathcal{S}) = 1$, so to complete the probability assignment, it remains only to determine $P(H)$ and $P(T)$. Since H and T are disjoint events and $H \cup T = \mathcal{S}$, Axiom 3 implies that $1 = P(\mathcal{S}) = P(H) + P(T)$. So $P(T) = 1 - P(H)$. Thus the only freedom allowed by the axioms in this experiment is the probability assigned to H. One possible assignment of probabilities is $P(H) = .5$, $P(T) = .5$, while another possible assignment is $P(H) = .75$, $P(T) = .25$. In fact, letting p represent any fixed number between 0 and 1, $P(H) = p$ and $P(T) = 1 - p$ is an assignment consistent with the axioms. ∎

Example 2.12   Consider the experiment in Example 2.4, in which batteries coming off an assembly line are tested one by one until one having a voltage within prescribed limits is found. The simple events are $E_1 = \{S\}$, $E_2 = \{FS\}$, $E_3 = \{FFS\}$, $E_4 = \{FFFS\}, \ldots$. Suppose that the probability of any particular battery being satisfactory is .99. Then it can be shown that $P(E_1) = .99$, $P(E_2) = (.01)(.99)$, $P(E_3) = (.01)^2(.99), \ldots$ is an assignment of probabilities to the simple events that satisfies the axioms. In particular, because the E_i's are disjoint and $\mathscr{S} = E_1 \cup E_2 \cup E_3 \cup \ldots$, we must have $1 = P(\mathscr{S}) = P(E_1) + P(E_2) + P(E_3) + \cdots = .99[1 + .01 + (.01)^2 + (.01)^3 + \cdots]$. The validity of this equality is a consequence of a mathematical result concerning the sum of a geometric series.

However, another legitimate (according to the axioms) probability assignment of the same "geometric" type is obtained by replacing .99 by any other number p between 0 and 1 (and .01 by $1 - p$). Thus there are an infinite number of legitimate probability assignments of this type.   ■

Interpreting Probability

Examples 2.11 and 2.12 show that the axioms do not completely determine an assignment of probabilities to events. The axioms only serve to rule out assignments inconsistent with our intuitive notions of probability. In the coin-tossing experiment of Example 2.11, two particular assignments were suggested. The appropriate or correct assignment depends on the manner in which the experiment is carried out and also on one's interpretation of probability. The interpretation that is most frequently used and most easily understood is based on the notion of relative frequencies.

Consider an experiment that can be repeatedly performed in an identical and independent fashion, and let A be an event consisting of a fixed set of outcomes of the experiment. Simple examples of such repeatable experiments include the coin-tossing and die-tossing experiments previously discussed. If the experiment is performed n times, on some of the replications the event A will occur (the outcome will be in the set A), and on others, A will not occur. Let $n(A)$ denote the number of replications on which A does occur. Then the ratio $n(A)/n$ is called the *relative frequency* of occurrence of the event A in the sequence of n replications. Empirical evidence, based on the results of many of these sequences of repeatable experiments, indicates that as n grows large, the relative frequency $n(A)/n$ stabilizes, as pictured in Figure 2.2. That is, as n gets

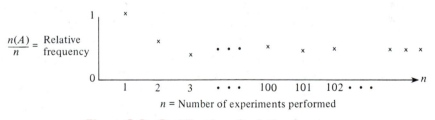

Figure 2.2   Stabilization of relative frequency

arbitrarily large, the relative frequency approaches a limiting value we refer to as the *limiting relative frequency* of the event A. The objective interpretation of probability identifies this limiting relative frequency with $P(A)$.

If probabilities are assigned to events in accordance with their limiting relative frequencies, then we can interpret a statement such as "The probability of that coin landing with the head facing up when it is tossed is .5" to mean that in a large number of such tosses, a head will appear on approximately half the tosses and a tail on the other half.

This relative frequency interpretation of probability is referred to as an objective interpretation because it rests on a property of the experiment rather than any particular individual concerned with the experiment. For example, two different observers of a sequence of coin tosses should both use the same probability assignments since the observers have nothing to do with limiting relative frequency. In practice, this interpretation is not as objective as it might seem, since the limiting relative frequency of an event will not be known. Thus, we will have to assign probabilities based on our beliefs about the limiting relative frequency of events under study. Fortunately, there are many experiments for which there will be a consensus with respect to probability assignments. When we speak of a fair coin, we shall mean $P(H) = P(T) = .5$, and a fair die is one for which limiting relative frequencies of the six outcomes are all $\frac{1}{6}$, suggesting probability assignments $P(\{1\}) = \cdots = P(\{6\}) = \frac{1}{6}$.

Because the objective interpretation of probability is based on the notion of limiting frequency, its applicability is limited to experimental situations that are repeatable. Yet the language of probability is often used in connection with situations that are inherently unrepeatable. Examples include: "The chances are good for a peace agreement"; "It is quite likely that our company will be awarded the contract"; and "Because their best quarterback is injured, I expect them to score no more than 10 points against us." In such situations we would like, as before, to assign numerical probabilities to various outcomes and events (for example, the probability is .9 that we will get the contract). We must therefore adopt an alternative interpretation of these probabilities. Because different observers may have different prior information and opinions concerning such experimental situations, probability assignments may now differ from individual to individual. Interpretations in such situations are thus referred to as subjective. The book by Winkler listed in the chapter references gives a very readable survey of several subjective interpretations.

Properties of Probability

Proposition

> For any event A, $P(A) = 1 - P(A')$.

Proof.   In Axiom 3a, let $n = 2$, $A_1 = A$, and $A_2 = A'$. Since by definition of A', $A \cup A' = \mathcal{S}$ while A and A' are disjoint, $1 = P(\mathcal{S}) = P(A \cup A') = P(A) + P(A')$, from which the desired result follows.

This proposition is surprisingly useful, since there are many situations in which $P(A')$ is more easily obtained by direct methods than is $P(A)$.

Example 2.13 Of all Americans, 85% have an agglutinating factor in their blood which classifies them as Rh positive, while 15% lack the factor, so are classified as Rh negative. Suppose that a doctor wishes to do an analysis of blood from a newborn Rh negative infant, so he examines blood-typing results from a sequence of newborn infants until he finds an Rh negative infant. What is the probability that at least two infants must be typed before the search is terminated?

Let $A = \{$at least two infants must be typed$\}$, the event whose probability is to be determined. One approach involves noting that $A = A_2 \cup A_3 \cup A_4 \cup \ldots$, where $A_i = \{$exactly i typings are necessary$\}$. Because the A_i's are disjoint, $P(A) = \sum_{i=2}^{\infty} P(A_i)$. However, it is more straightforward here to work with the complement of A. Since $A' = \{$exactly one typing is necessary$\} = \{$the first infant selected is Rh positive$\}$, the given information implies that $P(A') = .15$ and thus that $P(A) = 1 - P(A') = 1 - .15 = .85$. ■

In general, the foregoing proposition is useful when the event of interest can be expressed as "at least . . .", since then the complement "less than . . ." is often easier to work with (in some problems, "more than . . ." is easier to deal with than "at most . . .").

Proposition

> If A and B are mutually exclusive, then $P(A \cap B) = 0$.

Proof. Because $A \cap B$ contains no outcomes, $(A \cap B)' = \mathscr{S}$. Thus $1 = P[(A \cap B)'] = 1 - P(A \cap B)$. ■

When events A and B are mutually exclusive, Axiom 3 gives $P(A \cup B) = P(A) + P(B)$. When A and B are not mutually exclusive, the probability of the union is obtained from the following result.

Proposition

> For any two events A and B,
>
> $$P(A \cup B) = P(A) + P(B) - P(A \cap B)$$

Notice that the proposition is valid even if A and B are mutually exclusive, since then $P(A \cap B) = 0$. The key idea is that, in adding $P(A)$ and $P(B)$, the intersection $A \cap B$ is actually counted twice, so $P(A \cap B)$ must be subtracted out.

Proof. Note first that $A \cup B = A \cup (B \cap A')$, as illustrated in Figure 2.3.

Since A and $(B \cap A')$ are mutually exclusive, $P(A \cup B) = P(A) + P(B \cap A')$. But $B = (B \cap A) \cup (B \cap A')$ (the union of that part of B in A

and that part of B not in A), with $(B \cap A)$ and $(B \cap A')$ mutually exclusive, so that $P(B) = P(B \cap A) + P(B \cap A')$. This all gives

$$P(A \cup B) = P(A) + P(B \cap A') = P(A) + [P(B) - P(A \cap B)]$$
$$= P(A) + P(B) - P(A \cap B) \quad \blacksquare$$

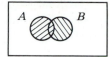

Figure 2.3 $A \cup B = A \cup (B \cap A')$

Example **2.14**   In a certain residential suburb, 60% of all households subscribe to the metro-politan newspaper published in a nearby city, 80% subscribe to the local after-noon paper, and 50% of all households subscribe to both papers. If a household is selected at random, what is the probability that it subscribes to (a) at least one of the two newspapers and (b) exactly one of the two newspapers?

With $A = \{$subscribes to the metropolitan paper$\}$ and $B = \{$subscribes to the local paper$\}$, the given information implies that $P(A) = .6$, $P(B) = .8$, and $P(A \cap B) = .5$. The above proposition then applies to give

P(subscribes to at least one of the two newspapers) $=$
$$P(A \cup B) = P(A) + P(B) - P(A \cap B) = .6 + .8 - .5 = .9$$

Because $\{$exactly one$\}$ and $\{$both$\}$ are mutually exclusive with $\{$exactly one$\}$ $\cup$ $\{$both$\} = \{$at least one$\}$

$$P(\text{exactly one}) + P(\text{both}) = P(\text{at least one}) = .9$$

giving

$$P(\text{exactly one}) = .9 - P(\text{both}) = .9 - .5 = .4. \quad \blacksquare$$

The probability of a union of more than two events can be computed analogously. For three events A, B, and C, the result is

$$P(A \cup B \cup C) = P(A) + P(B) + P(C) - P(A \cap B) - P(A \cap C)$$
$$- P(B \cap C) + P(A \cap B \cap C)$$

This can be seen by examining a Venn diagram of $A \cup B \cup C$, which is shown in Figure 2.4. When $P(A)$, $P(B)$, and $P(C)$ are added, certain intersections are counted twice, so must be subtracted out, but this results in $P(A \cap B \cap C)$ being subtracted once too often.

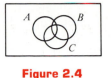

Figure 2.4

Determining Probabilities Systematically

When the number of possible outcomes (simple events) is large, there will be many compound events. A simple way to determine probabilities for these events that avoids violating the axioms and derived properties is to first determine probabilities $P(E_i)$ for all simple events. These should satisfy $P(E_i) \geq 0$ and $\sum_{\text{all } i} P(E_i) = 1$. Then the probability of any compound event A is computed by adding together the $P(E_i)$'s for all E_i's in A:

$$P(A) = \sum_{\text{all } E_i\text{'s in } A} P(E_i)$$

Example 2.15 Denote the six elementary events $\{1\}, \ldots, \{6\}$ associated with tossing a six-sided die once by $E_1, \ldots, E_6$. If the die is constructed so that any of the three even outcomes is twice as likely to occur as any of the three odd outcomes, then an appropriate assignment of probabilities to elementary events is $P(E_1) = P(E_3) = P(E_5) = \frac{1}{9}$, $P(E_2) = P(E_4) = P(E_6) = \frac{2}{9}$. Then for the event $A = \{\text{outcome is even}\} = E_2 \cup E_4 \cup E_6$, $P(A) = P(E_2) + P(E_4) + P(E_6) = \frac{6}{9} = \frac{2}{3}$; for $B = \{\text{outcome} \leq 3\} = E_1 \cup E_2 \cup E_3$, $P(B) = \frac{1}{9} + \frac{2}{9} + \frac{1}{9} = \frac{4}{9}$. ∎

Equally Likely Outcomes

In many experiments consisting of N outcomes, it is reasonable to assign equal probabilities to all N simple events. These include such obvious examples as tossing a fair coin or fair die once or twice (or any fixed number of times), or selecting one or several cards from a well-shuffled deck of 52. With $p = P(E_i)$ for every i,

$$1 = \sum_{i=1}^{N} P(E_i) = \sum_{i=1}^{N} p = p \cdot N, \quad \text{so } p = \frac{1}{N}$$

That is, if there are N possible outcomes, then the probability assigned to each is $1/N$.

Now consider an event A, with $N(A)$ denoting the number of outcomes contained in A. Then

$$P(A) = \sum_{E_i \text{ in } A} P(E_i) = \sum_{E_i \text{ in } A} \frac{1}{N} = \frac{N(A)}{N}$$

Once we have counted the number of outcomes N in the sample space, to compute the probability of any event we must count the number of outcomes contained in that event and take the ratio of the two numbers. Thus when outcomes are equally likely, computing probabilities reduces to counting.

Example **2.16**   When two dice are rolled separately as in Example 2.3, there are 36 possible outcomes ($N = 36$). If the dice are both fair, all 36 outcomes are equally likely, so $P(E_i) = \frac{1}{36}$. Then the event $A = \{\text{sum of two numbers} = 7\}$ consists of the six outcomes $(1, 6), (2, 5), (3, 4), (4, 3), (5, 2),$ and $(6, 1)$, so

$$P(A) = \frac{N(A)}{N} = \frac{6}{36} = \frac{1}{6}$$

■

Exercises / Section 2.2 [11–24]

11. A family that owns two automobiles is selected at random. Let $A_1 = \{$the older car is American$\}$ and $A_2 = \{$the newer car is American$\}$. If $P(A_1) = .7$, $P(A_2) = .5$, and $P(A_1 \cap A_2) = .4$, compute
  a. $P(A_1 \cup A_2)$ (the probability that at least one car is American)
  b. the probability that neither car is American
  c. the probability that exactly one of the two cars is American

12. A computer consulting firm presently has bids out on three projects. Let $A_i = \{$awarded project $i\}$ for $i = 1, 2, 3$, and suppose that $P(A_1) = .22$, $P(A_2) = .25$, $P(A_3) = .28$, $P(A_1 \cap A_2) = .11$, $P(A_1 \cap A_3) = .05$, $P(A_2 \cap A_3) = .07$, $P(A_1 \cap A_2 \cap A_3) = .01$. Express in words each of the following events and compute the probability of each event.
  a. $A_1 \cup A_2$
  b. $A_1' \cap A_2'$ *Hint:* $(A_1 \cup A_2)' = A_1' \cap A_2'$
  c. $A_1 \cup A_2 \cup A_3$ **d.** $A_1' \cap A_2' \cap A_3'$
  e. $A_1' \cap A_2' \cap A_3$ **f.** $(A_1' \cap A_2') \cup A_3$

13. A utility company offers a lifeline rate to any household whose electricity usage falls below 240 kWh during a particular month. Let A denote the event that a randomly selected household in a certain community does not exceed the lifeline usage during January, and let B be the analogous event for the month of July (A and B refer to the same household). Suppose that $P(A) = .8$, $P(B) = .7$, and $P(A \cup B) = .9$. Compute
  a. $P(A \cap B)$.

  b. the probability that the lifeline usage amount is exceeded in exactly one of the two months. Describe this event in terms of A and B.

14. In the experiment in which two fair dice (one red and one green) are rolled, let $S_2 = \{$the sum of the two resulting numbers is divisible by 2$\}$, $S_3 = \{$the sum is divisible by 3$\}$, $S_4 = \{$the sum is divisible by 4$\}$, and $Q = \{$the square root of the sum is an integer$\}$. Compute
  a. $P(S_2)$, $P(S_3)$, $P(S_4)$, and $P(Q)$, justifying each computation by identifying any axiom, rule, or property of probability used.
  b. $P(S_4 \cup Q)$ first by listing outcomes in $S_4 \cup Q$ and then by using the results of (a) along with an appropriate axiom or property.
  c. $P(S_2 \cap S_3)$ by listing outcomes in $S_2 \cap S_3$.
  d. $P(S_2 \cup S_3)$ by using the results of (a) and (c).
  e. $P(S_2 \cap S_4)$. Have you already computed this probability?
  f. $P(S_2 \cup S_3 \cup S_4)$.
  g. $P((S_2 \cup S_3)')$ without listing outcomes. *Hint:* See Exercise 9.

15. In a school machine shop, 60% of all machine breakdowns occur on lathes and 15% on drills. Let $A = \{$the next machine breakdown is a lathe$\}$, and $B = \{$the next machine breakdown is a drill$\}$ (so that A and B are mutually exclusive). With $P(A) = .60$ and $P(B) = .15$, calculate
  a. $P(A')$ **b.** $P(A \cup B)$ **c.** $P(A' \cap B')$

16. The route used by a certain motorist in commuting to work contains two intersections with traffic

signals. The probability that he must stop at the first signal is .4, the analogous probability for the second signal is .5, and the probability that he must stop at at least one of the two signals is .6. What is the probability that he must stop
a. at both signals?
b. at the first signal but not at the second one?
c. at exactly one signal?

17. A student has a box containing 25 computer disks, of which 15 are blank and the other 10 are not. If she randomly selects disks one by one, what is the probability that at least two must be selected in order to find one that is blank?

18. A state park has two campgrounds. Let A denote the event that there is still space in the first campground at 5:00 P.M. on any given day, and define B analogously for the second campground. If $P(A) = .3$, $P(B) = .2$, and $P(A \cap B) = .1$, what is the probability that a camper who shows up at precisely 5:00 P.M. will be able to camp
a. in at least one of the campgrounds?
b. in neither of the campgrounds?
c. in exactly one of the two campgrounds?

19. When I visit the local library, the probability that someone is reading the current issue of *Sports Illustrated* is .4, the probability that someone is reading *Time* is .3, and the probability that at least one of these two magazines is being read by someone is .5. What is the probability that
a. both of the magazines are being read?
b. neither of the two is being read?
c. exactly one is being read?

20. Use the axioms to show that if one event A is contained in another event B (A is a subset of B), then $P(A) \leq P(B)$. *Hint:* For such A and B, $B = A \cup (B \cap A')$, draw a Venn diagram.

21. If a standard deck of 52 cards (spades, hearts, diamonds, and clubs; with ace, 2, . . . , 10, jack, queen, and king in each suit) is well mixed before a single card is drawn, all 52 outcomes are equally likely. Define events by $A = \{\text{heart}\}$, $B = \{\text{black card}\}$, $C = \{\text{jack, queen, or king}\}$, and $D = \{\text{selected card is at most a five}\}$, and use elementary counting along with the axioms and properties to compute the following probabilities.
a. $P(A)$, $P(B)$, $P(C)$, and $P(D)$
b. $P(A \cup B)$ and $P(C \cup D)$
c. $P(A \cup C)$ and $P(B \cup D)$
d. $P(A \cup B \cup C \cup D)$

22. The three most popular options on a certain type of new car are automatic transmission (A), power steering (B), and a radio (C). If 70% of all purchasers request A, 80% request B, 75% request C, 85% request A or B, 90% request A or C, 95% request B or C, and 98% request A or B or C, compute the probabilities of the following events. *Note:* "A or B" could mean that both were requested. *Hint:* Try drawing a Venn diagram and labeling all regions.
a. The next purchaser will select at least one of the three options ($A \cup B \cup C$).
b. The next purchaser will select none of the three options.
c. The next purchaser will select only a radio.
d. The next purchaser will select exactly one of the three options.

23. An academic department with five faculty members—Anderson, Box, Cox, Cramer, and Fisher—must select two of its members to serve on a personnel review committee. Because the work will be time consuming, no one is anxious to serve, so it is decided that the representative will be selected by putting five slips of paper in a box, mixing them, and selecting two.
a. What is the probability that Anderson and Box are both selected? *Hint:* There are 20 equally likely outcomes for this experiment.
b. What is the probability that at least one of the two members whose name begins with C is selected?
c. If the five faculty members have taught for 3, 6, 7, 10, and 14 years, respectively, at the university, what is the probability that the two chosen representatives have at least 15 years' teaching experience at the university?

24. In Exercise 3, suppose that any incoming individual is equally likely to be assigned to any of the three stations irrespective of where other individuals have been assigned. What is the probability that
a. all three family members are assigned to the same station?
b. at most two family members are assigned to the same station?
c. every family member is assigned to a different station?

2.3   **Counting Techniques**

When the various outcomes of an experiment are equally likely (the same probability is assigned to each simple event), then the task of computing probabilities reduces to counting. In particular, if N is the number of outcomes in a sample space and $N(A)$ is the number of outcomes contained in an event A, then

$$P(A) = \frac{N(A)}{N} \tag{2.1}$$

If a list of the outcomes is available or easy to construct and N is small, then the numerator and denominator of (2.1) can be obtained without the benefit of any general counting principles.

There are, however, many experiments for which the effort involved in constructing such a list is prohibitive because N is quite large. By exploiting some general counting rules, it is possible to compute probabilities of the form (2.1) without a listing of outcomes. These rules are also useful in many problems involving outcomes that are not equally likely. Several of the rules developed here will be used in studying probability distributions in the next chapter.

The Product Rule for Ordered Pairs

Our first counting rule applies to any situation in which a set (event) consists of ordered pairs of objects, and we wish to count the number of such pairs. By an ordered pair we mean that, if O_1 and O_2 are objects, then the pair (O_1, O_2) is different from the pair (O_2, O_1). For example, if an individual selects one airline for a trip from L.A. to Chicago and (after transacting business in Chicago) a second one for continuing on to New York, one possibility is (American, United), another is (United, American), and still another is (United, United).

Proposition

> If the first element or object of an ordered pair can be selected in n_1 ways, and for each of these n_1 ways the second element of the pair can be selected in n_2 ways, then the number of pairs is $n_1 n_2$.

Example 2.17   A student wishes to commute first to a junior college for two years and then to a state college campus. Within commuting range there are four junior colleges and three state college campuses. How many choices of junior college and state college are available to her? Numbering junior colleges by 1, 2, 3, 4 and state colleges by a, b, c, choices are $(1, a)$, $(1, b)$, $\ldots$, $(4, c)$, a total of 12 choices. With $n_1 = 4$ and $n_2 = 3$, $N = n_1 n_2 = 12$ without a list.   ■

Example 2.18   A homeowner doing some remodeling requires the services of both a plumbing contractor and an electrical contractor. If there are 12 plumbing contractors and nine electrical contractors available in the area, in how many ways can the contractors be chosen? If we denote the plumbers by $P_1, \ldots, P_{12}$ and the electricians by $Q_1, \ldots, Q_9$, then we wish the number of pairs of the form (P_i, Q_j). With $n_1 = 12$ and $n_2 = 9$, the product rule yields $N = (12)(9) = 108$ possible ways of choosing the two types of contractors.   ■

In both Examples 2.17 and 2.18, the choice of the second element of the pair did not depend on which first element was chosen or occurred. As long as there are the same number of choices of the second element for each first element, the product rule is valid even when the set of possible second elements depends on the first element.

Example 2.19   A family has just moved to a new city and requires the services of both an obstetrician and a pediatrician. There are two easily accessible medical clinics, each having two obstetricians and three pediatricians. The family will obtain maximum health insurance benefits by joining a clinic and selecting both doctors from that clinic. In how many ways can this be done? Denote the obstetricians by O_1, O_2, O_3, and O_4 and the pediatricians by $P_1, \ldots, P_6$. Then we wish the number of pairs (O_i, P_j) for which O_i and P_j are associated with the same clinic. Because there are four obstetricians, $n_1 = 4$, and for each there are three choices of pediatrician, so $n_2 = 3$. Applying the product rule gives $N = n_1 n_2 = 12$ possible choices.   ■

Tree Diagrams

In problems in which the product rule can be applied, a configuration called a tree diagram can be used to represent pictorially all the possibilities. The tree diagram associated with Example 2.19 appears in Figure 2.5. Starting from a point on the left side of the diagram, for each possible first element of a pair a straight-line segment emanates rightward. Each of these lines is referred to as a first-generation branch. Now for any given first-generation branch we construct another line segment emanating from the tip of the branch for each possible choice of a second element of the pair. Each such line segment is a

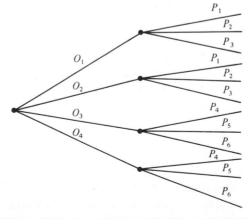

Figure 2.5   Tree diagram for Example 2.19

second-generation branch. Because there are four obstetricians, there are four first-generation branches, and three pediatricians for each obstetrician yields three second-generation branches emanating from each first-generation branch.

In the general case there are n_1 first-generation branches, and for each first-generation branch there are n_2 second-generation branches. The total number of second-generation branches is, therefore, $n_1 n_2$. Since each second-generation branch corresponds to exactly one possible pair (choosing a first element and then a second puts us at the end of exactly one second-generation branch), there are $n_1 n_2$ pairs, so the product rule is verified.

The construction of a tree diagram does not depend on having the same number of second-generation branches emanating from each first-generation branch. If the second clinic had four pediatricians then there would be only three branches emanating from two of the first-generation branches and four emanating from each of the other two first-generation branches. A tree diagram can thus be used to represent pictorially experiments other than those to which the product rule applies.

A More General Product Rule

If a six-sided die is tossed five times in succession rather than just twice, then each possible outcome is an ordered collection of five numbers such as (1, 3, 1, 2, 4) or (6, 5, 2, 2, 2). We shall call an ordered collection of k objects a **k-tuple** (so a pair is a 2-tuple and a triple is a 3-tuple). Each outcome of the die-tossing experiment is then a 5-tuple.

Product rule for k-tuples

> Suppose that a set consists of ordered collections of k elements (k-tuples) and that there are n_1 possible choices for the first element; for each choice of first element there are n_2 possible choices of the second element; . . . ; for each possible choice of the first $k - 1$ elements, there are n_k choices of the kth element. Then there are $n_1 n_2 \ldots n_k$ possible k-tuples.

This more general rule can also be illustrated by a tree diagram; simply construct a more elaborate diagram by adding third-generation branches emanating from the tip of each second generation, then fourth-generation branches, and so on, until finally kth-generation branches are added.

Example 2.20
(Example 2.18 continued)

Suppose that the home remodeling job involves first purchasing several kitchen appliances. They will all be purchased from the same dealer, and there are five dealers in the area. With the dealers denoted by $D_1, \ldots, D_5$ there are $N = n_1 n_2 n_3 = (5)(12)(9) = 540$ 3-tuples of the form (D_i, P_j, Q_k), so there are 540 ways to choose first an appliance dealer, then a plumbing contractor, and finally an electrical contractor. ∎

Example 2.21
(Example 2.19 continued)

If each clinic has both three specialists in internal medicine and two general surgeons, there are $n_1 n_2 n_3 n_4 = (4)(3)(3)(2) = 72$ ways to select one doctor of each type such that all doctors practice at the same clinic. ∎

Permutations

So far the successive elements of a k-tuple were selected from entirely different sets (for example, appliance dealers, then plumbers, and finally electricians). In several tosses of a die, the set from which successive elements are chosen is always $\{1, 2, 3, 4, 5, 6\}$, but the choices are made "with replacement," so that the same element can appear more than once. We now consider a fixed set consisting of n distinct elements and suppose that a k-tuple is formed by selecting successively from this set *without replacement,* so that an element can appear in at most one of the k positions.

Definition

> Any ordered sequence of k objects taken from a set of n distinct objects is called a **permutation** of size k of the objects. The number of permutations of size k that can be constructed from the n objects is denoted by $P_{k,n}$.

The number of permutations of size k is obtained immediately from the general product rule. The first element can be chosen in n ways, for each of these n ways the second element can be chosen in $n - 1$ ways, and so on; finally, for each way of choosing the first $k - 1$ elements, the kth element can be chosen in $n - (k - 1) = n - k + 1$ ways, so

$$P_{k,n} = n(n - 1)(n - 2) \cdots (n - k + 2)(n - k + 1)$$

Example 2.22

There are eight teaching assistants available for grading papers in a particular course. The first exam consists of four questions, and the professor wishes to select a different assistant to grade each question (only one assistant per question). In how many ways can assistants be chosen to grade the exam? Here n = the number of assistants = 8 and k = the number of questions = 4. The number of different grading assignments is then $P_{4,8} = (8)(7)(6)(5) = 1680$. ■

The use of factorial notation allows $P_{k,n}$ to be expressed more compactly.

Definition

> For any positive integer m, $m!$ is read "m factorial" and is defined by $m! = m(m - 1) \cdots (2)(1)$. Also, $0! = 1$.

Using factorial notation yields

$$P_{k,n} = n(n - 1) \cdots (n - k + 1)$$
$$= \frac{n(n - 1) \cdots (n - k + 1)(n - k)(n - k - 1) \cdots (2)(1)}{(n - k)(n - k - 1) \cdots (2)(1)}$$

which becomes

$$P_{k,n} = \frac{n!}{(n - k)!}$$

For example, $P_{3,9} = 9!/(9-3)! = 9!/6! = 9 \cdot 8 \cdot 7 \cdot 6!/6! = 9 \cdot 8 \cdot 7$. Note also that because $0! = 1$, $P_{n,n} = n!/(n-n)! = n!/0! = n!/1 = n!$, as it should.

Combinations

There are many counting problems in which one is given a set of n distinct objects and wishes to count the number of unordered subsets of size k. For example, in bridge it is only the 13 cards in a hand and not the order in which they are dealt that is important; in the formation of a committee, the order in which committee members are listed is frequently unimportant.

Definition

Given a set of n distinct objects, any unordered subset of size k of the objects is called a **combination.** The number of combinations of size k that can be formed from n distinct objects will be denoted by $\binom{n}{k}$. (This notation is more common in probability than $C_{k,n}$, which would be analogous to notation for permutations.)

The number of combinations of size k from a particular set is smaller than the number of permutations since, when order is disregarded, a number of permutations correspond to the same combination. Consider, for example, the set $\{A, B, C, D, E\}$ consisting of 5 elements. We know that there are $5!/(5-3)! = 60$ permutations of size 3. There are 6 permutations of size 3 consisting of the elements A, B, and C since these 3 can be ordered $3 \cdot 2 \cdot 1 = 3! = 6$ ways: (A, B, C), (A, C, B), (B, A, C), (B, C, A), (C, A, B), and (C, B, A). These 6 permutations are equivalent to the single combination $\{A, B, C\}$. Similarly for any other combination of size 3, there are 3! permutations, each obtained by ordering the 3 objects. Thus,

$$60 = P_{3,5} = \binom{5}{3} \cdot 3!; \quad \text{so} \quad \binom{5}{3} = \frac{60}{3!} = 10$$

These 10 combinations are

$$\{A, B, C\}, \{A, B, D\}, \{A, B, E\}, \{A, C, D\}, \{A, C, E\}, \{A, D, E\}, \{B, C, D\}, \{B, C, E\},$$
$$\{B, D, E\}, \{C, D, E\}$$

When there are n distinct objects, any permutation of size k is obtained by ordering the k unordered objects of a combination in one of $k!$ ways, so the number of permutations is the product of $k!$ and the number of combinations. This gives

$$\binom{n}{k} = \frac{P_{k,n}}{k!} = \frac{n!}{k!(n-k)!}$$

Notice that $\binom{n}{n} = 1$ and $\binom{n}{0} = 1$ since there is only one way to choose a set of (all) n elements or of no elements, and $\binom{n}{1} = n$ since there are n subsets of size 1.

Example 2.23 A bridge hand consists of any 13 cards selected from a 52-card deck without regard to order. There are $\binom{52}{13} = \dfrac{52!}{13!39!}$ different bridge hands, which works out to approximately 635 billion. Since there are 13 cards in each suit, the number of hands consisting entirely of clubs and/or spades (no red cards) is $\binom{26}{13} = \dfrac{26!}{13!13!} = 10{,}400{,}597$. One of these $\binom{26}{13}$ hands consists entirely of spades and one consists entirely of clubs, so there are $\left[\binom{26}{13} - 2\right]$ hands that consist entirely of clubs and spades with both suits represented in the hand. Suppose that a bridge hand is dealt from a well-shuffled deck (that is, 13 cards are randomly selected from among the 52 possibilities), and let

$A = \{$the hand consists entirely of spades and clubs with both suits represented$\}$

$B = \{$the hand consists of exactly two suits$\}$

The $N = \binom{52}{13}$ possible outcomes are equally likely, so

$$P(A) = \frac{N(A)}{N} = \frac{\binom{26}{13} - 2}{\binom{52}{13}} = .0000164$$

Since there are $\binom{4}{2} = 6$ combinations consisting of two suits, of which spades and clubs is one such combination,

$$P(B) = \frac{6\left[\binom{26}{13} - 2\right]}{\binom{52}{13}} = .0000984$$

That is, a hand consisting entirely of cards from exactly two of the four suits will occur roughly once in every 10,000 hands. If you play bridge only once a month, it is likely that you will never be dealt such a hand. ∎

Example 2.24 A rental car service facility has 10 foreign cars and 15 domestic cars waiting to be serviced on a particular Saturday morning. Because there are so few mechanics working on Saturday, only 6 can be serviced. If the 6 are chosen at random, what is the probability that 3 of the cars selected are domestic and the other 3 are foreign?

Let $D_3 = \{$exactly 3 of the 6 cars chosen are domestic$\}$. Assuming that any particular set of 6 cars is as likely to be chosen as is any other set of 6, we have equally likely outcomes, so $P(D_3) = N(D_3)/N$, where N is the number of ways of choosing 6 cars from the 25 and $N(D_3)$ is the number of ways of choosing 3 domestic cars and 3 foreign cars. We have immediately that $N = \binom{25}{6}$. To obtain $N(D_3)$, think of first choosing 3 of the 15 domestic cars and then 3 of the foreign cars. There are $\binom{15}{3}$ ways of choosing the 3 domestic cars, and there are $\binom{10}{3}$ ways of choosing the 3 foreign cars; $N(D_3)$ is now the product of these two numbers (visualize a tree diagram—we are really using a product rule argument here), so

$$P(D_3) = \frac{N(D_3)}{N} = \frac{\binom{15}{3}\binom{10}{3}}{\binom{25}{6}} = \frac{\frac{15!}{3!12!} \cdot \frac{10!}{3!7!}}{\frac{25!}{6!19!}} = .3083$$

Let $D_4 = \{$exactly 4 of the 6 cars chosen are domestic$\}$, and define D_5 and D_6 in an analogous manner. Then the probability that at least 3 domestic cars are selected is

$$P(D_3 \cup D_4 \cup D_5 \cup D_6) = P(D_3) + P(D_4) + P(D_5) + P(D_6)$$
$$= \frac{\binom{15}{3}\binom{10}{3}}{\binom{25}{6}} + \frac{\binom{15}{4}\binom{10}{2}}{\binom{25}{6}} + \frac{\binom{15}{5}\binom{10}{1}}{\binom{25}{6}} + \frac{\binom{15}{6}\binom{10}{0}}{\binom{25}{6}} = .8530$$

This is also the probability that at most 3 foreign cars are selected.   ■

Exercises / Section 2.3 (25–37)

25. The Student Engineers Council at a certain college has one student representative from each of the five engineering majors (civil, electrical, industrial, materials, and mechanical). In how many ways can
  a. both a council president and a vice president be selected?
  b. a president, a vice president, and a secretary be selected?
  c. two members be selected for the President's Council?

26. A real estate agent is showing homes to a prospective buyer. There are 10 homes in the desired price range listed in the area. The buyer only has time to visit three of them.

  a. In how many ways could the three homes be chosen if the order of visiting is considered?
  b. In how many ways could the three homes be chosen if the order is unimportant?
  c. If four of the homes are new and six have previously been occupied, and if the three homes to visit are randomly chosen, what is the probability that all three are new? (The same answer results whether or not order is considered.)

27. a. Beethoven wrote 9 symphonies and Mozart wrote 27 piano concertos. If a university radio station announcer wishes to play first a Beethoven symphony and then a Mozart concerto, in how many ways can this be done?

b. The station manager has decided that on each successive night (seven days per week), a Beethoven symphony will be played, followed by a Mozart piano concerto, followed by a Schubert string quartet (of which there are 15). For roughly how many years could this policy be continued before exactly the same program would have to be repeated?

28. During the fall quarter of an academic year, Statistics 1A will be offered at 12 different times, while during the winter quarter Statistics 1B will be offered at 10 different times.
   a. If a student wishes to take both 1A and 1B during the fall and winter quarters, respectively, how many different time choices for the two courses are possible?
   b. If each section of each course can accommodate up to 35 students, what is the maximum number of students that could be enrolled in these courses?
   c. If six of the 1A sections are given in the morning and seven of the 1B sections are given in the morning, what is the probability that a student will end up with two morning sections if the student selects one of the time choices of (a) in a completely random fashion?
   d. If Professor I. N. Coherent is scheduled to teach three sections of 1A and three sections of 1B, what is the probability that a student who randomly selects sections will end up having Professor Coherent for both quarters? For neither quarter?

29. A city is serviced by a cable television company that provides programs on seven different channels for its subscribers. From 5:30 to 6:00 P.M. on weekdays three channels show news, from 6:00 to 6:30 five channels show news, from 6:30 to 7:00 four channels show news, and from 7:00 to 7:30 two channels show news.
   a. If a viewing sequence consists of four half-hour programs between 5:30 and 7:30, how many viewing sequences are there?
   b. If a viewing sequence is chosen at random from the set of all possible sequences, what is the probability that all four programs selected are news programs?
   c. Choosing a sequence as in (b), what is the probability that at least two programs are news programs?

d. Choosing a sequence as in (b), what is the probability that a news program is selected both at 6:00 P.M. and at 6:30 P.M.?

30. A chain of stereo stores is offering a special price on a complete set of components (receiver, turntable, speakers, cassette deck). A purchaser is offered a choice of manufacturer for each component:

Receiver: Kenwood, Pioneer, Sansui, Sony, Sherwood
Turntable: BSR, Dual, Sony, Technics
Speakers: AR, KLH, JBL
Cassette Deck: Advent, Sony, Teac, Technics

A switchboard display in the store allows a customer to hook together any selection of components (consisting of one of each type). Use the product rules to answer the following questions.
   a. In how many ways can one component of each type be selected?
   b. In how many ways can components be selected if both the receiver and the turntable are to be Sony?
   c. In how many ways can components be selected if none is to be Sony?
   d. In how many ways can a selection be made if at least one Sony component is to be included?
   e. If someone flips switches on the selection in a completely random fashion, what is the probability that the system selected contains at least one Sony component? Exactly one Sony component?

31. Shortly after being put into service, some buses manufactured by a certain company have developed cracks on the underside of the main frame. Suppose that a particular city has 20 of these buses, and cracks have actually appeared in eight of them.
   a. How many ways are there to select a sample of five buses from the 20 for a thorough inspection?
   b. In how many ways can a sample of five buses contain exactly five with visible cracks?
   c. If a sample of five buses is chosen at random, what is the probability that at least four of the five have visible cracks?

32. A consumer group is concerned about the possibility that there is systematic underweighing in the meat department of a local supermarket. A

representative is sent to purchase five 1-lb packages of ground meat. If there are currently 25 such packages on display and eight of them are actually underweight, what is the probability that at most two of the five packages purchased are underweight?

33. Three molecules of type A, three of type B, three of type C, and three of type D are to be linked together to form a chain molecule. One such chain molecule is $ABCDABCDABCD$, and another is $BCDDAAABDBCC$.

   a. How many such chain molecules are there? *Hint:* If the three A's were distinguishable from one another—A_1, A_2, A_3—and the B's, C's, and D's were also, how many molecules would there be? How is this number reduced when the subscripts are removed from the A's?

   b. Suppose that a chain molecule of the type described is randomly selected. What is the probability that all three molecules of each type end up next to one another (such as in $BBBAAADDDCCC$)?

34. A mathematics professor wishes to schedule an appointment with each of her eight teaching assistants, four male and four female, to discuss her calculus course. Suppose that all possible orderings of appointments are equally likely to be selected.

   a. What is the probability that at least one female assistant is among the first three the professor meets with?

   b. What is the probability that after the first five appointments she has met with all female assistants?

   c. Suppose that the professor has the same eight assistants the following semester and again schedules appointments without regard to the ordering during the first semester. What is the probability that the orderings of appointments are different?

35. Three married couples have purchased theater tickets and are seated in a row consisting of just six seats. If they take their seats in a completely random fashion (random order), what is the probability that Jim and Paula (husband and wife) sit in the two seats on the far left? What is the probability that Jim and Paula end up sitting next to one another? What is the probability that at least one of the wives ends up sitting next to her husband?

36. In five-card poker, a straight consists of five cards with adjacent denominations (for example, 9 of clubs, 10 of hearts, jack of hearts, queen of spades, and king of clubs). Assuming that aces can be high or low, if you are dealt a five-card hand, what is the probability that it will be a straight with high card 10? What is the probability that it will be a straight? What is the probability that it will be a straight flush (all cards in the same suit)?

37. Show that $\binom{n}{k} = \binom{n}{n-k}$. Give an interpretation involving subsets.

2.4   Conditional Probability

The probabilities assigned to various events depend upon what is known about the experimental situation when the assignment is made. Subsequent to the initial assignment, partial information about or relevant to the outcome of the experiment may become available, and this information may cause us to revise some of our probability assignments. For a particular event A, we have used $P(A)$ to represent the probability assigned to A; we now think of $P(A)$ as the original or unconditional probability of the event A.

In this section we examine how the information "an event B has occurred" affects the probability assigned to A. For example, A might refer to an individual having a particular disease in the presence of certain symptoms. If a blood test is performed on the individual and the result is negative (B = negative blood test), then the probability of having the disease will change (it should decrease, but not usually to zero, since blood tests are not infallible). We will use the notation $P(A \mid B)$ to represent the **conditional probability of A given that the event B has occurred.**

Example 2.25

Complex components are assembled in a plant that uses two different assembly lines, A and A'. Line A uses older equipment than A', so it is somewhat slower and less reliable. Suppose that on a given day line A has assembled 8 components, of which 2 have been identified as defective (B) and 6 as nondefective (B'), while A' has produced 1 defective and 9 nondefective components. This information is summarized in the accompanying table.

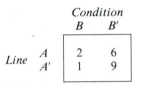

Unaware of this information, the sales manager randomly selects 1 of these 18 components for a demonstration. Prior to the demonstration

$$P(\text{line } A \text{ component selected}) = P(A) = \frac{N(A)}{N} = \frac{8}{18} = .44$$

However, if the chosen component turns out to be defective, then the event B has occurred, so the component must have been 1 of the 3 in the B column of the table. Since these 3 components are equally likely among themselves after B has occurred

$$P(A \mid B) = \frac{2}{3} = \frac{\dfrac{2}{18}}{\dfrac{3}{18}} = \frac{P(A \cap B)}{P(B)} \tag{2.2}$$

In (2.2) the conditional probability is expressed as a ratio of unconditional probabilities: The numerator is the probability of the intersection of the two events, while the denominator is the probability of the conditioning event B. A Venn diagram illuminates this relationship (Figure 2.6).

Given that B has occurred, the relevant sample space is no longer $\mathscr{S}$ but consists of outcomes in B; A has occurred if and only if one of the outcomes in the intersection occurred, so that the conditional probability of A given B is proportional to $P(A \cap B)$. The proportionality constant $1/P(B)$ is used to ensure that the probability $P(B \mid B)$ of the new sample space B equals one.

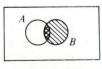

Figure 2.6

The Definition of Conditional Probability

Example 2.25 demonstrates that when outcomes are equally likely, computation of conditional probabilities can be based on intuition. When experiments are more complicated, though, intuition may fail us, so we want to have a general definition of conditional probability which will yield intuitive answers in simple problems. The Venn diagram and (2.2) suggest the appropriate definition.

Definition

> For any two events A and B with $P(B) > 0$, the **conditional probability of A given that B has occurred** is defined by
>
> $$P(A \mid B) = \frac{P(A \cap B)}{P(B)} \tag{2.3}$$

Example 2.26

Suppose that at a particular self-service gas station 10% of all customers check their oil level, 2% check their tire pressure, and 1% check both oil level and tire pressure. Consider selecting a customer in a completely random fashion, and let $A = \{$the customer checks oil$\}$ and $B = \{$the customer checks tires$\}$. Then $P(A) = .10$, $P(B) = .02$, and $P($customer checks both oil and tire pressure$) = P(A \cap B) = .01$. If the selected customer checked his or her tire pressure, the probability that oil level was also checked is

$$P(A \mid B) = \frac{P(A \cap B)}{P(B)} = \frac{.01}{.02} = .50$$

That is, if we focus only on customers who have checked tire pressure, 50% of those also check oil level. Similarly,

$$P(\text{checked tires} \mid \text{checked oil}) = P(B \mid A) = \frac{P(A \cap B)}{P(A)} = \frac{.01}{.10} = .10$$

Notice that $P(A \mid B) \neq P(A)$ and that $P(B \mid A) \neq P(B)$. ■

Example 2.27

A news magazine publishes three columns entitled "Art" (A), "Books" (B), and "Cinema" (C). Reading habits of a randomly selected reader with respect to these columns are

Read regularly	A	B	C	$A \cap B$	$A \cap C$	$B \cap C$	$A \cap B \cap C$
Probability	.14	.23	.37	.08	.09	.13	.05

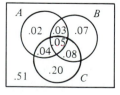

Figure 2.7

We thus have

$$P(A \mid B) = \frac{P(A \cap B)}{P(B)} = \frac{.08}{.23} = .348$$

$$P(A \mid B \cup C) = \frac{P(A \cap (B \cup C))}{P(B \cup C)} = \frac{.04 + .05 + .03}{.47} = \frac{.12}{.47} = .255$$

$$P(A \mid \text{reads at least one}) = P(A \mid A \cup B \cup C) = \frac{P(A \cap (A \cup B \cup C))}{P(A \cup B \cup C)}$$

$$= \frac{P(A)}{P(A \cup B \cup C)} = \frac{.14}{.49} = .286$$

and

$$P(A \cup B \mid C) = \frac{P((A \cup B) \cap C)}{P(C)} = \frac{.04 + .05 + .08}{.37} = .459 \qquad \blacksquare$$

The Multiplication Rule for $P(A \cap B)$

The definition of conditional probability yields the following result, obtained by multiplying both sides of (2.3) by $P(B)$.

**The multiplica-
tion rule**

$$P(A \cap B) = P(A \mid B) \cdot P(B)$$

This rule is important because it is often the case that $P(A \cap B)$ is desired while both $P(B)$ and $P(A \mid B)$ can be specified from the problem description.

Example 2.28   Four individuals have responded to a request by a blood bank for blood donations. None of them has donated before, so their blood types are unknown. Suppose that only type A positive is desired, and that only one of the four actually has this type. If the potential donors are selected in random order for typing, what is the probability that at least three individuals must be typed to obtain the desired type?

Making the identification $B = \{1\text{st type not A}+\}$ and $A = \{2\text{nd type not A}+\}$, $P(B) = \frac{3}{4}$. Given that the first type is not A$+$, two of the three individuals left are not A$+$, so $P(A \mid B) = \frac{2}{3}$. The multiplication rule now gives

$$P(\text{at least three individuals are typed}) = P(A \cap B)$$

$$= P(A \mid B) \cdot P(B) = \frac{3}{4} \cdot \frac{2}{3} = \frac{6}{12} = .5$$

$\blacksquare$

The multiplication rule is most useful when the experiment consists of several stages in succession. The conditioning event B then describes the outcome of the first stage and A the outcome of the second, so that $P(A \mid B)$—conditioning on what occurs first—will often be known. The rule is easily extended to experiments involving more than two stages. For example

$$P(A_1 \cap A_2 \cap A_3) = P(A_3 \mid A_1 \cap A_2) \cdot P(A_1 \cap A_2)$$
$$= P(A_3 \mid A_1 \cap A_2) \cdot P(A_2 \mid A_1) \cdot P(A_1) \qquad (2.4)$$

where A_1 occurs first, followed by A_2, and finally A_3.

Example 2.29 For the blood typing experiment of Example 2.28,

$$P(\text{third type is A}+) = P(\text{third is} \mid \text{first isn't} \cap \text{second isn't})$$
$$\cdot P(\text{second isn't} \mid \text{first isn't}) \cdot P(\text{first isn't})$$
$$= \frac{1}{2} \cdot \frac{2}{3} \cdot \frac{3}{4} = \frac{1}{4} = .25 \qquad \blacksquare$$

When the experiment of interest consists of a sequence of several stages, it is convenient to represent these with a tree diagram. Once we have an appropriate tree diagram, probabilities and conditional probabilities can be entered on the various branches; this will make repeated use of the multiplication rule quite straightforward.

Example 2.30 A chain of video stores sells three different brands of videocassette recorders (VCR's). Fifty percent of its sales are brand 1 VCR's (the least expensive), 30% are brand 2, and 20% are brand 3. Each manufacturer offers a one-year warranty on parts and labor. It is known that 25% of brand 1's VCR's require warranty repair work, whereas the corresponding percentages for brands 2 and 3 are 20% and 10%, respectively.

a. What is the probability that a randomly selected purchaser has bought a brand 1 VCR that will need repair while under warranty?
b. What is the probability that a randomly selected purchaser has a VCR that will need repair while under warranty?
c. If a customer returns to the store with a VCR that needs warranty repair work, what is the probability that it is a brand 1 VCR? A brand 2 VCR? A brand 3 VCR?

The first stage of the problem involves a customer selecting one of the three brands of VCR. Let $A_i = \{\text{brand } i \text{ is purchased}\}$ for $i = 1$, 2, and 3. Then $P(A_1) = .50$, $P(A_2) = .30$, and $P(A_3) = .20$. Once a brand of VCR is selected, the second stage involves observing whether or not the selected VCR needs warranty repair. With $B = \{\text{needs repair}\}$ and $B' = \{\text{doesn't need repair}\}$, the given information implies that $P(B \mid A_1) = .25$, $P(B \mid A_2) = .20$, and $P(B \mid A_3) = .10$.

The tree diagram representing this experimental situation appears in Figure 2.8. The initial branches correspond to different brands of VCR's and there are two second-generation branches emanating from the tip of each initial branch, one for "needs repair" and the other for "doesn't need repair." The probability $P(A_i)$ appears on the ith initial branch, whereas the conditional probabilities $P(B \mid A_i)$ and $P(B' \mid A_i)$ appear on the second-generation branches. To the right of each second-generation branch corresponding to the occurrence of B, we display the product of probabilities on the branches leading out to that point. This is simply the multiplication rule in action. The answer to the question posed in (a) is thus $P(A_1 \cap B) = P(B \mid A_1) \cdot P(A_1) = .125$. The answer to the question of (b) is

$$P(B) = P((\text{brand 1 and repair}) \text{ or } (\text{brand 2 and repair})$$
$$\text{or } (\text{brand 3 and repair}))$$
$$= P(A_1 \cap B) + P(A_2 \cap B) + P(A_3 \cap B)$$
$$= .125 + .060 + .020 = .205$$

Finally,

$$P(A_1 \mid B) = \frac{P(A_1 \cap B)}{P(B)} = \frac{.125}{.205} = .61$$

$$P(A_2 \mid B) = \frac{P(A_2 \cap B)}{P(B)} = \frac{.060}{.205} = .29$$

and

$$P(A_3 \mid B) = 1 - P(A_1 \mid B) - P(A_2 \mid B) = .10$$

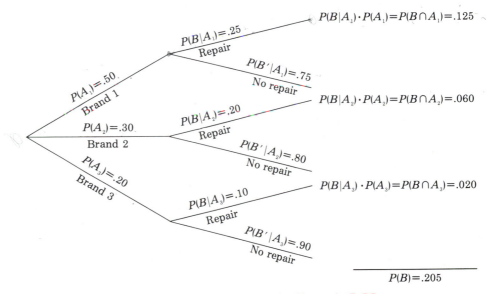

$$P(B) = .205$$

Figure 2.8 Tree diagram for Example 2.30

Notice that the initial or *prior* probability of brand 1 is .50, whereas once it is known that the selected VCR needed repair, the *posterior* probability of brand 1 increases to .61. This is because brand 1 VCR's are more likely to need warranty repair than are the other brands. The posterior probability of brand 3 is $P(A_3 | B) = .10$, which is much less than the prior probability $P(A_3) = .20$.

■

Bayes' Theorem

The computation of a posterior probability $P(A_k | B)$ from given prior probabilities $P(A_i)$ and conditional probabilities $P(B | A_i)$ occupies a central position in elementary probability. The general rule for such computations, which is really just a simple application of the multiplication rule, goes back to Reverend Thomas Bayes, who lived in the eighteenth century. To state it we first need another result. Recall events $A_1, \ldots, A_n$ are mutually exclusive if no two have any common outcomes. The events are *exhaustive* if one A_i must occur, so that $A_1 \cup \cdots \cup A_n = \mathscr{S}$.

The law of total probability

> Let $A_1, \ldots, A_n$ be mutually exclusive and exhaustive events. Then for any other event B,
>
> $$P(B) = \sum_{i=1}^{n} P(B | A_i) P(A_i) \tag{2.5}$$

Proof.   Because the A_i's are mutually exclusive and exhaustive, if B occurs it must be in conjunction with exactly one of the A_i's. That is, $B = (A_1$ and $B)$ or ... or $(A_n$ and $B) = (A_1 \cap B) \cup \cdots \cup (A_n \cap B)$, where the events $(A_i \cap B)$ are exclusive. This "partitioning of B" is illustrated in Figure 2.9. Thus

$$P(B) = \sum_{i=1}^{n} P(A_i \cap B) = \sum_{i=1}^{n} P(B | A_i) P(A_i)$$

as desired.

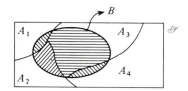

Figure 2.9   Partition of B by mutually exclusive and exhaustive A_i's

■

An example of the use of (2.5) appeared in answering question (b) of Example 2.30, where $A_1 = \{$brand 1$\}$, $A_2 = \{$brand 2$\}$, $A_3 = \{$brand 3$\}$, and $B = \{$repair$\}$.

Bayes' theorem

Let $A_1, A_2, \ldots, A_n$ be a collection of n mutually exclusive and exhaustive events with $P(A_i) > 0$ for $i = 1, \ldots, n$. Then for any other event B for which $P(B) > 0$

$$P(A_k \mid B) = \frac{P(A_k \cap B)}{P(B)} = \frac{P(B \mid A_k)P(A_k)}{\displaystyle\sum_{i=1}^{n} P(B \mid A_i) \cdot P(A_i)}; \, k = 1, \ldots, n \qquad (2.6)$$

The transition from the second to the third expression in (2.6) rests on using the multiplication rule in the numerator and the law of total probability in the denominator.

While the right-hand side expression in (2.6) enables one to compute posterior probabilities by substituting prior and conditional probabilities in the appropriate places, the proliferation of events and subscripts can often intimidate a newcomer to probability. Example 2.30 shows that the posterior probabilities can be computed from a tree diagram by using the middle expression in (2.6) without ever referring explicitly to Bayes' theorem.

Example 2.31 *Incidence of a rare disease.* Only one in 1000 adults is afflicted with a rare disease for which a diagnostic test has been developed. The test is such that, when an individual actually has the disease, a positive result will occur 99% of the time, while an individual without the disease will show a positive test result only 2% of the time. If a randomly selected individual is tested and the result is positive, what is the probability that the individual has the disease?

To use Bayes' theorem, let $A_1 = \{$individual has the disease$\}$, $A_2 = \{$individual does not have the disease$\}$, and $B = \{$positive test result$\}$. Then $P(A_1) = .001$, $P(A_2) = .999$, $P(B \mid A_1) = .99$, and $P(B \mid A_2) = .02$. The tree diagram for this problem appears in Figure 2.10.

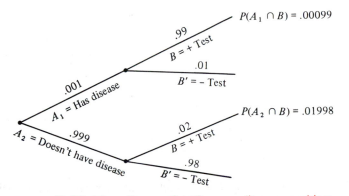

Figure 2.10 Tree diagram for the rare disease problem

Next to each branch corresponding to a positive test result, the multiplication rule yields the recorded probabilities. Therefore $P(B) = .00099 + .01998 = .02097$, from which we have

$$P(A_1 \mid B) = \frac{P(A_1 \cap B)}{P(B)} = \frac{.00099}{.02097} = .047$$

This result seems counter-intuitive; the diagnostic test appears so accurate we expect someone with a positive test result to be highly likely to have the disease, whereas the computed conditional probability is only .047. The reason for this seemingly paradoxical result is that, because the disease is rare and the test only moderately reliable, most positive test results arise from errors rather than from diseased individuals. The probability of having the disease has increased by a multiplicative factor of 47 (from prior .001 to posterior .047); but to get a further increase in the posterior probability, a diagnostic test with much smaller error rates is needed. If the disease were not so rare (for example, 25% incidence in the population), then the error rates for the present test would provide good diagnoses. ∎

Exercises / Section 2.4 (38–54)

38. A study of the relationship among adult drivers between income level (L = low, M = medium, and H = high) and preference for one of the "big three" automobile manufacturers (denoted by A, B, and C here) yielded the accompanying table of joint probabilities.

		Income			
		L	M	H	
	A	.10	.13	.02	.25
Preference	B	.20	.12	.08	.40
	C	.10	.15	.10	.35
		.40	.40	.20	

This table shows, for example, that P(low income and prefer A) = $P(L \cap A)$ = .10, P(low income) = $P(L)$ = .40, and P(prefer A) = $P(A)$ = .25. Use this table to compute the following conditional probabilities.

a. $P(B \mid H)$ **b.** $P(M \mid C)$
c. $P(A' \mid M)$ **d.** $P(M \mid A')$
e. $P(M \mid B \cup C)$ **f.** $P(L \cup M \mid C)$
g. What is the probability that a randomly selected adult driver will either prefer A or have a high income?

39. A mathematics professor is teaching both a morning and an afternoon section of introductory calculus. Let A = {the professor gives a bad morning lecture} and B = {the professor gives a bad afternoon lecture}. If $P(A) = .3$, $P(B) = .2$, and $P(A \cap B) = .1$, calculate the following probabilities (a Venn diagram might help):

a. $P(B \mid A)$ **b.** $P(B' \mid A)$
c. $P(B \mid A')$ **d.** $P(B' \mid A')$
e. If at the conclusion of the afternoon class, the professor is heard to mutter "what a rotten lecture," what is the probability that the morning lecture was also bad?

40. When a roulette wheel is spun once, there are 38 possible outcomes—18 red, 18 black, and two green (if the outcome is green, the house wins everything). If a wheel is spun twice, all (38)(38) outcomes are equally likely. If you are told that in two spins, at least one resulted in a green outcome, what is the probability that both outcomes were green?

41. Refer to Exercise 11, where A_1 = {older car is American}, A_2 = {newer car is American}, $P(A_1) = .7$, $P(A_2) = .5$, and $P(A_1 \cap A_2) = .4$. Compute the following probabilities:

a. $P(A_2 \mid A_1)$ **b.** $P(A_1 \mid A_2)$ **c.** $P(A_2' \mid A_1')$
d. $P(A_1 \cap A_2 \mid$ at least one car is American)

42. In Exercise 12, $A_i = \{$awarded project $i\}$ for $i = 1, 2, 3$. Use the probabilities given there to compute the following probabilities:
  a. $P(A_2 | A_1)$ **b.** $P(A_2 \cap A_3 | A_1)$
  c. $P(A_2 \cup A_3 | A_1)$
  d. $P(A_1 \cap A_2 \cap A_3 | A_1 \cup A_2 \cup A_3)$. Express in words the event whose probability you have calculated.

43. In Exercise 35, six people (three married couples) chose seats at random in a row consisting of six seats.
  a. Use the multiplication rule to compute the probability that Jim and Paula sit together on the far left (event A) and that John and Mary Lou (husband and wife) sit together in the middle (event B).
  b. Given that John and Mary Lou sit together in the middle, what is the probability that the two other husbands sit next to their wives?
  c. Given that John and Mary Lou sit together, what is the probability that all husbands sit next to their wives?

44. For any events A and B with $P(B) > 0$, show that $P(A | B) + P(A' | B) = 1$.

45. Show that for any events A, B, and C with $P(C) > 0$, $P(A \cup B | C) = P(A | C) + P(B | C) - P(A \cap B | C)$.

46. At a certain gas station 40% of the customers request regular gas (A_1), 35% request unleaded gas (A_2), and 25% request premium gas (A_3). Of those customers requesting regular gas, only 30% fill their tanks (event B). Of those customers requesting unleaded gas, 60% fill their tanks, while of those requesting premium, 50% fill their tanks.
  a. What is the probability that the next customer will request unleaded gas and fill the tank ($A_2 \cap B$)?
  b. What is the probability that the next customer fills the tank?
  c. If the next customer fills the tank, what is the probability that regular gas is requested? Unleaded gas? Premium gas?

47. Seventy percent of the light aircraft that disappear while in flight in a certain country are subsequently discovered. Of the aircraft that are discovered, 60% have an emergency locator, whereas 90% of the aircraft not discovered do not have such a locator. Suppose that a light aircraft has disappeared.
  a. If it has an emergency locator, what is the probability that it will not be discovered?
  b. If it doesn't have an emergency locator, what is the probability that it will be discovered?

48. Components of a certain type are shipped to a supplier in batches of ten. Suppose that 50% of all such batches contain no defective components, 30% contain one defective component, and 20% contain two defective components. Two components from a batch are randomly selected and tested. What are the probabilities associated with 0, 1, and 2 defective components being in the batch under each of the following conditions?
  a. Neither tested component is defective.
  b. One of the two tested components is defective.

49. A company employing 10,000 workers offers deluxe medical coverage, standard medical coverage, and economy medical coverage. Of the employees 30% have deluxe coverage, 60% have standard coverage, and 10% have economy coverage. From past experience, the probability that an employee with deluxe coverage will submit no claims during the next year is .1, the probability of an employee with standard coverage submitting no claim is .4, and the probability of an employee with economy coverage submitting no claim is .7. If an employee is selected at random, draw an appropriate tree diagram and answer the following questions.
  a. What is the probability that the selected employee has standard coverage and will submit no claims?
  b. What is the probability that the selected employee will submit no claims?
  c. If the selected employee submits no claims during the next year, what is the probability that the employee had standard coverage? Economy coverage?

50. For customers purchasing a full set of tires at a particular tire store, consider the events

$A = \{$tires purchased were made in the United States$\}$
$B = \{$purchaser has tires balanced immediately$\}$
$C = \{$purchaser requests front end alignment$\}$

along with A', B', and C'. Assume the following unconditional and conditional probabilities:

$P(A) = .75, P(B \mid A) = .9, P(B \mid A') = .8,$
$P(C \mid A \cap B) = .8, P(C \mid A \cap B') = .6,$
$P(C \mid A' \cap B) = .7, P(C \mid A' \cap B') = .3$

a. Construct a tree diagram consisting of first-, second-, and third-generation branches, and place an event label and appropriate probability next to each branch.
b. Compute $P(A \cap B \cap C)$.
c. Compute $P(B \cap C)$.
d. Compute $P(C)$.
e. Compute $P(A \mid B \cap C)$, the probability of a purchase of U.S. tires given that both balancing and an alignment were requested.

51. In Example 2.31, suppose that the incidence rate for the disease is 1 in 25 rather than 1 in 1000. What then is the probability of a positive test result? Given that the test result is positive, what is the probability that the individual has the disease? Given a negative test result, what is the probability that the individual does not have the disease?

52. At a large university, in the never-ending quest for a satisfactory textbook, the Statistics Department has tried a different text during each of the last three quarters. During the fall quarter 500 students used the text by Professor Mean, during the winter quarter 300 students used the text by Professor Median, and during the spring quarter 200 students used the text by Professor Mode. A survey at the end of each quarter showed that 200 students were satisfied with Mean's book, 150 were satisfied with Median's book, and 160 were satisfied with Mode's book. If a student who took statistics during one of these quarters is selected at random and admits to having been satisfied with the text, is the student most likely to have used the book by Mean, Median, or Mode?

Who is the least likely author? *Hint:* Draw a tree diagram or use Bayes' theorem.

53. A friend who works in a big city owns two cars, one small and one large. Three-quarters of the time he drives the small car to work, and one-quarter of the time he takes the large car. If he takes the small car, he usually has little trouble parking, and so is at work on time with probability .9. If he takes the large car, he is on time to work with probability .6. Given that he was on time on a particular morning, what is the probability that he drove the small car?

54. In Exercise 46, consider the following additional information on credit card usage:

70% of all regular fill-up customers use a credit card,

50% of all regular non-fill-up customers use a credit card,

60% of all unleaded fill-up customers use a credit card,

50% of all unleaded non-fill-up customers use a credit card,

50% of all premium fill-up customers use a credit card,

40% of all premium non-fill-up customers use a credit card.

Compute the probability of each of the following events for the next customer to arrive (a tree diagram might help).
a. {unleaded and fill-up and credit card}
b. {premium and non-fill-up and credit card}
c. {premium and fill-up and credit card}
d. {premium and credit card}
e. {fill-up and credit card}
f. {credit card}
g. If the next customer uses a credit card, what is the probability that premium was requested?

2.5   Independence

The definition of conditional probability enables us to revise the probability $P(A)$ originally assigned to A when we are subsequently informed that another event, B, has occurred; the new probability of A is $P(A \mid B)$. In our examples, it was frequently the case that $P(A \mid B)$ was unequal to the unconditional probability $P(A)$, indicating that the information "B has occurred" resulted in a change in the chance of A occurring. There are other situations, though, in which the chance that A will occur or has occurred is not affected by knowledge that B has occurred, so that $P(A \mid B) = P(A)$. It is then natural to think of A and B as independent events, meaning that the occurrence or nonoccurrence of one event has no bearing on the chance that the other will occur.

Definition

> Two events A and B are **independent** if $P(A \mid B) = P(A)$ and are **dependent** otherwise.

The definition of independence might seem "unsymmetric," since we do not demand that $P(B \mid A) = P(B)$ also. However, using the definition of conditional probability and the multiplication rule,

$$P(B \mid A) = \frac{P(A \cap B)}{P(A)} = \frac{P(A \mid B)P(B)}{P(A)} \tag{2.7}$$

The right-hand side of (2.7) is $P(B)$ if and only if $P(A \mid B) = P(A)$ (independence), so the equality in the definition implies the other equality (and vice versa).

Example 2.32

Consider tossing a fair six-sided die once and define events $A = \{2, 4, 6\}$, $B = \{1, 2, 3\}$, and $C = \{1, 2, 3, 4\}$. We then have $P(A) = \frac{1}{2}$, $P(A \mid B) = \frac{1}{3}$, and $P(A \mid C) = \frac{1}{2}$. That is, events A and B are dependent while events A and C are independent. Intuitively, if such a die is tossed and we are informed that the outcome was either 1, 2, 3, or 4 (C has occurred), then the probability that A occurred is $\frac{1}{2}$, as it originally was, since two of the four relevant outcomes are even and the outcomes are still equally likely. ∎

Example 2.33

Let A and B be any two mutually exclusive events with $P(A) > 0$. For example, for a randomly chosen automobile, let $A = \{$the car has four cylinders$\}$ and $B = \{$the car has six cylinders$\}$. Since the events are mutually exclusive, if B occurs, then A cannot possibly have occurred, so $P(A \mid B) = 0 \neq P(A)$. The message here is that *if two events are mutually exclusive, they cannot be independent.* When A and B are mutually exclusive, the information that A occurred says something about B (it cannot have occurred), so independence is precluded. ∎

$P(A \cap B)$ When Events Are Independent

There are a number of experimental situations in which, rather than having to verify the independence of A and B by using the definition, we wish to build the independence of A and B into our probability assignments. For example, if a card is selected from a deck of 52, then replaced and the deck reshuffled before a second card is drawn, then it is reasonable to assume that any event A defined with respect to the first card selected (red card, picture card, five, and so on) is independent of an event defined with respect to the second card. The next proposition tells us how to compute $P(A \cap B)$ when the events are independent.

Proposition

> A and B are independent if and only if
>
> $$P(A \cap B) = P(A) \cdot P(B) \qquad (2.8)$$

To paraphrase the proposition, A and B are independent events iff* the probability that they both occur ($A \cap B$) is the product of the two individual probabilities. The verification is as follows:

$$P(A \cap B) = P(A \mid B) \cdot P(B) = P(A) \cdot P(B) \qquad (2.9)$$

where the second equality in (2.9) is valid iff A and B are independent. Because of the equivalence of independence with (2.8), the latter can be used as a definition of independence.

Example 2.34   Two different record companies, X and Y, both produce classical music recordings. Label X is a "budget" label, and 5% of X's new records exhibit a significant degree of warpage. Label Y is manufactured under more stringent quality-control conditions (and sold at a higher price) than X, so only 2% of its new pressings are warped. If you purchase one label X recording and one label Y recording at your local record store, what is the probability that both records are warped?

To answer this question, define events A and B by $A = \{$X record is warped$\}$ and $B = \{$Y record is warped$\}$. Then $P(A) = .05$, $P(B) = .02$, and we wish $P(A \cap B)$. Because labels X and Y have no relationship to one another, it is reasonable to assume that A and B are independent events, whence $P(A \cap B) = P(A) \cdot P(B) = .001$.   ■

Example 2.35   A minicomputer salesperson makes either one or two sales contacts on any given day, with probabilities .6 and .4, respectively. If only one contact is made, the probability is .2 that a sale will result and .8 that no sale will result. If two contacts are made, the two customers will make their purchase decisions independently of one another, each purchasing with probability .2 and not purchasing with probability .8. What is the probability that the salesperson has made two sales at the end of the day? To begin,

$$P(\text{two sales}) = P(\text{two contacts and both purchase})$$

$$= P(\text{both purchase} \mid \text{two contacts}) \cdot P(\text{two contacts})$$

Since, given that two contacts are made, the two purchase decisions are independent, $P(\text{both purchase} \mid \text{two contacts}) = (.2)(.2) = .04$, while $P(\text{two contacts}) = .4$. This gives $P(\text{two sales}) = (.04)(.4) = .016$.   ■

---

*iff is an abbreviation for "if and only if."

Independence of More Than Two Events

The notion of independence of two events can be extended to collections of more than two events. While it is possible to extend the definition for two independent events by working in terms of conditional and unconditional probabilities, it is more direct and less cumbersome to proceed along the lines of the last proposition.

Definition

> Events $A_1, \ldots, A_n$ are **mutually independent** if for every k, $k = 2, 3, \ldots, n$, and every subset of indices $i_1, i_2, \ldots, i_k$,
> $$P(A_{i_1} \cap A_{i_2} \cap \cdots \cap A_{i_k}) = P(A_{i_1}) \cdot P(A_{i_2}) \cdots P(A_{i_k}).$$

Paraphrasing the definition, the events are mutually independent if the probability of the intersection of any subset of the n events is equal to the product of the individual probabilities. When $P(A_i \cap A_j) = P(A_i) \cdot P(A_j)$ for every pair i, j with $i \neq j$, the events are said to be **pairwise independent.** Intuitively, pairwise independence means that occurrence of any single one of the A_i's does not affect the chance that any other single one might occur. The next example shows that mutual independence is a more general notion than pairwise independence.

Example 2.36

The winner of a contest will be rewarded with one or more of three different prizes in the following manner. Four slips of paper marked 1, 2, 3, and (1, 2, 3), respectively, will be placed in a box. After the box is shaken, one slip will be withdrawn, and the contest winner will receive the prize (or prizes) whose number (numbers) is (are) on the slip. Let A_i ($i = 1, 2, 3$) denote the event that the ith prize is awarded. Notice that the ith prize will be awarded if either the slip with the number i on it is selected or if the slip with (1, 2, 3) on it is selected. Assigning probability $\frac{1}{4}$ to each of the four slips then gives $P(A_1) = P(A_2) = P(A_3) = \frac{1}{2}$; $P(A_1 \cap A_2) = P$ (win both prizes 1 and 2) $= P[(1, 2, 3)$ is selected$] = \frac{1}{4}$; and similarly $P(A_1 \cap A_3) = P(A_2 \cap A_3) = \frac{1}{4}$. Since $P(A_i \cap A_j) = P(A_i) \cdot P(A_j)$ for all i, j with $i \neq j$, the three events are pairwise independent. However, $P(A_1 \cap A_2 \cap A_3) = P[(1, 2, 3)$ selected$] = \frac{1}{4} \neq P(A_1) \cdot P(A_2) \cdot P(A_3) = \frac{1}{8}$, so the events are not mutually independent. That is, while the unconditional probability of winning prize 1 is $\frac{1}{2}$, if we are told that A_2 and A_3 both occurred, it must have been the case that the slip marked (1, 2, 3) was selected, so that the conditional probability of A_1 is 1. Given only that A_2 occurred, both the slip marked 2 and the one marked (1, 2, 3) are possible, so the conditional probability equals the unconditional probability. To summarize, mutual independence implies pairwise independence, but the reverse implication is not correct. ∎

As was the case with two events, we frequently specify at the outset of a problem the independence of certain events. The definition can then be used to calculate the probability of an intersection.

Example **2.37**   A system consists of four components as illustrated in Figure 2.11. The entire system will work if either the 1–2 subsystem works or if the 3–4 subsystem works (since the two subsystems are connected in parallel). Since the two components in each subsystem are connected in series, a subsystem will work only if

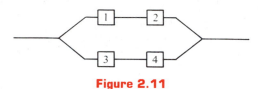

Figure 2.11

both its components work. If components work or fail independently of one another, and if each works with probability .9, what is the probability that the entire system will work (the system reliability coefficient)? Letting A_i ($i = 1, 2, 3, 4$) be the event that the ith component works, the A_i's are mutually independent. The event that the 1–2 subsystem works is $A_1 \cap A_2$, and similarly $A_3 \cap A_4$ denotes the event that the 3–4 subsystem works. The event that the system works is $(A_1 \cap A_2) \cup (A_3 \cap A_4)$, so

$$P[(A_1 \cap A_2) \cup (A_3 \cap A_4)]$$
$$= P(A_1 \cap A_2) + P(A_3 \cap A_4) - P[(A_1 \cap A_2) \cap (A_3 \cap A_4)]$$
$$= P(A_1) \cdot P(A_2) + P(A_3) \cdot P(A_4) - P(A_1) \cdot P(A_2) \cdot P(A_3) \cdot P(A_4)$$
$$= (.9)(.9) + (.9)(.9) - (.9)(.9)(.9)(.9) = .9639$$

Letting $x = P(A_i)$ for $i = 1, 2, 3, 4$, what value of x would yield a system reliability of .99? Proceeding analogously, $P(\text{system works}) = x^2 + x^2 - x^4 = .99$ or $y^2 - 2y + .99 = 0$ where $y = x^2$. Solving this quadratic gives $y = .9$, so $x = \sqrt{.9} \approx .95$ (recall that $\approx$ means approximately equal). To achieve a system reliability of .99 without introducing more components or changing the system configuration, the reliability of each component would have to be increased from .9 to .95. ∎

Exercises / Section 2.5 (55–68)

55. If $P(A_1) = .7$, $P(A_2) = .5$, and $P(A_1 \cap A_2) = .4$ as in Exercise 11, show that A_1 and A_2 are dependent first by using the definition of independence and then by verifying that the multiplication property does not hold.

56. In Exercise 12, is any A_i independent of any other A_j? Answer using the multiplication property for independent events.

57. An executive has both a morning and an afternoon meeting on a particular day. Let $A = \{$late to the morning meeting$\}$ and $B = \{$late to the afternoon meeting$\}$.
   a. If $P(A) = .4$, $P(B) = .5$, and $P(A \cap B) = .25$, are A and B independent events?
   b. If A and B are independent events with $P(A) = .4$ and $P(B) = .5$, what is the probability that the executive is on time to both meetings? To exactly one meeting?

58. The probability that a grader will make a marking error on any particular question of a multiple choice exam is .1. If there are n questions and questions are marked independently, what is the probability that no errors are made? That at least one error is made? If the probability of a marking error is p rather than .1, answer these two questions.

59. Referring back to Example 2.35, calculate the probability that the salesperson has made exactly one sale at the end of the day. *Hint:* One sale can be made with either one contact or with two contacts; if there are two contacts, the first could result in a purchase and the second in no purchase or vice versa. A tree diagram might be helpful.

60. Consider the system of components connected as in the accompanying picture. Components 1 and 2 are connected in parallel, so that subsystem works iff either 1 or 2 works; since 3 and 4 are connected in series, that subsystem works iff both 3 and 4 work. If components work independently of one another and P(component works) = .9, calculate P(system works).

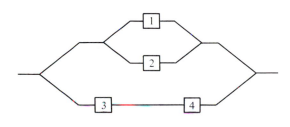

61. Components arriving at a distributor are checked for defects by two different inspectors (each component is checked by both inspectors). The first inspector detects 90% of all defectives that are present, and the second inspector does likewise. At least one inspector does not detect a defect on 20% of all defective components. What is the probability that
   a. a defective component will be detected only by the first inspector? By exactly one of the two inspectors?
   b. all three defective components in a batch escape detection by both inspectors (assuming inspections of different components are independent of one another)?

62. Suppose that with probability $\frac{2}{3}$, a randomly selected vehicle will pass inspection at a headlight inspection station. Assuming that successive vehicles pass or fail independently of one another, calculate the following probabilities.
   a. P(all of the next three vehicles inspected pass).
   b. P(at least one of the next three inspected fails).
   c. P(exactly one of the next three inspected passes). *Hint:* With A = {exactly one passes}, $A = E_1 \cup E_2 \cup E_3$ where E_i = {ith passes and the other two don't}.
   d. P(at most one of the next three vehicles inspected passes).
   e. Given that at least one of the next three vehicles passes inspection, what is the probability that all three pass (a conditional probability)?

63. A quality-control inspector is inspecting newly produced items for faults. The inspector searches an item for faults in a series of independent fixations, each of a fixed duration. Given that a flaw is actually present, let p denote the probability that the flaw is detected during any one fixation (this model is discussed in "Human Performance in Sampling Inspection," *Human Factors,* 1979, pp. 99–105).
   a. Assuming that an item has a flaw, what is the probability that it is detected by the end of the second fixation (once a flaw has been detected, the sequence of fixations terminates). *Hint:* {flaw detected in at most 2 fixations} = {flaw detected on the first fixation} ∪ {flaw undetected on the first and detected on the second}.
   b. Give an expression for the probability that a flaw will be detected by the end of the nth fixation.
   c. If when a flaw has not been detected in three fixations, the item is passed, what is the probability that a flawed item will pass inspection?
   d. Suppose that 10% of all items contain a flaw [P(randomly chosen item is flawed) = .1]. With the assumption of (c), what is the probability that a randomly chosen item will pass inspection (it will automatically pass if it is not flawed, but could also pass if it is flawed)?
   e. Given that an item has passed inspection (no flaws in three fixations), what is the probability that it is actually flawed? Calculate for p = .5.

64. a. A lumber company has just taken delivery on a lot of 10,000 2×4 boards. Suppose that 20% of these boards (2000) are actually too green to be used in first-quality construction. Two boards are selected at random, one after the other. Let $A =$ {the first board is green} and $B =$ {the second board is green}. Compute $P(A)$, $P(B)$, and $P(A \cap B)$ (a tree diagram might help). Are A and B independent?

b. With A and B independent and $P(A) = P(B) = .2$, what is $P(A \cap B)$? How much difference is there between this answer and $P(A \cap B)$ in (a)? For purposes of calculating $P(A \cap B)$, can we assume that A and B of (a) are independent to obtain essentially the correct probability?

c. Suppose that the lot consists of ten boards, of which two are green. Does the assumption of independence now yield approximately the correct answer for $P(A \cap B)$? What is the critical difference between the situation here and that of (a)? When do you think that an independence assumption would be valid in obtaining an approximately correct answer to $P(A \cap B)$?

65. Refer back to the assumptions stated in Exercise 60, and answer the question posed there for the system in the accompanying picture. How would the probability change if this were a subsystem connected in parallel to the subsystem pictured in Example 2.37?

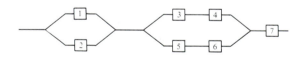

66. Professor Stan der Deviation can take one of two routes on his way home from work. On the first route, there are four railroad crossings. The prob-ability that he will be stopped by a train at any particular one of the crossings is .1, and trains operate independently at the four crossings. The other route is longer but there are only two cross-ings, independent of one another, with the same stoppage probability for each as on the first route. On a particular day, Professor Deviation has a meeting scheduled at home for a certain time. Whichever route he takes, he calculates that he will be late if he is stopped by trains at at least half the crossings encountered.

a. Which route should he take to minimize the probability of being late to the meeting?

b. If he tosses a fair coin to decide on a route and he is late, what is the probability that he took the four-crossing route?

67. In Exercise 6, suppose that Jack Twain is at loca-tion A on the diagram and that his sister Jill is at location I. At exactly the same instant, Jack and Jill decide to leave their respective locations and journey toward one another (Jack traveling north and east only and Jill traveling south and west). If they walk at exactly the same pace and toss a fair coin independently at each corner location at which there is a choice of direction, what is the probability that the Twains meet at G? What is the probability that the Twains do not meet?

68. Suppose that identical tags are placed on both the left ear and the right ear of a fox. The fox is then let loose for a period of time. Consider the two events $C_1 =$ {left ear tag is lost} and $C_2 =$ {right ear tag is lost}. Let $\pi = P(C_1) = P(C_2)$, and assume that C_1 and C_2 are independent events. Derive an expression (involving π) for the probability that exactly one tag is lost given that at most one is lost ("Ear Tag Loss in Red Foxes," *J. Wildlife Mgmt.,* 1976, pp. 164–167). *Hint:* Draw a tree diagram in which the two initial branches refer to whether or not the left ear tag was lost.

Supplementary Exercises / Chapter 2 (69–86)

69. A small manufacturing company will start oper-ating a night shift. There are 20 machinists em-ployed by the company.

a. If a night crew consists of three machinists, how many different crews are possible?

b. If the machinists are ranked 1, 2, . . . , 20 in order of competence, how many of these crews would not have the best machinist?

c. How many of the crews would have at least one of the best 10 machinists?

d. If one of these crews is selected at random to work on a particular night, what is the probability that the best machinist will not work that night?

70. A claims officer at a Social Security office will have time to examine six claims during a particular day. There are 10 claims on her desk, of which four concern disability and six concern old-age benefits. If claims are selected at random from the 10, what is the probability that all disability claims will have been examined by the end of the day? What is the probability that by the end of the day, only one of the two types of claims will remain on her desk?

71. Let A be the event that a certain book on fluid mechanics is presently checked out of the university library, and let B be the event that another book on the same subject is checked out.
a. If $P(A) = .5$, $P(B) = .4$, and $P(A \cup B) = .65$, compute $P(A \cap B)$.
b. Using the probabilities of (a), compute P(exactly one of the two books is checked out). *Hint:* Venn diagram.
c. If $P(A \cup B) = .7$, $P(A \cap B) = .2$, and P(only the first book is checked out) $= .4$, compute $P(A)$ and $P(B)$. *Hint:* Venn diagram.
d. If $P(A \cup B) = .7$, $P(A \cap B) = .2$, and P(exactly one book is checked out) $= .5$, can $P(A)$ and $P(B)$ be determined?

72. Suppose that I fly to New York on airline X and return to Los Angeles on airline Y. Let $A = \{$X loses my baggage$\}$ and $B = \{$Y loses my baggage$\}$. If A and B are independent events with $P(A) > P(B)$, $P(A \cap B) = .0002$, and $P(A \cup B) = .03$, determine $P(A)$ and $P(B)$.

73. Individual A has a circle of five close friends (B, C, D, E, and F). A has heard a certain rumor from outside the circle and has invited the five friends to a party in order to circulate the rumor. To begin, A selects one of the five at random and tells the rumor to the chosen individual. That individual then selects at random one of the four remaining individuals and repeats the rumor. Continuing, a new individual is selected from those not already having heard the rumor by the individual who has just heard it, until everyone has been told. What is the probability that the rumor is repeated in the order B, C, D, E, and F?

74. a. In Exercise 73, what is the probability that F is the third person at the party to be told the rumor?
b. What is the probability that F is the last person to hear the rumor?

75. Referring to Exercise 73, if at each stage the person who currently "has" the rumor does not know who has already heard it and selects the next recipient at random from all five possible individuals, what is the probability that F has still not heard the rumor after it has been told 10 times at the party?

76. An automobile insurance company classifies each driver as a good risk (A_1), a medium risk (A_2), or a poor risk (A_3). Of those currently insured, 30% are good risks, 50% are medium risks, and 20% are poor risks. In any given year the probability that a driver will have at least one accident is .1 for a good risk, .3 for a medium risk, and .5 for a poor risk. If a randomly selected driver insured by this company has an accident during the next year, what is the probability that the driver was actually a good risk? A medium risk?

77. In Exercise 76, suppose that each insuree has a three-year policy. If years are independent and a randomly selected driver reports no accidents during the three years, what is the probability that the driver was actually a good risk?

78. A chemical engineer is interested in determining if a certain trace impurity is present in a product. An experiment has a probability of .80 of detecting the impurity if it is present. The probability of not detecting the impurity if it is absent is .90. The prior probabilities of the impurity being present and being absent are .40 and .60, respectively. Three separate experiments result in only two detections. What is the posterior probability that the impurity is present?

79. Each contestant on a quiz show is asked to specify one of six possible categories from which questions will be asked. Suppose that P(contestant requests category i) $= \frac{1}{6}$ and that successive contestants choose their categories independently of one another. If there are three contestants on each show and all three contestants on a particular show select different categories, what is the probability that exactly one has selected category 1?

80. One method used to distinguish between granitic (G) and basaltic (B) rocks is to examine a portion of the infrared spectrum of the sun's energy reflected from the rock surface. Let R_1, R_2, and R_3 denote measured spectrum intensities at three different wavelengths; typically, for granite $R_1 < R_2 < R_3$ while for basalt $R_3 < R_1 < R_2$. When measurements are made remotely (using aircraft), various orderings of the R_i's may arise whether the rock is basalt or granite.

Flights over regions of known composition have yielded the following information.

	Granite	Basalt
$R_1 < R_2 < R_3$	60%	10%
$R_1 < R_3 < R_2$	25%	20%
$R_3 < R_1 < R_2$	15%	70%

Suppose that for a randomly selected rock in a certain region, $P(\text{granite}) = .25$ and $P(\text{basalt}) = .75$.

a. Show that $P(\text{granite} \mid R_1 < R_2 < R_3) > P(\text{basalt} \mid R_1 < R_2 < R_3)$. If measurements yielded $R_1 < R_2 < R_3$, would you classify the rock as granite or basalt?

b. If measurements yielded $R_1 < R_3 < R_2$, how would you classify the rock? Same question for $R_3 < R_1 < R_2$.

c. Using the classification rule indicated in (a) and (b), when selecting a rock from this region, what is the probability of an erroneous classification? *Hint:* Either G could be classified as B or B as G, and $P(B)$ and $P(G)$ are known.

d. If $P(\text{granite}) = p$ rather than .25, are there values of p (other than 1) for which one would always classify a rock as granite?

81. A subject is allowed a sequence of glimpses to detect a target. Let $G_i = \{$the target is detected on the ith glimpse$\}$, with $p_i = P(G_i)$. Suppose that the G_i's are independent events, and write an expression for the probability that the target has been detected by the end of the nth glimpse. *Note:* This model is discussed in "Predicting Aircraft Detectability," *Human Factors* (1979, pp. 277–291).

82. In a Little League baseball game, team A's pitcher throws a strike 50% of the time and a ball 50% of the time, successive pitches are independent of one another, and the pitcher never hits a batter. Knowing this, team B's manager has instructed the first batter not to swing at anything. Calculate the probability that

a. the batter walks on the fourth pitch.

b. the batter walks on the sixth pitch (so two of the first five must be strikes), using a counting argument or constructing a tree diagram.

c. the batter walks.

d. What is the probability that the first batter up scores while no one is out (assuming that each batter pursues a no-swing strategy)?

83. Four engineers, A, B, C, and D, have been scheduled for job interviews at 10 A.M. on Friday, January 13, at Random Sampling, Inc. The personnel manager has scheduled the four for interview rooms 1, 2, 3, and 4, respectively. However, the manager's secretary does not know this, so assigns them to the four rooms in a completely random fashion (what else!). What is the probability that

a. all four end up in the correct rooms?

b. none of the four ends up in the correct room?

84. A particular airline has 10 A.M. flights from Chicago to New York, Atlanta, and Los Angeles. Let A denote the event that the New York flight is full, and define events B and C analogously for the other two flights. Suppose that $P(A) = .6$, $P(B) = .5$, $P(C) = .4$, and that the three events are independent. What is the probability that

a. all three flights are full? That at least one flight is not full?

b. only the New York flight is full? That exactly one of the three flights is full?

85. A personnel manager is to interview four candidates for a job. These are ranked 1, 2, 3, and 4 in order of preference, and will be interviewed in random order. However, at the conclusion of each interview, the manager will know only how the current candidate compares to those previously interviewed. For example, the interview order 3, 4, 1, 2 generates no information after the first interview, shows that the second candidate is worse than the first, and that the third is better than the first two. However, the order 3, 4, 2, 1 would generate the same information after each of the first three interviews. The manager wants to hire the best candidate, but must make an irrevocable hire–no hire decision after each interview. Consider the strategy: automatically reject the first s candidates, and then hire the first subsequent candidate who is best among those already interviewed (if no such candidate appears, the last one interviewed is hired).

For example, with $s = 2$ the order 3, 4, 1, 2 would result in the best being hired, while 3, 1, 2, 4 would not. Of the four possible s values (0, 1, 2, and 3), which one maximizes P(best is hired)?

Hint: Write out the 24 equally likely interviewing orderings; $s = 0$ means that the first candidate is automatically hired.

86. If A and B are independent events, show that A' and B' are also. *Hint:* $A' \cap B' = (A \cup B)'$. Use this and the addition rule for the probability of a union.

Bibliography

Derman, Cyrus, Gleser, Leon, and Olkin, Ingram, *Probability Models and Applications,* Macmillan, New York, 1980. A comprehensive introduction to probability, written at a slightly higher mathematical level than this text but containing many good examples.

Larsen, Richard, and Marx, Morris, *Introduction to Mathematical Statistics* (2nd ed.), Prentice-Hall, Englewood Cliffs, N. J., 1985. Includes a nice concise exposition of probability written at a relatively modest mathematical level.

Mosteller, Frederick, Rourke, Robert, and Thomas, George, *Probability with Statistical Applications* (2nd ed.), Addison-Wesley, Reading, Mass., 1970. A very good precalculus introduction to probability, with many entertaining examples; especially good on counting rules and their application.

Ross, Sheldon, *A First Course in Probability* (3rd ed.), Macmillan, New York, 1985. Rather tightly written and more mathematically sophisticated than this text, but contains a wealth of interesting examples and exercises.

Winkler, Robert, *Introduction to Bayesian Inference and Decision,* Holt, Rinehart & Winston, New York, 1972. A very good introduction to subjective probability.

CHAPTER 3

Discrete Random Variables and Probability Distributions

Introduction

Whether an experiment yields qualitative or quantitative outcomes, methods of statistical analysis require that we focus on certain numerical aspects of the data (such as a sample proportion x/n, mean $\bar{x}$, or standard deviation s). The concept of a random variable allows us to pass from the experimental outcomes themselves to a numerical function of the outcomes. There are two fundamentally different types of random variables—discrete random variables and continuous random variables. In this chapter we examine the basic properties and discuss the most important examples of discrete variables, and we will study continuous variables in Chapter 4.

3.1   Random Variables

In any experiment there are numerous characteristics that can be observed or measured, but in most cases an experimenter will focus on some specific aspect or aspects of a sample. For example, in a study of commuting patterns in a metropolitan area, each individual in a sample might be asked about commuting distance and the number of people commuting in the same vehicle, but not about IQ, income, family size, and other such characteristics. Alternatively, a researcher may test a sample of components and record only the number that have failed within 1000 hours, rather than recording the individual failure times.

In general, each outcome of an experiment can be associated with a number by specifying a rule of association (for example, the number among the sample of 10 components that fail to last 1000 hours, or the total weight of baggage for a sample of 25 airline passengers). Such a rule of association is called a **random variable**—a variable because different numerical values are possible, and random because the observed value depends on which of the possible experimental outcomes results.

<table>
<tr><td>Definition</td><td>For a given sample space $\mathcal{S}$ of some experiment, a **random variable** is any rule that associates a number with each outcome in $\mathcal{S}$.</td></tr>
</table>

Figure 3.1   A random variable

We will often use the abbreviation r.v. in place of random variable. As is customary in probability and statistics, capital letters such as X or Y will be used to denote random variables. The notation $X(s) = x$ means that x is the number associated with the outcome s by the random variable X, so x is called the value of the variable associated with s.

Example 3.1   When a student attempts to log on to a computer time-sharing system, either all ports could be busy (F), in which case the student will fail to obtain access, or else there will be at least one port free (S), in which case the student will be successful in accessing the system. With $\mathcal{S} = \{S, F\}$, define a random variable X by

$$X(S) = 1, \quad X(F) = 0$$

The r.v. X indicates whether (1) or not (0) the student can log on.   ■

In Example 3.1, the r.v. X was specified by explicitly listing each element of $\mathcal{S}$ and the associated number. If $\mathcal{S}$ contains more than a few outcomes, such a listing is tedious, but it can frequently be avoided.

Example 3.2   Consider the experiment in which an automobile with California license plates is selected, and define a random variable Y by

$$Y = \begin{cases} 1 & \text{if the selected automobile uses unleaded gas} \\ 0 & \text{if the selected automobile does not use unleaded gas} \end{cases}$$

For example, if California license UZF002 uses unleaded gas, then $Y(\text{UZF002}) = 1$, while $Y(\text{AXJ375}) = 0$ tells us that the car with license AXJ375 does not use unleaded gas. A word description of this sort is more economical than a complete listing, so wherever possible we shall use such a description.   ■

In Examples 3.1 and 3.2, the only possible values of the random variable were 0 and 1. Such a random variable arises frequently enough to be given a special name, after the individual who first studied it.

Definition

> Any random variable whose only possible values are 0 and 1 is called a **Bernoulli random variable.**

We will often want to define and study several different random variables from the same sample space.

Example **3.3**

In the experiment in which a red and a green die are each rolled once (Example 2.3), there are 36 outcomes. Define random variables X, Y, and U by

X = sum of the two resulting numbers,

Y = difference between the number on the red die and the number on the green die, and

U = the maximum of the two resulting numbers.

If this experiment is performed and $s = (2, 3)$ results, then $X((2, 3)) = 2 + 3 = 5$, so we say that the observed value of X was $x = 5$. Similarly, the observed value of Y would be $y = 2 - 3 = -1$, and the observed value of U would be $u = \max(2, 3) = 3$. ∎

An Infinite Set of Possible *X* Values

Each of the random variables of Examples 3.1–3.3 can assume only a finite number of possible values. This need not be the case.

Example **3.4**

In Example 2.4 we considered the experiment in which batteries coming off an assembly line were examined until a good one (S) was obtained. The sample space was $\mathscr{S} = \{S, FS, FFS, \ldots\}$. Define a r.v. X by

X = the number of batteries examined before the experiment terminates

Then $X(S) = 1$, $X(FS) = 2$, $X(FFS) = 3, \ldots, X(FFFFFFS) = 7$, and so on. Any positive integer is a possible value of X, so the set of possible values is infinite. ∎

Example **3.5**

Suppose that in some random fashion, a location (latitude and longitude) in the continental United States is selected, and define a r.v. Y by

Y = the height above sea level at the selected location

For example, if the selected location were $(39°50' \text{ N}, 98°35' \text{ W})$, then we might have $Y((39°50' \text{ N}, 98°35' \text{ W})) = 1748.26$ ft. The largest possible value of Y is 14,494 (Mt. Whitney) and the smallest possible value is -282 (Death

Valley). The set of all possible values of Y is the set of all numbers in the interval between -282 and $14{,}494$—that is,

$$\{y: y \text{ is a number}, -282 \leq y \leq 14{,}494\}$$

and there are an infinite number of numbers in this interval.   ■

Discrete Random Variables

Although both X of Example 3.4 and Y of Example 3.5 can assume any one of an infinite number of possible values, the two infinite sets are actually quite different from one another. The variable X is more closely related to the finite valued variables of Examples 3.1–3.3 than to Y above.

Definition

> A set is **discrete** either if it consists of a finite number of elements, or if its elements can be listed so that there is a first element, a second element, a third element, and so on, in the list.

A discrete set that is infinite is sometimes called countably infinite, since according to the definition we can count $(1, 2, 3, \ldots)$ its elements. Most of the infinite discrete sets we shall encounter will consist either of the nonnegative integers $(\{0, 1, 2, \ldots\})$ or some subset of these integers. The set of possible values of the r.v. X in Example 3.4 is $D = \{1, 2, 3, \ldots\}$, which is clearly discrete.

That there are infinite sets that are not discrete is a profound and important result from pure mathematics. If D is any interval of real numbers, such as $\{x: -282 \leq x \leq 14{,}494\}$ of Example 3.5, then it can be shown that there is no way to list (or count) the elements of this set. More will be said about this at the outset of Chapter 4.

Definition

> A random variable is said to be **discrete** if its set of possible values is a discrete set.

Example **3.6**

All random variables of Examples 3.1–3.4 are discrete. As another example, suppose that we select married couples at random and do a blood test on each person until we find a husband and wife who both have the same Rh factor. With $X = $ the number of blood tests to be performed, possible values of X are $D = \{2, 4, 6, 8, \ldots\}$. Since the possible values have been listed in sequence, D is a discrete set, so X is a discrete random variable.   ■

To study basic properties of discrete random variables, only the tools of discrete mathematics—summation and differences—are required, while the study of continuous variables requires the continuous mathematics of the calculus—integrals and derivatives. This is the reason for making the distinction between the two types of variables.

Exercises / Section 3.1 [1–10]

1. Three automobiles are selected at random, and each is categorized as having a diesel (S) or nondiesel (F) engine (so outcomes are SSS, SSF, and so on). If $X =$ the number of cars among the three with diesel engines, list each outcome in $\mathscr{S}$ and its associated X value.

2. Give three examples of Bernoulli random variables (other than those in the text).

3. Using the experiment in Example 3.3, define two more random variables and list the possible values of each.

4. Let $X =$ the number of nonzero digits in a randomly selected zip code. What are the possible values of X? Give three possible outcomes and their associated X values.

5. If the sample space $\mathscr{S}$ is an infinite set, does this necessarily imply that any r.v. X defined from $\mathscr{S}$ will have an infinite set of possible values? If yes, say why. If no, give an example.

6. Starting at a fixed time, each car entering an intersection is observed to see whether it turns left (L), right (R), or goes straight ahead (A). The experiment terminates as soon as a car is observed to turn left. Let $X =$ the number of cars observed. What are possible X values? List five outcomes and their associated X values.

7. For each random variable defined below, describe the set of possible values for the variable and state whether or not the variable is discrete.
   a. $X =$ the number of unbroken eggs in a randomly chosen standard egg carton.
   b. $Y =$ the number of students on a class list for a particular course who are absent on the first day of classes.
   c. $U =$ the number of times that a duffer has to swing at a golf ball before hitting it.
   d. $X =$ the length of a randomly selected rattlesnake.
   e. $Z =$ the amount of royalties earned from the sale of a first edition of 10,000 textbooks.
   f. $Y =$ the pH of a randomly chosen soil sample.
   g. $X =$ the tension (p.s.i.) at which a randomly selected tennis racket has been strung.
   h. $X =$ the total number of coin tosses required for three individuals to obtain a match (HHH or TTT).

8. Each time a component is tested, the trial is a success (S) or failure (F). Suppose that the component is tested repeatedly until a success occurs on three consecutive trials. Let Y denote the number of trials necessary to achieve this. List all outcomes corresponding to the five smallest possible values of Y, and state which Y value is associated with each one.

9. Consider an experiment in which an individual named Claudius is located at the point 0 in the diagram below.

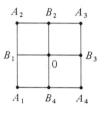

Using an appropriate randomization device (such as a tetrahedral die, one having four sides), Claudius first moves to one of the four locations B_1, B_2, B_3, B_4. Once at one of these locations, another randomization device is used to decide whether Claudius next returns to 0 or next visits one of the two other adjacent points. The experiment then continues in this fashion; after each move, another move to one of the (new) adjacent points is determined by tossing an appropriate die or coin.
   a. Let $X =$ the number of moves that Claudius makes before first returning to 0. What are possible values of X? Is X discrete or continuous?
   b. If moves are allowed also along the diagonal paths connecting 0 to A_1, A_2, A_3, and A_4, respectively, answer the questions in part (a).

10. If first an ordinary six-sided die is tossed, after which a four-sided die (tetrahedral, with faces marked 1, 2, 3, and 4) is tossed, list the possible values for the following random variables:
    a. $T =$ the total of the two numerical outcomes.
    b. $X =$ the absolute value of the difference between the two outcomes.

c. $Y =$ the number of tosses for which the outcome is even.

d. $Z =$ the number of tosses for which the outcome is six.

3.2 Probability Distributions for Discrete Random Variables

When probabilities are assigned to various outcomes in $\mathscr{S}$, these in turn determine probabilities associated with the values of any particular random variable X. The *probability distribution* of X says how the total probability of 1 is distributed among (allocated to) the various possible X values.

Example 3.7   Two friends, A and B, have made appointments at the student health center for 2 P.M. on a certain day. Two other students, C and D, also have appointments scheduled at that time. There is only one doctor on duty. Suppose that the students are seen by the doctor in random order, so that any one of the $4! = 24$ possible orderings has the same probability. Let $X =$ the number of students examined before A and B are both examined and can leave together. Then X is a discrete random variable with possible values 2, 3, and 4. If, for example, the ordering is $ACBD$, then $X = 3$, whereas if students are seen in the order $CDBA$, then $X = 4$. By listing all experimental outcomes (orderings), it can be verified that $X = 2$ for four outcomes, $X = 3$ for eight outcomes, and $X = 4$ for 12 outcomes. Since the outcomes are equally likely, the probabilities associated with these three X values are

$$p(2) = P(X = 2) = \frac{4}{24} = \frac{1}{6}$$

$$p(3) = P(X = 3) = \frac{8}{24} = \frac{1}{3}$$

$$p(4) = P(X = 4) = \frac{12}{24} = \frac{1}{2}$$

$$p(x) = P(X = x) = 0 \text{ for } x \neq 2, 3, \text{ or } 4$$

The values of X along with their probabilities collectively specify the probability distribution or *probability mass function* of X. If this experiment were repeated over and over again, in the long run $X = 2$ would occur one-sixth of the time, $X = 3$ one-third of the time, and $X = 4$ half of the time. ■

Definition

> The **probability distribution** or **probability mass function** (p.m.f.) of a discrete random variable is defined for every number x by $p(x) = P(X = x) = P(\text{all } s \in \mathscr{S}: X(s) = x).$*

---

*$P(X = x)$ is read "the probability that the r.v. X assumes the value x." For example, $P(X = 2)$ denotes the probability that the resulting X value is 2.

In words, for every possible value x of the random variable, the p.m.f. specifies the probability of observing that value when the experiment is performed. The conditions $p(x) \geq 0$ and $\underset{\text{all possible } x}{\Sigma p(x)} = 1$ are required of any p.m.f.

Example 3.8   Suppose we go to a large tire store during a particular week and observe whether the next customer to purchase tires purchases a radial or a bias-ply tire. Let

$$X = \begin{cases} 1 & \text{if the customer purchases a radial tire} \\ 0 & \text{if the customer purchases a bias-ply tire} \end{cases}$$

If 60% of all purchasers during that week select radials, the p.m.f. for X is

$$p(0) = P(X = 0) = P(\text{next customer purchases a bias-ply tire}) = .4$$
$$p(1) = P(X = 1) = P(\text{next customer purchases a radial tire}) = .6$$
$$p(x) = P(X = x) = 0 \text{ for } x \neq 0 \text{ or } 1$$

An equivalent description is

$$p(x) = \begin{cases} .4 & \text{if} \quad x = 0, \\ .6 & \text{if} \quad x = 1, \\ 0 & \text{if} \quad x \neq 0 \text{ or } 1. \end{cases}$$

A picture of this p.m.f., called a line graph, appears in Figure 3.2.

Figure 3.2   The line graph for the p.m.f. in Example 3.8   ■

Example 3.9   Consider a group of five potential blood donors—A, B, C, D, and E—of whom only A and B have type O+ blood. Five blood samples, one from each individual, will be typed in random order until an O+ individual is identified. Let the r.v. Y = the number of typings necessary to identify an O+ individual. Then the p.m.f. of Y is

$$p(1) = P(Y = 1) = P(\text{A or B typed first}) = \frac{2}{5} = .4$$

$$p(2) = P(Y = 2) = P(\text{C, D, or E first, and then A or B})$$
$$= \left(\frac{3}{5}\right)\left(\frac{2}{4}\right) = .3$$

$$p(3) = P(Y = 3) = P(\text{C, D, or E first and second, and then A or B})$$

$$= \left(\frac{3}{5}\right)\left(\frac{2}{4}\right)\left(\frac{2}{3}\right) = .2$$

$$p(4) = P(Y = 4) = P(\text{C, D, and E all done first}) = \left(\frac{3}{5}\right)\left(\frac{2}{4}\right)\left(\frac{1}{3}\right) = .1$$

$$p(y) = 0 \quad \text{if} \quad y \neq 1, 2, 3, 4$$

The p.m.f. can be presented nicely in tabular form:

y	1	2	3	4
$p(y)$	.4	.3	.2	.1

where any y value not listed receives zero probability. This p.m.f. can also be displayed in a line graph (Figure 3.3).

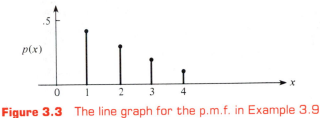

Figure 3.3   The line graph for the p.m.f. in Example 3.9   ■

The name "probability mass function" is suggested by a model used in physics for a system of "point masses." In this model masses are distributed at various locations x along a one-dimensional axis. Our p.m.f. describes how the total probability mass of 1 is distributed at various points along the axis of possible values of the random variable (where, and how much mass at each x).

Another useful pictorial representation of a p.m.f., called a **probability histogram,** is similar to histograms discussed in Chapter 1. Above each y with $p(y) > 0$, construct a rectangle centered at y. The height of each rectangle is proportional to $p(y)$, and the base is the same for all rectangles. When possible values are equally spaced, the base is frequently chosen as the distance between successive y values (though it could be smaller). Figure 3.4 shows two probability histograms.

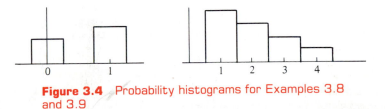

Figure 3.4   Probability histograms for Examples 3.8 and 3.9

A Parameter of a Probability Distribution

In Example 3.8 we were able to obtain $p(x)$ because we had specific information about the probabilities of various outcomes (radial or bias ply). Even without this specific information, we can write the general form of the p.m.f. Suppose that

P(the next customer purchases a radial tire) $= \alpha$

and

P(the next customer purchases a bias-ply tire) $= 1 - \alpha$

where $0 < \alpha < 1$ but α is otherwise unspecified. Then the p.m.f. can be written

$p(0) = P(X = 0) = P(\text{bias ply}) = 1 - \alpha$

$p(1) = P(X = 1) = P(\text{radial}) = \alpha$

$p(x) = 0$ for $x \neq 0$ or 1

Because the p.m.f. depends on the particular value of α, we often write $p(x; \alpha)$ rather than just $p(x)$:

$$p(x; \alpha) = \begin{cases} 1 - \alpha & \text{if } x = 0 \\ \alpha & \text{if } x = 1 \\ 0 & \text{otherwise} \end{cases} \tag{3.1}$$

Then each choice of α in (3.1) yields a different p.m.f. If the two events {radial} and {bias ply} are equally likely, then $\alpha = .5$.

Definition

> Suppose that $p(x)$ depends on a quantity that can be assigned any one of a number of possible values, with each different value determining a different probability distribution. Such a quantity is called a **parameter** of the distribution. The collection of all probability distributions for different values of the parameter is called a **family** of probability distributions.

The quantity α in (3.1) is a parameter. Each different number α between 0 and 1 determines a different member of a family of distributions; two such members are

$$p(x; .6) = \begin{cases} .4 & \text{if } x = 0 \\ .6 & \text{if } x = 1 \\ 0 & \text{otherwise} \end{cases} \quad \text{and} \quad p(x; .5) = \begin{cases} .5 & \text{if } x = 0 \\ .5 & \text{if } x = 1 \\ 0 & \text{otherwise} \end{cases}$$

Every probability distribution for a Bernoulli r.v. has the form (3.1), so (3.1) is called the family of Bernoulli distributions.

Example 3.10   Starting at a fixed time, we observe the sex of each newborn child at a certain hospital until a boy (B) is born. Let $p = P(B)$, assume that successive births are independent, and define the r.v. X by $X =$ number of births observed. Then

$$p(1) = P(X = 1) = P(B) = p$$
$$p(2) = P(X = 2) = P(GB) = P(G) \cdot P(B) = (1 - p)p$$

and

$$p(3) = P(X = 3) = P(GGB) = P(G) \cdot P(G) \cdot P(B) = (1 - p)^2 p$$

Continuing in this way, a general formula emerges:

$$p(x) = \begin{cases} (1 - p)^{x-1}p & x = 1, 2, 3, \ldots \\ 0 & \text{otherwise} \end{cases} \tag{3.2}$$

The quantity p in (3.2) represents a number between 0 and 1 and is a parameter of the probability distribution. In the sex example, $p = .51$ might be appropriate, but if we were looking for the first child with Rh positive blood, then we might let $p = .85$. ■

The Cumulative Distribution Function

For some fixed value x, we often wish to compute the probability that the observed value of X will be at most x. For example, the p.m.f. in Example 3.7 was

$$p(x) = \begin{cases} 1/6 & x = 2 \\ 1/3 & x = 3 \\ 1/2 & x = 4 \\ 0 & \text{otherwise} \end{cases}$$

The probability that X is at most 3 is then

$$P(X \leq 3) = p(2) + p(3) = \frac{1}{6} + \frac{1}{3} = \frac{1}{2}$$

In this example, $X \leq 3.5$ iff $X \leq 3$, so $P(X \leq 3.5) = P(X \leq 3) = \frac{1}{2}$. Similarly, $P(X \leq 2) = P(X = 2) = \frac{1}{6}$, and $P(X \leq 2.75) = \frac{1}{6}$ also. Since 2 is the smallest possible value of X, $P(X \leq 1.7) = 0$, $P(X \leq 1.999) = 0$, and so on. The largest possible X value is 4, so $P(X \leq 4) = 1$, and if x is any number larger than 4, $P(X \leq x) = 1$—that is, $P(X \leq 5) = 1$, $P(X \leq 10.23) = 1$, and so on. Notice that $P(X < 4) = \frac{1}{2} \neq P(X \leq 4)$, since the probability of the X value 4 is included in the latter probability but not in the former. When X is a discrete random variable and x is a possible value of X, $P(X < x) < P(X \leq x)$.

Definition

> The **cumulative distribution function** (c.d.f.) $F(x)$ of a discrete random variable X with p.m.f. $p(x)$ is defined for every number x by
>
> $$F(x) = P(X \leq x) = \sum_{y: y \leq x} p(y)$$
>
> For any number x, $F(x)$ is the probability that the observed value of X will be at most x.

Example **3.11**   The p.m.f. of Y for Example 3.9 was

y	1	2	3	4
$p(y)$	.4	.3	.2	.1

We first determine $F(y)$ for each value in the set $\{1, 2, 3, 4\}$ of possible values:

$$F(1) = P(Y \leq 1) = P(Y = 1) = p(1) = .4$$
$$F(2) = P(Y \leq 2) = P(Y = 1 \text{ or } 2) = p(1) + p(2) = .7$$
$$F(3) = P(Y \leq 3) = P(Y = 1 \text{ or } 2 \text{ or } 3) = p(1) + p(2) + p(3) = .9$$
$$F(4) = P(Y \leq 4) = P(Y = 1 \text{ or } 2 \text{ or } 3 \text{ or } 4) = 1$$

Now for any other number y, $F(y)$ will equal the value of F at the closest possible value of Y to the left of y. For example, $F(2.7) = P(Y \leq 2.7) = P(Y \leq 2) = .7$, and $F(3.999) = F(3) = .9$. The c.d.f. is thus

$$F(y) = \begin{cases} 0 & \text{if } y < 1 \\ .4 & \text{if } 1 \leq y < 2 \\ .7 & \text{if } 2 \leq y < 3 \\ .9 & \text{if } 3 \leq y < 4 \\ 1 & \text{if } 4 \leq y \end{cases} \tag{3.3}$$

A graph of $F(y)$ appears in Figure 3.5.

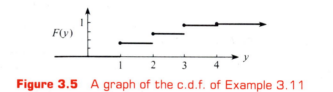

Figure 3.5   A graph of the c.d.f. of Example 3.11   ■

For X a discrete random variable, the graph of $F(x)$ will have a jump at every possible value of X and will be flat between possible values. Such a graph is called a **step function.**

Example **3.12**   In Example 3.10 any positive integer was a possible X value, and the p.m.f. was

$$p(x) = \begin{cases} (1 - p)^{x-1}p & x = 1, 2, 3, \ldots \\ 0 & \text{otherwise} \end{cases}$$

For any positive integer x,

$$F(x) = \sum_{y \leq x} p(y) = \sum_{y=1}^{x} (1 - p)^{y-1}p = p \sum_{y=0}^{x-1} (1 - p)^y \tag{3.4}$$

To evaluate this sum, we use the fact that the partial sum of a geometric series is

$$\sum_{y=0}^{k} a^y = \frac{1 - a^{k+1}}{1 - a}$$

Using this in (3.4) with $a = 1 - p$ and $k = x - 1$ gives

$$F(x) = p \cdot \frac{1 - (1 - p)^x}{1 - (1 - p)} = 1 - (1 - p)^x, \ x \text{ a positive integer}$$

Since F is constant in between positive integers,

$$F(x) = \begin{cases} 0 & x < 1 \\ 1 - (1 - p)^{[x]} & x \geq 1 \end{cases} \tag{3.5}$$

where $[x]$ is the largest integer $\leq x$ (such as $[2.7] = 2$). For example, if $p = .51$ as in the birth example, then the probability of having to examine at most five births to see the first boy is $F(5) = 1 - (.49)^5 = 1 - .0282 = .9718$, while $F(10) \approx 1.0000$. This c.d.f. is graphed in Figure 3.6.

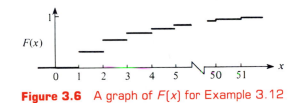

Figure 3.6 A graph of $F(x)$ for Example 3.12 ■

In examples thus far, the c.d.f. has been derived from the p.m.f. It is possible to reverse this procedure and obtain the p.m.f. from the c.d.f. whenever the latter function is available. Suppose, for example, that X represents the number of defective components in a shipment consisting of six components, so that possible X values are $0, 1, \ldots, 6$. Then $p(3) = P(X = 3) = [p(0) + p(1) + p(2) + p(3)] - [p(0) + p(1) + p(2)] = P(X \leq 3) - P(X \leq 2) = F(3) - F(2)$. More generally, the probability that X falls in a specified interval is easily obtained from the c.d.f. For example, $P(2 \leq X \leq 4) = p(2) + p(3) + p(4) = [p(0) + \cdots + p(4)] - [p(0) + p(1)] = P(X \leq 4) - P(X \leq 1) = F(4) - F(1)$. Notice that $P(2 \leq X \leq 4) \neq F(4) - F(2)$. This is because the X value 2 is included in $2 \leq X \leq 4$, so we don't want to subtract out its probability. It *is* true that $P(2 < X \leq 4) = F(4) - F(2)$, because $X = 2$ is not included in the interval $2 < X \leq 4$.

Proposition

For any two numbers a and b with $a \leq b$,

$$P(a \leq X \leq b) = F(b) - F(a-)$$

where "$a-$" represents the largest possible X value that is strictly less than a. In particular, if the only possible values are integers and if a and b are integers,

$$P(a \leq X \leq b) = P(X = a \text{ or } a + 1 \text{ or } \ldots \text{ or } b) = F(b) - F(a - 1)$$

Taking $a = b$ yields $P(X = a) = F(a) - F(a - 1)$ in this case.

Example **3.13**   Let X = the number of days of sick leave taken by a randomly selected employee of a large company during 1986. If the maximum number of allowable sick days per year is 14, possible values of X are 0, 1, . . . , 14. With $F(0) = .58$, $F(1) = .72$, $F(2) = .76$, $F(3) = .81$, $F(4) = .88$, and $F(5) = .94$,

$$P(2 \leq X \leq 5) = P(X = 2, 3, 4, \text{ or } 5) = F(5) - F(1) = .22$$

and

$$P(X = 3) = F(3) - F(2) = .05$$ ■

The reason for subtracting $F(a-)$ rather than $F(a)$ is that we want to include $P(X = a)$; $F(b) - F(a)$ gives $P(a < X \leq b)$. This proposition will be used extensively when computing binomial and Poisson probabilities in Sections 3.4 and 3.6.

Another View of p.m.f.'s

It is often helpful to think of a p.m.f. as specifying a mathematical model for a discrete population.

Example **3.14**   Consider selecting at random a student who is among the 15,000 registered for the current term at Mega University. Let X = the number of courses for which the selected student is registered, and suppose that X has p.m.f.

x	1	2	3	4	5	6	7
$p(x)$	.01	.03	.13	.25	.39	.17	.02

One way to view this situation is to think of the population as consisting of 15,000 individuals, each having his or her own X value; the proportion with each X value is given by $p(x)$ above. An alternative viewpoint is to forget about the students and think of the population itself as consisting of the X values: there are some 1's in the population, some 2's, . . . , and finally some 7's. The population then consists of the numbers 1, 2, . . . , 7 (so is discrete), and $p(x)$ gives a model for the distribution of population values. ■

Once we have such a mathematical model for a population, we will use it to compute values of population characteristics (such as the mean μ) and make inferences about such characteristics.

Exercises / Section 3.2 [11–25]

11. An automobile service facility specializing in engine tuneups knows that 25% of all tuneups are done on four-cylinder automobiles, 40% on six-cylinder automobiles, and 35% on eight-cylinder automobiles. Let X = the number of cylinders on the next car to be tuned.

a. What is the probability mass function of X?
b. Draw both a line graph and a probability histogram for the p.m.f. of (a).
c. Compute the cumulative distribution function of X, and then graph it.

12. Let X = the number of tires on a randomly selected automobile that are underinflated.
  a. Which of the following three $p(x)$ functions is a legitimate p.m.f. for X, and why are the other two not allowed?

x	0	1	2	3	4
$p(x)$	.3	.2	.1	.05	.05
$p(x)$	.4	.1	.1	.1	.3
$p(x)$	.4	.1	.2	.1	.3

  b. For the legitimate p.m.f. of (a), compute $P(2 \leq X \leq 4)$, $P(X \leq 2)$, and $P(X \neq 0)$.
  c. If $p(x) = c \cdot (5 - x)$ for $x = 0, 1, \ldots, 4$, what is the value of c? *Hint:* $\sum_{x=0}^{4} p(x) = 1$.

13. A mail-order computer business has six telephone lines. Let X denote the number of lines in use at a specified time. Suppose that the probability mass function of X is as given in the accompanying table.

x	0	1	2	3	4	5	6
$p(x)$	.10	.15	.20	.25	.20	.06	.04

Calculate the probability of each of the following events.
  a. {at most 3 lines are in use}
  b. {fewer than 3 lines are in use}
  c. {at least 3 lines are in use}
  d. {between 2 and 5 lines, inclusive, are in use}
  e. {between 2 and 4 lines, inclusive, are not in use}
  f. {at least 4 lines are not in use}

14. A contractor is required by a county planning department to submit one, two, three, four, or five forms (depending on the nature of the project) in applying for a building permit. Let Y = the number of forms required of the next applicant. The probability that y forms are required is known to be proportional to y—that is, $p(y) = ky$ for $y = 1, \ldots, 5$.
  a. What is the value of k? *Hint:* $\sum_{y=1}^{5} p(y) = 1$.
  b. What is the probability that at most three forms are required?
  c. What is the probability that between two and four forms (inclusive) are required?
  d. Could $p(y) = y^2/50$ for $y = 1, \ldots, 5$ be the p.m.f. of Y?

15. Two fair six-sided dice are tossed independently. Let M = the maximum of the two tosses.
  a. What is the p.m.f. of M? *Hint:* First determine $p(1)$, then $p(2)$, and so on.
  b. Compute the c.d.f. of M and graph it.

16. In Example 3.9 suppose that there are only four potential blood donors, of whom only one has type O+ blood. Compute the p.m.f. of Y.

17. A library subscribes to two different weekly news magazines, each of which is supposed to arrive in Wednesday's mail. In actuality, each one may arrive on Wednesday, Thursday, Friday, or Saturday. Suppose that the two arrive independently of one another, and that for each one $P(\text{Wed.}) = .4$, $P(\text{Thurs.}) = .3$, $P(\text{Fri.}) = .2$, and $P(\text{Sat.}) = .1$. Let Y = the number of days beyond Wednesday that it takes for both magazines to arrive (so possible Y values are 0, 1, 2, or 3). Compute the p.m.f. of Y.

18. Refer back to Exercise 13, and calculate and graph the cumulative distribution function $F(x)$. Then use it to calculate the probabilities of the events given in (a), (b), (c), and (d) of that problem.

19. A small town situated on a main highway has two gas stations, A and B. Station A sells regular, unleaded, and premium gas for 114.9, 117.6, and 119.9 cents per gallon, respectively, while B sells no regular, but sells unleaded and premium for 115.9 and 119.9 cents per gallon, respectively. Of the cars that stop at Station A, 50% buy regular, 30% buy unleaded, and 20% buy premium. Of the cars stopping at B, 60% buy unleaded and 40% buy premium. Suppose that A gets 60% of the cars that stop for gas in this town, while 40% go to B. Let V = the price per gallon paid by the next car that stops for gas in this town.
  a. Compute the p.m.f. of V. Then draw a probability histogram of the p.m.f.
  b. Compute and graph the c.d.f. of V.

20. An insurance company offers its policyholders a number of different premium payment options. For a randomly selected policyholder, let X = the number of months between successive payments. The c.d.f. of X is

$$F(x) = \begin{cases} 0 & \text{if} \quad x < 1 \\ .3 & \text{if} \quad 1 \le x < 3 \\ .4 & \text{if} \quad 3 \le x < 4 \\ .45 & \text{if} \quad 4 \le x < 6 \\ .60 & \text{if} \quad 6 \le x < 12 \\ 1 & \text{if} \quad 12 \le x \end{cases}$$

a. What is the p.m.f. of X?

b. Using just the c.d.f., compute $P(3 \le X \le 6)$ and $P(4 \le X)$.

21. In Example 3.10 let $Y =$ the number of girls born before the experiment terminates. With $p = P(B)$ and $1 - p = P(G)$, what is the p.m.f. of Y? *Hint:* First list the possible values of Y, starting with the smallest, and proceed until you see a general formula.

22. In Example 3.10 suppose that the experiment does not terminate until two boys are born, and let $Y =$ the number of births examined before the experiment terminates. Then possible Y values are 2, 3, 4, ... Again assuming independent births and $P(B) = p$, derive the p.m.f. of Y. *Hint:* To have $Y = y$, the yth birth must be a boy, and one of the first $y - 1$ must also be a boy, while the others are girls; there are $y - 1$ different ways for this latter event to occur, so first write down the probability of each such way. It might help to select, for example, $y = 5$ and list all corresponding outcomes.

23. Alvie Singer lives at 0 in the diagram below, and has four friends who live at A, B, C, and D. One day Alvie decides to go visiting, so he tosses a fair coin twice to decide which of the four to visit. Once at a friend's house, he will either return home or else proceed to one of the two adjacent houses (such as 0, A, or C when at B), with each of the three possibilities having probability $\frac{1}{3}$. In

this way Alvie continues to visit friends until he returns home.

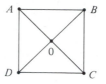

a. Let $X =$ the number of times that Alvie visits a friend. Derive the p.m.f. of X.

b. Let $Y =$ the number of straight-line segments that Alvie traverses (including those leading to and from 0). What is the p.m.f. of Y?

c. Suppose that female friends live at A and C and male friends at B and D. If $Z =$ the number of visits to female friends, what is the p.m.f. of Z?

24. After all students have left the classroom, a statistics professor notices that four copies of the text were left under desks. At the beginning of the next lecture, the professor distributes the four books in a completely random fashion to each of the four students (1, 2, 3, and 4) who claim to have left books. One possible outcome is that 1 receives 2's book, 2 receives 4's book, 3 receives his or her own book, and 4 receives 1's book. This outcome can be abbreviated as (2, 4, 3, 1).

a. List the other 23 possible outcomes.

b. Let X denote the number of students who receive their own book. Determine the p.m.f. of X.

25. Show that the c.d.f. $F(x)$ is a nondecreasing function; that is, $x_1 < x_2$ implies that $F(x_1) \le F(x_2)$. Under what condition will $F(x_1) = F(x_2)$?

3.3   Expected Values of Discrete Random Variables

In Example 3.14 we considered a university having 15,000 students and let $X =$ the number of courses for which a randomly selected student is registered. The probability mass function of X appears below. Since $p(1) = .01$, we know that $(.01) \cdot (15,000) = 150$ of the students are registered for one course, and similarly for the other x values.

x	1	2	3	4	5	6	7	
$p(x)$	.01	.03	.13	.25	.39	.17	.02	(3.6)
Number registered	150	450	1950	3750	5850	2550	300	

To compute the average number of courses per student, or the average value of X in the population, we should compute the total number of courses and divide by the total number of students. Since each of 150 students is taking one course, these 150 contribute 150 courses to the total. Similarly, 450 students contribute 2(450) courses, and so on. The population average value of X is then

$$\frac{1(150) + 2(450) + 3(1950) + \cdots + 7(300)}{15,000} = 4.57 \qquad (3.7)$$

Since $150/15,000 = .01 = p(1)$, $450/15,000 = .03 = p(2)$, and so on, an alternative expression for (3.7) is

$$1 \cdot p(1) + 2 \cdot p(2) + \cdots + 7 \cdot p(7) \qquad (3.8)$$

Expression (3.8) shows that to compute the population average value of X, we need only the possible values of X along with their probabilities (proportions). In particular the population size is irrelevant as long as the probability mass function is given by (3.6). The average or mean value of X is then a weighted average of the possible values $1, \ldots, 7$, where the weights are the probabilities of those values.

The Expected Value of X

Definition

> Let X be a discrete random variable with set of possible values D and probability mass function $p(x)$. The **expected value** or **mean value** of X, denoted by $E(X)$ or μ_X, is
>
> $$E(X) = \mu_X = \sum_{x \in D} x \cdot p(x)$$

When it is clear to which X the expected value refers, μ rather than μ_X is often used.

Example 3.15   For the p.m.f. in (3.6)

$$\begin{aligned}
\mu &= 1 \cdot p(1) + 2 \cdot p(2) + \cdots + 7 \cdot p(7) \\
&= (1)(.01) + 2(.03) + \cdots + (7)(.02) \\
&= .01 + .06 + .39 + 1.00 + 1.95 + 1.02 + .14 = 4.57
\end{aligned}$$

If we think of the population as consisting of the X values $1, 2, \ldots, 7$, then $\mu = 4.57$ is the population mean. In the sequel we shall often refer to μ as the population mean rather than the mean of X in the population. ∎

In the above example, the expected value μ was 4.57, which is not a possible value of X. The word "expected" should be interpreted with caution, since one would not expect to see an X value of 4.57 when a single student is selected.

Example 3.16   Just after birth, each newborn child is rated on a scale called the Apgar scale. The possible ratings are 0, 1, ..., 10, with the child's rating determined by color, muscle tone, respiratory effort, heartbeat, and reflex irritability (the best possible score is 10). Let X be the Apgar score of a randomly selected child born at a certain hospital during the next year, and suppose that the p.m.f. of X is

x	0	1	2	3	4	5	6	7	8	9	10
$p(x)$	.002	.001	.002	.005	.02	.04	.18	.37	.25	.12	.01

Then the mean value of X is

$$E(X) = \mu = 0(.002) + 1(.001) + 2(.002) + \cdots + 8(.25) + 9(.12)$$
$$+ 10(.01) = 7.15$$

Again μ is not a possible value of the variable X. Also, because the variable refers to a future child, there is no concrete existing population to which μ refers. Instead, we think of the p.m.f. as a model for a conceptual population consisting of the values 0, 1, 2, ..., 10. The mean value of this conceptual population is then $\mu = 7.15$.   ■

Example 3.17   Let X be a Bernoulli random variable with p.m.f.

$$p(x) = \begin{cases} 1 - p & x = 0 \\ p & x = 1 \\ 0 & x \neq 0, 1 \end{cases}$$

Then $E(X) = 0 \cdot p(0) + 1 \cdot p(1) = 0(1 - p) + 1(p) = p$. That is, the expected value of X is just the probability that X takes on the value 1. If we conceptualize a population consisting of 0's in proportion $1 - p$ and 1's in proportion p, then the population average is $\mu = p$.   ■

Example 3.18   The general form for the p.m.f. of $X =$ number of children born up to and including the first boy is

$$p(x) = \begin{cases} p(1 - p)^{x-1} & x = 1, 2, 3, \ldots \\ 0 & \text{otherwise} \end{cases}$$

From the definition,

$$E(X) = \sum_D x \cdot p(x) = \sum_{x=1}^{\infty} xp(1 - p)^{x-1} = p\sum_{x=1}^{\infty}\left[-\frac{d}{dp}(1 - p)^x \right] \quad (3.9)$$

If we exchange the order of taking the derivative and the summation, the sum is that of a geometric series. After computing it, the derivative is taken, and the final result is $E(X) = 1/p$. If p is near 1, we expect to see a boy very soon, while if p is near 0, we expect many births before the first boy. For $p = .5$, $E(X) = 2$.   ■

There is another frequently used interpretation of μ. Consider the p.m.f.

$$p(x) = \begin{cases} (.5) \cdot (.5)^{x-1} & \text{if} \quad x = 1, 2, 3, \ldots \\ 0 & \text{otherwise} \end{cases}$$

This is the p.m.f. of $X =$ the number of tosses of a fair coin necessary to obtain the first H (a special case of Example 3.18). Suppose that we observe a value x from this p.m.f. (toss a coin until an H appears), then observe independently another value (keep tossing), then another, and so on. If after observing a very large number of x values, we average them, the resulting sample average will be very near to $\mu = 2$. That is, μ can be interpreted as the long-run average observed value of X when the experiment is performed repeatedly.

Example 3.19   Let X have p.m.f.

$$p(x) = \begin{cases} \dfrac{k}{x^2} & x = 1, 2, 3, \ldots \\ 0 & \text{otherwise} \end{cases}$$

where k is chosen so that $\sum\limits_{x=1}^{\infty} (k/x^2) = 1$ (in a mathematics course on infinite series, it is shown that $\sum\limits_{x=1}^{\infty} (1/x^2) < \infty$, which implies that such a k exists, but its exact value need not concern us). The expected value of X is

$$\mu = E(X) = \sum_{x=1}^{\infty} x \cdot \frac{k}{x^2} = k \sum_{x=1}^{\infty} \frac{1}{x} \tag{3.10}$$

The sum on the right of (3.10) is the famous harmonic series of mathematics, and can be shown to equal ∞. $E(X)$ is not finite here because $p(x)$ does not decrease sufficiently fast as x increases; statisticians say that the probability distribution of X has "a heavy tail." If a sequence of X values is chosen using this distribution, the sample average will not settle down to some finite number but will tend to grow without bound.

Statisticians use the phrase "heavy tails" in connection with any distribution having a large amount of probability far from μ (so heavy tails does not require $\mu = \infty$). Such heavy tails make it difficult to make inferences about μ. ∎

The Expected Value of a Function
Often we will be interested in the expected value of some function $h(X)$ rather than X itself.

Example 3.20   Suppose that a bookstore purchases 10 copies of a book at \$6.00 each, to sell at \$12.00 with the understanding that at the end of a three-month period, any unsold copies can be redeemed for \$2.00. If $X =$ the number of copies purchased, then net revenue $= h(X) = 12X + 2(10 - X) - 60 = 10X - 40$. ∎

An easy way of computing the expected value of $h(X)$ is suggested by the following example.

Example 3.21 Let X be the outcome when a fair six-sided die is tossed once, and let $h(X) = X^2$. The p.m.f. of X is

x	1	2	3	4	5	6
$p(x)$	$\frac{1}{6}$	$\frac{1}{6}$	$\frac{1}{6}$	$\frac{1}{6}$	$\frac{1}{6}$	$\frac{1}{6}$

If we let $Y = X^2$, then Y is a random variable with set of possible values $D^* = \{1, 4, 9, 16, 25, 36\}$ and the p.m.f. of Y is

y	1	4	9	16	25	36
$p(y)$	$\frac{1}{6}$	$\frac{1}{6}$	$\frac{1}{6}$	$\frac{1}{6}$	$\frac{1}{6}$	$\frac{1}{6}$

Having obtained the p.m.f. of Y, its expected value (that is, the expected value of X^2) is

$$E(Y) = \sum_{D^*} y \cdot p(y)$$

$$= (1) \cdot \frac{1}{6} + (4) \cdot \frac{1}{6} + \cdots + (25) \cdot \frac{1}{6} + (36) \cdot \frac{1}{6} = \frac{91}{6} \qquad (3.11)$$

$$= (1)^2 \cdot \frac{1}{6} + (2)^2 \cdot \frac{1}{6} + \cdots + (5)^2 \cdot \frac{1}{6} + (6)^2 \cdot \frac{1}{6} = \sum_{D} x^2 \cdot p(x)$$

According to (3.11) it was not necessary to compute the p.m.f. of $Y = X^2$ to obtain $E(Y)$; instead the desired expected value is a weighted average of the possible $h(x)$ (rather than x) values. ∎

Proposition

> If the random variable X has set of possible values D and p.m.f. $p(x)$, then the expected value of any function $h(X)$, denoted by $E[h(X)]$ or $\mu_{h(X)}$, is computed by
>
> $$E[h(X)] = \sum_{D} h(x) \cdot p(x)$$

According to this proposition, $E[h(X)]$ is computed in the same way that $E(X)$ itself is, except that $h(x)$ is substituted in place of x.

Example 3.22 A computer store has purchased three computers of a certain type at $500 apiece. It will sell them for $1000 apiece. The manufacturer has agreed to repurchase any computers still unsold after a specified period at $200 apiece. Let X denote the number of computers sold, and suppose that $p(0) = .1$, $p(1) = .2$, $p(2) = .3$, and $p(3) = .4$. With $h(X)$ denoting the profit associated with selling X units, the given information implies that $h(X) = $ revenue −

cost $= 1000X + 200(3 - X) - 1500 = 800X - 900$. The expected profit is then

$$E[h(X)] = h(0) \cdot p(0) + h(1) \cdot p(1) + h(2) \cdot p(2) + h(3) \cdot p(0)$$
$$= (-900)(.1) + (-100)(.2) + (700)(.3) + (1500)(.4) = \$700$$

■

Rules of Expected Value

The $h(X)$ function of interest is quite frequently a linear function $aX + b$. In this case $E[h(X)]$ is easily computed from $E(X)$.

Proposition

$E(aX + b) = a \cdot E(X) + b$ (or, using alternative notation, $\mu_{aX + b} = a \cdot \mu_X + b$).

To paraphrase, the expected value of a linear function equals the linear function of the expected value $E(X)$. Since $h(X)$ in Example 3.22 is linear and $E(X) = 2$, $E(h(X)) = 800(2) - 900 = \$700$ as before.

Proof. $E(aX + b) = \sum_D (ax + b) \cdot p(x) = a \sum_D x \cdot p(x) + b \sum_D p(x)$

$$= aE(X) + b$$

■

Two special cases of the proposition yield two important rules of expected value.

1. For any constant a, $E(aX) = a \cdot E(X)$ (take $b = 0$)
2. For any constant b, $E(X + b) = E(X) + b$ (take $a = 1$)   (3.12)

Multiplication of X by a constant a changes the units of measurement (from dollars to cents, where $a = 100$, inches to cm, where $a = 2.54$, and so on). Rule (1) says that the expected value in the new units equals the expected value in the old units multiplied by the conversion factor a. Similarly, if a constant b is added to each possible value of X, then the expected value will be shifted by that same constant amount.

The Variance of X

The expected value of X measures where the probability distribution is centered. Using the physical analogy of placing point mass $p(x)$ at the value x on a one-dimensional axis, if the axis were then supported by a fulcrum placed at μ, there would be no tendency for the axis to tilt. This is illustrated for two different distributions in Figure 3.7.

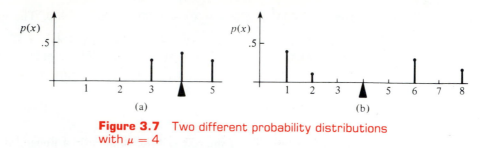

Figure 3.7   Two different probability distributions with $\mu = 4$

Although both distributions pictured in Figure 3.7 have the same center μ, the distribution of (*b*) has greater spread or variability or dispersion than does that of (*a*). We will use the variance of X to measure the amount of variability in (the distribution of) X, just as s^2 was used in Chapter 1 to measure variability in a sample.

Definition

> Let X have probability mass function $p(x)$ and expected value μ. Then the **variance** of X, denoted by $V(X)$ or σ_X^2, or just σ^2, is
>
> $$V(X) = \sum_D (x - \mu)^2 \cdot p(x) = E[(X - \mu)^2]$$
>
> The **standard deviation** of X is
>
> $$\sigma_X = \sqrt{\sigma_X^2}$$

The quantity $h(X) = (X - \mu)^2$ is the squared deviation of X from its mean, and σ^2 is the expected squared deviation. If most of the probability distribution is close to μ, then σ^2 will be relatively small, while if there are x values far from μ that have large $p(x)$, then σ^2 will be quite large.

Example 3.23   If X has p.m.f.

x	3	4	5
$p(x)$	.3	.4	.3

(as in Figure 3.7a) with mean $\mu = 4$, then

$$V(X) = \sigma^2 = \sum_{x=3}^{5} (x - 4)^2 \cdot p(x)$$

$$= (3 - 4)^2 \cdot (.3) + (4 - 4)^2 \cdot (.4) + (5 - 4)^2 \cdot (.3) = .6$$

The standard deviation of X is $\sigma = \sqrt{.6} = .77$.

If X has the p.m.f. of Figure 3.7b,

x	1	2	6	8
$p(x)$	.4	.1	.3	.2

then

$$\sigma^2 = (1 - 4)^2 \cdot (.4) + (2 - 4)^2 \cdot (.1) + (6 - 4)^2 \cdot (.3) + (8 - 4)^2 \cdot (.2)$$
$$= 8.4$$

and

$$\sigma = 2.90 \qquad\qquad\qquad\qquad\qquad\qquad\qquad\qquad\qquad\qquad\quad \blacksquare$$

When the p.m.f. $p(x)$ specifies a mathematical model for the distribution of population values, both σ^2 and σ measure the spread of values in the population; σ^2 is the population variance and σ is the population standard deviation.

A Shortcut Formula for σ^2

The number of arithmetic operations necessary to compute σ^2 can be reduced by using an alternative computing formula.

Proposition

$$V(X) = \sigma^2 = \Big[\sum_D x^2 \cdot p(x)\Big] - \mu^2 = E(X^2) - [E(X)]^2$$

In using this formula, $E(X^2)$ is computed first without any subtraction; then $E(X)$ is computed, squared, and subtracted (once) from $E(X^2)$.

Example 3.24 If X is the outcome of a single roll of a fair die, then

$$E(X^2) = \sum_D x^2 \cdot p(x) = \sum_{x=1}^{6} x^2 \cdot \frac{1}{6} = \frac{91}{6}$$

and

$$E(X) = \mu = \sum_D x \cdot p(x) = \sum_{x=1}^{6} x \cdot \frac{1}{6} = \frac{7}{2}$$

The variance is then

$$\sigma^2 = \frac{91}{6} - \left(\frac{7}{2}\right)^2 = \frac{91}{6} - \frac{49}{4} = \frac{70}{24} = 2.92,$$

and $\sigma = \sqrt{2.92} = 1.71.$ $\qquad\qquad\qquad\qquad\qquad\qquad\qquad\qquad\quad \blacksquare$

Proof of the shortcut formula.   Expand $(x - \mu)^2$ in the definition of σ^2 to obtain $x^2 - 2\mu x + \mu^2$, and then carry Σ through to each of the three terms:

$$\sigma^2 = \sum_D x^2 \cdot p(x) - 2\mu \cdot \sum_D x \cdot p(x) + \mu^2 \sum_D p(x)$$
$$= E(X^2) - 2\mu \cdot \mu + \mu^2 = E(X^2) - \mu^2 \qquad\qquad\qquad\qquad \blacksquare$$

Rules of Variance

The variance of $h(X)$ is the expected value of the squared difference between $h(X)$ and its expected value:

$$V[h(X)] = \sigma^2_{h(X)} = \sum_D \{h(x) - E[h(X)]\}^2 \cdot p(x) \tag{3.13}$$

When $h(X)$ is a linear function, $V[h(X)]$ is easily related to $V(X)$.

Proposition

$$V(aX + b) = \sigma^2_{aX+b} = a^2 \cdot \sigma^2_X, \text{ and } \sigma_{aX+b} = |a| \cdot \sigma_X$$

This result says that the addition of the constant b does not affect the variance, which is intuitive, since the addition of b changes the location (mean value) but not the spread of values. In particular,

$$\begin{aligned}
&\textbf{1. } \sigma^2_{aX} = a^2 \cdot \sigma^2_X, \qquad \sigma_{aX} = |a| \cdot \sigma_X \\
&\textbf{2. } \sigma^2_{X+b} = \sigma^2_X
\end{aligned} \tag{3.14}$$

The reason for the absolute value in σ_{aX} is that a may be negative, while a standard deviation cannot be negative; a^2 results when a is brought outside the term being squared in (3.13).

Example 3.25

In the computer sales problem of Example 3.22, $E(X) = 2$ and

$$E(X^2) = (0)^2(.1) + (1)^2(.2) + (2)^2(.3) + (3)^2(.4) = 5$$

so $V(X) = 5 - (2)^2 = 1$. The profit function $h(X) = 800X - 900$ then has variance $(800)^2 \cdot V(X) = (640,000)(1) = 640,000$ and standard deviation 800. ∎

Exercises / Section 3.3 (26–40)

26. The probability mass function for $X =$ the number of cars owned by a randomly selected family in Suburbanville is

x	0	1	2	3	4
$p(x)$	.08	.15	.45	.27	.05

Compute
a. $E(X)$
b. $V(X)$ directly from the definition.
c. the standard deviation of X.
d. $V(X)$ using the shortcut formula.

27. An instructor in a technical writing class has asked that a certain report be turned in the following week, adding the restriction that any report exceeding four pages will not be accepted. Let $Y =$ the number of pages in a randomly chosen student's report, and suppose that Y has p.m.f.

y	1	2	3	4
$p(y)$	.01	.19	.35	.45

a. Compute $E(Y)$.
b. Suppose that the instructor spends $\sqrt{Y}$ minutes grading a paper consisting of Y pages. What is the expected amount of time $[E(\sqrt{Y})]$ spent grading a randomly selected paper?
c. Compute the variance and standard deviation of Y.

28. An appliance dealer sells three different models of upright freezers having 13.5, 15.9, and 19.1 cubic feet of storage space respectively. Let $X =$ the amount of storage space purchased by the next customer to buy a freezer. Suppose that X has p.m.f.

x	13.5	15.9	19.1
$p(x)$	.2	.5	.3

  a. Compute $E(X)$, $E(X^2)$, and $V(X)$.
  b. If the price of a freezer having capacity X cubic feet is $25X - 8.5$, what is the expected price paid by the next customer to buy a freezer?
  c. What is the variance of the price $25X - 8.5$ paid by the next customer?
  d. Suppose that while the rated capacity of a freezer is X, the actual capacity is $h(X) = X - .01X^2$. What is the expected actual capacity of the freezer purchased by the next customer?

29. Let X be a Bernoulli r.v. with p.m.f. as in Example 3.17.
  a. Compute $E(X^2)$.
  b. Show that $V(X) = p(1 - p)$.
  c. Compute $E(X^{79})$.

30. Suppose that the number of plants of a particular type found in a rectangular region (called a quadrat by ecologists) in a certain geographical area is a r.v. X with p.m.f.

$$p(x) = \begin{cases} c/x^3 & x = 1, 2, 3, \ldots \\ 0 & \text{otherwise} \end{cases}$$

Is $E(X)$ finite? Justify your answer (this is another distribution that statisticians would call heavy-tailed).

31. A small drugstore orders copies of a news magazine for its magazine rack each week. Let $X =$ demand for the magazine, with p.m.f.

x	1	2	3	4	5	6
$p(x)$	$\frac{1}{15}$	$\frac{2}{15}$	$\frac{3}{15}$	$\frac{4}{15}$	$\frac{3}{15}$	$\frac{2}{15}$

Suppose that the store owner actually pays $.25 for each copy of the news magazine, and the price to customers is $1.00. If magazines left at the end of the week have no salvage value, is it better to order three or four copies of the magazine? *Hint:* For both three and four copies or-

dered, express net revenue as a function of demand X, and then compute the expected revenue.

32. You are offered the choice of playing one of two games. In game A, you roll a fair die once and receive the amount $X =$ the resulting number (in dollars). In game B, you roll a fair die twice and receive $Y =$ the maximum of the two outcomes. If the entry fee for game A is $3.00 and for game B is $3.50, which game should you play?

33. The n candidates for a job have been ranked $1, 2, 3, \ldots, n$. Let $X =$ the rank of a randomly selected candidate, so that X has p.m.f.

$$p(x) = \begin{cases} 1/n & x = 1, 2, \ldots, n \\ 0 & \text{otherwise} \end{cases}$$

(this is called the discrete uniform distribution). Compute $E(X)$ and $V(X)$ using the shortcut formula. *Hint:* The sum of the first n positive integers is $n(n + 1)/2$, while the sum of their squares is $n(n + 1)(2n + 1)/6$.

34. Let $X =$ the outcome when a fair die is rolled once. If before the die is rolled you are offered either $(1/3.5)$ dollars or $h(X) = 1/X$ dollars, would you accept the guaranteed amount or would you gamble? *Note:* It is not generally true that $1/E(X) = E(1/X)$.

35. A chemical supply company currently has in stock 100 lbs. of a certain chemical, which it sells to customers in 5-lb. lots. Let $X =$ the number of lots ordered by a randomly chosen customer, and suppose that X has p.m.f.

x	1	2	3	4
$p(x)$	.2	.4	.3	.1

Compute $E(X)$ and $V(X)$. Then compute the expected number of pounds left after the next customer's order is shipped, and the variance of the number of pounds left. *Hint:* The number of pounds left is a linear function of X.

36. a. Draw a line graph of the p.m.f. of X in Exercise 31. Then determine the p.m.f. of $-X$ and draw its line graph. From these two pictures, what can you say about $V(X)$ and $V(-X)$?
  b. Use the proposition involving $V(aX + b)$ to establish a general relationship between $V(X)$ and $V(-X)$.

37. Use the definition in Expression (3.13) to prove that $V(aX + b) = a^2 \cdot \sigma_X^2$. *Hint:* With $h(X) = aX + b$, $E[h(X)] = a\mu + b$ where $\mu = E(X)$.

38. Suppose that $E(X) = 5$ and $E[X(X - 1)] = 27.5$. What is
  a. $E(X^2)$? *Hint:* $E[X(X - 1)] = E[X^2 - X] =. E(X^2) - E(X)$.
  b. $V(X)$?
  c. the general relationship between $E(X)$, $E[X(X - 1)]$, and $V(X)$?

39. Write a general rule for $E(X - c)$ where c is a constant. What happens when you let $c = \mu$, the expected value of X?

40. A result called **Chebyshev's inequality** states that for any probability distribution of a random variable X and any number k that is at least 1, $P(|X - \mu| \geq k\sigma) \leq 1/k^2$. In words, the probability that the value of X lies at least k standard deviations from its mean is at most $1/k^2$.
  a. What is the value of the upper bound for $k = 2$? $k = 3$? $k = 4$? $k = 5$? $k = 10$?
  b. Compute μ and σ for the distribution of Exercise 3.13. Then evaluate $P(|X - \mu| \geq k\sigma)$ for the values of k given in (a). What does this suggest about the upper bound relative to the corresponding probability?
  c. Let X have three possible values, -1, 0, and 1, with probabilities $\frac{1}{18}$, $\frac{8}{9}$, and $\frac{1}{18}$, respectively. What is $P(|X - \mu| \geq 3\sigma)$, and how does it compare to the corresponding bound?
  d. Give a distribution for which $P(|X - \mu| \geq 5\sigma) = .04$.

3.4   The Binomial Probability Distribution

There are many experiments that conform either exactly or approximately to the following list of requirements:

1. The experiment consists of a sequence of n trials, where n is fixed in advance of the experiment.
2. The trials are identical, and each trial can result in one of the same two possible outcomes, which we denote by success (S) or failure (F).
3. The trials are independent, so that the outcome on any particular trial does not influence the outcome on any other trial.
4. The probability of success is constant from trial to trial; we denote this probability by p.

Definition

> An experiment for which conditions 1–4 are satisfied is called a **binomial experiment.**

Example 3.26   The same coin is tossed successively and independently n times. We arbitrarily use S to denote the outcome H (heads) and F to denote the outcome T (tails). Then this experiment satisfies 1–4. ■

Many experiments involve a sequence of independent trials for which there are more than two possible outcomes on any one trial. A binomial experiment can then be created by dividing the possible outcomes into two groups.

Example 3.27   The color of pea seeds is determined by a single genetic locus. If the two alleles at this locus are AA or Aa (the genotype), then the pea will be yellow (the phenotype), and if the allele is aa, the pea will be green. Suppose we pair off 20 Aa seeds and cross the two seeds in each of the 10 pairs to obtain 10 new genotypes. Call each new genotype a success S if it is aa and a failure otherwise. Then with this identification of S and F, the experiment is binomial with $n = 10$ and $p = P(\text{aa genotype})$. If each member of the pair is equally likely to contribute a or A, $p = P(a) \cdot P(a) = (\frac{1}{2})(\frac{1}{2}) = \frac{1}{4}$. ■

Example 3.28 Suppose that a certain city has 50 licensed restaurants, of which 15 currently have at least one serious health code violation and the other 35 have no serious violations. There are five inspectors, each of whom will inspect one restaurant during the coming week. The name of each restaurant is written on a different slip of paper, and after the slips are thoroughly mixed, each inspector in turn draws one of the slips *without replacement*. Label the ith trial as a success if the ith restaurant selected ($i = 1, \ldots, 5$) has no serious violations. Then

$$P(S \text{ on first trial}) = \frac{35}{50} = .70$$

and

$$
\begin{aligned}
P(S \text{ on second trial}) &= P(SS) + P(FS) = P(\text{second } S \mid \text{first } S)\, P(\text{first } S) \\
&\quad + P(\text{second } S \mid \text{first } F)\, P(\text{first } F) \\
&= \frac{34}{49} \cdot \frac{35}{50} + \frac{35}{49} \cdot \frac{15}{50} = \frac{35}{50} \left[\frac{34}{49} + \frac{15}{49} \right] = \frac{35}{50} = .70
\end{aligned}
$$

Similarly, it can be shown that $P(S \text{ on } i\text{th trial}) = .70$ for $i = 3, 4, 5$. However,

$$P(S \text{ on fifth trial} \mid SSSS) = \frac{31}{46} = .67$$

$$\neq .76 = \frac{35}{46}$$

$$= P(S \text{ on fifth trial} \mid FFFF)$$

so the experiment is not binomial because the trials are not independent. In general, if sampling is without replacement, the experiment will not yield independent trials. If each slip had been replaced after being drawn, then trials would be independent, but this might result in the same restaurant being inspected by more than one inspector. ■

Example 3.29 Suppose that a certain state has 500,000 registered automobiles, of which 400,000 were manufactured domestically. A sample of 10 cars is chosen without replacement. The ith trial is labeled S if the ith car chosen is a domestic model. Although this situation would seem identical to that of Example 3.28, the important difference is that the size of the population being sampled is very large relative to the sample size. In this case

$$P(S \text{ on } 2 \mid S \text{ on } 1) = \frac{399{,}999}{499{,}999} = .80000$$

and

$$P(S \text{ on } 10 \mid S \text{ on first } 9) = \frac{399{,}991}{499{,}991} = .799996 \approx .80000$$

These calculations indicate that while the trials are not exactly independent, the conditional probabilities differ so slightly from one another that for practical purposes the trials can be regarded as independent with constant $P(S) = .8$. Thus to a very good approximation, the experiment is binomial with $n = 10$ and $p = .8$. ■

We will use the following rule of thumb in deciding whether a "without replacement" experiment can be treated as a binomial experiment.

Rule

> Suppose that each trial of an experiment can result in S or F, but the sampling is without replacement from a population of size N. If the sample size (number of trials) n is at most 5% of the population size, the experiment can be analyzed as though it is exactly a binomial experiment.

By "analyzed," we mean that probabilities based on the binomial experiment assumptions will be quite close to the actual "without replacement" probabilities, which are typically more difficult to calculate. In Example 3.28 $n/N = 5/50 = .1 > .05$, so the binomial experiment is not a good approximation, but in Example 3.29 $n/N = 10/500{,}000 < .05$.

The Binomial Random Variable and Distribution

In most binomial experiments it is the total number of S's, rather than knowledge of exactly which trials yielded S's, that is of interest.

Definition

> Given a binomial experiment consisting of n trials, the **binomial random variable X** associated with this experiment is defined as
>
> $X =$ the number of S's among the n trials

Suppose, for example, that $n = 3$. Then there are eight possible outcomes for the experiment:

$$SSS, \; SSF, \; SFS, \; SFF, \; FSS, \; FSF, \; FFS, \; FFF$$

From the definition of X, $X(SSF) = 2$, $X(SFF) = 1$, and so on. Possible values for X in an n trial experiment are $x = 0, 1, 2, \ldots, n$. We will often write $X \sim \text{Bin}(n, p)$ to indicate that X is a binomial random variable based on n trials with success probability p.

Notation

> Because the p.m.f. of a binomial r.v. X depends on the two parameters n and p, we denote the p.m.f. by $b(x; n, p)$.

To gain insight into $b(x; n, p)$ for general n, consider the case $n = 4$ for which each outcome, its probability, and corresponding x value are listed in Table 3.1. For example,

$$P(SSFS) = P(S) \cdot P(S) \cdot P(F) \cdot P(S) \quad \text{(independent trials)}$$
$$= p \cdot p \cdot (1 - p) \cdot p \quad \text{(constant } P(S))$$
$$= p^3 \cdot (1 - p)$$

Table 3.1 Outcomes and probabilities for a binomial experiment with four trials

Outcome	Observed x	Probability	Outcome	Observed x	Probability
SSSS	4	p^4	FSSS	3	$p^3(1 - p)$
SSSF	3	$p^3(1 - p)$	FSSF	2	$p^2(1 - p)^2$
SSFS	3	$p^3(1 - p)$	FSFS	2	$p^2(1 - p)^2$
SSFF	2	$p^2(1 - p)^2$	FSFF	1	$p(1 - p)^3$
SFSS	3	$p^3(1 - p)$	FFSS	2	$p^2(1 - p)^2$
SFSF	2	$p^2(1 - p)^2$	FFSF	1	$p(1 - p)^3$
SFFS	2	$p^2(1 - p)^2$	FFFS	1	$p(1 - p)^3$
SFFF	1	$p(1 - p)^3$	FFFF	0	$(1 - p)^4$

In this special case we wish $b(x; 4, p)$ for $x = 0, 1, 2, 3,$ and 4. For $b(3; 4, p)$, we identify which of the 16 outcomes yields an x value of 3 and add up the probabilities associated with each such outcome:

$$b(3; 4, p) = P(FSSS) + P(SFSS) + P(SSFS) + P(SSSF)$$
$$= 4p^3(1 - p)$$

Since the number of outcomes with $x = 3$ is four, and each of the four has probability $p^3(1 - p)$ (the order of S's and F's is not important, but only the number of S's),

$$b(3; 4, p) = \left\{ \begin{matrix} \text{number of outcomes} \\ \text{with } X = 3 \end{matrix} \right\} \cdot \left\{ \begin{matrix} \text{probability of any particular} \\ \text{outcome with } X = 3 \end{matrix} \right\}$$

Similarly, $b(2; 4, p) = 6p^2(1 - p)^2$, which is also the product of the number of outcomes with $X = 2$ and the probability of any such outcome.

In general,

$$b(x; n, p) = \left\{ \begin{matrix} \text{number of sequences of} \\ \text{length } n \text{ consisting of } x \text{ } S\text{'s} \end{matrix} \right\} \cdot \left\{ \begin{matrix} \text{probability of any} \\ \text{particular such sequence} \end{matrix} \right\}$$

Since the ordering of S's and F's is not important, the second factor above is $p^x(1 - p)^{n - x}$ (for example, the first x trials resulting in S and the last $n - x$

in F). The first factor is the number of ways of choosing x of the n trials to be S's—that is, the number of combinations of size x that can be constructed from n distinct objects (trials here).

Theorem

$$b(x; n, p) = \begin{cases} \binom{n}{x}p^x(1 - p)^{n - x} & x = 0, 1, 2, \ldots, n \\ 0 & \text{otherwise} \end{cases}$$

Example 3.30   Each of six randomly selected beer drinkers is given a glass containing beer S and one containing beer F. The glasses are identical in appearance except for a code on the bottom to identify the beer. Suppose that there is actually no tendency among beer drinkers to prefer one beer to the other. Then $p = P($a selected individual prefers $S) = .5$, so with $X = $ the number among the six who prefer S, $X \sim$ Bin$(6, .5)$.

Thus

$$P(X = 3) = b(3; 6, .5) = \binom{6}{3}(.5)^3(.5)^3 = 20\,(.5)^6 = .313$$

The probability that at least three prefer S is

$$P(3 \le X) = \sum_{x = 3}^{6} b(x; 6, .5) = \sum_{x = 3}^{6} \binom{6}{x}(.5)^x(.5)^{6 - x} = .656$$

while the probability that at most one prefers S is

$$P(X \le 1) = \sum_{x = 0}^{1} b(x; 6, .5) = .109$$ ∎

Using Binomial Tables

Even for a relatively small value of n, the computation of binomial probabilities can be tedious. Appendix Table A.1 tabulates the cumulative distribution function $F(x) = P(X \le x)$ for $n = 5, 10, 15, 20, 25$ in combination with selected values of p. Various other probabilities can then be calculated using the proposition on c.d.f.'s from Section 3.2.

Notation

For $X \sim$ Bin(n, p), the cumulative distribution function will be denoted by

$$P(X \le x) = B(x; n, p) = \sum_{y = 0}^{x} b(y; n, p) \qquad x = 0, 1, \ldots, n$$

Example 3.31   A multiple-choice test consists of 15 questions, and for each question there are five possible answers. If for each question an answer is selected in a completely random fashion, what is the probability that (a) at most eight of the questions are answered correctly? (b) Exactly eight? (c) At least eight? (d) Between four and seven inclusive?

The binomial r.v. here is X = the number of correctly answered questions, with $n = 15$ and $p = \frac{1}{5} = .2$. We wish first $P(X \le 8) = \sum_{x=0}^{8} b(x; 15, .2) = B(8; 15, .2)$. This is the $x = 8$ entry (row $x = 8$) in the $p = .2$ column of the $n = 15$ table; the tabulated probability is $B(8; 15, .2) = .999$. Similarly $P(X = 8) = P(X \le 8) - P(X \le 7) = B(8; 15, .2) - B(7; 15, .2) = .999 - .996 = .003$, while $P(X \ge 8) = 1 - P(X \le 7) = 1 - B(7; 15, .2) = 1 - .996 = .004$. Finally, for question (d), $P(4 \le X \le 7) = P(X = 4, 5, 6,$ or $7) = P(X \le 7) - P(X \le 3) = B(7; 15, .2) - B(3; 15, .2) = .996 - .648 = .348$. ∎

Example 3.32   Cars coming to a dead-end intersection can turn either left or right. Suppose that successive cars choose a turning direction independently of one another and that $P(\text{left turn}) = .7$. Among the next 15 cars, what is the probability that at least 10 turn left? Among the next 15 cars, what is the probability that at least 10 turn in the same direction?

With X = the number among the next 15 cars that turn left, $X \sim$ Bin(15, .7), so

$$
\begin{aligned}
P(\text{at least 10 left turns}) = P(10 \le X) &= 1 - P(X \le 9) \\
&= 1 - B(9; 15, .7) = 1 - .278 \\
&= .722
\end{aligned}
$$

where .278 is the $n = 15$ binomial table entry in the $p = .7$ column and $x = 9$ row. Similarly

$$
\begin{aligned}
P\binom{\text{at least 10 turns in}}{\text{the same direction}} &= P\binom{\text{at least 10 left turns or}}{\text{at least 10 right turns}} \\
&= P(10 \le X \text{ or } X \le 5) \\
&= P(X \le 5) + P(10 \le X) \\
&= B(5; 15, .7) + 1 - B(9; 15, .7) \\
&= .004 + .722 = .726
\end{aligned}
$$
∎

Example 3.33   An electronics manufacturer claims that at most 10% of its power supply units need service during the warranty period. To investigate this claim, technicians at a testing laboratory purchase 20 units and subject each one to accelerated testing to simulate use during the warranty period. Let p denote the probability that a power supply unit needs repair during the period (the proportion of all such units that need repair). The laboratory technicians must decide whether or not the data resulting from the experiment supports the claim that $p \le .10$. Let X denote the number among the 20 sampled that need repair, so $X \sim$ Bin(20, p). Consider the decision rule

reject the claim that $p \le .10$ in favor of the conclusion that $p > .10$ if $x \ge 5$ (where x is the observed value of X), and consider the claim plausible if $x \le 4$.

The probability that the claim is rejected when $p = .10$ (an incorrect conclusion) is

$$P(X \geq 5 \text{ when } p = .10) = 1 - B(4; 20, .1) = 1 - .957 = .043$$

The probability that the claim is not rejected when $p = .20$ (a different type of incorrect conclusion) is

$$P(X \leq 4 \text{ when } p = .2) = B(4; 20, .2) = .630$$

The first probability is rather small, but the second is intolerably large. When $p = .20$, so that the manufacturer has grossly understated the percentage of units that need service, and the stated decision rule is used, 63% of all samples will result in the manufacturer's claim being judged plausible!

One might think that the probability of this second type of erroneous conclusion could be made smaller by changing the cutoff value 5 in the decision rule to something else. However, while replacing 5 by a smaller number would yield a probability smaller than .630, the other probability would then increase. The only way to make both "error probabilities" small is to base the decision rule on an experiment involving many more units.   ■

Note that a table entry of 0 signifies only that a probability is 0 to three significant digits, for all entries in the table are actually positive. Much more extensive tables of binomial probabilities have been published. In Chapter 4 we present a method for obtaining quick and accurate approximations to binomial probabilities when n is large.

The Mean and Variance of *X*

For $n = 1$, the binomial distribution becomes the Bernoulli distribution. From Example 3.17 the mean value of a Bernoulli variable is $\mu = p$, so the expected number of S's on any single trial is p. Since a binomial experiment consists of n trials, intuition suggests that for $X \sim \text{Bin}(n, p)$, $E(X) = np$, the product of the number of trials and the probability of success on a single trial. The expression for $V(X)$ is not so intuitive.

Proposition

> If $X \sim \text{Bin}(n, p)$, then $E(X) = np$, $V(X) = np(1 - p) = npq$, and $\sigma_X = \sqrt{npq}$ (where $q = 1 - p$).

Thus, calculating the mean and variance of a binomial random variable does not necessitate evaluating summations. The proof of the result for $E(X)$ is sketched in Exercise 52.

Example 3.34   If 75% of all consultations handled by student consultants at a computing center involve programs with syntax errors, and X is the number of programs with syntax errors in 10 randomly chosen consultations, then $X \sim \text{Bin}(10, .75)$. Thus $E(X) = np = (10)(.75) = 7.5$, $V(X) = npq = 10(.75)(.25) = 1.875$, and $\sigma = \sqrt{1.875}$. Again, even though X can take on only integer values, $E(X)$ need not be an integer. If we perform a large number of independent binomial experiments, each with $n = 10$ trials and $p = .75$, then the average number of S's per experiment will be close to 7.5.   ■

Exercises / Section 3.4* (41–55)

41. Compute the following binomial probabilities directly from the formula for $b(x; n, p)$.
   a. $b(3; 8, .6)$ **b.** $b(5; 8, .6)$
   c. $P(3 \leq X \leq 5)$ when $n = 8$ and $p = .6$
   d. $P(1 \leq X)$ when $n = 12$ and $p = .1$

42. Use Appendix Table A.1 to obtain the following probabilities:
   a. $B(4; 10, .3)$ **b.** $b(4; 10, .3)$ **c.** $b(6; 10, .7)$
   d. $P(2 \leq X \leq 4)$ when $X \sim \text{Bin}(10, .3)$
   e. $P(2 \leq X)$ when $X \sim \text{Bin}(10, .3)$
   f. $P(X \leq 1)$ when $X \sim \text{Bin}(10, .7)$
   g. $P(2 < X < 6)$ when $X \sim \text{Bin}(10, .3)$

43. A large supermarket stocks both national brands of coffee and its own house brand. Consider a single randomly selected customer purchasing coffee and let success = {the customer purchases a national brand}. Assume that $p = .75$ and that customers make coffee purchase decisions independently of one another. Then with $X =$ the number among 20 randomly selected coffee purchasers who select a national brand, $X \sim \text{Bin}(20, .75)$.
   a. Compute $P(X \leq 15)$.
   b. Compute $P(X = 15)$.
   c. Compute the probability that at most 17 of the 20 customers select a national brand.
   d. Compute the probability that at least 12 customers select a national brand.
   e. Compute the probability that between 14 and 18 customers inclusive select a national brand.
   f. Compute the probability that fewer than half the customers select a national brand.
   g. To increase the sales of its own brand, the store instituted a promotional campaign. After the campaign, if fewer than half of 20 randomly selected purchasers had selected the national brand, would this be strong evidence that p was now less than .75?

h. With $X \sim \text{Bin}(20, .75)$, compute $E(X)$ and $V(X)$. Why is $V(X)$ here larger than in Example 3.34?

44. Suppose that only 20% of all drivers come to a complete stop at an intersection having flashing red lights in all directions when no other cars are visible. What is the probability that, of 20 randomly chosen drivers coming to an intersection under these conditions,
   a. at most 5 will come to a complete stop
   b. exactly 5 will come to a complete stop
   c. at least 5 will come to a complete stop
   d. How many of the next 20 drivers do you expect to come to a complete stop?

45. If 90% of all students taking a beginning computer programming course fail to get their first program to run on first submission, what is the probability that among 15 randomly chosen such students,
   a. at least 12 fail on first submission
   b. between 10 and 13 inclusive fail on first submission
   c. at most 2 get their program to run properly on first submission?

46. A particular type of tennis racket comes strung with either nylon or gut strings. The local tennis shop currently has 10 such rackets strung with nylon and 10 strung with gut. If 60% of all customers who request this racket choose nylon strings, what is the probability that
   a. more than 10 of the next 15 customers request nylon
   b. all of the next 15 customers who ask for this racket can be given their choice of strings without the store having to reorder?

47. A very large batch of components has arrived at a distributor. The batch can be characterized as acceptable only if the proportion of defective components is at most .10. The distributor decides to randomly select 10 components and ac-

---

*"Between a and b inclusive" is equivalent to $(a \leq X \leq b)$.

cept the batch only if the number of defective components in the sample is at most 2.

a. What is the probability that the batch will be accepted when the actual proportion of defectives is .01? .05? .10? .20? .25?

b. Let p denote the actual proportion of defectives in the batch. A graph of P(lot is accepted) as a function of p with p on the horizontal axis and P(lot is accepted) on the vertical axis is called the *operating characteristic curve* for the lot acceptance sampling plan. Use the results of (a) to sketch this curve for $0 \le p \le 1$.

c. Repeat (a) and (b) with "1" replacing "2" in the lot acceptance sampling plan.

d. Repeat (a) and (b) with 15 replacing 10 in the lot acceptance sampling plan.

e. Which of the three sampling plans, that of (a), (c), or (d), appears most satisfactory, and why?

48. An ordinance requiring that a smoke detector be installed in all previously constructed houses has been in effect in a particular city for one year. The fire department is concerned that many houses remain without detectors. Let $p =$ the true proportion of such houses having detectors, and suppose that a random sample of 25 homes is inspected. If the sample strongly indicates that fewer than 80% of all houses have a detector, the fire department will campaign for a mandatory inspection program. Because of the costliness of the program, the department prefers not to call for such inspections unless sample evidence strongly argues for their necessity. Let X denote the number of homes with detectors among the 25 sampled. Consider rejecting the claim that $p \ge .8$ if $x \le 15$, where x is the observed value of X.

a. What is the probability that the claim is rejected when the actual value of p is .8?

b. What is the probability of not rejecting the claim when $p = .7$? When $p = .6$?

c. How do the "error probabilities" of (a) and (b) change if the value 15 in the decision rule is replaced by 14?

49. A student who is trying to write a paper for a course has a choice of two topics, A and B. If topic A is chosen, the student will order two books through interlibrary loan, while if topic B is chosen, the student will order four books. The student feels that a good paper necessitates receiving and using at least half the books ordered for either topic chosen. If the probability that a book ordered through interlibrary loan actually arrives in time is .9 and books arrive independently of one another, which topic should the student choose to maximize the probability of writing a good paper? What if the arrival probability is only .5 instead of .9?

50. a. For fixed n, are there values of $p(0 \le p \le 1)$ for which $V(X) = 0$? Explain why this is so.

b. For what value of p is $V(X)$ maximized? *Hint:* Either graph $V(X)$ as a function of p or else take a derivative.

51. a. Show that $b(x; n, 1 - p) = b(n - x; n, p)$.

b. Show that $B(x; n, 1 - p) = 1 - B(n - x - 1; n, p)$. *Hint:* At most x S's is equivalent to at least $n - x$ F's.

c. What do (a) and (b) imply about the necessity of including values of p greater than .5 in Appendix Table A.1?

52. Show that $E(X) = np$ when X is a binomial random variable. *Hint:* First express $E(X)$ as a sum with lower limit $x = 1$. Then factor out np, let $y = x - 1$ so that the sum is from $y = 0$ to $y = n - 1$, and show that the sum equals 1.

53. Customers at a gas station select either regular (A), premium (B), or diesel fuel (C). Assume that successive customers make independent choices, with $P(A) = .3$, $P(B) = .2$, and $P(C) = .5$.

a. Among the next 100 customers, what are the mean and variance of the number who select regular fuel? Explain your reasoning.

b. Answer (a) for the number among the 100 who select a nondiesel fuel.

54. An airport limousine can accommodate up to four passengers on any one trip. The company will accept a maximum of six reservations for a trip and a passenger must have a reservation. From previous records, 20% of all those making reservations do not appear for the trip. Answer the following questions, assuming independence wherever appropriate.

a. If six reservations are made, what is the probability that at least one individual with a reservation cannot be accommodated on the trip?

b. If six reservations are made, what is the expected number of available places when the limousine departs?

c. Suppose that the probability distribution of the number of reservations made is given in the accompanying table.

Number of reservations	3	4	5	6
Probability	.1	.2	.3	.4

Let X denote the number of passengers on a randomly selected trip. Obtain the probability mass function of X.

55. Refer back to Chebyshev's inequality given in Exercise 40. Calculate $P(|X - \mu| \geq k\sigma)$ for $k = 2$ and $k = 3$ when $X \sim \text{Bin}(20, .5)$, and compare to the corresponding upper bound. Repeat for $X \sim \text{Bin}(20, .75)$.

3.5   The Hypergeometric and Negative Binomial Distributions

The hypergeometric and negative binomial distributions are both closely related to the binomial distribution. Whereas the binomial distribution is the approximate probability model for sampling without replacement from a finite dichotomous (S–F) population, the hypergeometric distribution is the exact probability model for the number of S's in the sample. The binomial random variable X is the number of S's when the number of trials n is fixed, whereas the negative binomial distribution arises from fixing the number of S's and letting the number of trials be random.

The Hypergeometric Distribution

The assumptions leading to the hypergeometric distribution are:

1. The population or set to be sampled consists of N individuals, objects, or elements (a *finite* population).
2. Each individual can be characterized as a success (S) or a failure (F), and there are M successes in the population.
3. A sample of n individuals is drawn in such a way that each subset of size n is equally likely to be chosen.

The random variable of interest is $X =$ the number of S's in the sample. The probability distribution of X depends on the parameters n, M, and N, so we wish to obtain $P(X = x) = h(x; n, M, N)$.

Example 3.35   An undergraduate library has 20 copies of a certain introductory economics text, of which 8 are first printings and 12 are second printings (containing corrections of some minor errors that appeared in the first printing). The course instructor has requested that five copies be put on two-hour reserve. If the copies are selected in a completely random fashion, so that every subset of size five has the same probability of being selected, what is the probability that x ($x = 0, 1, 2, 3, 4,$ or 5) of those selected are second printings?

In this example the population size is $N = 20$, the sample size is $n = 5$, and the number of S's (second printing $= S$) and F's in the population are $M = 12$ and $N - M = 8$, respectively. Consider the value $x = 2$. Because all outcomes (each one consisting of five particular books) are equally likely,

$$P(X = 2) = h(2; 5, 12, 20) = \frac{\text{number of outcomes having } X = 2}{\text{number of possible outcomes}}$$

The number of possible outcomes in the experiment is the number of ways of selecting five from the 20 objects without regard to order—that is, $\binom{20}{5}$. To count the number of outcomes having $X = 2$, note that there are $\binom{12}{2}$ ways of selecting two of the second printings, and for each such way there are $\binom{8}{3}$ ways of selecting the three first printings to fill out the sample. The product rule from Chapter 2 then gives $\binom{12}{2}\binom{8}{3}$ as the number of outcomes with $X = 2$, so

$$h(2; 5, 12, 20) = \frac{\binom{12}{2}\binom{8}{3}}{\binom{20}{5}} = \frac{77}{323} = .238 \qquad \blacksquare$$

In general, if the sample size n is smaller than the number of successes in the population (M), then the largest possible X value is n. However, if $M < n$ (for example, a sample size of 25 and only 15 successes in the population), then X can be at most M. Similarly, whenever the number of population failures $N - M$ exceeds the sample size, the smallest possible X value is zero (since all sampled individuals might then be failures). However, if $N - M < n$, the smallest possible X value is $n - (N - M)$. Summarizing, possible values of X satisfy $\max(0, n - (N - M)) \le x \le \min(n, M)$. An argument parallel to that of the previous example gives the p.m.f. of X.

Proposition

> If X is the number of S's in a completely random sample of size n drawn from a population consisting of M S's and $(N - M)$ F's, then the probability distribution of X, called the **hypergeometric distribution,** is given by
>
> $$P(X = x) = h(x; n, M, N) = \frac{\binom{M}{x}\binom{N - M}{n - x}}{\binom{N}{n}} \qquad (3.15)$$
>
> for x an integer satisfying $\max(0, n - N + M) \le x \le \min(n, M)$.

In Example 3.35, $n = 5$, $M = 12$, and $N = 20$, so $h(x; 5, 12, 20)$ for $x = 0, 1, 2, 3, 4, 5$ can be obtained by substituting these numbers into (3.15).

Example 3.36   Five individuals from an animal population thought to be near extinction in a certain region have been caught, tagged, and released to mix into the population. After they have had an opportunity to mix in, a random sample of 10 of these animals is selected. Let $X = $ the number of tagged animals in the second sample. If there are actually 25 animals of this type in the region, what is the probability that (a) $X = 2$? (b) $X \le 2$?

The parameter values are $n = 10$, $M = 5$ (five tagged animals in the population), and $N = 25$, so

$$h(x; 10, 5, 25) = \frac{\binom{5}{x}\binom{20}{10-x}}{\binom{25}{10}} \quad x = 0, 1, 2, 3, 4, 5$$

For (a), $P(X = 2) = h(2; 10, 5, 25) = \dfrac{\binom{5}{2}\binom{20}{8}}{\binom{25}{10}} = .385$

For (b), $P(X \le 2) = P(X = 0, 1, \text{or } 2) = \displaystyle\sum_{x=0}^{2} h(x; 10, 5, 25)$

$$= .057 + .257 + .385 = .699 \quad \blacksquare$$

Comprehensive tables of the hypergeometric distribution are available, but because the distribution has three parameters, these tables require much more space than tables for the binomial distribution.

The Mean and Variance of *X*

As in the binomial case, there are simple expressions for $E(X)$ and $V(X)$.

Proposition

> The mean and variance of the hypergeometric random variable X having p.m.f. $h(x; n, M, N)$ are
>
> $$E(X) = n \cdot \frac{M}{N} \quad \text{and} \quad V(X) = \left(\frac{N-n}{N-1}\right) \cdot n \cdot \frac{M}{N} \cdot \left(1 - \frac{M}{N}\right)$$

The ratio M/N is the proportion of *S*'s in the population. If we replace M/N by p in $E(X)$ and $V(X)$, we get

$$E(X) = np$$

$$V(X) = \left(\frac{N-n}{N-1}\right) \cdot np(1-p) \tag{3.16}$$

Expression (3.16) shows that the means of the binomial and hypergeometric random variables are equal, while the variances of the two r.v.'s differ by the factor $(N - n)/(N - 1)$, often called the **finite population correction factor.** This factor is less than one, so the hypergeometric variable has smaller variance than does the binomial r.v. The correction factor can be written $(1 - n/N)/(1 - 1/N)$, which is approximately 1 when n is small relative to N.

Example **3.37**
(Example 3.36
continued)

In the animal-tagging example, $n = 10$, $M = 5$, and $N = 25$, so $p = \frac{5}{25} = .2$ and

$$E(X) = 10(.2) = 2$$

$$V(X) = \frac{15}{24}\,(10)(.2)(.8) = (.625)(1.6) = 1$$

If the sampling was carried out with replacement, $V(X) = 1.6$. Suppose that the population size N is not actually known, so the value x is observed and we wish to estimate N. It is reasonable to equate the observed sample proportion of S's x/n with the population proportion M/N, giving the estimate

$$\hat{N} = \frac{M \cdot n}{x}$$

If $M = 100$, $n = 40$, and $x = 16$, then $\hat{N} = 250$. ■

Approximating Hypergeometric Probabilities

Our general rule of thumb in Section 3.4 stated that if sampling was without replacement but n/N was at most .05, then the binomial distribution could be used to compute approximate probabilities involving the number of S's in the sample. A more precise statement is as follows: Let the population size N and number of population S's M get large with the ratio M/N remaining fixed at p. Then $h(x; n, M, N)$ approaches $b(x; n, p)$, so for n/N small, the two are approximately equal provided that p is not too near either 0 or 1. This is the rationale for our rule of thumb.

The Negative Binomial Distribution

The negative binomial random variable and distribution are based on an experiment satisfying the following conditions:

1. The experiment consists of a sequence of independent trials.
2. Each trial can result in either a success (S) or a failure (F).
3. The probability of success is constant from trial to trial, so $P(S$ on trial $i) = p$ for $i = 1, 2, 3 \ldots$.
4. The experiment continues (trials are performed) until a total of r successes have been observed, where r is a specified positive integer.

The random variable of interest is $X =$ the number of failures that precede the rth success; X is called a **negative binomial** random variable because, in contrast to the binomial r.v., the number of successes is fixed and the number of trials is random.

Possible values of X are $0, 1, 2, \ldots$. Let $nb(x; r, p)$ denote the p.m.f. of X. The event $\{X = x\}$ is equivalent to $\{r - 1$ S's in the first $x + r - 1$ trials and an S on the $(x + r)$th trial$\}$ (if, for example, $r = 5$ and $x = 10$, then there must be four S's in the first 14 trials and the trial 15 must be an S). Since trials are independent,

$$nb(x; r, p) = P(X = x)$$
$$= P(r - 1 \ S\text{'s on the first } x + r - 1 \text{ trials}) \cdot P(S) \qquad (3.17)$$

The first probability on the far right of (3.17) is the binomial probability

$$\binom{x + r - 1}{r - 1} p^{r-1}(1 - p)^x, \text{ while } P(S) = p$$

Proposition

> The p.m.f. of the negative binomial r.v. X with parameters $r = $ number of S's and $p = P(S)$ is
>
> $$nb(x; r, p) = \binom{x + r - 1}{r - 1} p^r (1 - p)^x \quad x = 0, 1, 2, \ldots$$

Example 3.38

A pediatrician wishes to recruit five couples, each of whom is expecting their first child, to participate in a new natural childbirth regimen. Let $p = P(\text{a randomly selected couple agrees to participate})$. If $p = .2$, what is the probability that 15 couples must be asked before five are found who agree to participate? That is, with $S = \{\text{agrees to participate}\}$, what is the probability that 10 F's occur before the fifth S? Substituting $r = 5$, $p = .2$, and $x = 10$ into $nb(x; r, p)$ gives

$$nb(10; 5, .2) = \binom{14}{4}(.2)^5(.8)^{10} = .034$$

The probability that at most 10 F's are observed (at most 15 couples are asked) is

$$P(X \leq 10) = \sum_{x=0}^{10} nb(x; 5, .2) = (.2)^5 \sum_{x=0}^{10} \binom{x + 4}{4}(.8)^x = .164 \qquad \blacksquare$$

In some sources the negative binomial r.v. is taken to be the number of trials $X + r$ rather than the number of failures.

In the special case $r = 1$, the p.m.f. is

$$nb(x; 1, p) = (1 - p)^x p \quad x = 0, 1, 2, \ldots \qquad (3.18)$$

In Example 3.10 we derived the p.m.f. for the number of trials necessary to obtain the first S, and the p.m.f. there is quite similar to (3.18). Both $X = $ number of F's and $Y = $ number of trials ($= 1 + X$) are referred to in the literature as **geometric** random variables, and the p.m.f. (3.18) is called the **geometric distribution.**

In Example 3.18 the expected number of trials until the first S was shown to be $1/p$, so that the expected number of F's until the first S is $(1/p) - 1 = (1 - p)/p$. Intuitively, we would expect to see $r \cdot (1 - p)/p$ F's before the rth S, and this is indeed $E(X)$. There is also a simple formula for $V(X)$.

Proposition

> If X is a negative binomial random variable with p.m.f. $nb(x; r, p)$, then
>
> $$E(X) = \frac{r(1 - p)}{p} \quad \text{and} \quad V(X) = \frac{r(1 - p)}{p^2}$$

Lastly, by expanding the binomial coefficient in front of $p^r(1 - p)^x$ and doing some cancellation, it can be seen that $nb(x; r, p)$ is well defined even when r is not an integer. This *generalized negative binomial distribution* has been found to fit observed data quite well in a wide variety of applications.

Exercises / Section 3.5 (56–67)

56. A shipment of 15 concrete cylinders has been received by a contractor, five for a small project and the other ten for a larger project. Suppose that six of the 15 have a crushing strength below the specified minimum. If the five for the smaller project are randomly selected from the 15 and $X =$ the number among the five that have a below minimum crushing strength, then X has a hypergeometric distribution with parameters $n = 5$, $M = 6$, and $N = 15$. Compute
   a. $P(X = 2)$ **b.** $P(X \leq 2)$ **c.** $P(X \geq 1)$
   d. $E(X)$ and $V(X)$.

57. Each of 12 refrigerators of a certain type has been returned to a distributor because of the presence of a high-pitched oscillating noise when the refrigerator is running. Suppose that four of these 12 have defective compressors and the other eight have less serious problems. If they are examined in random order, let $X =$ the number among the first six examined that have a defective compressor. Compute
   a. $P(X = 1)$ **b.** $P(X \geq 4)$ **c.** $P(1 \leq X \leq 3)$

58. Professor Samuel P. Ling, a new faculty member in the Statistics Department at Hypergeometric State University, is to be assigned five student advisees. The department has 15 advisees to assign, of whom 10 are honor students, and will assign them in such a way that any five of the 15 are equally likely to be assigned to Professor Ling. Let $X =$ the number of honor students assigned to Professor Ling. Write the probability mass function of X.
   a. What is the probability that all five of Professor Ling's advisees are honor students?

b. What is the probability that exactly three of the five are honor students?
   c. Compute $E(X)$ and $V(X)$.

59. A tennis instructor has 12 cans of Wilson tennis balls and 15 cans of Penn tennis balls. If there are 20 students in the class and after pairing off, each of the 10 pairs is given a can chosen at random from those available, write down the p.m.f. of the random variable $X =$ the number of cans of Wilson balls distributed to the class. What is the probability that five cans of each type of ball are used by the class?

60. A geologist has collected 10 specimens of basaltic rock and 10 specimens of granite. If the geologist instructs a laboratory assistant to randomly select 15 of the specimens for analysis, what is the p.m.f. of the number of basalt specimens selected for analysis? What is the probability that all specimens of one of the two types of rock are selected for analysis?

61. A personnel director interviewing 11 senior engineers for four job openings has scheduled six interviews for the first day and five for the second day of interviewing. Assume that the candidates are interviewed in random order.
   a. What is the probability that x of the top four candidates are interviewed on the first day?
   b. How many of the top four candidates can be expected to be interviewed on the first day?

62. Twenty pairs of individuals playing in a bridge tournament have been seeded $1, \ldots, 20$. In the first part of the tournament, the 20 are randomly divided into 10 east–west pairs and 10 north–south pairs.

a. What is the probability that x of the top 10 pairs end up playing east–west?

b. What is the probability that all of the top five pairs end up playing the same direction?

c. If there are $2n$ pairs, what is the p.m.f. of $X =$ the number among the top n pairs who end up playing east–west, and what are $E(X)$ and $V(X)$?

63. A second-stage smog alert has been called in a certain area of Los Angeles County in which there are 50 industrial firms. An inspector will visit 10 randomly selected firms to check for violations of regulations.

a. If 15 of the firms are actually violating at least one regulation, what is the p.m.f. of the number of firms visited by the inspector that are in violation of at least one regulation?

b. If there are 500 firms in the area, of which 150 are in violation, approximate the p.m.f. of (a) by a simpler p.m.f.

c. For $X =$ the number among the 10 visited that are in violation, compute $E(X)$ and $V(X)$ both for the exact p.m.f. and the approximating p.m.f. in (b).

64. Suppose that $p = P$(male birth) $= .5$. A couple wishes to have exactly two female children in their family. They will have children until this condition is fulfilled.

a. What is the probability that the family has x male children?

b. What is the probability that the family has four children?

c. What is the probability that the family has at most four children?

d. How many male children would you expect this family to have? How many children would you expect this family to have?

65. A family decides to have children until it has three children of the same sex. Assuming $P(B) = P(G) = .5$, what is the p.m.f. of $X =$ the number of children in the family?

66. Three brothers and their wives decide to have children until each family has two female children. What is the p.m.f. of $X =$ the total number of male children born to the brothers? What is $E(X)$, and how does it compare to the expected number of male children born to each brother?

67. Individual A has a red die and B has a green die (both fair). If they each roll until they obtain five "doubles" (1–1, ..., 6–6), what is the p.m.f. of $X =$ the total number of times a die is rolled? What are $E(X)$ and $V(X)$?

3.6 The Poisson Probability Distribution

The binomial, hypergeometric, and negative binomial distributions were all derived by starting with an experiment consisting of trials or draws and applying the laws of probability to various outcomes of the experiment. There is no simple experiment on which the Poisson distribution is based, though we will shortly describe how it can be obtained by certain limiting operations.

Definition

A random variable X is said to have a **Poisson distribution** if the probability mass function of X is

$$p(x; \lambda) = \frac{e^{-\lambda}\lambda^x}{x!} \quad x = 0, 1, 2, \ldots$$

for some $\lambda > 0$.

The value of λ is frequently a rate per unit time or per unit area. The letter e in $p(x; \lambda)$ represents the base of the natural logarithm system, and its numerical value is approximately 2.71828. Because λ must be positive, $p(x; \lambda) > 0$ for all possible x values. The fact that $\sum\limits_{x=0}^{\infty} p(x; \lambda) = 1$ is a consequence of the Maclaurin infinite series expansion of e^λ, which appears in most calculus texts:

$$e^\lambda = 1 + \lambda + \frac{\lambda^2}{2!} + \frac{\lambda^3}{3!} + \cdots = \sum_{x=0}^{\infty} \frac{\lambda^x}{x!} \tag{3.19}$$

If the two extreme terms in (3.19) are multiplied by $e^{-\lambda}$, and then $e^{-\lambda}$ is placed inside the summation, the result is

$$1 = \sum_{x=0}^{\infty} e^{-\lambda} \frac{\lambda^x}{x!}$$

which shows that $p(x; \lambda)$ fulfills the second condition necessary for specifying a p.m.f.

Example 3.39 Let X denote the number of flaws on the surface of a randomly selected boiler of a certain type. Suppose that X has a Poisson distribution with $\lambda = 5$. Then the probability that a randomly selected boiler has exactly two flaws is

$$P(X = 2) = \frac{e^{-5}(5)^2}{2!} = .084$$

The probability that a boiler contains at most two flaws is

$$P(X \le 2) = \sum_{x=0}^{2} \frac{e^{-5}(5)^x}{x!} = e^{-5}\left(1 + 5 + \frac{25}{2}\right) = .125 \qquad \blacksquare$$

The Poisson Distribution as a Limit

The rationale for using the Poisson distribution in many situations is provided by the following proposition.

Proposition

> Suppose that in the binomial p.m.f. $b(x; n, p)$, we let $n \to \infty$ and $p \to 0$ in such a way that np remains fixed at value $\lambda > 0$. Then $b(x; n, p) \to p(x; \lambda)$.

According to this proposition, *in any binomial experiment in which n is large and p is small, $b(x; n, p) \approx p(x; \lambda)$ where $\lambda = np$.* As a rule of thumb, this approximation can safely be applied if $n \ge 100$, $p \le .01$, and $np \le 20$.

Example 3.40 If a publisher of nontechnical books takes great pains to ensure that its books are free of typographical errors, so that the probability of any given page containing at least one such error is .005 and errors are independent from page to page, what is the probability that one of its 400-page novels will contain exactly one page with errors? At most three pages with errors?

With S denoting a page containing at least one error and F an error-free page, the number of pages X containing at least one error is a binomial r.v. with $n = 400$ and $p = .005$, so $np = 2$. We wish

$$P(X = 1) = b(1; 400, .005) \approx p(1; 2) = \frac{e^{-2}(2)^1}{1!} = .271$$

Similarly,

$$P(X \leq 3) \approx \sum_{x=0}^{3} p(x; 2) = \sum_{x=0}^{3} e^{-2} \frac{2^x}{x!} = .135 + .271 + .271 + .180$$

$$= .857 \qquad \blacksquare$$

Appendix Table A.2 exhibits the cumulative distribution function $F(x; \lambda)$ for $\lambda = .1, .2, \ldots, 1, 2, \ldots, 10, 15$, and 20. For example, if $\lambda = 2$, then $P(X \leq 3) = F(3; 2) = .857$ as in Example 3.40, while $P(X = 3) = F(3; 2) - F(2; 2) = .180$.

The Mean and Variance of *X*

Since $b(x; n, p) \to p(x; \lambda)$ as $n \to \infty$, $p \to 0$, $np = \lambda$, the mean and variance of a binomial variable should approach those of a Poisson variable. These limits are $np = \lambda$ and $np(1 - p) \to \lambda$.

Proposition

> If X has a Poisson distribution with parameter λ, then $E(X) = V(X) = \lambda$.

These results can also be derived directly from the definitions of mean and variance.

Example 3.41
(Example 3.39 continued)

Both the expected number of flaws and the variance of the number of flaws equal five, and $\sigma_X = \sqrt{\lambda} = \sqrt{5} = 2.24$. $\qquad \blacksquare$

The Poisson Process

A very important application of the Poisson distribution arises in connection with the occurrence of events of a particular type over time. As an example, suppose that starting from a time point that we label $t = 0$, we are interested in counting the number of radioactive pulses recorded by a Geiger counter. We make the following assumptions about the way in which pulses occur:

1. There exists a parameter $\alpha > 0$ such that for any short time interval of length Δt, the probability that exactly one pulse is received is $\alpha \cdot \Delta t + o(\Delta t)$.*

---

*A quantity is $o(\Delta t)$ (read "little o of delta t") if as Δt approaches 0, so does $o(\Delta t)/\Delta t$. That is, $o(\Delta t)$ is even more negligible than Δt itself. The quantity $(\Delta t)^2$ has this property, but $\sin(\Delta t)$ does not.

2. The probability of more than one pulse being received during Δt is $o(\Delta t)$ [which, along with (1), implies that the probability of no pulses during Δt is $1 - \alpha \cdot \Delta t - o(\Delta t)$].

3. The number of pulses received during the time interval Δt is independent of the number received prior to this time interval.

Informally, Assumption 1 says that for a short interval of time, the probability of receiving a single pulse is approximately proportional to the length of the time interval, where α is the constant of proportionality. Now let $P_k(t)$ denote the probability that k pulses have been received by the counter during any particular time interval of length t.

Proposition

> $P_k(t) = e^{-\alpha t} \cdot (\alpha t)^k / k!$, so that the number of pulses during a time interval of length t is a Poisson random variable with parameter $\lambda = \alpha t$. The expected number of pulses during any such time interval is then αt, so the expected number during a unit interval of time is α.

Example 3.42

Suppose that pulses arrive at the counter at an average rate of six per minute, so that $\alpha = 6$. To find the probability that in a half-minute interval at least one pulse is received, note that the number of pulses in such an interval has a Poisson distribution with parameter $\alpha t = 6(.5) = 3$ (.5 min is used because α is expressed as a rate per minute). Then with $X =$ the number of pulses received in the 30-sec interval,

$$P(1 \le X) = 1 - P(X = 0) = 1 - \frac{e^{-3}(3)^0}{0!} = .950 \qquad \blacksquare$$

If in (1), (2), and (3) we replace "pulse" by "event," then the number of events occurring during a fixed time interval of length t has a Poisson distribution with parameter αt. Any process that has this distribution is called a **Poisson process,** and α is called the *rate of the process*. Other examples of situations giving rise to a Poisson process include monitoring the status of a computer system over time, with breakdowns constituting the events of interest, recording the number of accidents in an industrial facility over time, answering calls at a telephone switchboard, and observing the number of cosmic-ray showers from a particular observatory over time.

Instead of observing events over time, consider observing events of some type that occur in a two- or three-dimensional region. For example, we might select on a map a certain region R of a forest, go to that region and count the number of trees. Each tree would represent an event occurring at a particular point in space. Under assumptions similar to (1)–(3), it can be shown that the number of events occurring in a region R has a Poisson distribution with parameter $\alpha \cdot a(R)$, where $a(R)$ is the area of R. The quantity α is the expected number of events per unit area or volume.

Exercises / Section 3.6 (68–81)

68. Let X have a Poisson distribution with parameter $\lambda = 5$, and use Appendix Table A.2 to compute the following probabilities.
   a. $P(X \le 8)$ **b.** $P(X = 8)$ **c.** $P(9 \le X)$
   d. $P(5 \le X \le 8)$ **e.** $P(5 < X < 8)$

69. Suppose that the number of tornadoes X observed in a particular region during a one-year period has a Poisson distribution with $\lambda = 8$.
   a. Compute $P(X \le 5)$.
   b. Compute $P(6 \le X \le 9)$.
   c. Compute $P(10 \le X)$.
   d. How many tornadoes can be expected to be observed during the one-year period, and what is the standard deviation of the number of observed tornadoes?

70. Let $X =$ the number of automobiles of a particular year and model that will at some time in the future suffer a catastrophic failure in the steering mechanism, causing complete loss of control at high speed. Suppose that X has a Poisson distribution with parameter $\lambda = 10$.
   a. What is the probability that at most 10 cars suffer such a failure?
   b. What is the probability that between 10 and 15 (inclusive) cars will suffer such a failure?
   c. What are $E(X)$ and σ_X?

71. Consider writing onto a computer disk and then sending it through a certifier that counts the number of missing pulses. Suppose that this number X has a Poisson distribution with parameter $\lambda = .2$. (Suggested in "Average Sample Number for Semi-Curtailed Sampling Using the Poisson Distribution," *J. Quality Technology*, 1983, pp. 126–129.)
   a. What is the probability that a disk has exactly one missing pulse?
   b. What is the probability that a disk has at least two missing pulses?
   c. If two disks are independently selected, what is the probability that neither contains a missing pulse?

72. A library employee shelves 1000 books on a particular day. If the probability that any particular book is misshelved is .001 and books are shelved independently of one another, what is the approximate probability distribution of the number of books misshelved on that day? Use this distribution to calculate the probability that
   a. exactly three books are misshelved on that day
   b. at least one book is misshelved on that day

73. A notice is sent to all owners of a certain type of automobile asking them to bring their cars to a dealer to check for the presence of a particular manufacturing defect. Suppose that only .05% of such cars have the defect. Consider a random sample of 10,000 cars.
   a. What are the expected value and standard deviation of the number of cars in the sample that have the defect?
   b. What is the (approximate) probability that at least 10 sampled cars have the defect?
   c. What is the (approximate) probability that no sampled cars have the defect?

74. Suppose that small aircraft arrive at a certain airport according to a Poisson process with rate $\alpha = 8$ per hour, so that the number of arrivals during a time period of t hours is a Poisson random variable with parameter $\lambda = 8t$.
   a. What is the probability that exactly five small aircraft arrive during a one-hour period? At least 5? At least 10?
   b. What are the expected value and standard deviation of the number of small aircraft that arrive during a 90-min period?
   c. What is the probability that at least 20 small aircraft arrive during a two-and-one-half hour period? That at most 10 arrive during this period?

75. The number of tickets issued by a meter reader for parking meter violations can be modeled by a Poisson process with a rate parameter of 5 per hour.
   a. What is the probability that exactly five tickets are given out during a particular hour?
   b. What is the probability that at least three tickets are given out during a particular hour?
   c. How many tickets do you expect to be given during a 45-min period?

76. The number of requests for assistance received by a towing service is a Poisson process with rate $\alpha = 4$ per hour.

a. Compute the probability that exactly 10 requests are received during a particular two-hour period.

b. If the operators of the towing service take a 30-minute break for lunch, what is the probability that they do not miss any calls for assistance?

c. How many calls would you expect during their break?

77. Reconsider the librarian introduced in Exercise 72. Suppose that what occurs on any one day is independent of what happens on any other day. What is the probability that the librarian mis-shelves exactly one book during a five-day work-week (assuming that 1000 books/day are shelved)?

78. Let X have a Poisson distribution with parameter λ. Show that $E(X) = \lambda$ directly from the definition of expected value. *Hint:* The first term in the sum equals zero, and then x can be canceled. Now factor out λ, and show that what is left sums to one.

79. Suppose that trees are distributed in a forest according to a two-dimensional Poisson process with parameter α = the expected number of trees per acre = 80.

a. What is the probability that in a certain quarter-acre plot, there will be at most 16 trees?

b. If the forest covers 85,000 acres, what is the expected number of trees in the forest?

c. Suppose that you select a point in the forest and construct a circle of radius .1 mile. Let X = the number of trees within that circular region. What is the p.m.f. of X? *Hint:* 1 sq. mile = 640 acres.

80. Automobiles arrive at a vehicle equipment inspection station according to a Poisson process with rate $\alpha = 10$ per hour. Suppose that with probability .5, an arriving vehicle will have no equipment violations.

a. What is the probability that exactly 10 arrive during the hour and all 10 have no violations?

b. For any fixed $y \geq 10$, what is the probability that y arrive during the hour, of which 10 have no violations?

c. What is the probability that 10 "no violation" cars arrive during the next hour? *Hint:* Sum the probabilities in (b) from $y = 10$ to ∞.

81. a. In a Poisson process, what has to happen in both the time interval $(0, t)$ and the interval $(t, t + \Delta t)$ in order that there be no events occurring in the entire interval $(0, t + \Delta t)$? Use this and assumptions (1)–(3) to write a relationship between $P_0(t + \Delta t)$ and $P_0(t)$.

b. Use the result of (a) to write an expression for $P_0(t + \Delta t) - P_0(t)$. Then divide by Δt and let $\Delta t \to 0$ to obtain an equation involving d/dt $P_0(t)$, the derivative of $P_0(t)$ with respect to t.

c. Verify that $P_0(t) = e^{-\alpha t}$ satisfies the equation of (b).

d. It can be shown in a manner similar to (a) and (b) that the $P_k(t)$'s must satisfy the system of differential equations

$$\frac{d}{dt} P_k(t) = \alpha P_{k-1}(t) - \alpha P_k(t)$$
$$k = 1, 2, 3, \ldots$$

Verify that $P_k(t) = e^{-\alpha t}(\alpha t)^k / k!$ satisfies the system. (This is actually the only solution.)

Supplementary Exercises / Chapter 3 (82–105)

82. Consider a deck consisting of seven cards, marked $1, 2, \ldots, 7$. Three of these cards are selected at random. Define a r.v. W by W = the sum of the resulting numbers, and compute the p.m.f. of W. Then compute μ and σ^2. *Hint:* Consider outcomes as unordered, so that $(1, 3, 7)$ and $(3, 1, 7)$ are not different outcomes. Then there are 35 outcomes, and they can be listed (this type of r.v. actually arises in connection with a hypothesis test called Wilcoxon's rank sum test, in which there is an x sample and a y sample, and W is the sum of the ranks of the x's in the combined sample).

83. After shuffling a deck of 52 cards, a dealer deals out five. Let X = the number of suits represented in the five-card hand.

a. Show that the p.m.f. of X is

x	1	2	3	4
$p(x)$	.002	.146	.588	.264

Hint: $p(1) = 4P$(all are spades), $p(2) = 6P$(only spades and hearts with at least one of each), and $p(4) = 4P$(2 spades $\cap$ one of each other suit).

b. Compute μ, σ^2, and σ.

84. The negative binomial random variable X was defined as the number of F's preceding the rth S. Let $Y =$ the number of trials necessary to obtain the rth S. In the same manner in which the p.m.f. of X was derived, derive the p.m.f. of Y.

85. Of all customers purchasing automatic garage door openers, 75% purchase a chain-driven model. Let $X =$ the number among the next 15 purchasers who select the chain-driven model.

a. What is the p.m.f. of X?
b. Compute $P(X \leq 12)$.
c. Compute $P(9 \leq X \leq 12)$.
d. Compute μ and σ^2.
e. If the store currently has in stock 10 of the chain-driven models and eight shaft-driven models, what is the probability that the requests of these 15 customers can all be met from existing stock?

86. A friend recently planned a camping trip. He had two flashlights, one that required a single 6-V battery and another that used two size-D batteries. He had previously packed two 6-V and four size-D batteries in his camper. Suppose the probability that any particular battery works is p and that batteries work or fail independently of one another. Our friend wants to take just one flashlight. For what values of p should he take the 6-V flashlight?

87. A manufacturer of flashlight batteries wishes to control the quality of its product by rejecting any lot in which the proportion of batteries having unacceptable voltage appears to be too high. To this end, out of each large lot (10,000 batteries), 25 will be selected and tested. If at least five of these generate an unacceptable voltage, the entire lot will be rejected. What is the probability that a lot will be rejected if

a. 5% of the batteries in the lot have unacceptable voltage
b. 10% of the batteries in the lot have unacceptable voltage
c. 20% of the batteries in the lot have unacceptable voltage

d. What would happen to the above probabilities if the critical rejection number were increased from five to six?

88. Of the people passing through an airport metal detector, .1% activate it; let $X =$ the number among a randomly selected group of 500 who activate the detector.

a. What is the (approximate) p.m.f. of X?
b. Compute $P(X = 2)$. **c.** Compute $P(2 \leq X)$.

89. An educational consulting firm is trying to decide whether high school students who have never before used a hand-held calculator can solve a certain type of problem more easily with a calculator that uses reverse Polish logic or one that doesn't use this logic. A sample of 25 students is selected and allowed to practice on both calculators. Then each student is asked to work one problem on the reverse Polish calculator and a similar problem on the other. Let $p = P(S)$ where S denotes a student who worked the problem more quickly using reverse Polish logic than without, and let $X =$ number of S's.

a. If $p = .5$, what is $P(7 \leq X \leq 18)$?
b. If $p = .8$, what is $P(7 \leq X \leq 18)$?
c. If the claim that $p = .5$ is to be rejected when either $X \leq 7$ or $X \geq 18$, what is the probability of rejecting the claim when it is actually correct?
d. If the decision to reject the claim $p = .5$ is made as in (c), what is the probability that the claim is not rejected when $p = .6$? When $p = .8$?
e. What decision rule would you choose for rejecting the claim $p = .5$ if you wanted the probability in (c) to be at most .01?

90. Consider a disease whose presence can be identified by carrying out a blood test. Let p denote the probability that a randomly selected individual has the disease. Suppose that n individuals are independently selected for testing. One way to proceed is to carry out a separate test on each of the n blood samples. A potentially more economical approach, group testing, was introduced during WWII to identify syphilitic men among army inductees. First take a part of each blood sample, combine these specimens, and carry out a single test. If no one has the disease, the result will be negative, and only the one test is required. If at least one individual is diseased, the test on the combined sample will yield a positive result, in which case the n individual tests are then carried out. If $p = .1$ and $n = 3$, what is the expected number of tests using this procedure? What is the expected number when $n = 5$? (The paper "Random Multiple-Access Communication and Group Testing," *IEEE Trans. on Commun.,* 1984, pp. 769–774, applied these ideas to a communication system in which the dichotomy was active/idle user rather than diseased/non-diseased.)

91. Let p_1 denote the probability that any particular code symbol is erroneously transmitted through a communication system. Assume that on different symbols, errors occur independently of one another. Suppose also that with probability p_2, an erroneous symbol is corrected upon receipt. Let X denote the number of correct symbols in a message block consisting of n symbols (after the correction process has ended). What is the probability distribution of X?

92. The purchaser of a power generating unit requires c consecutive successful startups before the unit will be accepted. Assume that the outcomes of individual startups are independent of one another, and let p denote the probability that any particular startup is successful. The random variable of interest is $X =$ the number of startups that must be made prior to acceptance. Give the probability mass function of X for the case $c = 2$. If $p = .9$, what is $P(X \le 8)$? *Hint:* For $x \ge 5$, express $p(x)$ "recursively" in terms of the p.m.f. evaluated at the smaller values $x - 3$, $x - 4$, ..., 2. (This problem was suggested by the paper "Evaluation of a Start-Up Demonstration Test," *J. Quality Technology,* 1983, pp. 103–106.)

93. Of all repair work done on TV sets by a certain store, 80% is done on sets that are no longer under warranty (so if $S = \{$not under warranty$\}$, $P(S) = .8$).
    a. Among 20 sets brought in for repair work, what is the expected number not under warranty?
    b. Among these 20 sets, what is the expected number under warranty?
    c. Among these 20 sets, what is the probability that at least 75% are not under warranty?

d. Suppose that there are 12 sets now in the shop of which four are under warranty. If the 12 were brought in in completely random order and are serviced in the same order, what is the p.m.f. of $X =$ the number of sets under warranty among the first five serviced? What are $E(X)$ and $V(X)$?

94. A trial has just resulted in a hung jury because eight of the jury were in favor of a guilty verdict while the other four were for acquittal. If the jurors leave the jury room in random order and the first four leaving the room are each accosted by a reporter in quest of an interview, what is the p.m.f. of $X =$ the number of jurors favoring acquittal among those interviewed? How many of those favoring acquittal do you expect to be interviewed?

95. A telephone company employs five information operators who receive requests for information independently of one another, each according to a Poisson process with rate $\alpha = 2$ per minute.
    a. What is the probability that during a given one-minute period, the first operator receives no requests?
    b. What is the probability that during a given one-minute period, exactly four of the five operators receive no requests?
    c. Write an expression for the probability that during a given one-minute period, all of the operators receive exactly the same number of requests.

96. Grasshoppers are distributed at random in a large field according to a Poisson distribution with parameter $\alpha = 2$ per square yard. How large should the radius R of a circular sampling region be taken so that the probability of finding at least one in the region equals .99?

97. A newsstand has ordered five copies of a certain issue of a photography magazine. Let $X =$ the number of individuals who come in to purchase this magazine. If X has a Poisson distribution with parameter $\lambda = 4$, what is the expected number of copies that are sold?

98. Individuals A and B begin to play a sequence of chess games. Let $S = \{$A wins a game$\}$, and suppose that outcomes of successive games are independent with $P(S) = p$ and $P(F) = 1 - p$ (they

never draw). They will play until one of them wins 10 games. Let X = the number of games played (with possible values 10, 11, ..., 19).

a. For $x = 10, 11, ..., 19$, obtain an expression for $p(x) = P(X = x)$.

b. If a draw is possible, with $p = P(S)$, $q = P(F)$, $1 - p - q = P(\text{draw})$, what are the possible values of X? What is $P(20 \le X)$? *Hint:* $P(20 \le X) = 1 - P(X < 20)$.

99. Two drugs are being considered for treatment of the same disease. Let $p_1 = P(S \text{ using drug A})$ and $p_2 = P(S \text{ using drug B})$. Patients will be paired off and then one patient from each pair will be treated with drug A and the other with drug B. This clinical trial experiment will continue until one of the drugs first accumulates 10 S's. Let X = the number of pairs of patients treated.

a. What is $P(X = 10)$?

b. What is the probability that both drugs achieve 10 S's on the same pair? *Hint:* Both can achieve 10 S's on, say, the fifteenth pair if each of them has nine S's on the first 14 and an S on the fifteenth.

c. Derive an expression for $F(x) = P(X \le x)$, $x = 10, 11, ...$.

100. The generalized negative binomial p.m.f. is given by

$$nb(x; r, p) = k(r, x) \cdot p^r (1 - p)^x$$

$$x = 0, 1, 2, ...$$

Let X, the number of plants of a certain species found in a particular region, have this distribution with $p = .3$ and $r = 2.5$. What is $P(X = 4)$? What is the probability that at least one plant is found?

101. Define a function $p(x; \lambda, \mu)$ by

$p(x; \lambda, \mu)$

$$= \begin{cases} \dfrac{1}{2} e^{-\lambda} \dfrac{\lambda^x}{x!} + \dfrac{1}{2} e^{-\mu} \dfrac{\mu^x}{x!} & x = 0, 1, 2, ... \\ 0 & \text{otherwise} \end{cases}$$

a. Show that $p(x; \lambda, \mu)$ satisfies the two conditions necessary for specifying a p.m.f. *Note:* If a firm employs two typists, one of whom makes typographical errors at the rate of λ per page and the other at rate μ per page, and they each do half the firm's typing, then

$p(x; \lambda, \mu)$ is the p.m.f. of X = the number of errors on a randomly chosen page.

b. If the first typist (rate λ) types 60% of all pages, what is the p.m.f. of X of part (a)?

c. What is $E(X)$ for $p(x; \lambda, \mu)$ given by the expression just above?

d. What is σ^2 for $p(x; \lambda, \mu)$ given by that expression?

102. The *mode* of a discrete r.v. X with p.m.f. $p(x)$ is that value x^* for which $p(x)$ is largest (the most probable x value).

a. Let $X \sim \text{Bin}(n, p)$. By considering the ratio $b(x + 1; n, p)/b(x; n, p)$, show that $b(x; n, p)$ increases with x as long as $x < np - (1 - p)$. Conclude that the mode x^* is the integer satisfying $(n + 1)p - 1 \le x^* \le (n + 1)p$.

b. Show that if X has a Poisson distribution with parameter λ, the mode is the largest integer $< \lambda$. If λ is an integer, show that both $\lambda - 1$ and λ are modes.

103. A computer disk storage device has 10 concentric tracks, numbered 1, 2, ..., 10 from outermost to innermost, and a single access arm. Let p_i = the probability that any particular requests for data will take the arm to track i ($i = 1, ...,$ 10). Assume that the tracks accessed in successive seeks are independent, and let X = the number of tracks over which the access arm passes during two successive requests (excluding the track that the arm has just left, so possible X values are $x = 0, 1, ..., 9$). Compute the p.m.f. of X. *Hint:* $P(\text{the arm is now on track } i \text{ and } X = j) = P(X = j \mid \text{arm now on } i) \cdot p_i$. After writing the conditional probability in terms of $p_1, ..., p_{10}$, by the law of total probability, the desired probability is obtained by summing over i.

104. If X is a hypergeometric r.v., show directly from the definition that $E(X) = nM/N$ (consider only the case $n < M$). *Hint:* Factor nM/N out of the sum for $E(X)$, and show that the terms inside the sum are of the form $h(y; n - 1, M - 1, N - 1)$ where $y = x - 1$.

105. Use the fact that

$$\sum_{\text{all } x} (x - \mu)^2 p(x) \ge \sum_{x:|x - \mu| \ge k} (x - \mu)^2 p(x)$$

to prove Chebyshev's inequality given in Exercise 40.

Bibliography

Derman, Cyrus, Gleser, Leon, and Olkin, Ingram, *Probability Models and Applications,* Macmillan, New York, 1980. Contains an in-depth discussion of both general properties of discrete and continuous distributions and results for specific distributions.

Johnson, Norman, and Kotz, Samuel, *Distributions in Statistics: Discrete Distributions,* Houghton Mifflin, Boston, 1969. An encyclopedia of information on discrete distributions.

Ross, Sheldon, *Introduction to Applied Probability Models* (3rd ed.), Academic Press, New York, 1985. A good source of material on the Poisson process and generalizations, and a nice introduction to other topics in applied probability.

CHAPTER 4

Continuous Random Variables and Probability Distributions

Introduction

As mentioned at the beginning of Chapter 3, not all random variables are discrete. In this chapter we study the second general type of random variable that arises in many applied problems. Sections 4.1 and 4.2 present the basic definitions and properties of continuous random variables and their probability distributions. In Section 4.3 we study in detail the normal random variable and distribution, unquestionably the most important and useful in probability and statistics. Sections 4.4 and 4.6 discuss some other continuous distributions that are often used in applied work. In Section 4.6 we introduce a method for assessing whether given sample data is consistent with a specified distribution.

4.1 Continuous Random Variables and Probability Density Functions

A discrete random variable is one whose possible values either constitute a finite set or else can be listed in an infinite sequence (a list in which there is a first element, a second element, and so on). A random variable whose set of possible values is an entire interval of numbers is not discrete.

Continuous Random Variables

Definition

> A random variable X is said to be **continuous** if its set of possible values is an entire interval of numbers—that is, if for some $A < B$, any number x between A and B is possible.

Example **4.1**   If in the study of the ecology of a lake, we make depth measurements at randomly chosen locations, then $X = $ the depth at such a location is a continuous random variable. Here A is the minimum depth in the region being sampled and B is the maximum depth.   ■

Example **4.2**   If a chemical compound is randomly selected and its pH X is determined, then X is a continuous random variable, because any pH value between 0 and 14 is possible. If more is known about the compound selected for analysis, then the set of possible values might be a subinterval of [0, 14] such as $5.5 \leq x \leq 6.5$, but X would still be continuous.   ■

If the measurement scale of X can be subdivided to any extent desired, then the variable is continuous; if it cannot, the variable is discrete. For example, if the variable is height or length, then it can be measured in kilometers, meters, centimeters, millimeters, and so on, so the variable is continuous. If, however, $X = $ the billing amount on a randomly chosen monthly natural gas statement, then the smallest unit of measurement is pennies, so any value of X is a multiple of \$.01 and X is a discrete variable.

One might argue that although in principle variables such as height, weight, and temperature are continuous, in practice the limitations of our measuring instruments restrict us to a discrete (though sometimes very finely subdivided) world. However, continuous models often approximate real-world situations very well, and continuous mathematics (the calculus) is frequently easier to work with than mathematics of discrete variables and distributions.

Probability Distributions for Continuous Variables

Suppose that the variable of interest X is the depth of a lake at a randomly chosen point on the surface. Let $M = $ the maximum depth (in meters), so that any number in the interval [0, M] is a possible value of X. If we "discretize" X by measuring depth to the nearest meter, then possible values are nonnegative integers $\leq M$. The resulting discrete distribution of depth can be pictured using a probability histogram. If we draw the histogram so that the area of the rectangle above any possible integer k is the proportion of the lake whose depth is (to the nearest meter) k, then the total area of all rectangles is one. A possible histogram appears in Figure 4.1(a).

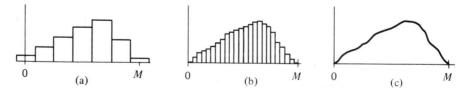

Figure 4.1   (a) Probability histogram of depth measured to the nearest meter; (b) probability histogram of depth measured to the nearest centimeter; (c) a limit of a sequence of discrete histograms

If depth is measured to the nearest centimeter and the same measurement axis as in Figure 4.1(*a*) is used, each rectangle in the resulting probability histogram is much narrower, though the total area of all rectangles is still one. A possible histogram is pictured in Figure 4.1(*b*); it has a much smoother appearance than the histogram in Figure 4.1(*a*). If we continue in this way to measure depth more and more finely, the resulting sequence of histograms approaches a smooth curve such as is pictured in Figure 4.1(*c*). Because for each histogram the total area of all rectangles equals one, the total area under the smooth curve is also one. The probability that the depth at a randomly chosen point is between *a* and *b* is just the area under the smooth curve between *a* and *b*. It is exactly a smooth curve of the type pictured in Figure 4.1(*c*) that specifies a continuous probability distribution.

Definition

Let X be a continuous random variable. Then a **probability distribution** or **probability density function** (p.d.f.) of X is a function $f(x)$ such that for any two numbers a and b with $a \leq b$,

$$P(a \leq X \leq b) = \int_a^b f(x)\, dx$$

That is, the probability that X takes on a value in the interval $[a, b]$ is the area under the graph of the density function, as illustrated in Figure 4.2.

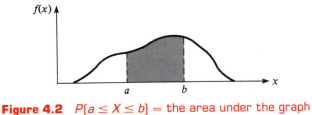

Figure 4.2 $P(a \leq X \leq b)$ = the area under the graph of $f(x)$ between a and b

In order that $f(x)$ be a legitimate p.d.f., it must satisfy the two conditions

1. $f(x) \geq 0$ for all x

2. $\int_{-\infty}^{\infty} f(x)\, dx$ = area under the entire graph of $f(x) = 1$

Example 4.3

Suppose that I take a bus to work, and that every five minutes a bus arrives at my stop. Because of variation in the time that I leave my house, I don't always arrive at the bus stop at the same time, so my waiting time X for the next bus is a continuous random variable. The set of possible values of X is the interval $[0, 5]$. One possible probability density function for X is

$$f(x) = \begin{cases} \dfrac{1}{5} & 0 \le x \le 5 \\ 0 & \text{otherwise} \end{cases}$$

The p.d.f. $f(x)$ is graphed in Figure 4.3. Clearly $f(x) \ge 0$, and the total area under the graph is $5 \cdot (\frac{1}{5}) = 1$. The probability that I wait between one and three minutes is

$$P(1 \le X \le 3) = \int_1^3 f(x)\, dx = \int_1^3 \frac{1}{5}\, dx = \frac{x}{5}\Big|_{x=1}^{x=3} = \frac{2}{5}$$

Similarly, $P(2 \le X \le 4) = \frac{2}{5}$. The probability that I wait at least four minutes is

$$P(4 \le X) = \int_4^\infty f(x)\, dx = \int_4^5 \frac{1}{5}\, dx + \int_5^\infty 0\, dx = \frac{x}{5}\Big|_{x=4}^{x=5} = \frac{1}{5}$$

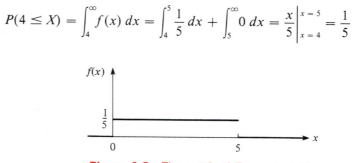

Figure 4.3   The p.d.f. of Example 4.3   ■

Because whenever $0 \le a \le b \le 5$ in Example 4.3, the probability $P(a \le X \le b)$ depends only on the length $b - a$ of the interval, X is said to have a uniform distribution.

Definition

A continuous r.v. X is said to have a **uniform distribution** on the interval $[A, B]$ if the p.d.f. of X is

$$f(x; A, B) = \begin{cases} \dfrac{1}{B - A} & A \le x \le B \\ 0 & \text{otherwise} \end{cases}$$

The graph of any uniform p.d.f. looks like the graph in Figure 4.3 except that the interval of positive density is $[A, B]$ rather than $[0, 5]$.

In the discrete case a probability mass function tells us how little "blobs" of probability mass of various magnitudes are distributed along the measurement axis. In the continuous case, probability density is "smeared" in a continuous fashion along the interval of possible values. When density is smeared uniformly over the interval, a uniform p.d.f., as in Figure 4.3, results.

When X is a discrete r.v., each possible value was assigned positive probability, but this is not the case when X is continuous.

Proposition

> If X is a continuous r.v., then for any number c, $P(X = c) = 0$. Furthermore, for any two numbers a and b with $a < b$,
>
> $$P(a \leq X \leq b) = P(a < X \leq b)$$
> $$= P(a \leq X < b) = P(a < X < b) \qquad (4.1)$$

In words, the probability assigned to any particular value is zero, and the probability of an interval does not depend on whether or not either of its end points is included. These properties follow from the fact that the area under the graph of $f(x)$ and above the single value c is zero, and that the area under the graph above an interval is unaffected by exclusion or inclusion of the end points of the interval. If X is discrete and there is positive probability mass at both $X = a$ and $X = b$, then all four probabilities in (4.1) will differ from one another.

The fact that a continuous distribution assigns probability zero to each value has a physical analogue. Consider a solid circular rod with cross-sectional area $= 1$ in.^{2}. Place the rod alongside a measurement axis and suppose that the density of the rod at any point x is given by the value $f(x)$ of a density function. Then if the rod is sliced at points a and b and this segment is removed, the amount of mass removed is $\int_a^b f(x)\, dx$, while if the rod is sliced just at the point c, no mass is removed. Mass is assigned to interval segments of the rod but not to individual points.

Example 4.4

"Time headway" in traffic flow is the elapsed time between the time that one car finishes passing a fixed point and the instant that the next car begins to pass that point. Let $X =$ the time headway for two randomly chosen consecutive cars on a freeway during a period of heavy flow. The following p.d.f. of X is essentially the one suggested in "The Statistical Properties of Freeway Traffic," *Transportation Research* (vol. 11, pp. 221–228):

$$f(x) = \begin{cases} .15e^{-.15(x - .5)} & x \geq .5 \\ 0 & \text{otherwise} \end{cases}$$

The graph of $f(x)$ appears in Figure 4.4; there is no density associated with headway times less than .5, and headway density decreases rapidly (exponentially fast) as x increases from .5. Clearly $f(x) \geq 0$; to show that $\int_{-\infty}^{\infty} f(x)\, dx = 1$ we use the calculus result $\int_a^{\infty} e^{-kx}\, dx = (1/k)\, e^{-k \cdot a}$. Then

$$\int_{-\infty}^{\infty} f(x)\, dx = \int_{.5}^{\infty} .15e^{-.15(x - .5)}\, dx = .15e^{.075} \int_{.5}^{\infty} e^{-.15x}\, dx$$

$$= .15e^{.075} \cdot \frac{1}{.15}\, e^{-(.15)(.5)} = 1$$

The probability that headway time is at most five seconds is

$$P(X \leq 5) = \int_{-\infty}^{5} f(x)\, dx = \int_{.5}^{5} .15e^{-.15(x - .5)}\, dx$$

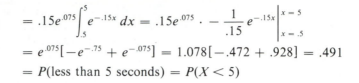

$$= .15e^{.075} \int_{.5}^{5} e^{-.15x} \, dx = .15e^{.075} \cdot \left. -\frac{1}{.15} e^{-.15x} \right|_{x=.5}^{x=5}$$

$$= e^{.075}[-e^{-.75} + e^{-.075}] = 1.078[-.472 + .928] = .491$$

$$= P(\text{less than 5 seconds}) = P(X < 5)$$

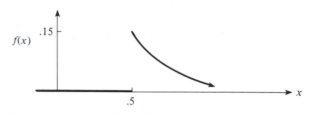

Figure 4.4   The p.d.f. of headway time in Example 4.4   ■

Unlike discrete distributions such as the binomial, hypergeometric, and negative binomial, the distribution of any given continuous random variable cannot usually be derived using simple probabilistic arguments. Instead, one must make a judicious choice of p.d.f. based on prior knowledge and available data. Fortunately there are some general families of p.d.f.'s that have been found to fit well in a wide variety of experimental situations; several of these are discussed later in the chapter.

Just as in the discrete case, it is often helpful to think of the population of interest as consisting of X values rather than individuals or objects. The probability density function is then a model for the distribution of values in this numerical population, and from this model various population characteristics (such as the mean) can be calculated.

Exercises / Section 4.1 (1–10)

1. Let X denote the amount of time for which a book on two-hour reserve at a college library is checked out by a randomly selected student, and suppose that X has density function

$$f(x) = \begin{cases} \frac{1}{2}x & 0 \le x \le 2 \\ 0 & \text{otherwise} \end{cases}$$

Calculate
a. $P(X \le 1)$
b. $P(.5 \le X \le 1.5)$
c. $P(1.5 < X)$

2. Suppose that the reaction temperature X (in °C) in a certain chemical process has a uniform distribution with $A = -5$ and $B = 5$.

a. Compute $P(X < 0)$.
b. Compute $P(-2 < X < 2)$.
c. Compute $P(-2 \le X \le 3)$.
d. For k satisfying $-5 < k < k + 4 < 5$, compute $P(k < X < k + 4)$.

3. Suppose that the distance X between a point target and a shot aimed at the point in a coin-operated target game is a continuous random variable with p.d.f.

$$f(x) = \begin{cases} \frac{3}{4}(1 - x^2) & -1 \le x \le 1 \\ 0 & \text{otherwise} \end{cases}$$

a. Sketch the graph of $f(x)$.
b. Compute $P(X > 0)$.

c. Compute $P(-.5 < X < .5)$.

d. Compute $P(X < -.25 \text{ or } X > .25)$.

4. Let X denote the vibratory stress (psi) on a wind turbine blade at a particular wind speed in a wind tunnel. The paper "Blade Fatigue Life Assessment with Application to VAWTS" (*J. Solar Energy Engr.*, 1982, pp. 107–111) proposes the Rayleigh distribution, with probability density function

$$f(x; \theta) = \begin{cases} \dfrac{x}{\theta^2} \cdot e^{-x^2/2\theta^2} & x > 0 \\ 0 & \text{otherwise} \end{cases}$$

as a model for the X distribution.

a. Verify that $f(x; \theta)$ is a legitimate p.d.f.

b. Suppose that $\theta = 100$ (a value suggested by a graph in the paper). What is the probability that X is at most 200? Less than 200? At least 200?

c. What is the probability that X is between 100 and 200?

d. Give an expression for $P(X \leq x)$.

5. A college professor never finishes his lecture before the bell rings to end the period, and always finishes his lectures within one minute after the bell rings. Let $X =$ the time that elapses between the bell and the end of the lecture, and suppose that the p.d.f. of X is

$$f(x) = \begin{cases} kx^2 & 0 \leq x \leq 1 \\ 0 & \text{otherwise} \end{cases}$$

a. Find the value of k. *Hint:* Total area under the graph of $f(x)$ is 1.

b. What is the probability that the lecture ends within $\frac{1}{2}$ minute of the bell ringing?

c. What is the probability that the lecture continues beyond the bell for between 15 and 30 seconds?

d. What is the probability that the lecture continues for at least 40 seconds beyond the bell?

6. The actual tracking weight of a stereo cartridge that is set to track at 3 gm on a particular changer can be regarded as a continuous r.v. X with p.d.f.

$$f(x) = \begin{cases} k[1 - (x - 3)^2] & 2 \leq x \leq 4 \\ 0 & \text{otherwise} \end{cases}$$

a. Sketch the graph of $f(x)$.

b. Find the value of k.

c. What is the probability that the actual tracking weight is greater than the prescribed weight?

d. What is the probability that the actual weight is within .25 gm of the prescribed weight?

e. What is the probability that the actual weight differs from the prescribed weight by more than .5 gm?

7. At a well-known restaurant the preparation time for a house specialty dish is listed on the menu as 30 minutes. Assume the actual preparation time X has a uniform distribution with $A = 25$ and $B = 35$.

a. Write the p.d.f. of X and sketch its graph.

b. What is the probability that actual preparation time exceeds 33 minutes?

c. What is the probability that actual preparation time is within two minutes of the time advertised?

d. For any a such that $25 < a < a + 2 < 35$, what is the probability that actual preparation time is between a and $a + 2$ minutes?

8. In commuting to work I must first get on a bus near my house and then transfer to a second bus. If the waiting time at each stop has a uniform distribution with $A = 0$ and $B = 5$, then it can be shown that my total waiting time Y has the p.d.f.

$$f(y) = \begin{cases} \dfrac{1}{25}y & 0 \leq y < 5 \\ \dfrac{2}{5} - \dfrac{1}{25}y & 5 \leq y \leq 10 \\ 0 & y < 0 \text{ or } y > 10 \end{cases}$$

a. Sketch a graph of the p.d.f. of Y.

b. Verify that $\int_{-\infty}^{\infty} f(y)\, dy = 1$.

c. What is the probability that total waiting time is at most three minutes?

d. What is the probability that total waiting time is at most eight minutes?

e. What is the probability that total waiting time is between three and eight minutes?

f. What is the probability that total waiting time is either less than two minutes or more than six minutes?

9. Consider again the p.d.f. of $X =$ time headway given in Example 4.4. What is the probability that time headway
   a. is at most six seconds?
   b. is more than six seconds? At least six seconds?
   c. is between five and six seconds?

10. A family of p.d.f.'s that has been used to approximate the distribution of income, city population size, and size of firms is the Pareto family. The family has two parameters, k and θ, both > 0, and the p.d.f. is

$$f(x; k, \theta) = \begin{cases} \dfrac{k \cdot \theta^k}{x^{k+1}} & x \geq \theta \\ 0 & x < \theta \end{cases}$$

   a. Sketch the graph of $f(x; k, \theta)$.
   b. Verify that the total area under the graph equals 1.
   c. If the r.v. X has p.d.f. $f(x; k, \theta)$, for any fixed $b > \theta$ obtain an expression for $P(X \leq b)$.
   d. For $\theta < a < b$, obtain an expression for $P(a \leq X \leq b)$.

4.2 Cumulative Distribution Functions and Expected Values

Several of the most important concepts introduced in the study of discrete distributions also play an important role for continuous distributions. Definitions analogous to those in Chapter 3 involve replacing summation by integration.

The Cumulative Distribution Function

The cumulative distribution function $F(x)$ for a discrete random variable X gives, for any specified number x, the probability $P(X \leq x)$. It is obtained by summing the p.m.f. $p(y)$ over all possible values y satisfying $y \leq x$. The c.d.f. of a continuous random variable gives the same probabilities $P(X \leq x)$, and is obtained by integrating the p.d.f. $f(y)$ between the limits $-\infty$ and x.

Definition

> The **cumulative distribution function** (c.d.f.) $F(x)$ for a continuous r.v. X is defined for every number x by
>
> $$F(x) = P(X \leq x) = \int_{-\infty}^{x} f(y)\, dy$$

Example 4.5

Let X have a uniform distribution on $[A, B]$. The density function appears in Figure 4.5. For $x < A$, $F(x) = 0$, since there is no area under the graph of the

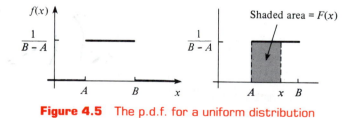

Figure 4.5 The p.d.f. for a uniform distribution

density function to the left of such an x. For $x \geq B$, $F(x) = 1$, since all the area is accumulated to the left of such an x. Finally, for $A \leq x \leq B$,

$$F(x) = \int_{-\infty}^{x} f(y)\, dy = \int_{A}^{x} \frac{1}{B-A}\, dy = \frac{1}{B-A} \cdot y \Big|_{y=A}^{y=x} = \frac{x-A}{B-A}$$

The entire c.d.f. is

$$F(x) = \begin{cases} 0 & x < A \\ \dfrac{x-A}{B-A} & A \le x < B \\ 1 & x \ge B \end{cases}$$

The graph of this c.d.f. appears in Figure 4.6.

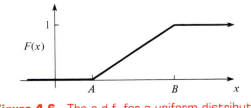

Figure 4.6   The c.d.f. for a uniform distribution   ∎

Using *F(x)* to Compute Probabilities

The importance of the c.d.f. here, just as for discrete random variables, is that probabilities of various intervals can be computed from a formula for or table of $F(x)$.

Proposition

Let X be a continuous random variable with p.d.f. $f(x)$ and c.d.f. $F(x)$. Then for any two numbers a, b with $a < b$,

$$P(a \le X \le b) = F(b) - F(a)$$

Figure 4.7 illustrates this proposition; the desired probability is the shaded area under the density curve between a and b and equals the difference between the two shaded cumulative areas.

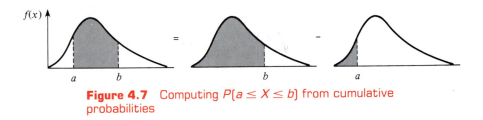

Figure 4.7   Computing $P(a \le X \le b)$ from cumulative probabilities

Example 4.6

Suppose that the p.d.f. of the magnitude X of a dynamic load on a bridge (in Newtons) is given by

$$f(x) = \begin{cases} \dfrac{1}{8} + \dfrac{3}{8}x & 0 \le x \le 2 \\ 0 & \text{otherwise} \end{cases}$$

The c.d.f. $F(x)$ is 0 for $x < 0$, 1 for $x > 2$, and

$$F(x) = \int_{-\infty}^{x} f(y)\, dy = \int_{0}^{x} \left(\frac{1}{8} + \frac{3}{8}y \right) dy = \frac{x}{8} + \frac{3}{16}x^2 \quad 0 \le x \le 2$$

The graphs of $f(x)$ and $F(x)$ appear in Figure 4.8. The probability that the load is between 1 and 1.5 is

$$P(1 \le X \le 1.5) = F(1.5) - F(1)$$

$$= \left[\frac{1}{8}(1.5) + \frac{3}{16}(1.5)^2 \right] - \left[\frac{1}{8}(1) + \frac{3}{16}(1)^2 \right]$$

$$= \frac{19}{64} = .297$$

The probability that the load exceeds 1 is

$$P(X > 1) = 1 - P(X \le 1) = 1 - F(1) = 1 - \left[\frac{1}{8}(1) + \frac{3}{16}(1)^2 \right]$$

$$= \frac{11}{16} = .688$$

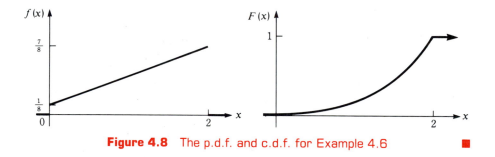

Figure 4.8   The p.d.f. and c.d.f. for Example 4.6   ■

In the previous example $P(1 \le X \le 1.5)$ could just as easily have been obtained by integrating $f(x)$, but when the integration must be done numerically (as with the normal and gamma distributions to be encountered shortly), tabulating $F(x)$ enables probabilities to be computed quickly.

Obtaining *f(x)* from *F(x)*

For X discrete, the p.m.f. is obtained from the c.d.f. by taking the difference between two $F(x)$ values. The continuous analogue of a difference is a derivative. The following result is a consequence of the Fundamental Theorem of Calculus.

Proposition

If X is a continuous r.v. with p.d.f. $f(x)$ and c.d.f. $F(x)$, then at every x at which the derivative $F'(x)$ exists, $F'(x) = f(x)$.

Example 4.7
(Example 4.5 continued)

When X has a uniform distribution, $F(x)$ is differentiable except at $x = A$ and $x = B$, where the graph of $F(x)$ has sharp corners. Since $F(x) = 0$ for $x < A$ and $F(x) = 1$ for $x > B$, $F'(x) = 0 = f(x)$ for such x. For $A < x < B$,

$$F'(x) = \frac{d}{dx}\left(\frac{x - A}{B - A}\right) = \frac{1}{B - A} = f(x)$$ ∎

Percentiles of a Continuous Distribution

When we say that an individual's test score was at the eighty-fifth percentile of the population, we mean that 85% of all population scores were below that score and 15% were above. Similarly, the 40th percentile is the score that exceeds 40% of all scores and is exceeded by 60% of all scores.

Definition

Let p be a number between 0 and 1. The **($100p$)th percentile** of the distribution of a continuous r.v. X, denoted by $\eta(p)$, is defined by

$$p = F(\eta(p)) = \int_{-\infty}^{\eta(p)} f(y)\, dy \qquad (4.2)$$

According to (4.2), $\eta(p)$ is that value on the measurement axis such that $100p\%$ of the area under the graph of $f(x)$ lies to the left of $\eta(p)$ and $100(1 - p)\%$ lies to the right. Thus $\eta(.75)$, the 75th percentile, is such that the area under the graph of $f(x)$ to the left of $\eta(.75)$ is .75. Figure 4.9 illustrates the definition.

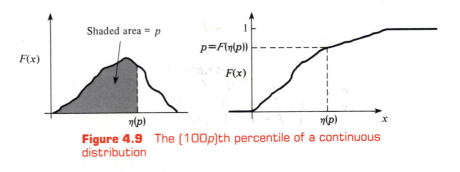

Figure 4.9   The ($100p$)th percentile of a continuous distribution

Example 4.8

The distribution of the amount of gravel (in tons) sold by a particular construction supply company in a given week is a continuous r.v. X with p.d.f.

$$f(x) = \begin{cases} \dfrac{3}{2}(1 - x^2) & 0 \le x \le 1 \\ 0 & \text{otherwise} \end{cases}$$

The cumulative distribution function of sales is then, for $0 < x < 1$,

$$F(x) = \int_0^x \frac{3}{2}(1 - y^2)\, dy = \frac{3}{2}\left(y - \frac{y^3}{3}\right)\bigg|_{y=0}^{y=x} = \frac{3}{2}\left(x - \frac{x^3}{3}\right)$$

The graphs of both $f(x)$ and $F(x)$ appear in Figure 4.10. The $(100p)$th percentile of this distribution satisfies the equation

$$p = F(\eta(p)) = \frac{3}{2}\left(\eta(p) - \frac{(\eta(p))^3}{3}\right)$$

that is,

$$(\eta(p))^3 - 3\eta(p) + 2p = 0$$

For the fiftieth percentile, $p = .5$ and the equation to be solved is $\eta^3 - 3\eta + 1 = 0$; the solution is $\eta = \eta(.5) = .347$. If the distribution remains the same from week to week, then in the long run 50% of all weeks will result in sales of less than .347 ton and 50% in more than .347 ton.

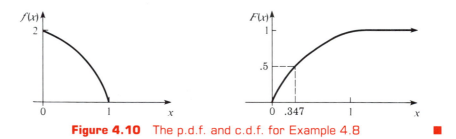

Figure 4.10 The p.d.f. and c.d.f. for Example 4.8 ■

Definition

> The **median** of a continuous distribution, denoted by $\tilde{\mu}$, is the fiftieth percentile, so $\tilde{\mu}$ satisfies $.5 = F(\tilde{\mu})$. That is, half the area under the density curve is to the left of $\tilde{\mu}$ and half is to the right of $\tilde{\mu}$.

Example **4.9**

A continuous distribution whose p.d.f. is **symmetric**—which means that the graph of the p.d.f. to the left of some point is a mirror image of the graph to the right of that point—has median $\tilde{\mu}$ equal to the point of symmetry, since half the area under the curve lies to either side of this point. Figure 4.11 gives several examples. The amount of error in a measurement of a physical quantity is often assumed to have a symmetric distribution.

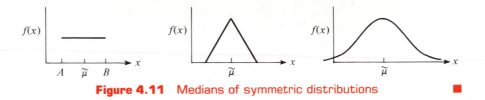

Figure 4.11 Medians of symmetric distributions ■

Expected Values for Continuous Random Variables

For a discrete r.v. X, $E(X)$ was obtained by summing $x \cdot p(x)$ over possible X values. Here we replace summation by integration and the p.m.f. by the p.d.f. to get a continuous weighted average.

Definition

> The **expected** or **mean value** of a continuous r.v. X with p.d.f. $f(x)$ is
>
> $$\mu_X = E(X) = \int_{-\infty}^{\infty} x \cdot f(x)\, dx$$

Example 4.10
(Example 4.8 continued)

The p.d.f. of weekly gravel sales X was

$$f(x) = \begin{cases} \dfrac{3}{2}(1 - x^2) & 0 \le x \le 1 \\ 0 & \text{otherwise} \end{cases}$$

so

$$E(X) = \int_{-\infty}^{\infty} x \cdot f(x)\, dx = \int_0^1 x \cdot \frac{3}{2}(1 - x^2)\, dx$$

$$= \frac{3}{2}\int_0^1 (x - x^3)\, dx = \frac{3}{2}\left(\frac{x^2}{2} - \frac{x^4}{4}\right)\Bigg|_{x=0}^{x=1} = \frac{3}{8}$$

■

When the p.d.f. $f(x)$ specifies a model for the distribution of values in a numerical population, then μ is the population mean, which is the most frequently used measure of population location or center.

Often we wish to compute the expected value of some function $h(X)$ of the r.v. X. If we think of $h(X)$ as a new r.v. Y, techniques from mathematical statistics can be used to derive the p.d.f. of Y, and $E(Y)$ can be computed from the definition. Fortunately, as in the discrete case, there is an easier way to compute $E[h(X)]$.

Proposition

> If X is a continuous r.v. with p.d.f. $f(x)$ and $h(X)$ is any function of X, then
>
> $$E[h(X)] = \mu_{h(X)} = \int_{-\infty}^{\infty} h(x) \cdot f(x)\, dx$$

Example 4.11

Two species are competing in a region for control of a limited amount of a certain resource. Let X = the proportion of the resource controlled by species 1, and suppose that X has p.d.f.

$$f(x) = \begin{cases} 1 & 0 \le x \le 1 \\ 0 & \text{otherwise} \end{cases}$$

which is a uniform distribution on $[0, 1]$ (in her book *Ecological Diversity*, E. C. Pielou calls this the "broken stick" model for resource allocation, since it is analogous to breaking a stick at a randomly chosen point). Then the species that controls the majority of this resource controls the amount

$$h(X) = \max(X, 1 - X) = \begin{cases} 1 - X & \text{if} \quad 0 \le X < \dfrac{1}{2} \\ X & \text{if} \quad \dfrac{1}{2} \le X \le 1 \end{cases}$$

The expected amount controlled by the species having majority control is then

$$E[h(X)] = \int_{-\infty}^{\infty} \max(x, 1 - x) \cdot f(x)\, dx = \int_{0}^{1} \max(x, 1 - x) \cdot 1\, dx$$

$$= \int_{0}^{1/2} (1 - x) \cdot 1\, dx + \int_{1/2}^{1} x \cdot 1\, dx = \frac{3}{4} \qquad \blacksquare$$

For $h(X)$ a linear function, $E[h(X)] = E(aX + b) = aE(X) + b$.

The Variance of a Continuous Random Variable

Definition

> The **variance** of a continuous r.v. X with p.d.f. $f(x)$ and mean value μ is
>
> $$\sigma_X^2 = V(X) = \int_{-\infty}^{\infty} (x - \mu)^2 \cdot f(x)\, dx = E[(X - \mu)^2]$$
>
> The **standard deviation** of X is $\sigma_X = \sqrt{V(X)}$.

As in the discrete case, σ_X^2 is the expected or average squared deviation about the mean μ, and gives a measure of how much spread there is in the distribution or population of x values. The easiest way to compute σ^2 is to again use a shortcut formula.

Proposition

> $$V(X) = E(X^2) - [E(X)]^2$$

Example 4.12
(Example 4.10 continued)

For X = weekly gravel sales, we computed $E(X) = \frac{3}{8}$. Since

$$E(X^2) = \int_{-\infty}^{\infty} x^2 \cdot f(x)\, dx = \int_{0}^{1} x^2 \cdot \frac{3}{2}(1 - x^2)\, dx = \int_{0}^{1} \frac{3}{2}(x^2 - x^4)\, dx = \frac{1}{5}$$

$$V(X) = \frac{1}{5} - \left(\frac{3}{8}\right)^2 = \frac{19}{320} = .059 \quad \text{and} \quad \sigma_X = .244$$ ∎

When $h(X)$ is a linear function and $V(X) = \sigma^2$, $V[h(X)] = V(aX + b) = a^2 \cdot \sigma^2$ and $\sigma_{aX+b} = |a| \cdot \sigma$.

Exercises / Section 4.2 (11–24)

11. The cumulative distribution function of checkout duration X as described in Exercise 1 is

$$F(x) = \begin{cases} 0 & x < 0 \\ \dfrac{x^2}{4} & 0 \le x < 2 \\ 1 & 2 \le x \end{cases}$$

Use this to compute
a. $P(X \le 1)$ **b.** $P(.5 \le X \le 1)$ **c.** $P(X > .5)$
d. the median checkout duration $\tilde{\mu}$ [solve $.5 = F(\tilde{\mu})$]
e. $F'(x)$ to obtain the density function $f(x)$.

12. The cumulative distribution function for X ($=$ shot-to-target distance) of Exercise 3 is

$$F(x) = \begin{cases} 0 & x < -1 \\ \dfrac{1}{2} + \dfrac{3}{4}\left(x - \dfrac{x^3}{3}\right) & -1 \le x < 1 \\ 1 & 1 \le x \end{cases}$$

a. Compute $P(X < 0)$.
b. Compute $P(-.5 < X < .5)$.
c. Compute $P(.25 < X)$.
d. Verify that $f(x)$ is as given in Exercise 3 by obtaining $F'(x)$.
e. Verify that $\tilde{\mu} = 0$.

13. Let X denote checkout time duration with p.d.f. given in Exercise 1.
a. Compute $E(X)$. **b.** Compute $V(X)$ and σ_X.
c. If the borrower is charged an amount $h(X) = X^2$ when checkout duration is X, compute the expected charge $E[h(X)]$.

14. The article "Modeling Sediment and Water Column Interactions for Hydrophobic Pollutants" (*Water Research*, 1984, pp. 1169–1174) suggests the uniform distribution on the interval (7.5, 20) as a model for depth (cm) of the bioturbation layer in sediment in a certain region.
a. What are the mean and variance of depth?
b. What is the cumulative distribution function of depth?

c. What is the probability that observed depth is at most 10? Between 10 and 15?
d. What is the probability that the observed depth is within one standard deviation of the mean value? Within two standard deviations?

15. Suppose that the p.d.f. of weekly gravel sales X (in tons) is

$$f(x) = \begin{cases} 2(1 - x) & 0 \le x \le 1 \\ 0 & \text{otherwise} \end{cases}$$

a. Obtain the c.d.f. of X and graph it.
b. What is $P(X \le .5)$ [that is, $F(.5)$]?
c. Using (a), what is $P(.25 < X \le .5)$? $P(.25 \le X \le .5)$?
d. What is the seventy-fifth percentile of the sales distribution?
e. What is the median $\tilde{\mu}$ of the sales distribution?
f. Compute $E(X)$ and σ_X.

16. Answer (a)–(f) of Exercise 15 with $X =$ lecture time past the bell given in Exercise 5.

17. Consider the p.d.f. of $X =$ actual tracking weight given in Exercise 6.
a. Obtain and graph the c.d.f. of X.
b. From the graph of $f(x)$, what is $\tilde{\mu}$?
c. Compute $E(X)$ and $V(X)$.

18. Let X have a uniform distribution on the interval $[A, B]$.
a. Obtain an expression for the $(100p)$th percentile.
b. Compute $E(X)$, $V(X)$, and σ_X.
c. For n a positive integer, compute $E(X^n)$.

19. Let X be a continuous random variable with cumulative distribution function

$$F(x) = \begin{cases} 0 & x \le 0 \\ \dfrac{x}{4}\left[1 + \ln\left(\dfrac{4}{x}\right)\right] & 0 < x \le 4 \\ 1 & x > 4 \end{cases}$$

(This type of c.d.f. is suggested in the paper "Variability in Measured Bedload-Transport Rates," *Water Resources Bull.*, 1985, pp. 39–48, as a model for a certain hydrologic variable.) What is

a. $P(X \leq 1)$?

b. $P(1 \leq X \leq 3)$?

c. the p.d.f. of X?

20. Consider the p.d.f. for total waiting time Y for two buses

$$f(y) = \begin{cases} \dfrac{1}{25}y & 0 \leq y < 5 \\ \dfrac{2}{5} - \dfrac{1}{25}y & 5 \leq y \leq 10 \\ 0 & \text{otherwise} \end{cases}$$

introduced in Exercise 8.

a. Compute and sketch the c.d.f. of Y. *Hint:* Consider separately $0 \leq y < 5$ and $5 \leq y \leq 10$ in computing $F(y)$. A picture of the p.d.f. should be helpful.

b. Obtain an expression for the $(100p)$th percentile. *Hint:* Consider separately $0 < p < .5$ and $.5 < p < 1$.

c. Compute $E(Y)$ and $V(Y)$. How do these compare with the expected waiting time and variance for a single bus when the time is uniformly distributed on $[0, 5]$?

21. An ecologist wishes to mark off a circular sampling region having radius 10 meters. However, the radius of the resulting region is actually a random variable R with p.d.f.

$$f(r) = \begin{cases} \dfrac{3}{4}[1 - (10 - r)^2] & 9 \leq r \leq 11 \\ 0 & \text{otherwise} \end{cases}$$

What is the expected area of the resulting circular region?

22. The weekly demand for propane gas (in thousands of gallons) from a particular facility is a r.v. X with p.d.f.

$$f(x) = \begin{cases} 2\left(1 - \dfrac{1}{x^2}\right) & 1 \leq x \leq 2 \\ 0 & \text{otherwise} \end{cases}$$

a. Compute the c.d.f. of X.

b. Obtain an expression for the $(100p)$th percentile. What is the value of $\tilde{\mu}$?

c. Compute $E(X)$ and $V(X)$.

d. If 1.5 thousand gallons is in stock at the beginning of the week and no new supply is due in during the week, how much of the 1.5 thousand gallons is expected to be left at the end of the week?

23. If the temperature of a compound at a particular time during a chemical reaction is a r.v. with mean value 120 °C and standard deviation 2 °C, what are the mean temperature and standard deviation measured in °F? *Hint:* $°F = \frac{9}{5} °C + 32$.

24. Let X have the Pareto p.d.f.

$$f(x; k, \theta) = \begin{cases} \dfrac{k \cdot \theta^k}{x^{k+1}} & x \geq \theta \\ 0 & x < \theta \end{cases}$$

introduced in Exercise 10.

a. If $k > 1$, compute $E(X)$.

b. What can you say about $E(X)$ if $k = 1$?

c. If $k > 2$, show that $V(X) = k\theta^2(k-1)^{-2}(k-2)^{-1}$.

d. If $k = 2$, what can you say about $V(X)$?

e. What conditions on k are necessary to ensure that $E(X^n)$ is finite?

4.3 The Normal Distribution

Definition

> A continuous random variable X is said to have a **normal distribution with parameters μ and σ (or μ and σ^2)**, where $-\infty < \mu < \infty$ and $0 < \sigma$, if the probability density function of X is
>
> $$f(x; \mu, \sigma) = \frac{1}{\sqrt{2\pi}\,\sigma} e^{-(x-\mu)^2/2\sigma^2} \qquad -\infty < x < \infty \qquad (4.3)$$

Again e denotes the base of the natural logarithm system and equals approximately 2.71828, while π represents the familiar mathematical constant with approximate value 3.14159. The statement that X is normally distributed with parameters μ and σ^2 is often abbreviated $X \sim N(\mu, \sigma^2)$.

Clearly $f(x; \mu, \sigma) \geq 0$ for any number x, but techniques from multivariate calculus must be used to show that $\int_{-\infty}^{\infty} f(x; \mu, \sigma)\, dx = 1$. It can be shown that $E(X) = \mu$ and $V(X) = \sigma^2$, so the parameters are the mean and standard deviation of X. Figure 4.12 presents graphs of $f(x; \mu, \sigma)$ for several different (μ, σ) pairs. Each graph is symmetric about μ and bell-shaped, so the center of the bell (point of symmetry) is both the mean of the distribution and the median. The value of σ is the distance from μ to the inflection points of the curve (the points at which the curve changes from turning downward to turning upward). Large values of σ yield graphs that are quite spread out about μ, whereas small values of σ yield graphs with a high peak above μ and most of the area under the graph quite close to μ. Thus a large σ implies that a value of X far from μ may well be observed, while such a value is quite unlikely when σ is small.

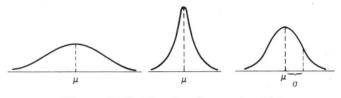

Figure 4.12   Graphs of normal p.d.f.'s

The normal distribution is the most important one in all of probability and statistics. Many numerical populations have distributions that can be fit very closely by a normal curve with appropriate values of μ and σ. Examples include heights, weights, and other physical characteristics (the famous 1903 *Biometrika* paper "On the Laws of Inheritance in Man" discussed many examples of this sort), measurement errors in scientific experiments, anthropometric measurements on fossils, reaction times in psychological experiments, measurements of intelligence and aptitude, scores on various tests, and numerous economic measures and indicators. Even when the underlying distribution is discrete, the normal curve often gives an excellent approximation. In addition, even when individual variables themselves are not normally distributed, sums and averages of the variables will under suitable conditions have approximately a normal distribution; this is the content of the Central Limit Theorem discussed in the next chapter.

The Standard Normal Distribution

To compute $P(a \leq X \leq b)$ when X is a normal r.v. with parameters μ and σ, we must evaluate

$$\int_a^b \frac{1}{\sqrt{2\pi}\,\sigma} e^{-(x-\mu)^2/2\sigma^2}\, dx \tag{4.4}$$

None of the standard integration techniques can be used to evaluate (4.4). Instead, for $\mu = 0$ and $\sigma = 1$, (4.4) has been numerically evaluated and tabulated for certain values of a and b. Using this table, probabilities can also be computed for any other values of μ and σ under consideration.

Definition

> The normal distribution with parameter values $\mu = 0$ and $\sigma = 1$ is called a **standard normal distribution.** A random variable that has a standard normal distribution is called a **standard normal random variable,** and will be denoted by Z. The p.d.f. of Z is
>
> $$ f(z; 0, 1) = \frac{1}{\sqrt{2\pi}} e^{-z^2/2} \quad -\infty < z < \infty $$
>
> The cumulative distribution function of Z is $P(Z \leq z) = \int_{-\infty}^{z} f(y; 0, 1)\, dy$, which we shall denote by $\Phi(z)$.

The standard normal distribution does not frequently serve as a model for a naturally arising population. Instead it is a reference distribution from which information about other normal distributions can be obtained. Appendix Table A.3 gives $\Phi(z) = P(Z \leq z)$, the area under the graph of the standard normal p.d.f. to the left of z, for $z = -3.49, -3.48, \ldots, 3.48, 3.49$. Figure 4.13 illustrates the type of cumulative area (probability) tabulated in Table A.3. From this table various other probabilities involving Z can be calculated.

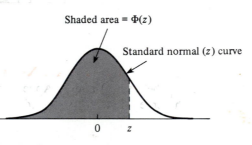

Figure 4.13 Standard normal curve areas tabulated in Appendix Table A.3

Example 4.13   Compute the following probabilities:

    (a) $P(Z \leq 1.25)$ (b) $P(Z > 1.25)$ (c) $P(Z \leq -1.25)$
    (d) $P(-.38 \leq Z \leq 1.25)$

  1. $P(Z \leq 1.25) = \Phi(1.25)$, a probability that is tabulated in Appendix Table A.3 at the intersection of the row marked 1.2 and the column marked .05. The number there is .8944, so $P(Z \leq 1.25) = .8944$. (See Figure 4.14(a).)

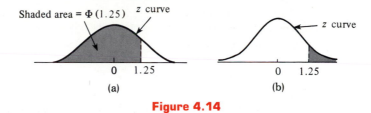

Figure 4.14

2. $P(Z > 1.25) = 1 - P(Z \leq 1.25) = 1 - \Phi(1.25)$, the area under the standard normal curve to the right of 1.25 (an upper-tail area). Since $\Phi(1.25) = .8944$, $P(Z > 1.25) = .1056$. Since Z is a continuous r.v., $P(Z \geq 1.25)$ also equals .1056. (See Figure 4.14(b).)

3. $P(Z \leq -1.25) = \Phi(-1.25)$, a lower-tail area. Directly from Table A.3, $\Phi(-1.25) = .1056$. By symmetry of the normal curve, this is the same answer as in (b).

4. $P(-.38 \leq Z \leq 1.25)$ is the area under the standard normal curve above the interval whose left end point is $-.38$ and whose right end point is 1.25. From Section 4.2 if X is a continuous r.v. with c.d.f. $F(x)$, then $P(a \leq X \leq b) = F(b) - F(a)$. This gives $P(-.38 \leq Z \leq 1.25) = \Phi(1.25) - \Phi(-.38) = .8944 - .3520 = .5424$. (See Figure 4.15.)

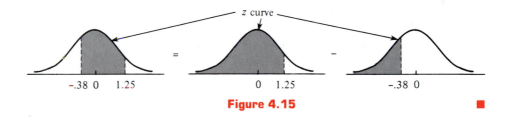

Figure 4.15 ■

Percentiles of the Standard Normal Distribution

For any p between 0 and 1, Appendix Table A.3 can be used to obtain the $(100p)$th percentile of the standard normal distribution.

Example **4.14**   The ninety-ninth percentile of the standard normal distribution is that point on the axis such that the area under the curve to the left of the point is .9900. Now Appendix Table A.3 gives for fixed z the area under the standard normal curve to the left of z, whereas here we have the area and want the value of z. This is the "inverse" problem to "$P(Z \leq z) = ?$" so the table is used in an inverse fashion: Find in the middle of the table .9900, and then the row and column in which it lies identify the ninety-ninth percentile z. Here .9901 lies in the row marked 2.3 and column marked .03, so the ninety-ninth percentile is (approximately) $z = 2.33$. (See Figure 4.16.) By symmetry, the first percentile is the negative of the ninety-ninth percentile, so equals -2.33 (1% lies below the first and above the ninety-ninth). (See Figure 4.17.)

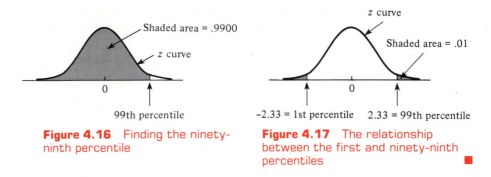

Figure 4.16 Finding the ninety-ninth percentile

Figure 4.17 The relationship between the first and ninety-ninth percentiles ∎

In general, the $(100p)$th percentile is identified by the row and column of Appendix Table A.3 in which the entry p is found (for example, the sixty-seventh percentile is obtained by finding .6700 in the body of the table, which gives $z = 0.44$). If p does not appear, the number closest to it is often used, though linear interpolation gives a more accurate answer. For example, to find the ninety-fifth percentile, we look for .9500 inside the table. While .9500 does not appear, both .9495 and .9505 do, corresponding to $z = 1.64$ and 1.65, respectively. Since .9500 is halfway between the two probabilities which do appear, we shall use 1.645 as the ninety-fifth percentile and -1.645 as the fifth percentile.

z_α Notation

In statistical inference we shall need the values on the measurement axis that capture certain small tail areas under the standard normal curve.

Notation

> z_α will denote the value on the measurement axis for which α of the area under the z curve lies to the right of z_α. (See Figure 4.18.)

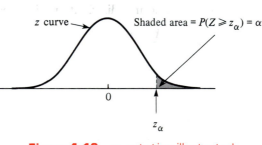

Figure 4.18 z_α notation illustrated

Since α of the area under the standard normal curve lies to the right of z_α, $1 - \alpha$ of the area lies to the left of z_α. Thus z_α is the $100(1 - \alpha)$th percentile of the standard normal distribution. Further, by symmetry the area under the standard normal curve to the left of $-z_\alpha$ is also α. The z_α's are usually referred to as **z critical values**. Table 4.1 lists the most useful standard normal percentiles and z_α values.

Table 4.1   Standard normal percentiles and critical values

Percentile	90	95	97.5	99	99.5	99.9	99.95
α (tail area)		.1	.05	.025	.01	.005	.001	.0005
$z_\alpha = 100(1 - \alpha)$th percentile		1.28	1.645	1.96	2.33	2.58	3.08	3.27

Example **4.15** $z_{.05}$ is the $100(1 - .05)$th $=$ ninety-fifth percentile of the standard normal distribution, so $z_{.05} = 1.645$. The area under the standard normal curve to the left of $-z_{.05}$ is also .05. (See Figure 4.19.)

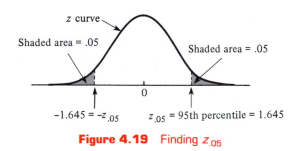

Figure 4.19   Finding $z_{.05}$ ■

Nonstandard Normal Distributions

When $X \sim N(\mu, \sigma^2)$, probabilities involving X are computed by "standardizing." The **standardized variable** is $(X - \mu)/\sigma$. Subtracting μ shifts the mean from μ to zero, and then dividing by σ scales the variable so that the standard deviation is one rather than σ.

Proposition

If X has a normal distribution with mean μ and standard deviation σ, then

$$Z = \frac{X - \mu}{\sigma}$$

is a standard normal random variable.

The key idea of the proposition is that by standardizing, any probability involving X can be expressed as a probability involving a standard normal r.v. Z, so that Appendix Table A.3 can be used. In particular,

$$P(X \leq x) = P[Z \leq (x - \mu)/\sigma] = \Phi[(x - \mu)/\sigma]$$

as illustrated in Figure 4.20. The proposition can be proved by writing the c.d.f. of $Z = (X - \mu)/\sigma$ as $P(Z \leq z) = P(X \leq \sigma z + \mu) = \int_{-\infty}^{\sigma z + \mu} f(x; \mu, \sigma) \, dx$. Using a result from calculus, this integral can be differentiated with respect to z to yield the desired p.d.f. $f(z; 0, 1)$.

Figure 4.20   Equality of nonstandard and standard normal curve areas

Example 4.16   The breakdown voltage X of a randomly chosen diode of a particular type is known to be normally distributed with $\mu = 40$ volts and $\sigma = 1.5$ volts. What is the probability that the breakdown voltage will be between 39 and 42 volts?

The desired probability is $P(39 \leq X \leq 42)$. To compute it we express the event $39 \leq X \leq 42$ in equivalent form by standardizing:

$$39 \leq X \leq 42 \quad \text{iff} \quad \frac{39 - 40}{1.5} \leq \frac{X - 40}{1.5} \leq \frac{42 - 40}{1.5}$$

that is, $-.67 \leq Z \leq 1.33$. Thus

$$P(39 \leq X \leq 42) = P(-.67 \leq Z \leq 1.33) = \Phi(1.33) - \Phi(-.67)$$
$$= .9082 - .2514 = .6568$$

Similarly,

$$P(40 \leq X \leq 43) = P\left(\frac{40 - 40}{1.5} \leq \frac{X - 40}{1.5} \leq \frac{43 - 40}{1.5}\right)$$
$$= P(0 \leq Z \leq 2) = .4772 \qquad \blacksquare$$

Example 4.17   Tissue respiration rate in diaphragms of rats (microliters per mg dry wt per hr) under standard temperature conditions is normally distributed with $\mu = 2.03$ and $\sigma = .44$. To compute the probability that a randomly chosen rat has rate X which is at least 2.5, we again standardize:

$$P(X \geq 2.5) = P\left(\frac{X - 2.03}{.44} \geq \frac{2.5 - 2.03}{.44}\right)$$
$$= P(Z \geq 1.07) = 1 - \Phi(1.07) = .1423$$

The probability that X falls outside the interval $(1.59, 2.47)$ is

$$P(X \leq 1.59 \text{ or } \geq 2.47) = P(X \leq 1.59) + P(X \geq 2.47)$$
$$= P\left(\frac{X - 2.03}{.44} \leq \frac{1.59 - 2.03}{.44}\right) + P\left(\frac{X - 2.03}{.44} \geq \frac{2.47 - 2.03}{.44}\right)$$
$$= P(Z \leq -1) + P(Z \geq 1) = .1587 + .1587 = .3174 \qquad \blacksquare$$

The difference $X - \mu$ gives the distance between X and its mean, so $Z = (X - \mu)/\sigma$ is this distance measured as a number of standard deviations. For example, if $\mu = 40$ and $\sigma = 1.5$ as in Example 4.16, then $x = 43$ corresponds to $z = (43 - 40)/1.5 = 2$, so the value 43 is two standard deviations above the mean. The value $x = 38.5$ is one standard deviation below the mean since $z = (38.5 - 40)/1.5 = -1$. Thus the probability that any normal r.v. falls within, say, one standard deviation of its mean is $P(-1 \leq Z \leq 1) = .6826$, independent of μ and σ. Similarly the probability of an observed value within two standard deviations of μ is $P(-2 \leq Z \leq 2) = .9544$, and within 3σ of μ is $P(-3 \leq Z \leq 3) = .9974$.

Approximately 68% of the values in any normal population lie within one standard deviation of the mean, approximately 95% lie within two standard deviations of μ, and approximately 99.7% lie within three standard deviations of μ.

It is indeed rare to observe a value from a normal population that is much further than two standard deviations from μ. These results will be important in testing hypotheses about μ.

Percentiles of an Arbitrary Normal Distribution

The $(100p)$th percentile of a normal distribution with mean μ and standard deviation σ is easily related to the $(100p)$th percentile of the standard normal distribution.

Proposition

$$\begin{array}{l} (100p)\text{th percentile} \\ \text{for normal } (\mu, \sigma) \end{array} = \mu + \left[\begin{array}{l} (100p)\text{th for} \\ \text{standard normal} \end{array}\right] \cdot \sigma$$

Another way of saying this is that if z is the desired percentile for the standard normal distribution, then the desired percentile for the normal (μ, σ) distribution is z standard deviations from μ.

Example 4.18

The amount of distilled water dispensed by a certain machine is normally distributed with mean value 64 oz and standard deviation .78 oz. What container size c will ensure that overflow occurs only one-half of one percent of the time? If X denotes the amount dispensed, the desired condition is that $P(X > c) = .005$, or equivalently that $P(X \leq c) = .995$. Thus c is the 99.5th percentile of the normal distribution with $\mu = 64$ and $\sigma = .78$. The 99.5th percentile of the standard normal distribution is 2.58, so

$$c = \eta(.995) = 64 + (2.58)(.78) = 64 + 2.0 = 66 \text{ oz}$$

This is illustrated in Figure 4.21.

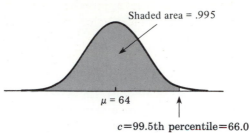

Shaded area = .995

$\mu = 64$

$c = $ 99.5th percentile$= 66.0$

Figure 4.21   Distribution of amount dispensed for Example 4.18

The Normal Distribution and Discrete Populations

The normal distribution is often used as an approximation to the distribution of values in a discrete population. In such situations extra care must be taken to ensure that probabilities are computed in an accurate manner.

Example 4.19   IQ in a particular population (as measured by a standard test) is known to be approximately normally distributed with $\mu = 100$ and $\sigma = 15$. What is the probability that a randomly selected individual has an IQ of at least 125? Letting $X = $ the IQ of a randomly chosen person, we wish $P(X \geq 125)$. The temptation here is to standardize $X \geq 125$ immediately as in previous examples. However, the IQ population is actually discrete, since IQ's are integer valued, so the normal curve is an approximation to a discrete probability histogram as pictured in Figure 4.22.

The rectangles of the histogram are *centered* at integers, so IQ's of at least 125 correspond to rectangles beginning at 124.5, as shaded in Figure 4.22. Thus we really want $P(X \geq 124.5)$, which can now be standardized to yield $P(Z \geq 1.63) = .0516$. If we had standardized $X \geq 125$, we would have obtained $P(Z \geq 1.67) = .0475$. The difference is not great, but the answer .0516 is more accurate. Similarly, $P(X = 125)$ would be approximated by the area between 124.5 and 125.5, since the area under the normal curve above the single value 125 is zero.

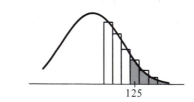

125

Figure 4.22   A normal approximation to a discrete distribution

The correction for discreteness of the underlying distribution in Example 4.19 is often called a **continuity correction.** It is useful in the following application of the normal distribution to the computation of binomial probabilities.

The Normal Approximation to the Binomial Distribution

Recall that the mean value and standard deviation of a binomial random variable X are $\mu_X = np$ and $\sigma_X = \sqrt{npq}$, respectively. Figure 4.23 displays a binomial probability histogram for the binomial distribution with $n = 20$, $p = .6$ (so $\mu = 20(.6) = 12$ and $\sigma = \sqrt{20(.6)(.4)} = 2.19$). A normal curve with mean

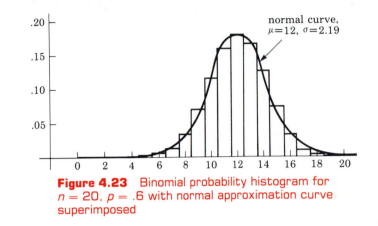

Figure 4.23 Binomial probability histogram for $n = 20$, $p = .6$ with normal approximation curve superimposed

value and standard deviation equal to the corresponding values for the binomial distribution has been superimposed on the probability histogram. Although the probability histogram is a bit skewed (because $p \neq .5$), the normal curve gives a very good approximation, especially in the middle part of the picture. The area of any rectangle (probability of any particular X value) except those in the extreme tails can be accurately approximated by the corresponding normal curve area. For example, $P(X = 10) = B(10; 20, .6) - B(9; 20, .6) = .117$, whereas the area under the normal curve between 9.5 and 10.5 is $P(-1.14 \leq Z \leq -.68) = .1212$.

More generally, as long as the binomial probability histogram is not too skewed, binomial probabilities can be well approximated by normal curve areas. It is then customary to say that X has approximately a normal distribution.

Proposition

Let X be a binomial random variable based on n trials with success probability p. Then if the binomial probability histogram is not too skewed, X has approximately a normal distribution with $\mu = np$ and $\sigma = \sqrt{npq}$. In particular, for $x = $ a possible value of X,

$$P(X \leq x) = B(x; n, p) \approx \left(\begin{array}{c} \text{area under the normal curve} \\ \text{to the left of } x + .5 \end{array} \right)$$

$$= \Phi\left(\frac{x + .5 - np}{\sqrt{npq}} \right)$$

In practice, the approximation is adequate provided that both $np \geq 5$ and $nq \geq 5$.

If either $np < 5$ or $nq < 5$, the binomial distribution is too skewed for the (symmetric) normal curve to give accurate approximations.

Example 4.20   *The Los Angeles Times* (Dec. 11, 1983) reported that of roughly 3 million gasoline and petroleum product storage tanks buried below ground, one in four is leaking. Suppose that $n = 50$ tanks are randomly selected for inspection. Assuming that the one-in-four figure is correct, calculate the probability that at most 10 of the selected tanks leak. Let a success be identified with a leaking tank, so that $p = .25$. Then $\mu = 12.5$ and $\sigma = 3.06$. Since $np = 50(.25) = 12.5 \geq 5$ and $nq = 37.5 \geq 5$, the approximation can safely be applied:

$$P(X \leq 10) = B(10; 50, .25) \approx \Phi\left(\frac{10 + .5 - 12.5}{3.06}\right) = \Phi(-.65) = .2578$$

Similarly, the probability that between 5 and 15 (inclusive) of the selected tanks are leaking is

$$P(5 \leq X \leq 15) = B(15; 50, .25) - B(4; 50, .25)$$
$$\approx \Phi\left(\frac{15.5 - 12.5}{3.06}\right) - \Phi\left(\frac{4.5 - 12.5}{3.06}\right) = .8320$$

The exact probabilities are .2622 and .8348, respectively, so the approximations are quite good. In the last calculation, the probability $P(5 \leq X \leq 15)$ is being approximated by the area under the normal curve between 4.5 and 15.5—the continuity correction is used for both the upper and lower limits.   ■

When the objective of our investigation is to make an inference about a population proportion p, interest will focus on the sample proportion of successes X/n rather than on X itself. Because this proportion is just X multiplied by the constant $1/n$, it will also have approximately a normal distribution (with mean $\mu = p$ and standard deviation $\sigma = \sqrt{pq/n}$) provided that both $np \geq 5$ and $nq \geq 5$. This normal approximation is the basis for several large-sample inferential procedures to be discussed in later chapters.

It is quite difficult to give a direct proof of the validity of this normal approximation (the first one goes back over 150 years to Laplace). In the next chapter we'll show that it is a consequence of an important general result called the Central Limit Theorem.

Exercises / Section 4.3 (25–46)

25. Let Z be a standard normal random variable and calculate the following probabilities, drawing pictures wherever appropriate:

   a. $P(0 \leq Z \leq 2.17)$ **b.** $P(0 \leq Z \leq 1)$
   c. $P(-2.50 \leq Z \leq 0)$
   d. $P(-2.50 \leq Z \leq 2.50)$
   e. $P(Z \leq 1.37)$ **f.** $P(-1.75 \leq Z)$
   g. $P(-1.50 \leq Z \leq 2.00)$
   h. $P(1.37 \leq Z \leq 2.50)$ **i.** $P(1.50 \leq Z)$
   j. $P(|Z| \leq 2.50)$

26. In each case below, determine the value of the constant c that makes the probability statement correct.

a. $\Phi(c) = .9838$ **b.** $P(0 \leq Z \leq c) = .291$
c. $P(c \leq Z) = .121$
d. $P(-c \leq Z \leq c) = .668$
e. $P(c \leq |Z|) = .016$

27. Find the following percentiles for the standard normal distribution. Interpolate where appropriate.
a. ninety-first **b.** ninth **c.** seventy-fifth
d. twenty-fifth **e.** sixth

28. Determine z_α for
a. $\alpha = .0055$ **b.** $\alpha = .09$ **c.** $\alpha = .663$

29. If X is a normal random variable with mean 80 and standard deviation 10, compute the following probabilities by standardizing:
a. $P(X \leq 100)$ **b.** $P(X \leq 80)$
c. $P(65 \leq X \leq 100)$ **d.** $P(70 \leq X)$
e. $P(85 \leq X \leq 95)$ **f.** $P(|X - 80| \leq 10)$

30. Suppose that the force acting on a column that helps to support a building is normally distributed with mean 15.0 kips and standard deviation 1.25 kips. What is the probability that the force
a. is at most 17 kips?
b. is between 12 and 17 kips?
c. differs from 15.0 kips by at most two standard deviations?

31. Assume that development time for a particular type of photographic printing paper when it is exposed to a light source for five seconds is normally distributed with a mean of 25 seconds and a standard deviation of 1.3 seconds. What is the probability that
a. a particular print will require more than 26.5 seconds to develop?
b. development time is at least 23 seconds?
c. development time differs from the expected time by more than 2.5 seconds?

32. A particular type of gasoline tank for a compact car is designed to hold 15 gallons. Suppose that the actual capacity X of a randomly chosen tank of this type is normally distributed with mean 15 and standard deviation .2 gallon.
a. What is the probability that a randomly selected tank will hold at most 14.8 gallons?
b. What is the probability that a randomly selected tank will hold between 14.7 and 15.1 gallons?
c. If the car on which a randomly chosen tank is mounted gets exactly 25 m.p.g., what is the probability that the car can travel 370 miles without refueling?

33. There are two machines available for cutting corks intended for use in wine bottles. The first produces corks with diameters that are normally distributed with mean 3 cm and standard deviation .1 cm. The second machine produces corks with diameters that have a normal distribution with mean 3.04 cm and standard deviation .02 cm. Acceptable corks have diameters between 2.9 cm and 3.1 cm. Which machine is more likely to produce an acceptable cork?

34. a. If a normal distribution has $\mu = 25$ and $\sigma = 5$, what is the ninety-first percentile of the distribution?
b. What is the sixth percentile of the distribution of (a)?
c. If bearing diameter is normally distributed with mean .25 in. and standard deviation .002 in., below what value will 95% of all bearings have diameters?
d. As in (c), above what value will 10% of all bearings have diameters?

35. Suppose that the pH of soil samples taken from a certain geographic region is normally distributed with mean pH 6.00 and standard deviation .10. If the pH of a randomly selected soil sample from this region is determined,
a. What is the probability that the resulting pH is between 5.90 and 6.15?
b. What is the probability that the resulting pH exceeds 6.10?
c. What is the probability that the resulting pH is at most 5.95?
d. What value will be exceeded by only 5% of all such pH's?

36. The Rockwell hardness of a metal is determined by impressing a hardened point into the surface of the metal and then measuring the depth of penetration of the point. Suppose that the Rockwell hardness of a particular alloy is normally distributed with mean 70 and standard deviation 3 (assume that Rockwell hardness is measured on a continuous scale).
a. If a specimen is acceptable only if its hardness is between 65 and 75, what is the probability that a randomly chosen specimen has an acceptable hardness?

b. If the acceptable range of hardness was $(70 - c, 70 + c)$, for what value of c would 95% of all specimens have acceptable hardness?

c. If the acceptable range is as in (a) and the hardness of each of 10 randomly selected specimens is independently determined, what is the expected number of acceptable specimens among the 10?

d. What is the probability that at most eight of 10 independently selected specimens have a hardness of less than 73.84? *Hint:* $Y =$ the number among the 10 specimens with hardness less than 73.84 is a binomial variable; what is p?

37. The weight distribution of parcels sent in a certain manner is normal with mean value 10 lb and standard deviation 2 lb. The parcel service wishes to establish a weight value c beyond which there will be a surcharge. What value of c is such that 99% of all parcels are at least one pound under the surcharge weight?

38. Suppose that Appendix Table A.3 contained $\Phi(z)$ only for $z \geq 0$. Explain how you could still compute
   a. $P(-1.72 \leq Z \leq -.55)$
   b. $P(-1.72 \leq Z \leq .55)$
   Is it necessary to table $\Phi(z)$ for z negative? What property of the standard normal curve justifies your answer?

39. Chebyshev's inequality, introduced in Exercise 40 (Chapter 3) is valid for continuous as well as discrete distributions. It states that for any number k satisfying $k \geq 1$, $P(|X - \mu| \geq k\sigma) \leq 1/k^2$ (see Exercise 40 in Chapter 3 for an interpretation). Obtain this probability in the case of a normal distribution for $k = 1$, 2, and 3, and compare to the upper bound.

40. If the diameter of bearings produced by a certain machine has a normal distribution with $\sigma = .002$ in., what is the probability that a randomly chosen bearing will have a diameter that differs from the mean diameter by at least .005 in.?

41. Let X represent the number of pages of text in a randomly chosen mathematics Ph.D. thesis. While X can assume only positive integer values,

suppose that it is approximately normally distributed with expected value 90 and standard deviation 15. What is the probability that a randomly chosen thesis contains
   a. at most 100 pages (using the continuity correction)?
   b. between 80 and 110 pages (using the continuity correction)?

42. Let X have a binomial distribution with parameters $n = 25$ and p. Calculate each of the following probabilities using the normal approximation (with the continuity correction) for the cases $p = .5$, .6, and .8, and compare to the exact probabilities calculated from Appendix Table A.1.
   a. $P(15 \leq X \leq 20)$ **b.** $P(X \leq 15)$
   c. $P(20 \leq X)$

43. Suppose that 70% of all consultations handled by student consultants at a computer center involve programs with syntax errors, and let X denote the number with such errors in a random sample of 100 consultations. What is the (approximate) probability that X is:
   a. between 60 and 80, inclusive?
   b. at most 75?   **c.** less than 75?

44. Suppose that only 40% of all drivers in a certain state regularly wear a seatbelt. A random sample of 500 drivers is selected. What is the probability that:
   a. between 180 and 230 (inclusive) of the drivers in the sample regularly wear a seatbelt?
   b. fewer than 175 of those in the sample regularly wear a seatbelt? Fewer than 150?

45. Show that the relationship between a general normal percentile and the corresponding z percentile is as stated in this section.

46. **a.** If X has a normal distribution with parameters μ and σ, show that $Y = aX + b$ (a linear function of X) also has a normal distribution. What are the parameters of the distribution of Y [that is, $E(Y)$ and $V(Y)$]? *Hint:* Write the c.d.f. of Y, $P(Y \leq y)$, as an integral involving the p.d.f. of X, and then differentiate with respect to y to get the p.d.f. of Y.
   b. If when measured in °C, temperature is normally distributed with mean 115 and standard deviation 2, what can be said about the distribution of temperature measured in °F?

4.4   The Gamma Distribution and Its Relatives

The graph of any normal p.d.f. is bell-shaped and thus symmetric. There are many practical situations in which the variable of interest to the experimenter might have a skewed distribution. A family of p.d.f.'s that yields a wide variety of skewed distributional shapes is the gamma family. To define the family of gamma distributions, we first need to introduce a function that plays an important role in many branches of mathematics.

Definition

> For $\alpha > 0$, the **gamma function** $\Gamma(\alpha)$ is defined by
>
> $$\Gamma(\alpha) = \int_0^\infty x^{\alpha - 1} e^{-x} \, dx \tag{4.5}$$

The most important properties of the gamma function are:

1. For any $\alpha > 1$, $\Gamma(\alpha) = (\alpha - 1) \cdot \Gamma(\alpha - 1)$
2. For any positive integer n, $\Gamma(n) = (n - 1)!$
3. $\Gamma(\frac{1}{2}) = \sqrt{\pi}$

By (4.5) if we let

$$f(x; \alpha) = \begin{cases} \dfrac{x^{\alpha - 1} e^{-x}}{\Gamma(\alpha)} & x \geq 0 \\ 0 & \text{otherwise} \end{cases} \tag{4.6}$$

then $f(x; \alpha) \geq 0$ and $\int_0^\infty f(x; \alpha) \, dx = \dfrac{\Gamma(\alpha)}{\Gamma(\alpha)} = 1$, so $f(x; \alpha)$ satisfies the two basic properties of a p.d.f.

The Family of Gamma Distributions

Definition

> A continuous random variable X will be said to have a **gamma distribution** if the p.d.f. of X is
>
> $$f(x; \alpha, \beta) = \begin{cases} \dfrac{1}{\beta^\alpha \Gamma(\alpha)} x^{\alpha - 1} e^{-x/\beta} & x \geq 0 \\ 0 & \text{otherwise} \end{cases} \tag{4.7}$$
>
> where the parameters α and β satisfy $\alpha > 0$, $\beta > 0$. The **standard gamma distribution** has $\beta = 1$, so the p.d.f. of a standard gamma r.v. is given by (4.6).

Figure 4.24(a) illustrates the graphs of the gamma p.d.f. (4.7) for several (α, β) pairs, while Figure 4.24(b) presents graphs of the standard gamma p.d.f. For the standard p.d.f., when $\alpha \leq 1$, $f(x; \alpha)$ is strictly decreasing as x increases from 0; when $\alpha > 1$, $f(x; \alpha)$ rises from 0 at $x = 0$ to a maximum and then decreases. The parameter β in (4.7) is called the scale parameter because values other than 1 either stretch or compress the p.d.f. in the x direction.

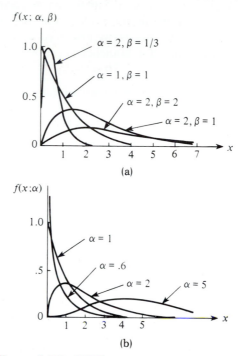

Figure 4.24 (*a*) Gamma density functions; (*b*) Standard gamma density functions

$E(X)$ and $E(X^2)$ can be obtained from a reasonably straightforward integration, and then $V(X) = E(X^2) - [E(X)]^2$.

Proposition

The mean and variance of a random variable X having the gamma distribution (4.7) are

$$E(X) = \mu = \alpha\beta \qquad V(X) = \sigma^2 = \alpha\beta^2$$

Computing Probabilities from the Gamma Distribution

When X is a standard gamma r.v., the c.d.f. of X,

$$F(x; \alpha) = \int_0^x \frac{y^{\alpha-1}e^{-y}}{\Gamma(\alpha)} \, dy \quad x > 0 \tag{4.8}$$

is called the **incomplete gamma function** [sometimes the incomplete gamma function refers to (4.8) without the denominator $\Gamma(\alpha)$ in the integrand]. There are extensive tables of $F(x; \alpha)$ available; in Appendix Table A.4 we present a small table for $\alpha = 1, 2, \ldots, 10$ and $x = 1, 2, \ldots, 15$.

Example 4.21 Suppose that the reaction time X of a randomly selected individual to a certain stimulus has a standard gamma distribution with $\alpha = 2$ seconds. Since

$$P(a \leq X \leq b) = F(b) - F(a) \text{ when } X \text{ is continuous,}$$
$$P(3 \leq X \leq 5) = F(5; 2) - F(3; 2) = .960 - .801 = .159$$

The probability that the reaction time is more than four seconds is

$$P(X > 4) = 1 - P(X \leq 4) = 1 - F(4; 2) = 1 - .908 = .092 \qquad \blacksquare$$

The incomplete gamma function can also be used to compute probabilities involving nonstandard gamma distributions.

Proposition

> Let X have a gamma distribution with parameters α and β. Then for any $x > 0$, the c.d.f. of X is given by
>
> $$P(X \leq x) = F\left(\frac{x}{\beta}; \alpha\right)$$
>
> where $F(\cdot; \alpha)$ is the incomplete gamma function.

Example 4.22 Suppose that the survival time X in weeks of a randomly selected male mouse exposed to 240 rads of gamma radiation has a gamma distribution with $\alpha = 8$ and $\beta = 15$ (data in *Survival Distributions: Reliability Applications in the Biomedical Services* by A. J. Gross and V. Clark, suggests $\alpha \approx 8.5$ and $\beta \approx 13.3$). The expected survival time is $E(X) = (8)(15) = 120$ weeks, while $V(X) = (8)(15)^2 = 1800$ and $\sigma_X = \sqrt{1800} = 42.43$ weeks. The probability that a mouse survives between 60 and 120 weeks is

$$P(60 \leq X \leq 120) = P(X \leq 120) - P(X \leq 60)$$
$$= F(120/15; 8) - F(60/15; 8)$$
$$= F(8; 8) - F(4; 8) = .547 - .051 = .496$$

The probability that a mouse survives at least 30 weeks is

$$P(X \geq 30) = 1 - P(X < 30) = 1 - P(X \leq 30)$$
$$= 1 - F(30/15; 8) = .999 \qquad \blacksquare$$

The Exponential Distribution

Definition

> X is said to have an **exponential distribution** if the p.d.f. of X is
>
> $$f(x; \lambda) = \begin{cases} \lambda e^{-\lambda x} & x \geq 0 \\ 0 & \text{otherwise} \end{cases} \quad \text{where} \quad \lambda > 0 \qquad (4.9)$$

The exponential p.d.f. is a special case of the general gamma p.d.f. (4.7) in which $\alpha = 1$ and β has been replaced by $1/\lambda$ [some authors use the form $(1/\beta)e^{-x/\beta}$]. The mean and variance of X are then

$$\mu = \alpha\beta = \frac{1}{\lambda} \qquad \sigma^2 = \alpha\beta^2 = \frac{1}{\lambda^2}$$

Both the mean and standard deviation of the exponential distribution equal $1/\lambda$. Graphs of several exponential p.d.f.'s appear in Figure 4.25.

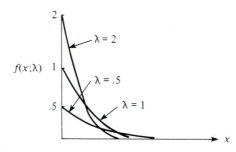

Figure 4.25   Exponential density functions

Unlike the general gamma p.d.f., the exponential p.d.f. can be easily integrated. In particular, the c.d.f. of X is

$$F(x; \lambda) = \begin{cases} 0 & x < 0 \\ 1 - e^{-\lambda x} & x \geq 0 \end{cases}$$

Example 4.23   Suppose that the response time X at a certain on-line computer terminal (the elapsed time between the end of a user's inquiry and the beginning of the system's response to that inquiry) has an exponential distribution with expected response time equal to 5 seconds. Then $E(X) = 1/\lambda = 5$, so $\lambda = .2$. The probability that the response time is at most 10 seconds is

$$P(X \leq 10) = F(10; .2) = 1 - e^{-(.2)(10)} = 1 - e^{-2} = 1 - .135 = .865$$

The probability that response time is between 5 and 10 seconds is

$$P(5 \leq X \leq 10) = F(10; 2) - F(5; .2) = (1 - e^{-2}) - (1 - e^{-1}) = .233$$

∎

Applications of the Exponential Distribution

The exponential distribution is frequently used as a model for the distribution of times between the occurrence of successive events such as customers arriving at a service facility or calls coming in to a switchboard. The reason for this is that the exponential distribution is closely related to the Poisson process discussed in Chapter 3.

Proposition

> Suppose that the number of events occurring in any time interval of length t has a Poisson distribution with parameter λt (where λ, the rate of the event process, is the expected number of events occurring in one unit of time) and that numbers of occurrences in nonoverlapping intervals are independent of one another. Then the distribution of elapsed time between the occurrence of two successive events is exponential with parameter λ.

While a complete proof is beyond the scope of the text, the result is easily verified for the time X_1 until the first event occurs:

$$P(X_1 \le t) = 1 - P(X_1 > t) = 1 - P[\text{no events in } (0, t)]$$

$$= 1 - \frac{e^{-\lambda t} \cdot (\lambda t)^0}{0!} = 1 - e^{-\lambda t}$$

which is exactly the c.d.f. of the exponential distribution.

Example 4.24

Suppose that calls are received at a 24-hour "suicide hotline" according to a Poisson process with rate $\lambda = .5$ call per day. Then the number of days X between successive calls has an exponential distribution with parameter $\lambda = .5$, so the probability that more than two days elapse between calls is

$$P(X > 2) = 1 - P(X \le 2) = 1 - F(2; .5) = e^{-(.5)(2)} = .368$$

The expected time between successive calls is $1/\lambda = 2$ days. ∎

Another important application of the exponential distribution is to model the distribution of component lifetime. A partial reason for the popularity of such applications is the **"memoryless" property** of the exponential distribution. Suppose that component lifetime is exponentially distributed with parameter λ. After putting the component into service, we leave for a period of t_0 hours and then return to find the component still working; what now is the probability that it lasts at least an additional t hours? In symbols we wish $P(X \ge t + t_0 \mid X \ge t_0)$. By the definition of conditional probability,

$$P(X \ge t + t_0 \mid X \ge t_0) = \frac{P[(X \ge t + t_0) \cap (X \ge t_0)]}{P(X \ge t_0)}$$

But the event $X \ge t_0$ in the numerator is redundant, since both events can occur if and only if $X \ge t + t_0$. Therefore

$$P(X \ge t + t_0 \mid X \ge t_0) = \frac{P(X \ge t + t_0)}{P(X \ge t_0)} = \frac{1 - F(t + t_0; \lambda)}{1 - F(t_0; \lambda)} = e^{-\lambda t}$$

This conditional probability is identical to the original probability $P(X \ge t)$ that the component lasted t hours. Thus *the distribution of additional lifetime is exactly the same as the original distribution of lifetime,* so at each point in time the component shows no effect of wear. A way to paraphrase this is to say that the distribution of remaining lifetime is independent of current age.

While the memoryless property can be justified at least approximately in many applied problems, in other situations components deteriorate with age or occasionally improve with age (at least up to a certain point). More general lifetime models are then furnished by the gamma, Weibull, or lognormal distributions (the latter two are discussed in the next section).

The Chi-Squared Distribution

Definition

> Let ν be a positive integer. Then a r.v. X is said to have a **chi-squared distribution** with parameter ν if the p.d.f. of X is the gamma density (4.7) with $\alpha = \nu/2$ and $\beta = 2$. The p.d.f. of a chi-squared r.v. is thus
>
> $$f(x; \nu) = \begin{cases} \dfrac{1}{2^{\nu/2}\Gamma(\nu/2)} x^{(\nu/2)-1} e^{-x/2} & x \geq 0 \\ 0 & x < 0 \end{cases} \qquad (4.10)$$
>
> The parameter ν is called the **number of degrees of freedom** (d.f.) of X. The symbol χ^2 is often used in place of "chi-squared."

The chi-squared distribution is important because it is the basis for a number of procedures in statistical inference. The reason for this is that chi-squared distributions are intimately related to normal distributions (see Exercise 59). We will discuss the chi-squared distribution in more detail in the chapters on inference.

Exercises / Section 4.4 (47–59)

47. Evaluate
  a. $\Gamma(6)$ **b.** $\Gamma(5/2)$
  c. $F(4; 5)$ (the incomplete gamma function)
  d. $F(5; 4)$ **e.** $F(0; 4)$

48. Let X have a standard gamma distribution with $\alpha = 7$. Evaluate
  a. $P(X \leq 5)$ **b.** $P(X < 5)$
  c. $P(X > 8)$ **d.** $P(3 \leq X \leq 8)$
  e. $P(3 < X < 8)$ **f.** $P(X < 4 \text{ or } X > 6)$

49. Suppose that the time (in hours) taken by a homeowner to mow his lawn is a r.v. X having a gamma distribution with parameters $\alpha = 2$ and $\beta = \frac{1}{2}$. What is the probability that it takes:
  a. at most 1 hour to mow the lawn?
  b. at least 2 hours to mow the lawn?
  c. between .5 and 1.5 hours to mow the lawn?

50. Suppose that the time spent by a randomly selected student who uses a terminal connected to a local timesharing computer facility has a gamma distribution with mean 20 min and variance 80 min².
  a. What are the values of α and β?
  b. What is the probability that a student uses the terminal for at most 24 minutes?
  c. What is the probability that a student spends between 20 and 40 minutes using the terminal?

51. Suppose that when a transistor of a certain type is subjected to an accelerated life test, the lifetime X (in weeks) has a gamma distribution with mean 24 weeks and standard deviation 12 weeks.

a. What is the probability that a transistor will last between 12 and 24 weeks?

b. What is the probability that a transistor will last at most 24 weeks? Is the median of the lifetime distribution less than 24? Why or why not?

c. What is the ninety-ninth percentile of the lifetime distribution?

d. Suppose that the test will actually be terminated after t weeks. What value of t is such that only one-half of 1% of all transistors would still be operating at termination?

52. Let X = the time between two successive arrivals at the drive-up window of a local bank. If X has an exponential distribution with $\lambda = 1$ (which is identical to a standard gamma distribution with $\alpha = 1$), compute

a. the expected time between two successive arrivals

b. the standard deviation of the time between successive arrivals

c. $P(X \le 4)$ **d.** $P(2 \le X \le 5)$

53. The time X (seconds) that it takes a librarian to locate a card in a file of records on checked-out books has an exponential distribution with expected time = 20 seconds. Calculate the following probabilities.

a. $P(X \le 30)$ **b.** $P(X \ge 20)$

c. $P(20 \le X \le 30)$

d. For what value of t is $P(X \le t) = .5$ (t is the fiftieth percentile of the distribution)?

54. Extensive experience with fans of a certain type used in diesel engines has suggested that the exponential distribution provides a good model for time until failure. Suppose that the mean time until failure is 25,000 hours. What is the probability

a. that a randomly selected fan will last at least 20,000 hours? At most 30,000 hours? Between 20,000 and 30,000 hours?

b. that the lifetime of a fan exceeds the mean value by more than two standard deviations? More than three standard deviations?

55. The special case of the gamma distribution in which α is a positive integer n is called an Erlang distribution. If we replace β by $1/\lambda$ in (4.7) the Erlang p.d.f. is

$$f(x; \lambda, n) = \begin{cases} \dfrac{\lambda(\lambda x)^{n-1}e^{-\lambda x}}{(n-1)!} & x \ge 0 \\ 0 & x < 0 \end{cases}$$

It can be shown that if the times between successive events are independent, each with an exponential distribution with parameter λ, then the total time X that elapses before all of the next n events occur has p.d.f. $f(x; \lambda, n)$.

a. What is the expected value of X? If the time (min) between arrivals of successive customers is exponentially distributed with $\lambda = .5$, how much time can be expected to elapse before the tenth customer arrives?

b. If customer interarrival time is exponentially distributed with $\lambda = .5$, what is the probability that the tenth customer (after the one who has just arrived) will arrive within the next 30 minutes?

c. The event $\{X \le t\}$ occurs iff at least n events occur in the next t units of time. Use the fact that the number of events occurring in an interval of length t has a Poisson distribution with parameter λt to write down an expression (involving Poisson probabilities) for the c.d.f. $F(t; \lambda, n) = P(X \le t)$ of the Erlang distribution.

56. A system consists of five identical components connected in series as shown:

As soon as one component fails the entire system will fail. Suppose that each component has a lifetime that is exponentially distributed with $\lambda = .01$, and that components fail independently of one another. Define events $A_i = \{i\text{th component lasts at least } t \text{ hours}\}$, $i = 1, \ldots, 5$, so that the A_i's are independent events. Let X = the time at which the system fails = the shortest (minimum) lifetime among the five components.

a. The event $\{X \ge t\}$ is equivalent to what event involving $A_1, \ldots, A_5$?

b. Using the independence of the A_i's, compute $P(X \ge t)$. Then obtain $F(t) = P(X \le t)$ and the p.d.f. of X. What type of distribution does X have?

c. Suppose that there are n components, each having exponential lifetime with parameter λ. What type of distribution does X have?

57. If X has an exponential distribution with parameter λ, derive a general expression for the $(100p)$th percentile of the distribution. Then specialize to obtain the median.

58. Show that $\int_0^\infty f(x; \alpha, \beta)\, dx = 1$ for the gamma p.d.f. of (4.7). *Hint:* Let $y = x/\beta$ to obtain an integrand like that of (4.6).

59. a. The event $\{X^2 \le y\}$ is equivalent to what event involving X itself?

b. If X has a standard normal distribution, use (a) to write the integral that equals $P(X^2 \le y)$. Then differentiate this with respect to y to obtain the p.d.f. of X^2 [the square of a $N(0, 1)$ variable]. Finally, show that X^2 has a chi-squared distribution with $\nu = 1$ degrees of freedom (see equation (4.10)).

Hint: $\dfrac{d}{dy}\left\{\displaystyle\int_{a(y)}^{b(y)} f(x)\, dx\right\}$

$$= f[b(y)] \cdot b'(y) - f[a(y)] \cdot a'(y)$$

4.5   Other Continuous Distributions

The normal, gamma (including exponential), and uniform families of distributions provide a wide variety of probability models for continuous variables, but there are many practical situations in which no member of these families fits a set of observed data very well. Statisticians and other investigators have developed other families of distributions that are often appropriate in practice.

The Weibull Distribution
The family of Weibull distributions was introduced by the Swedish physicist Waloddi Weibull in 1939; his 1951 paper "A Statistical Distribution Function of Wide Applicability" (*J. Applied Mechanics,* vol. 18, pp. 293–297) discusses a number of applications.

Definition

> A random variable X is said to have a **Weibull distribution** with parameters α and β $(\alpha > 0, \beta > 0)$ if the p.d.f. of X is
>
> $$f(x; \alpha, \beta) = \begin{cases} \dfrac{\alpha}{\beta^\alpha} x^{\alpha - 1} e^{-(x/\beta)^\alpha} & x \ge 0 \\[2mm] 0 & x < 0 \end{cases} \qquad (4.11)$$

While in some situations there are theoretical justifications for the appropriateness of the Weibull distribution, in many applications $f(x; \alpha, \beta)$ simply provides a good fit to observed data for particular values of α and β. When $\alpha = 1$, the p.d.f. reduces to the exponential distribution (with $\lambda = 1/\beta$), so the exponential distribution is a special case of both the gamma and Weibull distributions. However, there are gamma distributions which are not Weibull distributions and vice versa, so one family is not a subset of the other. By varying α and β a number of different distributional shapes can be obtained, as illustrated in Figure 4.26. β is a scale parameter, so different values stretch or compress the graph in the x direction.

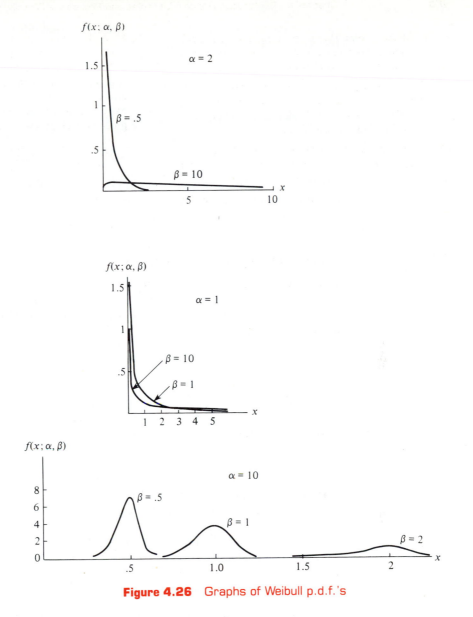

Figure 4.26   Graphs of Weibull p.d.f.'s

Integrating to obtain $E(X)$ and $E(X^2)$ yields

$$\mu = \beta\Gamma\left(1 + \frac{1}{\alpha}\right), \qquad \sigma^2 = \beta^2\left\{\Gamma\left(1 + \frac{2}{\alpha}\right) - \left[\Gamma\left(1 + \frac{1}{\alpha}\right)\right]^2\right\}$$

The computation of μ and σ^2 thus necessitates using a table of the gamma function.

The integration $\int_0^x f(y; \alpha, \beta)\, dy$ is easily carried out to obtain the cumulative distribution function of X.

Proposition

> The c.d.f. of a Weibull r.v. having parameters α and β is
>
> $$F(x; \alpha, \beta) = \begin{cases} 0 & x < 0 \\ 1 - e^{-(x/\beta)^\alpha} & x \geq 0 \end{cases} \qquad (4.12)$$

Example 4.25

Let $X =$ the ultimate tensile strength (ksi) at $-200°\mathrm{F}$ of a type of steel that exhibits "cold brittleness" at low temperatures. Suppose that X has a Weibull distribution with parameters $\alpha = 20$ and $\beta = 100$. Then

$$P(X \leq 105) = F(105; 20, 100) = 1 - e^{-(105/100)^{20}} = 1 - .070 = .930$$

and

$$P(98 \leq X \leq 102) = F(102; 20, 100) - F(98; 20, 100)$$
$$= e^{-(.98)^{20}} - e^{-(1.02)^{20}} = .513 - .226 = .287 \qquad \blacksquare$$

Frequently in practical situations a Weibull model may be reasonable except that the smallest possible X value may be some value γ not assumed to be zero (this would also apply to a gamma model). The quantity γ can then be regarded as a third parameter of the distribution, which is what Weibull did in his original work. For, say, $\gamma = 3$, all curves in Figure 4.26 would be shifted three units to the right. This is equivalent to saying that $X - \gamma$ has the p.d.f. (4.11), so that the c.d.f. of X is obtained by replacing x in (4.12) by $x - \gamma$.

Example 4.26

Let $X =$ the corrosion weight loss for a small square magnesium alloy plate immersed for seven days in an inhibited aqueous 20% solution of $MgBr_2$. Suppose that the minimum possible weight loss is $\gamma = 3$ and that the excess $X - 3$ over this minimum has a Weibull distribution with $\alpha = 2$ and $\beta = 4$ (this example was considered in "Practical Applications of the Weibull Distribution," *Industrial Quality Control*, August 1964, pp. 71–78; values for α and β were taken to be 1.8 and 3.67, respectively, though a slightly different choice of parameters was used in the article). The c.d.f. of X is then

$$F(x; \alpha, \beta, \gamma) = F(x; 2, 4, 3) = \begin{cases} 0 & x < 3 \\ 1 - e^{-[(x-3)/4]^2} & x \geq 3 \end{cases}$$

Therefore

$$P(X > 3.5) = 1 - F(3.5; 2, 4, 3) = e^{-.0156} = .985$$

and

$$P(7 \leq X \leq 9) = 1 - e^{-2.25} - (1 - e^{-1}) = .895 - .632 = .263 \qquad \blacksquare$$

The Lognormal Distribution

Definition

A nonnegative r.v. X is said to have a **lognormal distribution** if the r.v. $Y = \ln(X)$ has a normal distribution. The resulting p.d.f. of a lognormal r.v. when $\ln(X)$ is normally distributed with parameters μ and σ is

$$f(x; \mu, \sigma) = \begin{cases} \dfrac{1}{\sqrt{2\pi}\,\sigma x} e^{-\frac{1}{2\sigma^2}[\ln(x) - \mu]^2} & x \geq 0 \\[2mm] 0 & x < 0 \end{cases}$$

Be careful here; μ and σ are not the mean and standard deviation of X but of $\ln(X)$. The mean and variance of X can be shown to be

$$E(X) = e^{\mu + \sigma^2/2} \qquad V(X) = e^{2\mu + \sigma^2} \cdot (e^{\sigma^2} - 1)$$

In the next chapter we will present a theoretical justification for this distribution in connection with the Central Limit Theorem, but as with other distributions, the lognormal can be used as a model even in the absence of such justification. Figure 4.27 illustrates the graphs of the lognormal p.d.f.; although the normal curve is symmetric, a lognormal curve has a positive skew.

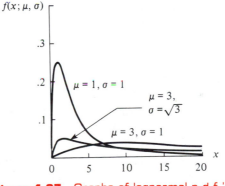

Figure 4.27   Graphs of lognormal p.d.f.'s

Because $\ln(X)$ has a normal distribution, the c.d.f. of X can be expressed in terms of the c.d.f. $\Phi(z)$ of a standard normal r.v. Z. For $x \geq 0$,

$$\begin{aligned} F(x; \mu, \sigma) &= P(X \leq x) = P[\ln(X) \leq \ln(x)] \\[2mm] &= P\left(Z \leq \frac{\ln(x) - \mu}{\sigma}\right) = \Phi\left(\frac{\ln(x) - \mu}{\sigma}\right) \end{aligned} \qquad (4.13)$$

Example **4.27** Let $X =$ the hourly median power (in decibels) of received radio signals transmitted between two cities. Because a decibel is a unit of measurement equal to the logarithm of the ratio I/I_0, where I is the intensity of the signal of interest and I_0 is the intensity of a standard signal or sound, it is reasonable to assume that X has a lognormal distribution ("Families of Distributions for Hourly Median Power and Instantaneous Power of Received Radio Signals," *J. Research National Bureau of Standards,* 1963, vol. 67D, pp. 753–762). If the parameter values are $\mu = 3.5$ and $\sigma = 1.2$, then

$$E(X) = e^{3.5 + .72} = 68.0, \qquad V(X) = e^{8.44}(e^{1.44} - 1) = 14{,}907.2$$

The probability that received power is between 50 and 250 db is

$$
\begin{aligned}
P(50 \le X \le 250) &= F(250; 3.5, 1.2) - F(50; 3.5, 1.2) \\
&= \Phi\!\left(\frac{\ln(250) - 3.5}{1.2}\right) - \Phi\!\left(\frac{\ln(50) - 3.5}{1.2}\right) \\
&= \Phi(1.68) - \Phi(.41) \\
&= .9535 - .6591 = .2944 \quad \text{(Appendix Table A.3)}
\end{aligned}
$$

The probability that X does not exceed its mean is

$$P(X \le 68.0) = \Phi\!\left(\frac{\ln(68.0) - 3.5}{1.2}\right) = \Phi(.60) = .7257$$

If the distribution were symmetric, this probability would equal .5; it is much larger because of the positive skew (long upper tail) of the distribution, which pulls μ outward past the median. ∎

The Beta Distribution

All families of continuous distributions discussed so far except for the uniform distribution have positive density over an infinite interval (though typically the density function decreases rapidly to zero beyond a few standard deviations from the mean). The beta distribution provides positive density only for X in an interval of finite length.

> A random variable X is said to have a **beta distribution** with parameters α, β (both positive), A, and B if the p.d.f. of X is
>
> $$
> f(x; \alpha, \beta, A, B)
> $$
> $$
> = \begin{cases} \dfrac{1}{B - A} \cdot \dfrac{\Gamma(\alpha + \beta)}{\Gamma(\alpha) \cdot \Gamma(\beta)} \left(\dfrac{x - A}{B - A}\right)^{\alpha - 1}\left(\dfrac{B - x}{B - A}\right)^{\beta - 1} & A \le x \le B \\ 0 & \text{otherwise} \end{cases}
> $$
>
> The case $A = 0$, $B = 1$ gives the **standard beta distribution.**

Figure 4.28 illustrates several standard beta p.d.f.'s. Graphs of the general p.d.f. are similar, except they are shifted and then stretched or compressed to fit over $[A, B]$. Unless α and β are integers, integration of the p.d.f. to calculate probabilities is difficult, so a table of the incomplete beta function is generally used. The mean and variance of X are

$$\mu = A + (B - A) \cdot \frac{\alpha}{\alpha + \beta}, \qquad \sigma^2 = \frac{(B - A)^2 \alpha \beta}{(\alpha + \beta)^2(\alpha + \beta + 1)}$$

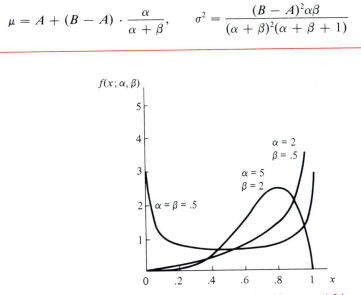

Figure 4.28 Graphs of standard beta p.d.f.'s

Example 4.28 Project managers often use a method labelled PERT—for program evaluation and review technique—to coordinate the various activities making up a large project (one successful application was in the construction of the Apollo spacecraft). A standard assumption in PERT analysis is that the time necessary to complete any particular activity once it has been started has a beta distribution with $A =$ the optimistic time (if everything goes well) and $B =$ the pessimistic time (if everything goes badly). Suppose that in constructing a single-family house, the time X (in days) necessary for laying the foundation has a beta distribution with $A = 2$, $B = 5$, $\alpha = 2$, and $\beta = 3$. Then $\alpha/(\alpha + \beta) = .4$, so $E(X) = 2 + (3)(.4) = 3.2$. For these values of α and β, the p.d.f. of X is a simple polynomial function. The probability that it takes at most 3 days to lay the foundation is

$$P(X \leq 3) = \int_2^3 \frac{1}{3} \cdot \frac{4!}{1!2!} \left(\frac{x - 2}{3}\right)\left(\frac{5 - x}{3}\right)^2 dx$$

$$= \frac{4}{27} \int_2^3 (x - 2)(5 - x)^2 \, dx = \frac{4}{27} \cdot \frac{11}{4} = \frac{11}{27} = .407 \qquad \blacksquare$$

The standard beta distribution is commonly used to model variation in the proportion or percentage of a quantity occurring in different samples, such as the proportion of a 24-hour day that an individual is asleep or the proportion of a certain element in a chemical compound.

Exercises / Section 4.5 (60–71)

60. The lifetime X (in hundreds of hours) of a certain type of vacuum tube has a Weibull distribution with parameters $\alpha = 2$ and $\beta = 3$. Compute:
  a. $E(X)$ and $V(X)$.   **b.** $P(X \leq 6)$.
  c. $P(1.5 \leq X \leq 6)$.

61. The authors of the paper "A Probabilistic Insulation Life Model for Combined Thermal-Electrical Stresses" (*IEEE Trans. on Elect. Insulation,* 1985, pp. 519–522) state that "the Weibull distribution is widely used in statistical problems relating to aging of solid insulating materials subjected to aging and stress." They propose the use of the distribution as a model for time to failure of solid insulating specimens subjected to AC voltage. The values of the parameters depend on the voltage and temperature; suppose that $\alpha = 2.5$ and $\beta = 200$ (values suggested by data in the paper).
  a. What is the probability that a specimen's lifetime is at most 200? Less than 200? More than 300?
  b. What is the probability that a specimen's lifetime is between 100 and 200?
  c. What value is such that exactly 50% of all specimens have lifetimes exceeding that value?

62. Let $X =$ the time (in 10^{-1} weeks) from shipment of a defective product until the customer returns the product. Suppose that the minimum return time is $\gamma = 3.5$ and that the excess $X - 3.5$ over the minimum has a Weibull distribution with parameters $\alpha = 2$ and $\beta = 1.5$ (see the *Industrial Quality Control* article referenced in Example 4.26).
  a. What is the c.d.f. of X?
  b. What are the expected return time and variance of return time? *Hint:* First obtain $E(X - 3.5)$ and $V(X - 3.5)$.
  c. Compute $P(X > 5)$.
  d. Compute $P(5 \leq X \leq 8)$.

63. Let X have a Weibull distribution with p.d.f. (4.11). Verify that $\mu = \beta \Gamma(1 + 1/\alpha)$. *Hint:* In

the integral for $E(X)$, make the change of variable $y = (x/\beta)^\alpha$, so that $x = \beta y^{1/\alpha}$.

64. a. In Exercise 60, what is the median lifetime of such tubes? *Hint:* Use expression (4.12).
  b. In Exercise 62, what is the median return time?
  c. If X has a Weibull distribution with c.d.f. (4.12), obtain a general expression for the $(100p)$th percentile of the distribution.
  d. In Exercise 62, the company wants to refuse to accept returns after t weeks. For what value of t will only 10% of all returns be refused?

65. In Example 4.27, in which X has a lognormal distribution with parameters $\mu = 3.5$ and $\sigma = 1.2$, compute
  a. $P(X \leq 100)$ **b.** $P(100 \leq X \leq 200)$

66. a. Use equation (4.13) to write a formula for the median $\tilde{\mu}$ of the lognormal distribution. What is the median for the power distribution of Example 4.27?
  b. Recalling that z_α is our notation for the $100(1 - \alpha)$ percentile of the standard normal distribution, write an expression for the $100(1 - \alpha)$ percentile of the lognormal distribution. In Example 4.27, above what value will median power be only 5% of the time?

67. A theoretical justification based on a certain material failure mechanism underlies the assumption that ductile strength X of a material has a lognormal distribution. Suppose that the parameters are $\mu = 5$ and $\sigma = .1$.
  a. Compute $E(X)$ and $V(X)$.
  b. Compute $P(X > 120)$.
  c. Compute $P(110 \leq X \leq 130)$.
  d. What is the value of median ductile strength?
  e. If 10 different samples of an alloy steel of this type were subjected to a strength test, how many would you expect to have strength at least 120?
  f. If the smallest 5% of strength values were unacceptable, what would the minimum acceptable strength be?

68. What condition on α and β is necessary for the standard beta p.d.f. to be symmetric?

69. Suppose that the proportion X of surface area in a randomly selected quadrat that is covered by a certain plant has a standard beta distribution with $\alpha = 5$ and $\beta = 2$.
  a. Compute $E(X)$ and $V(X)$.
  b. Compute $P(X \le .2)$.
  c. Compute $P(.2 \le X \le .4)$.
  d. What is the expected proportion of the sampling region not covered by the plant?

70. Let X have a standard beta density with parameters α and β.
  a. Verify the formula for $E(X)$ given in the section.

b. Compute $E[(1 - X)^m]$. If X represents the proportion of a substance consisting of a particular ingredient, what is the expected proportion that does not consist of this ingredient?

71. Stress is applied to a 20-in. steel bar that is clamped in a fixed position at each end. Let $Y =$ the distance from the left end at which the bar snaps. Suppose that $Y/20$ has a standard beta distribution with $E(Y) = 10$ and $V(Y) = \frac{100}{7}$.
  a. What are the parameters of the relevant standard beta distribution?
  b. Compute $P(8 \le Y \le 12)$.
  c. Compute the probability that the bar snaps more than two inches from where you expect it to.

4.6 Probability Plots

An investigator will often have obtained a numerical sample $x_1, x_2, \ldots, x_n$ and wish to know whether it is plausible that it came from a population distribution of some particular type (for example, from a normal distribution). For one thing, many formal procedures from statistical inference are based on the population distribution being of a specified type. The use of such a procedure is inappropriate if the actual underlying probability distribution differs greatly from the assumed type. Additionally, understanding the underlying distribution can sometimes give insight into the physical mechanisms involved in generating the data. An effective way to check a distributional assumption is to construct what statisticians call a **probability plot.** The essence of such a plot is that if the distribution on which the plot is based is correct, the points in the plot will fall close to a straight line. If the actual distribution is quite different from the one used to construct the plot, the points should depart substantially from a linear pattern.

Sample Percentiles

The details involved in constructing probability plots differ a bit from source to source. The basis for our construction is a comparison between percentiles of the sample data and the corresponding percentiles of the distribution under consideration. Recall that the $(100p)$th percentile of a continuous distribution with cumulative distribution function $F(\cdot)$ is the number $\eta(p)$ that satisfies $F(\eta(p)) = p$. That is, $\eta(p)$ is the number on the measurement scale such that the area under the density curve to the left of $\eta(p)$ is p. Thus the 50th percentile $\eta(.5)$ satisfies $F(\eta(.5)) = .5$ and the 90th percentile satisfies $F(\eta(.9)) = .9$. Consider as an example the standard normal distribution, for which we have denoted the c.d.f. by $\Phi(\cdot)$. From Appendix Table A.3, the 20th percentile is found by locating the row and column in which .2000 (or a number as close to it as possible) appears inside the table. Since .2005 appears at the intersection of the $-.8$ row and the .04 column, the 20th percentile is approximately $-.84$. Similarly, the 25th percentile of the standard normal distribution is (using linear interpolation) approximately $-.675$.

Roughly speaking, sample percentiles are defined in the same way that percentiles of a population distribution are defined. The 50th sample percentile should separate the smallest 50% of the sample from the largest 50%, the 90th percentile should be such that 90% of the sample lies below that value and 10% lies above, and so on. Unfortunately we run into problems when we actually try to compute the sample percentiles for a particular sample of n observations. If, for example, $n = 10$, we can split off 20% of these values or 30% of the data, but there is no value that will split off exactly 23% of these 10 observations. To proceed further, we need an operational definition of sample percentiles (this is one place where different people do slightly different things). Recall that when n is odd, the sample median or 50th sample percentile is the middle value in the ordered list—for example, the 6th largest value when $n = 11$. This amounts to regarding the middle observation as being half in the lower half of the data and half in the upper half. Similarly, suppose that $n = 10$. Then if we call the 3rd smallest value the 25th percentile, we are regarding that value as being half in the lower group (consisting of the two smallest observations) and half in the upper group (the seven largest observations). This leads to the following general definition of sample percentiles.

Definition

> Order the n sample observations from smallest to largest. Then the ith smallest observation in the list is taken to be the **$[100(i - .5)/n]$th sample percentile.**

Once the percent values $100(i - .5)/n$ $(i = 1, 2, \ldots, n)$ have been calculated, sample percentiles corresponding to intermediate percentages can be obtained by linear interpolation. For example, if $n = 10$ the percentages corresponding to the ordered sample observations are $100(1 - .5)/10 = 5\%$, $100(2 - .5)/10 = 15\%$, 25%, ..., and $100(10 - .5)/10 = 95\%$. The 10th percentile is then halfway between the 5th percentile (smallest sample observation) and the 15th percentile (second smallest observation). For our purposes such interpolation is not necessary because a probability plot will be based only on the percentages $100(i - .5)/n$ corresponding to the n sample observations.

A Probability Plot

Suppose now that for percentages $100(i - .5)/n$ $(i = 1, \ldots, n)$ the percentiles are determined for a specified population distribution whose plausibility is being investigated. If the sample was actually selected from the specified distribution, the sample percentiles (ordered sample observations) should be reasonably close to the corresponding population distribution percentiles. That is, for $i = 1, 2, \ldots, n$ there should be reasonable agreement between the ith smallest sample observation and the $[100(i - .5)/n]$th percentile for the specified distribution. Consider the (population percentile, sample percentile) pairs—that is, the pairs

$$\begin{pmatrix} [100(i - .5)/n]\text{th percentile} & i\text{th smallest} \\ \text{of the distribution} & , & \text{sample observation} \end{pmatrix}$$

for $i = 1, \ldots, n$. Each such pair can be plotted as a point on a two-dimensional coordinate system. If the sample percentiles are close to the corresponding population distribution percentiles, the first number in each pair will be roughly equal to the second number. The plotted points will then fall close to a 45° line. Substantial deviations of the plotted points from a 45° line cast doubt on the assumption that the distribution under consideration is the correct one.

Example 4.29 The value of a certain physical constant is known to an experimenter. The experimenter makes $n = 10$ independent measurements of this value using a particular measurement device and records the resulting measurement errors (error = observed value − true value). These observations appear in the accompanying table. Is it plausible that the random variable *measurement error* has a standard normal distribution? The needed standard normal (z) percentiles are also displayed in the table. Thus the points in the probability plot are $(-1.645, -1.91), (-1.037, -1.25), \ldots,$ and $(1.645, 1.56)$. Figure 4.29(a) shows the resulting plot. Although the points deviate a bit from the 45° line, the predominant impression is that this line fits the points very well. The plot suggests that the standard normal distribution is a reasonable probability model for measurement error.

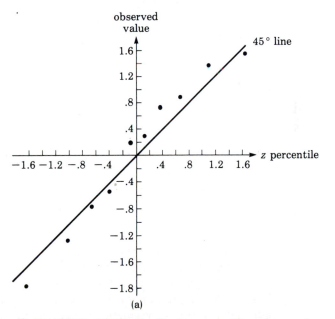

Figure 4.29 Plots of pairs (z percentile, observed value) for the data of Example 4.29: (*a*) first sample

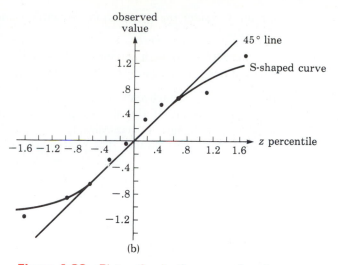

(b)

Figure 4.29   Plots of pairs (z percentile, observed value) for the data of Example 4.29: (*b*) second sample

Percentage:	5	15	25	35	45
z percentile:	−1.645	−1.037	−.675	−.385	−.126
Sample observation:	−1.91	−1.25	−.75	−.53	.20

Percentage:	55	65	75	85	95
z percentile:	.126	.385	.675	1.037	1.645
Sample observation:	.35	.72	.87	1.40	1.56

Figure 4.29(*b*) shows a plot of pairs (z percentile, observation) for a second sample of 10 observations. The 45° line gives a good fit to the middle part of the sample but not to the extremes. The plot has a well-defined s-shaped appearance. The two smallest sample observations are considerably larger than the corresponding z percentiles (the points on the far left of the plot are well above the 45° line). Similarly, the two largest sample observations are much smaller than the associated z percentiles. This plot indicates that the standard normal distribution would not be a plausible choice for the probability model that gave rise to these observed measurement errors.   ■

An investigator is typically not interested in knowing just whether a specified probability distribution, such as the standard normal distribution (normal with $\mu = 0$ and $\sigma = 1$) or the exponential distribution with $\lambda = .1$, is a plausible model for the population distribution from which the sample was selected. Instead the investigator will want to know whether *some* member of a family of probability distributions specifies a plausible model—the family of normal distributions, the family of exponential distributions, the family of Weibull distributions, and so on. The values of the parameters of a distribution are usually not specified at the outset. If the family of Weibull distributions is under consideration as a model for lifetime data, the issue is whether there are *any* values of the parameters α and β for which the corresponding Weibull distribution gives a good fit to the data. Fortunately, it is almost always the case that just one probability plot will suffice for assessing the plausibility of an entire family. If the plot deviates substantially from a straight line, no member of the family is plausible. When the plot is quite straight, further work is necessary to estimate values of the parameters (for example, find values for μ and σ) that yield the most reasonable distribution of the specified type.

Let's focus on a plot for checking normality. Such a plot can be very useful in applied work, because many formal statistical procedures are appropriate (give accurate inferences) only when the population distribution is at least approximately normal. These procedures should generally not be used if the normal probability plot shows a very pronounced departure from linearity. The key to constructing an omnibus normal probability plot is the relationship between standard normal (z) percentiles and those for any other normal distribution:

$$\begin{array}{l} \text{percentile for a normal} \\ (\mu, \sigma) \text{ distribution} \end{array} = \mu + \sigma \cdot (\text{corresponding } z \text{ percentile})$$

Consider first the case $\mu = 0$. Then if each observation is exactly equal to the corresponding normal percentile for a particular value of σ, the pairs ($\sigma \cdot$ [z percentile], observation) fall on a 45° line, which has slope 1. This implies that the pairs (z percentile, observation) fall on a line passing through $(0, 0)$ (that is, one with y intercept 0) but having slope σ rather than 1. The effect of a nonzero value of μ is simply to change the y intercept from 0 to μ.

A plot of the n pairs

($[100(i - .5)/n]$th z percentile, corresponding observation)

on a two-dimensional coordinate system is called a **normal probability plot.** If the sample observations are in fact drawn from a normal distribution with mean value μ and standard deviation σ, the points should fall close to a straight line with slope σ and intercept μ. Thus a plot for which the points fall close to some straight line suggests that the assumption of a normal population distribution is plausible.

Example 4.30   The article "Nonbloated Burned Clay Aggregate Concrete" (*J. Materials,* 1972, pp. 555–563) reported 7-day compressive strength observations for $n = 30$ concrete specimens. The data appears in the accompanying stem and leaf display (a stem with an ℓ includes observations with "low leaves," those between 000 and 499, whereas an h stem is for leaves between 500 and 999).

1ℓ	400
$1h$	932
2ℓ	000, 200, 200
$2h$	530, 630, 665, 735, 735, 800, 935
3ℓ	000, 000, 030, 065, 065, 065, 170, 200, 235, 260, 335, 365, 465
$3h$	500, 600, 600, 835
4ℓ	460

A normal probability plot of the data is given in Figure 4.30. Each 2 in the plot denotes two points that are very close to one another, and the 3 plays a similar role. The plot is clearly very straight, providing substantial support for the conclusion that compressive strength is a normally distributed variable.

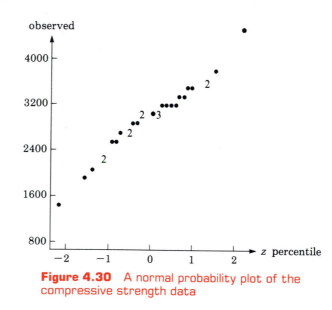

Figure 4.30   A normal probability plot of the compressive strength data

A nonnormal population distribution can often be placed in one of the following three categories:

1. It is symmetric and has "lighter tails" than does a normal distribution—that is, the density curve declines more rapidly out in the tails than does a normal curve.
2. It is symmetric and heavy tailed compared to a normal distribution.
3. It is skewed.

A uniform distribution is light tailed, since its density function drops to zero outside a finite interval. The density function $f(x) = 1/[\pi(1 + x^2)]$ for $-\infty < x < \infty$ is one example of a heavy-tailed distribution, since $1/(1 + x^2)$ declines much less rapidly than does $e^{-x^2/2}$. Lognormal and Weibull distributions are among those which are skewed. When the points in a normal probability plot do not adhere to a straight line, the pattern will frequently suggest that the population distribution is in a particular one of these three categories.

When the distribution from which the sample is selected is light tailed, the largest and smallest observations are usually not as extreme as would be expected from a normal random sample. Visualize a straight line drawn through the middle part of the plot; points on the far right tend to be below the line (observed value $< z$ percentile), whereas points on the left end of the plot tend to fall above the straight line (observed value $> z$ percentile). The result is an s-shaped pattern of the type pictured in the second plot of Figure 4.29.

A sample from a heavy-tailed distribution also tends to produce an s-shaped plot. However, in contrast to the light-tailed case, the left end of the plot curves downward (observed $< z$ percentile). An example appears in Figure 4.31(a). If the underlying distribution is positively skewed (a short left tail and a long right tail), the smallest sample observations will be larger than expected from a normal sample and so will the largest observations. In this case points

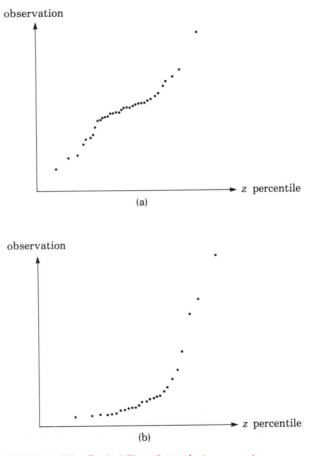

Figure 4.31   Probability plots that suggest a nonnormal distribution: (a) a plot consistent with a heavy-tailed distribution; (b) a plot consistent with a positively skewed distribution

on both ends of the plot will fall above a straight line through the middle part, yielding a curved pattern as illustrated in Figure 4.31(b). A sample from a lognormal distribution will usually produce such a pattern. A plot of (z percentile, $\ln(x)$) pairs should then resemble a straight line.

Even when the population distribution is normal, the sample percentiles will not coincide exactly with the theoretical percentiles because of sampling variability. How much can the points in the probability plot deviate from a straight line pattern before the assumption of population normality is no longer plausible? This is not an easy question to answer. Generally speaking, a small sample from a normal distribution is more likely to yield a plot with a nonlinear pattern than is a large sample. The book *Fitting Equations to Data* (see Chapter 13 bibliography) presents the results of a simulation study in which numerous samples of different sizes were selected from normal distributions. The authors concluded that there is typically great variation in the appearance of the probability plot for sample sizes smaller than 30, and only for much larger sample sizes does a linear pattern generally predominate. When a plot is based on a small sample size, only a very substantial departure from linearity should be taken as conclusive evidence of nonnormality. A similar comment applies to probability plots for checking the plausibility of other types of distributions.

Beyond Normality

Consider a family of probability distributions involving two parameters, θ_1 and θ_2, and let $F(x; \theta_1, \theta_2)$ denote the corresponding cumulative distribution functions. The family of normal distributions is one such family, with $\theta_1 = \mu$, $\theta_2 = \sigma$, and $F(x; \mu, \sigma) = \Phi((x - \mu)/\sigma)$. Another example is the Weibull family, with $\theta_1 = \alpha$, $\theta_2 = \beta$, and

$$F(x; \alpha, \beta) = 1 - e^{-(x/\beta)^\alpha}$$

Still another family of this type is the gamma family, for which the c.d.f. is an integral involving the incomplete gamma function that cannot be expressed in any simpler form.

The parameters θ_1 and θ_2 are said to be **location** and **scale parameters,** respectively, if $F(x; \theta_1, \theta_2)$ is a function of $(x - \theta_1)/\theta_2$. The parameters μ and σ of the normal family are location and scale parameters, respectively. Changing μ shifts the location of the bell-shaped density curve to the right or left, and changing σ amounts to stretching or compressing the measurement scale (the scale on the horizontal axis when the density function is graphed). Another example is given by the c.d.f.

$$F(x; \theta_1, \theta_2) = 1 - e^{-e^{(x - \theta_1)/\theta_2}} \qquad -\infty < x < \infty$$

A random variable with this c.d.f. is said to have an *extreme value distribution*. It is used in applications involving component lifetime and material strength.

Although the form of the extreme value c.d.f. might at first glance suggest that θ_1 is the point of symmetry for the density function, and therefore the mean and median, this is not the case. Instead, $P(X \leq \theta_1) = F(\theta_1; \theta_1, \theta_2) =$

$1 - e^{-1} = .632$, and the density function $f(x; \theta_1, \theta_2) = F'(x; \theta_1, \theta_2)$ is negatively skewed (a long lower tail). Similarly, the scale parameter θ_2 is not the standard deviation ($\mu = \theta_1 - .5772\theta_2$ and $\sigma = 1.283\theta_2$). However, changing the value of θ_1 does change the location of the density curve, whereas a change in θ_2 rescales the measurement axis.

The parameter β of the Weibull distribution is a scale parameter, but α is not a location parameter. The parameter α is usually referred to as a **shape parameter.** A similar comment applies to the parameters α and β of the gamma distribution. In the usual form, the density function for any member of either the gamma or Weibull distribution is positive for $x > 0$ and zero otherwise. A location parameter can be introduced as a third parameter γ (we did this for the Weibull distribution) to shift the density function so that it is positive if $x > \gamma$ and zero otherwise.

When the family under consideration has only location and scale parameters, the issue of whether or not any member of the family is a plausible population distribution can be addressed via a single easily constructed probability plot. One first obtains the percentiles of the *standard distribution,* the one with $\theta_1 = 0$ and $\theta_2 = 1$, for percentages $100(i - .5)/n$ $(i = 1, \ldots, n)$. The n (standardized percentile, observation) pairs give the points in the plot. This is of course exactly what we did to obtain an omnibus normal probability plot. Somewhat surprisingly, this methodology can be applied to yield an omnibus Weibull probability plot. The key result is that if X has a Weibull distribution with shape parameter α and scale parameter β, then the transformed variable $\ln(X)$ has an extreme value distribution with location parameter $\theta_1 = \ln(\beta)$ and scale parameter α. Thus a plot of the (extreme value standardized percentile, $\ln(x)$) pairs that shows a strong linear pattern provides support for choosing the Weibull distribution as a population model.

Example 4.31   The accompanying observations are on lifetime (hours) of power apparatus insulation when thermal and electrical stress acceleration were fixed at particular values ("On the Estimation of Life of Power Apparatus Insulation Under Combined Electrical and Thermal Stress," *IEEE Trans. on Electrical Insulation,* 1985, pp. 70–78). A Weibull probability plot necessitates first computing the 5th, 15th, . . . , and 95th percentiles of the standard extreme value distribution. The $(100p)$th percentile $\eta(p)$ satisfies

$$p = F(\eta(p)) = 1 - e^{-e^{\eta(p)}}$$

from which $\eta(p) = \ln(-\ln(1 - p))$.

Percentile:	−2.97	−1.82	−1.25	−.84	−.51
x:	282	501	741	851	1072
ln(x):	5.64	6.22	6.61	6.75	6.98

Percentile:	−.23	.05	.33	.64	1.10
x:	1122	1202	1585	1905	2138
ln(x):	7.02	7.09	7.37	7.55	7.67

The pairs $(-2.97, 5.64)$, $(-1.82, 6.22)$, ..., $(1.10, 7.67)$ are plotted as points in Figure 4.32. The straightness of the plot argues strongly for using the Weibull distribution as a model for insulation life, a conclusion also reached by the author of the cited paper.

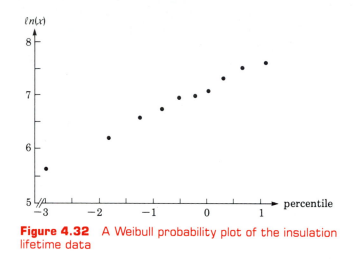

Figure 4.32   A Weibull probability plot of the insulation lifetime data

The gamma distribution is an example of a family involving a shape parameter for which there is no transformation $h(\cdot)$ such that $h(X)$ has a distribution depending only on location and scale parameters. Construction of a probability plot necessitates first estimating the shape parameter from sample data (some methods for doing this are described in Chapter 6). Sometimes an investigator wishes to know whether the transformed variable X^θ has a normal distribution for some value of θ (by convention, $\theta = 0$ is identified with the logarithmic transformation, in which case X has a lognormal distribution). The book *Graphical Methods of Data Analysis,* listed in the Chapter 1 bibliography, discusses this type of problem as well as other refinements of probability plotting.

Finally, construction of various probability plots can be facilitated by the use of commercially available special probability paper. The axes of such paper are scaled so that calculation of percentiles is unnecessary. The book *Applied Life Data Analysis,* listed in the bibliography, presents examples and related discussion. However, the ability of statistical computer packages to construct probability plots will eventually make probability paper as obsolete as the slide rule.

Exercises / Section 4.6 (72–78)

72. Consider the following 10 observations on bearing lifetime (in hours):

152.7, 172.0, 172.5, 173.3, 193.0, 204.7, 216.5, 234.9, 262.6, 422.6

Construct a normal probability plot, and comment on the plausibility of the normal distribution as a model for bearing lifetime (data from "Modified Moment Estimation for the Three-Parameter Lognormal Distribution," *J. Quality Technology,* 1985, pp. 92–99).

73. Construct a normal probability plot for the fatigue crack propagation data given in Exercise 20 (Chapter 1). Does it appear plausible that propagation life has a normal distribution? Explain.

74. The paper "The Load-Life Relationship for M50 Bearings with Silicon Nitride Ceramic Balls" (*Lubrication Engr.,* 1984, pp. 153–159) reported the accompanying data on bearing load life (million revs.) for bearings tested at a 6.45 kN load.

    47.1 68.1 68.1 90.8 103.6 106.0 115.0
    126.0 146.6 229.0 240.0 240.0 278.0 278.0
    289.0 289.0 367.0 385.9 392.0 505.0

    a. Construct a normal probability plot. Is normality plausible?
    b. Construct a Weibull probability plot. Is the Weibull distribution family plausible?

75. If you have access to a statistical computer package, use it to construct a normal probability plot for the shear strength data of Exercise 7 (Chapter 1). (If not, construct a plot using just the first 25 observations in the data set.) Comment.

76. Construct a probability plot that will allow you to assess the plausibility of the lognormal distribution as a model for the rainfall data of Exercise 44 (Chapter 1).

77. The accompanying observations are precipitation values during March over a 30-year period in Minneapolis–St. Paul.

.77	1.20	3.00	1.62	2.81	2.48
1.74	.47	3.09	1.31	1.87	.96
.81	1.43	1.51	.32	1.18	1.89
1.20	3.37	2.10	.59	1.35	.90
1.95	2.20	.52	.81	4.75	2.05

    a. Construct and interpret a normal probability plot for this data set.
    b. Calculate the square root of each value and then construct a normal probability plot based on this transformed data. Does it seem plausible that the square root of precipitation is normally distributed?
    c. Repeat (b) after transforming by cube roots.

78. The following failure time observations (1000's of hours) resulted from accelerated life testing of 16 integrated circuit chips of a certain type.

82.8	11.6	359.5	502.5	307.8	179.7
242.0	26.5	244.8	304.3	379.1	212.6
229.9	558.9	366.7	204.6		

    Use the corresponding percentiles of the exponential distribution with $\lambda = 1$ to construct a probability plot. Then explain why the plot assesses the plausibility of the sample having been generated from *any* exponential distribution.

Supplementary Exercises / Chapter 4 (79–97)

79. Let $X =$ the time that it takes a read/write head to locate a desired record on a computer disk memory device once the head has been positioned over the correct track. If the disks rotate once every 25 milliseconds, a reasonable assumption is that X is uniformly distributed on the interval $[0, 25]$.
    a. Compute $P(10 \leq X \leq 20)$.
    b. Compute $P(X \geq 10)$.
    c. Obtain the c.d.f. $F(X)$.
    d. Compute $E(X)$ and σ_X.

80. A 12-in. bar that is clamped at both ends is to be subjected to an increasing amount of stress until it snaps. Let $Y =$ the distance from the left end at which the break occurs, and suppose that Y has p.d.f.

$$f(y) = \begin{cases} \left(\dfrac{1}{24}\right) y \left(1 - \dfrac{y}{12}\right) & 0 \leq y \leq 12 \\ 0 & \text{otherwise} \end{cases}$$

Compute
a. the c.d.f. of Y and graph it.
b. $P(Y \leq 4)$, $P(Y > 6)$, and $P(4 \leq Y \leq 6)$.
c. $E(Y)$, $E(Y^2)$, and $V(Y)$.
d. the probability that the break point occurs more than 2 in. from the expected break point.
e. the expected length of the shorter segment when the break occurs.

81. The completion time X for a certain task has cumulative distribution function

$$F(x) = \begin{cases} 0 & x < 0 \\ \dfrac{x^3}{3} & 0 \le x < 1 \\ 1 - \dfrac{1}{2}\left(\dfrac{7}{3} - x\right)\left(\dfrac{7}{4} - \dfrac{3}{4}x\right) & 1 \le x < \dfrac{7}{3} \\ 1 & x \ge \dfrac{7}{3} \end{cases}$$

  a. Obtain the p.d.f. $f(x)$ and sketch its graph.
  b. Compute $P(.5 \le X \le 2)$.
  c. Compute $E(X)$.

82. Suppose that the time X necessary to process an application for a license plate for a newly purchased automobile is normally distributed with $\mu = 6$ and $\sigma = 1.5$ min.

  a. What is the probability that it takes at least 9.50 min to process a single application?
  b. If 10 such applications are processed independently of one another, what is the probability that all take less than 9.5 min?
  c. If 1000 such applications are processed and $Y =$ the number among the 1000 that take at least 10.62 min (a "success"), what is $P(Y \le 1)$ (approximately)?

83. The reaction time (in seconds) to a certain stimulus is a continuous random variable with p.d.f.

$$f(x) = \begin{cases} \dfrac{3}{2} \cdot \dfrac{1}{x^2} & 1 \le x \le 3 \\ 0 & \text{otherwise} \end{cases}$$

  a. Obtain the cumulative distribution function.
  b. What is the probability that reaction time is at most 2.5 sec? Between 1.5 and 2.5 sec?
  c. Compute the expected reaction time.
  d. Compute the standard deviation of reaction time.
  e. If an individual takes more than 1.5 sec to react, a light comes on and stays on either until one further second has elapsed or until the person reacts (whichever happens first). Determine the expected amount of time that the light remains lit. *Hint:* Let $h(X) =$ the time that the light is on as a function of reaction time X.

84. Let X denote the temperature at which a certain chemical reaction takes place, and suppose that X has p.d.f.

$$f(x) = \begin{cases} \dfrac{1}{9}(4 - x^2) & -1 \le x \le 2 \\ 0 & \text{otherwise} \end{cases}$$

  a. Sketch the graph of $f(x)$.
  b. Determine the cumulative distribution function and sketch it.
  c. Is zero the median temperature at which the reaction takes place? If not, is the median temperature smaller or larger than zero?
  d. Suppose that this reaction is independently carried out once in each of 10 different labs, and that the p.d.f. of reaction time in each lab is as given. Let $Y =$ the number among the 10 labs at which the temperature exceeds 1. What kind of distribution does Y have (give the name and values of any parameters)?

85. The paper "Determination of the MTF of Positive Photoresists Using the Monte Carlo Method" (*Photographic Sci. and Engr.,* 1983, pp. 254–260) proposes the exponential distribution with parameter $\lambda = .93$ as a model for the distribution of a photon's free path length (μm) under certain circumstances. Suppose that this is the correct model.

  a. What is the expected path length, and what is the standard deviation of path length?
  b. What is the probability that path length exceeds 3.0? What is the probability that path length is between 1.0 and 3.0?
  c. What value is exceeded by only 10% of all path lengths?

86. The paper "The Prediction of Corrosion by Statistical Analysis of Corrosion Profiles" (*Corrosion Science,* 1985, pp. 305–315) suggests the following cumulative distribution function for the depth X of the deepest pit in an experiment involving the exposure of carbon manganese steel to acidified sea water:

$$F(x; \alpha, \beta) = e^{-e^{-(x - \alpha)/\beta}} \qquad -\infty < x < \infty$$

The authors propose the values $\alpha = 150$, $\beta = 90$. Assume this to be the correct model.

  a. What is the probability that the depth of the deepest pit is at most 150? At most 300? Between 150 and 300?
  b. Below what value will the depth of the maximum pit be observed in 90% of all such experiments?

c. What is the density function of X?

d. The density function can be shown to be unimodal (a single peak). Above what value on the measurement axis does this peak occur (this value is the mode)?

e. It can be shown that $E(X) \approx .5772\beta - \alpha$. What is the mean for the given values of α and β, and how does it compare to the median and mode? Sketch the graph of the density function. *Note:* This is called the *largest extreme value distribution.*

87. A component has lifetime X which is exponentially distributed with parameter λ.

  a. If the cost of operation per unit time is c, what is the expected cost of operating this component over its lifetime?

  b. Instead of a constant cost rate c as in part (a), suppose that the cost rate is $c(1 - .5e^{\alpha x})$, so that the cost per unit time is less than c when the component is new and gets more expensive as the component ages. Now compute the expected cost of operation over the lifetime of the component.

88. The *mode* of a continuous distribution is that value x^* which maximizes $f(x)$.

  a. What is the mode of a normal distribution with parameters μ and σ?

  b. Does the uniform distribution with parameters A and B have a single mode? Why or why not?

  c. What is the mode of an exponential distribution with parameter λ? (Draw a picture here.)

  d. If X has a gamma distribution with parameters α and β, and $\alpha > 1$, find the mode. *Hint:* $\ln[f(x)]$ will be maximized iff $f(x)$ is, and it may be simpler to take the derivative of $\ln[f(x)]$.

  e. What is the mode of a chi-squared distribution having ν degrees of freedom?

89. The article "Error Distribution in Navigation" (*J. Institute of Navigation,* 1971, pp. 429–442) suggests that the frequency distribution of positive errors (magnitudes of errors) is well approximated by an exponential distribution. Let $X =$ the lateral position error (nautical miles), which can be either negative or positive, and suppose that the p.d.f. of X is

$$f(x) = \begin{cases} (.1)e^{-.2|x|} & -\infty < x < \infty \\ 0 & \text{otherwise} \end{cases}$$

  a. Sketch a graph of $f(x)$, and verify that $f(x)$ is a legitimate p.d.f. (show that it integrates to 1).

  b. Obtain the c.d.f. of X and sketch it.

  c. Compute $P(X \leq 0)$, $P(X \leq 2)$, $P(-1 \leq X \leq 2)$, and the probability that an error of more than two miles is made.

90. In some systems, a customer is allocated to one of two service facilities. If the service time for a customer served by facility i has an exponential distribution with parameter λ_i $(i = 1, 2)$ and p is the proportion of all customers served by facility 1, then the p.d.f. of $X =$ the service time of a randomly selected customer is

$$f(x; \lambda_1, \lambda_2, p)$$
$$= \begin{cases} p\lambda_1 e^{-\lambda_1 x} + (1 - p)\lambda_2 e^{-\lambda_2 x} & x \geq 0 \\ 0 & \text{otherwise} \end{cases}$$

This is often called the hyperexponential distribution. As an example, many computer systems process some programs using both a fast in-core compiler (FORTRAN WATFIV) and a slower compiler.

  a. Verify that $f(x; \lambda_1, \lambda_2, p)$ is indeed a p.d.f.

  b. What is the c.d.f. $F(x; \lambda_1, \lambda_2, p)$?

  c. If X has $f(x; \lambda_1, \lambda_2, p)$ as its p.d.f., what is $E(X)$?

  d. Using the fact that $E(X^2) = 2/\lambda^2$ when X has an exponential distribution with parameter λ, compute $E(X^2)$ when X has p.d.f. $f(x; \lambda_1, \lambda_2, p)$. Then compute $V(X)$.

  e. The coefficient of variation of a random variable (or distribution) is $CV = \sigma/\mu$. What is CV for an exponential r.v.? What can you say about the value of CV when X has a hyperexponential distribution?

  f. What is CV for an Erlang distribution with parameters λ and n as defined in Exercise 55? *Note:* In applied work, the sample CV is used to decide which of the three distributions might be appropriate.

91. Suppose that a particular state allows individuals filing tax returns to itemize deductions only if the total of all itemized deductions is at least $5,000.00, and let X (in thousands of dollars) be the total of itemized deductions on a randomly chosen form. Assume that X has the p.d.f.

$$f(x; \alpha) = \begin{cases} \dfrac{k}{x^\alpha} & x \ge 5 \\ 0 & \text{otherwise} \end{cases}$$

a. Find the value of k. What restriction on α is necessary?
b. What is the c.d.f. of X?
c. What is the expected total deduction on a randomly chosen form. What restriction on α is necessary for $E(X)$ to be finite?
d. Show that $\ln(X/5)$ has an exponential distribution with parameter $\alpha - 1$.

92. Let I_i be the input current to a transistor and I_0 be the output current. Then the current gain is proportional to $\ln(I_0/I_i)$. Suppose that the constant of proportionality is 1 (which amounts to choosing a particular unit of measurement), so that current gain $= X = \ln(I_0/I_i)$. Assume X is normally distributed with $\mu = 1$ and $\sigma = .05$.
a. What type of distribution does the ratio I_0/I_i have?
b. What is the probability that the output current is more than twice the input current?
c. What are the expected value and variance of the ratio of output to input current?

93. Let Z have a standard normal distribution, and define a new r.v. Y by $Y = \sigma Z + \mu$. Show that Y has a normal distribution with parameters μ and σ. *Hint:* $Y \le y$ iff $Z \le ?$ Use this to find the c.d.f. of Y and then differentiate it with respect to y.

94. a. Suppose that the lifetime X of a component, when measured in hours, has a gamma distribution with parameters α and β. Let $Y =$ the lifetime measured in minutes. Derive the p.d.f. of Y. *Hint:* $Y \le y$ iff $X \le y/60$. Use this to obtain the c.d.f. of Y and then differentiate to obtain the p.d.f.
b. If X has a gamma distribution with parameters α and β, what is the probability distribution of $Y = cX$?

95. In Exercises 93 and 94, as well as many other situations, one has the p.d.f. $f(x)$ of X and wishes the p.d.f. of $Y = h(X)$. Assume that $h(\cdot)$ is an invertible function, so that $y = h(x)$ can be solved for x to yield $x = k(y)$. Then it can be shown that the p.d.f. of Y is

$$g(y) = f[k(y)] \cdot |k'(y)|$$

a. If X has a uniform distribution with $A = 0$, $B = 1$, derive the p.d.f. of $Y = -\ln(X)$.
b. Work Exercise 93 using this result.
c. Work Exercise 94(b) using this result.

96. Let X denote the lifetime of a component, with $f(x)$ and $F(x)$ the p.d.f. and c.d.f. of X. The probability that the component fails in the interval $(x, x + \Delta x)$ is approximately $f(x) \cdot \Delta x$. The probability that it fails in $(x, x + \Delta x)$ given that it has lasted at least x is $f(x) \cdot \Delta x/[1 - F(x)]$. Dividing this by Δx produces the **failure rate function:**

$$r(x) = \frac{f(x)}{1 - F(x)}$$

An increasing failure rate function indicates that older components are increasingly likely to wear out, while a decreasing failure rate is evidence of increasing reliability with age. In practice a "bathtub-shaped" failure is often assumed.
a. If X is exponentially distributed, what is $r(x)$?
b. If X has a Weibull distribution with parameters α and β, what is $r(x)$? For what parameter values will $r(x)$ be increasing? For what parameter values will $r(x)$ decrease with x?
c. Since $r(x) = -d/dx \ln[1 - F(x)]$, $\ln[1 - F(x)] = -\int r(x) \, dx$. Suppose that

$$r(x) = \begin{cases} \alpha\left(1 - \dfrac{x}{\beta}\right) & 0 \le x \le \beta \\ 0 & \text{otherwise} \end{cases}$$

so that if a component lasts β hours, it will last forever (while seemingly unreasonable, this model can be used to study just "initial wearout"). What are the c.d.f. and p.d.f. of X?

97. Let U have a uniform distribution on the interval $[0, 1]$. Then observed values having this distribution can be obtained from a computer's random number generator. Let $X = -(1/\lambda)\ln(1 - U)$.
a. Show that X has an exponential distribution with parameter λ. *Hint:* The c.d.f. of X is $F(x) = P(X \le x)$; $X \le x$ is equivalent to $U \le ?$
b. How would you use (a) and a random number generator to obtain observed values from an exponential distribution with parameter $\lambda = 10$?

Bibliography

Bury, Karl, *Statistical Models in Applied Science,* Wiley, New York, 1975. Somewhat difficult to read, but the author presents a number of good engineering and science examples involving continuous probability models.

Derman, Cyrus, Gleser, Leon, and Olkin, Ingram, *Probability Models and Applications,* Macmillan, New York, 1980. Good coverage of general properties and specific distributions.

Johnson, Norman, and Kotz, Samuel, *Distributions in Statistics: Continuous Distributions 1 and 2,* Houghton Mifflin, Boston, 1970. These two volumes together present an exhaustive survey of various continuous distributions.

Nelson, Wayne, *Applied Life Data Analysis,* Wiley, New York, 1982. Gives a comprehensive discussion of distributions and methods that are used in the analysis of lifetime data.

CHAPTER 5

Joint Probability Distributions and Random Samples

Introduction

In Chapters 3 and 4 we studied probability models for a single random variable. Many problems in probability and statistics lead to models involving several random variables simultaneously. In the present chapter we first discuss probability models for the joint behavior of several random variables, putting special emphasis on the case in which the variables are independent of one another. We then study expected values of functions of several random variables, including covariance and correlation as measures of the degree of association between two variables.

Many statistical procedures are based on linear functions of random variables, especially totals $X_1 + X_2 + \cdots + X_n$ and averages $(X_1 + \cdots + X_n)/n$, so properties of such functions are presented next. The last section focuses on the relationship between linear functions and the normal distribution. The key result is the Central Limit Theorem (C.L.T.), which provides the justification for many large-sample statistical procedures.

5.1 Jointly Distributed Random Variables

There are many experimental situations in which more than one random variable will be of interest to an investigator. We shall first consider joint probability distributions for two discrete random variables, then for two continuous variables, and finally for more than two variables.

The Joint P.M.F. for Two Discrete Random Variables

The probability mass function of a single discrete random variable X specifies how much probability mass is placed on each possible X value. The joint p.m.f.

of two discrete random variables X and Y describes how much probability mass is placed on each possible pair of values (x, y).

Definition

> Let X and Y be two discrete random variables defined on the sample space $\mathcal{S}$ of an experiment. The **joint probability mass function** $p(x, y)$ is defined for each pair of numbers (x, y) by
>
> $$p(x, y) = P(X = x \text{ and } Y = y)$$
>
> For any set A consisting of pairs of (x, y) values, the probability $P[(X, Y) \in A]$ is obtained by summing the joint p.m.f. over pairs in A:
>
> $$P[(X, Y) \in A] = \sum\sum_{(x, y) \in A} p(x, y)$$

Example 5.1

A large insurance agency services a number of customers who have purchased both a homeowners policy and an automobile policy from the agency. For each type of policy, a deductible amount must be specified. For an automobile policy, the choices are \$100 and \$250, while for a homeowners policy the choices are 0, \$100, and \$200. Suppose that an individual with both types of policy is selected at random from the agency's files, and let $X =$ the deductible amount on the auto policy and $Y =$ the deductible amount on the homeowners policy. Possible (X, Y) pairs are then $(100, 0)$, $(100, 100)$, $(100, 200)$, $(250, 0)$, $(250, 100)$, and $(250, 200)$; the joint p.m.f. specifies the probability associated with each one of these pairs, with any other pair having probability zero. Suppose that the joint p.m.f. is given in the accompanying **joint probability table:**

		y		
$p(x, y)$		0	100	200
	100	.20	.10	.20
x				
	250	.05	.15	.30

Then $p(100, 100) = P(X = 100 \text{ and } Y = 100) = P(\$100 \text{ deductible on both policies}) = .10$. The probability $P(Y \geq 100)$ is computed by summing probabilities of all (x, y) pairs for which $y \geq 100$:

$$P(Y \geq 100) = p(100, 100) + p(250, 100) + p(100, 200) + p(250, 200)$$

$$= .75 \qquad \blacksquare$$

A function $p(x, y)$ can be used as a joint p.m.f. provided that $p(x, y) \geq 0$ for all x and y and $\sum_{x}\sum_{y} p(x, y) = 1$.

The p.m.f. of one of the variables alone is obtained by summing $p(x, y)$ over values of the other variable. The result is called a marginal p.m.f. because when the $p(x, y)$'s appear in a rectangular table, the sums are just marginal (row or column) totals.

Definition

> The **marginal probability mass functions** of X and of Y, denoted by $p_X(x)$ and $p_Y(y)$, respectively, are given by
>
> $$p_X(x) = \sum_y p(x, y), \qquad p_Y(y) = \sum_x p(x, y)$$

Thus to obtain the marginal p.m.f. of X evaluated at, say, $x = 100$, the probabilities $p(100, y)$ are added over all possible y values. Doing this for each possible X value gives the marginal p.m.f. of X alone (without reference to Y). From the marginal p.m.f.'s, probabilities of events involving only X or only Y can be computed.

Example 5.2
(Example 5.1 continued)

The possible X values are $x = 100$ and $x = 250$, so computing row totals in the joint probability table yields

$$p_X(100) = p(100, 0) + p(100, 100) + p(100, 200) = .50$$

and

$$p_X(250) = p(250, 0) + p(250, 100) + p(250, 200) = .50$$

The marginal p.m.f. of X is then

$$p_X(x) = \begin{cases} .5 & x = 100, 250 \\ 0 & \text{otherwise} \end{cases}$$

Similarly, the marginal p.m.f. of Y is obtained from column totals as

$$p_Y(y) = \begin{cases} .25 & y = 0, 100 \\ .50 & y = 200 \\ 0 & \text{otherwise} \end{cases}$$

so $P(Y \geq 100) = p_Y(100) + p_Y(200) = .75$ as before. ■

The Joint P.D.F. for Two Continuous Random Variables

The probability that the observed value of a continuous random variable X lies in a one-dimensional set A (such as an interval) is obtained by integrating the p.d.f. $f(x)$ over the set A. Similarly, the probability that the pair (X, Y) of continuous random variables falls in a two-dimensional set A (such as a rectangle) is obtained by integrating a function called the *joint density function* over A.

Definition

Let X and Y be continuous random variables. Then $f(x, y)$ is the **joint probability density function** for X and Y if for any two-dimensional set A

$$P[(X, Y) \in A] = \iint_A f(x, y) \, dx \, dy$$

In particular, if A is the two-dimensional rectangle $\{(x, y): a \le x \le b, c \le y \le d\}$, then

$$P[(X, Y) \in A] = P(a \le X \le b, c \le Y \le d) = \int_a^b \int_c^d f(x, y) \, dy \, dx$$

For $f(x, y)$ to be a candidate for a joint p.d.f., it must satisfy $f(x, y) \ge 0$ and $\int_{-\infty}^{\infty} \int_{-\infty}^{\infty} f(x, y) \, dx \, dy = 1$. We can think of $f(x, y)$ as specifying a surface at height $f(x, y)$ above the point (x, y) in a three-dimensional coordinate system. Then $P[(X, Y) \in A]$ is the volume underneath this surface and above the region A, analogous to the area under a curve in the one-dimensional case. This is illustrated in Figure 5.1.

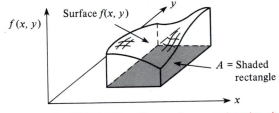

Figure 5.1 $P[(X, Y) \in A] =$ volume under density surface above A

Example 5.3

A bank operates both a drive-up facility and a walk-in facility. On a randomly selected day, let $X =$ the proportion of time that the drive-up facility is in use (at least one customer is being served or waiting to be served) and $Y =$ the proportion of time that the walk-in facility is in use. Then the set of possible values for (X, Y) is the rectangle $D = \{(x, y): 0 \le x \le 1, 0 \le y \le 1\}$. Suppose that the joint p.d.f. of (X, Y) is given by

$$f(x, y) = \begin{cases} \dfrac{6}{5}(x + y^2) & 0 \le x \le 1, 0 \le y \le 1 \\ 0 & \text{otherwise} \end{cases}$$

To verify that this is a legitimate p.d.f., note that $f(x, y) \ge 0$ and

$$\int_{-\infty}^{\infty} \int_{-\infty}^{\infty} f(x, y) \, dx \, dy = \int_0^1 \int_0^1 \frac{6}{5}(x + y^2) \, dx \, dy$$

$$= \int_0^1 \int_0^1 \frac{6}{5} x \, dx \, dy + \int_0^1 \int_0^1 \frac{6}{5} y^2 \, dx \, dy$$

$$= \int_0^1 \frac{6}{5} x \, dx + \int_0^1 \frac{6}{5} y^2 \, dy = \frac{6}{10} + \frac{6}{15} = 1$$

The probability that neither facility is busy more than one-quarter of the time is

$$P\left(0 \le X \le \frac{1}{4}, 0 \le Y \le \frac{1}{4}\right) = \int_0^{1/4} \int_0^{1/4} \frac{6}{5}(x + y^2) \, dx \, dy$$

$$= \frac{6}{5} \int_0^{1/4} \int_0^{1/4} x \, dx \, dy + \frac{6}{5} \int_0^{1/4} \int_0^{1/4} y^2 \, dx \, dy$$

$$= \frac{6}{20} \cdot \frac{x^2}{2} \Big|_{x=0}^{x=1/4} + \frac{6}{20} \cdot \frac{y^3}{3} \Big|_{y=0}^{y=1/4} = \frac{7}{640}$$

$$= .0109 \qquad \blacksquare$$

As with joint p.m.f.'s, from the joint p.d.f. of X and Y each of the two marginal density functions can be computed.

Definition

> The **marginal probability density functions** of X and Y, denoted by $f_X(x)$ and $f_Y(y)$, respectively, are given by
>
> $$f_X(x) = \int_{-\infty}^{\infty} f(x, y) \, dy \quad \text{for } -\infty < x < \infty$$
>
> $$f_Y(y) = \int_{-\infty}^{\infty} f(x, y) \, dx \quad \text{for } -\infty < y < \infty$$

Example 5.4
(Example 5.3
continued)

The marginal p.d.f. of X, which gives the probability distribution of busy time for the drive-up facility without reference to the walk-in facility, is

$$f_X(x) = \int_{-\infty}^{\infty} f(x, y) \, dy = \int_0^1 \frac{6}{5}(x + y^2) \, dy = \frac{6}{5}x + \frac{2}{5}$$

for $0 \le x \le 1$ and 0 otherwise. The marginal p.d.f. of Y is

$$f_Y(y) = \begin{cases} \dfrac{6}{5}y^2 + \dfrac{3}{5} & 0 \le y \le 1 \\ 0 & \text{otherwise} \end{cases}$$

Then

$$P\left(\frac{1}{4} \le Y \le \frac{3}{4}\right) = \int_{1/4}^{3/4} f_Y(y) \, dy = \frac{37}{80} = .4625 \qquad \blacksquare$$

In Example 5.3 the region of positive joint density was a rectangle, which made computation of the marginal p.d.f.'s relatively easy. Consider now an example in which the region of positive density is a more complicated figure.

Example 5.5

A nut company markets cans of deluxe mixed nuts containing almonds, cashews, and peanuts. Suppose that the net weight of each can is exactly one pound, but that the weight contribution of each type of nut is random. Because the three weights sum to one, a joint probability model for any two gives all necessary information about the weight of the third type. Let $X =$ the weight of almonds in a selected can and $Y =$ the weight of cashews. Then the region of positive density is $D = \{(x, y): 0 \le x \le 1, 0 \le y \le 1, x + y \le 1\}$, the shaded region pictured in Figure 5.2. Now let the joint p.d.f. for (X, Y) be

$$f(x, y) = \begin{cases} 24xy & 0 \le x \le 1, 0 \le y \le 1, x + y \le 1 \\ 0 & \text{otherwise} \end{cases}$$

For any fixed $x, f(x, y)$ increases with y, and for fixed $y, f(x, y)$ increases with x. This is appropriate since the word "deluxe" implies that most of the can should consist of almonds and cashews rather than peanuts, so that the density function should be large near the upper boundary and small near the origin. The surface determined by $f(x, y)$ slopes upward from zero as (x, y) moves away from either axis.

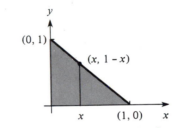

Figure 5.2 Region of positive density for Example 5.5

Clearly $f(x, y) \ge 0$. To verify the second condition on a joint p.d.f., recall that a double integral is computed as an iterated integral by holding one variable fixed (such as x as in Figure 5.2), integrating over values of the other variable lying along the straight line passing through the value of the fixed variable, and finally integrating over all possible values of the fixed variable. Thus

$$\int_{-\infty}^{\infty} \int_{-\infty}^{\infty} f(x, y) \, dy \, dx = \iint_D f(x, y) \, dy \, dx = \int_0^1 \left\{ \int_0^{1-x} 24xy \, dy \right\} dx$$

$$= \int_0^1 24x \left\{ \frac{y^2}{2} \Big|_{y=0}^{y=1-x} \right\} dx = \int_0^1 12x(1-x)^2 \, dx = 1$$

To compute the probability that the two types of nuts together make up at most 50% of the can, let $A = \{(x, y): 0 \le x \le 1, 0 \le y \le 1, \text{ and } x + y \le .5\}$ as shown in Figure 5.3. Then

$$P((X, Y) \in A) = \iint_A f(x, y) \, dx \, dy = \int_0^{.5} \int_0^{.5-x} 24xy \, dy \, dx = .125$$

The marginal p.d.f. for almonds is obtained by holding X fixed at x and integrating $f(x, y)$ along the vertical line through x:

$$f_X(x) = \int_{-\infty}^{\infty} f(x, y)\, dy = \begin{cases} \int_0^{1-x} 24xy\, dy = 12x(1-x)^2 & 0 \le x \le 1 \\ 0 & \text{otherwise} \end{cases}$$

By symmetry of $f(x, y)$ and the region D, the marginal p.d.f. of Y is obtained by replacing x and X in $f_X(x)$ by y and Y, respectively.

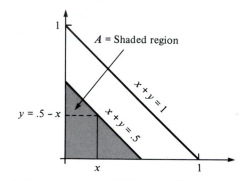

Figure 5.3   Computing $P[(X, Y) \in A]$ for Example 5.5   ■

Independent Random Variables

In many situations information about the observed value of one of the two variables X and Y gives information about the value of the other variable. In Example 5.1 the marginal probability of X at $x = 250$ was .5, as was the probability that $X = 100$. If, however, we are told that the selected individual had $Y = 0$, then $X = 100$ is four times as likely as $X = 250$. Thus there is a dependence between the two variables.

In Chapter 2 we pointed out that one way of defining independence of two events was to say that A and B are independent if $P(A \cap B) = P(A) \cdot P(B)$. We now use an analogous definition for the independence of two random variables.

Definition

> Two random variables X and Y are said to be **independent** if for every pair of x and y values,
>
> $$p(x, y) = p_X(x) \cdot p_Y(y) \quad \text{when } X \text{ and } Y \text{ are discrete}$$
>
> or   (5.1)
>
> $$f(x, y) = f_X(x) \cdot f_Y(y) \quad \text{when } X \text{ and } Y \text{ are continuous}$$
>
> If (5.1) is not satisfied for all (x, y), X and Y are said to be **dependent**.

The definition says that two variables are independent if their joint p.m.f. or p.d.f. is the product of the two marginal p.m.f.'s or p.d.f.'s.

Example 5.6   In the insurance problem of Example 5.1, $p(100, 100) = .10$, while $p_X(100) = .5$ and $p_Y(100) = .25$. Since $.10 \neq (.5)(.25)$, X and Y are not independent. Independence of X and Y requires that *every* entry in the joint probability table be the product of the corresponding row and column marginal probabilities.   ∎

Example 5.7
(Example 5.4
continued)
Because $f(x, y)$ has the form of a product, X and Y would appear to be independent. However, while $f_X(\frac{3}{4}) = f_Y(\frac{3}{4}) = (\frac{9}{16})^2$, $f(\frac{3}{4}, \frac{3}{4}) = 0 \neq (\frac{9}{16})^2 \cdot (\frac{9}{16})^2$, so the variables are not in fact independent. To be independent, $f(x, y)$ must have the form $g(x) \cdot h(y)$ and the region of positive density must be a rectangle whose sides are parallel to the coordinate axes.   ∎

Independence of two random variables is most useful when the description of the experiment under study tells us that X and Y have no effect on one another. Then once the marginal p.m.f.'s or p.d.f.'s have been specified, the joint p.m.f. or p.d.f. is simply the product of the two marginal functions. It follows that

$$P(a \leq X \leq b, c \leq Y \leq d) = P(a \leq X \leq b) \cdot P(c \leq Y \leq d)$$

Example 5.8   Suppose that the lifetimes of two components are independent of one another and that the first lifetime X_1 has an exponential distribution with parameter λ_1 while the second X_2 has an exponential distribution with parameter λ_2. Then the joint p.d.f. is

$$f(x_1, x_2) = f_{X_1}(x_1) \cdot f_{X_2}(x_2)$$
$$= \begin{cases} \lambda_1 e^{-\lambda_1 x_1} \cdot \lambda_2 e^{-\lambda_2 x_2} = \lambda_1 \lambda_2 e^{-\lambda_1 x_1 - \lambda_2 x_2} & x_1 > 0, x_2 > 0 \\ 0 & \text{otherwise} \end{cases}$$

Let $\lambda_1 = 1/1000$ and $\lambda_2 = 1/1200$, so that the expected lifetimes are 1000 hours and 1200 hours, respectively. The probability that both component lifetimes are at least 1500 hours is

$$P(1500 \leq X_1, 1500 \leq X_2) = P(1500 \leq X_1) \cdot P(1500 \leq X_2)$$
$$= e^{-\lambda_1(1500)} \cdot e^{-\lambda_2(1500)}$$
$$= (.2231)(.2865) = .0639 \qquad ∎$$

More Than Two Random Variables

To model the joint behavior of more than two random variables, we extend the concept of a joint distribution of two variables.

Definition

> If $X_1, X_2, \ldots, X_n$ are all discrete random variables, the joint p.m.f. of the variables is the function
>
> $$p(x_1, x_2, \ldots, x_n) = P(X_1 = x_1, X_2 = x_2, \ldots, X_n = x_n)$$
>
> If the variables are continuous, the joint p.d.f. of $X_1, \ldots, X_n$ is the function $f(x_1, x_2, \ldots, x_n)$ such that for any n intervals $[a_1, b_1], \ldots, [a_n, b_n]$,
>
> $$P(a_1 \leq X_1 \leq b_1, \ldots, a_n \leq X_n \leq b_n)$$
>
> $$= \int_{a_1}^{b_1} \cdots \int_{a_n}^{b_n} f(x_1, \ldots, x_n)\, dx_n \ldots dx_1$$

Example 5.9

In a binomial experiment each trial could result in one of only two possible outcomes. Consider now an experiment consisting of n independent and identical trials, in which each trial can result in any one of r possible outcomes. Let $p_i = P(\text{outcome } i \text{ on any particular trial})$, and define random variables by $X_i = $ the number of trials resulting in outcome i ($i = 1, \ldots, r$). Such an experiment is called a *multinomial experiment*, and the joint p.m.f. of $X_1, \ldots, X_r$ is called the *multinomial distribution*. By using a counting argument analogous to the one used in deriving the binomial distribution, the joint p.m.f. of $X_1, \ldots, X_r$ can be shown to be

$$p(x_1, \ldots, x_r) = \begin{cases} \dfrac{n!}{(x_1!)(x_2!) \cdots (x_r!)} p_1^{x_1} \cdots p_r^{x_r} & \begin{matrix} x_i = 0, 1, 2, \ldots \text{ with} \\ x_1 + \cdots + x_r = n \end{matrix} \\ 0 & \text{otherwise} \end{cases}$$

As an example, if the allele of each of 10 independently obtained pea sections is determined, and $p_1 = P(AA)$, $p_2 = P(Aa)$, $p_3 = P(aa)$, $X_1 = $ number of AA's, $X_2 = $ number of Aa's, $X_3 = $ number of aa's, then

$$p(x_1, x_2, x_3) = \frac{10!}{(x_1!)(x_2!)(x_3!)} p_1^{x_1} p_2^{x_2} p_3^{x_3} \quad \begin{matrix} x_i = 0, 1, \ldots \text{ and} \\ x_1 + x_2 + x_3 = 10 \end{matrix}$$

If $p_1 = p_3 = .25$, $p_2 = .5$, then

$$P(X_1 = 2, X_2 = 5, X_3 = 3) = p(2, 5, 3)$$

$$= \frac{10!}{2!\ 5!\ 3!}(.25)^2(.5)^5(.25)^3 = .0769$$

The case $r = 2$ gives the binomial distribution, with $X_1 = $ number of successes and $X_2 = n - X_1 = $ number of failures. ∎

Example 5.10

When a certain method is used to collect a fixed volume of rock samples in a region, there are four resulting rock types. Let X_1, X_2, and X_3 denote the proportion by volume of rock types 1, 2, and 3 in a randomly selected sample (the proportion of rock type 4 is $1 - X_1 - X_2 - X_3$, so a variable X_4 would be redundant). If the joint p.d.f. of X_1, X_2, X_3 is

$$f(x_1, x_2, x_3)$$
$$= \begin{cases} kx_1 x_2 (1 - x_3) & 0 \le x_1 \le 1, 0 \le x_2 \le 1, 0 \le x_3 \le 1, x_1 + x_2 + x_3 \le 1 \\ 0 & \text{otherwise} \end{cases}$$

then k is determined by

$$1 = \int_{-\infty}^{\infty} \int_{-\infty}^{\infty} \int_{-\infty}^{\infty} f(x_1, x_2, x_3) \, dx_3 \, dx_2 \, dx_1$$
$$= \int_0^1 \left\{ \int_0^{1-x_1} \left[\int_0^{1-x_1-x_2} kx_1 x_2 (1 - x_3) \, dx_3 \right] dx_2 \right\} dx_1$$

This iterated integral has value $k/144$, so $k = 144$. The probability that rocks of type 1 and 2 together account for at most 50% of the sample is

$$P(X_1 + X_2 \le .5) = \iiint_{\substack{\{0 \le x_i \le 1 \text{ for } i = 1, 2, 3 \\ x_1 + x_2 + x_3 \le 1, x_1 + x_2 \le .5\}}} f(x_1, x_2, x_3) \, dx_3 \, dx_2 \, dx_1$$
$$= \int_0^{.5} \left\{ \int_0^{.5-x_1} \left[\int_0^{1-x_1-x_2} 144 x_1 x_2 (1 - x_3) \, dx_3 \right] dx_2 \right\} dx_1$$

$$= .6066 \qquad \blacksquare$$

The notion of independence of more than two random variables is similar to the notion of independence of more than two events.

<table>
<tr><td>Definition</td><td>The random variables $X_1, X_2, \ldots, X_n$ are said to be **independent** if for *every* subset $X_{i_1}, X_{i_2}, \ldots, X_{i_k}$ of the variables (each pair, each triple, and so on), the joint p.m.f. or p.d.f. of the subset is equal to the product of the marginal p.m.f.'s or p.d.f.'s.</td></tr>
</table>

Thus if the variables are independent with $n = 4$, then the joint p.m.f. or p.d.f. of any two variables is the product of the two marginals, and similarly for any three variables and all four variables together. Most importantly, once we are told that n variables are independent, then the joint p.m.f. or p.d.f. is the product of the n marginals.

Example 5.11 If $X_1, \ldots, X_n$ represent the lifetimes of n components, the components operate independently of one another, and each lifetime is exponentially distributed with parameter λ, then

$$f(x_1, x_2, \ldots, x_n) = (\lambda e^{-\lambda x_1}) \cdot (\lambda e^{-\lambda x_2}) \cdots (\lambda e^{-\lambda x_n})$$

$$= \begin{cases} \lambda^n e^{-\lambda \Sigma x_i} & x_1 \geq 0, x_2 \geq 0, \ldots, x_n \geq 0 \\ 0 & \text{otherwise} \end{cases}$$

If these n components constitute a system that will fail as soon as a single component fails, then the probability that the system lasts past time t is

$$P(X_1 > t, \ldots, X_n > t) = \int_t^\infty \cdots \int_t^\infty f(x_1, \ldots, x_n) \, dx_1, \ldots, dx_n$$

$$= \left(\int_t^\infty \lambda e^{-\lambda x_1} \, dx_1 \right) \cdots \left(\int_t^\infty \lambda e^{-\lambda x_n} \, dx_n \right)$$

$$= (e^{-\lambda t})^n = e^{-n\lambda t}$$

Therefore

$$P(\text{system lifetime} \leq t) = 1 - e^{-n\lambda t} \quad \text{for} \quad t \geq 0$$

which shows that system lifetime has an exponential distribution with parameter λn. ∎

In many experimental situations to be considered in this book, independence is a reasonable assumption, so that specifying the joint distribution reduces to deciding on appropriate marginal distributions.

Exercises / Section 5.1 (1–14)

1. A service station has both self-service and full-service islands. On each island there is a single regular unleaded pump with two hoses. Let X denote the number of hoses being used on the self-service island at a particular time, and let Y denote the number of hoses on the full-service island in use at that time. The joint p.m.f. of X and Y appears in the accompanying tabulation.

$p(x, y)$	y 0	1	2
x 0	.10	.04	.02
1	.08	.20	.06
2	.06	.14	.30

a. What is $P(X = 1$ and $Y = 1)$?

b. Compute $P(X \leq 1$ and $Y \leq 1)$.

c. Give a word description of the event $\{X \neq 0$ and $Y \neq 0\}$, and compute the probability of this event.

d. Compute the marginal p.m.f. of X and of Y. Using $p_X(x)$, what is $P(X \leq 1)$?

e. Are X and Y independent random variables? Explain.

2. When an automobile is stopped by a roving safety patrol, each tire is checked for tire wear and each headlight is checked to see whether it is properly aimed. Let X denote the number of headlights that need adjustment and let Y denote the number of defective tires.

a. If X and Y are independent with $p_X(0) = .5$, $p_X(1) = .3$, $p_X(2) = .2$, and $p_Y(0) = .6$, $p_Y(1) = .1, p_Y(2) = p_Y(3) = .05, p_Y(4) = .2$, display the joint p.m.f. of (X, Y) in a joint probability table.

b. Compute $P(X \leq 1$ and $Y \leq 1)$ from the joint probability table and verify that it equals $P(X \leq 1) \cdot P(Y \leq 1)$.

c. What is $P(X + Y = 0)$ (the probability of no violations)?

d. Compute $P(X + Y \leq 1)$.

3. A fair four-sided (tetrahedral) die is rolled twice independently; possible outcomes on each toss are 1, 2, 3, and 4.
   a. Let X = the number of times that a one results, and Y = the number of times that a two results. What is the joint p.m.f. of X and Y?
   b. Let X = the outcome of the first toss, and Y = the outcome of the second toss. What is the joint p.m.f. of X and Y?

4. The joint probability distribution of the number of cars X and the number of buses Y per signal cycle at a proposed left turn lane is displayed in the accompanying joint probability table.

$p(x, y)$		0	1	2
			y	
	0	.025	.015	.010
	1	.050	.030	.020
	2	.125	.075	.050
x	3	.150	.090	.060
	4	.100	.060	.040
	5	.050	.030	.020

   a. What is the probability that there is exactly one car and exactly one bus during a cycle?
   b. What is the probability that there is at most one car and at most one bus during a cycle?
   c. What is the probability that there is exactly one car during a cycle? Exactly one bus?
   d. Suppose that the left turn lane is to have a capacity of five cars, and one bus is equivalent to three cars. What is the probability of an overflow during a cycle?
   e. Are X and Y independent random variables? Explain.

5. A college bookstore carries three different editions of Shakespeare's *Hamlet*. At the beginning of the term it has on the shelf eight copies of publisher A's edition, 10 copies of publisher B's, and 12 copies of publisher C's. Six students have signed up for a course in which *Hamlet* will be read (this is a technical college). Each enters the bookstore and randomly selects a book. Let X = the number of publisher A's edition selected, Y = the number of publisher B's edition selected, and $p(x, y)$ denote the joint p.m.f. of X and Y.
   a. What is $p(3, 2)$? *Hint:* Each sample of size six is equally likely to be selected. Therefore $p(3, 2)$ = (number of outcomes with $X = 3$

and $Y = 2$)/(total number of outcomes). Now use the product rule for counting to obtain the numerator and denominator.
   b. Using the logic of (a), obtain $p(x, y)$. (This can be thought of as a multivariate hypergeometric distribution—sampling without replacement from a finite population consisting of more than two categories.)

6. Each front tire on a particular type of automobile is supposed to be filled to a pressure of 26 p.s.i. Suppose that the actual air pressure in each tire is a r.v.—X for the right tire and Y for the left tire, with joint p.d.f.

$$f(x, y) = \begin{cases} K(x^2 + y^2) & 20 \leq x \leq 30, 20 \leq y \leq 30 \\ 0 & \text{otherwise} \end{cases}$$

   a. What is the value of K?
   b. What is the probability that both tires are underfilled?
   c. What is the probability that the difference in air pressure between the two tires is at most 2 p.s.i.?
   d. Determine the (marginal) distribution of air pressure in the right tire alone.
   e. Are X and Y independent random variables?

7. Annie and Alvie have agreed to meet between 5:00 P.M. and 6:00 P.M. for dinner at a local health food restaurant. Let X = Annie's arrival time and Y = Alvie's arrival time, and suppose that X and Y are independent with each uniformly distributed on the interval $[5, 6]$.
   a. What is the joint p.d.f. of X and Y?
   b. What is the probability that they both arrive between 5:15 and 5:45?
   c. If the first one to arrive will wait only 10 minutes before leaving to eat elsewhere, what is the probability that they have dinner at the health food restaurant? *Hint:* The event of interest is $A = \{(x, y): |x - y| \leq \frac{1}{6}\}$.

8. A professor has just given one long technical paper to one typist and a somewhat shorter paper to another typist. Let X = the number of typing errors on the first paper, and Y the number of typing errors on the second paper. Suppose that X has a Poisson distribution with parameter λ, Y

has a Poisson distribution with parameter μ, and that X and Y are independent.

a. What is the joint p.m.f. of X and Y?

b. What is the probability that at most one error is made on both papers combined?

c. Obtain a general expression for the probability that the total number of errors in the two papers is m (where m is a nonnegative integer). *Hint:* $A = \{(x, y): x + y = m\} = \{(m, 0), (m - 1, 1), \ldots, (1, m - 1), (0, m)\}$. Now sum the joint p.m.f. over $(x, y) \in A$ and use the binomial theorem, which says that for any a, b

$$\sum_{k=0}^{m} \binom{m}{k} a^k b^{m-k} = (a + b)^m$$

9. Two components of a mini-computer have the following joint p.d.f. for their useful lifetimes X and Y:

$$f(x, y) = \begin{cases} xe^{-x(1+y)} & x \geq 0 \text{ and } y \geq 0 \\ 0 & \text{otherwise} \end{cases}$$

a. What is the probability that the lifetime X of the first component exceeds 3?

b. What are the marginal p.d.f.'s of X and Y? Are the two lifetimes independent? Explain.

c. What is the probability that the lifetime of at least one component exceeds 3?

10. You have two light bulbs for a particular lamp. Let $X =$ the lifetime of the first bulb, and $Y =$ the lifetime of the second bulb (both in 1000's of hours). Suppose that X and Y are independent and that each has an exponential distribution with parameter $\lambda = 1$.

a. What is the joint p.d.f. of X and Y?

b. What is the probability that each bulb lasts at most 1000 hours (that is, $X \leq 1$ and $Y \leq 1$)?

c. What is the probability that the total lifetime of the two bulbs is at most 2? *Hint:* Draw a picture of the region $A = \{(x, y): x \geq 0, y \geq 0, x + y \leq 2\}$ before integrating.

d. What is the probability that the total lifetime is between 1 and 2?

11. Suppose that you have 10 light bulbs, that the lifetime of each is independent of all the other lifetimes, and that each lifetime has an exponential distribution with parameter λ.

a. What is the probability that all 10 bulbs fail before time t?

b. What is the probability that exactly k of the 10 bulbs fail before time t?

c. Suppose that 9 of the bulbs have lifetimes that are exponentially distributed with parameter λ, and that the remaining bulb has a lifetime that is exponentially distributed with parameter μ (it's made by another manufacturer). What is the probability that exactly 5 of the 10 bulbs fail before time t?

12. Consider a system consisting of three components as pictured. The system will continue to function as long as the first component functions and either component 2 or component 3 functions. Let X_1, X_2, and X_3 denote the lifetimes of components 1, 2, and 3, respectively. Suppose that the X_i's are independent of one another and that each X_i has an exponential distribution with parameter λ.

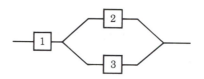

a. Let Y denote the system lifetime. Obtain the cumulative distribution function of Y and differentiate it to obtain the p.d.f. *Hint:* $F(y) = P(Y \leq y)$; express the event $\{Y \leq y\}$ in terms of unions and/or intersections of the three events $\{X_1 \leq y\}$, $\{X_2 \leq y\}$, and $\{X_3 \leq y\}$.

b. Compute the expected system lifetime.

13. a. For $f(x_1, x_2, x_3)$ as given in Example 5.10, compute the **joint marginal density function** of X_1 and X_3 alone (by integrating over x_2).

b. What is the probability that rocks of type 1 and 3 together make up at most 50% of the sample? *Hint:* Use the result of (a).

c. Compute the marginal p.d.f. of X_1 alone. *Hint:* Use the result of (a).

14. An ecologist wishes to select a point inside a circular sampling region according to a uniform distribution (in practice this could be done by first selecting a direction and then a distance from the center in that direction). Let $X =$ the x coordinate of the point selected, and $Y =$ the y coordinate of the point selected. If the circle is centered at $(0, 0)$ and has radius R, then the joint p.d.f. of X and Y is

$$f(x, y) = \begin{cases} \dfrac{1}{\pi R^2} & x^2 + y^2 \leq R^2 \\ 0 & \text{otherwise} \end{cases}$$

a. What is the probability that the selected point is within $R/2$ of the center of the circular re-

gion? *Hint:* Draw a picture of the region of positive density D. Because $f(x, y)$ is constant on D, computing a probability reduces to computing an area.

b. What is the probability that both X and Y differ from 0 by at most $R/2$?

c. Answer (b) for $R/\sqrt{2}$ replacing $R/2$.

d. What is the marginal p.d.f. of X? Of Y? are X and Y independent?

5.2 Expected Values, Covariance, and Correlation

We previously saw that any function $h(X)$ of a single r.v. X is itself a random variable. However, to compute $E[h(X)]$, it was not necessary to obtain the probability distribution of $h(X)$; instead $E[h(X)]$ was computed as a weighted average of $h(x)$ values, where the weight function was the p.m.f. $p(x)$ or p.d.f. $f(x)$ of X. A similar result holds for a function $h(X, Y)$ of two jointly distributed random variables.

Proposition

> Let X and Y be jointly distributed random variables with p.m.f. $p(x, y)$ or p.d.f. $f(x, y)$ according to whether the variables are discrete or continuous. Then the expected value of a function $h(X, Y)$, denoted by $E[h(X, Y)]$ or $\mu_{h(X, Y)}$, is given by
>
> $$E[h(X, Y)]$$
> $$= \begin{cases} \displaystyle\sum_x \sum_y h(x, y) \cdot p(x, y) & \text{if } X \text{ and } Y \text{ are discrete} \\ \displaystyle\int_{-\infty}^{\infty} \int_{-\infty}^{\infty} h(x, y) \cdot f(x, y) \, dx \, dy & \text{if } X \text{ and } Y \text{ are continuous} \end{cases}$$

Example 5.12

Five friends have purchased tickets to a certain concert. If the tickets are for seats 1–5 in a particular row and the tickets are randomly distributed among the five, what is the expected number of seats separating any particular two of the five? Let X and Y denote the seat number of the first and second individuals, respectively. Possible (X, Y) pairs are $\{(1, 2), (1, 3), \ldots, (5, 4)\}$, and the joint p.m.f. of (X, Y) is

$$p(x, y) = \begin{cases} \dfrac{1}{20} & x = 1, \ldots, 5; y = 1, \ldots, 5; x \neq y \\ 0 & \text{otherwise} \end{cases}$$

The number of seats separating the two individuals is $h(X, Y) = |X - Y| - 1$. The accompanying table gives $h(x, y)$ for each possible (x, y) pair.

$$x$$

$h(x, y)$	1	2	3	4	5
1	—	0	1	2	3
2	0	—	0	1	2
y 3	1	0	—	0	1
4	2	1	0	—	0
5	3	2	1	0	—

Thus

$$E[h(X, Y)] = \sum_{(x, y)} \sum h(x, y) \cdot p(x, y) = \sum_{\substack{x = 1 \\ x \neq y}}^{5} \sum_{y = 1}^{5} (|x - y| - 1) \cdot \frac{1}{20} = 1$$

■

Example 5.13   In Example 5.5 of the previous section, the joint p.d.f. of the amount X of almonds and amount Y of cashews in a one-pound can of nuts was

$$f(x, y) = \begin{cases} 24xy & 0 \leq x \leq 1, 0 \leq y \leq 1, x + y \leq 1 \\ 0 & \text{otherwise} \end{cases}$$

If a pound of almonds costs the company \$1.00, a pound of cashews costs \$1.50, and a pound of peanuts costs \$.50, then the total cost of the contents of a can is

$$h(X, Y) = (1)X + (1.5)Y + (.5)(1 - X - Y) = .5 + .5X + Y$$

(since $1 - X - Y$ of the weight consists of peanuts). The expected total cost is

$$E[h(X, Y)] = \int_{-\infty}^{\infty} \int_{-\infty}^{\infty} h(x, y) \cdot f(x, y) \, dx \, dy$$

$$= \int_{0}^{1} \int_{0}^{1 - x} (.5 + .5x + y) \cdot 24xy \, dy \, dx = \$1.10$$

■

The method of computing the expected value of a function $h(X_1, \ldots, X_n)$ of n random variables is similar to that for two random variables. If the X_i's are discrete, $E[h(X_1, \ldots, X_n)]$ is an n-dimensional sum, and if the X_i's are continuous, it is an n-dimensional integral.

Covariance

When two random variables X and Y are not independent, it is frequently of interest to measure how strongly they are related to one another.

Definition

The **covariance** between two random variables X and Y is

$$\text{Cov}(X, Y) = E[(X - \mu_X)(Y - \mu_Y)]$$

$$= \begin{cases} \sum_x \sum_y (x - \mu_X)(y - \mu_Y) p(x, y) & X, Y \text{ discrete} \\ \int_{-\infty}^{\infty} \int_{-\infty}^{\infty} (x - \mu_X)(y - \mu_Y) f(x, y) \, dx \, dy & X, Y \text{ continuous} \end{cases}$$

The rationale for the definition is as follows. Suppose that X and Y have a strong positive relationship to one another, by which we mean that large values of X tend to occur with large values of Y and small values of X with small values of Y. Then most of the probability mass or density will be associated with $(x - \mu_X)$ and $(y - \mu_Y)$ either both positive (both X and Y above their respective means) or both negative, so the product $(x - \mu_X)(y - \mu_Y)$ will tend to be positive. Thus for a strong positive relationship, $\text{Cov}(X, Y)$ should be quite positive. For a strong negative relationship, the signs of $(x - \mu_X)$ and $(y - \mu_Y)$ will tend to be opposite to one another, yielding a negative product. Thus for a strong negative relationship, $\text{Cov}(X, Y)$ should be quite negative. If X and Y are not strongly related, positive and negative products will tend to cancel one another, yielding a covariance near 0. Figure 5.4 illustrates the different possibilities. The covariance depends *both* on the set of possible pairs and on the probabilities. In Figure 5.4 the probabilities could be changed without altering the set of possible pairs, and this could drastically change the value of $\text{Cov}(X, Y)$.

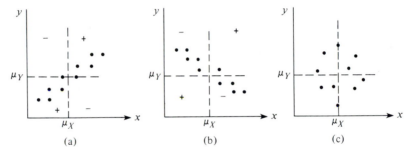

Figure 5.4 $p(x, y) = \frac{1}{10}$ for each of ten pairs corresponding to indicated points; (a) positive covariance; (b) negative covariance; (c) covariance near zero

Example 5.14   The joint and marginal p.m.f.'s for $X =$ automobile policy deductible amount and $Y =$ homeowner policy deductible amount in Example 5.1 were

$p(x, y)$		0	100	200
			y	
x	100	.20	.10	.20
	250	.05	.15	.30

x	100	250
$p_X(x)$	.5	.5

y	0	100	200
$p_Y(y)$	.25	.25	.5

from which $\mu_X = \Sigma x p_X(x) = 175$ and $\mu_Y = 125$. Therefore

$$\text{Cov}(X, Y) = \sum\sum_{(x,y)} (x - 175)(y - 125)p(x, y)$$

$$= (100 - 175)(0 - 125)(.20) + \cdots$$
$$+ (250 - 175)(200 - 125)(.30) = 1875$$

∎

The following shortcut formula for $\text{Cov}(X, Y)$ simplifies the computations.

Proposition

$$\text{Cov}(X, Y) = E(XY) - \mu_X \cdot \mu_Y$$

According to this formula, no intermediate subtractions are necessary; only at the end of the computation is $\mu_X \cdot \mu_Y$ subtracted from $E(XY)$. The proof involves expanding $(X - \mu_X)(Y - \mu_Y)$ and then taking the expected value of each term separately. Note that $\text{Cov}(X, X) = E(X^2) - \mu_X^2 = V(X)$.

Example 5.15
(Example 5.9 continued)

The joint and marginal p.m.f.'s of $X =$ amount of almonds and $Y =$ amount of cashews were

$$f(x, y) = \begin{cases} 24xy & 0 \leq x \leq 1, 0 \leq y \leq 1, x + y \leq 1 \\ 0 & \text{otherwise} \end{cases}$$

$$f_X(x) = \begin{cases} 12x(1 - x)^2 & 0 \leq x \leq 1 \\ 0 & \text{otherwise} \end{cases}$$

with $f_Y(y)$ obtained by replacing x by y in $f_X(x)$. It is easily verified that $\mu_X = \mu_Y = \frac{2}{5}$, and

$$E(XY) = \int_{-\infty}^{\infty} \int_{-\infty}^{\infty} xy f(x, y) \, dx \, dy = \int_0^1 \int_0^{1-x} xy \cdot 24xy \, dy \, dx$$

$$= 8 \int_0^1 x^2 (1 - x)^3 \, dx = \frac{2}{15}$$

Thus $\text{Cov}(X, Y) = \frac{2}{15} - (\frac{2}{5})(\frac{2}{5}) = \frac{2}{15} - \frac{4}{25} = -\frac{2}{75}$. A negative covariance is reasonable here, since more almonds in the can implies fewer cashews. ∎

It would appear that the relationship in the insurance example is quite strong, since $\text{Cov}(X, Y) = 1875$, while $\text{Cov}(X, Y) = -\frac{2}{75}$ in the nut example would seem to imply quite a weak relationship. Unfortunately the covariance has a serious defect that makes it impossible to interpret a computed value of the covariance. In the insurance example, suppose that we had measured deductible amount in cents rather than dollars. Then $100X$ would replace X, $100Y$ would replace Y, and the resulting covariance would be $\text{Cov}(100X, 100Y) = (100)(100)\text{Cov}(X, Y) = 18{,}750{,}000$. If, on the other hand, the deductible amount had been measured in 100's of dollars, the computed covariance would have been $(.01)(.01)(1875) = .1875$. *The defect of covariance is that its computed value depends critically on the units of measurement.* Ideally the choice of units should have no effect on a measure of strength of relationship. This is achieved by scaling the covariance.

Correlation

Definition

> The **correlation coefficient** of X and Y, denoted by $\text{Corr}(X, Y)$, $\rho_{X,Y}$, or just ρ, is defined by
>
> $$\rho_{X,Y} = \frac{\text{Cov}(X, Y)}{\sigma_X \cdot \sigma_Y}$$

Example 5.16

It is easily verified that in the insurance problem of Example 5.14, $E(X^2) = 36{,}250$, $\sigma_X^2 = 36{,}250 - (175)^2 = 5625$, $\sigma_X = 75$, $E(Y^2) = 22{,}500$, $\sigma_Y^2 = 6875$ and $\sigma_Y = 82.92$. This gives

$$\rho = \frac{1875}{(75)(82.92)} = .301 \qquad\blacksquare$$

The following proposition shows that ρ remedies the defect of $\text{Cov}(X, Y)$ and also suggests how to recognize the existence of a strong relationship.

Proposition

> **1.** If a and c are either both positive or both negative
>
> $$\text{Corr}(aX + b, cY + d) = \text{Corr}(X, Y)$$
>
> **2.** For any two r.v.'s X and Y, $-1 \le \text{Corr}(X, Y) \le 1$.

Statement 1 says precisely that the correlation coefficient is not affected by a linear change in the units of measurement (if, say, $X = $ temperature in $°C$, then $9X/5 + 32 = $ temperature in $°F$). According to statement 2 the strongest possible positive relationship is evidenced by $\rho = +1$, while the strongest possible negative relationship corresponds to $\rho = -1$. The proof of the first statement is sketched out in Exercise 5.27, while that of the second appears in a supplementary exercise at the end of the chapter. For descriptive purposes, the relationship will be described as strong if $|\rho| \ge .8$, moderate if $.5 < |\rho| < .8$, and weak if $|\rho| \le .5$.

If we think of $p(x, y)$ or $f(x, y)$ as prescribing a mathematical model for how the two numerical variables X and Y are distributed in some population (height and weight, verbal SAT score and quantitative SAT score, and the like), then ρ is a population characteristic or parameter that measures how strongly X and Y are related in the population. In Chapter 12 we will consider taking a sample of pairs $(x_1, y_1), \ldots, (x_n, y_n)$ from the population. The sample correlation coefficient r will then be defined and used to make inferences about ρ.

The correlation coefficient ρ is actually not a completely general measure of the strength of a relationship.

Proposition

> **1.** If X and Y are independent, then $\rho = 0$, but $\rho = 0$ does not imply independence.
> **2.** $\rho = 1$ or -1 iff $Y = aX + b$ for some numbers a and b with $a \neq 0$.

This proposition says that ρ is a measure of the degree of **linear** relationship between X and Y, and only when the two variables are perfectly related in a linear manner will ρ be as positive or negative as it can be. A ρ less than 1 in absolute value indicates only that the relationship is not completely linear, but there may still be a very strong nonlinear relation. Also, $\rho = 0$ does not imply that X and Y are independent, but only that there is complete absence of a linear relationship. When $\rho = 0$, X and Y are said to be **uncorrelated.** Two variables could be uncorrelated yet highly dependent because there is a strong nonlinear relationship, so be careful not to conclude too much from knowing that $\rho = 0$.

Example 5.17   Let X and Y be discrete r.v.'s with joint p.m.f.

$$p(x, y) = \begin{cases} \dfrac{1}{4} & (x, y) = (-4, 1), (4, -1), (2, 2), (-2, -2) \\ 0 & \text{otherwise} \end{cases}$$

The points that receive positive probability mass are identified on the (x, y) coordinate system in Figure 5.5. It is evident from the figure that the value of X is completely determined by the value of Y and vice versa, so the two variables are completely dependent. However, by symmetry $\mu_X = \mu_Y = 0$ and $E(XY) = (-4)\frac{1}{4} + (-4)\frac{1}{4} + (4)\frac{1}{4} + (4)\frac{1}{4} = 0$, so $\text{Cov}(X, Y) = E(XY) - \mu_X \cdot \mu_Y = 0$ and thus $\rho_{X, Y} = 0$. Although there is perfect dependence, there is also complete absence of any linear relationship!

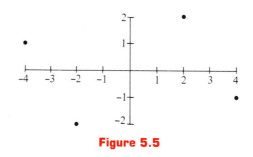

Figure 5.5   ■

A value of ρ near 1 does not necessarily imply that increasing the value of X *causes* Y to increase. It only implies that large X values are *associated* with large Y values. For example, in the population of children, vocabulary size and number of cavities are quite positively correlated, but it is certainly not true that cavities cause vocabulary to grow. Instead, the values of both these variables tend to increase as the value of age, a third variable, increases. For children of a fixed age, there is probably a very low correlation between number of cavities and vocabulary size. In summary, association (a high correlation) is not the same as causation.

Exercises / Section 5.2 (15–28)

15. An instructor has given a short quiz consisting of two parts. For a randomly selected student, let X = the number of points earned on the first part, Y = the number of points earned on the second part, and suppose that the joint p.m.f. of X and Y is given in the accompanying tabulation.

			y		
$p(x, y)$		0	5	10	15
	0	.02	.06	.02	.10
x	5	.04	.15	.20	.10
	10	.01	.15	.14	.01

a. If the score recorded in the grade book is the total number of points earned on the two parts, what is the expected recorded score $E(X + Y)$?

b. If the maximum of the two scores is recorded, what is the expected recorded score?

16. Six individuals, including A and B, take seats around a circular table in a completely random fashion. Suppose that the seats are numbered $1, \ldots, 6$ and let X = A's seat number and Y = B's seat number. If A sends a written message around the table to B in the direction in which they are closest, how many individuals (including A and B) would you expect to handle the message?

17. A surveyor wishes to lay out a square region with each side having length L. However, because of measurement error, he instead lays out a rectangle in which the north/south sides both have length X and the east/west sides both have length Y. Suppose that X and Y are independent and that each is uniformly distributed on the interval $[L - A, L + A]$ (where $0 < A < L$). What is the expected area of the resulting rectangle?

18. Consider a small ferry that can accommodate cars and buses. The toll for cars is $3 and the toll for buses is $10. Let X and Y denote the number of cars and buses, respectively, carried on a single trip, and suppose that the joint distribution of

X and Y is as given in the table of Exercise 4. Compute the expected revenue from a single trip.

19. Annie and Alvie have agreed to meet for lunch between noon (0:00 P.M.) and 1:00 P.M. Denote Annie's arrival time by X, Alvie's by Y, and suppose that X and Y are independent with p.d.f.'s

$$f_X(x) = \begin{cases} 3x^2 & 0 \le x \le 1 \\ 0 & \text{otherwise} \end{cases}$$

$$f_Y(y) = \begin{cases} 2y & 0 \le y \le 1 \\ 0 & \text{otherwise} \end{cases}$$

What is the expected amount of time that the one who arrives first must wait for the other person? *Hint:* $h(X, Y) = |X - Y|$.

20. Show that if X and Y are independent random variables, then $E(XY) = E(X) \cdot E(Y)$.

21. Compute the correlation coefficient ρ for X and Y of Example 5.15 (the covariance has already been computed).

22. a. Compute the covariance for X and Y in Exercise 15.

b. Compute ρ for X and Y in the same exercise.

23. a. Compute the covariance between X and Y in Exercise 6.

b. Compute the correlation coefficient ρ for this X and Y.

24. Compute the value of the correlation coefficient for lifetimes of the two mini-computer components as described in Exercise 9.

25. Use the result of Exercise 20 to show that when X and Y are independent, $\text{Cov}(X, Y) = \text{Corr}(X, Y) = 0$.

26. a. Recalling the definition of σ^2 for a single r.v. X, write a formula that would be appropriate for computing the variance of a function $h(X, Y)$ of two r.v.'s. *Hint:* Remember that variance is just a special expected value.

b. Use this formula to compute the variance of the recorded score $h(X, Y)$ [$= \max(X, Y)$] in (b) of Exercise 15.

27. a. Use the rules of expected value to show that $\text{Cov}(aX + b, cY + d) = ac \, \text{Cov}(X, Y)$.
   b. Use (a) along with the rules of variance and standard deviation to show that $\text{Corr}(aX + b,$

$cY + d) = \text{Corr}(X, Y)$ when a and c have the same sign.
   c. What happens if a and c have opposite signs?

28. Show that if $Y = aX + b$ $(a \neq 0)$, then $\text{Corr}(X, Y) = +1$ or -1. Under what conditions will $\rho = +1$?

5.3   Sums and Averages of Random Variables

In many statistics problems the observations in a sample can be thought of as observed values of a sequence of random variables. In this section we let $X_1, X_2, \ldots, X_n$ be such a sequence and define and study several new random variables derived from the sequence.

Definition

> Let $X_1, X_2, \ldots, X_n$ be a collection of n random variables. These random variables are said to constitute a **random sample** of size n if (a) the X_i's are independent random variables, and (b) every X_i has the same probability distribution.

When (a) and (b) are satisfied, the X_i's are said to be **independent and identically distributed,** abbreviated **i.i.d.** If sampling is with replacement or from an infinite conceptual population, (a) and (b) are satisfied exactly. If the sample is taken without replacement from a finite population with the population size N much larger than the sample size n, (a) and (b) are approximately satisfied, so for all practical purposes the X_i's can be regarded as a random sample. In most problems we will proceed with our analysis by making the assumption that we have a random sample. A random sample will thus describe a collection of observations taken from the same population. The observed values of $X_1, X_2, \ldots, X_n$ will be denoted by $x_1, x_2, \ldots, x_n$.

The Distribution of the Sample Mean and Sample Total

Definition

> The random variables
> $$\overline{X} = \frac{1}{n} \sum_{i=1}^{n} X_i \quad \text{and} \quad T_o = \sum_{i=1}^{n} X_i$$
> are referred to as the **sample mean** and **sample total,** respectively, of the sample $X_1, \ldots, X_n$.

In Chapter 1 the sample mean was an average of a fixed set of numbers. Now we are thinking of the sample mean $\overline{X}$ before the X_i's are observed, so that the observed value $\overline{x}$ is not yet known. Information about the probability distribution of $\overline{X}$ will be used to specify inferential procedures in subsequent chapters.

Example 5.18 A large automobile service center charges \$40, \$45, and \$50 for a tuneup of 4-, 6-, and 8-cylinder cars, respectively. If 20% of its tuneups are done on 4-cylinder cars, 30% on 6-cylinder cars, and 50% on 8-cylinder cars, then the probability distribution of revenue from a single randomly selected tuneup is given by

x	40	45	50
$p(x)$	.2	.3	.5

with $\mu = 46.5$, $\sigma^2 = 15.25$ (5.2)

Suppose that on a particular day only two servicing jobs involve tuneups. Let X_1 = the revenue from the first tuneup, X_2 = the revenue from the second, and suppose that X_1 and X_2 are independent, each having probability distribution shown in (5.2) [so that X_1 and X_2 constitute a random sample from the distribution (5.2)]. Table 5.1 lists possible (x_1, x_2) pairs, the probability of each (computed using (5.2) and the assumption of independence), and the resulting t (total) and $\overline{x}$ values. Now to obtain the probability distribution of $\overline{X}$, the sample average revenue per tuneup, we must consider each possible value $\overline{x}$ and compute its probability. For example, $\overline{x} = 45$ occurs three times in the table with probabilities .10, .09, and .10, so $P(\overline{X} = 45) = .10 + .09 + .10 = .29$. Proceeding in this manner, we obtain the p.m.f. $p_{\overline{X}}(\overline{x})$ of $\overline{X}$ and $p_{T_o}(t)$ of the sample total revenue T_o.

Table 5.1

x_1	x_2	$p(x_1, x_2)$	t	$\overline{x}$
40	40	.04	80	40
40	45	.06	85	42.5
40	50	.10	90	45
45	40	.06	85	42.5
45	45	.09	90	45
45	50	.15	95	47.5
50	40	.10	90	45
50	45	.15	95	47.5
50	50	.25	100	50

$\overline{x}$	40	42.5	45	47.5	50
$p_{\overline{X}}(\overline{x})$	.04	.12	.29	.30	.25

(5.3)

t	80	85	90	95	100
$p_{T_o}(t)$	.04	.12	.29	.30	.25

(5.4)

Figure 5.6 pictures a probability histogram both for the original distribution (5.2) and the $\overline{X}$ distribution (5.3). Figure 5.6 suggests first that the mean (expected value) of the $\overline{X}$ distribution is equal to the mean 46.5 of the original distribution, since both histograms appear to be centered at the same place.

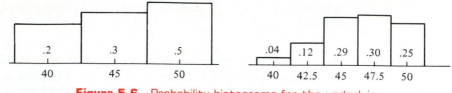

Figure 5.6   Probability histograms for the underlying distribution and $\overline{X}$ distribution in Example 5.18

From (5.3)

$$\mu_{\overline{X}} = E(\overline{X}) = \Sigma \overline{x}\, p_{\overline{X}}(\overline{x}) = (40)(.04) + \cdots + (50)(.25) = 46.5 = \mu$$

Second, it appears that the $\overline{X}$ distribution has smaller spread (variability) than the original distribution, since probability mass has moved in toward the mean. Again from (5.3)

$$\sigma_{\overline{X}}^2 = V(\overline{X}) = \Sigma \overline{x}^2 \cdot p_{\overline{X}}(\overline{x}) - \mu_{\overline{X}}^2$$

$$= (40)^2(.04) + \cdots + (50)^2(.25) - (46.5)^2$$

$$= 7.625 = \frac{(15.25)}{2} = \frac{\sigma^2}{2}$$

The variance of $\overline{X}$ is precisely half that of the original variance. Similar calculations with $p_{T_o}(t)$ yield $\mu_{T_o} = E(T_o) = 93 = 2\mu$ and $\sigma_{T_o}^2 = 30.5 = 2\sigma^2$, so both the expected value and variance of T_o are twice the mean and variance of the original distribution.

If four tuneups had been done on the day of interest, the sample average revenue $\overline{X}$ would be based on a random sample of four X_i's, each having distribution (5.2). More calculation eventually yields the p.m.f. of $\overline{X}$ for $n = 4$ as

$\overline{x}$	40	41.25	42.5	43.75	45	46.25	47.5	48.75	50
$p_{\overline{X}}(\overline{x})$	.0016	.0096	.0376	.0936	.1761	.2340	.2350	.1500	.0625

From this, for $n = 4$, $\mu_{\overline{X}} = 46.50 = \mu$ and $\sigma_{\overline{X}}^2 = 3.8125 = \sigma^2/4$. A probability histogram of this p.m.f. appears in Figure 5.7.

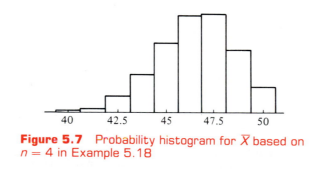

Figure 5.7   Probability histogram for $\overline{X}$ based on $n = 4$ in Example 5.18   ■

Example 5.18 should suggest first of all that the computation of $p_{\overline{X}}(\overline{x})$ and $p_{T_o}(t)$ can be tedious. If the original distribution (5.2) had allowed for more than the three possible values 40, 45, and 50, then even for $n = 2$ the computations would have been more involved. The example should also suggest, however, that there are some general relationships between $E(\overline{X})$, $V(\overline{X})$, $E(T_o)$, $V(T_o)$, and the mean μ and variance σ^2 of the original distribution. Before developing these relationships, consider an example in which the random sample is drawn from a continuous distribution.

Example 5.19 The time that it takes to serve a customer at the cash register in a mini-market is a random variable having an exponential distribution with parameter λ. Suppose that X_1 and X_2 are service times for two different customers, assumed independent of each other. The c.d.f. of the total service time $T_o = X_1 + X_2$ for the two customers is, for $t \geq 0$,

$$F_{T_o}(t) = P(X_1 + X_2 \leq t) = \iint\limits_{\{(x_1, x_2):\, x_1 + x_2 \leq t\}} f(x_1, x_2)\, dx_1\, dx_2$$

$$= \int_0^t \int_0^{t-x_1} \lambda e^{-\lambda x_1} \cdot \lambda e^{-\lambda x_2}\, dx_2\, dx_1 = \int_0^t [\lambda e^{-\lambda x_1} - \lambda e^{-\lambda t}]\, dx_1$$

$$= 1 - e^{-\lambda t} - \lambda t e^{-\lambda t}$$

The p.d.f. of T_o is obtained by differentiating $F_{T_o}(t)$:

$$f_{T_o}(t) = \begin{cases} \lambda^2 t e^{-\lambda t} & t \geq 0 \\ 0 & t < 0 \end{cases} \tag{5.5}$$

This is a gamma p.d.f. ($\alpha = 2$ and $\beta = 1/\lambda$). The p.d.f. of $\overline{X} = T_o/2$ is obtained from the relation $\{\overline{X} \leq \overline{x}\}$ iff $\{T_o \leq 2\overline{x}\}$ as

$$f_{\overline{X}}(\overline{x}) = \begin{cases} 4\lambda^2 \overline{x} e^{-2\lambda \overline{x}} & \overline{x} \geq 0 \\ 0 & \overline{x} < 0 \end{cases} \tag{5.6}$$

The mean and variance of the underlying exponential distribution are $\mu = 1/\lambda$ and $\sigma^2 = 1/\lambda^2$. From (5.5) and (5.6) it can be verified that $E(\overline{X}) = 1/\lambda$, $V(\overline{X}) = 1/2\lambda^2$, $E(T_o) = 2/\lambda$, and $V(T_o) = 2/\lambda^2$. These results again suggest some general relationships between means and variances of $\overline{X}$, T_o, and the underlying distribution.

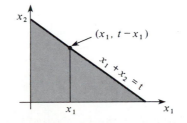

Figure 5.8 Region of integration to obtain c.d.f. of T_o in Example 5.19 ■

Rules for Expected Value

Examples 5.18 and 5.19 demonstrate that the derivation of the distribution of $\overline{X}$ or T_o can be difficult (if, say, the integration is difficult) and time consuming (though methods of mathematical statistics can sometimes be used to obtain the distributions). Given the individual p.m.f.'s or p.d.f.'s of the X_i's, and in particular $E(X_i)$ and $V(X_i)$ for each i, what can we conclude about the distributions of $\overline{X}$ and T_o? To answer this, we first note that $\overline{X}$ and T_o are special cases of a general type of function of X_i's.

Definition

> Given a collection of n random variables $X_1, \ldots, X_n$ and n numerical constants $a_1, \ldots, a_n$, the random variable
>
> $$Y = a_1 X_1 + \cdots + a_n X_n = \sum_{i=1}^{n} a_i X_i \qquad (5.7)$$
>
> is called a **linear combination** of the X_i's.

If we take $a_1 = \cdots = a_n = 1$ in (5.7), $Y = T_o$, while $a_1 = \cdots = a_n = 1/n$ produces $Y = \overline{X}$. Thus both T_o and $\overline{X}$ are special linear combinations.

Proposition

> For any linear combination of $X_1, \ldots, X_n$, whether or not the X_i's are independent,
>
> $$E(a_1 X_1 + \cdots + a_n X_n) = a_1 E(X_1) + \cdots + a_n E(X_n) \qquad (5.8)$$

To paraphrase (5.8), the expected value of a linear combination is the same linear combination of expected values.

Proof for $n = 2$. Assuming that X_1 and X_2 are continuous with joint p.d.f. $f(x_1, x_2)$,

$$E(a_1 X_1 + a_2 X_2) = \int_{-\infty}^{\infty} \int_{-\infty}^{\infty} (a_1 x_1 + a_2 x_2) f(x_1, x_2) \, dx_1 \, dx_2$$

$$= a_1 \int_{-\infty}^{\infty} \int_{-\infty}^{\infty} x_1 f(x_1, x_2) \, dx_2 \, dx_1$$

$$+ a_2 \int_{-\infty}^{\infty} \int_{-\infty}^{\infty} x_2 f(x_1, x_2) \, dx_1 \, dx_2$$

$$= a_1 \int_{-\infty}^{\infty} x_1 f_{X_1}(x_1) \, dx_1 + a_2 \int_{-\infty}^{\infty} x_2 f_{X_2}(x_2) \, dx_2$$

$$= a_1 E(X_1) + a_2 E(X_2)$$

The general case is proved in a similar manner, with summation replacing integration when the X_i's are discrete. ∎

Example 5.20 A bookstore stocks three different hardbound dictionaries selling for $10, $12, and $15, respectively. Let X_1, X_2, and X_3 refer to the number of $10, $12, and $15 dictionaries, respectively, sold during a given period. Then the total revenue from dictionary sales is $Y = 10X_1 + 12X_2 + 15X_3$. If $E(X_1) = 13$, $E(X_2) = 10$, and $E(X_3) = 6$, then

$$E(Y) = E(10X_1 + 12X_2 + 15X_3)$$
$$= 10E(X_1) + 12E(X_2) + 15E(X_3) = 340$$

∎

Corollary

When $E(X_i) = \mu$ for $i = 1, \ldots, n$ (as would be the case for a random sample),

$$E(\overline{X}) = \mu_{\overline{X}} = \mu \qquad (5.9)$$

and

$$E(T_o) = n\mu \qquad (5.10)$$

Expression (5.9) comes from letting $a_1 = \cdots = a_n = 1/n$ in (5.8), yielding $E(\overline{X})$ on the left and $(1/n)\,\mu + \cdots + (1/n)\,\mu = \mu$ on the right. Similarly (5.10) comes from $a_i = 1$ for all i in (5.8).

Example 5.21 In a notched tensile fatigue test on a titanium specimen, the expected number of cycles to first acoustic emission (used to indicate crack initiation) is 28,000. Let $X_1, \ldots, X_{10}$ be a random sample of size 10, where each X_i is the number of cycles on a different specimen. Then the expected average number of cycles per specimen until first emission is $E(\overline{X}) = \mu = 28{,}000$, while the expected total number of cycles on all specimens is $E(T_o) = 10\mu = 280{,}000$. ∎

Rules for Variance

The expected value of $a_1 X_1 + \cdots + a_n X_n$ is the same linear combination of the individual expected values whether or not the X_i's are independent. However, while the variance of a linear combination involves only the individual variances in the case of independence, it is substantially more complicated when there is dependence among the X_i's.

Proposition

Let $X_1, \ldots, X_n$ be a collection of n random variables with $V(X_i) = \sigma_i^2$.

1. If $X_1, \ldots, X_n$ are independent, then

$$V(a_1 X_1 + \cdots + a_n X_n) = a_1^2 \sigma_1^2 + \cdots + a_n^2 \sigma_n^2$$

$$= \sum_{i=1}^{n} a_i^2 \cdot V(X_i) \tag{5.11}$$

2. For any $X_1, \ldots, X_n$,

$$V(a_1 X_1 + \cdots + a_n X_n) = \sum_{i=1}^{n} \sum_{j=1}^{n} a_i a_j \, \mathrm{Cov}(X_i, X_j) \tag{5.12}$$

The first part of this proposition is a special case of the second part; when the X_i's are independent, $\mathrm{Cov}(X_i, X_j) = 0$ for $i \neq j$ and $= V(X_i)$ for $i = j$, so (5.12) reduces to (5.11) (this simplification actually occurs when the X_i's are uncorrelated, a weaker condition than independence).

Corollary

Let $X_1, \ldots, X_n$ be a random sample (independent and identically distributed) from a distribution with variance σ^2 [so $V(X_i) = \sigma^2$, $i = 1, \ldots, n$]. Then

$$V(\overline{X}) = \sigma_{\overline{X}}^2 = \frac{\sigma^2}{n}, \quad \sigma_{\overline{X}} = \frac{\sigma}{\sqrt{n}} \tag{5.13}$$

and

$$V(T_o) = n\sigma^2, \quad \sigma_{T_o} = \sqrt{n}\,\sigma \tag{5.14}$$

Result (5.13) is extremely important and useful. It says that if σ^2 is the variance of the distribution or population being sampled and n is the size of the random sample, then the variance of the sample mean is smaller than that of the original variance by the factor $1/n$. This result along with $E(\overline{X}) = \mu$ is illustrated in Figure 5.9.

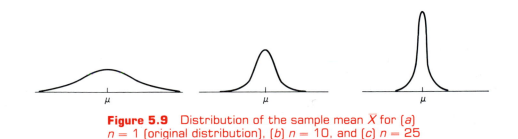

Figure 5.9   Distribution of the sample mean $\overline{X}$ for (a) $n = 1$ (original distribution), (b) $n = 10$, and (c) $n = 25$

For every value of n, the distribution is centered at μ, the mean of the population being sampled, but as n increases, more and more probability is "squeezed in" close to μ. Intuitively this is quite plausible, since the average value in a large sample should be close to the population average.

The story is different for T_o. If μ is positive, $E(T_o) = n\mu$ increases with n, so the distribution of T_o is centered further and further to the right of zero, and since $V(T_o) = n\sigma^2$, the distribution spreads out as n increases. This is because a sum is more variable than any single component in the sum.

Example 5.22
(Example 5.21 continued)

If the standard deviation of the number of cycles to first acoustic emission is $\sigma = 5000$ cycles, then for a random sample consisting of $n = 25$ observations, $\sigma_{\bar{X}} = 5000/\sqrt{25} = 1000$ and $\sigma_{T_o} = \sqrt{25}(5000) = 25{,}000$. ∎

Another special linear combination is obtained by taking $n = 2$, $a_1 = 1$, and $a_2 = -1$, so that $Y = X_1 - X_2$. Then (5.11) gives

Corollary

When X_1 and X_2 are independent, $V(X_1 - X_2) = \sigma_1^2 + \sigma_2^2$.

That is, *the variance of a difference is the sum, not the difference, of the individual variances.* If we write $X_1 - X_2$ as $X_1 + (-X_2)$, this is plausible since $-X_2$ has exactly the same amount of variability as X_2 itself.

Example 5.23

My commute time X_1 to work has expected value 25 min and standard deviation 5 min, and my commute time X_2 home from work has expected value 20 min and standard deviation 4 min. The expected difference in the two times is then

$$E(X_1 - X_2) = E(X_1) - E(X_2) = 25 - 20 = 5 \text{ min}$$

Assuming that X_1 and X_2 are independent, the variance of the difference in times is

$$V(X_1 - X_2) = \sigma_1^2 + \sigma_2^2 = (5)^2 + (4)^2 = 41$$

so

$$\sigma_{X_1 - X_2} = \sqrt{41} = 6.40 \text{ min}$$

These are also the variance and standard deviation of total commute time $T_o = X_1 + X_2$. ∎

Exercises / Section 5.3 (29–39)

29. Let X_1 and X_2 be independent, each having the probability distribution

x	1	2	3	4
$p(x)$	.2	.3	.4	.1

a. List all possible (x_1, x_2) pairs, the probability of each one, and the corresponding values of $x_1 + x_2$ and $\bar{x}$. Then write the p.m.f. of $T_o = X_1 + X_2$ and $\bar{X}$.

b. Draw probability histograms of the distributions of X_1, T_o, and $\overline{X}$. What can you say about the centers of the histograms relative to one another? Which has the smallest spread and which has the largest?

c. Compute $E(X_1)$, and then use the rules of expected value to compute $E(T_o)$ and $E(\overline{X})$.

d. Compute $V(X_1)$, and then use the rules of variance to obtain $V(T_o)$, σ_{T_o}, $V(\overline{X})$, and $\sigma_{\overline{X}}$.

30. A local clothing store carries three different brands of blue oxford cloth shirts, priced at \$12, \$15, and \$20, respectively. Based on past experience, the probability distribution of the price X paid by a randomly selected customer purchasing such a shirt is known to be

x	12	15	20
$p(x)$	.5	.2	.3

a. Let X_1 and X_2 denote the prices paid by two different randomly selected individuals for such shirts, and suppose that X_1, X_2 constitute a random sample of size $n = 2$ from the distribution. Compute the p.m.f. of $T_o = X_1 + X_2$ and of $\overline{X}$.

b. Compute $\mu = E(X)$. Then compute $\mu_{\overline{X}}$ and μ_{T_o} from the distributions derived in (a), and verify that $\mu_{\overline{X}} = \mu$ and $\mu_{T_o} = 2\mu$.

c. Compute $\sigma^2 = V(X)$. Then compute $\sigma_{\overline{X}}^2$ and $\sigma_{T_o}^2$ from the distributions derived in (a), and verify that $\sigma_{\overline{X}}^2 = \sigma^2/2$ and $\sigma_{T_o}^2 = 2\sigma^2$.

d. In the course of a week, 20 different customers purchase (independently of one another) one shirt according to the probability distribution. What is the expected total revenue from these 20 sales? What is the expected average revenue per sale?

e. As in (d), what is the variance of total revenue, what is the variance of average revenue, and what is the standard deviation of average revenue?

31. A friend of mine keeps a supply of yogurt in both his apartment refrigerator for breakfast and his office refrigerator for lunch. His favorite flavors are plain, vanilla, and strawberry, listed as containing 160, 200, and 240 calories, respectively. Suppose that he selects a container at random for breakfast and (independently) selects another for lunch, and let X_1 and X_2 denote the number

of calories in the breakfast and lunch container, respectively.

a. Compute the p.m.f. of $T_o = X_1 + X_2$ and $\overline{X}$ if each refrigerator contains two containers of plain, two of vanilla, and one of strawberry yogurt.

b. Compute the p.m.f. of T_o if the apartment refrigerator is stocked as in (a), but the office refrigerator contains two plain, one vanilla, and two strawberry. Is X_1, X_2 a random sample in this case? Why or why not?

c. Use the p.m.f. derived in (b) to compute μ_{T_o} and $\sigma_{T_o}^2$. Then verify that $\mu_{T_o} = \mu_1 + \mu_2$, where $\mu_i = E(X_i)$, and that $\sigma_{T_o}^2 = \sigma_1^2 + \sigma_2^2$, where $\sigma_i^2 = V(X_i)$.

d. Suppose that the refrigerators are stocked as in (a), but now after having selected at random for breakfast, my friend selects at random a different flavor for lunch. Are X_1 and X_2 still independent? Why or why not? Compute the p.m.f. of T_o. Hint: A tree diagram might help here.

e. In the situation described in (d), compute $\mu_{T_o} = E(X_1 + X_2)$ from the distribution that you have derived. Then compute $\mu_1 = E(X_1)$ and $\mu_2 = E(X_2)$, and verify that $\mu_{T_o} = \mu_1 + \mu_2$. Hint: You must be careful in writing down the distribution of X_2. It's best to use a tree diagram.

32. Exercise 18 introduced random variables X and Y, the number of cars and buses, respectively, carried by a ferry on a single trip. The joint p.m.f. of X and Y is given in the table in Exercise 4. It is readily verified that X and Y are independent.

a. Compute the expected value, variance, and standard deviation of the total number of vehicles on a single trip.

b. If each car is charged \$3 and each bus \$10, compute the expected value, variance, and standard deviation of the revenue resulting from a single trip.

33. Suppose that your waiting time for a bus in the morning is uniformly distributed on [0, 5], while waiting time in the evening is uniformly distributed on [0, 10] independent of morning waiting time.

a. If you take the bus each morning and evening for a week, what is your total expected waiting time? *Hint:* Define random variables $X_1, \ldots, X_{10}$ and use a rule of expected value.

b. What is the variance of your total waiting time?

c. What are the expected value and variance of the difference between morning and evening waiting time on a given day?

d. What are the expected value and variance of the difference between total morning waiting time and total evening waiting time for a particular week?

34. A store owner has purchased 20 light bulbs of type A and 25 of type B. Let $X_1, \ldots, X_{20}$ represent the 20 type-A lifetimes, assumed independent with mean and standard deviation 1000 hours and 100 hours, respectively. Similarly, let $Y_1, \ldots, Y_{25}$ denote the type-B lifetimes, assumed independent with mean and standard deviation 1200 hours and 150 hours, respectively. Use appropriate rules of expected value and variance to obtain the expected value, variance, and standard deviation of the difference $\overline{X} - \overline{Y}$ between the sample average lifetimes for the two types of bulbs.

35. Although a lecture period at a certain university lasts exactly 50 min, the actual lecture time of a statistics instructor on any particular day is a random variable (what else!) with expected value 52 min and standard deviation 2 min. If times of different lectures are independent of one another and the instructor gives 25 lectures in a particular course, compute

a. the expected total lecture time, variance of total lecture time, and standard deviation of total lecture time in the course.

b. the expected average lecture time per day, variance of average lecture time per day, and standard deviation of average lecture time per day for the 25-day course.

36. Let X_1 denote the time that you must wait before the bus that you take to work arrives, and let X_2 denote your waiting time for the bus that takes you home. Suppose that X_1 and X_2 are independent, each distributed uniformly on $[0, 5]$, and let $T_0 = X_1 + X_2$ be your total waiting time.

a. Compute the c.d.f. $F_{T_0}(t)$ of T_0. Then differentiate it to obtain $f_{T_0}(t)$. *Hint:* The region of positive density is a square with side length $= 5$, and the joint density of (X_1, X_2) is $f(x_1, x_2) = 1/25$ on this square. In computing $F_{T_0}(t) = P(T_0 \le t)$, consider separately $t \le 5$ and $t > 5$.

b. If X_1 and X_2 are independent with X_1 uniform on $[0, 5]$ and X_2 uniform on $[0, 10]$, compute $f_{T_0}(t)$.

37. In Sections 5.1 and 5.2 we considered the following joint p.d.f. for $X =$ amount of almonds and $Y =$ amount of cashews in a one-pound can.

$$f(x, y) = \begin{cases} 24\,xy & x \ge 0, y \ge 0, x + y \le 1 \\ 0 & \text{otherwise} \end{cases}$$

Compute the expected value of the total amount $X + Y$ of almonds and cashews in the can, and compute the variance of this total. *Hint:* The marginal p.d.f. of both X and Y as well as $\text{Cov}(X, Y)$ were previously computed.

38. Three different roads feed into a particular freeway entrance. Suppose that during a fixed time period, the number of cars coming from each road onto the freeway is a random variable, with expected value and standard deviation as given.

	Road 1	Road 2	Road 3
Expected value	800	1000	600
Standard deviation	16	25	18

a. What is the expected total number of cars entering the freeway at this point during the period? *Hint:* Let $X_i =$ the number from road i.

b. What is the variance of the total number of entering cars? Have you made any assumptions about the relationship between the number of cars on the different roads?

c. With X_i denoting the number of cars entering from road i during the period, suppose that $\text{Cov}(X_1, X_2) = 80$, $\text{Cov}(X_1, X_3) = 90$, and $\text{Cov}(X_2, X_3) = 100$ (so that the three streams of traffic are not independent). Compute the expected total number of entering cars and the standard deviation of the total.

39. Suppose we take a random sample of size n from a continuous distribution having median 0, so that the probability of any one observation being positive is .5. We now disregard the signs of the observations, rank them from smallest to largest in absolute value, and then let W = the sum of the ranks of the observations having positive signs. For example, if the observations are $-.3$, $+.7$, $+2.1$, and -2.5, then the ranks of positive observations are 2 and 3, so $W = 5$. In Chapter 15, W will be called *Wilcoxon's signed rank statistic*. W can be represented as follows:

$$W = 1 \cdot Y_1 + 2 \cdot Y_2 + 3 \cdot Y_3 + \cdots + n \cdot Y_n$$

$$= \sum_{i=1}^{n} i \cdot Y_i$$

where the Y_i's are independent Bernoulli random variables, each with $p = .5$ ($Y_i = 1$ corresponds to the observation with rank i being positive). Compute

a. $E(Y_i)$ and then $E(W)$ using the equation for W. *Hint:* The first n positive integers sum to $n(n + 1)/2$.

b. $V(Y_i)$ and then $V(W)$. *Hint:* The sum of the squares of the first n positive integers is $n(n + 1)(2n + 1)/6$.

5.4   The Central Limit Theorem

In the previous section we gave several examples in which the probability distributions of the sample mean $\overline{X}$ and sample total T_o were computed. In most cases these distributions are quite difficult to obtain, though we did obtain information about the expected values and variances of these variables. There is, though, one very important special case in which the distribution of any linear combination of the X_i's is easily obtained.

Linear Combinations of Normal Random Variables

Proposition

> If $X_1, X_2, \ldots, X_n$ are independent normal random variables, then any linear combination
>
> $$Y = a_1 X_1 + a_2 X_2 + \cdots + a_n X_n$$
>
> also has a normal distribution with mean $\mu_Y = \Sigma a_i \mu_i$ and variance $\sigma_Y^2 = \Sigma a_i^2 \sigma_i^2$. In particular if each X_i has the same distribution (a normal random sample) with mean μ and variance σ^2, then $\overline{X}$ is normally distributed with mean μ and variance σ^2/n, and T_o is normally distributed with mean $n\mu$ and variance $n\sigma^2$.

Figure 5.9 on p. 206 illustrates this result for $\overline{X}$. When the distribution being sampled is normal, then for any n the p.d.f. of $\overline{X}$ is also normal and centered at μ; the larger the sample size, the smaller the spread of the p.d.f. about μ.

A proof of the proposition for the linear combination $Y = a_1 X_1 + a_2 X_2$ is possible using the method in Example 5.19 but even here the details are messy. The general result is usually proved using a theoretical tool called a moment generating function. A good discussion of moment generating functions appears in both of the books listed in the chapter bibliography.

Once we know that a linear combination of interest is normally distributed, probabilities involving it are computed by standardizing exactly as before.

Example 5.24

The time that it takes a randomly selected rat of a certain subspecies to find its way through a maze is a normally distributed random variable with $\mu = 1.5$ min and $\sigma = .35$ min. Suppose that five rats are selected and let $X_1, \ldots, X_5$ denote their times in the maze. Assuming the X_i's to be a random sample from this normal distribution, what is the probability that the total time $T_o = X_1 + \cdots + X_5$ for the five is between 6 and 8 min? By the proposition, T_o has a normal distribution with $\mu_{T_o} = n\mu = 5(1.5) = 7.5$ and variance $\sigma_{T_o}^2 = n\sigma^2 = 5(.1225) = .6125$, so $\sigma_{T_o} = .783$. To standardize T_o, subtract μ_{T_o} and divide by σ_{T_o}:

$$P(6 \leq T_o \leq 8) = P\left(\frac{6 - 7.5}{.783} \leq Z \leq \frac{8 - 7.5}{.783}\right)$$
$$= P(-1.92 \leq Z \leq .64) = \Phi(.64) - \Phi(-1.92) = .7115$$

Determination of the probability that the sample average time $\overline{X}$ (a normally distributed variable) is at most 2.0 min requires $\mu_{\overline{X}} = \mu = 1.5$ and $\sigma_{\overline{X}} = \sigma/\sqrt{n} = .35/\sqrt{5} = .1565$. Then

$$P(\overline{X} \leq 2.0) = P\left(Z \leq \frac{2.0 - 1.5}{.1565}\right) = P(Z \leq 3.19) = \Phi(3.19) = .9993$$

Example 5.25

Pistons used in a certain type of engine have diameters that are known to follow a normal distribution with mean value 22.40 and standard deviation .03. Cylinders for this type of engine have normally distributed diameters with mean 22.50 and standard deviation .04. What is the probability that a randomly chosen piston will fit inside a randomly chosen cylinder?

Let X_1 and X_2 denote the piston diameter and cylinder diameter, respectively. The piston will fit inside the cylinder if $X_1 < X_2$, or equivalently if $X_1 - X_2 < 0$. The difference $X_1 - X_2$, being a linear combination of independent normal random variables, has a normal distribution with

$$E(X_1 - X_2) = E(X_1) - E(X_2) = 22.40 - 22.50 = -.10$$
$$V(X_1 - X_2) = V(X_1) + V(X_2) = (.03)^2 + (.04)^2 = .0025$$

and

$$\sigma_{X_1 - X_2} = \sqrt{.0025} = .05$$

The desired probability is now obtained by standardizing:

$$P(X_1 - X_2 < 0) = P\left(Z < \frac{0 - (-.1)}{.05}\right) = P(Z < 2.00) = \Phi(2.00) = .9772$$

Suppose that the supplier of pistons can set the mean piston diameter μ_1 to any desired value. What value will ensure that 99% of all piston-cylinder pairings yield a fit? The desired condition is

$$P(X_1 - X_2 < 0) = P\left(Z < \frac{0 - (\mu_1 - 22.50)}{.05}\right) = .99$$

which implies that

$$z_{.01} = 2.33 = \frac{-(\mu_1 - 22.50)}{.05}$$

Solving this equation gives $\mu_1 = 22.3835$. ■

Example 5.26

On Mondays, Wednesdays, and Fridays I drive to work by myself, while on Tuesdays and Thursdays I carpool with another individual by driving every other week. Assume that all driving times are independent of one another and normally distributed as follows: my house directly to work: $\mu = 30$ min, $\sigma = 8$ min; my house to carpooler's house: $\mu = 15$ min, $\sigma = 3$ min; carpooler's house to work: $\mu = 20$ min, $\sigma = 6$ min. What is the probability that my total driving time to work exceeds 200 min during a week in which I drive every day? To answer this, define random variables by $X_1, X_2, X_3 =$ M, W, F direct driving times; $X_4, X_5 =$ T, Th times to carpooler's house; $X_6, X_7 =$ T, Th times from carpooler's house to work.

The total driving time is $T_o = X_1 + \cdots + X_7$, and we wish $P(T_o \geq 200)$. To standardize, we compute

$$\mu_{T_o} = E(X_1 + \cdots + X_7) = \sum_{i=1}^{7} \mu_i$$

$$= 30 + 30 + 30 + 15 + 15 + 20 + 20 = 160$$

$$\sigma_{T_o}^2 = V(X_1 + \cdots + X_7) = \sum_{i=1}^{7} \sigma_i^2$$

$$= 64 + 64 + 64 + 9 + 9 + 36 + 36 = 282$$

and

$$\sigma_{T_o} = \sqrt{282} = 16.79$$

Therefore, since T_o has a normal distribution

$$P(T_o \geq 200) = P\left(Z \geq \frac{200 - 160}{16.79}\right) = P(Z \geq 2.38) = .0087 \qquad ■$$

The Central Limit Theorem

When the X_i's are normally distributed, so is $\overline{X}$ for every value of n. Figures 5.6 and 5.7 of the previous section suggest that even when the distribution of each X_i is highly nonnormal, averaging produces a distribution more symmetric and bell-shaped than the distribution being sampled. Notice that even when $n = 4$, probability has begun to pile up around μ in a bell-shaped fashion, though the original p.m.f. in Figure 5.6 is quite skewed. A reasonable conjecture is that if n is large, a suitable normal curve will approximate the actual distribution of $\overline{X}$. The formal statement of this result is the most important theorem of probability.

The Central Limit Theorem (C.L.T.)

Let $X_1, X_2, \ldots, X_n$ be a random sample from a distribution with mean μ and variance σ^2. Then if n is sufficiently large, $\overline{X}$ has approximately a normal distribution with $\mu_{\overline{X}} = \mu$ and $\sigma_{\overline{X}}^2 = \sigma^2/n$, and T_o also has approximately a normal distribution with $\mu_{T_o} = n\mu$, $\sigma_{T_o}^2 = n\sigma^2$. The larger the value of n, the better the approximation.

According to the C.L.T., when n is large and we wish to calculate, say, $P(a \leq \overline{X} \leq b)$, we need only "pretend" that $\overline{X}$ is normal, standardize it, and use the normal table. The resulting answer will be approximately correct. The exact answer could be obtained only by first finding the distribution of $\overline{X}$, so the C.L.T. provides a truly impressive shortcut. The proof of the theorem involves much advanced mathematics.

Example 5.27

The nicotine content in a single cigarette of a particular brand is a random variable with mean $\mu = .8$ mg and standard deviation .1 mg. If an individual smokes five packs of these cigarettes per week, what is the probability that the total amount of nicotine consumed in a week is at least 82 mg?

Notice that there is no assumption of normality here, but five packs consist of $n = 100$ cigarettes. Letting $X_1, \ldots, X_{100}$ be the nicotine contents of the 100 cigarettes, the X_i's constitute a random sample from a distribution with $\mu = .8$ and $\sigma = .1$. Assuming that $n = 100$ is sufficiently large for the C.L.T. to apply, T_o is approximately normal with mean $(100)(.8) = 80$ and standard deviation $(.1) \cdot \sqrt{100} = 1$. Therefore $(T_o - 80)/1$ is *approximately* standard normal, so that

$$P(T_o \geq 82) \approx P[Z \geq (82 - 80)/1] = P(Z \geq 2) = .0228$$

Thus the probability that at least 82 mg is consumed is approximately .0228. ∎

$N(.8, .1)$

$N(80, 1)$

Example 5.28

When a batch of a certain chemical product is prepared, the amount of a particular impurity in the batch is a random variable with mean value 4.0 g and standard deviation 1.5 g. If 50 batches are independently prepared, what is the (approximate) probability that the sample average amount of impurity $\overline{X}$ is between 3.5 and 3.8 g? According to the rule of thumb to be stated shortly, $n = 50$ is large enough for the C.L.T. to be applicable. $\overline{X}$ then has approximately a normal distribution with mean value $\mu_{\overline{X}} = 4.0$ and $\sigma_{\overline{X}} = 1.5/\sqrt{50} = .2121$, so

$$P(3.5 \leq \overline{X} \leq 3.8) \approx P\left(\frac{3.5 - 4.0}{.2121} \leq Z \leq \frac{3.8 - 4.0}{.2121}\right)$$

$$= \Phi(-.94) - \Phi(-2.36) = .1645$$

∎

Example 5.29

A certain consumer organization customarily reports the number of major defects for each new automobile that it tests. Suppose that the number of such defects for a certain model is a random variable with mean value 3.2 and standard deviation 2.4. Among 100 randomly selected cars of this model, how likely is it that the sample average number of major defects exceeds 4? Let X_i denote the number of major defects for the ith car in the random sample. Notice that X_i is a discrete random variable, but the C.L.T. is applicable whether the variable of interest is discrete or continuous. Also, although the fact that the standard deviation of this nonnegative variable is quite large relative to the mean value suggests that its distribution is positively skewed, the large sample size implies that $\overline{X}$ does have approximately a normal distribution. Using $\mu_{\overline{X}} = 3.2$ and $\sigma_{\overline{X}} = .24$,

$$P(\overline{X} > 4) \approx P\left(Z > \frac{4 - 3.2}{.24}\right) = 1 - \Phi(3.33) = .0004 \qquad \blacksquare$$

Understanding the C.L.T.

The Central Limit Theorem provides insight into why many random variables have probability distributions that are approximately normal. For example, the measurement error in a scientific experiment can be thought of as a sum of a number of underlying perturbations and errors of small magnitude.

Although the usefulness of the C.L.T. for inference will soon be apparent, the intuitive content of the result gives many beginning students difficulty. Again looking back to Figure 5.6, the probability histogram on the left is a picture of the distribution being sampled. It is discrete and quite skewed, so does not look at all like a normal distribution. The distribution of $\overline{X}$ for $n = 2$ starts to exhibit some symmetry, and this is even more pronounced for $n = 4$ in Figure 5.7. Figure 5.10 contains the probability distribution of $\overline{X}$ for $n = 8$, as well as a probability histogram for this distribution. With $\mu_{\overline{X}} = \mu = 46.5$ and $\sigma_{\overline{X}} = \sigma/\sqrt{n} = 3.905/\sqrt{8} = 1.38$, if we fit a normal curve with this mean and standard deviation through the histogram of $\overline{X}$, the areas of rectangles in the probability histogram are reasonably well approximated by the normal curve areas, at least in the central part of the distribution. The picture for T_o is similar except that the horizontal scale is much more spread out, with T_o ranging between 320 ($\overline{x} = 40$) and 360 ($\overline{x} = 50$).

$\overline{x}$	40	40.625	41.25	41.875	42.5	43.125
$p(\overline{x})$	.0000	.0000	.0003	.0012	.0038	.0112
$\overline{x}$	43.75	44.375	45	45.625	46.25	46.875
$p(\overline{x})$	.0274	.0556	.0954	.1378	.1704	.1746
$\overline{x}$	47.5	48.125	48.75	49.375	50	
$p(\overline{x})$	.1474	.0998	.0519	.0188	.0039	

(a)

Figure 5.10 (a) Probability distribution of $\overline{X}$ for $n = 8$, and (b) probability histogram and normal approximation to the distribution of X, when the original distribution is as in Example 5.18

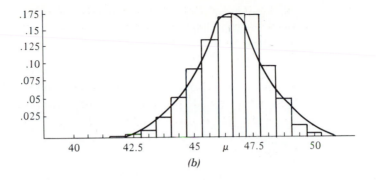

(b)

A better understanding of the C.L.T. can also be obtained by performing a simulation experiment of the following sort: Select a probability distribution and a reasonably large value of n, such as $n = 50$. Then obtain 50 observed x_i's from this distribution and compute the sample average $\bar{x}$ for the 50. To do this with the p.m.f.

x	40	45	50
$p(x)$	.2	.3	.5

that we have been working with, one could put two slips marked 40, three marked 45, and five marked 50 into a box and then sample with replacement. Alternatively, there are computer program packages such as MINITAB or GPSS that allow for such simulation. Next obtain another 50 x_i's and compute $\bar{x}$, and so on until 1000 $\bar{x}$'s, each an average of 50 observations, have been obtained. Now construct a sample histogram of the 1000 $\bar{x}$'s. This histogram will have the appearance of the true distribution of $\bar{X}$ *and* will be very close to bell-shaped. If $n = 100$ x_i's had been averaged to obtain each $\bar{x}$, the bell-shaped appearance would be even more pronounced, since a larger n improves the approximation.

A practical difficulty in applying the C.L.T. is in knowing when n is sufficiently large. The problem is that the accuracy of the approximation for a particular n depends on the shape of the original underlying distribution being sampled. If the underlying distribution is close to bell-shaped, then the approximation will be good even for a small n, while if it is far from bell-shaped then a large n will be required. We will use the following rule of thumb, which is frequently somewhat conservative.

Rule of thumb

If $n > 30$, the C.L.T. can be used.

Other Applications of the C.L.T.

The C.L.T. can be used to justify the normal approximation to the binomial distribution discussed in Chapter 4. Recall that a binomial variable X is the number of successes in a binomial experiment consisting of n independent success-failure trials with $p = P(S)$ for any particular trial. Define new random variables $X_1, X_2, \ldots, X_n$ by

$$X_i = \begin{cases} 1 & \text{if the } i\text{th trial results in a success} \\ 0 & \text{if the } i\text{th trial results in a failure} \end{cases} \quad (i = 1, \ldots, n)$$

Because the trials are independent and $P(S)$ is constant from trial to trial, the X_i's are i.i.d. (a random sample from a Bernoulli distribution). The C.L.T. then implies that if n is sufficiently large, both the sum and the average of the X_i's have approximately normal distributions. When the X_i's are summed, a 1 is added in for every S that occurs and a 0 for every F, so $X_1 + \cdots + X_n = X$. The sample mean of the X_i's is X/n, the sample proportion of successes. That is, both X and X/n are approximately normal when n is large. The necessary sample size for this approximation depends on the value of p. This is because when p is close to .5, the distribution of each X_i is reasonably symmetric (Figure 5.11), whereas the distribution is quite skewed when p is near 0 or 1. Using the approximation only if both $np \geq 5$ and $n(1 - p) \geq 5$ ensures that n is large enough to overcome any skewness in the underlying Bernoulli distribution.

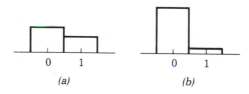

Figure 5.11   Two Bernoulli distributions: (a) $p = .4$ (reasonably symmetric); (b) $p = .1$ (very skewed)

Recall from Section 4.5 that a r.v. X has a lognormal distribution if $\ln(X)$ has a normal distribution.

Proposition

> Let $X_1, X_2, \ldots, X_n$ be a random sample from a distribution for which only positive values are possible $[P(X_i > 0) = 1]$. Then if n is sufficiently large, the product $Y = X_1 X_2 \cdots X_n$ has approximately a lognormal distribution.

To verify this, note that

$$\ln(Y) = \ln(X_1) + \ln(X_2) + \cdots + \ln(X_n)$$

Since $\ln(Y)$ is a sum of independent and identically distributed r.v.'s [the $\ln(X_i)$'s], it is approximately normal when n is large, so Y itself has approximately a lognormal distribution. As an example of the applicability of this result, Bury (*Statistical Models in Applied Science,* p. 590) argues that the damage process in plastic flow and crack propagation is a multiplicative process, so that variables such as percentage elongation and rupture strength have approximately lognormal distributions.

Exercises / Section 5.4 (40–55)

40. Let X_1, X_2, and X_3 be three independent normal random variables with expected values μ_1, μ_2, and μ_3 and variances σ_1^2, σ_2^2, and σ_3^2, respectively.
  a. If $\mu_1 = \mu_2 = \mu_3 = 100$ and $\sigma_1^2 = \sigma_2^2 = \sigma_3^2 = 12$, calculate $P(X_1 + X_2 + X_3 \le 309)$ and $P(288 \le X_1 + X_2 + X_3 \le 312)$.
  b. Using the μ_i's and σ_i's of (a), calculate $P(105 \le \overline{X})$ and $P(98 \le \overline{X} \le 102)$.
  c. Using the μ_i's and σ_i's of (a), calculate $P(-10 \le X_1 - .5X_2 - .5X_3 \le 5)$.
  d. If $\mu_1 = 90$, $\mu_2 = 100$, $\mu_3 = 110$, $\sigma_1^2 = 10$, $\sigma_2^2 = 12$, and $\sigma_3^2 = 14$, calculate $P(X_1 + X_2 + X_3 \le 306)$ and $P(98 \le \overline{X} \le 102)$.

41. Five automobiles of the same type are to be driven on a 300-mile trip. The first two will use an economy brand of gasoline and the other three will use a name brand. Let X_1, X_2, X_3, X_4, and X_5 be the observed numbers of miles per gallon for the five cars, and suppose that these variables are independent and normally distributed with $\mu_1 = \mu_2 = 20$, $\mu_3 = \mu_4 = \mu_5 = 21$, and $\sigma^2 = 4$ for all five variables. Define a r.v. Y by

$$Y = \frac{X_1 + X_2}{2} - \frac{X_3 + X_4 + X_5}{3}$$

so that Y is a measure of the difference in efficiency between economy gas and name brand gas. Compute $P(0 \le Y)$ and $P(-1 \le Y \le 1)$.

42. Let X_1, X_2, . . . , X_{100} denote the actual net weights of 100 randomly selected 50-lb bags of fertilizer.
  a. If the expected weight of each bag is 50 and the variance is 1, calculate $P(49.75 \le \overline{X} \le 50.25)$ (approximately) using the C.L.T.
  b. If the expected weight is 49.8 lbs rather than 50 lbs so that on average bags are underfilled, calculate $P(49.75 \le \overline{X} \le 50.25)$.

43. There are 40 students in an elementary statistics class. On the basis of years of experience, the instructor knows that the time needed to grade a randomly chosen first examination paper is a random variable with an expected value of six minutes and a standard deviation of six minutes.
  a. If grading times are independent and the instructor begins grading at 6:50 P.M. and grades continuously, what is the (approximate) probability that he is through grading before the 11:00 P.M. TV news begins?
  b. If the sports report begins at 11:10, what is the probability that he misses part of the report if he waits until grading is done before turning on the TV?

44. Suppose that when the pH of a certain chemical compound is 5.00, the pH measured by a randomly selected beginning chemistry student is a random variable with mean 5.00 and standard deviation .3. A large batch of the compound is subdivided and a sample given to each student in a morning lab and each student in an afternoon lab. Let $\overline{X}$ = the average pH as determined by the morning students and $\overline{Y}$ = the average pH as determined by the afternoon students.
  a. If pH is a normal variable and there are 25 students in each lab, compute $P(-.1 \le \overline{X} - \overline{Y} \le .1)$. *Hint:* $\overline{X} - \overline{Y}$ is a linear combination of normal variables, so is normally distributed. Compute $\mu_{\overline{X} - \overline{Y}}$ and $\sigma_{\overline{X} - \overline{Y}}$.
  b. If there are 36 students in each lab, but pH determinations are not assumed normal, calculate (approximately) $P(-.1 \le \overline{X} - \overline{Y} \le .1)$.

45. The first assignment in an introductory computer programming class involves running a short program. If past experience indicates that 40% of all beginning students will make no typographical errors, compute the (approximate) probability that in a class of 50 students
  a. at least 25 will make no errors. *Hint:* normal approximation to the binomial.
  b. between 15 and 25 will make no errors.

46. If two loads are applied to a cantilever beam as shown in the accompanying drawing, the bending moment at 0 due to the loads is $a_1 X_1 + a_2 X_2$.

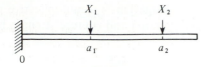

a. Suppose that X_1 and X_2 are independent random variables with means 2 and 4 kips, respectively, and standard deviations .5 and 1.0 kip, respectively. If $a_1 = 5$ ft and $a_2 = 10$ ft, what is the expected bending moment and what is the standard deviation of the bending moment?

b. If X_1 and X_2 are normally distributed, what is the probability that the bending moment will exceed 75 kip-ft?

c. Suppose that the positions of the two loads are random variables. Denoting them by A_1 and A_2, assume that these variables have means of 5 and 10 ft, respectively, that each has a standard deviation of .5, and that all A_i's and X_i's are independent of one another. What is the expected moment now?

d. For the situation of (c), what is the variance of the bending moment?

e. If the situation is as described in (a) except that $\text{Corr}(X_1, X_2) = .5$ (so that the two loads are not independent), what is the variance of the bending moment?

47. Suppose that the sediment density (g/cm) of a randomly selected specimen from a certain region is normally distributed with mean 2.65 and standard deviation .85 (suggested in "Modeling Sediment and Water Column Interactions for Hydrophobic Pollutants," *Water Research*, 1984, pp. 1169–1174.

a. If a random sample of 25 specimens is selected, what is the probability that the sample average sediment density is at most 3.00? Between 2.65 and 3.00?

b. How large a sample size would be required to ensure that the first probability in (a) is at least .99?

48. One piece of PVC pipe is to be inserted inside another piece. The length of the first piece is normally distributed with mean value 20 in. and standard deviation .5 in. The length of the second piece is a normal random variable with mean and standard deviation 15 in. and .4 in., respectively.

The amount of overlap is normally distributed with mean value 1 in. and standard deviation .1 in. Assuming that the lengths and amount of overlap are independent of one another, what is the probability that the total length after insertion is between 34.5 in. and 35 in.?

49. Two airplanes are flying in the same direction in adjacent parallel corridors. At time $t = 0$, the first airplane is 10 km ahead of the second one. Suppose that the speed of the first plane (km/hr) is normally distributed with mean 520 and standard deviation 10, and the second plane's speed is also normally distributed with mean and standard deviation 500 and 10, respectively.

a. What is the probability that after two hours of flying, the second plane has not caught up to the first plane?

b. Determine the probability that the planes are separated by at most 10 km after two hours.

50. In Exercise 46, the weight of the beam itself contributes to the bending moment. Assume that the beam is of uniform thickness and density, so that the resulting load is uniformly distributed on the beam. If the weight of the beam is random, the resulting load from the weight is also random; denote this load by W (kips/ft).

a. If the beam is 12 ft long, W has mean 1.5 and standard deviation .25, and the fixed loads are as described in (a) of the previous problem, what are the expected value and variance of the bending moment? *Hint:* If the load due to the beam were w kips/ft, the contribution to the bending moment would be $w \int_0^{12} x \, dx$.

b. If all three variables (X_1, X_2, and W) are normally distributed, what is the probability that the bending moment will be at most 200 kip-ft?

51. I have three errands to take care of in the Administration Building. Let X_i = the time that it takes for the ith errand ($i = 1, 2, 3$) and let X_4 = the total time that I spend walking to and from the building and between each errand. Suppose that the X_i's are independent, normally distributed, with the following means and standard deviations: $\mu_1 = 15$ min, $\sigma_1 = 4$, $\mu_2 = 5$, $\sigma_2 = 1$, $\mu_3 = 8$, $\sigma_3 = 2$, $\mu_4 = 12$, $\sigma_4 = 3$. I plan to leave my office at precisely 10:00 A.M. and wish to post a note on my door which reads, "I will return by t

A.M." What time t should I write down if I want the probability of my arriving after t to be .01?

52. Suppose that the expected tensile strength of type-A steel is 106 k.s.i. and the standard deviation of tensile strength is 8 k.s.i. For type-B steel suppose that the expected tensile strength and standard deviation of tensile strength are 104 k.s.i. and 6 k.s.i., respectively. Let $\bar{X}$ = the sample average tensile strength of a random sample of 40 type-A specimens, and let $\bar{Y}$ = the sample average tensile strength of a random sample of 35 type-B specimens.
   a. What is the approximate distribution of $\bar{X}$? Of $\bar{Y}$?
   b. What is the approximate distribution of $\bar{X} - \bar{Y}$? Justify your answer.
   c. Calculate (approximately) $P(-1 \le \bar{X} - \bar{Y} \le 1)$.
   d. Calculate $P(\bar{X} - \bar{Y} \ge 6)$. If you actually observed $\bar{X} - \bar{Y} \ge 6$, would you doubt that $\mu_1 - \mu_2 = 2$?

53. In an area having sandy soil, 50 small trees of a certain type were planted, and another 50 trees were planted in an area having clay soil. Let X = the number of trees planted in sandy soil that survive one year, and Y = the number of trees planted in clay soil that survive one year. If the probability that a tree planted in sandy soil will survive one year is .7 and the probability of one-year survival in clay soil is .6, compute (approximately) $P(-5 \le X - Y \le 5)$ (do not bother with the continuity correction).

54. The number of parking tickets issued in a certain city on any given weekday has a Poisson distribution with parameter $\lambda = 50$. What is the approximate probability that:
   a. between 35 and 70 tickets are given out on a particular day? *Hint:* When λ is large, a Poisson random variable has approximately a normal distribution.
   b. the total number of tickets given out during a five-day week is between 225 and 275?

55. Suppose that the distribution of the time X (hours) spent by students at a certain university on their senior projects is gamma with parameters $\alpha = 50$ and $\beta = 2$. Because α is large, it can be shown that X has approximately a normal distribution. Use this fact to compute the probability that a randomly selected student spends at most 125 hours on the senior project.

Supplementary Exercises / Chapter 5 (56–71)

56. A restaurant serves three fixed-price dinners costing \$7, \$9, and \$10. For a randomly selected couple dining at this restaurant, let X = the cost of the man's dinner and Y = the cost of the woman's dinner. Suppose that the joint p.m.f. of X and Y is given in the following table:

$p(x, y)$		y	
	7	9	10
7	.05	.05	.10
x 9	.05	.10	.35
10	0	.20	.10

   a. Compute the marginal p.m.f. of X and of Y.
   b. What is the probability that the man's and the woman's dinner cost at most \$9 each?
   c. Are X and Y independent? Justify your answer.
   d. What is the expected total cost of the dinner for the two people?

   e. Suppose that when a couple opens fortune cookies at the conclusion of the meal, they find the message "You will receive as a refund the difference between the cost of the more expensive and the less expensive meal that you have chosen." How much does the restaurant expect to refund?

57. A health food store stocks two different brands of a certain type of grain. Let X = the amount (lbs) of brand A on hand, and Y = the amount of brand B on hand. Suppose that the joint p.d.f. of X and Y is

$$f(x, y) = \begin{cases} kxy & x \ge 0, y \ge 0, 20 \le x + y \le 30 \\ 0 & \text{otherwise} \end{cases}$$

   a. Draw the region of positive density and determine the value of k.

b. Are X and Y independent? Answer by first deriving the marginal p.d.f. of each variable.

c. Compute $P(X + Y \leq 25)$.

d. What is the expected total amount of this grain on hand?

e. Compute $\text{Cov}(X, Y)$ and $\text{Corr}(X, Y)$.

f. What is the variance of the total amount of grain on hand?

58. Let $X_1, X_2, \ldots, X_n$ be random variables denoting n independent bids for an item that is for sale. Suppose that each X_i is uniformly distributed on the interval $[100, 200]$. If the seller sells to the highest bidder, how much can he expect to earn on the sale? *Hint:* Let $Y = \max (X_1, X_2, \ldots, X_n)$. First find $F_Y(y)$ by noting that $Y \leq y$ iff each X_i is $\leq y$. Then obtain the p.d.f. and $E(Y)$.

59. Suppose that my calorie intake at breakfast is a random variable with expected value 500 and standard deviation 50, my calorie intake at lunch is random with expected value 800 and standard deviation 100, and my calorie intake at dinner is a random variable with expected value 1700 and standard deviation 200. Assuming that intakes on different meals are independent of one another and that I eat three meals per day each day, what is the probability that my average calorie intake per day over the next (365-day) year is between 2950 and 3050? *Hint:* Let X_i, Y_i, and Z_i denote the three calorie intakes on day i. Then total intake $= \Sigma (X_i + Y_i + Z_i)$.

60. Suppose that the proportion of rural voters in a certain state who favor a particular gubernatorial candidate is .45, and the proportion of suburban and urban voters favoring the candidate is .60. If a sample of 200 rural voters and 300 urban and suburban voters is obtained, what is the approximate probability that at least 250 of these voters favor this candidate?

61. Let μ denote the true pH of a chemical compound. A sequence of n independent sample pH determinations will be made. Suppose that each sample pH is a random variable with expected value μ and standard deviation .1. How many determinations are required if we wish the probability that the sample average is within .02 of the

true pH to be at least .95? What theorem justifies your probability calculation?

62. If the amount of soft drink that I consume on any given day is independent of consumption on any other day and is normally distributed with $\mu = 13$ oz and $\sigma = 2$, and if I currently have two six-packs of 16-oz bottles, what is the probability that I still have some soft drink left at the end of two weeks (14 days)?

63. A student has a class that is supposed to end at 9:00 A.M. and another that is supposed to begin at 9:10 A.M. Suppose that the actual ending time of the 9 A.M. class is a normally distributed r.v. X_1 with mean 9:02 and standard deviation 1.5 min, and that the starting time of the next class is also a normally distributed r.v. X_2 with mean 9:10 and standard deviation 1 min. Suppose also that the time necessary to get from one classroom to the other is a normally distributed r.v. X_3 with mean 6 min and standard deviation 1 min. What is the probability that the student makes it to the second class before the lecture starts? (Assume independence of X_1, X_2, and X_3, which is reasonable if the student pays no attention to the finishing time of the first class.)

64. a. Use the general formula for the variance of a linear combination to write an expression for $V(aX + Y)$. Then let $a = \sigma_Y/\sigma_X$ and show that $\rho \geq -1$. *Hint:* Variance is always ≥ 0, and $\text{Cov}(X, Y) = \sigma_X \cdot \sigma_Y \cdot \rho$.

b. By considering $V(aX - Y)$, conclude that $\rho \leq 1$.

c. Use the fact that $V(W) = 0$ only if W is a constant to show that $\rho = 1$ only if $Y = aX + b$.

65. Suppose that a randomly chosen individual's verbal score X and quantitative score Y on a nationally administered aptitude examination have joint p.d.f.

$$f(x, y) = \begin{cases} \dfrac{2}{5}(2x + 3y) & 0 \leq x \leq 1, 0 \leq y \leq 1 \\ 0 & \text{otherwise} \end{cases}$$

You are asked to provide a prediction t of the individual's total score $X + Y$. The error of prediction is the mean square error $E[(X +$

$Y - t)^2]$. What value of t minimizes the error of prediction?

66. a. Let X_1 have a chi-squared distribution with parameter ν_1 (see Section 4.4), and let X_2 be independent of X_1 and have a chi-squared distribution with parameter ν_2. Use the technique of Example 5.13 to show that $X_1 + X_2$ has a chi-squared distribution with parameter $\nu_1 + \nu_2$.

b. In Exercise 59 (Chapter 4), you were asked to show that if Z is a standard normal r.v., then Z^2 has a chi-squared distribution with $\nu = 1$. Let $Z_1, Z_2, \ldots, Z_n$ be n independent standard normal r.v.'s. What is the distribution of $Z_1^2 + \cdots + Z_n^2$? Justify your answer.

c. Let $X_1, \ldots, X_n$ be a random sample from a normal distribution with mean μ and variance σ^2. What is the distribution of the sum $Y = \sum_{i=1}^{n} [(X_i - \mu)/\sigma]^2$? Justify your answer.

67. a. Show that $\text{Cov}(X, Y + Z) = \text{Cov}(X, Y) + \text{Cov}(X, Z)$.

b. Let X_1 and X_2 be quantitative and verbal scores on one aptitude exam, and Y_1 and Y_2 be corresponding scores on another exam. If $\text{Cov}(X_1, Y_1) = 5$, $\text{Cov}(X_1, Y_2) = 1$, $\text{Cov}(X_2, Y_1) = 2$, and $\text{Cov}(X_2, Y_2) = 8$, what is the covariance between the two total scores $X_1 + X_2$ and $Y_1 + Y_2$?

68. A rock specimen from a particular area is randomly selected and weighed two different times. Let W denote the actual weight, and X_1 and X_2 the two measured weights. Then $X_1 = W + E_1$ and $X_2 = W + E_2$, where E_1 and E_2 are the two measurement errors. Suppose that the E_i's are independent of one another and of W, and that $V(E_1) = V(E_2) = \sigma_E^2$.

a. Express ρ, the correlation coefficient between the two measured weights X_1 and X_2, in terms of σ_W^2, the variance of actual weight, and σ_X^2, the variance of measured weight.

b. Compute ρ when $\sigma_W = 1$ kg and $\sigma_E = .01$ kg.

69. Let A denote the percentage of one constituent in a randomly selected rock specimen, and let B denote the percentage of a second constituent in that same specimen. Suppose that D and E are measurement errors in determining the values of A and B, so that measured values are $X = A + D$ and $Y = B + E$, respectively. Assume that measurement errors are independent of one another and of actual values.

a. Show that $\text{Corr}(X, Y) = \text{Corr}(A, B) \cdot \sqrt{\text{Corr}(X_1, X_2)} \cdot \sqrt{\text{Corr}(Y_1, Y_2)}$, where X_1 and X_2 are replicate measurements on the value of A, and Y_1 and Y_2 are defined analogously with respect to B. What effect does the presence of measurement error have on the correlation?

b. What is the maximum value of $\text{Corr}(X, Y)$ when $\text{Corr}(X_1, X_2) = .8100$ and $\text{Corr}(Y_1, Y_2) = .9025$? Is this disturbing?

70. Let $X_1, \ldots, X_n$ be independent random variables with mean values $\mu_1, \ldots, \mu_n$ and variances $\sigma_1^2, \ldots, \sigma_n^2$. Consider a function $h(x_1, \ldots, x_n)$, and use it to define a new random variable $Y = h(X_1, \ldots, X_n)$. Under rather general conditions on the h function, if the σ_i's are all small relative to the corresponding μ_i's, it can be shown that $E(Y) \approx h(\mu_1, \ldots, \mu_n)$ and

$$V(Y) \approx \left(\frac{\partial h}{\partial x_1}\right)^2 \cdot \sigma_1^2 + \cdots + \left(\frac{\partial h}{\partial x_n}\right)^2 \cdot \sigma_n^2$$

where each partial derivative is evaluated at $(x_1, \ldots, x_n) = (\mu_1, \ldots, \mu_n)$. Suppose that three resistors with resistances X_1, X_2, X_3 are connected in parallel across a battery with voltage X_4. Then by Ohm's law, the current is

$$Y = X_4 \left[\frac{1}{X_1} + \frac{1}{X_2} + \frac{1}{X_3}\right]$$

Let $\mu_1 = 10$ ohms, $\sigma_1 = 1.0$ ohm, $\mu_2 = 15$ ohms, $\sigma_2 = 1.0$ ohm, $\mu_3 = 20$ ohms, $\sigma_3 = 1.5$ ohms, $\mu_4 = 120$ volts, $\sigma_4 = 4.0$ volts. Calculate the (approximate) expected value and standard deviation of the current (suggested by "Random Samplings," *CHEMTECH*, 1984, pp. 696–697).

71. A more accurate approximation to $E(h(X_1, \ldots, X_n))$ in Exercise 70 is

$$h(\mu_1, \ldots, \mu_n) + \frac{1}{2}\sigma_1^2\left(\frac{\partial^2 h}{\partial x_1^2}\right) + \cdots + \frac{1}{2}\sigma_n^2\left(\frac{\partial^2 h}{\partial x_n^2}\right)$$

Compute this for $Y = h(X_1, X_2, X_3, X_4)$ given in Exercise 70, and compare it to the leading term $h(\mu_1, \ldots, \mu_n)$.

Bibliography

Derman, Cyrus, Gleser, Leon, and Olkin, Ingram, *Probability Models and Applications,* Macmillan, New York, 1980. Contains a careful and comprehensive exposition of joint distributions, rules of expectation, and limit theorems.

Larsen, Richard, and Marx, Morris, *Introduction to Mathematical Statistics* (2nd ed.), Prentice-Hall, Englewood Cliffs, N.J., 1985. More limited coverage than in the book by Derman et al., but well written and readable.

CHAPTER 6

Point Estimation

Introduction

Given a parameter of interest, such as a population mean μ or population proportion p, the objective of point estimation is to use a sample to compute a number that represents in some sense a good guess for the true value of the parameter. The resulting number is called a *point estimate*. In Section 6.1 we present some general concepts of point estimation. Section 6.2 describes and illustrates two important methods for obtaining point estimates, the method of moments and the method of maximum likelihood.

6.1 Some General Concepts of Point Estimation

Statistical inference is almost always directed toward drawing some type of conclusion about one or more parameters (population characteristics). To do so requires that an investigator obtain sample data from each of the populations under study. Conclusions can then be based on the computed values of various sample quantities. For example, let μ (a parameter) denote the true average breaking strength of wire connections used in bonding semiconductor wafers. A random sample of $n = 10$ connections might be made and the breaking strength of each one determined, resulting in observed strengths $x_1, x_2, \ldots, x_{10}$. The sample mean breaking strength $\overline{x}$ could then be used to draw a conclusion about the value of μ. Similarly, if σ^2 is the variance of the breaking strength distribution (population variance, another parameter), the value of the sample variance s^2 can be used to infer something about σ^2.

Consider selecting a sample of size n from a single population. Before the sample is obtained, we are uncertain about what the value of each observation will be. Thus the first observation must be considered a random variable X_1, the second observation another random variable X_2, and so on. That is, prior to

223

actually removing the sample from the population, we denote the sample observations by $X_1, X_2, \ldots, X_n$. After sampling, each observation is a number, and these numbers are denoted by $x_1, x_2, \ldots, x_n$. Similarly, prior to sampling, any function of the sample observations, such as the sample mean $\Sigma X_i/n$ or the sample variance $\Sigma(X_i - \overline{X})^2/(n-1)$, is a random variable. Thus before sampling, we denote the sample mean by $\overline{X}$, the sample variance by S^2, the sample median by $\tilde{X}$, and so on—an upper case letter to emphasize the status of each as a random variable. Once specific sample values $x_1, x_2, \ldots, x_n$ have been obtained, $\overline{x}$, s^2, and $\tilde{x}$ will represent the computed (or observed) values of the sample mean, variance, and median, respectively.

A parameter, such as μ, σ^2, or $\tilde{\mu}$, is a characteristic of the population (or probability distribution used to model the population). Quantities such as $\overline{X}$, S^2, and $\tilde{X}$ are sample characteristics—the value of each one varies from sample to sample, whereas a parameter has a fixed numerical value (usually unknown to the investigator). The word *statistic* is used to distinguish a sample characteristic from a parameter.

Definition

> A **statistic** is any function of the random variables constituting one or more samples, provided that the function does not depend on any unknown parameter values.

Thus the sample mean $\overline{X}$ is a statistic, with computed value denoted by $\overline{x}$, as is the sample variance S^2, with computed value given by s^2. As another example, an investigator might wish to compare true average lifetimes μ_1 and μ_2 for two different types of light bulbs. Let $X_1, X_2, \ldots, X_m$ denote a random sample of m type I lifetimes and $Y_1, Y_2, \ldots, Y_n$ denote a second random sample of n type II lifetimes. Then $\overline{X} - \overline{Y}$, the difference between the two sample means, is a statistic. Its computed value $\overline{x} - \overline{y}$ can be used to infer something about the parameter $\mu_1 - \mu_2$, the difference between the two population means. The random variable $\Sigma(X_i - \mu)^2/n$, the average squared deviation from the population mean, is not a statistic unless the value of μ is known. Similarly, the standardized random variable $(\overline{X} - \mu)/(\sigma/\sqrt{n})$ is not a statistic unless the values of μ and σ are known.

When discussing general concepts and methods of inference, it is convenient to have a generic symbol for the parameter of interest. We shall use the Greek letter θ for this purpose. The objective of point estimation is to select a single number, based on sample data, that represents the most plausible value of θ. Suppose, for example, that the parameter of interest is μ, the true average lifetime of calculator batteries of a certain type. A random sample of $n = 3$ batteries might yield observed lifetimes (hours) $x_1 = 5.0$, $x_2 = 6.4$, $x_3 = 5.9$. The computed value of the sample mean lifetime is $\overline{x} = 5.77$, and it is reasonable to regard 5.77 as the most plausible value of μ—our "best guess" for the value of μ based on the available sample information.

Definition

> A **point estimate** of a parameter θ is a single number that can be regarded as the most plausible value of θ. A point estimate is obtained by selecting a suitable statistic and computing its value from the given sample data. The selected statistic is called the **point estimator** of θ.

In the calculator battery example just given, the estimator used to obtain the point estimate of μ was $\overline{X}$, and the point estimate of μ was 5.77. If the three observed lifetimes had instead been $x_1 = 5.6$, $x_2 = 4.5$, and $x_3 = 6.1$, use of the estimator $\overline{X}$ would have resulted in the estimate $\overline{x} = (5.6 + 4.5 + 6.1)/3 = 5.40$. The symbol $\hat{\theta}$ ("theta hat") is customarily used to denote both the estimator of θ and the point estimate resulting from a given sample. Thus $\hat{\mu} = \overline{X}$ is read as "the point estimator of μ is the sample mean $\overline{X}$," and the statement "the point estimate of μ is 5.77" can be written concisely as $\hat{\mu} = 5.77$. Notice that in writing $\hat{\theta} = 72.5$, there is no indication of how this point estimate was obtained (what statistic was used). It is recommended that both the estimator and the resulting estimate be reported.

Example **6.1**

An automobile manufacturer has developed a new type of bumper, which is supposed to absorb impacts with less damage than previous bumpers. The manufacturer has used this bumper in a sequence of 25 controlled crashes against a wall, each at 10 m.p.h., using one of its compact car models. Let $X =$ the number of crashes that result in no visible damage to the automobile. The parameter to be estimated is $p =$ the proportion of all such crashes that result in no damage [alternatively, $p = P(\text{no damage in a single crash})$]. If X is observed to be $x = 15$, the most reasonable estimator and estimate are

$$\text{estimator } \hat{p} = \frac{X}{n}, \quad \text{estimate} = \frac{x}{n} = \frac{15}{25} = .60 \qquad \blacksquare$$

If for each parameter of interest there were only one reasonable point estimator, there would not be much to point estimation. In most problems, though, there will be more than one reasonable estimator.

Example **6.2**

A consumer organization wishes to obtain a point estimate for the true average lifetime (hours of writing) for an inexpensive brand of ball-point pen. Suppose that the distribution of lifetime is normal with parameters μ and σ, so that μ is both the expected lifetime and median lifetime (since the normal p.d.f. is symmetric about μ). The organization purchases $n = 10$ such pens and inserts each in a specially constructed machine that will cause the pen to write continuously until the ink runs out. Let $X_1, X_2, \ldots, X_{10}$ denote the lifetimes, assumed to be a random sample from the normal distribution with parameters μ and σ. Suppose that the observed lifetimes are $x_1 = 26.3$, $x_2 = 35.1$, $x_3 = 23.0$, $x_4 = 28.4$, $x_5 = 31.6$, $x_6 = 30.9$, $x_7 = 25.2$, $x_8 = 28.0$, $x_9 = 27.3$, and $x_{10} = 29.2$. Consider the following estimators and resulting estimates for μ:

a. estimator $= \overline{X}$, estimate $= \overline{x} = \Sigma x_i / 10 = 28.50$
b. estimator $= \tilde{X}$, estimate $= \tilde{x} = (28.0 + 28.4)/2 = 28.20$
c. estimator $= [\min(X_i) + \max(X_i)]/2 =$ the average of the two extreme lifetimes, estimate $= [\min(x_i) + \max(x_i)]/2 = (23.0 + 35.1)/2 = 29.05$
d. estimator $= \overline{X}_{tr(10)}$, the 10% trimmed mean (discard the smallest and largest 10% of the sample and then average), estimate $=$

$$\overline{x}_{tr(10)} = \frac{25.2 + 26.3 + 27.3 + 28.0 + 28.4 + 29.2 + 30.9 + 31.6}{8}$$

$$= 28.36$$

Each one of the estimators (a)–(d) uses a different measure of the center of the sample to estimate μ. Which of the estimates is closest to the true value? We can't answer this without knowing the true value. A question which can be answered is, "Which estimator, when used on other samples of X_i's, will tend to produce estimates closest to the true value?" We will shortly consider this type of question. ∎

Example 6.3 Officials of a paint company have been concerned about the variability in drying time of the company's top-quality interior latex paint. Let X be the drying time of a paint sample on a test board, and let $\sigma^2 = V(X)$ (the variance of the population of all such drying times). If n test boards are set up and the drying times are $X_1, X_2, \ldots, X_n$, then one estimator of σ^2 is the sample variance:

$$\hat{\sigma}^2 = S^2 = \frac{\Sigma(X_i - \overline{X})^2}{n-1} = \frac{\Sigma X_i^2 - (\Sigma X_i)^2/n}{n-1}$$

If $n = 10$ and the observed x_i's are as in Example 6.2 (with minutes replacing hours), then the corresponding estimate is

$$\hat{\sigma}^2 = s^2 = \frac{\Sigma x_i^2 - (\Sigma x_i)^2/10}{9} = \frac{(26.3)^2 + \cdots + (29.2)^2 - (285)^2/10}{9}$$

$$= 11.90$$

The estimate of σ would then be $\hat{\sigma} = s = \sqrt{11.90} = 3.45$.

An alternative estimator would result from using divisor n instead of $n - 1$ (that is, the average squared deviation):

$$\hat{\sigma}^2 = \frac{\Sigma(X_i - \overline{X})^2}{n}, \quad \text{estimate} = \frac{107.10}{10} = 10.71$$

We will shortly indicate why many statisticians prefer S^2 to the estimator with divisor n. ∎

In the best of all possible worlds, we could find an estimator $\hat{\theta}$ for which $\hat{\theta} = \theta$ always. However, $\hat{\theta}$ is a function of the sample X_i's, so is itself a random variable. For some samples $\hat{\theta}$ will yield a value larger than θ, while for other samples $\hat{\theta}$ will underestimate θ. If we write

$$\hat{\theta} = \theta + \text{error of estimation}$$

then an accurate estimator would be one resulting in small estimation errors, so that estimated values will be near the true value. An estimator that has the properties of unbiasedness and minimum variance will often be accurate in this sense.

Unbiased Estimators

Suppose that we have two measuring instruments; one instrument has been accurately calibrated, but the other systematically gives readings under the true value being measured. When each instrument is used repeatedly on the same object, because of measurement error the observed measurements will not be identical. However, the measurements produced by the first instrument will be distributed about the true value in such a way that on average this instrument measures what it purports to measure, so is called an unbiased instrument. The second instrument yields observations that have a systematic error component or bias.

Definition

> A point estimator $\hat{\theta}$ is said to be an **unbiased estimator** of θ if $E(\hat{\theta}) = \theta$ for every possible value of θ. If $\hat{\theta}$ is not unbiased, the difference $E(\hat{\theta}) - \theta$ is called the **bias** of θ.

Thus $\hat{\theta}$ is unbiased if its distribution is "centered" at the true value of the parameter. Figure 6.1 pictures the distributions of several biased and unbiased estimators. Note that "centered" here means that the expected value, not the median, of the distribution of $\hat{\theta}$ is equal to θ.

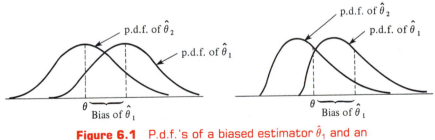

Figure 6.1 P.d.f.'s of a biased estimator $\hat{\theta}_1$ and an unbiased estimator $\hat{\theta}_2$ for a parameter θ

In Example 6.1 the sample proportion X/n was used as an estimator of p, where X, the number of sample successes, had a binomial distribution with parameters n and p. Thus

$$E(\hat{p}) = E\left(\frac{X}{n}\right) = \frac{1}{n}E(X) = \frac{1}{n}(np) = p$$

Proposition

> When X is a binomial random variable with parameters n and p, the sample proportion $\hat{p} = X/n$ is an unbiased estimator of p.

No matter what the true value of p is, the distribution of the estimator $\hat{p}$ will be centered at the true value.

Example 6.4   Suppose that X, the reaction time to a certain stimulus, has a uniform distribution on the interval from 0 to an unknown upper limit θ (so the density function of X is rectangular in shape with height $1/\theta$ for $0 \le x \le \theta$). It is desired to estimate θ on the basis of a random sample $X_1, X_2, \ldots, X_n$ of reaction times. Since θ is the largest possible time in the entire population of reaction times, consider as a first estimator the largest sample reaction time: $\hat{\theta}_1 = \max(X_1, \ldots, X_n)$. If $n = 5$ and $x_1 = 4.2$, $x_2 = 1.7$, $x_3 = 2.4$, $x_4 = 3.9$, $x_5 = 1.3$, the point estimate of θ is $\hat{\theta}_1 = \max(4.2, 1.7, 2.4, 3.9, 1.3) = 4.2$.

Unbiasedness implies that some samples will yield estimates that exceed θ and other samples will yield estimates smaller than θ—otherwise θ couldn't possibly be the center (balance point) of $\hat{\theta}_1$'s distribution. However, our proposed estimator will never overestimate θ (the largest sample value can't exceed the largest population value) and will underestimate θ unless the largest sample value equals θ. This intuitive argument shows that $\hat{\theta}_1$ is a biased estimator. More precisely, it can be shown (see Exercise 30) that $E(\hat{\theta}_1) = n\theta/(n + 1)$. Since $n/(n + 1) < 1$, $E(\hat{\theta}_1) < \theta$—there is a tendency for $\hat{\theta}_1$ to underestimate θ. The bias of $\hat{\theta}_1$ is $n\theta/(n + 1) - \theta = -\theta/(n + 1)$, which approaches 0 as n gets large.

It is easy to modify $\hat{\theta}_1$ to obtain an unbiased estimator of θ. Consider the estimator

$$\hat{\theta}_2 = \frac{n + 1}{n} \cdot \max(X_1, \ldots, X_n)$$

Using this estimator on the above data gives the estimate $(6/5)(4.2) = 5.04$. The fact that $(n + 1)/n > 1$ implies that $\hat{\theta}_2$ will overestimate θ for some samples and underestimate it for others. The mean value of this estimator is

$$E(\hat{\theta}_2) = E\left[\frac{n + 1}{n} \max(X_1, \ldots, X_n)\right] = \frac{n + 1}{n} \cdot E[\max(X_1, \ldots, X_n)]$$

$$= \frac{n + 1}{n} \cdot \frac{n}{n + 1} \theta = \theta$$

Thus $\hat{\theta}_2$ is an unbiased estimator. If it is used repeatedly on different samples to estimate θ, some estimates will be too large and others will be too small, but in the long run there will be no systematic tendency to underestimate or overestimate θ.   ∎

Principle of unbiased estimation

> When choosing between several different estimators of θ, select one that is unbiased.

According to this principle, the unbiased estimator $\hat{\theta}_2$ in Example 6.4 should be preferred to the biased estimator $\hat{\theta}_1$. Consider now the problem of estimating σ^2.

Proposition

> Let $X_1, X_2, \ldots, X_n$ be a random sample from a distribution with mean μ and variance σ^2. Then the estimator
>
> $$\hat{\sigma}^2 = S^2 = \frac{\Sigma(X_i - \bar{X})^2}{n - 1}$$
>
> is an unbiased estimator of σ^2.

Proof.   For any r.v. Y, $V(Y) = E(Y^2) - [E(Y)]^2$, so $E(Y^2) = V(Y) + [E(Y)]^2$. Applying this to

$$S^2 = \frac{1}{n - 1}\left[\Sigma X_i^2 - \frac{(\Sigma X_i)^2}{n}\right]$$

gives

$$E(S^2) = \frac{1}{n - 1}\left\{\Sigma E(X_i^2) - \frac{1}{n} E[(\Sigma X_i)^2]\right\}$$

$$= \frac{1}{n - 1}\left\{\Sigma(\sigma^2 + \mu^2) - \frac{1}{n}\{V(\Sigma X_i) + [E(\Sigma X_i)]^2\}\right\}$$

$$= \frac{1}{n - 1}\left\{n\sigma^2 + n\mu^2 - \frac{1}{n} n\sigma^2 - \frac{1}{n}(n\mu)^2\right\}$$

$$= \frac{1}{n - 1}\{n\sigma^2 - \sigma^2\} = \sigma^2 \quad \text{as desired} \qquad \blacksquare$$

The estimator which uses divisor n can be expressed as $(n - 1)S^2/n$, so

$$E\left[\frac{(n - 1)S^2}{n}\right] = \frac{n - 1}{n} E(S^2) = \frac{n - 1}{n} \sigma^2$$

This estimator is therefore not unbiased. The bias is $(n - 1)\sigma^2/n - \sigma^2 = -\sigma^2/n$. Because the bias is negative, the estimator with divisor n tends to underestimate σ^2, and this is why the $n - 1$ divisor is preferred by many statisticians (though when n is large, the bias is small and there is little difference between the two).

Although S^2 is unbiased for σ^2, S is a biased estimator of σ (its bias is small unless n is quite small). However, there are other good reasons to use S as an estimator, especially when the population distribution is normal. These will become more apparent when we discuss confidence intervals and hypothesis testing in the next several chapters.

In Example 6.2 we proposed several different estimators for the mean μ of a normal distribution. If there were a unique unbiased estimator for μ, the estimation problem would be resolved by using that estimator. Unfortunately, this is not the case.

Proposition

> If $X_1, X_2, \ldots, X_n$ is a random sample from a distribution with mean μ, then $\overline{X}$ is an unbiased estimator of μ. If in addition the distribution is continuous and symmetric, then $\tilde{X}$ and any trimmed mean are also unbiased estimators of μ.

The fact that $\overline{X}$ is unbiased is just a restatement of one of our rules of expected value: $E(\overline{X}) = \mu$ for every possible value of μ (for discrete as well as continuous distributions). The unbiasedness of the other estimators is more difficult to verify.

According to this proposition, the principle of unbiasedness by itself does not always allow us to select a single estimator. When the underlying population is normal, even the third estimator in Example 6.2 is unbiased, and there are many other unbiased estimators. What we now need is a way of selecting among unbiased estimators.

Estimators with Minimum Variance

Suppose that $\hat{\theta}_1$ and $\hat{\theta}_2$ are two estimators of θ that are both unbiased. Then although the distribution of each estimator is centered at the true value of θ, the spreads of the distributions about the true value may be different.

Principle of minimum variance unbiased estimation

> Among all estimators of θ that are unbiased, choose the one that has minimum variance. The resulting $\hat{\theta}$ is called the minimum variance unbiased estimator (MVUE) of θ.

Figure 6.2 pictures the p.d.f.'s of two unbiased estimators, with $\hat{\theta}_1$ having smaller variance than $\hat{\theta}_2$. Then $\hat{\theta}_1$ is more likely than $\hat{\theta}_2$ to produce an estimate close to the true θ. The MVUE is in a certain sense the most likely among all unbiased estimators to produce an estimate close to the true θ.

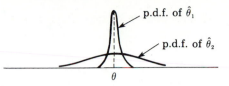

Figure 6.2 Graphs of the p.d.f.'s of two different unbiased estimators

Example 6.5

We argued in Example 6.4 that when $X_1, \ldots, X_n$ is a random sample from a uniform distribution on $[0, \theta]$, the estimator

$$\hat{\theta}_1 = \frac{n+1}{n} \cdot \max(X_1, \ldots, X_n)$$

is unbiased for θ (we previously denoted this estimator by $\hat{\theta}_2$). This is not the only unbiased estimator of θ. The expected value of a uniformly distributed random variable is just the midpoint of the interval of positive density, so $E(X_i) = \theta/2$. This implies that $E(\overline{X}) = \theta/2$, from which $E(2\overline{X}) = \theta$. That is, the estimator $\hat{\theta}_2 = 2\overline{X}$ is unbiased for θ.

If X is uniformly distributed on the interval $[A, B]$, $V(X) = \sigma^2 = (B - A)^2/12$. Thus in our situation $V(X_i) = \theta^2/12$, $V(\overline{X}) = \sigma^2/n = \theta^2/12n$, and $V(\hat{\theta}_2) = V(2\overline{X}) = 4V(\overline{X}) = \theta^2/3n$. The results of Exercise 30 can be used to show that $V(\hat{\theta}_1) = \theta^2/[n(n + 2)]$. The estimator $\hat{\theta}_1$ has smaller variance than does $\hat{\theta}_2$ if $3n < n(n + 2)$; that is, if $0 < n^2 - n = n(n - 1)$. As long as $n > 1$, $V(\hat{\theta}_1) < V(\hat{\theta}_2)$, so $\hat{\theta}_1$ is a better estimator than is $\hat{\theta}_2$. More advanced methods can be used to show that $\hat{\theta}_1$ is the MVUE of θ—every other unbiased estimator of θ has variance that exceeds $\theta^2/[n(n + 2)]$. ∎

One of the triumphs of mathematical statistics has been the development of methodology for identifying the MVUE in a wide variety of situations. The most important result of this type for our purposes concerns estimating the mean μ of a normal distribution.

Theorem

> Let $X_1, \ldots, X_n$ be a random sample from a normal distribution with parameters μ and σ. Then the estimator $\hat{\mu} = \overline{X}$ is the MVUE for μ.

The proof of this result can be found in the book *Introduction to Mathematical Statistics* by Hogg and Craig. Whenever we are convinced that the population being sampled is normal, the result says that $\overline{X}$ should be used to estimate μ. In Example 6.2, then, our estimate would be $\bar{x} = 28.50$.

In some situations it is possible to obtain an estimator with small bias that would be preferred to the best unbiased estimator. This is illustrated in Figure 6.3. However, MVUE's are often easier to obtain than the type of biased estimator whose distribution is pictured.

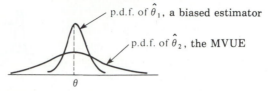

Figure 6.3   A biased estimator that is preferable to the MVUE

Some Complications

The last theorem does not say that in estimating a population mean μ, the estimator $\overline{X}$ should be used irrespective of the distribution being sampled.

Example 6.6   Suppose that we wish to estimate the thermal conductivity μ of a certain material. Using standard measurement techniques, we will obtain a random sample $X_1, \ldots, X_n$ of n thermal conductivity measurements. Let us assume that the p.d.f. of each of these measurements belongs to one of the following families:

$$f(x) = \frac{1}{\sqrt{2\pi\sigma^2}} e^{-(1/2\sigma^2)(x - \mu)^2} \quad -\infty < x < \infty \tag{6.1}$$

$$f(x) = \frac{1}{\pi[1 + (x - \mu)^2]} \quad -\infty < x < \infty \tag{6.2}$$

$$f(x) = \begin{cases} \dfrac{1}{2c} & -c \le x - \mu \le c \\ 0 & \text{otherwise} \end{cases} \tag{6.3}$$

The p.d.f. (6.1) is the normal distribution, (6.2) is called the Cauchy distribution, and (6.3) is a uniform distribution. All three of the distributions are symmetric about μ, and in fact the Cauchy distribution is bell-shaped but with much heavier tails (more probability further out) than the normal curve. The uniform distribution has no tails. The four estimators for μ considered earlier are $\overline{X}$, $\tilde{X}$, $\overline{X}_e$ (the average of the two extreme observations), and $\overline{X}_{\text{tr}(10)}$, a trimmed mean.

The very important moral here is that the best estimator for μ depends crucially on which distribution is being sampled. In particular

1. If the random sample comes from a normal distribution, then $\overline{X}$ is the best of the four estimators, since it has minimum variance among all unbiased estimators.

2. If the random sample comes from a Cauchy distribution, then $\overline{X}$ and $\overline{X}_e$ are terrible estimators for μ, while $\tilde{X}$ is quite good (the MVUE is not known); $\overline{X}$ is bad because it is very sensitive to outlying observations, and the heavy tails of the Cauchy distribution make a few such observations likely to appear in any sample.

3. If the underlying distribution is uniform, the best estimator is $\overline{X}_e$; this estimator is greatly influenced by outlying observations, but the lack of tails makes such observations impossible.

4. *The trimmed mean is best in none of these three situations, but works reasonably well in all three.* That is, $\overline{X}_{tr(10)}$ does not suffer too much in comparison with the best procedure in any of the three situations. ∎

More generally, recent research in statistics has established that when estimating a point of symmetry μ of a continuous probability distribution, a trimmed mean with trimming proportion 10% or 20% (from each end of the sample) produces reasonably behaved estimates over a very wide range of possible models. For this reason a trimmed mean with small trimming percentage is said to be a **robust estimator.**

In some situations the choice is not between two different estimators constructed from the same sample, but instead between estimators based on two different experiments.

Example 6.7 Suppose that a certain type of component has a lifetime distribution that is exponential with parameter λ, so that expected lifetime is $\mu = 1/\lambda$. A sample of n such components is selected, and each is put into operation. If the experiment is continued until all n lifetimes $X_1, \ldots, X_n$ have been observed, then $\overline{X}$ is an unbiased estimator of μ.

In some experiments, though, the components are left in operation only until the time of the rth failure, where $r < n$. This procedure is referred to as **censoring.** Let Y_1 denote the time of the first failure (the minimum lifetime among the n components), Y_2 denote the time at which the second failure occurs (the second smallest lifetime), and so on. Since the experiment terminates at time Y_r, the total accumulated lifetime at termination is

$$T_r = \sum_{i=1}^{r} Y_i + (n - r)Y_r$$

We now demonstrate that $\hat{\mu} = T_r/r$ is an unbiased estimator for μ. To do so, we need two properties of exponential variables:

1. The memoryless property (see Section 4.4), which says that at any time point, remaining lifetime has the same exponential distribution as original lifetime.

2. If $X_1, \ldots, X_k$ are independent, each exponential with parameter λ, then $\min(X_1, \ldots, X_k)$ is exponential with parameter $k\lambda$ and has expected value $1/k\lambda$.

Since all n components last until Y_1, $n - 1$ last an additional $Y_2 - Y_1$, $n - 2$ an additional $Y_3 - Y_2$ amount of time, and so on, another expression for T_r is

$$T_r = nY_1 + (n - 1)(Y_2 - Y_1) + (n - 2)(Y_3 - Y_2) + \cdots$$
$$+ (n - r + 1)(Y_r - Y_{r-1})$$

But Y_1 is the minimum of n exponential variables, so $E(Y_1) = 1/n\lambda$. Similarly, $Y_2 - Y_1$ is the smallest of the $n-1$ remaining lifetimes, each exponential with parameter λ (by the memoryless property), so $E(Y_2 - Y_1) = 1/(n-1)\lambda$. Continuing, $E(Y_{i+1} - Y_i) = 1/(n-i)\lambda$, so

$$E(T_r) = nE(Y_1) + (n-1)\,E(Y_2 - Y_1) + \cdots + (n-r+1)\,E(Y_r - Y_{r-1})$$

$$= n \cdot \frac{1}{n\lambda} + (n-1) \cdot \frac{1}{(n-1)\lambda} + \cdots + (n-r-1) \cdot \frac{1}{(n-r-1)\lambda}$$

$$= \frac{r}{\lambda}$$

Therefore, $E(T_r/r) = (1/r)\,E(T_r) = 1/r \cdot r/\lambda = 1/\lambda = \mu$ as claimed.

As an example, suppose that 20 components are put on test and $r = 10$. Then if the first 10 failure times are 11, 15, 29, 33, 35, 40, 47, 55, 58, and 72, the estimate of μ is

$$\hat{\mu} = \frac{11 + 15 + \cdots + 72 + (10)(72)}{10} = 111.5$$

The advantage of the experiment with censoring is that it terminates more quickly than the uncensored experiment. However, it can be shown that $V(T_r/r) = 1/\lambda^2 r$, which is larger than $1/\lambda^2 n$, the variance of $\overline{X}$ in the uncensored experiment.   ■

Reporting a Point Estimate: The Standard Error

In addition to reporting the value of a point estimate, some indication of its precision should be given. The usual measure of precision is the standard error of the estimator used.

Definition

> The **standard error** of an estimator $\hat{\theta}$ is its standard deviation $\sigma_{\hat{\theta}} = \sqrt{V(\hat{\theta})}$. If the standard error itself involves unknown parameters whose values can be estimated, substitution of these estimates into $\sigma_{\hat{\theta}}$ yields the **estimated standard error** (estimated standard deviation) of the estimator. The estimated standard error can be denoted either by $\hat{\sigma}_{\hat{\theta}}$ (the ^ over σ emphasizes that $\sigma_{\hat{\theta}}$ is being estimated) or by $s_{\hat{\theta}}$.

Example **6.8**
(Example 6.2
continued)

Assuming that writing lifetime is normally distributed, $\hat{\mu} = \overline{X}$ is the best estimator of μ. If the value of σ is known to be 3.5, the standard error of $\overline{X}$ is $\sigma_{\overline{X}} = \sigma/\sqrt{n} = 3.5/\sqrt{10} = 1.11$. If, as is usually the case, the value of σ is unknown, the estimate $\hat{\sigma} = s = 3.45$ is substituted into $\sigma_{\overline{X}}$ to obtain the estimated standard error $\hat{\sigma}_{\overline{X}} = s_{\overline{X}} = s/\sqrt{n} = 3.45/\sqrt{10} = 1.09$.   ■

Example **6.9**
(Example 6.1
continued)

The standard error of $\hat{p} = X/n$ is

$$\sigma_{\hat{p}} = \sqrt{V(X/n)} = \sqrt{\frac{V(X)}{n^2}} = \sqrt{\frac{npq}{n^2}} = \sqrt{\frac{pq}{n}}$$

Since p (and $q = 1 - p$) are unknown (else why estimate?), we substitute $\hat{p} = x/n$ and $\hat{q} = 1 - x/n$ into $\sigma_{\hat{p}}$, yielding the estimated standard error $\hat{\sigma}_{\hat{p}} = \sqrt{\hat{p}\hat{q}/n} = \sqrt{(.6)(.4)/25} = .098$. Alternatively, since the largest value of pq is attained when $p = q = .5$, an upper bound on the standard error is $\sqrt{1/4n} = .10$. ■

When the point estimator $\hat{\theta}$ has approximately a normal distribution, which will often be the case when n is large, then we can be reasonably confident that the true value of θ lies within approximately two standard errors (standard deviations) of $\hat{\theta}$. Thus in the pen lifetime example, if $n = 36$, $\hat{\mu} = \bar{x} = 28.50$, and $s = 3.60$, then $s/\sqrt{n} = .60$, so within two estimated standard errors of $\hat{\mu}$ translates to the interval $28.50 \pm (2)(.60) = (27.30, 29.70)$.

If $\hat{\theta}$ is not necessarily approximately normal but is unbiased, then it can be shown that the estimate will deviate from θ by as much as four standard errors at most 6% of the time. We would then expect the true value to lie within four standard errors of $\hat{\theta}$ (and this is a very conservative statement, since it applies to *any* unbiased $\hat{\theta}$).

Summarizing, the standard error tells us roughly within what distance of $\hat{\theta}$ we can expect the true value of θ to lie.

Exercises / Section 6.1 (1–15)

1. The interpupillary distance was measured for each of 10 randomly selected adult males, yielding the following measurements (mm): 64.9, 62.1, 70.8, 65.3, 68.7, 57.2, 62.5, 69.8, 64.3, 71.6.
   a. Assuming that interpupillary distance is normally distributed, use the sample mean $\bar{X}$ and the sample standard deviation S to obtain estimates of μ and σ.
   b. Supposing only that the distribution is continuous, use the sample median $\tilde{X}$ to obtain an estimate of the population median $\tilde{\mu}$. Why in this situation wouldn't you use $\bar{X}$ to estimate $\tilde{\mu}$?
   c. Assuming that interpupillary distance has a symmetric, but not necessarily normal, distribution, use a 10% trimmed mean to obtain an estimate of μ.
   d. Let p denote the proportion of all males having interpupillary distance between 60 and 70. Determine the number of sample x_i's that satisfy $60 \le x_i \le 70$, and use it to estimate the population proportion p.

2. A sample of 20 students who had recently taken elementary statistics yielded the following information on brand of calculator owned (T = Texas Instruments, H = Hewlett-Packard, C = Casio, S = Sharp): T, T, H, T, C, T, T, S, C, H, S, S, T, H, C, T, T, T, H, T
   a. Estimate the true proportion of all such students who own a Texas Instruments calculator.
   b. Among the brands owned, only Hewlett-Packard uses reverse Polish logic. Estimate the proportion of all such students who own a calculator that does not use reverse Polish logic.
   c. Three of the four Hewlett-Packard calculators in the sample were programmable. Estimate the proportion of all students who own a calculator that is both made by H-P and is programmable.

3. Twelve samples of a certain brand of white bread were analyzed and the carbohydrate content (percentage of nitrogen-free extract) of each sample was determined, yielding the following data:

   76.93, 76.88, 77.07, 76.68, 76.39, 75.09,
   76.88, 77.67, 78.15, 76.50, 77.16, 76.42

a. Assuming that the distribution of carbohydrate content is normal, estimate the true average carbohydrate content for this brand. What estimator did you use?

b. Making no assumption whatsoever about the distribution of carbohydrate content, estimate the value below which (and above which) half of all carbohydrate contents will fall. What estimator did you use?

c. Again making no assumption about the underlying distribution, estimate the true proportion of samples whose carbohydrate content would not exceed 76%.

d. What is the estimated standard error of the estimator that you used in (a)?

4. In Exercise 3, let $X_1, \ldots, X_m$ denote the random variables observed, assumed to be a random sample from a distribution with mean μ_1 and variance σ_1^2. For n samples of a second brand, let $Y_1, \ldots, Y_n$ denote the carbohydrate contents, assumed to be a random sample from a distribution with mean μ_2 and variance σ_2^2.

a. Show using rules of expected value that $\overline{X} - \overline{Y}$ is an unbiased estimator of $\mu_1 - \mu_2$. If the observed $\overline{Y}$ is $\overline{y} = 74.28$ for $n = 14$ samples, what is the point estimate?

b. Use the rules of variance from Section 5.3 to obtain an expression for the variance and standard deviation (standard error) of $\overline{X} - \overline{Y}$.

c. If σ_1 and σ_2 are unknown, but $s_1 = .75$, $s_2 = .70$, compute the estimated standard error of $\overline{X} - \overline{Y}$.

5. Consider the accompanying observations on streamflow (1000's of acre feet) recorded at a station in Colorado for the period April 1–August 31 over a 31-year span (from an article in the 1974 volume of *Water Resources Research*).

127.96	210.07	203.24	108.91	178.21
285.37	100.85	89.59	185.36	126.94
200.19	66.24	247.11	299.87	109.64
125.86	114.79	109.11	330.33	85.54
117.64	302.74	280.55	145.11	95.36
204.91	311.13	150.58	262.09	477.08
94.33				

An appropriate probability plot supports the use of the lognormal distribution (see Section 4.5) as a reasonable model for streamflow.

a. Estimate the parameters of the distribution. *Hint:* Remember that X has a lognormal distribution with parameters μ and σ^2 if $\ln(X)$ is normally distributed with mean μ and variance σ^2.

b. Use the estimates of (a) to calculate an estimate of the expected value of streamflow. *Hint:* What is $E(X)$?

6. Each of 150 newly manufactured items is examined and the number of scratches per item is recorded (the items are supposed to be free of scratches), yielding the following data.

Number of scratches per item	0	1	2	3	4	5	6	7
Observed frequency	18	37	42	30	13	7	2	1

Let X = the number of scratches on a randomly chosen item, and assume that X has a Poisson distribution with parameter λ.

a. Find an unbiased estimator of λ and compute the estimate for the above data. *Hint:* $E(X) = \lambda$ for X Poisson, so $E(\overline{X}) = ?$

b. What is the standard deviation (standard error) of your estimator? Compute the estimated standard error. *Hint:* $\sigma_X^2 = \lambda$ for X Poisson.

7. Using a long rod that has length μ, you are going to lay out a square plot in which the length of each side is μ. Thus the area of the plot will be μ^2. However, you don't know the value of μ, so you decide to make n independent measurements $X_1, X_2, \ldots, X_n$ of the length. Assume that each X_i has mean μ (unbiased measurements) and variance σ^2.

a. Show that $\overline{X}^2$ is not an unbiased estimator for μ^2. *Hint:* For any r.v. Y, $E(Y^2) = V(Y) + [E(Y)]^2$. Apply this with $Y = \overline{X}$.

b. For what value of k is the estimator $\overline{X}^2 - kS^2$ unbiased for μ^2? *Hint:* Compute $E(\overline{X}^2 - kS^2)$.

8. Of n_1 randomly selected male smokers, X_1 smoked filter cigarettes, while of n_2 randomly selected female smokers, X_2 smoked filter cigarettes. Let p_1 and p_2 respectively denote the probabilities that a randomly selected male and female smoke filter cigarettes.

a. Show that $(X_1/n_1) - (X_2/n_2)$ is an unbiased estimator for $p_1 - p_2$. *Hint:* $E(X_i) = n_i p_i$ for $i = 1, 2$.

b. What is the standard error of the estimator in (a)?

c. How would you use the observed values x_1 and x_2 to estimate the standard error of your estimator?

d. If $n_1 = n_2 = 200$, $x_1 = 127$, $x_2 = 176$, use the estimator of (a) to obtain an estimate of $p_1 - p_2$.

e. Use the result of (c) and the data of (d) to estimate the standard error of the estimator.

9. Suppose that a certain type of fertilizer has an expected yield per acre of μ_1 with variance σ^2, while the expected yield for a second type of fertilizer is μ_2 with the same variance σ^2. Let S_1^2 and S_2^2 denote the sample variances of yields based on sample sizes n_1 and n_2, respectively, of the two fertilizers. Show that the pooled (combined) estimator

$$\hat{\sigma}^2 = \frac{(n_1 - 1)S_1^2 + (n_2 - 1)S_2^2}{n_1 + n_2 - 2}$$

is unbiased for σ^2 (this estimator will be used in Chapter 9).

10. A sample of n captured Pandamonian jet fighters results in serial numbers $x_1, x_2, x_3, \ldots, x_n$. The CIA knows that the aircraft were numbered consecutively at the factory starting with α and ending with β, so that the total number of planes manufactured is $\beta - \alpha + 1$ (that is, if $\alpha = 17$ and $\beta = 29$, then $29 - 17 + 1 = 13$ planes having serial numbers 17, 18, 19, ..., 28, 29 were manufactured). However, the CIA does not know the values of α or β. A CIA statistician suggests using the estimator $\max(X_i) - \min(X_i) + 1$ to estimate the total number of planes manufactured.

a. If $n = 5$, $x_1 = 237$, $x_2 = 375$, $x_3 = 202$, $x_4 = 525$, and $x_5 = 418$, what is the corresponding estimate?

b. Under what conditions on the sample will the value of the estimate be exactly equal to the true total number of planes? Will the estimate ever be larger than the true total? Do you think the estimator is unbiased for estimating $\beta - \alpha + 1$? Explain in one or two sentences.

11. Let $X_1, X_2, \ldots, X_n$ represent a random sample from a Rayleigh distribution with p.d.f.

$$f(x; \theta) = \frac{x}{\theta} e^{-x^2/2\theta} \quad x > 0$$

a. It can be shown that $E(X^2) = 2\theta$. Use this fact to construct an unbiased estimator of θ (and use rules of expected value to show that it is unbiased).

b. Estimate θ from the accompanying $n = 10$ observations on vibratory stress of a turbine blade under specified conditions:

16.88	10.23	4.59	6.66	13.68
14.23	19.87	9.40	6.51	10.95

12. Suppose that the true average growth μ of one type of plant during a one-year period is identical to that of a second type, but the variance of growth for the first type is σ^2 while for the second type the variance is $4\sigma^2$. Let $X_1, \ldots, X_m$ be m independent growth observations on the first type [so $E(X_i) = \mu$, $V(X_i) = \sigma^2$], and let $Y_1, \ldots, Y_n$ be n independent growth observations on the second type [$E(Y_i) = \mu$, $V(Y_i) = 4\sigma^2$].

a. Show that for any δ between 0 and 1, $\hat{\mu} = \delta \overline{X} + (1 - \delta)\overline{Y}$ is unbiased for μ.

b. For fixed m and n, compute $V(\hat{\mu})$ and then find the value of δ which minimizes $V(\hat{\mu})$. *Hint:* Differentiate $V(\hat{\mu})$ with respect to δ.

13. In Chapter 3 we defined a negative binomial r.v. as the number of failures that occur before the rth success in a sequence of independent and identical success/failure trials. The p.m.f. of X is

$$nb(x; r, p)$$

$$= \begin{cases} \binom{x + r - 1}{x} p^r (1 - p)^x & x = 0, 1, 2, \ldots \\ 0 & \text{otherwise} \end{cases}$$

a. Suppose that $r \geq 2$. Show that

$$\hat{p} = (r - 1)/(X + r - 1)$$

is an unbiased estimator for p. *Hint:* Write out $E(\hat{p})$ and cancel $x + r - 1$ inside the sum.

b. A reporter wishes to interview five individuals who support a certain candidate, so begins asking people whether (S) or not (F) they support the candidate. If the sequence of responses is $SFFSFFFSSS$, estimate p = the true proportion who support the candidate.

14. Let $X_1, X_2, \ldots, X_n$ be a random sample from a p.d.f. $f(x)$ that is symmetric about μ, so that $\tilde{X}$ is an unbiased estimator of μ. If n is large, it can be shown that $V(\tilde{X}) \approx 1/4n[f(\mu)]^2$.
   a. Compare $V(\tilde{X})$ to $V(\overline{X})$ when the underlying distribution is normal.
   b. When the underlying p.d.f. is Cauchy (see Example 6.6), $V(\overline{X}) = \infty$, so $\overline{X}$ is a terrible estimator. What is $V(\tilde{X})$ in this case when n is large?

15. An investigator wishes to estimate the proportion of students at a certain university who have violated the honor code. Having obtained a random sample of n students, she realizes that asking each, "Have you violated the honor code?" will probably result in some untruthful responses. Consider the following scheme, called a **randomized response** technique. The investigator makes up a deck of 100 cards, of which 50 are of type I and 50 are of type II.

   Type I: Have you violated the honor code (yes or no)?

   Type II: Is the last digit of your telephone number a 0, 1, or 2 (yes or no)?

   Each student in the random sample is asked to mix the deck, draw a card, and answer the resulting question truthfully. Because of the irrelevant question on type-II cards, a yes response no longer stigmatizes the respondent, so we assume that responses are truthful. Let p denote the proportion of honor code violators (that is, the probability of a randomly selected student being a violator), and let $\lambda = P(\text{yes response})$. Then λ and p are related by $\lambda = .5p + (.5)(.3)$.
   a. Let Y denote the number of yes responses, so $Y \sim \text{Bin}(n, \lambda)$. Thus Y/n is an unbiased estimator of λ. Derive an estimator for p based on Y. If $n = 80$ and $y = 20$, what is your estimate? *Hint:* Solve $\lambda = .5p + .15$ for p and then substitute Y/n in place of λ.
   b. Use the fact that $E(Y/n) = \lambda$ to show that your estimator $\hat{p}$ is unbiased.
   c. If there were 70 type-I and 30 type-II cards, what would your estimator for p be?

6.2   Methods of Point Estimation

There are many estimation problems in which an unbiased estimator cannot be obtained. We now discuss two "constructive" methods for obtaining point estimators, the method of moments and method of maximum likelihood. By constructive we mean that the general definition of each type of estimator suggests explicitly how to obtain the estimator in any specific problem. While maximum likelihood estimators are generally preferable to moment estimators because of certain efficiency properties, they often require significantly more computation than do moment estimators.

The Method of Moments

The basic idea of this method is to equate certain sample characteristics, such as the mean, to the corresponding population expected values. Then solving these equations for unknown parameter values yields the estimators.

Definition

> Let $X_1, \ldots, X_n$ be a random sample from a p.m.f. or p.d.f. $f(x)$. For $k = 1, 2, 3, \ldots$ the **kth population moment,** or **kth moment of the distribution $f(x)$,** is $E(X^k)$. The **kth sample moment** is $(1/n) \sum_{i=1}^{n} X_i^k$.

Definition

> Let $X_1, X_2, \ldots, X_n$ be a random sample from a distribution with p.m.f. or p.d.f. $f(x; \theta_1, \ldots, \theta_m)$, where $\theta_1, \ldots, \theta_m$ are parameters whose values are unknown. Then the **moment estimators** $\hat{\theta}_1, \ldots, \hat{\theta}_m$ are obtained by equating the first m sample moments to the corresponding first m population moments and solving for $\theta_1, \ldots, \theta_m$.

If, for example, $m = 2$, $E(X)$ and $E(X^2)$ will be functions of θ_1 and θ_2. Setting $E(X) = (1/n) \Sigma X_i (= \overline{X})$ and $E(X^2) = (1/n) \Sigma X_i^2$ gives two equations in θ_1 and θ_2. The solution then defines the estimators. For estimating a population mean μ, the method gives $\mu = \overline{X}$, so the estimator is the sample mean.

Example 6.10

Let $X_1, X_2, \ldots, X_n$ represent a random sample of service times of n customers at a certain facility, where the underlying distribution is assumed exponential with parameter λ. Since there is only one parameter to be estimated, the estimator is obtained by equating $E(X)$ to $\overline{X}$. Since $E(X) = 1/\lambda$ for an exponential distribution, this gives $1/\lambda = \overline{X}$ or $\lambda = 1/\overline{X}$. The moment estimator of λ is then $\hat{\lambda} = 1/\overline{X}$. ∎

Example 6.11

Let $X_1, \ldots, X_n$ be a random sample from a gamma distribution with parameters α and β. From Chapter 4, $E(X) = \alpha\beta$ and $E(X^2) = \beta^2 \Gamma(\alpha + 2)/\Gamma(\alpha) = \beta^2(\alpha + 1)\alpha$. The moment estimators of α and β are obtained from solving

$$\overline{X} = \alpha\beta, \quad \frac{1}{n} \Sigma X_i^2 = \alpha(\alpha + 1)\beta^2$$

Since $\alpha(\alpha + 1)\beta^2 = \alpha^2\beta^2 + \alpha\beta^2$ and the first equation implies $\alpha^2\beta^2 = \overline{X}^2$, the second equation becomes

$$\frac{1}{n} \Sigma X_i^2 - \overline{X}^2 = \alpha\beta^2$$

Now dividing each side of this second equation by the corresponding side of the first equation and substituting back gives the estimators

$$\hat{\alpha} = \frac{\overline{X}^2}{(1/n) \Sigma X_i^2 - \overline{X}^2}, \quad \hat{\beta} = \frac{(1/n) \Sigma X_i^2 - \overline{X}^2}{\overline{X}}$$

To illustrate, the survival time data mentioned in Example 4.22 is

152, 115, 109, 94, 88, 137, 152, 77, 160, 165, 125, 40, 128, 123, 136, 101, 62, 153, 83, 69

with $\overline{x} = 113.5$ and $(1/20) \Sigma x_i^2 = 14{,}087.8$. The estimates are

$$\hat{\alpha} = \frac{(113.5)^2}{14{,}087.8 - (113.5)^2} = 10.7, \quad \hat{\beta} = \frac{14{,}087.8 - (113.5)^2}{113.5} = 10.6$$

These estimates of α and β differ from the values suggested by Gross and Clark because they used a different estimation technique.   ∎

Example 6.12   Let $X_1, \ldots, X_n$ be a random sample from a generalized negative binomial distribution with parameters r and p (Section 3.5). Since $E(X) = r(1 - p)/p$ and $V(X) = r(1 - p)/p^2$, $E(X^2) = V(X) + [E(X)]^2 = r(1 - p)(r - rp + 1)/p^2$. Equating $E(X)$ to $\overline{X}$ and $E(X^2)$ to $(1/n) \Sigma X_i^2$ gives eventually

$$\hat{p} = \frac{\overline{X}}{(1/n) \, \Sigma X_i^2 - \overline{X}^2}, \quad \hat{r} = \frac{\overline{X}^2}{(1/n) \, \Sigma X_i^2 - \overline{X}^2 - \overline{X}}$$

As an illustration, Reep, Pollard, and Benjamin ("Skill and Chance in Ball Games," *J. Royal Stat. Soc.*, 1971, pp. 623–629) consider the negative binomial distribution as a model for the number of goals per game scored by National Hockey League teams. The data for 1966–1967 follows (420 games):

Goals	0	1	2	3	4	5	6	7	8	9	10
Freq.	29	71	82	89	65	45	24	7	4	1	3

Then

$$\bar{x} = \Sigma x_i / 420 = [(0)(29) + (1)(71) + \cdots + (10)(3)]/420 = 2.98$$

and

$$\Sigma x_i^2 / 420 = [(0)^2(29) + (1)^2(71) + \cdots + (10)^2(3)]/420 = 12.40$$

Thus

$$\hat{p} = \frac{2.98}{12.40 - (2.98)^2} = .85, \quad \hat{r} = \frac{(2.98)^2}{12.40 - (2.98)^2 - 2.98} = 16.5$$

Although r by definition must be positive, the denominator of $\hat{r}$ could be negative, indicating that the negative binomial distribution is not appropriate (or that the moment estimator is flawed).   ∎

Maximum Likelihood Estimation

The method of maximum likelihood was first introduced by R. A. Fisher, a geneticist and statistician, in the 1920s. Most statisticians recommend this method, at least when the sample size is large, since the resulting estimators have certain desirable efficiency properties (see the proposition on page 245).

Example 6.13   A sample of 10 new phonograph records pressed by a certain record company is obtained. Upon opening each record jacket, it is found that the first, third, and tenth records are warped while the others are not. Let $p = P(\text{warped record})$ and define $X_1, \ldots, X_{10}$ by $X_i = 1$ if the ith record is warped and 0 otherwise. Then the observed x_i's are 1, 0, 1, 0, 0, 0, 0, 0, 0, 1, so the joint p.m.f. of the sample is

$$f(x_1, x_2, \ldots, x_{10}; p) = p(1 - p)p \cdots p = p^3(1 - p)^7 \tag{6.4}$$

We now ask "for what value of p is the observed sample most likely to have occurred?" That is, we wish to find the value of p that maximizes (6.4), or equivalently maximizes the natural log of (6.4).* Since

$$\ln[f(x_1, \ldots, x_{10}; p)] = 3\ln(p) + 7\ln(1 - p) \tag{6.5}$$

which is a differentiable function of p, equating the derivative of (6.5) to zero gives the maximizing value:[†]

$$\frac{d}{dp}\ln[f(x_1, \ldots, x_{10}; p)] = \frac{3}{p} - \frac{7}{1 - p} = 0 \Rightarrow p = \frac{3}{10} = \frac{x}{n}$$

where x is the observed number of successes (warped records). The estimate of p is now $\hat{p} = \frac{3}{10}$. It is called the maximum likelihood estimate because for fixed $x_1, \ldots, x_{10}$, it is the parameter value that maximizes the likelihood (joint p.m.f.) of the observed sample.

Note that if we had been told only that among the 10 records there were 3 that were warped, (6.4) would be replaced by the binomial p.m.f. $\binom{10}{3}p^3(1 - p)^7$, which is also maximized for $\hat{p} = \frac{3}{10}$. ■

Definition

> Let $X_1, X_2, \ldots, X_n$ have joint p.m.f. or p.d.f.
>
> $$f(x_1, x_2, \ldots, x_n; \theta_1, \ldots, \theta_m), \tag{6.6}$$
>
> where the parameters $\theta_1, \ldots, \theta_m$ have unknown values. When $x_1, \ldots, x_n$ are the observed sample values and (6.6) is regarded as a function of $\theta_1, \ldots, \theta_m$, it is called the **likelihood function**. The maximum likelihood estimates (m.l.e.'s) $\hat{\theta}_1, \ldots, \hat{\theta}_m$ are those values of the θ_i's that maximize the likelihood function, so that
>
> $$f(x_1, \ldots, x_n; \hat{\theta}_1, \ldots, \hat{\theta}_m) \geq f(x_1, \ldots, x_n; \theta_1, \ldots, \theta_m) \text{ for all } \theta_1, \ldots, \theta_m.$$
>
> When the X_i's are substituted in place of the x_i's, the **maximum likelihood estimators** result.

The likelihood function tells us how likely the observed sample is as a function of the possible parameter values. Maximizing the likelihood gives the parameter values for which the observed sample is most likely to have been generated—that is, the parameter values which "agree most closely" with the observed data.

---

*Since $\ln[g(x)]$ is a monotonic function of $g(x)$, finding x to maximize $\ln[g(x)]$ is equivalent to maximizing $g(x)$ itself. In statistics, taking the logarithm frequently changes a product to a sum, which is easier to work with.
[†]This conclusion requires checking the second derivative, but the details are omitted.

Example **6.14** Suppose that $X_1, X_2, \ldots, X_n$ is a random sample from an exponential distribution with parameter λ. Because of independence, the likelihood function is a product of the individual p.d.f.'s:

$$f(x_1, \ldots, x_n; \lambda) = (\lambda e^{-\lambda x_1}) \cdots (\lambda e^{-\lambda x_n}) = \lambda^n e^{-\lambda \Sigma x_i}$$

The ln(likelihood) is

$$\ln[f(x_1, \ldots, x_n; \lambda)] = n \ln(\lambda) - \lambda \Sigma x_i$$

Equating $d/d\lambda$ [ln(likelihood)] to zero results in $n/\lambda - \Sigma x_i = 0$, or $\lambda = n/\Sigma x_i = 1/\bar{x}$. Thus the maximum likelihood estimator is $\hat{\lambda} = 1/\bar{X}$; it is identical to the method of moments estimator [but is not an unbiased estimator, since $E(1/\bar{X}) \neq 1/E(\bar{X})$]. ∎

Example **6.15** Let $X_1, \ldots, X_n$ be a random sample from a normal distribution. The likelihood function is

$$f(x_1, \ldots, x_n; \mu, \sigma^2) = \frac{1}{\sqrt{2\pi\sigma^2}} e^{-(1/2\sigma^2)(x_1 - \mu)^2} \cdots \frac{1}{\sqrt{2\pi\sigma^2}} e^{-(1/2\sigma^2)(x_n - \mu)^2}$$

$$= \left(\frac{1}{2\pi\sigma^2}\right)^{n/2} e^{-(1/2\sigma^2)\Sigma(x_i - \mu)^2}$$

so

$$\ln[f(x_1, \ldots, x_n; \mu, \sigma^2)] = -\frac{n}{2}\ln(2\pi\sigma^2) - \frac{1}{2\sigma^2}\Sigma(x_i - \mu)^2$$

To find the maximizing values of μ and σ^2, we must take the partial derivatives of $\ln(f)$ with respect to μ and σ^2, equate them to zero, and solve the resulting two equations. Omitting the details, the resulting m.l.e.'s are

$$\hat{\mu} = \bar{X}, \quad \hat{\sigma}^2 = \frac{\Sigma(X_i - \bar{X})^2}{n}$$

The m.l.e. of σ^2 is not the unbiased estimator, so two different principles of estimation (unbiasedness and maximum likelihood) yield two different estimators. ∎

Example **6.16** In Chapter 3 we discussed the use of the Poisson distribution for modeling the number of "events" that occur in a two-dimensional region. Assume that when the region R being sampled has area $a(R)$, the number of events X occurring in R has a Poisson distribution with parameter $\lambda a(R)$ (where λ is the expected number of events per unit area), and also that nonoverlapping regions yield independent X's.

Suppose that an ecologist selects n nonoverlapping regions $R_1, \ldots, R_n$ and counts the number of plants of a certain species found in each region. The joint p.m.f. (likelihood) is then

$$p(x_1, \ldots, x_n; \lambda) = \frac{[\lambda \cdot a(R_1)]^{x_1} e^{-\lambda \cdot a(R_1)}}{x_1!} \cdots \frac{[\lambda \cdot a(R_n)]^{x_n} e^{-\lambda \cdot a(R_n)}}{x_n!}$$

$$= \frac{[a(R_1)]^{x_1} \cdots [a(R_n)]^{x_n} \cdot \lambda^{\Sigma x_i} \cdot e^{-\lambda \Sigma a(R_i)}}{x_1! \cdots x_n!}$$

The ln(likelihood) is

$$\ln[p(x_1, \ldots, x_n; \lambda)] = \Sigma x_i \cdot \ln[a(R_i)] + \ln(\lambda) \cdot \Sigma x_i - \lambda \Sigma a(R_i) - \Sigma \ln(x_i!)$$

Taking $\dfrac{d}{d\lambda} \ln(p)$ and equating it to zero yields

$$\frac{\Sigma x_i}{\lambda} - \Sigma a(R_i) = 0, \quad \text{so} \quad \lambda = \frac{\Sigma x_i}{\Sigma a(R_i)}$$

The maximum likelihood estimator is then $\hat{\lambda} = \Sigma X_i / \Sigma a(R_i)$. This is intuitively reasonable because λ is the true density (plants per unit area), while $\hat{\lambda}$ is the sample density since $\Sigma a(R_i)$ is just the total area sampled. Because $E(X_i) = \lambda \cdot a(R_i)$, the estimator is unbiased.

Sometimes an alternative sampling procedure is used. Instead of fixing regions to be sampled, the ecologist will select n points in the entire region of interest and let $y_i =$ the distance from the ith point to the nearest plant. The c.d.f. of $Y =$ distance to the nearest plant is

$$F_Y(y) = P(Y \le y) = 1 - P(Y > y) = 1 - P\left(\begin{matrix}\text{no plants in a} \\ \text{circle of radius } y\end{matrix}\right)$$

$$= 1 - \frac{e^{-\lambda \pi y^2}(\lambda \pi y^2)^0}{0!} = 1 - e^{-\lambda \cdot \pi y^2}$$

Taking the derivative of $F_Y(y)$ with respect to y yields

$$f_Y(y; \lambda) = \begin{cases} 2\pi\lambda y e^{-\lambda \pi y^2} & y \ge 0 \\ 0 & \text{otherwise} \end{cases}$$

If we now form the likelihood $f_Y(y_1; \lambda) \cdots f_Y(y_n; \lambda)$, differentiate ln(likelihood), and so on, the resulting m.l.e. is

$$\hat{\lambda} = \frac{n}{\pi \Sigma Y_i^2} = \frac{\text{number of plants observed}}{\text{total area sampled}}$$

which is also a sample density. It can be shown that in a sparse environment (small λ), the distance method is in a certain sense better, while in a dense environment the first sampling method is better. ■

Example 6.17 Let $X_1, \ldots, X_n$ be a random sample from a Weibull p.d.f.

$$f(x; \alpha, \beta) = \begin{cases} \dfrac{\alpha}{\beta^\alpha} \cdot x^{\alpha - 1} \cdot e^{-(x/\beta)^\alpha} & x \ge 0 \\ 0 & \text{otherwise} \end{cases}$$

Writing the likelihood and ln(likelihood), then setting $\partial/\partial\alpha[\ln(f)] = 0$, $\partial/\partial\beta[\ln(f)] = 0$ yields the equations

$$\alpha = \left[\frac{\Sigma x_i^\alpha \cdot \ln(x_i)}{\Sigma x_i^\alpha} - \frac{\Sigma \ln(x_i)}{n}\right]^{-1}, \qquad \beta = \left(\frac{\Sigma x_i^\alpha}{n}\right)^{1/\alpha}$$

These two equations cannot be solved explicitly to give general formulas for the m.l.e.'s $\hat{\alpha}$ and $\hat{\beta}$. Instead, for each sample $x_1, \ldots, x_n$, the equations must be solved using an iterative numerical procedure. Even moment estimators of α and β are somewhat complicated (see Exercise 17). ■

Estimating Functions of Parameters

In Example 6.15 we obtained the maximum likelihood estimator of σ^2 when the underlying distribution was normal. The m.l.e. of $\sigma = \sqrt{\sigma^2}$, as well as many other m.l.e.'s, can be easily derived using the following proposition.

Proposition

> Let $\hat{\theta}_1, \hat{\theta}_2, \ldots, \hat{\theta}_m$ be the m.l.e.'s of the parameters $\theta_1, \theta_2, \ldots, \theta_m$. Then the m.l.e. of any function $h(\theta_1, \theta_2, \ldots, \theta_m)$ of these parameters is the function $h(\hat{\theta}_1, \hat{\theta}_2, \ldots, \hat{\theta}_m)$ of the m.l.e.'s.

Example 6.18
(Example 6.15 continued)

In the normal case, the m.l.e.'s of μ and σ^2 were $\hat{\mu} = \overline{X}$ and $\hat{\sigma}^2 = \Sigma(X_i - \overline{X})^2/n$. To obtain the m.l.e. of the function $h(\mu, \sigma^2) = \sqrt{\sigma^2} = \sigma$, substitute the m.l.e.'s into the function:

$$\hat{\sigma} = \sqrt{\hat{\sigma}^2} = \left[\frac{1}{n}\Sigma(X_i - \overline{X})^2\right]^{1/2}$$

The m.l.e. of σ is not the sample standard deviation S, though they are close unless n is quite small. ■

Example 6.19
(Example 6.17 continued)

The mean value of a r.v. X that has a Weibull distribution is

$$\mu = \beta \cdot \Gamma(1 + 1/\alpha)$$

The m.l.e. of μ is therefore $\hat{\mu} = \hat{\beta}\Gamma(1 + 1/\hat{\alpha})$, where $\hat{\alpha}$ and $\hat{\beta}$ are the m.l.e.'s of α and β. In particular, $\overline{X}$ is not the m.l.e. of μ, though it is an unbiased estimator. At least for large n, $\hat{\mu}$ is a better estimator than $\overline{X}$. ■

The proof of the above proposition can be found in the mathematical statistics book by DeGroot. The proposition itself is usually called the **invariance principle** for m.l.e.'s.

A Desirable Property of the M.L.E.

While the principle of maximum likelihood estimation has considerable intuitive appeal, the following proposition provides additional rationale for the use of m.l.e.'s.

Proposition

> Under very general conditions on the joint distribution of the sample, when the sample size n is large, the maximum likelihood estimator of any parameter θ is approximately unbiased $[E(\hat{\theta}) \approx \theta]$ and has variance that is nearly as small as can be achieved by any estimator. Stated another way, the m.l.e. $\hat{\theta}$ is approximately the MVUE of θ.

Because of this result and the fact that calculus-based techniques can usually be used to derive the m.l.e.'s (though often numerical methods, such as Newton's method, are necessary), maximum likelihood estimation is the most widely used estimation technique among statisticians. Many of the estimators used in the remainder of the book are m.l.e.'s. Obtaining an m.l.e., however, does require that the underlying distribution be specified.

Some Complications

Sometimes calculus cannot be used to obtain m.l.e.'s.

Example 6.20

Suppose that my waiting time for a bus is uniformly distributed on $[0, \theta]$, and that the results $x_1, \ldots, x_n$ of a random sample from this distribution have been observed. Since $f(x; \theta) = 1/\theta$ for $0 \le x \le \theta$ and 0 otherwise,

$$
f(x_1, \ldots, x_n; \theta) = \begin{cases} \dfrac{1}{\theta^n} & 0 \le x_1 \le \theta, \ldots, 0 \le x_n \le \theta \\ 0 & \text{otherwise} \end{cases}
$$

As long as $\max(x_i) \le \theta$, the likelihood is $1/\theta^n$, which is positive, but as soon as $\theta < \max(x_i)$, the likelihood drops to 0. This is illustrated in Figure 6.4. Calculus won't work because the maximum of the likelihood occurs at a point of discontinuity, but the figure shows that $\hat{\theta} = \max(X_i)$. Thus if my waiting times are 2.3, 3.7, 1.5, .4, and 3.2, then the maximum likelihood estimate is $\hat{\theta} = 3.7$.

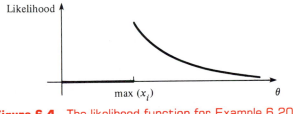

Figure 6.4 The likelihood function for Example 6.20 ■

Example 6.21

A method that is often used to estimate the size of a wildlife population involves performing a capture/recapture experiment. In this experiment an initial sample of M animals is captured, each of these animals is tagged, and the animals are then returned to the population. After allowing enough time for the tagged individuals to mix into the population, another sample of size n is captured. With $X =$ the number of tagged animals in the second sample, the objective is to use the observed x to estimate the population size N.

The parameter of interest is $\theta = N$, which can assume only integer values, so even after determining the likelihood function (p.m.f. of X here), using calculus to obtain N would present difficulties. If we think of a success as a previously tagged animal being recaptured, then sampling is without replacement from a population containing M successes and $N - M$ failures, so that X is a hypergeometric r.v. and the likelihood function is

$$p(x; N) = h(x; n, M, N) = \frac{\binom{M}{x} \cdot \binom{N - M}{n - x}}{\binom{N}{n}}$$

The integer-valued nature of N notwithstanding, it would be quite difficult to take the derivative of $p(x; N)$. However, if we consider the ratio of $p(x, N)$ to $p(x; N - 1)$, we have

$$\frac{p(x; N)}{p(x; N - 1)} = \frac{(N - M) \cdot (N - n)}{N(N - M - n + x)}$$

This ratio is larger than 1 if and only if $N < Mn/x$. The value of N for which $p(x; N)$ is maximized is therefore the largest integer less than Mn/x. If we use standard mathematical notation $[r]$ for the largest integer less than r, the maximum likelihood estimate of N is $\hat{N} = [Mn/x]$. As an illustration, if $M = 200$ fish are taken from a lake and tagged, subsequently $n = 100$ fish are recaptured, and among the 100 there are $x = 11$ tagged fish, then $\hat{N} = [(200)(100)/11] = [1818.18] = 1818$. The estimate is actually rather intuitive; x/n is the proportion of the recaptured sample that is tagged, while M/N is the proportion of the entire population that is tagged. The estimate is obtained by equating these two proportions (estimating a population proportion by a sample proportion). ■

Suppose $X_1, X_2, \ldots, X_n$ is a random sample from a p.d.f. $f(x; \theta)$ that is symmetric about θ, but that the investigator is unsure of the form of the f function. It is then desirable to use an estimator $\hat{\theta}$ that is *robust*—that is, one that performs well for a wide variety of underlying p.d.f.'s. One such estimator is a trimmed mean. In recent years statisticians have proposed another type of estimator, called an *M-estimator,* based on a generalization of maximum likelihood estimation. Instead of maximizing the log likelihood $\Sigma \ln[f(x_i; \theta)]$ for a specified f, one maximizes $\Sigma \rho(x_i; \theta)$. The "objective function" ρ is selected to yield an estimator with good robustness properties. The book by Hoaglin et al. contains a good exposition.

Exercises / Section 6.2 (16-26)

16. Suppose that I select n records manufactured by a particular company, and let X = the number among the n that are warped, $p = P$(warped record). Assume that only X is observed, rather than the sequence of S's and F's.
 a. Derive the maximum likelihood estimator of p. If $n = 20$ and $x = 3$, what is the estimate?
 b. Is the estimator of (a) unbiased?
 c. If $n = 20$ and $x = 3$, what is the m.l.e. of the probability $(1 - p)^5$ that none of the next five records examined is warped? *Hint:* the invariance principle.

17. Let X have a Weibull distribution with parameters α and β, so

$$E(X) = \beta \cdot \Gamma(1 + 1/\alpha)$$

$$V(X) = \beta^2 \{\Gamma(1 + 2/\alpha) - [\Gamma(1 + 1/\alpha)]^2\}$$

 a. Based on a random sample $X_1, \ldots, X_n$, write equations for the method of moments estimators of β and α. Show that once the estimate of α has been obtained, the estimate of β can be found from a table of the gamma function, and that the estimate of α is the solution to a complicated equation involving the gamma function.
 b. If $n = 20$, $\bar{x} = 28.0$, and $\Sigma x_i^2 = 16,500$, compute the estimates. *Hint:*

$$[\Gamma(1.2)]^2/\Gamma(1.4) = .95$$

18. Let X denote the proportion of allotted time that a randomly selected student spends working on a certain aptitude test, and suppose that the p.d.f. of X is

$$f(x; \theta) = \begin{cases} (\theta + 1)x^\theta & 0 \le x \le 1 \\ 0 & \text{otherwise} \end{cases}$$

where $-1 < \theta$. A random sample of 10 students yielded data $x_1 = .92$, $x_2 = .79$, $x_3 = .90$, $x_4 = .65$, $x_5 = .86$, $x_6 = .47$, $x_7 = .73$, $x_8 = .97$, $x_9 = .94$, $x_{10} = .77$.
 a. Use the method of moments to obtain an estimator of θ, and then compute the estimate for this data.
 b. Obtain the maximum likelihood estimator of θ, and then compute the estimate for the given data.

19. Two different computer systems are monitored for a total of n weeks. Let X_i denote the number of breakdowns of the first system during the ith week, and suppose that the X_i's are independent and drawn from a Poisson distribution with parameter λ_1. Similarly let Y_i denote the number of breakdowns of the second system during the ith week, and assume independence with each Y_i Poisson with parameter λ_2. Derive the m.l.e.'s of λ_1, λ_2, and $\lambda_1 - \lambda_2$. *Hint:* Using independence, write the joint p.m.f. (likelihood) of the X_i's and Y_i's together.

20. In the warped record problem (Exercise 16), instead of selecting $n = 20$ records to examine, suppose that I examine records in succession until I have found $r = 3$ warped ones. If the twentieth record is the third warped one (so that the number of records examined that were not warped is $x = 17$), what is the maximum likelihood estimate of p? Is this the same as the estimate in Exercise 16? Why or why not? Is it the same as the estimate computed from the unbiased estimator of Exercise 13?

21. The shear strength of each of 10 test spot welds is determined, yielding the following data (p.s.i.):

392, 376, 401, 367, 389, 362, 409, 415, 358, 375

 a. Assuming that shear strength is normally distributed, estimate the true average shear strength and standard deviation of shear strength using the method of maximum likelihood.
 b. Again assuming a normal distribution, estimate the strength value below which 95% of all welds will have their strengths. *Hint:* What is the ninety-fifth percentile in terms of μ and σ? Now use the invariance principle.

22. Referring back to Exercise 21, suppose that we decide to examine another test spot weld. Let X = shear strength of the weld. Use the given data to obtain the m.l.e. of $P(X \le 400)$. *Hint:* $P(X \le 400) = \Phi((400 - \mu)/\sigma)$.

23. Let $X_1, \ldots, X_n$ be a random sample from a gamma distribution with parameters α and β.

a. Derive the equations whose solution yields the maximum likelihood estimators of α and β. Do you think that they can be solved explicitly?

b. Show that the m.l.e. of $\mu = \alpha\beta$ is $\hat{\mu} = \overline{X}$.

24. Let $X_1, X_2, \ldots, X_n$ represent a random sample from the Rayleigh distribution with density function given in Exercise 10. Determine

a. the maximum likelihood estimator of θ, and then calculate the estimate for the vibratory stress data given in that exercise. Is this estimator the same as the unbiased estimator suggested in Exercise 10?

b. the maximum likelihood estimate of the median of the vibratory stress distribution. *Hint:* First express the median in terms of θ.

25. Consider a random sample $X_1, X_2, \ldots, X_n$ from the shifted exponential p.d.f.

$$f(x; \lambda, \theta) = \begin{cases} \lambda e^{-\lambda(x-\theta)} & x \geq \theta \\ 0 & \text{otherwise} \end{cases}$$

Taking $\theta = 0$ gives the p.d.f. of the exponential distribution considered previously (with positive density to the right of zero). An example of the shifted exponential distribution appeared in Ex-

ample 4.4, in which the variable of interest was time headway in traffic flow and $\theta = .5$ was the minimum possible time headway.

a. Obtain the maximum likelihood estimators of θ and λ.

b. If $n = 10$ time headway observations are made, resulting in the values 3.11, .64, 2.55, 2.20, 5.44, 3.42, 10.39, 8.93, 17.82, and 1.30, calculate the estimates of θ and λ.

26. At time $t = 0$, 20 identical components are put on test. The lifetime distribution of each is exponential with parameter λ. The experimenter then leaves the test facility unmonitored. On his return 24 hours later, the experimenter immediately terminates the test after noticing that $y = 15$ of the 20 components are still in operation (so five have failed). Derive the maximum likelihood estimate of λ. *Hint:* Let $Y = $ the number that survive 24 hours. Then $Y \sim \text{Bin}(n, p)$. What is the m.l.e. of p? Now notice that $p = P(X_i \geq 24)$ where X_i is exponentially distributed. This relates λ to p, so the former can be estimated once the latter has been.

Supplementary Exercises / Chapter 6 (27–33)

27. An estimator $\hat{\theta}$ is said to be **consistent** if for any $\epsilon > \theta$, $P(|\hat{\theta} - \theta| \geq \epsilon) \to 0$ as $n \to \infty$. That is, $\hat{\theta}$ is consistent if, as the sample size gets larger, it is less and less likely that $\hat{\theta}$ will be further than ϵ from the true value of θ. Show that $\overline{X}$ is a consistent estimator of μ when $\sigma^2 < \infty$ by using Chebyshev's inequality from Exercise 40 (Chapter 3). *Hint:* The inequality can be rewritten in the form

$$P(|Y - \mu_Y| \geq \epsilon) \leq \sigma_Y^2/\epsilon.$$

Now identify Y with $\overline{X}$.

28. a. Let $X_1, \ldots, X_n$ be a random sample from a uniform distribution on $[0, \theta]$. Then the m.l.e. of θ is $\hat{\theta} = Y = \max(X_i)$. Use the fact that $Y \leq y$ iff each $X_i \leq y$ to derive the c.d.f. of Y. Then show that the p.d.f. of $Y = \max(X_i)$ is

$$f_Y(y) = \begin{cases} \dfrac{ny^{n-1}}{\theta^n} & 0 \leq y \leq \theta \\ 0 & \text{otherwise} \end{cases}$$

b. Use the result of (a) to show that the m.l.e. is biased but that $(n + 1) \max(X_i)/n$ is unbiased.

29. At time $t = 0$ there is one individual alive in a certain population. **A pure birth process** then unfolds as follows. The time until the first birth is exponentially distributed with parameter λ. After the first birth there are two individuals alive. The time until the first gives birth again is exponential with parameter λ, and similarly for the second individual. Therefore the time until the next birth is the minimum of two exponential (λ) variables, which is exponential with parameter 2λ. Similarly, once the second birth has occurred, there are three individuals alive, so the time until the next birth is an exponential r.v. with parameter 3λ, and so on (the memoryless property of the exponential distribution is being used here). Suppose that the process is observed until the sixth birth has occurred, and the successive birth times are 25.2, 41.7, 51.2, 55.5, 59.5, 61.8 (from which you should calculate the times between successive births). Derive the m.l.e. of λ. *Hint:* The likelihood is a product of exponential terms.

30. The **mean square error** of an estimator $\hat{\theta}$ is $\text{MSE}(\hat{\theta}) = E(\hat{\theta} - \theta)^2$. If $\hat{\theta}$ is unbiased, then $\text{MSE}(\hat{\theta}) = V(\hat{\theta})$, but in general $\text{MSE}(\hat{\theta}) = V(\hat{\theta}) + (\text{bias})^2$. Consider the estimator $\hat{\sigma}^2 = KS^2$, where $S^2 = $ sample variance. What value of K minimizes the mean square error of this estimator when the population distribution is normal? *Hint:* It can be shown that

$$E[(S^2)^2] = (n + 1)\sigma^4/(n - 1)$$

    [in general it is quite difficult to find $\hat{\theta}$ to minimize $\text{MSE}(\hat{\theta})$, which is why we look only at unbiased estimators and minimize $V(\hat{\theta})$].

31. Let $X_1, \ldots, X_n$ be a random sample from a p.d.f. that is symmetric about μ. An estimator for μ that has been found to perform well for a variety of underlying distributions is the *Hodges-Lehmann estimator*. To define it, first compute for each $i \leq j$ and each $j = 1, 2, \ldots, n$ the pairwise average $\overline{X}_{i,j} = (X_i + X_j)/2$. Then the estimator is $\hat{\mu} = $ the median of the $\overline{X}_{i,j}$'s. Compute the value of this estimate using the data of Exercise 25 (Chapter 1). *Hint:* Construct a square table with the x_i's listed on the left margin and on top. Then compute averages on and above the diagonal.

32. When the population distribution is normal, the statistic median $\{|X_1 - \tilde{X}|, \ldots, |X_n - \tilde{X}|\}/.6745$ can be used to estimate σ. This estimator is more resistant to the effects of outliers (observations far from the bulk of the data) than is the sample standard deviation. Compute both the corresponding point estimate and s for the data of Example 6.2.

33. When the sample standard deviation S is based on a random sample from a normal population distribution, it can be shown that

$$E(S) = \sqrt{2/(n - 1)}\; \Gamma(n/2)\, \sigma/\Gamma((n - 1)/2)$$

    Use this to obtain an unbiased estimator for σ of the form cS. What is c when $n = 20$?

Bibliography

DeGroot, Morris, *Probability and Statistics* (2nd ed.), Addison-Wesley, Reading, Mass., 1986. Includes an excellent discussion of both general properties and methods of point estimation; of particular interest are examples showing how general principles and methods can yield unsatisfactory estimators in particular situations.

Hoaglin, David, Mosteller, Frederick, and Tukey, John, *Understanding Robust and Exploratory Data Analysis,* Wiley, New York, 1983. Contains several good chapters on robust point estimation, including one on *M*-estimation.

Hogg, Robert, and Craig, Allen, *Introduction to Mathematical Statistics* (4th ed.), Macmillan, New York, 1978. A good discussion of unbiasedness.

Larsen, Richard, and Marx, Morris, *Introduction to Mathematical Statistics* (2nd ed.), Prentice-Hall, Englewood Cliffs, N.J., 1985. A very good discussion of point estimation from a slightly more mathematical perspective than the present text.

CHAPTER 7

Interval Estimation
Based on a Single Sample

Introduction

Almost any parameter that we might wish to estimate has as its set of possible values an entire interval of numbers. If, for example, we wish to estimate the true average net weight μ of fertilizer bags having nominal weight 50 lb, then μ might be any number between, say, 45 and 55. Even if the variable of interest is discrete, as with $X =$ the number of phonograph records purchased by a randomly chosen customer leaving a certain record store, the true average number of records purchased per customer (μ) could be any number between, say, .5 and 4.0, and the true proportion p of customers purchasing at least one record might be any number between 0 and 1.

The fact that the set of possible parameter values is a continuum implies that a point estimate, though it will represent our best guess for the true value of the parameter, may be close to that true value but will virtually never actually equal it. Because of this, reporting only the estimated value is generally unsatisfactory; some measure of how close the point estimate is likely to be to the true value is required. One way to do this, as suggested in Section 6.1, is to report both the estimate and its standard deviation (standard error). Then, if the estimator has at least approximately a normal distribution, we can be quite confident that the true value lies within two or three standard deviations of the estimated value. This amounts to replacing the point estimate, a single number, by an entire interval of plausible values, and that is exactly what an interval estimate or confidence interval is—an interval of plausible values for the parameter being estimated. The degree of plausibility will be specified by a confidence level, so that we will speak of a 95% confidence interval (confidence level 95%) or a 99% interval. If the confidence level is high and the resulting interval is narrow, the investigator has reasonably precise knowledge of the parameter's value.

7.1 Basic Properties of Confidence Intervals

The basic concepts and properties of confidence intervals are most easily introduced by first focusing on a simple, albeit somewhat unrealistic, problem situation. Suppose that the parameter of interest is a population mean μ and that

1. the population distribution is normal, and
2. the value of the population standard deviation σ is known.

Normality of the population distribution is often a reasonable assumption. However, if the value of μ is unknown, it is implausible that the value of σ would be available (knowledge of a population's center typically precedes information concerning spread). In later sections we will develop methods based on less restrictive assumptions.

Example 7.1 Industrial engineers who specialize in ergonomics are concerned with designing workspace and devices operated by workers so as to achieve high productivity and comfort. The paper "Studies on Ergonomically Designed Alphanumeric Keyboards" (*Human Factors,* 1985, pp. 175–187) reported on a study of preferred height for an experimental keyboard with large forearm–wrist support. A sample of $n = 31$ trained typists was selected, and the preferred keyboard height was determined for each typist. The resulting sample average preferred height was $\bar{x} = 80.0$ cm. Assuming that preferred height is normally distributed with $\sigma = 2.0$ cm (a value suggested by data in the paper), obtain a confidence interval for μ, the true average preferred height for the population of all experienced typists. ∎

The actual sample observations $x_1, x_2, \ldots, x_n$ are assumed to be the result of a random sample $X_1, \ldots, X_n$ from a normal distribution with mean value μ and standard deviation σ. This implies that irrespective of the sample size n, the sample mean $\bar{X}$ is normally distributed with expected value μ and standard deviation $\sigma/\sqrt{n}$. Standardizing $\bar{X}$ by first subtracting its expected value and then dividing by its standard deviation yields the variable

$$Z = \frac{\bar{X} - \mu}{\sigma/\sqrt{n}} \tag{7.1}$$

which has a standard normal distribution. Because the area under the standard normal curve between -1.96 and 1.96 is .95, the following probability statement is valid:

$$P\left(-1.96 < \frac{\bar{X} - \mu}{\sigma/\sqrt{n}} < 1.96\right) = .95 \tag{7.2}$$

The next step in the development is to manipulate the inequalities inside the parentheses in (7.2) so that they appear in the equivalent form $\ell < \mu < u$, where the endpoints ℓ and u involve $\bar{X}$ and $\sigma/\sqrt{n}$. This is achieved through the following sequence of operations, each one yielding inequalities equivalent to those we started with:

1. Multiply through by $\sigma/\sqrt{n}$ to obtain

$$-1.96 \cdot \frac{\sigma}{\sqrt{n}} < \overline{X} - \mu < 1.96 \cdot \frac{\sigma}{\sqrt{n}}$$

2. Subtract $\overline{X}$ from each term to obtain

$$-\overline{X} - 1.96 \cdot \frac{\sigma}{\sqrt{n}} < -\mu < -\overline{X} + 1.96 \cdot \frac{\sigma}{\sqrt{n}}$$

3. Multiply through by -1 to eliminate the minus sign in front of μ (which reverses the direction of each inequality) to obtain

$$\overline{X} + 1.96 \cdot \frac{\sigma}{\sqrt{n}} > \mu > \overline{X} - 1.96 \cdot \frac{\sigma}{\sqrt{n}}$$

that is,

$$\overline{X} - 1.96 \cdot \frac{\sigma}{\sqrt{n}} < \mu < \overline{X} + 1.96 \cdot \frac{\sigma}{\sqrt{n}}$$

Because each set of inequalities in the sequence is equivalent to the original one, the probability associated with each is .95. That is, for the last set

$$P\left(\overline{X} - 1.96\,\frac{\sigma}{\sqrt{n}} < \mu < \overline{X} + 1.96\,\frac{\sigma}{\sqrt{n}}\right) = .95 \qquad (7.3)$$

The event inside the parentheses in (7.3) has a somewhat unfamiliar appearance; always before the random quantity has appeared in the middle with constants on both ends, as in $a \le Y \le b$. In (7.3) the random quantity appears on the two ends while the unknown constant μ appears in the middle. To interpret (7.3), think of a **random interval** having left endpoint $\overline{X} - 1.96 \cdot \sigma/\sqrt{n}$ and right endpoint $\overline{X} + 1.96 \cdot \sigma/\sqrt{n}$, which in interval notation is

$$\left(\overline{X} - 1.96 \cdot \frac{\sigma}{\sqrt{n}}, \quad \overline{X} + 1.96 \cdot \frac{\sigma}{\sqrt{n}}\right) \qquad (7.4)$$

The interval (7.4) is random because the two endpoints of the interval involve a random variable. Note that the interval is centered at the sample mean $\overline{X}$, and extends $1.96\sigma/\sqrt{n}$ to either side of $\overline{X}$. Thus the length of the interval is $2 \cdot (1.96) \cdot \sigma/\sqrt{n}$, which is not random; only the location of the interval (its midpoint $\overline{X}$) is random (Figure 7.1). Now (7.3) can be paraphrased by *"the*

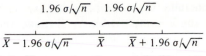

Figure 7.1 The random interval (7.4) centered at $\overline{X}$

probability is .95 that the random interval (7.4) includes or covers the true value of μ." Before any experiment is performed and any data is gathered, it is quite likely (probability .95) that μ will lie inside (7.4).

Definition

> If after observing $X_1 = x_1$, $X_2 = x_2$, ..., $X_n = x_n$, we compute the observed sample mean $\overline{x}$ and then substitute $\overline{x}$ into (7.4) in place of $\overline{X}$, the resulting fixed interval is called a **95% confidence interval for μ.** This confidence interval can be expressed either as
>
> $$\left(\overline{x} - 1.96 \cdot \frac{\sigma}{\sqrt{n}}, \quad \overline{x} + 1.96 \cdot \frac{\sigma}{\sqrt{n}}\right) \quad \text{is a 95\% confidence interval for } \mu$$
>
> or as
>
> $$\overline{x} - 1.96 \cdot \frac{\sigma}{\sqrt{n}} < \mu < \overline{x} + 1.96 \cdot \frac{\sigma}{\sqrt{n}} \quad \text{with 95\% confidence}$$
>
> A concise expression for the interval is $\overline{x} \pm 1.96 \cdot \sigma/\sqrt{n}$, where $-$ gives the left endpoint (lower limit) and $+$ gives the right endpoint (upper limit).

Example 7.2
(Example 7.1 continued)

The quantities needed for computation of the 95% confidence interval for true average preferred height are $\sigma = 2.0$, $n = 31$, and $\overline{x} = 80.0$. The resulting interval is

$$\overline{x} \pm 1.96 \cdot \frac{\sigma}{\sqrt{n}} = 80.0 \pm (1.96)\frac{2.0}{\sqrt{31}} = 80.0 \pm .7 = (79.3, 80.7)$$

That is, we can be highly confident that $79.3 < \mu < 80.7$. This interval is relatively narrow, indicating that μ has been rather precisely estimated. ∎

Interpreting a Confidence Interval

The confidence level 95% for the interval defined above was inherited from the probability .95 for the random interval (7.4). Intervals having other levels of confidence will be introduced shortly. For now, though, consider how 95% confidence can be interpreted.

Because we started with an event whose probability was .95—that the random interval (7.4) would capture the true value of μ—and then used the data in Example 7.1 to compute the fixed interval (79.3, 80.7), it is tempting to conclude that μ is within this fixed interval with probability .95. But by substituting $\bar{x} = 80.0$ in for $\bar{X}$, all randomness disappears; the interval (79.3, 80.7) is not a random interval, and μ is a constant (unfortunately unknown to us), so it is *incorrect* to write $P(\mu$ lies in (79.3, 80.7)$) = .95$.

If a probability statement involving the fixed interval is not appropriate, how can we give meaning to "95% confidence"? The answer lies in recalling the long-run frequency interpretation of probability: To say that an event has probability .95 is to say that if the experiment on which an event A is defined is performed over and over again, in the long run A will occur 95% of the time. Suppose that we obtain another sample of typists' preferred heights, and compute another 95% interval. Then suppose we consider repeating this for a third sample, a fourth sample, and so on. Let A be the event that $\bar{X} - 1.96 \cdot \sigma/\sqrt{n} < \mu < \bar{X} + 1.96 \cdot \sigma/\sqrt{n}$. Since $P(A) = .95$, in the long run 95% of our computed confidence intervals will contain μ. This is illustrated in Figure 7.2, where the vertical line cuts the measurement axis at the true (but unknown) value of μ. Notice that of the 11 intervals pictured, only intervals 3 and 11 fail to contain μ. In the long run, only 5% of the intervals so constructed would fail to contain μ.

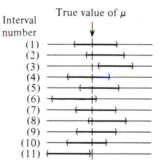

Figure 7.2   Repeated construction of 95% confidence intervals

According to this interpretation, the confidence level 95% is not so much a statement about any particular interval such as (79.3, 80.7), but pertains to what would happen if a very large number of like intervals were to be constructed. While this may seem unsatisfactory, the root of the difficulty lies with our interpretation of probability—it applies to a long sequence of replications of an experiment rather than just a single replication. There is another approach to the construction of and interpretation of confidence intervals that uses the notion of subjective probability and Bayes' theorem, but the technical details are beyond the scope of this text; the book by Winkler (see the Chapter 2 bibliography) is a good source. The interval presented here (as well as each interval presented subsequently) is called a "classical" confidence interval because its interpretation rests on the classical notion of probability (though the main ideas were developed as recently as the 1930s).

Other Levels of Confidence

Suppose that we wish a 99% confidence interval rather than a 95% interval. Then rather than starting with a probability of .95, we must begin with a probability of .99. Since the area under the standard normal curve between -2.58 and 2.58 equals .99, replacing 1.96 with 2.58 in the definition yields a 99% interval.

This suggests that any desired level of confidence can be achieved by replacing 1.96 or 2.58 with the appropriate standard normal critical value. As Figure 7.3 shows, a probability of $1 - \alpha$ is achieved by using $z_{\alpha/2}$ in place of 1.96.

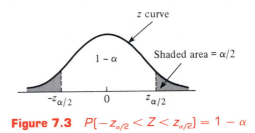

Figure 7.3 $P(-z_{\alpha/2} < Z < z_{\alpha/2}) = 1 - \alpha$

Definition

> A **100$(1 - \alpha)$% confidence interval** for the mean μ of a normal population when the value of σ is known is given by
>
> $$\left(\overline{x} - z_{\alpha/2} \cdot \frac{\sigma}{\sqrt{n}}, \quad \overline{x} + z_{\alpha/2} \cdot \frac{\sigma}{\sqrt{n}} \right) \qquad (7.5)$$
>
> or equivalently, by $\overline{x} \pm z_{\alpha/2} \cdot \sigma/\sqrt{n}$.

Example 7.3

A savings and loan association that finances a large number of home purchases in a particular region wanted information on the extent to which mortgage payments reduced the amount of disposable income during the initial year of occupancy for first-time home buyers. A sample of 35 recently granted mortgage applications was obtained, and for each the amount of the monthly mortgage payment as a percentage of take-home income $(100 \cdot \text{payment}/\text{income})$ was computed. The sample average percentage was found to be 24.7. Assuming that σ is known to be 3.0 and that percentage is a normally distributed variable (among all successful applicants), compute a 90% confidence interval for the true average percentage μ.

From the definition above, a 90% interval requires that $100(1 - \alpha) = 90$, so that $\alpha = .10$; then $z_{\alpha/2} = z_{.05} = 1.645$. Substituting this into (7.5) along with $\overline{x} = 24.7$, $\sigma = 3.0$, and $n = 35$ yields the interval

$$24.7 \pm (1.645)\frac{(3.0)}{\sqrt{35}} = 24.7 \pm .8 = (23.9, 25.5) \qquad \blacksquare$$

Although (7.5) can be used to obtain an interval with any desired degree of confidence, there are only three levels of confidence that appear at all frequently in the literature. These are 99%, 95%, and 90%, and the intervals are obtained by using 2.58, 1.96, and 1.645, respectively, in place of $z_{\alpha/2}$.

Confidence Level, Precision, and Choice of Sample Size

In deciding whether to compute a 95% confidence interval or a 99% interval, it may seem ridiculous to settle for a lower level of confidence when a higher level can be obtained. However, the old adage "you must give up something to get something" remains true here. What do we give up by using the 99% interval? Recall that because the 95% interval extends $1.96 \cdot \sigma/\sqrt{n}$ to either side of $\bar{x}$, the length of the interval is $2(1.96) \cdot \sigma/\sqrt{n} = 3.92 \cdot \sigma/\sqrt{n}$ Similarly, the length of the 99% interval is $2(2.58) \cdot \sigma/\sqrt{n} = 5.16 \cdot \sigma/\sqrt{n}$. That is, we have more confidence in the 99% interval precisely because it is longer. The higher the desired degree of confidence, the longer the resulting interval. In fact the only 100% confidence interval for μ is $(-\infty, \infty)$, which is not terribly informative since even before sampling, we knew that this interval covered μ.

If we think of the length of the interval as specifying its precision or accuracy, then the confidence level (or reliability) of the interval is inversely related to its precision. A highly reliable interval estimate may be quite imprecise in that the endpoints of the interval may be far apart, while a precise interval may entail relatively low reliability. Thus it cannot be said unequivocally that a 99% interval is to be preferred to a 95% interval; the gain in reliability entails a loss in precision.

The astute reader may by this time have observed that there is the possibility of "having one's cake and eating it too." Until now we have proceeded as though the sample size n was fixed before consideration of interval length. But if we wish our 95% interval to have a prescribed length, the sample size can be chosen to achieve this.

Example **7.4**   Extensive monitoring of a computer time-sharing system has suggested that response time to a particular editing command is normally distributed with standard deviation 25 milliseconds. A new operating system has been installed, and it is desired to estimate the true average response time μ for the new environment. Assuming that response times are still normally distributed with $\sigma = 25$, what sample size is necessary to ensure that the resulting 95% confidence interval has a length of (at most) 10? The sample size n must satisfy

$$10 = 2 \cdot (1.96)(25/\sqrt{n})$$

Rearranging this equation gives

$$\sqrt{n} = 2 \cdot (1.96)(25)/10 = 9.80, \text{ so } n = (9.80)^2 = 96.04$$

Since n must be an integer, a sample size of 97 is required.   ■

The general formula for the sample size n necessary to ensure an interval length L is obtained from $L = 2 \cdot z_{\alpha/2} \cdot \sigma/\sqrt{n}$ as

$$n = \left(2z_{\alpha/2} \cdot \frac{\sigma}{L}\right)^2$$

The smaller the desired length L, the larger n must be. In addition, n is an increasing function of σ (more population variability necessitates a larger sample size) and of the confidence level $100(1 - \alpha)$ (as α decreases, $z_{\alpha/2}$ increases).

Deriving a Confidence Interval

Let $X_1, X_2, \ldots, X_n$ denote the sample on which the confidence interval for a parameter θ is to be based. Suppose that a random variable satisfying the following two properties can be found:

1. The variable depends functionally both on $X_1, \ldots, X_n$ and on θ.
2. The probability distribution of the variable does not depend on θ or on any other unknown parameters.

Let $h(X_1, X_2, \ldots, X_n; \theta)$ denote this random variable. For example, if the population distribution is normal with known σ and $\theta = \mu$, then $h(X_1, \ldots, X_n; \mu) = (\overline{X} - \mu)/(\sigma/\sqrt{n})$ satisfies both properties; it clearly depends functionally on μ, yet has the standard normal probability distribution, which does not depend on μ. In general, the form of the h function is usually suggested by examining the distribution of an appropriate estimator $\hat{\theta}$.

For any α between 0 and 1, constants a and b can be found to satisfy

$$P(a < h(X_1, \ldots, X_n; \theta) < b) = 1 - \alpha \qquad (7.6)$$

Because of the second property, a and b do not depend on θ. In the normal example, $a = -z_{\alpha/2}$ and $b = z_{\alpha/2}$. Now suppose that the inequalities in (7.6) can be manipulated to isolate θ, giving the equivalent probability statement

$$P(\ell(X_1, X_2, \ldots, X_n) < \theta < u(X_1, X_2, \ldots, X_n)) = 1 - \alpha$$

Then $\ell(x_1, x_2, \ldots, x_n)$ and $u(x_1, \ldots, x_n)$ are the lower and upper confidence limits, respectively, for a $100(1 - \alpha)\%$ confidence interval. In the normal example, we saw that $\ell(X_1, \ldots, X_n) = \overline{X} - z_{\alpha/2} \cdot \sigma/\sqrt{n}$ and $u(X_1, \ldots, X_n) = \overline{X} + z_{\alpha/2} \cdot \sigma/\sqrt{n}$.

Example 7.5   A theoretical model suggests that the time to breakdown of an insulating fluid between electrodes at a particular voltage has an exponential distribution with parameter λ (see Section 4.4). A random sample of $n = 10$ breakdown times yielded the following sample data (in min.): $x_1 = 41.53$, $x_2 = 18.73$, $x_3 = 2.99$, $x_4 = 30.34$, $x_5 = 12.33$, $x_6 = 117.52$, $x_7 = 73.02$, $x_8 = 223.63$, $x_9 = 4.00$, $x_{10} = 26.78$. A 95% confidence interval for λ and for the true average breakdown time are desired.

Let $h(X_1, X_2, \ldots, X_n; \lambda) = 2\lambda\Sigma X_i$. It can be shown that this random variable has a probability distribution called a chi-squared distribution with $2n$ degrees of freedom ($\nu = 2n$, where ν is the parameter of a chi-squared distribution as discussed in Section 4.4). Appendix Table 6 pictures a typical chi-squared density curve, and tabulates critical values that capture specified tail areas. The relevant number of d.f. here is $2(10) = 20$. The $\nu = 20$ row of the table shows that 34.170 captures upper-tail area .025 and 9.591 captures lower-tail area .025 (upper-tail area .975). Thus for $n = 10$,

$$P(9.591 < 2\lambda\Sigma X_i < 34.170) = .95$$

Division by $2\Sigma X_i$ isolates λ, yielding

$$P(9.591/2\Sigma X_i < \lambda < 34.170/2\Sigma X_i) = .95$$

The lower limit of the 95% confidence interval for λ is $9.591/2\Sigma x_i$ and the upper limit is $34.170/2\Sigma x_i$. For the given data, $\Sigma x_i = 550.87$, giving the interval $(.00871, .03101)$.

The expected value of an exponential random variable is $\mu = 1/\lambda$. Since

$$P(2\Sigma X_i/34.170 < 1/\lambda < 2\Sigma X_i/9.591) = .95$$

the 95% confidence interval for true average breakdown time is $(2\Sigma x_i/34.170, 2\Sigma x_i/9.591) = (32.24, 114.87)$. This interval is obviously quite wide, reflecting substantial variability in breakdown times and a small sample size. ■

In general, the upper and lower confidence limits result from replacing each $<$ in (7.6) by $=$ and solving for θ. In the insulating fluid example just considered, $2\lambda\Sigma x_i = 34.170$ gives $\lambda = 34.170/2\Sigma x_i$ as the upper confidence limit, and the lower limit is obtained from the other equation. Notice that the two interval limits are not equidistant from the point estimate, since the interval is not of the form $\hat{\theta} \pm c$.

Exercises / Section 7.1 (1–9)

1. Consider a normal population distribution with the value of σ known.
   a. What is the confidence level for the interval $\bar{x} \pm 2.81\sigma/\sqrt{n}$?
   b. What is the confidence level for the interval $\bar{x} \pm 1.44\sigma/\sqrt{n}$?
   c. What value of $z_{\alpha/2}$ in the confidence interval formula (7.5) results in a confidence level of 99.7%?
   d. Answer the question posed in (c) for a confidence level of 75%.

2. A confidence interval is desired for the true average stray load loss μ (watts) for a certain type of induction motor when the line current is held at 10 amps for a speed of 1500 rpm. Assume that stray load loss is normally distributed with $\sigma = 3.0$.
   a. Compute a 95% confidence interval for μ when $n = 25$ and $\bar{x} = 58.3$.

   b. Compute a 95% confidence interval for μ when $n = 100$ and $\bar{x} = 58.3$.
   c. Compute a 99% confidence interval for μ when $n = 100$ and $\bar{x} = 58.3$.
   d. Compute an 82% confidence interval for μ when $n = 100$ and $\bar{x} = 58.3$.
   e. How large must n be if the length of the 99% interval for μ is to be 1.0?

3. Assume that the helium porosity (in percentage) of coal samples taken from any particular seam is normally distributed with true standard deviation .75.
   a. Compute a 95% confidence interval for the true average porosity of a certain seam if the average porosity for 20 specimens from the seam was 4.85.
   b. Compute a 98% confidence interval for true average porosity of another seam based on 16

specimens with a sample average porosity of 4.56.

c. How large a sample size is necessary if the length of the 95% interval is to be .40?

4. On the basis of extensive tests, the yield point of a particular type of mild steel reinforcing bar is known to be normally distributed with $\sigma = 100$. The composition of the bar has been slightly modified, but the modification is not believed to have affected either the normality or the value of σ.

a. Assuming this to be the case, if a sample of 25 modified bars resulted in a sample average yield point of 8439 lb, compute a 90% confidence interval for the true average yield point of the modified bar.

b. How would you modify the interval in (a) to obtain a confidence level of 92%?

5. By how much must the sample size n be increased if the length of the confidence interval (7.5) is to be halved? If the sample size is increased by a factor of 25, what effect will this have on the length of the interval? Justify your assertions.

6. Let $\alpha_1 > 0$, $\alpha_2 > 0$, with $\alpha_1 + \alpha_2 = \alpha$. Then

$$P\left(-z_{\alpha_1} < \frac{\overline{X} - \mu}{\sigma/\sqrt{n}} < z_{\alpha_2}\right) = 1 - \alpha$$

a. Use this equation to derive a more general expression for a $100(1 - \alpha)\%$ confidence interval for μ of which the interval (7.5) is a special case.

b. Let $\alpha = .05$ and $\alpha_1 = \alpha/4$, $\alpha_2 = 3\alpha/4$. Does this result in a shorter or longer interval than the interval (7.5)?

7. a. Under the same conditions as those leading to the interval (7.5), $P[(\overline{X} - \mu)/(\sigma/\sqrt{n}) < z_\alpha] = 1 - \alpha$. Use this to derive a one-sided interval for μ that has infinite length and provides a lower bound on μ. What is this 95% interval for the data in Exercise 3(a)?

b. What is an analogous interval to that of (a) which provides an upper bound on μ? Compute this 99% interval for the data of Exercise 2(a).

8. A random sample of $n = 15$ heat pumps of a certain type yielded the following observations on lifetime (in years): 2.0, 1.3, 6.0, 1.9, 5.1, .4, 1.0, 5.3, 15.7, .7, 4.8, .9, 12.2, 5.3, .6.

a. Assume that the lifetime distribution is exponential, and use an argument parallel to that of Example 7.5 to obtain a 95% confidence interval for expected (true average) lifetime.

b. How should the interval of (a) be altered to achieve a confidence level of 99%?

c. What is a 95% confidence interval for the standard deviation of the lifetime distribution? *Hint:* What is the standard deviation of an exponential random variable?

9. Consider the next 1000 95% confidence intervals for μ that a statistical consultant will obtain for various clients. Suppose that the data sets on which the intervals are based are selected independently of one another. How many of these 1000 intervals do you expect to capture the corresponding value of μ? What is the probability that between 940 and 960 of these intervals contain the corresponding value of μ? *Hint:* Let $Y = $ the number among the 1000 intervals that contain μ. What kind of random variable is Y?

7.2   Large-Sample Confidence Intervals for a Population Mean and Proportion

The confidence interval for μ given in the previous section assumed that the population distribution was normal and that the value of σ was known. We now present a large-sample confidence interval whose validity does not require making these assumptions. After showing how the argument leading to this interval generalizes to yield other large-sample intervals, we focus on an interval for a population proportion p.

A Large-Sample Interval for μ

Let $X_1, X_2, \ldots, X_n$ be a random sample from a population having mean μ and standard deviation σ. Provided that n is large, the Central Limit Theorem implies that $\overline{X}$ has approximately a normal distribution whatever the nature of the population distribution. It then follows that $Z = (\overline{X} - \mu)/(\sigma/\sqrt{n})$ has approximately a standard normal distribution, so that

$$P\left(-z_{\alpha/2} < \frac{\overline{X} - \mu}{\sigma/\sqrt{n}} < z_{\alpha/2}\right) \approx 1 - \alpha$$

An argument parallel to that given in Section 7.1 yields $\overline{x} \pm z_{\alpha/2} \cdot \sigma/\sqrt{n}$ as a large-sample confidence interval for μ with a confidence level of *approximately* $100(1 - \alpha)\%$. That is, when n is large, the confidence interval for μ given earlier remains valid whatever the population distribution, provided that the qualifier "approximately" is inserted in front of the confidence level.

A practical difficulty with the above development is that computation of the interval requires the value of σ, which will almost never be known. Consider the standardized variable

$$Z = \frac{\overline{X} - \mu}{S/\sqrt{n}}$$

in which the sample standard deviation S replaces σ. Previously there was randomness only in the numerator of Z (by virtue of $\overline{X}$). Now there is randomness both in the numerator and in the denominator—the values of both $\overline{X}$ and S vary from sample to sample. However, when n is large, the use of S rather than σ adds very little extra variability to Z. This is because the computed value s will be quite close to σ for almost all samples. More specifically, in this case the new Z also has approximately a standard normal distribution. Manipulation of the inequalities in a probability statement involving this new Z yields a general large-sample interval for μ.

Proposition

> If n is sufficiently large,
>
> $$Z = \frac{\overline{X} - \mu}{S/\sqrt{n}}$$
>
> has approximately a standard normal distribution. This implies that
>
> $$\overline{x} \pm z_{\alpha/2} \cdot \frac{s}{\sqrt{n}} \tag{7.7}$$
>
> is a **large-sample confidence interval for μ** with confidence level approximately $100(1 - \alpha)\%$.

Generally speaking, $n > 30$ will be sufficient to justify the use of this interval.

Example 7.6 A sample of 56 research cotton samples resulted in a sample average percentage elongation of 8.17 and a sample standard deviation of 1.42 ("An Apparent Relation Between the Spiral Angle ϕ, the Percent Elongation E_1, and the

Dimensions of the Cotton Fiber," *Textile Research J.,* 1978, pp. 407–410). A 95% large-sample confidence interval for the true average percent elongation μ is then

$$\bar{x} \pm \frac{1.96s}{\sqrt{n}} = 8.17 \pm \frac{(1.96)(1.42)}{\sqrt{56}}$$

$$= 8.17 \pm .37 = (7.80, 8.54)$$ ∎

Example 7.7   In the article "Petitioners Under Chapter XIII of the Bankruptcy Act" (*J. Consumer Affairs,* vol. 3, Summer 1969), it was reported that in a sample of 250 petitioners who filed for nonbusiness bankruptcy between January 1964 and December 1966 (out of a total of 31,956 such petitioners), the sample average number of months taken to acquire the debts listed on the petitions was 35.41 with a sample standard deviation of 21.34. Compute a 99% confidence interval for the true average number of months taken by bankruptcy petitioners to acquire debts listed on their petitions.

A 99% interval requires $100(1 - \alpha) = 99$, so $\alpha = .01$ and $z_{\alpha/2} = z_{.005} = 2.58$. Substituting into (7.7) gives

$$35.41 \pm \frac{(2.58)(21.34)}{\sqrt{250}} = (31.93, 38.89)$$

Note that if we go two estimated standard deviations ($2s$) to the left of $\bar{x}$, we are below zero. Since zero is a lower bound for number of months while there is effectively no upper bound, the actual distribution of number of months would appear to be at least somewhat positively skewed rather than normal. But of course the large-sample interval does not require that the underlying distribution be normal. ∎

Unfortunately the choice of sample size to yield a desired interval length is not as straightforward here as it was for the case of known σ. This is because the length of (7.7) is $2 \cdot z_{\alpha/2} \cdot s/\sqrt{n}$; since s is not known before the data has been gathered, the length of the interval cannot be determined solely by choice of n. The only option for an investigator who wishes to specify a desired length is to make a guess as to what the value of s (or σ) might be—by being conservative and guessing at a large value of s, an n larger than necessary will be chosen.

A General Large-Sample Confidence Interval

The large-sample intervals $\bar{x} \pm z_{\alpha/2} \cdot \sigma/\sqrt{n}$ and $\bar{x} \pm z_{\alpha/2} \cdot s/\sqrt{n}$ are special cases of a general large-sample confidence interval for a parameter θ. Suppose that $\hat{\theta}$ is an estimator satisfying the following properties: (a) it has approximately a normal distribution, (b) it is (at least approximately) unbiased, and (c) an expression for $\sigma_{\hat{\theta}}$, the standard deviation of $\hat{\theta}$, is available. For example, in the case $\theta = \mu$, $\hat{\mu} = \bar{X}$ is an unbiased estimator whose distribution is approximately normal when n is large, and $\sigma_{\hat{\mu}} = \sigma_{\bar{X}} = \sigma/\sqrt{n}$. Standardizing $\hat{\theta}$ yields the random variable $Z = (\hat{\theta} - \theta)/\sigma_{\hat{\theta}}$, which has approximately a standard normal distribution. This justifies the probability statement

$$P\left(-z_{\alpha/2} < \frac{\hat{\theta} - \theta}{\sigma_{\hat{\theta}}} < z_{\alpha/2}\right) \approx 1 - \alpha \tag{7.8}$$

Suppose first that $\sigma_{\hat{\theta}}$ doesn't involve any unknown parameters (for example, known σ in the case $\theta = \mu$). Then replacing each $<$ in (7.8) by $=$ yields $\theta = \hat{\theta} \pm z_{\alpha/2} \cdot \sigma_{\hat{\theta}}$, so the lower and upper confidence limits are $\hat{\theta} - z_{\alpha/2} \cdot \sigma_{\hat{\theta}}$ and $\hat{\theta} + z_{\alpha/2} \cdot \sigma_{\hat{\theta}}$, respectively. Now suppose that $\sigma_{\hat{\theta}}$ does not involve θ but does involve at least one other unknown parameter. Let $s_{\hat{\theta}}$ be the estimate of $\sigma_{\hat{\theta}}$ obtained by using estimates in place of the unknown parameters (for example, $s/\sqrt{n}$ estimates $\sigma/\sqrt{n}$). Under general conditions (essentially that $s_{\hat{\theta}}$ be close to $\sigma_{\hat{\theta}}$ for most samples), a valid confidence interval is $\hat{\theta} \pm z_{\alpha/2} \cdot s_{\hat{\theta}}$. The interval $\bar{x} \pm z_{\alpha/2} \cdot s/\sqrt{n}$ is an example.

Finally, suppose that $\sigma_{\hat{\theta}}$ does involve the unknown θ. This is the case, for example, when $\theta = p$, a population proportion. Then $(\hat{\theta} - \theta)/\sigma_{\hat{\theta}} = z_{\alpha/2}$ can be difficult to solve. An approximate solution can often be obtained by replacing θ in $\sigma_{\hat{\theta}}$ by its estimate $\hat{\theta}$. This results in an estimated standard deviation $s_{\hat{\theta}}$, and the corresponding interval is again $\hat{\theta} \pm z_{\alpha/2} \cdot s_{\hat{\theta}}$. We now apply this in the case $\theta = p$.

A Large-Sample Confidence Interval for a Population Proportion

Let p denote the proportion of "successes" in a population, where *success* identifies an individual or object that has a specified property. A random sample of n individuals is to be selected, and X is the number of successes in the sample. Provided that n is small compared to the population size, X can be regarded as a binomial random variable with $E(X) = np$ and $\sigma_X = \sqrt{np(1-p)}$. Furthermore, if n is large, X has approximately a normal distribution.

The natural estimator of p is $\hat{p} = X/n$, the sample fraction of successes. Since $\hat{p}$ is just X multiplied by the constant $1/n$, $\hat{p}$ also has approximately a normal distribution. As shown in Section 6.1, $E(\hat{p}) = p$ (unbiasedness) and $\sigma_{\hat{p}} = \sqrt{p(1-p)/n}$. The standard deviation $\sigma_{\hat{p}}$ does involve the unknown parameter p. Proceeding as indicated in the previous subsection on a general confidence interval, estimate $\sigma_{\hat{p}}$ by $\sqrt{\hat{p}(1-\hat{p})/n}$. This estimated standard deviation then appears in both the lower and upper confidence limits.

Proposition

> A large-sample **$100(1 - \alpha)\%$ confidence interval for a population proportion p** is
>
> $$\hat{p} \pm z_{\alpha/2}\sqrt{\hat{p}\hat{q}/n} \tag{7.9}$$
>
> where $\hat{p} = x/n$, n is the sample size, x is the observed number of successes, and $\hat{q} = 1 - \hat{p}$. This interval can be used whenever $n\hat{p} \geq 5$ and $n\hat{q} \geq 5$.

An alternative derivation of the confidence interval starts with the probability statement

$$P\left(-z_{\alpha/2} < \frac{\hat{p} - p}{\sqrt{p(1-p)/n}} < z_{\alpha/2}\right) \approx 1 - \alpha$$

Replacing each $<$ by $=$ yields a quadratic equation in p whose solutions are

$$p = \frac{\hat{p} + \dfrac{z_{\alpha/2}^2}{2n} \pm z_{\alpha/2}\sqrt{\dfrac{\hat{p}\hat{q}}{n} + \dfrac{z_{\alpha/2}^2}{4n^2}}}{1 + (z_{\alpha/2}^2)/n}$$

The $+$ sign goes with the upper endpoint of the interval and the $-$ sign with the lower endpoint. Since n is large, the three terms involving $z_{\alpha/2}^2$ are negligible compared to other terms. Ignoring them gives (7.9).

Example 7.8   There is controversy concerning the proposed Strategic Defense Initiative both in the scientific community and in the general public. One survey (San Luis Obispo *Telegram-Tribune,* March 14, 1984) reported that of 1010 randomly selected Californians, 750 favored development of a satellite system to defend against nuclear attack. Let p denote the proportion of all Californians who favored such development at the time of the survey. A point estimate of p is $\hat{p} = x/n = 750/1010 = .743$, and the estimated standard deviation (standard error) of $\hat{p}$ is $\sqrt{\hat{p}\hat{q}/n} = \sqrt{(.743)(.257)/1010} = .0137$. A 95% confidence interval for p is

$$\hat{p} \pm z_{.025}\sqrt{\hat{p}\hat{q}/n} = .743 \pm (1.96)(.0137) = .743 \pm .027 = (.716, .770)$$

We can be highly confident that between 71.6% and 77.0% of all Californians favored such development at the time of the survey. The interval is relatively narrow, indicating that rather precise information about the value of p is available. ■

It is also possible to obtain a confidence interval for p when n is small. The interval is based directly on the binomial distribution. The book by Johnson and Leone (see the Chapter 11 bibliography) has details.

Precision and Sample Size
The interval (7.9) extends a distance $z_{\alpha/2}\sqrt{\hat{p}\hat{q}/n}$ both to the left and to the right of the point estimate $\hat{p}$, so the length of the interval is

$$L = 2z_{\alpha/2}\sqrt{\frac{\hat{p}\hat{q}}{n}} \tag{7.10}$$

Because the right-hand side of (7.10) involves the unknown $\hat{p}$, the length is not determined solely by specification of n. Solving for n yields

$$n = \frac{4z_{\alpha/2}^2 \cdot \hat{p}\hat{q}}{L^2} = \frac{4z_{\alpha/2}^2 \hat{p}(1-\hat{p})}{L^2} \qquad (7.11)$$

There are now several possible approaches to a choice of n. One approach is to specify approximately what you expect the value of $\hat{p}$ to be, and use that in (7.11). Alternatively, a choice of n can be made by taking advantage of the fact that $\hat{p}(1-\hat{p})$ is maximized for $\hat{p} = \frac{1}{2}$ and decreases as $\hat{p}$ moves away from $\frac{1}{2}$ in either direction. Suppose, for example, that p_0 represents an upper bound on what we believe the value of $\hat{p}$ will be, and that $p_0 \leq \frac{1}{2}$. Then use of p_0 in (7.11) in place of $\hat{p}$ will give the maximum value of n necessary to arrive at length L. *The most conservative approach is to use $\hat{p} = \frac{1}{2}$, for then the length will be $\leq L$ no matter what $\hat{p}$ is actually observed.*

Example 7.9

The article "An Evaluation of Football Helmets Under Impact Conditions" (*Amer. J. Sports Medicine,* 1984, pp. 233–237) reported that when each football helmet in a random sample of 37 suspension-type helmets was subjected to a certain impact test, 24 showed damage. Let p denote the proportion of all helmets of this type that would show damage when tested in the prescribed manner. With $\hat{p} = 24/37 = .649$ and $\sqrt{\hat{p}\hat{q}/n} = .0785$, a 95% confidence interval for p based on this data has length $2(1.96)(.0785) \approx .31$. This length indicates that p would be rather imprecisely estimated. The sample size n required to yield a 95% confidence interval whose length is at most .10, whatever the resulting value of $\hat{p}$, is

$$n = \frac{4(1.96)^2(.5)^2}{(.10)^2} = 384.16.$$

It would be necessary to test 385 helmets in order to fulfill the requirement. ■

Exercises / Section 7.2 [10–19]

10. A random sample of 110 lightning flashes in a certain region resulted in a sample average radar echo duration of .81 sec and a sample standard deviation of .34 sec ("Lightning Strikes to an Airplane in a Thunderstorm," *J. Aircraft,* 1984, pp. 607–611). Compute a 99% confidence interval for the true average echo duration μ.

11. A random sample of 200 First National Bank customers having free checking accounts (which require a $300 minimum balance) revealed that during a particular month, the sample average number of checks written was 38.7 with a sample standard deviation of 8.3. Construct a 95% confidence interval for μ, the true average number of checks written during the month by all customers having such accounts. What assumptions are you making about the distribution of the number of checks written?

12. The article "Song Dialects and Colonization in the House Finch" (*Condor,* 1975, pp. 407–422) reported that the sample average duration for $n = 175$ songs of black house finches was 1.47 sec with a sample standard deviation of .95 sec. Compute a 90% confidence interval for the true average duration for black house finch songs. Does the data suggest that an assumption of normality for the distribution of song duration is plausible?

13. The sample average ultimate tensile strength for a sample of 35 high-strength magnetic alloy steel rings used in turbine generators was 152.3 ksi, while the sample standard deviation was 4.8 ksi. Obtain a 99% confidence interval for the true average ultimate tensile strength of such rings.

14. The paper "Extravisual Damage Detection? Defining the Standard Normal Tree" (*Photogrammetric Engineering and Remote Sensing,* 1981, pp. 515–522) discussed the use of color infrared photography in identification of normal trees in Douglas fir stands. Among data reported were summary statistics for green-filter analytical optical densitometric measurements on samples of both healthy and diseased trees. For a sample of 69 healthy trees, the sample average dye layer density was 1.028, with a sample standard deviation of .163. Compute a 95% confidence interval for the true average dye layer density μ for all such trees.

15. In a survey of 277 randomly selected adult female shoppers, 69 stated that whenever an advertised item is unavailable at their local supermarket, they request a raincheck ("Consumer Attitudes Toward Unavailability and Mispricing of Advertised Items by Grocery Stores," *J. Consumer Affairs,* 1977, no. 1, pp. 158–166). Obtain a 99% confidence interval for the true proportion p of adult female shoppers who request a raincheck in such situations.

16. An article in the December 12, 1977, *Los Angeles Times* reported that a new technique, graphic stress telethermometry (GST), accurately detected 23 out of 29 known breast cancer cases. Construct a 90% confidence interval for the true proportion of breast cancers that would be detected by the GST technique (because n is small, the interval will be quite wide).

17. A state legislator wishes to survey residents of her district to see what proportion of the electorate is aware of her position on using state funds to pay for abortions.

a. What sample size is necessary if the 95% confidence interval for p is to have length at most .10 irrespective of p (that is, what n is necessary for the legislator to be 95% confident that $\hat{p}$ is within .05 of the true proportion)?

b. If the legislator has strong reason to believe that at least $\frac{2}{3}$ of the electorate know of her position, how large a sample size would you recommend?

18. A manufacturer of video recorder tapes sells tapes labeled as giving 6 hours of playing time. Sixty-four of these tapes are selected and the actual playing time for each is determined. If the mean and standard deviation of the 64 observed playing times are 352 min and 8 min, respectively, construct a 99% confidence interval for μ, the true average playing time for 6-h tapes made by this manufacturer. Based on your interval, do you think that the manufacturer could be accused of false advertising? Explain.

19. The superintendent of a large school district, having once had a course in probability and statistics, believes that the number of teachers absent on any given day has a Poisson distribution with parameter λ. Use the accompanying data on absences for 50 days to derive a large-sample confidence interval for λ. *Hint:* The mean and variance of a Poisson variable both equal λ, so

$$Z = \frac{\overline{X} - \lambda}{\sqrt{\lambda/n}}$$

has approximately a standard normal distribution. Now proceed as in the derivation of the interval for p by making a probability statement (with probability $1 - \alpha$) and solving the resulting inequalities for λ (see the argument just after (7.9)).

Number of absences	0	1	2	3	4	5	6	7	8	9	10
Frequency	1	4	8	10	8	7	5	3	2	1	1

7.3   A Confidence Interval for the Mean of a Normal Population

The confidence interval for μ presented in Section 7.2 is valid provided that n is large enough for the Central Limit Theorem to apply. The resulting interval can be used whatever the nature of the population distribution. The Central Limit Theorem cannot be invoked, however, when n is small. In this case, one way to proceed is to make a specific assumption about the form of the population distribution and then derive a confidence interval tailored to that assumption. For example, we could develop a confidence interval for μ when the population is described by a gamma distribution, another interval for the case of a Weibull population, and so on. Statisticians have indeed carried out this program for a number of different distributional families. Because the normal distribution is more frequently appropriate as a population model than is any other type of distribution, we shall focus here on a confidence interval for this situation.

Assumption

> The population of interest is normal, so that $X_1, \ldots, X_n$ constitutes a random sample from a normal distribution with both μ and σ unknown.

The key result underlying the interval in Section 7.2 was that for large n, the r.v. $Z = (\overline{X} - \mu)/(S/\sqrt{n})$ has approximately a standard normal distribution. When n is small, S is no longer likely to be close to σ, so the variability in the distribution of Z arises from randomness both in the numerator and in the denominator. This implies that the probability distribution of $(\overline{X} - \mu)/(S/\sqrt{n})$ will be more spread out than the standard normal distribution. The result on which inferences are based introduces a new family of probability distributions called the family of t distributions.

Theorem

> When $\overline{X}$ is the mean of a random sample of size n from a normal distribution with mean μ, the random variable
>
> $$T = \frac{\overline{X} - \mu}{S/\sqrt{n}} \qquad (7.12)$$
>
> has a probability distribution called a t distribution with $n - 1$ degrees of freedom.

Properties of t Distributions

Before applying this theorem, we must first discuss properties of t distributions. While the variable of interest is still $(\overline{X} - \mu)/(S/\sqrt{n})$, we now denote it by T to emphasize that it does not have a standard normal distribution when n is small. Recall that a normal distribution is governed by two parameters, the mean μ and the standard deviation σ. A t distribution is governed by only one parameter, called the **number of degrees of freedom** of the distribution, abbreviated by d.f. We denote this parameter by the Greek letter ν. Possible values of ν are the positive integers $1, 2, 3, \ldots$. Each different value of ν corresponds to a different t distribution.

For any fixed value of the parameter ν, the density function that specifies the associated t curve has an even more complicated appearance than the normal density function. Fortunately we need concern ourselves only with several of the more important features of these curves.

Properties of *t* **distributions**

Let t_ν denote the density function curve for ν degrees of freedom.

1. Each t_ν curve is bell-shaped and centered at 0.
2. As ν increases, the spread of the corresponding t_ν curve decreases.
3. Each t_ν curve is more spread out than the standard normal (z) curve.
4. As $\nu \to \infty$, the sequence of t_ν curves approaches the standard normal curve (so the z curve is often called the t curve with d.f. $= \infty$).

Figure 7.4 illustrates several of these properties for selected values of ν.

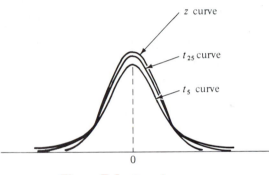

Figure 7.4 t_ν and z curves

The number of d.f. for T in (7.12) is $n - 1$ because although S is based on the n deviations $X_1 - \overline{X}, \ldots, X_n - \overline{X}$, $\Sigma(X_i - \overline{X}) = 0$ implies that only $n - 1$ of these are "freely determined." In general, the number of d.f. for a t variable is the number of freely determined deviations on which the estimated standard deviation in the denominator of T is based.

Since we want to use T to obtain a confidence interval in the same way that Z was previously used, it is necessary to establish notation analogous to z_α for the t distribution.

Notation

Let $t_{\alpha, \nu}$ = the number on the measurement axis for which the area under the t curve with ν degrees of freedom to the right of $t_{\alpha, \nu}$ is α; $t_{\alpha, \nu}$ is called a **t critical value.**

This notation is illustrated in Figure 7.5. Table A.5 in the Appendix gives $t_{\alpha, \nu}$ for selected values of α and ν. The columns of the table correspond to different values of α. To obtain $t_{.05, 15}$, go to the $\alpha = .05$ column, look down to the $\nu = 15$

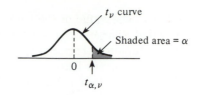

Figure 7.5   A pictorial definition of $t_{\alpha,\,\nu}$

row, and read $t_{.05,\,15} = 1.753$. Similarly, $t_{.05,\,22} = 1.717$ (.05 column, $\nu = 22$ row), and $t_{.01,\,22} = 2.508$.

The values of $t_{\alpha,\,\nu}$ exhibit regular behavior as we move across a row or down a column. For fixed ν, $t_{\alpha,\,\nu}$ increases as α decreases, since we must move further to the right of zero to capture area α in the tail. For fixed α, as ν is increased (that is, as we look down any particular column of the t table) the value of $t_{\alpha,\,\nu}$ decreases. This is because a larger value of ν implies a t distribution with smaller spread, so it is not necessary to go so far from zero to capture tail area α. Furthermore, $t_{\alpha,\,\nu}$ decreases more slowly as ν increases. Consequently, the table skips from $\nu = 30$ to 40 to 60 to ∞; many tables go directly from $\nu = 30$ to ∞. Because t_{∞} is the standard normal curve, the familiar z_{α} values appear in the last row of the table. The rule of thumb suggested earlier for deciding whether the Central Limit Theorem could be used (if $n > 30$) comes from the approximate equality of the standard normal and t distributions for $\nu \geq 30$.

The One-Sample t Confidence Interval

The standardized variable T has a t distribution with $n - 1$ d.f., and the area under the corresponding t density curve between $-t_{\alpha/2,\,n-1}$ and $t_{\alpha/2,\,n-1}$ is $1 - \alpha$ (area $\alpha/2$ lies in each tail), so

$$P(-t_{\alpha/2,\,n-1} < T < t_{\alpha/2,\,n-1}) = 1 - \alpha \qquad (7.13)$$

Expression (7.13) differs from expressions in earlier sections in that T and $t_{\alpha/2,\,n-1}$ are used in place of Z and $z_{\alpha/2}$, but it can be manipulated in the same manner to obtain a confidence interval for μ.

Proposition

> Let $\bar{x}$ and s be the sample mean and sample standard deviation computed from the results of a random sample from a normal population with mean μ. Then a **100(1 − α)% confidence interval for μ** is
>
> $$\left(\bar{x} - t_{\alpha/2,\,n-1} \cdot \frac{s}{\sqrt{n}}, \quad \bar{x} + t_{\alpha/2,\,n-1} \cdot \frac{s}{\sqrt{n}} \right) \qquad (7.14)$$
>
> or, more compactly, $\bar{x} \pm t_{\alpha/2,\,n-1} \cdot s/\sqrt{n}$.

Example **7.10**   The results of a Wagner turbidity test performed on 15 samples of standard Ottawa testing sand were (in microamperes)

26.7, 25.8, 24.0, 24.9, 26.4, 25.9, 24.4, 21.7, 24.1, 25.9, 27.3, 26.9, 27.3, 24.8, 23.6

("Influence of Washing on Properties of Standard Ottawa Testing Sand," *J. Materials,* 1971, pp. 218–233). Figure 7.6 displays a normal probability plot of the data. Only the lowest point in the plot departs at all from linearity. In our view, the magnitude of the departure is not substantial enough to cast doubt on the assumption that turbidity has a normal distribution. To obtain a 95% confidence interval for μ = the true average Wagner turbidity of all such sand samples, we first compute $\bar{x} = 25.31$ and $s = 1.58$. With $n = 15$ and $\alpha = .05$, $t_{\alpha/2, n-1} = t_{.025, 14} = 2.145$. The desired interval is then

$$\bar{x} \pm t_{\alpha/2, n-1}\frac{s}{\sqrt{n}} = 25.31 \pm \frac{(2.145)(1.58)}{\sqrt{15}} = 25.31 \pm .88$$

$$= (24.43, 26.19)$$

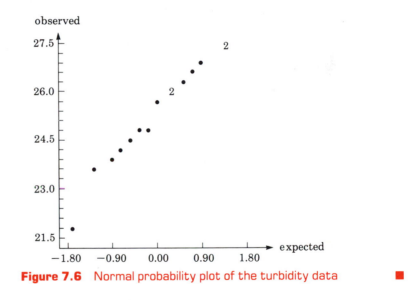

Figure 7.6   Normal probability plot of the turbidity data ■

Unfortunately it is not easy to select n to control the length of the t interval. This is because the length involves the unknown (before the data is collected) s, and because n enters not only through $1/\sqrt{n}$ but also through $t_{\alpha/2, n-1}$. As a result, an appropriate n can be obtained only by trial and error.

Chapter 15 discusses a small-sample confidence interval for μ which is valid provided only that the population distribution is symmetric, a weaker assumption than normality. However, when the population distribution is normal, the t interval tends to be shorter than would be any other interval with the same confidence level. If n is small and the investigator believes that the population distribution is quite skewed, a statistician should be consulted.

Exercises / Section 7.3 (20–26)

20. Determine the values of the following quantities:
    a. $t_{.1, 15}$ **b.** $t_{.05, 15}$ **c.** $t_{.05, 25}$ **d.** $t_{.05, 40}$ **e.** $t_{.005, 40}$

21. a. What is the 90th percentile of the t distribution with 22 d.f.?
    b. What is the 10th percentile of the t distribution with 22 d.f.?
    c. What is the 5th percentile of the t distribution with 16 d.f.?
    d. What is $P(T \leq 3.365)$ when T has a t distribution with 5 d.f.?
    e. What is $P(-2.306 < T < 2.306)$ when T has a t distribution with 8 d.f.?
    f. What is $P(|T| \geq 3.435)$ when T has a t distribution with 26 d.f.?

22. A random sample of $n = 8$ E-glass fiber test specimens of a certain type yielded a sample average interfacial shear yield stress of 30.2 and a sample standard deviation of 3.1 ("On Interfacial Failure in Notched Unidirectional Glass/Epoxy Composites," *J. Composite Materials,* 1985, pp. 276–286). Assuming that interfacial shear yield stress is normally distributed, compute a 95% confidence interval for true average stress (as did the authors of the paper).

23. In a study of the flammability of material used in children's sleepwear, the char length (inches) for five samples of washed acetate/nylon brushed tricot fabric was measured. The resulting observations were

8.1, 10.4, 9.5, 8.9, 10.7

Assuming that char length is a normally distributed variable, compute a 95% confidence interval for the true average char length for this fabric.

24. Trisodium carboxymethylaxysuccinate (Na-CMOS) is a chemical compound being considered for inclusion in various detergent products. Because it would eventually become a component

of wastewater, an experiment was performed to assess its effects on aquatic life. Following exposure to 100 mg NaCMOS/liter for a 16-hour period, the NaCMOS content was determined for each of six different goldfish. The resulting sample mean and sample standard deviation were 9.1 and 3.2, respectively. Compute a 90% confidence interval for the true average NaCMOS content for goldfish undergoing such exposure ("Acute Fish Toxicity and Absorption Tests of an Experimental Detergent Builder, Trisodium Carboxymethylaxysuccinate," *J. Testing and Evaluation,* 1979, pp. 16–17).

25. For each of 18 preserved cores from oil-wet carbonate reservoirs, the amount of residual gas saturation after a solvent injection was measured at water flood-out. Observations, in percent of pore volume, were

23.5, 31.5, 34.0, 46.7, 45.6, 32.5, 41.4, 37.2, 42.5, 46.9, 51.5, 36.4, 44.5, 35.7, 33.5, 39.3, 22.0, 51.2

Compute a 98% confidence interval for the true average amount of residual gas saturation ("Relative Permeability Studies of Gas-Water Flow Following Solvent Injection in Carbonate Rocks," *Soc. Petroleum Engineers J.,* 1976, pp. 23–30).

26. A more extensive tabulation of t critical values than that given in Appendix Table A.5 shows that $t_{.25, 20} = .687$, $t_{.20, 20} = .860$, and $t_{.15, 20} = 1.064$. What is the confidence level for each of the following three confidence intervals for the mean μ of a normal population distribution? Which of the three intervals would you recommend be used, and why?
    a. $(\bar{x} - .687s/\sqrt{21}, \bar{x} + 1.725s/\sqrt{21})$
    b. $(\bar{x} - .860s/\sqrt{21}, \bar{x} + 1.325s/\sqrt{21})$
    c. $(\bar{x} - 1.064s/\sqrt{21}, \bar{x} + 1.064s/\sqrt{21})$

7.4

Confidence Intervals for the Variance and Standard Deviation of a Normal Population

Although inferences concerning a population variance σ^2 or standard deviation σ are usually of less interest than those about a mean or proportion, there are occasions when such procedures are needed. In the case of a normal population distribution, inferences are based on the following result concerning the sample variance S^2.

Theorem

> Let $X_1, X_2, \ldots, X_n$ be a random sample from a normal distribution with parameters μ and σ^2. Then the random variable
>
> $$\frac{(n-1)S^2}{\sigma^2} = \frac{\Sigma(X_i - \overline{X})^2}{\sigma^2}$$
>
> has a chi-squared (χ^2) probability distribution with $n - 1$ degrees of freedom (d.f.).

As discussed in Sections 4.4 and 7.1, the chi-squared distribution is a continuous probability distribution with a single parameter ν, called the number of degrees of freedom, with possible values 1, 2, 3, ... The graphs of several χ^2 p.d.f.'s are illustrated in Figure 7.7. Each p.d.f. $f(x; \nu)$ is positive only for $x > 0$, and each has a positive skew (long upper tail), though the distribution moves rightward and becomes more symmetric as ν increases. To specify inferential procedures that use the chi-squared distribution, we need notation analogous to that for a t critical value $t_{\alpha, \nu}$.

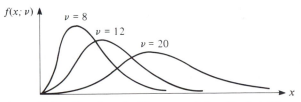

Figure 7.7 Graphs of chi-squared density functions

Notation

> Let $\chi^2_{\alpha, \nu}$, called a **chi-squared critical value,** denote the number on the measurement axis such that α of the area under the chi-squared curve with ν d.f. lies to the right of $\chi^2_{\alpha, \nu}$.

Figure 7.8 illustrates the notation for chi-squared critical values.

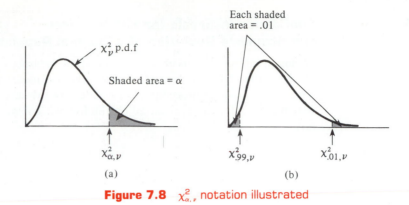

Figure 7.8 $\chi^2_{\alpha, \nu}$ notation illustrated

Because the t distribution is symmetric, it was necessary to tabulate only upper-tail critical values ($t_{\alpha, \nu}$ for small values of α). The chi-squared distribution is not symmetric, so Appendix Table A.6 contains values of $\chi^2_{\alpha, \nu}$ both for α near 0 and near 1, as illustrated in Figure 7.8(b). For example, $\chi^2_{.025, 14} = 26.119$ and $\chi^2_{.95, 20}$ (the fifth percentile) $= 10.851$.

The random variable $(n - 1)S^2/\sigma^2$ satisfies the two properties on which the general method for obtaining a confidence interval is based: it is a function of the parameter of interest σ^2, yet its probability distribution (chi-squared) does not depend on this parameter. The area under a chi-squared curve with ν d.f. to the right of $\chi^2_{\alpha/2, \nu}$ is $\alpha/2$, as is the area to the left of $\chi^2_{1 - \alpha/2, \nu}$. Thus the area captured between these two critical values is $1 - \alpha$. As a consequence of this and the theorem just stated,

$$P\left(\chi^2_{1 - \alpha/2, n - 1} < \frac{(n - 1)S^2}{\sigma^2} < \chi^2_{\alpha/2, n - 1}\right) = 1 - \alpha \qquad (7.15)$$

The inequalities in (7.15) are equivalent to

$$\frac{(n - 1)S^2}{\chi^2_{\alpha/2, n - 1}} < \sigma^2 < \frac{(n - 1)S^2}{\chi^2_{1 - \alpha/2, n - 1}}$$

Substitution of the computed value s^2 into the limits gives a confidence interval for σ^2, and taking square roots gives an interval for σ.

A 100(1 − α)% confidence interval for the variance σ^2 of a normal population has lower limit

$$(n - 1)s^2/\chi^2_{\alpha/2, n - 1}$$

and upper limit

$$(n - 1)s^2/\chi^2_{1 - \alpha/2, n - 1}$$

A confidence interval for σ has lower and upper limits that are the square roots of the corresponding limits in the interval for σ^2.

Example 7.11   The accompanying data on breakdown voltage of electrically stressed circuits was read from a normal probability plot that appeared in the paper "Damage of Flexible Printed Wiring Boards Associated with Lightning Induced Voltage Surges" (*IEEE Transactions on Components, Hybrids, and Manuf. Tech.,* 1985, pp. 214–220). The straightness of the plot gave strong support to the assumption that breakdown voltage is approximately normally distributed.

1470, 1510, 1690, 1740, 1900, 2000, 2030, 2100, 2190, 2200, 2290, 2380, 2390, 2480, 2500, 2580, 2700

Let σ^2 denote the variance of the breakdown voltage distribution. The computed value of the sample variance is $s^2 = 137{,}324.3$, the point estimate of σ^2. With d.f. $= n - 1 = 16$, a 95% confidence interval requires $\chi^2_{.975, 16} = 6.908$ and $\chi^2_{.025, 16} = 28.845$. The interval is

$$\left(\frac{16(137{,}324.3)}{28.845}, \frac{16(137{,}324.3)}{6.908} \right) = (76{,}172.3, 318{,}064.4)$$

Taking the square root of each endpoint yields (276.0, 564.0) as the 95% confidence interval for σ. These intervals are quite wide, reflecting substantial variability in breakdown voltage in combination with a small sample size.   ∎

Confidence intervals for σ^2 and σ when the population distribution is not normal can be difficult to obtain, even when the sample size is large. For such cases, please consult a knowledgeable statistician.

Exercises / Section 7.4 (27–30)

27. Determine the values of the following quantities:
    a. $\chi^2_{.1, 15}$
    b. $\chi^2_{.1, 25}$
    c. $\chi^2_{.01, 25}$
    d. $\chi^2_{.005, 25}$
    e. $\chi^2_{.99, 25}$
    f. $\chi^2_{.995, 25}$

28. Determine
    a. the ninety-fifth percentile of the chi-squared distribution with $\nu = 10$.
    b. the fifth percentile of the chi-squared distribution with $\nu = 10$.
    c. $P(10.98 \le \chi^2 \le 36.78)$ where χ^2 is a chi-squared r.v. with $\nu = 22$.
    d. $P(\chi^2 < 14.611$ or $\chi^2 > 37.652)$ where χ^2 is a chi-squared random variable with $\nu = 25$.

29. The amount of lateral expansion (mils) was determined for a sample of $n = 9$ pulsed-power gas metal arc welds used in LNG ship containment

tanks. The resulting sample standard deviation was $s = 2.81$ mils. Assuming normality, derive a 95% confidence interval for σ^2 and for σ.

30. The following observations were made on fracture toughness of base plate of 18% nickel maraging steel ["Fracture Testing of Weldments," *ASTM Special Publ. No. 381,* 1965, pp. 328–356 (in ksi $\sqrt{\text{in.}}$, given in increasing order)]:

    69.5, 71.9, 72.6, 73.1, 73.3, 73.5, 75.5, 75.7, 75.8, 76.1, 76.2, 76.2, 77.0, 77.9, 78.1, 79.6, 79.7, 79.9, 80.1, 82.2, 83.7, 93.3

    Calculate a 99% confidence interval for the standard deviation of the fracture toughness distribution. Is this interval valid whatever the nature of the distribution? Explain.

Supplementary Exercises / Chapter 7 (31–44)

31. Television advertisers are becoming concerned over the use of videocassette recorders (VCR's)

to tape television shows, since many viewers fast-forward through the commercials when viewing the taped shows. A survey conducted by A. C. Nielsen Co. of 1100 VCR owners found that 715 used the fast-forward feature to avoid commercials on taped programs (*Los Angeles Times,* September 2, 1984). Construct and interpret a 95% confidence interval for the proportion of all VCR owners who use the fast-forward feature to avoid advertisements.

32. Analysis of the venom of seven eight-day-old worker bees yielded the following observations on histamine content (nanograms):

    649, 832, 418, 530, 384, 899, 755

    Compute a 90% confidence interval for the true average histamine content for all worker bees of this age. Were any assumptions necessary to justify your computation?

33. A sample of 40 transistors of a certain type yielded a sample average saturation current of 9.46 milliamps with a sample standard deviation of .58. Compute a 95% confidence interval for the true average saturation current for transistors of this type. What assumptions, if any, are you making?

34. The financial manager of a large department store chain selected a random sample of 200 of its credit card customers and found that 136 had incurred an interest charge during the previous year because of an unpaid balance.
    a. Compute a 90% confidence interval for the true proportion of credit card customers who incurred an interest charge during the previous year.
    b. If the desired length of the 90% interval is .05, what sample size is necessary to ensure this?
    c. Compute an 82% confidence interval for the true proportion.

35. The reaction time (RT) to a stimulus is the interval of time commencing with stimulus presentation and ending with the first discernible movement of a certain type. The paper "Relationship of Reaction Time and Movement Time in a Gross Motor Skill" (*Perceptual and Motor Skills,* 1973, pp. 453–454) reported that the sample average RT for 16 experienced swimmers to a pistol start was .214 second with a sample standard deviation of .036 second. Making any necessary assumptions, derive a 90% confidence interval for true average RT for all experienced swimmers.

36. Aphid infestation of fruit trees can be controlled either by spraying with pesticide or by inundation with ladybugs. In a particular area, four different groves of fruit trees are selected for experimentation. The first three groves are sprayed with pesticides 1, 2, and 3, respectively, and the fourth is treated with ladybugs, with the following results on yield:

Treatment	$n_i =$ number of trees	$\bar{x}_i$ (bushels/tree)	s_i
1	100	10.5	1.5
2	90	10.0	1.3
3	100	10.1	1.8
4	120	10.7	1.6

Let $\mu_i =$ the true average yield (bushels/tree) after receiving the ith treatment. Then

$$\theta = \tfrac{1}{3}(\mu_1 + \mu_2 + \mu_3) - \mu_4$$

measures the difference in true average yields between treatment with pesticides and treatment with ladybugs. When n_1, n_2, n_3, and n_4 are all large, the estimator $\hat{\theta}$ obtained by replacing each μ_i by $\bar{X}_i$ is approximately normal. Use this to derive a large-sample $100(1 - \alpha)\%$ confidence interval for θ, and compute the 95% interval for the given data.

37. It is important that face masks used by firefighters be able to withstand high temperatures, since firefighters commonly work in temperatures of 200 to 500° F. In a test of one type of mask, 11 of 35 had their lenses pop off at 250°. Construct a 90% confidence interval for the true proportion of masks of this type whose lenses would pop out at 250°.

38. A manufacturer of college textbooks is interested in estimating the strength of the bindings produced by a particular binding machine. Strength can be measured by recording the force required to pull the pages from the binding. If this force is measured in pounds, how many books should be tested in order to estimate the average force required to break the binding to within .1 lb with 95% confidence? Assume that σ is known to be .8.

39. Chronic exposure to asbestos fiber is a well-known health hazard. The paper "The Acute Effects of Chrysotile Asbestos Exposure on Lung Function" (*Environ. Research*, 1978, pp. 360–372) reported results of a study based on a sample of construction workers who had been exposed to asbestos over a prolonged period. Among the data given in the article were the following (ordered) values of pulmonary compliance ($cm^3/cm\ H_2O$) for each of 16 subjects 8 months after the exposure period (pulmonary compliance is a measure of lung elasticity, or how effectively the lungs are able to inhale and exhale):

167.9   180.8   184.8   189.8   194.8   200.2
201.9   206.9   207.2   208.4   226.3   227.7
228.5   232.4   239.8   258.6

a. Is it plausible that the population distribution is normal?
b. Compute a 95% confidence interval for the true average pulmonary compliance after such exposure.

40. In Example 6.7 we introduced the concept of a censored experiment in which n components are put on test and the experiment terminates as soon as r of the components have failed. Suppose that component lifetimes are independent, each having an exponential distribution with parameter λ. Let Y_1 denote the time at which the first failure occurs, Y_2 the time at which the second failure occurs, and so on, so that $T_r = Y_1 + \cdots + Y_r + (n - r)Y_r$ is the total accumulated lifetime at termination. Then it can be shown that $2\lambda T_r$ has a chi-squared distribution with $2r$ d.f. Use this fact to develop a $100(1 - \alpha)\%$ confidence interval formula for true average lifetime $1/\lambda$. Compute a 95% confidence interval from the data in the aforementioned example.

41. Let $X_1, X_2, \ldots, X_n$ be a random sample from a continuous probability distribution having median $\tilde{\mu}$ (so that $P(X_i \le \tilde{\mu}) = P(X_i \ge \tilde{\mu}) = .5$).
a. Show that

$$P(\min(X_i) < \tilde{\mu} < \max(X_i)) = 1 - (\tfrac{1}{2})^{n-1}$$

so that $(\min(x_i), \max(x_i))$ is a $100(1 - \alpha)\%$ confidence interval for $\tilde{\mu}$ with $\alpha = (\tfrac{1}{2})^{n-1}$.
Hint: The complement of the event $\{\min(X_i) < \tilde{\mu} < \max(X_i)\}$ is $\{\max(X_i) \le \tilde{\mu}\} \cup \{\min(X_i) \ge \tilde{\mu}\}$. But $\max(X_i) \le \tilde{\mu}$ iff $X_i \le \tilde{\mu}$ for all i.

b. For each of six normal male infants, the amount of the amino acid alanine (mg/100 ml) was determined while on an isoleucine-free diet, resulting in the following data:

2.84, 3.54, 2.80, 1.44, 2.94, 2.70

Compute a 97% confidence interval for the true median amount of alanine for infants on such a diet ("The Essential Amino Acid Requirements of Infants," *Amer. J. Nutrition*, 1964, pp. 322–330).
c. Let $x_{(2)}$ denote the second smallest of the x_i's and $x_{(n-1)}$ denote the second largest of the x_i's. What is the confidence coefficient of the interval $(x_{(2)}, x_{(n-1)})$ for $\tilde{\mu}$?

42. Let $X_1, X_2, \ldots, X_n$ be a random sample from a uniform distribution on the interval $[0, \theta]$, so that

$$f(x) = \begin{cases} \dfrac{1}{\theta} & 0 \le x \le \theta \\ 0 & \text{otherwise} \end{cases}$$

Then if $Y = \max(X_i)$, it can be shown that the random variable $U = Y/\theta$ has density function

$$f_U(u) = \begin{cases} nu^{n-1} & 0 \le u \le 1 \\ 0 & \text{otherwise} \end{cases}$$

a. Use $f_U(u)$ to verify that

$$P\left((\alpha/2)^{1/n} \le \frac{Y}{\theta} \le (1 - \alpha/2)^{1/n}\right) = 1 - \alpha$$

and use this to derive a $100(1 - \alpha)\%$ confidence interval for θ.
b. Verify that $P(\alpha^{1/n} \le Y/\theta \le 1) = 1 - \alpha$, and derive a $100(1 - \alpha)\%$ confidence interval for θ based on this probability statement.
c. Which of the two intervals derived above is shorter? If my waiting time for a morning bus is uniformly distributed and observed waiting times are $x_1 = 4.2$, $x_2 = 3.5$, $x_3 = 1.7$, $x_4 = 1.2$, and $x_5 = 2.4$, derive a 95% confidence interval for θ by using the shorter of the two intervals.

43. Let $0 \le \gamma \le \alpha$. Then a $100(1 - \alpha)\%$ confidence interval for μ when n is large is

$$\left(\bar{x} - z_\gamma \cdot \frac{s}{\sqrt{n}}, \ \bar{x} + z_{\alpha - \gamma} \cdot \frac{s}{\sqrt{n}}\right)$$

The choice $\gamma = \alpha/2$ yields the usual interval derived in Section 7.2; if $\gamma \ne \alpha/2$, the above interval is not symmetric about $\bar{x}$. The length of this interval is $L = s(z_\gamma + z_{\alpha-\gamma})/\sqrt{n}$. Show that L is minimized for the choice $\gamma = \alpha/2$, so that the symmetric interval is the shortest. *Hints:* (a) By definition of z_α, $\Phi(z_\alpha) = 1 - \alpha$, so that $z_\alpha = \Phi^{-1}(1 - \alpha)$; (b) The relationship between the derivative of a function $y = f(x)$ and its inverse $x = f^{-1}(y)$ is $(d/dy)f^{-1}(y) = 1/f'(x)$.

44. Suppose that $x_1, x_2, \ldots, x_n$ are observed values resulting from a random sample from a symmetric but possibly heavy-tailed distribution. Let $\tilde{x}$ and f_s denote the sample median and fourth spread, respectively. Chapter 11 of *Understanding Robust and Exploratory Data Analysis* (see the bibliography in Chapter 6) suggests the following robust 95% confidence interval for the population mean (point of symmetry):

$$\tilde{x} \pm \left(\frac{\text{conservative } t \text{ critical value}}{1.075}\right) \cdot \frac{f_s}{\sqrt{n}}$$

The value of the quantity in parentheses is 2.10 for $n = 10$, 1.94 for $n = 20$, and 1.91 for $n = 30$. Compute this confidence interval for the data of Exercise 30 and compare to the t confidence interval appropriate for a normal population distribution.

Bibliography

DeGroot, Morris, *Probability and Statistics* (2nd ed.), Addison-Wesley, Reading, Mass., 1986. A very good exposition of the general principles of statistical inference.

Larsen, Richard, and Marx, Morris, *Introduction to Mathematical Statistics* (2nd ed.), Prentice-Hall, Englewood Cliffs, N.J., 1985. Similar to DeGroot's presentation, but slightly less mathematical.

McClave, James, and Dietrich, Frank, *Statistics* (3rd ed.), Dellen, San Francisco, 1985. A very good introduction to methods of statistical inference in a book written at a precalculus level.

CHAPTER 8

Tests of Hypotheses Based on a Single Sample

Introduction

As we have now seen, a parameter can be estimated from sample data either by a single number (a point estimate) or an entire interval of plausible values (a confidence interval). Frequently, however, the objective of an investigation is not to estimate a parameter, but instead to decide which of two contradictory claims about the parameter is correct. Methods for accomplishing this comprise the part of statistical inference called *hypothesis testing*. In this chapter we first discuss some of the basic concepts and terminology in hypothesis testing, and then develop decision procedures for the most frequently encountered testing problems based on a sample from a single population.

8.1   Hypotheses and Test Procedures

A **statistical hypothesis,** or hypothesis, is a claim either about the value of a single population characteristic or about the values of several population characteristics. One example of a hypothesis is the claim $\mu = .75$, where μ is the true average inside diameter of a certain type of PVC pipe. Another example is the statement $p < .10$, where p is the proportion of defective circuit boards among all circuit boards produced by a certain manufacturer. If μ_1 and μ_2 denote the true average breaking strengths of two different types of twine, one hypothesis is the assertion that $\mu_1 - \mu_2 = 0$, and another is the statement $\mu_1 - \mu_2 > 5$.

In any hypothesis testing problem, there are two contradictory hypotheses under consideration. One hypothesis might be the claim $\mu = .75$ and the other $\mu \neq .75$, or the two contradictory statements might be $p \geq .10$ and $p < .10$. The objective is to decide, based on sample information, which of the two hypotheses is correct. There is a familiar analogy to this in a criminal trial. One claim is the assertion that the accused individual is innocent. In the American judicial system, this is the claim that is initially believed to be true. Only in the face of strong evidence to the contrary should the jury reject this claim in favor of the alternative assertion that the accused is guilty. In this sense, the claim of innocence is the favored or protected hypothesis, and the burden of proof is placed on those who believe in the alternative claim.

Similarly, in testing statistical hypotheses, the problem will be formulated so that one of the claims is initially favored. This initially favored claim will not be rejected in favor of the alternative claim unless sample evidence contradicts it and provides strong support for the alternative assertion. The claim initially favored or believed to be true is called the **null hypothesis** and is denoted by H_0. The other claim in a hypothesis testing problem is called the **alternative hypothesis** and is denoted by H_a. Thus we might test $H_0 : \mu = .75$ against the alternative $H_a : \mu \neq .75$. Only if sample data strongly suggests that μ is something other than .75 should the null hypothesis be rejected. In the absence of such evidence, H_0 should not be rejected, since it is still quite plausible.

Sometimes an investigator does not want to accept a particular assertion unless and until data can provide strong support for the assertion. As an example, suppose that a company is considering putting a new type of coating on bearings that it produces. The true average wear life with the current coating is known to be 1000 hours. With μ denoting the true average life for the new coating, the company would not want to make a change unless evidence strongly suggested that μ exceeds 1000. An appropriate problem formulation would involve testing $H_0 : \mu = 1000$ against $H_a : \mu > 1000$. The conclusion that a change is justified is identified with H_a, and it would take conclusive evidence to justify rejecting H_0 and switching to the new coating.

Scientific research often involves trying to decide whether a current theory should be replaced by a more plausible and satisfactory explanation of the phenomena under investigation. A conservative approach is to identify the current theory with H_0 and the researcher's alternative explanation with H_a. Rejection of the current theory will then occur only when evidence is much more consistent with the new theory. In many situations H_a is referred to as the "researcher's hypothesis," since it is the claim that the researcher would really like to validate. The word *null* means "of no value, effect, or consequence," which suggests that H_0 should be identified with the hypothesis of no change (from current opinion), no difference, no improvement, and so on. Suppose, for example, that 10% of all circuit boards produced by a certain manufacturer during a recent period were defective. An engineer has suggested a change in the production process in the belief that it will result in a reduced defective rate. Let p denote the true proportion of defective boards resulting from the changed process. Then the research hypothesis, on which the burden of proof is placed, is the assertion that $p < .10$. Thus the alternative hypothesis is $H_a : p < .10$.

In our treatment of hypothesis testing, H_0 will always be stated as an equality claim. If θ denotes the parameter of interest, the null hypothesis will have the form $H_0 : \theta = \theta_0$, where θ_0 is a specified number called the *null value* of the parameter (value claimed for θ by the null hypothesis). As an example, consider the circuit board example just discussed. The suggested alternative hypothesis was $H_a : p < .10$, the claim that the defective rate is reduced by the process modification. A natural choice of H_0 in this situation is the claim that $p \geq .10$, according to which the new process is either no better *or* worse than the one currently used. We shall instead consider $H_0 : p = .10$ versus $H_a : p < .10$. The rationale for using this simplified null hypothesis is that any reasonable decision procedure for deciding between $H_0 : p = .10$ and $H_a : p < .10$ will also be reasonable for deciding between the claim that $p \geq .10$ and H_a. The use of a simplified H_0 is preferred because it has certain technical benefits, which will be apparent shortly.

The alternative to the null hypothesis $H_0 : \theta = \theta_0$ will look like one of the following three assertions: (a) $H_a : \theta > \theta_0$ (in which case the implicit null hypothesis is $\theta \leq \theta_0$); (b) $H_a : \theta < \theta_0$ (so the implicit null hypothesis states that $\theta \geq \theta_0$); or (c) $H_a : \theta \neq \theta_0$. For example, let σ denote the standard deviation of the distribution of inside diameters (inches) for a certain type of metal sleeve. If the decision was made to use the sleeve unless sample evidence conclusively demonstrated that $\sigma > .001$, the appropriate hypotheses would be $H_0 : \sigma = .001$ versus $H_a : \sigma > .001$.

Test Procedures

A test procedure is a rule, based on sample data, for deciding whether or not to reject H_0. A test of $H_0 : p = .10$ versus $H_a : p < .10$ in the circuit board problem might be based on examining a random sample of $n = 200$ boards. Let X denote the number of defective boards in the sample, a binomial random variable; x represents the observed value of X. If H_0 is true, $E(X) = np = 200(.10) = 20$, whereas we can expect fewer than 20 defective boards if H_a is true. A value x just a bit below 20 does not strongly contradict H_0, so it is reasonable to reject H_0 only if x is substantially less than 20. One such test procedure is to reject H_0 if $x \leq 15$ and not reject H_0 otherwise. This procedure has two constituents: (a) a *test statistic* or function of the sample data used to make a decision, and (b) a *rejection region* consisting of those x values for which H_0 will be rejected in favor of H_a. For the rule just suggested, the rejection region consists of $x = 0, 1, 2, \ldots,$ and 15. H_0 will not be rejected if $x = 16, 17, \ldots, 199,$ or 200.

A test procedure is specified by

1. a **test statistic,** a function of the sample data on which the decision (reject H_0 or don't reject H_0) is to be based, and
2. a **rejection region,** the set of all test statistic values for which H_0 will be rejected.

The null hypothesis will then be rejected if and only if the observed or computed test statistic value falls in the rejection region.

As another example, suppose that a cigarette manufacturer claims that the average nicotine content μ of brand B cigarettes is (at most) 1.5 mg. It would be unwise to reject the manufacturer's claim without strong contradictory evidence, so an appropriate problem formulation is to test $H_0 : \mu = 1.5$ versus $H_a : \mu > 1.5$. Consider a decision rule based on analyzing a random sample of 32 cigarettes. Let $X_1, \ldots, X_{32}$ denote the nicotine contents of cigarettes in the sample, so that $\overline{X}$ is the sample average nicotine content. If H_0 is true, $E(\overline{X}) = \mu = 1.5$; whereas if H_0 is false, we expect $\overline{X}$ to exceed 1.5. Strong evidence against H_0 is provided by a value $\overline{x}$ that considerably exceeds 1.5. Thus we might use $\overline{X}$ as a test statistic along with the rejection region $\overline{x} \geq 1.6$.

In both the circuit board and nicotine examples, the choice of test statistic and form of the rejection region make sense intuitively. However, the choice of cutoff value used to specify the rejection region is somewhat arbitrary. Instead of rejecting $H_0 : p = .10$ in favor of $H_a : p < .10$ when $x \leq 15$, we could use the rejection region $x \leq 14$. For this region, H_0 would not be rejected if 15 defective boards are observed, whereas this would lead to rejection of H_0 if the initially suggested region is employed. Similarly, the rejection region $\overline{x} \geq 1.55$ might be used in the nicotine problem in place of the region $\overline{x} \geq 1.60$.

Errors in Hypothesis Testing

Before satisfactory test procedures can be obtained, the consequences of using one rejection region as opposed to another must be understood. The basis for choosing a particular region lies in an understanding of the errors that one might be faced with in drawing a conclusion. Consider the rejection region $x \leq 15$ in the circuit board problem. Even when $H_0 : p = .10$ is true, it might happen that an unusual sample results in $x = 13$, so that H_0 is erroneously rejected. On the other hand, even when $H_a : p < .10$ is true, an unusual sample might yield $x = 20$, in which case H_0 would not be rejected, again an incorrect conclusion. Thus it is possible that H_0 may be rejected when it is true or that H_0 may not be rejected when it is false. These possible errors are not consequences of a foolishly chosen rejection region. Either one of these two errors might result when the region $x \leq 14$ is employed, or indeed when any other region is used.

Definition

> A **type I error** consists of rejecting the null hypothesis H_0 when it is true. A **type II error** involves not rejecting H_0 when H_0 is false.

In the nicotine problem, a type I error consists of rejecting the manufacturer's claim that $\mu = 1.5$ when it is actually true. If the rejection region $\overline{x} \geq 1.6$ is employed, it might happen that $\overline{x} = 1.63$ even when $\mu = 1.5$, resulting in a type I error. Alternatively, it may be that H_0 is false and yet $\overline{x} = 1.52$ is observed, leading to H_0 not being rejected (a type II error).

In the best of all possible worlds, test procedures for which neither type of error is possible could be developed. However, this ideal can be achieved only by basing a decision on an examination of the entire population, which is almost always impractical. The difficulty with using a procedure based on sample data is that because of sampling variability, an unrepresentative sample may result. Even though $E(\overline{X}) = \mu$, the observed value $\overline{x}$ may differ substantially from μ (at least if n is small). Thus when $\mu = 1.5$ in the nicotine situation, $\overline{x}$ may be much larger than 1.5, resulting in erroneous rejection of H_0. Alternatively, it may be that $\mu = 1.6$ yet an $\overline{x}$ much smaller than this is observed, leading to a type II error.

Instead of demanding error-free procedures, we must look for procedures for which either type of error is unlikely to occur. That is, a good procedure is one for which the probability of making either type of error is small. The choice of a particular rejection region cutoff value fixes the probabilities of type I and type II errors. These error probabilities are traditionally denoted by α and β, respectively. Because H_0 specifies a unique value of the parameter, there is a single value of α. However, there is a different value of β for each value of the parameter consistent with H_a.

Example **8.1** A certain type of automobile is known to sustain no visible damage 25% of the time in 10 mile per hour crash tests. A modified bumper design has been proposed in an effort to increase this percentage. Let p denote the proportion of all 10 mph crashes with this new bumper that result in no visible damage. The hypotheses to be tested are $H_0 : p = .25$ (no improvement) versus $H_a : p > .25$. The test will be based on an experiment involving $n = 20$ independent crashes with prototypes of the new design. Intuitively, H_0 should be rejected if a substantial number of the crashes show no damage. Consider the following test procedure:

> Test statistic: $X =$ the number of crashes with no visible damage (a binomial random variable)
> Rejection region: $R_8 = \{8, 9, 10, \ldots, 19, 20\}$; that is, reject H_0 if $x \geq 8$, where x is the observed value of the test statistic

This rejection region is called *upper-tailed,* because it consists only of large values of the test statistic.

When H_0 is true, X has a binomial probability distribution with $n = 20$ and $p = .25$. Thus

$$\alpha = P(\text{type I error}) = P(H_0 \text{ is rejected when it is true})$$
$$= P(X \geq 8 \text{ when } X \sim \text{Bin}(20, .25)) = 1 - B(7; 20, .25)$$
$$= 1 - .898 = .102$$

That is, when H_0 is actually true, roughly 10% of all experiments consisting of 20 crashes would result in H_0 being incorrectly rejected (a type I error).

In contrast to α, there is not a single β. Instead there is a different β for each different p that exceeds .25. Thus there is a value of β for $p = .3$ (in which case $X \sim \text{Bin}(20, .3)$), another value of β for $p = .5$, and so on. For example,

$$\beta(.3) = P(\text{type II error when } p = .3)$$
$$= P(H_0 \text{ is not rejected when it is false because } p = .3)$$
$$= P(X \le 7 \text{ when } X \sim \text{Bin}(20, .3)) = B(7; 20, .3) = .772$$

When p is actually .3 rather than .25 (a "small" departure from H_0), roughly 77% of all experiments of this type would result in H_0 being incorrectly not rejected! Similarly,

$$\beta(.5) = P(H_0 \text{ is not rejected when it is false because } p = .5)$$
$$= P(X \le 7 \text{ when } X \sim \text{Bin}(20, .5)) = B(7; 20, .5) = .132$$

This β is much smaller than $\beta(.3)$ because $p = .5$ represents a much more substantial departure from H_0 than does $p = .3$. The accompanying table displays β for these and other selected values of p (each calculated for the rejection region R_8). Clearly β decreases as the value of p moves further to the right of the null value .25. Intuitively, the greater the departure from H_0, the less likely it is that such a departure will not be detected.

p	.3	.4	.5	.6	.7	.8
$\beta(p)$	.772	.416	.132	.021	.001	.000

The proposed test procedure is still reasonable for testing the more realistic null hypothesis that $p \le .25$. In this case, there is no longer a single α, but instead an α for each p that is at most .25—$\alpha(.25)$, $\alpha(.23)$, $\alpha(.20)$, $\alpha(.15)$, and so on. It is easily verified, though, that $\alpha(p) < \alpha(.25) = .102$ if $p < .25$. That is, the largest value of α occurs for the boundary value .25 between H_0 and H_a. Thus if α is small for the simplified null hypothesis, it will also be as small or smaller for the more realistic H_0. ∎

Example 8.2   The drying time of a certain type of paint under specified test conditions is known to be normally distributed with mean value 75 min and standard deviation 9 min. Chemists have proposed a new additive designed to decrease average drying time. It is believed that drying times with this additive will remain normally distributed with $\sigma = 9$. Because of the expense associated with the additive, evidence should strongly suggest an improvement in average drying time before such a conclusion is adopted. Let μ denote the true average drying time when the additive is used. The appropriate hypotheses are $H_0: \mu = 75$ versus $H_a: \mu < 75$. Only if H_0 can be rejected will the additive be declared successful and used.

Experimental data is to consist of drying times from $n = 25$ test specimens. Let $X_1, \ldots, X_{25}$ denote the 25 drying times—a random sample of size 25 from a normal distribution with mean value μ and standard deviation $\sigma = 9$. The sample mean drying time $\overline{X}$ then has a normal distribution with expected value $\mu_{\overline{X}} = \mu$ and standard deviation $\sigma_{\overline{X}} = \sigma/\sqrt{n} = 9/\sqrt{25} = 1.80$. When H_0 is true, $\mu_{\overline{X}} = 75$, so an $\overline{x}$ value somewhat less than 75 would not strongly contradict H_0. A reasonable rejection region has the form $\overline{x} \le c$, where the cutoff value c is suitably chosen. Consider the choice $c = 70.8$, so that the test procedure consists of test statistic $\overline{X}$ and rejection region $\overline{x} \leq 70.8$. Because the rejection region consists only of small values of the test statistic, the test is said to be *lower-tailed*. Calculation of α and β now involves a routine standardization of $\overline{X}$ followed by reference to the standard normal probabilities of Appendix Table A.3:

$$\alpha = P(\text{type I error}) = P(H_0 \text{ is rejected when it is true})$$
$$= P(\overline{X} \leq 70.8 \text{ when } \overline{X} \sim \text{normal with } \mu_{\overline{x}} = 75, \sigma_{\overline{x}} = 1.8)$$
$$= \Phi\left(\frac{70.8 - 75}{1.8}\right) = \Phi(-2.33) = .01$$

$$\beta(72) = P(\text{type II error when } \mu = 72)$$
$$= P(H_0 \text{ is not rejected when it is false because } \mu = 72)$$
$$= P(\overline{X} > 70.8 \text{ when } \overline{X} \sim \text{normal with } \mu_{\overline{x}} = 72 \text{ and } \sigma_{\overline{x}} = 1.8)$$
$$= 1 - \Phi\left(\frac{70.8 - 72}{1.8}\right) = 1 - \Phi(-.67) = 1 - .2514 = .7486$$

$$\beta(70) = 1 - \Phi\left(\frac{70.8 - 70}{1.8}\right) = .3300$$

$$\beta(67) = 1 - \Phi\left(\frac{70.8 - 67}{1.8}\right) = .0174$$

For the specified test procedure, only 1% of all experiments carried out as described will result in H_0 being rejected when it is actually true. However, the chance of a type II error is very large if $\mu = 72$ (only a small departure from H_0), somewhat less if $\mu = 70$, and quite small when $\mu = 67$ (a very substantial departure from H_0). These error probabilities are illustrated in Figure 8.1.

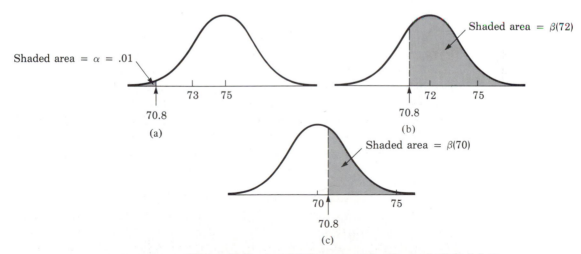

Figure 8.1 α and β illustrated for Example 8.2: (a) the sampling distribution of $\overline{X}$ when $\mu = 75$ (H_0 true); (b) the sampling distribution of $\overline{X}$ when $\mu = 72$ (H_0 false); (c) the sampling distribution of $\overline{X}$ when $\mu = 70$ (H_0 false)

Notice that α is computed using the probability distribution of the test statistic when H_0 is true, whereas determination of β requires knowing the test statistic's distribution when H_0 is false.

As in Example 8.1, if the more realistic null hypothesis $\mu \geq 75$ is considered, there is an α for each parameter value for which H_0 is true—$\alpha(75)$, $\alpha(75.8)$, $\alpha(76.5)$, and so on. It is easily verified, though, that $\alpha(75)$ is the largest of all these type I error probabilities. Focusing on the boundary value amounts to working explicitly with the "worst case." ∎

The specification of a cutoff value for the rejection region in the examples just considered was somewhat arbitrary. Use of the rejection region $R_8 = \{8, 9, \ldots, 20\}$ in Example 8.1 resulted in $\alpha = .102$, $\beta(.3) = .772$, and $\beta(.5) = .132$. Many would think these error probabilities intolerably large. Perhaps they can be decreased by changing the cutoff value.

Example 8.3
(Example 8.1 continued)

Let us use the same experiment and test statistic X as previously described in the automobile bumper problem, but now consider the rejection region $R_9 = \{9, 10, \ldots, 20\}$. Since X still has a binomial distribution with parameters $n = 20$ and p,

$$\alpha = P(H_0 \text{ is rejected when } p = .25)$$
$$= P(X \geq 9 \text{ when } X \sim \text{Bin}(20, .25)) = 1 - B(8; 20, .25) = .041$$

The type I error probability has been decreased by using the new rejection region. However, a price has been paid for this decrease:

$$\beta(.3) = P(H_0 \text{ is not rejected when } p = .3)$$
$$= P(X \leq 8 \text{ when } X \sim \text{Bin}(20, .3)) = B(8; 20, .3) = .887$$
$$\beta(.5) = B(8; 20, .5) = .252$$

Both these β's are larger than the corresponding error probabilities .772 and .132 for the region R_8. In retrospect, this is not surprising; α is computed by summing over probabilities of test statistic values *in the rejection region,* whereas β is the probability that X falls in the *complement* of the rejection region. Making the rejection region smaller must therefore decrease α while increasing β for any fixed alternative value of the parameter. ∎

Example 8.4
(Example 8.2 continued)

The use of cutoff value $c = 70.8$ in the paint drying example resulted in a very small value of α (.01) but rather large β's. Consider the same experiment and test statistic $\overline{X}$ with the new rejection region $\overline{x} \leq 72$. Because $\overline{X}$ is still normally distributed with mean value $\mu_{\overline{X}} = \mu$ and $\sigma_{\overline{X}} = 1.8$,

$$\alpha = P(H_0 \text{ is rejected when it is true})$$
$$= P(\overline{X} \le 72 \text{ when } \overline{X} \text{ is a normal r.v. with mean 75 and standard}$$
$$\text{deviation 1.8})$$
$$= \Phi\left(\frac{72 - 75}{1.8}\right) = \Phi(-1.67) = .0475 \approx .05$$

$$\beta(72) = P(H_0 \text{ is not rejected when } \mu = 72)$$
$$= P(\overline{X} > 72 \text{ when } \overline{X} \text{ is a normal r.v. with mean 72 and}$$
$$\text{standard deviation 1.8})$$
$$= 1 - \Phi\left(\frac{72 - 72}{1.8}\right) = 1 - \Phi(0) = .5$$

$$\beta(70) = 1 - \Phi\left(\frac{72 - 70}{1.8}\right) = .1335$$

$$\beta(67) = 1 - \Phi\left(\frac{72 - 67}{1.8}\right) = .0027$$

The change in cutoff value has made the rejection region larger (it includes more $\overline{x}$ values), resulting in a decrease in β for each fixed μ less than 75. However, α for this new region has increased from the previous value .01 to approximately .05. If a type I error probability this large can be tolerated, though, the second region ($c = 72$) is preferable to the first ($c = 70.8$) because of the smaller β's. ∎

The results of these examples can be generalized in the following manner.

Proposition

> Suppose that an experiment and a sample size are fixed, and a test statistic is chosen. Then decreasing the size of the rejection region to obtain a smaller value of α results in a larger value of β for any particular parameter value consistent with H_a.

This proposition says that once the test statistic and n are fixed, there is no rejection region that will simultaneously make both α and all β's small. A region must be chosen to effect a compromise between α and β.

Because of the suggested guidelines for specifying H_0 and H_a, a type I error is usually more serious than a type II error (this can always be achieved by proper choice of the hypotheses). The approach adhered to by most statistical practitioners is then to specify the largest value of α that can be tolerated and find a rejection region having that value of α rather than anything smaller. This makes β as small as possible subject to the bound on α. The resulting value of α is often referred to as the **significance level** of the test. Traditional levels of significance are .10, .05, and .01, though the level in any particular problem will depend on the seriousness of a type I error—the more serious this error, the smaller should be the significance level. The corresponding test procedure is called a **level α test** (for example, a level .05 test or a level .01 test). A test with significance level α is one for which the type I error probability is controlled at the specified level.

Example 8.5   Consider the situation mentioned earlier in which μ was the true average nicotine content of brand B cigarettes. The objective is to test $H_0: \mu = 1.5$ versus $H_a: \mu > 1.5$ based on a random sample $X_1, X_2, \ldots, X_{32}$ of nicotine contents. Suppose that the distribution of nicotine content is known to be normal with $\sigma = .20$. Then $\overline{X}$ is normally distributed with mean value $\mu_{\overline{X}} = \mu$ and standard deviation $\sigma_{\overline{X}} = .20/\sqrt{32} = .0354$. The appropriate rejection region in this situation has the form $\overline{x} \geq c$, where c is chosen to yield the desired significance level. For a level .05 test, we must have

$$\alpha = .05 = P(H_0 \text{ is rejected when } \mu = 1.5)$$
$$= P(\overline{X} \geq c \text{ when } \overline{X} \text{ is normally distributed with mean 1.5 and}$$
$$\text{standard deviation .0354})$$
$$= P\left(Z \geq \frac{c - 1.5}{.0354} \text{ when } Z \text{ is a standard normal r.v.}\right)$$

Thus c must be such that $(c - 1.5)/.0354$ captures upper-tail area .05 under the standard normal curve. From Section 4.3 or from Appendix Table A.3, the z critical value that captures this tail area is 1.645, so $(c - 1.5)/.0354 = 1.645$. This gives $c = 1.5 + .0354(1.645) = 1.56$. Thus the rejection region $\overline{x} \geq 1.56$ yields a test with significance level .05.

This test can be described in a slightly different way. Instead of using $\overline{X}$ as the test statistic, consider the standardized statistic $Z = (\overline{X} - 1.5)/.0354$ (where the standardization is done assuming that H_0 is true, so that $\mu_{\overline{X}} = 1.5$). Then rejecting H_0 if $\overline{x} \geq 1.56$ is equivalent to rejecting H_0 if $z \geq 1.645$. In succeeding sections and chapters, many test statistics will be obtained by standardizing in a similar manner. ∎

Example 8.6   A manufacturer of sprinkler systems used for fire protection in office buildings claims that the true average system activation temperature is 130°. A sample of $n = 9$ systems, when tested, yields a sample average activation temperature of 131.08° F. If the distribution of activation times is normal with standard deviation 1.5° F, does the data contradict the manufacturer's claim at significance level $\alpha = .05$?

Here μ is the true average activation temperature, and the appropriate null hypothesis is $H_0: \mu = 130$. Because a departure from the claimed value in either direction is of concern (it is dangerous to have either too high or too low an activation temperature), the relevant alternative hypothesis is $H_a: \mu \neq 130$. A reasonable test procedure uses the sample mean $\overline{X}$ as the test statistic and rejects H_0 if $\overline{x}$ deviates too much from 130. That is, the rejection region should have the form *reject H_0 either if $\overline{x} \leq 130 - c$ or if $\overline{x} \geq 130 + c$*. This region is *two-tailed*, since it consists of both large and small test statistic values.

The constant c must now be determined to yield $\alpha = .05$. Because $\sigma_{\overline{X}} = \sigma/\sqrt{n} = 1.5/\sqrt{9} = .5$ and, when H_0 is true, $\mu_{\overline{X}} = \mu = 130$, the standardized statistic $Z = (\overline{X} - 130)/.5$ has a standard normal distribution when H_0 is true. The desired condition is

$$\alpha = .05 = P(H_0 \text{ is rejected when it is true})$$
$$= P(\overline{X} \le 130 - c \text{ or } \overline{X} \ge 130 + c \text{ when } H_0 \text{ is true})$$
$$= P(Z \le -c/.5 \text{ or } Z \ge c/.5 \text{ when } Z \sim N(0, 1))$$
$$= 2P(Z \ge c/.5 \text{ when } Z \sim N(0, 1))$$

Thus c is such that $P(Z \ge c/.5) = \alpha/2 = .025$. From Section 4.3 or Appendix Table A.3, the z critical value that captures upper-tail area .025 is 1.96, so $c/.5 = 1.96$ and $c = .98$. The desired rejection region is then $\overline{x} \ge 130.98$ or $\overline{x} \le 129.02$. Since the observed $\overline{x}$ value of 131.08 does fall in the upper-tail of this rejection region, H_0 is rejected at level .05. We conclude that the manufacturer's claim is incorrect.

The test used to reach this conclusion was one for which the type I error probability was controlled at level .05. How likely is it that this test will result in a type II error for any particular μ different from 130? Since H_0 is not rejected if $129.02 < \overline{x} < 130.98$,

$$\beta(131) = P(\text{type II error when } \mu = 131)$$
$$= P(H_0 \text{ is not rejected when } \mu = 131)$$
$$= P(129.02 < \overline{X} < 130.98 \text{ when } \overline{X} \text{ is normal with mean } 131$$
$$\text{and standard deviation } .5)$$
$$= \Phi[(130.98 - 131)/.5] - \Phi[(129.02 - 131)/.5]$$
$$= \Phi(-.04) - \Phi(-3.96) = .4840$$

By symmetry, $\beta(129) = .4840$ also. A similar calculation yields $\beta(132) = \beta(128) = .0207$, which is much smaller than those values for $\mu = 131$ and 129 since 132 and 128 are further from the hypothesized value. A type II error is very unlikely to occur when μ differs from 130 by 2° F (or more).

Finally, notice that the test procedure phrased in terms of $\overline{X}$ can be stated in an equivalent form using the standardized statistic $Z = (\overline{X} - 130)/.5$: rejecting H_0 if either $\overline{x} \ge 130.98$ or $\overline{x} \le 129.02$ is identical to rejecting H_0 if either $z \ge 1.96$ or $z \le -1.96$. ◼

Exercises / Section 8.1 [1–12]

1. For each of the following assertions, say whether or not it is a legitimate statistical hypothesis and why.
   a. $H: \sigma > 100$ b. $H: \overline{x} = 45$ c. $H: s \le .20$
   d. $H: \sigma_1/\sigma_2 < 1$ e. $H: \overline{X} - \overline{Y} = 5$
   f. $H: \lambda \le .01$, where λ is the parameter of an exponential distribution used to model component lifetime.

2. For the following pairs of assertions, indicate which don't comply with our rules for setting up hypotheses and why (the subscripts 1 and 2 differentiate between quantities for two different populations or samples).
   a. $H_0: \mu = 100, H_a: \mu > 100$
   b. $H_0: \sigma > 20, H_a: \sigma \le 20$
   c. $H_0: p \ne .25, H_a: p = .25$

d. $H_0: \mu_1 - \mu_2 = 25, H_a: \mu_1 - \mu_2 > 100$
e. $H_0: S_1^2 = S_2^2, H_a: S_1^2 \neq S_2^2$
f. $H_0: \mu = 120, H_a: \mu = 150$
g. $H_0: \sigma_1/\sigma_2 = 1, H_a: \sigma_1/\sigma_2 \neq 1$
h. $H_0: p_1 - p_2 = -.1, H_a: p_1 - p_2 < -.1$

3. In order to determine whether the pipe welds in a nuclear power plant meet specifications, a random sample of welds is selected, and tests are conducted on each weld in the sample. Weld strength is measured as the force required to break the weld. Suppose that the specifications state that mean strength of welds should exceed 100 lb/in.^{2}; the inspection team decides to test $H_0: \mu = 100$ versus $H_a: \mu > 100$. Explain why it might be preferable to use this H_a rather than $\mu < 100$.

4. Let μ denote the true average radioactivity level (picocuries per liter). The value 5 picocuries per liter is considered the dividing line between safe and unsafe water. Would you recommend testing $H_0: \mu = 5$ versus $H_a: \mu > 5$ or $H_0: \mu = 5$ versus $H_a: \mu < 5$? Explain your reasoning. *Hint:* Think about the consequences of a type I and type II error for each possibility.

5. Before agreeing to purchase a large order of polyethylene sheaths for a particular type of high-pressure oil-filled submarine power cable, a company wants to see conclusive evidence that the true standard deviation of sheath thickness is less than .05 mm. What hypotheses should be tested, and why?

6. Many older homes have electrical systems that use fuses rather than circuit breakers. A manufacturer of 40-A fuses wants to make sure that the mean amperage at which its fuses burn out is in fact 40. If the mean amperage is lower than 40, customers will complain because the fuses require replacement too often. If the mean amperage is higher than 40, the manufacturer might be liable for damage to an electrical system due to fuse malfunction. In order to verify the amperage of the fuses, a sample of fuses is to be selected and inspected. If a hypothesis test were to be performed on the resulting data, what null and alternative hypotheses would be of interest to the manufacturer?

7. Let p denote the proportion of all registered voters in a certain area who favor candidate A over candidate B in the race for county supervisor. Consider testing $H_0: p = .5$ versus $H_a: p \neq .5$ based on a random sample of 25 registered voters. Let X denote the number in the sample who favor A, and x represent the observed value of X.
a. Which of the following rejection regions is most appropriate and why?

$R_1 = \{x : x \leq 7 \text{ or } x \geq 18\}, R_2 = \{x : x \leq 8\},$
$R_3 = \{x : x \geq 17\}$

b. In the context of this problem situation describe what a type I and type II error are.
c. What is the probability distribution of the test statistic X when H_0 is true? Use it to compute the probability of a type I error.
d. Compute the probability of a type II error for the selected region when $p = .3$, again when $p = .4$, and also for both $p = .6$ and $p = .7$.
e. Using the selected region, what would you conclude if six of the 25 queried favored A?

8. A mixture of pulverized fuel ash and Portland cement to be used for grouting should have a compressive strength of more than 1300 KN/m^2. The mixture will not be used unless experimental evidence indicates conclusively that the strength specification has been met. Suppose that compressive strength for specimens of this mixture is normally distributed with $\sigma = 60$. Let μ denote the true average compressive strength.
a. What are the appropriate null and alternative hypotheses?
b. Let $\overline{X}$ denote the sample average compressive strength for $n = 20$ randomly selected specimens. Consider the test procedure with test statistic $\overline{X}$ and rejection region $\overline{x} \geq 1331.26$. What is the probability distribution of the test statistic when H_0 is true? What is the probability of a type I error for the test procedure?
c. What is the probability distribution of the test statistic when $\mu = 1350$? Using the test procedure of (b), what is the probability that the mixture will be judged unsatisfactory when in fact $\mu = 1350$ (a type II error)?
d. How would you change the test procedure of (b) to obtain a test with significance level .05? What impact would this change have on the error probability of (c)?

e. Consider the standardized test statistic $Z = (\overline{X} - 1300)/(\sigma/\sqrt{n}) = (\overline{X} - 1300)/13.42$. What are the values of Z corresponding to the rejection region of (b)?

9. The calibration of a scale is to be checked by weighing a 10 kg test specimen 25 times. Suppose that the results of different weighings are independent of one another and that the weight on each trial is normally distributed with $\sigma = .200$ kg. Let μ denote the true average weight reading on the scale.

   a. What hypotheses should be tested?

   b. Suppose that the scale is to be recalibrated if either $\overline{x} \geq 10.1032$ or if $\overline{x} \leq 9.8968$. What is the probability that recalibration is carried out when it is actually unnecessary?

   c. What is the probability that recalibration is judged unnecessary when in fact $\mu = 10.1$? When $\mu = 9.8$?

   d. What change in the procedure of (b) yields a test with significance level .001?

   e. If the sample size were only 10 rather than 20, how should the procedure of (b) be altered so that $\alpha = .01$?

   f. Using the test of (e), what would you conclude from the following sample data:

   9.981, 10.006, 9.857, 10.107, 9.888, 9.793, 9.728, 10.439, 10.214, 10.190

   g. Re-express the test procedure of (b) in terms of the standardized test statistic $Z = (\overline{X} - 10)/(\sigma/\sqrt{n})$.

10. A new design for the braking system on a certain type of car has been proposed. For the current system the true average braking distance at 40 mph under specified conditions is known to be 120 ft. It is proposed that the new design be implemented only if sample data strongly indicates a reduction in true average braking distance for the new design.

   a. Define the parameter of interest and state the relevant hypotheses.

   b. Suppose that braking distance for the new system is normally distributed with $\sigma = 10$. Let $\overline{X}$ denote the sample average braking distance for a random sample of 36 observations. Which of the following rejection regions is appropriate: $R_1 = \{\overline{x} : \overline{x} \geq 124.80\}$, $R_2 = \{\overline{x} : \overline{x} \leq 115.20\}$, $R_3 = \{\overline{x} : \text{either } \overline{x} \geq 125.13$ or $\overline{x} \leq 114.87\}$?

   c. What is the significance level for the appropriate region of (b)? How would you change the region to obtain a test with $\alpha = .001$?

   d. What is the probability that the new design is not implemented when its true average braking distance is actually 115 ft and the appropriate region from (b) is used?

   e. Let $Z = (\overline{X} - 120)/(\sigma/\sqrt{n})$. What is the significance level for the rejection region $\{z : z \leq -2.33\}$? For the region $\{z : z \leq -2.88\}$?

11. Let $X_1, \ldots, X_n$ denote a random sample from a normal population distribution with a known value of σ.

   a. For testing $H_0 : \mu = \mu_0$ versus $H_a : \mu > \mu_0$ (where μ_0 is a fixed number), show that the test with test statistic $\overline{X}$ and rejection region $\overline{x} \geq \mu_0 + 2.33\sigma/\sqrt{n}$ has significance level .01.

   b. Suppose that the procedure of (a) is used to test $H_0 : \mu \leq \mu_0$ versus $H_a : \mu > \mu_0$. If $\mu_0 = 100$, $n = 25$, and $\sigma = 5$, what is the probability of committing a type I error when $\mu = 99$? When $\mu = 98$? In general, what can be said about the probability of a type I error when the actual value of μ is less than μ_0? Verify your assertion.

12. Reconsider the situation of Exercise 9, and suppose that the rejection region is $\{\overline{x} : \overline{x} \geq 10.1004$ or $\overline{x} \leq 9.8940\}$.

   a. What is α for this procedure?

   b. What is β when $\mu = 10.1$? When $\mu = 9.9$? Is this desirable?

8.2   Tests about a Population Mean

The general discussion in Chapter 7 of confidence intervals for a population mean μ focused on three different cases. We now develop test procedures for these same three cases.

Case I: A Normal Population with Known σ

Although the assumption that the value of σ is known is rarely met in practice, this case provides a good starting point because of the ease with which general procedures and their properties can be developed. The null hypothesis in all three cases will state that μ has a particular numerical value, the *null value*, which we shall denote by μ_0. Let $X_1, \ldots, X_n$ represent a random sample of size n from the normal population. Then the sample mean $\overline{X}$ has a normal distribution with expected value $\mu_{\overline{X}} = \mu$ and standard deviation $\sigma_{\overline{X}} = \sigma/\sqrt{n}$. When H_0 is true, $\mu_{\overline{X}} = \mu_0$. Consider now the statistic Z obtained by standardizing $\overline{X}$ under the assumption that H_0 is true:

$$Z = \frac{\overline{X} - \mu_0}{\sigma/\sqrt{n}}$$

Substitution of the computed sample mean $\overline{x}$ gives z, the distance between $\overline{x}$ and μ_0 expressed in "standard deviation units." For example, if the null hypothesis is $H_0 : \mu = 100$, $\sigma_{\overline{x}} = \sigma/\sqrt{n} = 10/\sqrt{25} = 2.0$ and $\overline{x} = 103$, then $z = (103 - 100)/2.0 = 1.5$. That is, the observed value of $\overline{x}$ is 1.5 standard deviations (of $\overline{X}$) above what we expect it to be when H_0 is true. The statistic Z is a natural measure of the distance between $\overline{X}$, the estimator of μ, and its expected value when H_0 is true. If this distance is too great in a direction consistent with H_a, the null hypothesis should be rejected.

Suppose first that the alternative hypothesis has the form $H_a : \mu > \mu_0$. Then an $\overline{x}$ value less than μ_0 certainly does not provide support for H_a. Such an $\overline{x}$ corresponds to a negative value of z (since $\overline{x} - \mu_0$ is negative and the divisor $\sigma/\sqrt{n}$ is positive). Similarly, an $\overline{x}$ value that exceeds μ_0 by only a small amount (corresponding to z which is positive but small) does not suggest that H_0 should be rejected in favor of H_a. The rejection of H_0 is appropriate only when $\overline{x}$ considerably exceeds μ_0—that is, when the z value is positive and large. In summary, the appropriate rejection region, based on the test statistic Z rather than $\overline{X}$, has the form $z \geq c$.

As discussed in Section 8.1, the cutoff value c should be chosen to control the probability of a type I error at the desired level α. This is easily accomplished because the distribution of the test statistic Z when H_0 is true is the standard normal distribution (that's why μ_0 was subtracted in standardizing). The required cutoff c is the z critical value that captures upper-tail area α under the standard normal curve. As an example, let $c = 1.645$, the value that captures tail area .05 ($z_{.05} = 1.645$). Then

$$\alpha = P(\text{type I error}) = P(H_0 \text{ is rejected when } H_0 \text{ is true})$$
$$= P(Z \geq 1.645 \text{ when } Z \text{ has a standard normal distribution})$$
$$= 1 - \Phi(1.645) = .05$$

More generally, the rejection region $z \geq z_\alpha$ has type I error probability α. The test procedure is upper-tailed because the rejection region consists only of large values of the test statistic.

Analogous reasoning for the alternative hypothesis $H_a : \mu < \mu_0$ suggests a rejection region of the form $z \le c$, where c is a suitably chosen negative number ($\bar{x}$ is far below μ_0 if and only if z is quite negative). Because Z has a standard normal distribution when H_0 is true, taking $c = -z_\alpha$ yields $P(\text{type I error}) = \alpha$. This is a lower-tailed test. For example, $z_{.10} = 1.28$ implies that the rejection region $z \le -1.28$ specifies a test with significance level .10.

Finally, when the alternative hypothesis is $H_a : \mu \ne \mu_0$, H_0 should be rejected if $\bar{x}$ is too far to either side of μ_0. This is equivalent to rejecting H_0 either if $z \ge c$ or if $z \le -c$. Suppose that we desire $\alpha = .05$. Then

$$.05 = P(Z \ge c \text{ or } Z \le -c \text{ when } Z \text{ has a standard normal distribution})$$
$$= \Phi(-c) + 1 - \Phi(c) = 2[1 - \Phi(c)]$$

Thus c is such that $1 - \Phi(c)$, the area under the standard normal curve to the right of c, is .025 (and not .05!). From Section 4.3 or Appendix Table A.3, $c = 1.96$ and the rejection region is $z \ge 1.96$ or $z \le -1.96$. For any α, the two-tailed rejection region $z \ge z_{\alpha/2}$ or $z \le -z_{\alpha/2}$ has type I error probability α (since area $\alpha/2$ is captured under each of the two tails of the z curve). Again, the key reason for using the standardized test statistic Z is that, because Z has a known distribution when H_0 is true (standard normal), a rejection region with desired type I error probability is easily obtained by using an appropriate critical value.

The test procedure for case I is summarized in the accompanying box, and the corresponding rejection regions are illustrated in Figure 8.2.

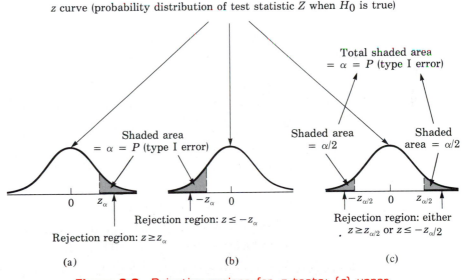

Figure 8.2 Rejection regions for z tests: (*a*) upper-tailed test; (*b*) lower-tailed test; (*c*) two-tailed test

Null hypothesis: $H_0: \mu = \mu_0$

Test statistic value: $z = \dfrac{\overline{x} - \mu_0}{\sigma/\sqrt{n}}$

Alternative hypothesis	Rejection region for level α test
$H_a: \mu > \mu_0$	$z \geq z_\alpha$
$H_a: \mu < \mu_0$	$z \leq -z_\alpha$
$H_a: \mu \neq \mu_0$	either $z \geq z_{\alpha/2}$ or $z \leq -z_{\alpha/2}$

Use of the following sequence of steps is recommended in any hypothesis testing analysis.

1. Identify the parameter of interest, and describe it in the context of the problem situation.
2. Determine the null value and state the null hypothesis.
3. State the appropriate alternative hypothesis.
4. Give the formula for the computed value of the test statistic (substituting the null value and the known values of any other parameters, but *not* those of any sample-based quantities).
5. State the rejection region for the selected significance level α.
6. Compute any necessary sample quantities, substitute into the formula for the test statistic value, and compute that value.
7. Decide whether or not H_0 should be rejected, and state this conclusion in the problem context.

The formulation of hypotheses (steps 2 and 3) should be done prior to examining the data.

Example 8.7 A company that produces bias-ply tires is considering a certain modification in the tread design. An economic feasibility study indicates that the modification can be justified only if true average tire life under standard test conditions exceeds 20,000 miles. A random sample of $n = 16$ prototype tires is manufactured and tested, resulting in a sample average tire life of $\overline{x} = 20{,}758$. Suppose that tire life is normally distributed with $\sigma = 1500$ (the value for the current version of the tire). Does this data suggest that the modification meets the condition required for changeover? Let's test the appropriate hypothesis using significance level .01.

1. Parameter of interest: $\mu =$ the true average tire life for tires with the new tread design.
2. Null hypothesis: $H_0: \mu = 20{,}000$ (null value $= \mu_0 = 20{,}000$).
3. Alternative hypothesis: $H_a: \mu > 20{,}000$ (so the changeover will take place only if H_0 is rejected in favor of H_a).
4. Test statistic value: $z = \dfrac{\overline{x} - \mu_0}{\sigma/\sqrt{n}} = \dfrac{\overline{x} - 20{,}000}{1500/\sqrt{n}}$

5. Rejection region: The form of H_a implies that an upper-tailed test with rejection region $z \geq z_\alpha$ should be used. For $\alpha = .01$, $z_\alpha = z_{.01} = 2.33$ (area .01 lies to the right of 2.33 under the z curve), so H_0 will be rejected if $z \geq 2.33$.

6. Substituting $n = 16$ and $\bar{x} = 20{,}758$ yields

$$z = \frac{20{,}758 - 20{,}000}{1500/\sqrt{16}} = \frac{758}{375} = 2.02$$

That is, the observed sample mean 20,758 is roughly two standard deviations above what would have been expected were H_0 true.

7. The computed value $z = 2.02$ does not fall in the rejection region ($2.02 \not\geq 2.33$), so H_0 cannot be rejected at significance level .01. The data does not give strong support to the claim that true average tire life for the new tread design exceeds 20,000 miles. Production of the current version should be continued.   ∎

β and Sample Size Determination

The z tests for case I are among the few in statistics for which there are simple formulas available for β, the probability of a type II error. Consider first the upper-tailed test with rejection region $z \geq z_\alpha$. This is equivalent to $\bar{x} \geq \mu_0 + z_\alpha \cdot \sigma/\sqrt{n}$, so H_0 will not be rejected if $\bar{x} < \mu_0 + z_\alpha \cdot \sigma/\sqrt{n}$. Now let μ' denote a particular value of μ that exceeds the null value μ_0. Then

$$
\begin{aligned}
\beta(\mu') &= P(H_0 \text{ is not rejected when } \mu = \mu') \\
&= P(\bar{X} < \mu_0 + z_\alpha \cdot \sigma/\sqrt{n} \text{ when } \mu = \mu') \\
&= P\left(\frac{\bar{X} - \mu'}{\sigma/\sqrt{n}} < z_\alpha + \frac{\mu_0 - \mu'}{\sigma/\sqrt{n}} \text{ when } \mu = \mu'\right) \\
&= \Phi\left(z_\alpha + \frac{\mu_0 - \mu'}{\sigma/\sqrt{n}}\right)
\end{aligned}
$$

As μ' increases, $\mu_0 - \mu'$ becomes more negative, so $\beta(\mu')$ will be small when μ' greatly exceeds μ_0 (because the value at which Φ is evaluated will then be quite negative). Error probabilities for the lower-tailed and two-tailed tests are derived in an analogous manner.

If σ is large, the probability of a type II error can be large at an alternative value μ' that is of particular concern to an investigator. Suppose that we fix α and also specify β for such an alternative value. In the tire example, company officials might view $\mu' = 21{,}000$ as a very substantial departure from $H_0: \mu = 20{,}000$, and therefore wish $\beta(21{,}000) = .10$ in addition to $\alpha = .01$. More generally, consider the two restrictions $P(\text{type I error}) = \alpha$ and $\beta(\mu') = \beta$ for specified α, μ', and β. Then for an upper-tailed test, the sample size n should be chosen to satisfy

$$\Phi\left(z_\alpha + \frac{\mu_0 - \mu'}{\sigma/\sqrt{n}}\right) = \beta$$

This implies that

$$-z_\beta = \frac{z \text{ critical value that}}{\text{captures lower-tail area } \beta} = z_\alpha + \frac{\mu_0 - \mu'}{\sigma/\sqrt{n}}$$

It is easy to solve this equation for the desired n. A parallel argument yields the necessary sample size for lower- and two-tailed tests as summarized in the accompanying box.

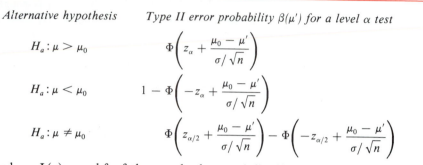

Alternative hypothesis   *Type II error probability $\beta(\mu')$ for a level α test*

$H_a : \mu > \mu_0$ $\Phi\left(z_\alpha + \dfrac{\mu_0 - \mu'}{\sigma/\sqrt{n}}\right)$

$H_a : \mu < \mu_0$ $1 - \Phi\left(-z_\alpha + \dfrac{\mu_0 - \mu'}{\sigma/\sqrt{n}}\right)$

$H_a : \mu \neq \mu_0$ $\Phi\left(z_{\alpha/2} + \dfrac{\mu_0 - \mu'}{\sigma/\sqrt{n}}\right) - \Phi\left(-z_{\alpha/2} + \dfrac{\mu_0 - \mu'}{\sigma/\sqrt{n}}\right)$

where $\Phi(z)$ = c.d.f. of the standard normal distribution.

The sample size n for which a level α test also has $\beta(\mu') = \beta$ at the alternative value μ' is

$$n = \begin{cases} \left[\dfrac{\sigma(z_\alpha + z_\beta)}{\mu_0 - \mu'}\right]^2 & \text{for a one-tailed} \\ & \text{(upper or lower) test} \\ \left[\dfrac{\sigma(z_{\alpha/2} + z_\beta)}{\mu_0 - \mu'}\right]^2 & \text{for a two-tailed test} \\ & \text{(an approximate solution)} \end{cases}$$

Example 8.8
(Example 8.7
continued)

In the tire problem we tested $H_0 : \mu = 20{,}000$ versus $H_a : \mu > 20{,}000$ using significance level .01. With $\sigma = 1500$, $n = 16$, and $z_\alpha = z_{.01} = 2.33$, the probability of making a type II error when $\mu = 21{,}000$ is

$$\beta(21{,}000) = \Phi\left(2.33 + \frac{20{,}000 - 21{,}000}{1500/\sqrt{16}}\right) = \Phi(-.34) = .3669$$

Since $z_{.1} = 1.28$, the requirement that the level .01 test also have $\beta(21{,}000) = .1$ necessitates

$$n = \left[\frac{1500(2.33 + 1.28)}{20{,}000 - 21{,}000}\right]^2 = (5.42)^2 = 29.32$$

The sample size must be an integer, so $n = 30$ prototype tires should be used. ■

Case II: Large-Sample Tests

When the sample size is large, the z tests for case I are easily modified to yield valid test procedures without requiring either a normal population distribution or known σ. The key result was used in Chapter 7 to justify large-sample confidence intervals: a large n implies that the sample standard deviation S will be close to σ for most samples, so that the standardized variable

$$Z = \frac{\overline{X} - \mu}{S/\sqrt{n}}$$

has *approximately* a standard normal distribution. Substitution of the null value μ_0 in place of μ yields the test statistic

$$Z = \frac{\overline{X} - \mu_0}{S/\sqrt{n}}$$

which has approximately a standard normal distribution when H_0 is true. The use of rejection regions given earlier for case I (for example, $z \geq z_\alpha$ when the alternative hypothesis is $H_a : \mu > \mu_0$) then results in test procedures for which the significance level is approximately (rather than exactly) α. The rule of thumb $n > 30$ will again be used to characterize a large sample size.

Example 8.9   A certain type of brick is being considered for use in a particular construction project. The brick will be used unless sample evidence strongly suggests that the true average compressive strength is below 3200 psi. A random sample of 36 bricks is selected and each is subjected to a compressive strength test. The resulting sample average compressive strength and sample standard deviation of compressive strength are 3109 psi and 156 psi respectively. State the relevant hypotheses and carry out a test to reach a decision using level of significance .05.

1. μ = true average compressive strength for this type of brick.
2. $H_0 : \mu = 3200$.
3. $H_a : \mu < 3200$ (so that the brick will be used unless H_0 is rejected in favor of H_a).

4. Test statistic value: $z = \dfrac{\overline{x} - 3200}{s/\sqrt{n}}$

5. Because $<$ appears in H_a, a lower-tailed test is used. For $\alpha = .05$, the appropriate one-tailed critical value is $z_\alpha = z_{.05} = 1.645$. H_0 will now be rejected if $z \leq -1.645$.
6. $n = 36$, $\overline{x} = 3109$, $s = 156$, so

$$z = \frac{3109 - 3200}{156/\sqrt{36}} = \frac{-91}{26} = -3.50$$

That is, $\overline{X}$ has been observed to fall 3.5 estimated standard deviations (of $\overline{X}$) below what would have been expected were H_0 true.
7. Since $-3.50 \leq -1.645$, H_0 is rejected at the 5% level of significance. We conclude that true average compressive strength is below 3200, so the brick

should not be used. Notice that in reaching this conclusion, we may have made a type I error (rejecting H_0 when it is true), but the low level of significance (.05) has made that unlikely.   ■

Example 8.10   An automobile manufacturer recommends that any purchaser of one of its new cars bring it in to a dealer for a 3000 mile checkup. The company wishes to know whether the true average mileage for initial servicing differs from 3000. A random sample of 50 recent purchasers resulted in a sample average mileage of 3208 and a sample standard deviation of 273 miles. Does the data strongly suggest that true average mileage for this checkup is something other than the recommended value? State and test the relevant hypotheses using level of significance .01.

1. μ = true average mileage of cars brought to the dealer for 3000 mile checkups.

2. $H_0 : \mu = 3000$.

3. $H_a : \mu \neq 3000$ (which says that μ differs from what the manufacturer recommends).

4. Test statistic value: $z = \dfrac{\bar{x} - 3000}{s/\sqrt{n}}$

5. The alternative hypothesis involves $\neq$, so a two-tailed test should be used. Since $z_{\alpha/2} = z_{.005} = 2.58$, H_0 should be rejected if either $z \geq 2.58$ or $z \leq -2.58$.

6. $n = 50$, $\bar{x} = 3208$, $s = 273$, which gives

$$z = \frac{3208 - 3000}{273/\sqrt{50}} = \frac{208}{38.61} = 5.39$$

7. Since 5.39 is in the upper tail of the two-tailed rejection region ($5.39 \geq 2.58$), H_0 is rejected using $\alpha = .01$. The data does strongly suggest that true average initial checkup mileage differs from the manufacturer's recommended value.   ■

Determination of β and the necessary sample size for these large-sample tests can be based either on specifying a plausible value of σ and using the case I formulas (even though s is used in the test) or on using the curves to be introduced shortly in connection with case III.

Case III: A Normal Population Distribution

When n is small, the Central Limit Theorem can no longer be invoked to justify the use of a large-sample test. We faced this same difficulty in obtaining a small-sample confidence interval for μ in Chapter 7. Our approach here will be the same one used there: we will assume that the population distribution is at least approximately normal, and describe test procedures whose validity rests on this assumption. If an investigator has good reason to believe that the population distribution is quite nonnormal, a distribution-free test from Chapter 15 can be used. Alternatively, a statistician can be consulted regarding procedures valid for specific families of population distributions other than the normal family.

The key result on which tests for a normal population mean are based was used in Chapter 7 to derive the t confidence interval: if $X_1, X_2, \ldots, X_n$ is a random sample from a normal distribution, the standardized variable

$$T = \frac{\overline{X} - \mu}{S/\sqrt{n}}$$

has a t distribution with $n - 1$ d.f. Consider testing $H_0 : \mu = \mu_0$ against $H_a : \mu > \mu_0$ by using the test statistic $T = (\overline{X} - \mu_0)/(S/\sqrt{n})$. That is, the test statistic results from standardizing $\overline{X}$ under the assumption that H_0 is true (using $S/\sqrt{n}$, the estimated standard deviation of $\overline{X}$, rather than $\sigma/\sqrt{n}$). When H_0 is true, the test statistic has a t distribution with $n - 1$ d.f. Knowledge of the test statistic's distribution when H_0 is true (the "null distribution") allows us to construct a rejection region for which the type I error probability is controlled at the desired level. In particular, use of the upper-tail t critical value $t_{\alpha, n-1}$ to specify the rejection region $t \geq t_{\alpha, n-1}$ implies that

$P(\text{type I error}) = P(H_0 \text{ is rejected when it is true})$
$= P(T \geq t_{\alpha, n-1} \text{ when } T \text{ has a } t \text{ distribution with } n - 1 \text{ d.f.})$
$= \alpha$

The test statistic is really the same here as in the large-sample case, but is labeled T to emphasize that its null distribution is a t distribution with $n - 1$ d.f. rather than the standard normal (z) distribution. The rejection region for the t test differs from that for the z test only in that a t critical value $t_{\alpha, n-1}$ replaces the z critical value z_α. Similar comments apply to alternatives for which a lower-tailed or two-tailed test is appropriate.

The one-sample t test

Null hypothesis: $H_0 : \mu = \mu_0$

Test statistic value: $t = \dfrac{\overline{x} - \mu_0}{s/\sqrt{n}}$

Alternative hypothesis	Rejection region for a level α test
$H_a : \mu > \mu_0$	$t \geq t_{\alpha, n-1}$
$H_a : \mu < \mu_0$	$t \leq -t_{\alpha, n-1}$
$H_a : \mu \neq \mu_0$	either $t \geq t_{\alpha/2, n-1}$ or $t \leq -t_{\alpha/2, n-1}$

Example 8.11

In order to test gasoline mileage performance for a new version of one of its compact cars, an automobile manufacturer selected six nonprofessional drivers to drive a car from Phoenix to Los Angeles. The miles per gallon values for the six cars at the conclusion of the trip were 27.2, 29.3, 31.5, 28.7, 30.2, and 29.6. The manufacturer wishes to advertise that cars of this type average (at least)

30 mpg on such long trips. Does the sample data contradict the validity of this claim?

The parameter of interest is

μ = true average mpg for cars of this type on such trips

The manufacturer will claim that $\mu \geq 30$ unless the data strongly suggests otherwise, so the relevant hypotheses are $H_0 : \mu = 30$ versus $H_a : \mu < 30$. The form of H_a dictates the use of a lower-tailed test. The formula for the test statistic value is

$$t = \frac{\bar{x} - 30}{s/\sqrt{n}}$$

At significance level .05, H_0 should be reflected if $t \leq -t_{\alpha, n-1} = -t_{.05, 5} = -2.015$ (this critical value is larger in magnitude than the corresponding z critical value -1.645, reflecting the extra variability associated with using S rather than σ in the test statistic). The data yields $\Sigma x_i = 176.5$ and $\Sigma x_i^2 = 5202.47$, so

$$\bar{x} = \frac{176.5}{6} = 29.42, \quad s^2 = \frac{5202.47 - (176.5)^2/6}{6 - 1} = 2.086, \quad s = 1.44,$$

and

$$t = \frac{29.42 - 30}{1.44/\sqrt{6}} = \frac{-.58}{.59} = -.99$$

Since $t = -.99$ is not in the rejection region ($t > -2.015$), H_0 is not rejected at level .05. The claim that the manufacturer wishes to make is not strongly contradicted by the data.   ■

β and Sample Size Determination

The calculation of β at the alternative value μ' in case I was carried out by expressing the rejection region in terms of $\bar{x}$ (for example, $\bar{x} \geq \mu_0 + z_\alpha \cdot \sigma/\sqrt{n}$) and then subtracting μ' to standardize correctly. An equivalent approach involves noting that when $\mu = \mu'$, the test statistic $Z = (\bar{X} - \mu_0)/(\sigma/\sqrt{n})$ still has a normal distribution with variance 1, but now the mean value is given by $(\mu' - \mu_0)/(\sigma/\sqrt{n})$. That is, when $\mu = \mu'$, the test statistic still has a normal distribution though not the standard normal distribution. Because of this, $\beta(\mu')$ is an area under the normal curve with mean value $(\mu' - \mu_0)/(\sigma/\sqrt{n})$ and variance 1. Both α and β involve working with normally distributed variables.

The calculation of $\beta(\mu')$ for the t test is much less straightforward. This is because the distribution of the test statistic $T = (\bar{X} - \mu_0)/(S/\sqrt{n})$ is quite complicated when H_0 is false and H_a is true. Thus for an upper-tailed test, determining

$$\beta(\mu') = P(T < t_{\alpha, n-1} \text{ when } \mu = \mu' \text{ rather than } \mu_0)$$

involves integrating a very unpleasant density function. This must be done numerically, but fortunately it has been done by research statisticians for both one- and two-tailed t tests. The results are summarized in graphs of β that appear in Appendix Table A.13. There are four sets of graphs, corresponding to one-tailed tests at level .05 and level .01 and two-tailed tests at the same levels.

To understand how these graphs are used, note first that both β and the necessary sample size n in case I are functions not just of the absolute difference $|\mu_0 - \mu'|$ but of $d = |\mu_0 - \mu'|/\sigma$. Suppose, for example, that $|\mu_0 - \mu'| = 10$. This departure from H_0 will be much easier to detect (smaller β) when $\sigma = 2$, in which case μ_0 and μ' are 5 population standard deviations apart, than when $\sigma = 10$. The fact that β for the t test depends upon d rather than just $|\mu_0 - \mu'|$ is unfortunate, since to use the graphs one must have some idea of the true value of σ. A conservative (large) guess for σ will yield a conservative (large) value of $\beta(\mu')$ and a conservative estimate of the sample size necessary for prescribed α and $\beta(\mu')$.

Once the alternative μ' and value of σ are selected, d is calculated and its value located on the horizontal axis of the relevant set of curves. The value of β is the height of the $n - 1$ d.f. curve above the value of d (visual interpolation is necessary if $n - 1$ is not a value for which the corresponding curve appears). Rather than fixing n (that is, $n - 1$, and thus the particular curve from which β is read), one might prescribe both α (.05 or .01 here) and a value of β for the chosen μ' and σ. After computing d, the point (d, β) is located on the relevant set of graphs. The curve below and closest to this point gives $n - 1$ and thus n (again, interpolation is often necessary).

Example 8.12 An engineering company has run out of space at its present facility, so is interested in obtaining space at a nearby location. The company is considering several existing buildings. To accommodate interaction between employees, a shuttle bus is to be operated between the two locations during working hours. It is decided that some trial runs be carried out to obtain data, and then any location for which there is strong evidence that μ, the true average midday travel time, exceeds 20 minutes should be eliminated from consideration. The hypotheses to be tested are then $H_0 : \mu = 20$ versus $H_a : \mu > 20$, and the selected significance level is $\alpha = .05$.

Suppose that prior evidence suggests $\sigma = 5$ as a plausible value, and that $n = 10$ runs are to be made. What is the chance of a type II error when $\mu = 25$ (the value 25 represents a substantial departure from H_0, so not rejecting H_0 when $\mu = 25$ is a serious error)? With $d = |20 - 25|/5 = 1$, the point on the $n - 1 = 9$ d.f. curve above $d = 1$ has a height of approximately .1, so $\beta \approx .1$. The company might think $\beta = .1$ too large for such an extreme departure from H_0, so might wish to use $\beta = .05$. Since $d = 1$, the point $(d, \beta) = (1, .05)$ must be located. This point is very close to the 14 d.f. curve, so using $n = 15$ will achieve $\alpha = .05$ and $\beta = .05$ when $\mu' = 25$, $\sigma = 5$. Of course, if σ is actually larger than 5, this value of n will not suffice. ∎

Exercises / Section 8.2 (13–31)

13. Let the test statistic Z have a standard normal distribution when H_0 is true. Give the significance level for each of the following situations.
    a. $H_a: \mu > \mu_0$, rejection region $z \geq 1.88$
    b. $H_a: \mu < \mu_0$, rejection region $z \leq -2.75$
    c. $H_a: \mu \neq \mu_0$, rejection region $z \geq 2.88$ or $z \leq -2.88$

14. Let the test statistic T have a t distribution when H_0 is true. Give the significance level for each of the following situations.
    a. $H_a: \mu > \mu_0$, d.f. $= 15$, rejection region $t \geq 3.733$
    b. $H_a: \mu < \mu_0$, $n = 24$, rejection region $t \leq -2.500$
    c. $H_a: \mu \neq \mu_0$, $n = 31$, rejection region $t \geq 1.697$ or $t \leq -1.697$

15. Answer the following questions for the tire problem in Example 8.7.
    a. If $\bar{x} = 20,960$ and a level $\alpha = .01$ test is used, what is the decision?
    b. If a level .01 test is used, what is $\beta(20,500)$?
    c. If a level .01 test is used and it is also required that $\beta(20,500) = .05$, what sample size n is necessary?
    d. If $\bar{x} = 20,960$, what is the smallest α at which H_0 can be rejected (based on $n = 16$)?

16. Reconsider the paint drying situation of Example 8.2, in which drying time for a test specimen is normally distributed with $\sigma = 9$. The hypotheses $H_0: \mu = 75$ versus $H_a: \mu < 75$ are to be tested using a random sample of $n = 25$ observations.
    a. How many standard deviations (of $\bar{X}$) below the null value is $\bar{x} = 72.3$?
    b. If $\bar{x} = 72.3$, what is the conclusion using $\alpha = .01$?
    c. What is α for the test procedure that rejects H_0 when $z \leq -2.88$?
    d. For the test procedure of (c), what is $\beta(70)$?
    e. If the test procedure of (c) is used, what n is necessary to ensure that $\beta(70) = .01$?
    f. If a level .01 test is used with $n = 100$, what is the probability of a type I error when $\mu = 76$?

17. The melting point of each of 16 samples of a certain brand of hydrogenated vegetable oil was determined, resulting in $\bar{x} = 94.32$. Assume that

the distribution of melting point is normal with $\sigma = 1.20$.
    a. Test $H_0: \mu = 95$ versus $H_a: \mu \neq 95$ using a two-tailed level .01 test.
    b. If a level .01 test is used, what is $\beta(94)$, the probability of a type II error when $\mu = 94$?
    c. What value of n is necessary to ensure that $\beta(94) = .1$ when $\alpha = .01$?

18. A television manufacturer claims that at most 250 microamperes of current are needed to attain a certain brightness level with a particular type of set. A sample of 20 sets yields a sample average current of $\bar{x} = 257.3$. Let μ denote the true average current necessary to achieve the desired brightness with sets of this type, and assume that μ is the mean of a normal population with $\sigma = 15$.
    a. Test at level .05 the null hypothesis that μ is at most 250 against the appropriate alternative.
    b. If $\mu = 260$, what is the probability of a type II error?
    c. Suppose that it was decided that the manufacturer's claim should not be accepted unless the data strongly supports it. Would you choose as your null hypothesis the H_0 suggested in (a), or would you select a different H_0? Explain. *Hint:* From a practical viewpoint, the claim that at most 250 microamps are required is the same as the claim that fewer than 250 microamps are required.

19. The desired percentage of SiO_2 in a certain type of aluminous cement is 5.5. To test whether the true average percentage is 5.5 for a particular production facility, 16 independently obtained samples are analyzed. Suppose that the percentage of SiO_2 in a sample is normally distributed with $\sigma = .3$, and that $\bar{x} = 5.25$.
    a. Does this indicate conclusively that the true average percentage differs from 5.5? Carry out the analysis using the sequence of steps suggested in the text.
    b. If the true average percentage is $\mu = 5.6$ and a level $\alpha = .01$ test based on $n = 16$ is used, what is the probability of detecting this departure from H_0?

c. What value of n is required to satisfy $\alpha = .01$ and $\beta(5.6) = .01$?

20. To obtain information on the corrosion resistance properties of a certain type of steel conduit, 35 specimens are buried in soil for a two-year period. The maximum penetration (in mils) for each specimen is then measured, yielding a sample average penetration of $\bar{x} = 52.7$ and a sample standard deviation of $s = 4.8$. The conduits were manufactured with the specification that true average penetration be at most 50 mils. To see whether experimental data indicates that specifications have not been met, test $H_0 : \mu = 50$ versus $H_a : \mu > 50$ using a large-sample test with $\alpha = .05$.

21. A sample of 40 speedometers of a particular brand is obtained, and each is calibrated to check for accuracy at 55 mph. The resulting sample average and sample standard deviation are $\bar{x} = 53.8$ and $s = 1.3$, respectively. Let $\mu =$ the true average reading when actual speed is 55 mph. Does the sample evidence suggest strongly that $\mu \neq 55$? Use a level .01 test.

22. The Charpy V-notch impact test is the basis for studying many material toughness criteria. This test was applied to 32 samples of a particular alloy at $110°$ F. The sample average amount of transverse lateral expansion was computed to be 73.1 mils, and the sample standard deviation was $s = 5.9$ mils. To be suitable for a particular application, the true average amount of expansion should be less than 75 mils. The alloy will not be used unless the sample provides strong evidence of this criteria having been met. Test the relevant hypotheses using $\alpha = .01$ to decide whether the alloy is suitable.

23. The relative conductivity of a semiconductor device is determined by the amount of impurity "doped" into the device during its manufacture. A silicon diode to be used for a specific purpose requires a cut-on voltage of .60 volt, and if this is not achieved then the mechanism governing the amount of impurity must be adjusted. A sample of 120 such diodes yielded a sample average voltage of .62 volt and sample standard deviation of .11. At level .001, does the data indicate that the true average cut-on voltage is something other than .60?

24. Minor surgery on horses under field conditions requires a reliable short-term anesthetic producing good muscle relaxation, minimal cardiovascular and respiratory changes, and a quick, smooth recovery with minimal aftereffects so that horses can be left unattended. The article "A Field Trial of Ketamine Anesthesia in the Horse" (*Equine Vet. J.*, 1984, pp. 176–179) reported that for a sample of $n = 73$ horses to which ketamine was administered under certain conditions, the sample average lateral recumbency (lying-down) time was 18.86 min and the standard deviation was 8.6 min. Does this data suggest that true average lateral recumbency time under these conditions is less than 20 min? Test the appropriate hypotheses at level of significance .10.

25. The amount of shaft wear (.0001 in.) after a fixed mileage was determined for each of $n = 8$ internal combustion engines having copper lead as a bearing material, resulting in $\bar{x} = 3.72$ and $s = 1.25$.
   a. Assuming that the distribution of shaft wear is normal with mean μ, use the t test at level .05 to test $H_0 : \mu = 3.50$ versus $H_a : \mu > 3.50$.
   b. Using $\sigma = 1.25$, what is the type II error probability $\beta(\mu')$ of the test for the alternative $\mu' = 4.00$?

26. The following radiation readings (milliroentgens per hour) were obtained from television display areas in a sample of 10 department stores ("Many Set Color TV Lounges Show Highest Radiation," *J. Environmental Health*, 1969, pp. 359–360):

.40, .48, .60, .15, .50, .80, .50, .36, .16, .89

The recommended limit for this type of radiation exposure is .5 mr/hr. Assuming that the observations come from a normal distribution with mean μ (the true average amount of radiation in television display areas in all department stores), test $H_0 : \mu = .5$ versus $H_a : \mu > .5$ using a level .1 test.

27. In an experiment designed to measure the time necessary for an inspector's eyes to become used to the reduced amount of light necessary for penetrant inspection, the sample average time for

$n = 9$ inspectors was 6.32 sec with a sample standard deviation of 1.65 sec. It has previously been assumed that the average adaptation time was at least 7 sec. Assuming adaptation time to be normally distributed, does the data contradict prior belief? Use the t test with $\alpha = .1$.

28. A certain type of soil was determined to have a natural mean pH value of 8.75. The authors of the paper "Effects of Brewery Effluent on Agricultural Soil and Crop Plants" (*Environ. Pollution*, 1984, pp. 341–351) treated soil samples with various dilutions of an acidic effluent. Five soil samples were treated with a solution of 25% water and 75% effluent. The mean and standard deviation of the 5 pH measurements were 8.00 and .05, respectively. Does this data indicate that at this concentration the effluent results in a mean pH that differs from the natural pH of the soil? Use a level .01 test.

29. The National Bureau of Standards had previously reported the value of Se content in NBS orchard leaves to be .080 ppm. The paper "A

Neutron Activation Method for Determining Submicrogram Selenium in Forage Grasses" (*Soil Science Soc. Amer. J.*, 1978, pp. 57–60) reported the following Se content for six different determinations:

.072, .073, .080, .078, .088, .080

Does the data contradict the previously reported value?

a. Test the relevant hypotheses using $\alpha = .01$.
b. Suppose that prior to the experiment, a value of $\sigma = .005$ had been assumed. How many determinations would then have been appropriate to obtain $\beta = .10$ for the alternative $\mu = .075$?

30. Show that for any $\Delta > 0$, when the population distribution is normal and σ is known, the two-tailed test satisfies $\beta(\mu_0 - \Delta) = \beta(\mu_0 + \Delta)$, so that $\beta(\mu')$ is symmetric about μ_0.

31. For a fixed alternative value μ', show that $\beta(\mu') \to 0$ as $n \to \infty$ for either a one-tailed or a two-tailed z test in the case of a normal population distribution with known σ.

8.3   Tests Concerning a Population Proportion

Let p denote the proportion of individuals or objects in a population who possess a specified property (for example, cars that were domestically manufactured, smokers who smoke a filter cigarette, and so on). If an individual or object with the property is labeled a success (S), then p is the population proportion of successes. Tests concerning p will be based on a random sample of size n from the population. Provided that n is small relative to the population size, X (the number of S's in the sample) has (approximately) a binomial distribution. Furthermore, if n itself is large, both X and the estimator $\hat{p} = X/n$ are approximately normally distributed. We first consider large-sample tests based on this latter fact, and then turn to the small-sample case that directly uses the binomial distribution.

Large-Sample Tests

Large-sample tests concerning p are a special case of the more general large-sample procedures for a parameter θ. Let $\hat{\theta}$ be an estimator of θ that is (at least approximately) unbiased and has approximately a normal distribution. The null hypothesis has the form $H_0 : \theta = \theta_0$, where θ_0 denotes a number (the null value) appropriate to the problem context. Suppose that when H_0 is true, the standard deviation of $\hat{\theta}$, $\sigma_{\hat{\theta}}$, involves no unknown parameters. For example, if $\theta = \mu$ and $\hat{\theta} = \overline{X}$, $\sigma_{\hat{\theta}} = \sigma_{\overline{X}} = \sigma/\sqrt{n}$, which involves no unknown parameters only if the value of σ is known. A large-sample test statistic results from standardizing $\hat{\theta}$ under the assumption that H_0 is true (so that $E(\hat{\theta}) = \theta_0$):

$$\text{test statistic} = Z = \frac{\hat{\theta} - \theta_0}{\sigma_{\hat{\theta}}}$$

If the alternative hypothesis is $H_a : \theta > \theta_0$, an upper-tailed test whose significance level is approximately α is specified by the rejection region $z \geq z_\alpha$. The other two alternatives, $H_a : \theta < \theta_0$ and $H_a : \theta \neq \theta_0$, are tested using a lower-tailed z test and a two-tailed z test, respectively.

In the case $\theta = p$, $\sigma_{\hat{\theta}}$ will not involve any unknown parameters when H_0 is true, but this is atypical. When $\sigma_{\hat{\theta}}$ does involve unknown parameters, it is often possible to use an estimated standard deviation $S_{\hat{\theta}}$ in place of $\sigma_{\hat{\theta}}$ and still have Z approximately normally distributed when H_0 is true (because when n is large, $s_{\hat{\theta}} \approx \sigma_{\hat{\theta}}$ for most samples). The large-sample test of the previous section furnishes an example of this: because σ is usually unknown, we use $s_{\hat{\theta}} = s_{\bar{X}} = s/\sqrt{n}$ in place of $\sigma/\sqrt{n}$ in the denominator of z.

The estimator $\hat{p} = X/n$ is unbiased ($E(\hat{p}) = p$), has approximately a normal distribution, and its standard deviation is $\sigma_{\hat{p}} = \sqrt{p(1-p)/n}$. These facts were used in Section 7.2 to obtain a large-sample confidence interval for p. When H_0 is true, $E(\hat{p}) = p_0$ and $\sigma_{\hat{p}} = \sqrt{p_0(1-p_0)/n}$, so $\sigma_{\hat{p}}$ does not involve any unknown parameters. It then follows that when n is large and H_0 is true, the test statistic

$$Z = \frac{\hat{p} - p_0}{\sqrt{p_0(1-p_0)/n}}$$

has approximately a standard normal distribution. If the alternative hypothesis is $H_a : p > p_0$ and the upper-tailed rejection region $z \geq z_\alpha$ is used, then

$$P(\text{type I error}) = P(H_0 \text{ is rejected when it is true})$$
$$= P(Z \geq z_\alpha \text{ when } Z \text{ has approximately a standard normal}$$
$$\text{distribution}) \approx \alpha$$

Thus the desired level of significance α is attained by using the critical value that captures area α in the upper-tail of the z curve. Rejection regions for the other two alternative hypotheses, lower-tailed for $H_a : p < p_0$ and two-tailed for $H_a : p \neq p_0$, are justified in an analogous manner.

Null hypothesis: $H_0 : p = p_0$

Test statistic value: $z = \dfrac{\hat{p} - p_0}{\sqrt{p_0(1-p_0)/n}}$

Alternative hypothesis	Rejection region
$H_a : p > p_0$	$z \geq z_\alpha$
$H_a : p < p_0$	$z \leq -z_\alpha$
$H_a : p \neq p_0$	either $z \geq z_{\alpha/2}$ or $z \leq -z_{\alpha/2}$

These test procedures are valid provided that both $np_0 \geq 5$ and $n(1 - p_0) \geq 5$.

Notice that the standard deviation in the denominator of z is $\sqrt{p_0(1 - p_0)/n}$, whereas the estimated standard deviation $\sqrt{\hat{p}(1 - \hat{p})/n}$ was used in the large-sample confidence interval for p.

Example 8.13   Environmental problems associated with leaded gasolines are well known. Many motorists have tampered with emission control devices to save money by purchasing leaded rather than unleaded gas. A recent *Los Angeles Times* article (March 17, 1984) reported that 15% of all California motorists have engaged in such tampering. Suppose that a random sample of 200 cars from a particular county is obtained, and the emission control devices of 21 are found to have been tampered with. Does this suggest that the proportion of cars in this county with tampered devices differs from the published statewide proportion? Use a test with level of significance .05.

1. p = the proportion of cars in this county whose emission control devices have been tampered with.
2. $H_0 : p = .15$.
3. $H_a : p \neq .15$.
4. Since $(200)(.15) \geq 5$ and $(200)(.85) \geq 5$, the z test can be used. The test statistic value is

   $$z = (\hat{p} - .15)/\sqrt{(.15)(.85)/n}$$

5. The form of H_a implies that a two-tailed test is appropriate. Because $z_{\alpha/2} = z_{.025} = 1.96$, H_0 will be rejected either if $z \geq 1.96$ or if $z \leq -1.96$.
6. $\sqrt{(.15)(.85)/200} = .0252$ and $\hat{p} = 21/200 = .105$, so substitution gives $z = (.105 - .150)/.0252 = -.045/.0252 = -1.79$.
7. Since -1.79 is neither ≥ 1.96 nor ≤ -1.96, H_0 cannot be rejected at level .05. The data does not suggest that the proportion of cars in this county with tampered devices differs from the statewide proportion.   ■

β and Sample Size Determination

When H_0 is true, the test statistic Z has approximately a standard normal distribution. Now suppose that H_0 is not true and that $p = p'$. Then Z still has approximately a normal distribution (because it is a linear function of $\hat{p}$), but its mean value and variance are no longer 0 and 1, respectively. Instead,

$$E(Z) = \frac{p' - p_0}{\sqrt{p_0(1 - p_0)/n}}, \quad V(Z) = \frac{p'(1 - p')/n}{p_0(1 - p_0)/n}$$

The probability of a type II error for an upper-tailed test is $\beta(p') = P(Z < z_\alpha$ when $p = p')$. This can be computed by using the above mean and variance to standardize and then referring to the standard normal c.d.f. In addition, if it is desired that the level α test also have $\beta(p') = \beta$ for a specified value of β, this equation can be solved for the necessary n as in Section 8.2. General expressions for $\beta(p')$ and n are given in the accompanying box.

Alternative hypothesis	$\beta(p')$
$H_a : p > p_0$	$\Phi\left[\dfrac{p_0 - p' + z_\alpha \sqrt{p_0(1 - p_0)/n}}{\sqrt{p'(1 - p')/n}}\right]$
$H_a : p < p_0$	$1 - \Phi\left[\dfrac{p_0 - p' - z_\alpha \sqrt{p_0(1 - p_0)/n}}{\sqrt{p'(1 - p')/n}}\right]$
$H_a : p \neq p_0$	$\Phi\left[\dfrac{p_0 - p' + z_{\alpha/2} \sqrt{p_0(1 - p_0)/n}}{\sqrt{p'(1 - p')/n}}\right]$ $\quad - \Phi\left[\dfrac{p_0 - p' - z_{\alpha/2} \sqrt{p_0(1 - p_0)/n}}{\sqrt{p'(1 - p')/n}}\right]$

The sample size n for which the level α test also satisfies $\beta(p') = \beta$ is

$$n = \begin{cases} \left[\dfrac{z_\alpha \sqrt{p_0(1 - p_0)} + z_\beta \sqrt{p'(1 - p')}}{p' - p_0}\right]^2 & \text{one-tailed test} \\[4mm] \left[\dfrac{z_{\alpha/2} \sqrt{p_0(1 - p_0)} + z_\beta \sqrt{p'(1 - p')}}{p' - p_0}\right]^2 & \begin{array}{l} \text{two-tailed test} \\ \text{(an approximate solution)} \end{array} \end{cases}$$

Example 8.14 A package delivery service advertises that at least 90% of all packages brought to its office by 9 A.M. for delivery in the same city are delivered by noon that day. Let p denote the true proportion of such packages that are delivered as advertised, and consider the hypotheses $H_0 : p = .9$ versus $H_a : p < .9$. If only 80% of the packages are delivered as advertised, how likely is it that a level .01 test based on $n = 225$ packages will detect such a departure from H_0? What should the sample size be in order to ensure that $\beta(.8) = .01$? With $\alpha = .01$, $p_0 = .9$, $p' = .8$, and $n = 225$,

$$\beta(.8) = 1 - \Phi\left(\frac{.9 - .8 - 2.33\sqrt{(.9)(.1)/225}}{\sqrt{(.8)(.2)/225}}\right) = 1 - \Phi(2.00) = .0228$$

Thus the probability that H_0 will be rejected using the test when $p = .8$ is .9772—roughly 98% of all samples will result in correct rejection of H_0.

Using $z_\alpha = z_\beta = 2.33$ in the sample-size formula yields

$$n = \left[\frac{2.33\sqrt{(.9)(.1)} + 2.33\sqrt{(.8)(.2)}}{.8 - .9}\right]^2 \approx 266 \qquad \blacksquare$$

Small-Sample Tests

Test procedures when the sample size n is small are based directly on the binomial distribution rather than the normal approximation. Consider the alternative hypothesis $H_a : p > p_0$, and again let X be the number of successes in the sample. Then X is the test statistic, and the upper-tailed rejection region has the form $x \geq c$. When H_0 is true, X has a binomial distribution with parameters n and p_0, so

$$
\begin{aligned}
P(\text{type I error}) &= P(H_0 \text{ is rejected when it is true}) \\
&= P(X \geq c \text{ when } X \sim \text{Bin}(n, p_0)) \\
&= 1 - P(X \leq c - 1 \text{ when } X \sim \text{Bin}(n, p_0)) \\
&= 1 - B(c - 1; n, p_0)
\end{aligned}
$$

As the critical value c decreases, more x values are included in the rejection region and $P(\text{type I error})$ increases. Because X has a discrete probability distribution, it is usually not possible to find a value of c for which $P(\text{type I error})$ is exactly the desired significance level α (for example, .05 or .01). Instead the largest rejection region of the form $\{c, c + 1, \ldots, n\}$ satisfying the condition $1 - B(c - 1; n, p_0) \leq \alpha$ is used.

Let p' denote an alternative value of p ($p' > p_0$). Then when $p = p'$, $X \sim \text{Bin}(n, p')$, so

$$
\begin{aligned}
\beta(p') &= P(\text{type II error when } p = p') \\
&= P(X < c \text{ when } X \sim \text{Bin}(n, p')) = B(c - 1; n, p')
\end{aligned}
$$

That is, $\beta(p')$ is the result of a straightforward binomial probability calculation. The sample size n necessary to ensure that a level α test also has specified β at a particular alternative value p' must be determined by trial and error using the binomial c.d.f.

Test procedures for $H_a: p < p_0$ and for $H_a: p \neq p_0$ are constructed in a similar manner. In the former case, the appropriate rejection region has the form $x \leq c$ (a lower-tailed test). The critical value c is the largest number satisfying $B(c; n, p_0) \leq \alpha$. The rejection region when the alternative hypothesis is $H_a: p \neq p_0$ consists of both large and small x values.

Example 8.15   A plastics manufacturer has developed a new type of plastic trash can and proposes to sell them with an unconditional six-year warranty. To see whether this is economically feasible, 20 prototype cans are subjected to an accelerated life test to simulate six years of use. The proposed warranty will be modified only if the sample data strongly suggests that fewer than 90% of such cans would survive the six-year period. Let p denote the proportion of all cans that survive the accelerated test. The relevant hypotheses are $H_0: p = .9$ versus $H_a: p < .9$. A decision will be based on the test statistic X, the number among the 20 that survive. If the desired significance level is $\alpha = .05$, c must satisfy $B(c; 20, .9) \leq .05$. From Appendix Table A.1, $B(15; 20, .9) = .043$, whereas $B(16; 20, .9) = .133$. The appropriate rejection region is therefore $x \leq 15$. If the accelerated test results in $x = 14$, H_0 would be rejected in favor of H_a, necessitating a modification of the proposed warranty. The probability of a type II error for the alternative value $p' = .8$ is

$$\beta(.8) = P(H_0 \text{ is not rejected when } X \sim \text{Bin}(20, .8))$$
$$= P(X \geq 16 \text{ when } X \sim \text{Bin}(20, .8))$$
$$= 1 - B(15; 20, .8) = 1 - .370 = .630$$

That is, when $p = .8$, 63% of all samples consisting of $n = 20$ cans would result in H_0 being incorrectly not rejected. This error probability is high because 20 is a small sample size and $p' = .8$ is close to the null value $p_0 = .9$. ■

Exercises / Section 8.3 (32–40)

32. Let p denote the true proportion of Budweiser drinkers who can distinguish their beer from Schlitz. If X denotes the number of correct identifications in a sample of 100 Bud drinkers and we observe $x = 57$, test $H_0 : p = .5$ versus $H_a : p \neq .5$ using a level .10 test.

33. The city council in a large city has become concerned about the trend toward exclusion of renters with children in apartments within the city. The housing coordinator has decided to select a random sample of 125 apartments and determine for each whether or not children would be permitted. Let $p =$ the true proportion of apartments that prohibit children. If $p > .75$ the council will consider appropriate legislation.

a. If 102 of the 125 sampled exclude renters with children, test $H_0 : p = .75$ versus $H_a : p > .75$ using a level .05 test.

b. What is the probability of a type II error when $p = .8$ (not rejecting H_0 when $p = .8$)?

c. How many apartments would have to be sampled to ensure that $\beta(.8) = .1$?

34. A telephone company is trying to decide whether some new lines in a large community should be installed underground. Because a small surcharge will be added to telephone bills to pay for the extra installation costs, the company has decided to survey customers and proceed only if the survey strongly indicates that more than 60% of all customers favor underground installation. If 118 of 160 customers surveyed favor underground installation in spite of the surcharge, what should the company do? Test the relevant hypotheses using $\alpha = .05$.

35. A university library ordinarily has a complete shelf inventory done once every year. Because of new shelving rules instituted the previous year, the head librarian feels that it may be possible to save money by postponing the inventory. The librarian decides to select at random 800 books from the library's collection and have them searched in a preliminary manner. If evidence indicates strongly that the true proportion of misshelved or unlocatable books is less than .02, then the inventory will be postponed.

a. Among the 800 books searched, 12 were misshelved or unlocatable. Test the relevant hypotheses and advise the librarian what to do (use $\alpha = .05$).

b. If the true proportion of misshelved and lost books is actually .01, what is the probability that the inventory will be (unnecessarily) taken?

c. If the true proportion is .05, what is the probability that the inventory will be postponed?

36. The article "Statistical Evidence of Discrimination" (*J. Amer. Stat. Assoc.*, 1982, pp. 773–783) discussed the court case *Swain v. Alabama* (1965), in which it was alleged that there was discrimination against blacks in grand jury selection. Census data suggested that 25% of those eligible for grand jury service were black, yet a random sample of 1050 called to appear for possible duty yielded only 177 blacks. Using a level .01 test, does this data argue strongly for a conclusion of discrimination?

37. A plan for an executive traveler's club has been developed by an airline on the premise that 5% of its current customers would qualify for membership. A random sample of 500 customers yielded 40 who would qualify.

a. Using this data, test at level .01 the null hypothesis that the company's premise is correct against the alternative that it is not correct.

b. What is the probability that when the test of (a) is used, the company's premise will be judged correct when in fact 10% of all current customers qualify?

c. What sample size would be sufficient to ensure that both α and the probability of (b) are .01?

38. A manufacturer of plumbing fixtures has developed a new type of washerless faucet. Let $p = P$(a randomly selected faucet of this type will develop a leak within two years under normal use). The manufacturer has decided to proceed with production unless it can be determined that p is too large; the borderline acceptable value of p is specified as .10. The manufacturer decides to subject n of these faucets to accelerated testing (approximating two years of normal use). With $X =$ the number among the n faucets that leak before the test concludes, production will commence unless the observed X is too large. It is decided that if $p = .10$, the probability of not proceeding should be at most .10, while if $p = .30$ the probability of proceeding should be at most .10. Can $n = 10$ be used? $n = 20$? $n = 25$? What is the appropriate rejection region for the chosen n, and what are the actual error probabilities when this region is used?

39. (Based on an article in the May 1978 issue of *Consumer Reports*.) To decide whether tennis enthusiasts strongly prefer gut strings to nylon strings in their tennis rackets, each of 20 randomly selected enthusiasts is given two tennis rackets that are identical except that one is strung with nylon and the other with gut. After several weeks of playing alternately with each racket, each of the enthusiasts will be asked to state a preference for one of the two rackets. Let $p =$ the true proportion of enthusiasts who prefer gut strings, and $X =$ the number of enthusiasts in the sample who prefer gut. Because gut is more expensive and breaks sooner than nylon, the null hypothesis is that at most 50% of all enthusiasts prefer gut; we simplify this to $H_0: p = .5$, planning to reject H_0 only if experimental evidence strongly favors gut strings.

a. Which of the rejection regions $\{15, 16, 17, 18, 19, 20\}$, $\{0, 1, 2, 3, 4, 5\}$, or $\{0, 1, 2, 3, 17, 18, 19, 20\}$ is most appropriate, and why are the other two not appropriate?

b. What is the probability of a type I error for the chosen region of (a)? Does the region specify a level .05 test? Is it the best level .05 test?

c. If 60% of all enthusiasts prefer gut, calculate the probability of a type II error using the appropriate region from (a). Repeat if 80% of all enthusiasts prefer gut.

d. If 13 out of the 20 players prefer gut, should H_0 be rejected using a significance level of .10?

40. Scientists think that robots will play a crucial role in factories in the next 20 years. Suppose that in an experiment to determine whether the use of robots to weave computer cables is feasible, a robot was used to assemble 500 cables. The cables were examined and there were 14 defectives. If human assemblers have a defect rate of .03 (3%), does this data support the hypothesis that the proportion of defectives is lower for robots than humans? Use a .01 significance level.

8.4   *P*-Values

One way to report the result of a hypothesis-testing analysis is to simply say whether or not the null hypothesis was rejected at a specified level of significance. Thus an investigator might state that H_0 was rejected at level of significance .05, or that use of a level .01 test resulted in not rejecting H_0. This type of statement is somewhat inadequate because it says nothing about whether the computed value of the test statistic just barely fell into the rejection region or whether it exceeded the critical value by a large amount. A related difficulty is that such a report imposes the specified significance level on other decision makers. In many decision situations individuals may have different views concerning the consequences of a type I or type II error. Each individuals would then want to select his or her own significance level—some selecting $\alpha = .05$, others .01 and so on—and reach a conclusion accordingly. This could result in some individuals rejecting H_0 while others conclude that the data does not show a strong enough contradiction of H_0 to justify its rejection.

Example 8.16

The true average time to initial relief of pain for the current best-selling pain reliever is known to be 10 minutes. Let μ denote the true average time to relief for a company's newly developed reliever. The company wishes to produce and market this reliever only if it provides quicker relief than does the current best-seller, so wishes to test $H_0 : \mu = 10$ versus $H_a : \mu < 10$. Only if experimental evidence leads to rejection of H_0 will the new reliever be introduced. After weighing the relative seriousness of each type of error, a single level of significance must be agreed upon and a decision—to reject H_0 and introduce the reliever or to not do so—made at that level.

Suppose that the new reliever has been introduced. The company supports its claim of quicker relief by stating that, based on an analysis of experimental data, $H_0 : \mu = 10$ was rejected in favor of $H_a : \mu < 10$ using level of significance $\alpha = .10$. Any individuals contemplating a switch to this new reliever would naturally want to reach their own conclusions concerning the validity of the claim. Individuals who are satisfied with the current best-seller would view a type I error (concluding that the new product provides quicker relief when it actually doesn't) as serious, so might wish to use $\alpha = .05$, .01, or even smaller levels. Unfortunately the nature of the company's statement prevents an individual decision maker from reaching a conclusion at such a level. The company has imposed its own choice of significance level on others. The report could have been done in a manner that allowed each individual flexibility in drawing a conclusion at a personally selected α. ∎

A *P*-value conveys much information about the strength of evidence against H_0 and allows an individual decision maker to draw a conclusion at any specified level α. Before we give a general definition, consider how the conclusion in a hypothesis-testing problem depends on the selected level α.

Example 8.17

The nicotine content problem discussed in Example 8.5 involved testing $H_0 : \mu = 1.5$ versus $H_a : \mu > 1.5$. Because of the inequality in H_a, the rejection region is upper-tailed, with H_0 rejected if $z \geq z_\alpha$. Suppose that $z = 2.10$. The accompanying table displays the rejection region for each of four different α's along with the resulting conclusion.

Level of significance α	Rejection region	Conclusion
.05	$z \geq 1.645$	reject H_0
.025	$z \geq 1.96$	reject H_0
.01	$z \geq 2.33$	don't reject H_0
.005	$z \geq 2.58$	don't reject H_0

For α relatively large, the z critical value z_α is not very far out in the upper tail; 2.10 exceeds the critical value, and so H_0 is rejected. However, as α decreases, the critical value increases. For small α the z critical value is large; 2.10 is less than z_α, and H_0 is not rejected.

Recall that for an upper-tailed z test, α is just the area under the z curve to the right of the critical value z_α. That is, once α is specified, the critical value is chosen to capture upper-tail area α. Appendix Table A.3 shows that the area to the right of 2.10 is .0179. Using an α larger than .0179 corresponds to $z_\alpha < 2.10$. An α less than .0179 necessitates using a z critical value that exceeds 2.10. The decision at a particular level α thus depends on how the selected α compares to the tail area captured by the computed z. This is illustrated in Figure 8.3. Notice in particular that .0179, the captured tail area, is the smallest level α at which H_0 would be rejected, because using any smaller α results in a z critical value that exceeds 2.10, so that 2.10 is not in the rejection region.

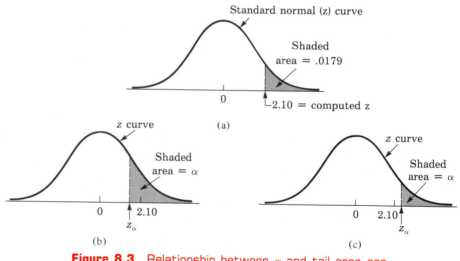

Figure 8.3   Relationship between α and tail area captured by computed z: (a) tail area captured by computed z; (b) when $\alpha > .0179$, $z_\alpha < 2.10$ and H_0 is rejected; (c) when $\alpha < .0179$, $z_\alpha > 2.10$ and H_0 is not rejected   ■

In general, suppose that the sampling distribution of a test statistic when H_0 is true has been determined. Then for specified α the rejection region is determined by finding a critical value or values that capture tail area α (upper, lower, or two-tailed, whichever is appropriate) under the sampling distribution curve. The smallest α for which H_0 would be rejected is the tail area captured by the computed value of the test statistic. This smallest α is the *P*-value.

Definition

> The ***P*-value** is the smallest level of significance at which H_0 would be rejected when a specified test procedure is used on a given data set. Once the *P*-value has been determined, the conclusion at any particular level α results from comparing the *P*-value to α:
>
> **a.** *P*-value $\leq \alpha \Rightarrow$ reject H_0 at level α.
> **b.** *P*-value $> \alpha \Rightarrow$ do not reject H_0 at level α.

It is customary to call the data *significant* when H_0 is rejected and *not significant* otherwise. The *P*-value is then the smallest level at which the data is significant. An easy way to visualize the comparison of the *P*-value with the chosen α is to draw a picture like that of Figure 8.4. The calculation of the *P*-value depends on whether the test is upper, lower, or two-tailed. However, once it has been calculated, the comparison with α does not depend on which type of test was used.

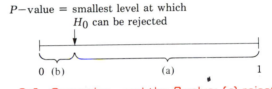

Figure 8.4   Comparing α and the *P*-value: (*a*) reject H_0 when α lies here; (*b*) do not reject H_0 when α lies here

Example 8.18
(Example 8.16
continued)

Suppose that when data from an experiment involving the new pain reliever was analyzed, the *P*-value for testing $H_0 : \mu = 10$ versus $H_a : \mu < 10$ was calculated as .0384. Since $\alpha = .05$ is larger than the *P*-value (.05 lies in the interval (*a*) of Figure 8.4), H_0 would be rejected by anyone carrying out the test at level .05. However, at level .01 H_0 would not be rejected, because .01 is smaller than the smallest level (.0384) at which H_0 can be rejected.   ■

Most standard computer packages for doing statistical analyses automatically calculate and print out a *P*-value when a hypothesis-testing analysis is performed. A conclusion can then be drawn directly from the output, without reference to a table of critical values.

The *P*-Value for a *z* Test
The *P*-value for a *z* test (one based on a test statistic whose distribution when H_0 is true is at least approximately standard normal) is easily determined from the information in Appendix Table A.3. Consider an upper-tailed test, and let *z*

denote the computed value of the test statistic Z. The null hypothesis is rejected if $z \geq z_\alpha$, and the P-value is the smallest α for which this is the case. Since z_α increases as α decreases, the P-value is the value of α for which $z = z_\alpha$. That is, the P-value is just the area captured by the computed value z in the upper-tail of the standard normal curve. The corresponding cumulative area is $\Phi(z)$, so in this case P-value $= 1 - \Phi(z)$.

An analogous argument for a lower-tailed test shows that the P-value is the area captured by the computed value z in the lower-tail of the standard normal curve. More care must be exercised in the case of a two-tailed test. Suppose first that z is positive. Then the P-value is the value of α satisfying $z = z_{\alpha/2}$ (that is, computed $z = $ upper-tail critical value). This says that the area captured in the upper-tail is half the value, so that P-value $= 2[1 - \Phi(z)]$. If z is negative, the P-value is the α for which $z = -z_{\alpha/2}$, or equivalently $-z = z_{\alpha/2}$, so P-value $= 2[1 - \Phi(-z)]$. Since $-z = |z|$ when z is negative, P-value $= 2[1 - \Phi(|z|)]$ for either positive or negative z.

$$
P\text{-value} = P = \begin{cases} 1 - \Phi(z) & \text{for an upper-tailed test} \\ \Phi(z) & \text{for a lower-tailed test} \\ 2[1 - \Phi(|z|)] & \text{for a two-tailed test} \end{cases}
$$

Example 8.19   The manufacturer's recommended air pressure for a certain type of tire is 32 psi. A random sample of $n = 35$ such tires from different cars operated by a large rental company yields a sample average pressure and sample standard deviation of 31.4 and 1.2, respectively. Does this data suggest that the average pressure for all such tires on the company's cars differs from the recommended value?

Let μ denote the population average pressure. The hypotheses of interest are $H_0 : \mu = 32$ versus $H_a : \mu \neq 32$. Because n is large, a large-sample z test can be used. The computed value of the test statistic is

$$
z = \frac{\bar{x} - \mu_0}{s / \sqrt{n}} = \frac{31.4 - 32}{1.2 / \sqrt{35}} = -2.96
$$

The P-value for a two-tailed test is then

$$
P\text{-value} = 2[1 - \Phi(|-2.96|)] = 2(1 - .9985) = .003
$$

At significance level .01, H_0 is rejected (because the P-value is smaller than this level, as in (a) of Figure 8.4) in favor of the conclusion that the population average pressure differs from the recommended value. If, however, a test with level $\alpha = .001$ is employed, H_0 cannot be rejected. ■

The *P*-Value for a *t* Test

Whereas the standard normal table gives critical values corresponding to a great many different tail areas, Appendix Table A.5 contains only seven critical values for each different t distribution. This implies that the exact P-value—a captured tail area for a one-tailed test or twice the captured area for a two-tailed test—cannot usually be obtained. In practice it is usually sufficient to say where the *P*-value lies with respect to the most frequently used significance levels.

Consider a *t* test based on 10 d.f., so that the relevant *t* critical values from Appendix Table A.5 are as follows:

Tail area:	.10	.05	.025	.01	.005	.001	.0005
Critical value:	1.372	1.812	2.228	2.764	3.169	4.144	4.587

Suppose first that the test is upper-tailed (rejection region $t \geq t_{\alpha, 10}$) and that $t = 2.53$. This computed value is between the two critical values 2.228 and 2.764 corresponding to $\alpha = .025$ and .01, respectively. Therefore, *P*-value < .025 (H_0 is rejected at level .025, so the smallest α at which rejection is possible is less than .025) and .01 < *P*-value (H_0 can't be rejected at level .01). This gives the bounds .01 < *P*-value < .025. If $t = .93$, then we can only say H_0 cannot be rejected at level .10, so .10 < *P*-value (a lower bound). Similarly, if $t = 4.78$, H_0 can be rejected at level .0005 ($4.78 \geq t_{.0005, 10}$), resulting in the upper bound *P*-value < .0005.

For a lower-tailed test, an upper and/or lower bound on the *P*-value results from changing the sign of *t* and proceeding as in the case of an upper-tailed test. Thus if $t = -1.98$ for a 10 d.f. test, .025 < *P*-value < .05 (H_0 can be rejected at level .05 but not at level .025). Finally, for a two-tailed test, remember that the level of significance associated with using a particular critical value is twice the corresponding upper-tail area. For example, $\alpha = .05$ for the rejection region $t \geq 2.228$ or $t \leq -2.228$. Suppose that $t = 2.41$, which lies between the two critical values 2.228 (corresponding to $\alpha = .05$) and 2.764 (for $\alpha = .02$). Then .02 < *P*-value < .05. That is, for a two-tailed test, bounds result from doubling the tail areas at the top of the bounding columns. Similarly, if $t = -1.15$, H_0 cannot be rejected at level $2(.10) = .20$, so .20 < *P*-value.

Several widely used statistical computer packages incorporate algorithms for computing captured *t* curve areas. Resulting computer output then displays an exact *P*-value rather than just the bounds. These packages have similar capabilities for other test procedures to be encountered shortly.

Exercises / Section 8.4 (41–52)

41. For which of the given *P*-values would the null hypothesis be rejected when performing a level .05 test?

a. .001 **b.** .021 **c.** .078 **d.** .047 **e.** .148

42. Pairs of *P*-values and significance levels, α, are given. For each pair, state whether the observed *P*-value would lead to rejection of H_0 at the given significance level.

a. *P*-value $= .084$, $\alpha = .05$
b. *P*-value $= .003$, $\alpha = .001$

c. *P*-value $= .498$, $\alpha = .05$
d. *P*-value $= .084$, $\alpha = .10$
e. *P*-value $= .039$, $\alpha = .01$
f. *P*-value $= .218$, $\alpha = .10$

43. Let μ denote the reaction time to a certain stimulus. For a large-sample *z* test of $H_0 : \mu = 5$ versus $H_a : \mu > 5$, find the *P*-value associated with each of the given values of the test statistic.

a. $z = 1.4$ **b.** .9 **c.** 1.9 **d.** 2.4 **e.** $-.1$

44. Newly purchased automobile tires of a certain type are supposed to be filled to a pressure of 30 lb/in.^{2} Let μ denote the true average pressure. Find the P-value associated with each given z statistic value for testing $H_0: \mu = 30$ versus $H_a: \mu \neq 30$.
a. 2.1   **b.** -1.7 **c.** $-.5$ **d.** 1.4   **e.** -5

45. A researcher collected data in order to test $H_0: \mu = 17$ versus $H_a: \mu > 17$. Place bounds on the P-value for each of the given t test statistic values and associated degrees of freedom.
a. $t = 1.84$, df $= 14$
b. $t = 3.74$, df $= 25$
c. $t = 2.42$, df $= 13$
d. $t = 1.32$, df $= 8$
e. $t = 2.67$, df $= 45$

46. Place bounds on the P-value for a two-tailed t test for each case.
a. $t = 2.3$, df $= 6$
b. $t = -3.0$, df $= 14$
c. $t = 4.2$, df $= 24$
d. $t = -1.3$, df $= 17$

47. An aspirin manufacturer fills bottles by weight rather than by count. Since each bottle should contain 100 tablets, the average weight per tablet should be five grains. Each of 100 tablets taken from a very large lot is weighed, resulting in a sample average weight per tablet of 4.87 grains and a sample standard deviation of .35 grain. Does this information provide strong evidence for concluding that the company is not filling its bottles as advertised? Test the appropriate hypotheses using $\alpha = .01$ by first computing the P-value and then comparing it to the specified significance level.

48. The owner of a gas station which makes headlight inspections for the state is trying to decide whether or not to discontinue the service. He figures that to make the operation profitable, the station must average in excess of 15 inspections per week. Unless data indicates strongly that this is the case, inspections will be discontinued. Data for the past year (52 weeks) yields a sample average of 16.7 inspections per week with a sample standard deviation of 4.5. Is this strong enough evidence to cause the owner to retain the inspection service? Test the relevant hypotheses using the P-value method.

49. Because of variability in the manufacturing process, the actual yielding point of a sample of mild steel subjected to increasing stress will usually differ from the theoretical yielding point. Let p denote the true proportion of samples which yield before their theoretical yielding point. If on the basis of a sample it can be concluded that more than 20% of all specimens yield before the theoretical point, the production process will have to be modified.
a. If 15 of 60 specimens yield before the theoretical point, what is the P-value when the appropriate test is used, and what would you advise the company to do?
b. If the true percentage of "early yields" is actually 50% (so that the theoretical point is the median of the yield distribution) and a level .01 test is used, what is the probability that the company concludes a modification of the process is necessary?

50. The times of first sprinkler activation for a series of tests with fire prevention sprinkler systems using an aqueous film-forming foam were (in sec)

27, 41, 22, 27, 23, 35, 30, 33, 24, 27, 28, 22, 24

(see "Use of AFFF in Sprinkler Systems," *Fire Technology,* 1976, p. 5). The system has been designed so that true average activation time is at most 25 sec under such conditions. Does the data strongly contradict the validity of this design specification?

Test the relevant hypotheses using $\alpha = .05$ by first obtaining an upper and lower bound for the P-value of the data. What assumption are you making about the activation time distribution?

51. A spectrophotometer used for measuring CO concentration [ppm (parts per million) by volume] is checked for accuracy by taking readings on a manufactured gas (called span gas) in which the CO concentration is very precisely controlled at 70 ppm. If the readings suggest that the spectrophotometer is not working properly, it will have to be recalibrated. Assume that if properly calibrated, measured concentration for span gas samples is normally distributed. On the basis of the six readings, 85, 77, 82, 68, 72, 69, is recalibration necessary? Test using $\alpha = .05$ by first obtaining as much information as you can about the *P*-value.

52. Federal officials are currently investigating the problems associated with disposal of hazardous wastes. One disposal site is the abandoned Stringfellow acid pits in Riverside County, California. The EPA had sampled water from 11 wells in nearby Glen Avon. Radiation levels from 38 to 67 pCi/L were observed (*Los Angeles* *Times*, May 31, 1984). The EPA standard for maximum allowable radiation level for drinking water is 15 pCi. Suppose that the sample of 11 wells has resulted in a sample mean radiation level of 52.5 pCi and a sample standard deviation of 8. Use a level .01 test to determine whether the data strongly suggests that the mean radiation level exceeds the EPA standard by first placing a bound or bounds on the *P*-value.

8.5   Some Comments on Selecting a Test Procedure

Once the experimenter has decided on the question of interest and the method for gathering data (the design of the experiment), construction of an appropriate test procedure consists of three distinct steps:

1. Specify a test statistic (the function of the observed values which will serve as the decision maker).
2. Decide on the general form of the rejection region (typically reject H_0 for suitably large values of the test statistic, reject for suitably small values, or reject for both small and large values).
3. Select the specific numerical critical value or values which will separate the rejection region from the acceptance region (by obtaining the distribution of the test statistic when H_0 is true, and then selecting a level of significance).

In the examples thus far, both steps 1 and 2 above were carried out in an ad hoc manner through intuition. For example, when the underlying population was assumed normal with mean μ and known σ, we were led from $\overline{X}$ to the standardized test statistic

$$Z = \frac{\overline{X} - \mu_0}{\sigma/\sqrt{n}}$$

For testing $H_0: \mu = \mu_0$ versus $H_a: \mu > \mu_0$, intuition then suggested rejecting H_0 when z was large. Finally the critical value was determined by specifying the level of significance α and using the fact that Z has a standard normal distribution when H_0 is true. The reliability of the test in reaching a correct decision can be assessed by studying type II error probabilities.

Issues which ought to be considered in carrying out steps 1–3 encompass the following questions:

1. What are the practical implications and consequences of choosing a particular level of significance once the other aspects of a test procedure have been determined?
2. Does there exist a general principle, not dependent just on intuition, which can be used to obtain best or good test procedures?

3. When there exist two or more tests which are appropriate in a given situation, how can the tests be compared to decide which should be used?
4. If a test is derived under specific assumptions about the distribution or population being sampled, how well will the test procedure work when the assumptions are violated?

Statistical versus Practical Significance

While the process of reaching a decision by using the methodology of classical hypothesis testing involves selecting a level of significance and then accepting or rejecting H_0 at that level α, simply reporting the α used and the decision reached conveys little of the information contained in the sample data. Especially when the results of an experiment are to be communicated to a large audience, rejection of H_0 at level .05 will be much more convincing if the observed value of the test statistic greatly exceeds the 5% critical value than if it barely exceeds that value. This is precisely what led to the notion of P-value as a way of reporting significance without imposing a particular α on others who might wish to draw their own conclusions.

Even if a P-value is included in a summary of results, however, there may be difficulty in interpreting this value and in making a decision. This is because a small P-value, which would ordinarily indicate **statistical significance** in that it would strongly suggest rejection of H_0 in favor of H_a, may be the result of a large sample size in combination with a departure from H_0 which has little **practical significance.** In many experimental situations, only departures from H_0 of large magnitude would be worthy of detection, whereas a small departure from H_0 would have little practical significance.

Consider as an example testing $H_0: \mu = 100$ versus $H_a: \mu > 100$ where μ is the mean of a normal population with $\sigma = 10$. Suppose that a true value of $\mu = 101$ would not represent a serious departure from H_0 in the sense that not rejecting H_0 when $\mu = 101$ would be a relatively inexpensive error. For a reasonably large sample size n, this μ would lead to an $\bar{x}$ value near 101, so we would not want this sample evidence to argue strongly for rejection of H_0 when $\bar{x} = 101$ is observed. For various sample sizes, Table 8.1 records both the P-value when $\bar{x} = 101$ and also the probability of not rejecting H_0 at level .01 when $\mu = 101$.

Table 8.1 An illustration of the effect of large sample size on P-values and β

n	P-value when $\bar{x} = 101$	$\beta(101)$ for level .01 test
25	.3085	.9664
100	.1587	.9082
400	.0228	.6293
900	.0013	.2514
1600	.0000335	.0475
2500	.000000297	.0038
10000	7.69×10^{-24}	.0000

The second column in Table 8.1 shows that even for moderately large sample sizes, the P-value of $\bar{x} = 101$ argues very strongly for rejection of H_0, whereas the observed $\bar{x}$ itself suggests that in practical terms the true value of μ differs little from the null value $\mu_0 = 100$. The third column points out that even when there is little practical difference between the true μ and the null value, for a fixed level of significance a large sample size will almost always lead to rejection of the null hypothesis at that level. To summarize, *one must be especially careful in interpreting evidence when the sample size is large, since any small departure from H_0 will almost surely be detected by a test, yet such a departure may have little practical significance.*

The Likelihood Ratio Principle

Let $x_1, x_2, \ldots, x_n$ be the observations in a random sample from a probability distribution $f(x; \theta)$. The joint distribution evaluated at these sample values is $f(x_1; \theta) \cdot f(x_2; \theta) \cdot \ldots \cdot f(x_n; \theta)$. As in the discussion of maximum likelihood estimation, the *likelihood function* is this joint distribution regarded as a function of θ. Consider testing $H_0: \theta$ is in Ω_0 versus $H_a: \theta$ is in Ω_a, where Ω_0 and Ω_a are disjoint (for example, $H_0: \theta \leq 100$ versus $H_a: \theta > 100$). The likelihood ratio principle for test construction proceeds as follows:

1. Find the largest value of the likelihood for any θ in Ω_0 (by finding the maximum likelihood estimate within Ω_0 and substituting back into the likelihood function).
2. Find the largest value of the likelihood for any θ in Ω_a.
3. Form the ratio

$$\lambda(x_1, \ldots, x_n) = \frac{\text{maximum likelihood for } \theta \text{ in } \Omega_0}{\text{maximum likelihood for } \theta \text{ in } \Omega_a}$$

The ratio $\lambda(x_1, \ldots, x_n)$ is called the *likelihood ratio statistic value*. The test procedure consists of rejecting H_0 when this ratio is small. That is, a constant $k \cdot$ is chosen and H_0 is rejected if $\lambda(x_1, \ldots, x_n) \leq k$. Thus H_0 is rejected when the denominator of λ greatly exceeds the numerator, indicating that the data is much more consistent with H_a than with H_0.

The constant k is selected to yield the desired type I error probability. Often the inequality $\lambda \leq k$ can be manipulated to yield a simpler equivalent condition. For example, for testing $H_0: \mu \leq \mu_0$ versus $H_a: \mu > \mu_0$ in the case of normality, $\lambda \leq k$ is equivalent to $t \geq c$. Thus with $c = t_{\alpha, n-1}$, the likelihood ratio test is the one-sample t test.

The likelihood ratio principle can also be applied when the X_i's have different distributions and even when they are dependent, though the likelihood function can be complicated in such cases. Many of the test procedures to be presented in subsequent chapters are obtained from the likelihood ratio principle. These tests often turn out to minimize β among all tests that have the desired α, so are truly best tests. For more details and some worked out examples, refer to the book by DeGroot or the one by Hogg and Craig listed in the Chapter 6 bibliography.

A practical limitation on the use of the likelihood ratio principle is that in order to construct the likelihood ratio test statistic, the form of the probability distribution from which the sample comes must be specified. To derive the t test from the likelihood ratio principle, the investigator must assume a normal p.d.f. If an investigator is willing to assume that the distribution is symmetric but does not want to be specific about its exact form (such as normal, uniform, or Cauchy), then the principle fails because there is no way to write a joint p.d.f. simultaneously valid for all symmetric distributions. In Chapter 15 we will present several **distribution-free** test procedures, so called because the probability of a type I error is controlled simultaneously for many different underlying distributions. These procedures are useful when the investigator has limited knowledge of the underlying distribution. We shall also say more about issues 3 and 4 listed at the outset of this section.

Exercises / Section 8.5 (53–54)

53. Reconsider the paint drying problem discussed in Example 8.2. The hypotheses were $H_0: \mu = 75$ versus $H_a: \mu < 75$, with σ assumed to have value 9.0. Consider the alternative value $\mu = 74$, which in the context of the problem would presumably not be a practically significant departure from H_0.

a. For a level .01 test, compute β at this alternative for sample sizes $n = 100$, 900, and 2500.

b. If the observed value of $\overline{X}$ is $\overline{x} = 74$, what can you say about the resulting P-value when $n = 2500$? Is the data statistically significant at any of the standard values of α?

c. Would you really want to use a sample size of 2500 along with a level .01 test (leaving aside the cost of such an experiment)? Explain.

54. Consider the large-sample level .01 test in Section 8.3 for testing $H_0: p = .2$ against $H_a: p > .2$.

a. For the alternative value $p = .21$, compute $\beta(.21)$ for sample sizes $n = 100$, 2500, 10,000, 40,000, and 90,000.

b. For $\hat{p} = x/n = .21$, compute the P-value when $n = 100$, 2500, 10,000, and 40,000.

c. In most situations, would it be reasonable to use a level .01 test in conjunction with a sample size of 40,000? Why or why not?

Supplementary Exercises / Chapter 8 (55–71)

55. A small dog food company being considered for acquisition by a large conglomerate asserts that at least half the grocery stores in its market area stock its brand. To see whether this is the case, a random sample of 200 stores is selected, of which 78 are found to carry the brand.

a. Does this data contradict the dog food company's claim? Test using $\alpha = .05$.

b. What is the P-value of the data?

56. A sample of 50 lenses used in eyeglasses yields a sample mean thickness of 3.05 mm and a sample standard deviation of .34 mm. The desired true average thickness of such lenses is 3.20 mm. Does the data strongly suggest that the true average thickness of such lenses is something other than what is desired? Test using $\alpha = .05$.

57. In Exercise 56, suppose that the experimenter had believed before collecting the data that the value of σ was approximately .30. If the experimenter wished the probability of a type II error to be .05 when $\mu = 3.00$, was a sample size 50 unnecessarily large?

58. In an experiment to test the effects of hormones on growth of beef cattle, 200 mg of progesterone and 20 mg of estradiol benzoate are implanted in the outer ear of 16 randomly selected steers, each weighing approximately 500 lb. The sample average weight gain per day for a certain number of days is found to be 2.79 lb with a sample standard deviation of .41 lb per day. Does this data strongly suggest that the true average daily weight gain for steers treated with the hormone implant exceeds 2.50? Making any necessary assumptions, test the relevant hypotheses using $\alpha = .001$.

59. The incidence of a certain type of chromosome defect in the U.S. adult male population is believed to be 1 in 80. A random sample of 600 individuals in U.S. penal institutions reveals 12 who have such defects. Can it be concluded that the incidence rate of this defect among prisoners differs from the presumed rate for the entire adult male population?

  a. State and test the relevant hypotheses using $\alpha = .05$. What type of error might you have made in reaching a conclusion?

  b. What P-value is associated with this test? Based on this P-value, could H_0 be rejected at significance level .20?

60. In an investigation of the toxin produced by a certain poisonous snake, a researcher prepared 26 different vials, each containing 1 gm of the toxin, and then determined the amount of antitoxin needed to neutralize the toxin. The sample average amount of antitoxin necessary was found to be 1.89 mg with a sample standard deviation of .42. Previous research had indicated that the true average neutralizing amount was 1.75 mg per gram of toxin. Does the new data contradict the value suggested by prior research?

  a. Test the relevant hypotheses at level .05.

  b. What can you say about the P-value of the data?

61. The sample unrestrained compressive strength for 45 specimens of a particular type of brick was computed to be 3107 psi with a sample standard deviation of 188. The distribution of unrestrained compressive strength may be somewhat skewed. Does the data strongly indicate that the true average unrestrained compressive strength is less than the design value of 3200? Test using $\alpha = .001$.

62. To test the ability of auto mechanics to identify simple engine problems, an automobile with a single such problem was taken in turn to 72 different car repair facilities. Only 42 of the 72 mechanics who worked on the car correctly identified the problem. Does this strongly indicate that the true proportion of mechanics who could identify this problem is less than .75? Compute the P-value and reach a conclusion accordingly.

63. When $X_1, X_2, \ldots, X_n$ are independent Poisson variables, each with parameter λ, and n is large, the sample mean $\overline{X}$ has approximately a normal distribution with $\mu = E(\overline{X}) = \lambda$ and $\sigma^2 = V(\overline{X}) = \lambda/n$. This implies that

$$Z = \frac{\overline{X} - \lambda}{\sqrt{\lambda/n}}$$

has approximately a standard normal distribution. For testing $H_0 : \lambda = \lambda_0$, we can replace λ by λ_0 in the equation for Z to obtain a test statistic. This statistic is actually preferred to the large-sample statistic with denominator $S/\sqrt{n}$ (when the X_i's are Poisson) because it is tailored explicitly to the Poisson assumption. If the number of requests for consulting received by a certain statistician during a five-day work week has a Poisson distribution and the total number of consulting requests during a 36-week period is 160, does this suggest that the true average number of weekly requests exceeds 4.0? Test using $\alpha = .02$.

64. A hot-tub manufacturer advertises that with its heating equipment, a temperature of 100° F can be achieved in at most 15 min. A random sample of 32 tubs is selected and the time necessary to achieve a 100° F temperature is determined for each tub. The sample average time and sample standard deviation are 17.5 min and 2.2 min, respectively. Does this data cast doubt on the company's claim? Compute the P-value, and use it to reach a conclusion at level .05.

65. The drug cibenzoline is currently being investigated for possible use in controlling cardiac arrhythmia. The paper "Quantification of Cibenzoline in Human Plasma by Gas Chromatography-Negative Ion Chemical-Ionization Mass

Spectrometry" (*J. Chromatography*, 1984, 403–409) describes a new method of determining the concentration of cibenzoline in a solution. After 5 ng of cibenzoline was added to a solution, the concentration was measured by the new method. This process was repeated three times, resulting in $n = 3$ concentration readings. The sample mean and standard deviation were reported to be 4.59 ng and .08 ng, respectively. Does this data suggest that the new method produces a mean concentration reading that is too small (less than 5 ng)? Use a .05 significance level and test the appropriate hypotheses.

66. Chapter 7 presented a confidence interval for the variance σ^2 of a normal population distribution. The key result there was that the random variable $\chi^2 = (n-1)S^2/\sigma^2$ has a chi-squared distribution with $n-1$ d.f. Consider the null hypothesis $H_0 : \sigma^2 = \sigma_0^2$ (equivalently, $\sigma = \sigma_0$). Then when H_0 is true, the test statistic $\chi^2 = (n-1)S^2/\sigma_0^2$ has a chi-squared distribution with $n-1$ d.f. If the relevant alternative is $H_a : \sigma^2 > \sigma_0^2$, rejecting H_0 if $(n-1)s^2/\sigma_0^2 \geq \chi^2_{\alpha, n-1}$ gives a test with significance level α. To ensure reasonably uniform characteristics for a particular application, it is desired that the true standard deviation of the softening point of a certain type of petroleum pitch be at most .50° C. The softening points of 10 different specimens were determined, yielding a sample standard deviation of .58° C. Does this strongly contradict the uniformity specification? Test the appropriate hypotheses using $\alpha = .01$.

67. Referring back to Exercise 66, suppose that an investigator wishes to test $H_0 : \sigma^2 = .04$ versus $H_a : \sigma^2 < .04$ based on a sample of 21 observations. The computed value of $20s^2/.04$ is 8.58. Place bounds on the *P*-value, and then reach a conclusion at level .01.

68. When the population distribution is normal and n is large, the sample standard deviation S has approximately a normal distribution with $E(S) \approx \sigma$, $V(S) \approx \sigma^2/2n$. We already know that in this case, for any n, $\overline{X}$ is normal with $E(\overline{X}) = \mu$ and $V(\overline{X}) = \sigma^2/n$.
   a. Assuming that the underlying distribution is normal, what is an approximately unbiased estimator of the ninety-ninth percentile $\theta = \mu + 2.33\sigma$?

b. When the X_i's are normal, it can be shown that $\overline{X}$ and S are independent r.v.'s (one measures location while the other measures spread). Use this to compute $V(\hat{\theta})$ and $\sigma_{\hat{\theta}}$ for the estimator $\hat{\theta}$ of (a). What is the estimated standard error $\hat{\sigma}_{\hat{\theta}}$?
   c. Write a test statistic for testing $H_0 : \theta = \theta_0$ which has approximately a standard normal distribution when H_0 is true. If soil pH is normally distributed in a certain region and 64 soil samples yield $\overline{x} = 6.33$, $s = .16$, does this provide strong evidence for concluding that at least 95% of all possible samples would have a pH of less than 6.75? Test using $\alpha = .01$.

69. Let $X_1, X_2, \ldots, X_n$ be a random sample from an exponential distribution with parameter λ. Then it can be shown that $2\lambda\Sigma X_i$ has a chi-squared distribution with $\nu = 2n$ (by first showing that $2\lambda X_i$ has a chi-squared distribution with $\nu = 2$).
   a. Use this fact to obtain a test statistic and rejection region which together specify a level α test for $H_0 : \mu = \mu_0$ versus each of the three commonly encountered alternatives. *Hint:* $E(X_i) = \mu = 1/\lambda$, so $\mu = \mu_0$ is equivalent to $\lambda = 1/\mu_0$.
   b. Suppose that 10 identical components, each having exponentially distributed time until failure, are tested. The resulting failure times are

   95, 16, 11, 3, 42, 71, 225, 64, 87, 123

   Use the test procedure of (a) to decide if the data strongly suggests that the true average lifetime is less than the previously claimed value of 75.

70. Suppose that the population distribution is normal with known σ. Let γ be such that $0 < \gamma < \alpha$. For testing $H_0 : \mu = \mu_0$ versus $H_a : \mu \neq \mu_0$, consider the test that rejects H_0 if either $z \geq z_\gamma$, or $z \leq -z_{\alpha - \gamma}$, where $Z = (\overline{X} - \mu_0)/(\sigma/\sqrt{n})$.
   a. Show that $P(\text{type I error}) = \alpha$.
   b. Derive an expression for $\beta(\mu')$. *Hint:* Express the test in the form "reject H_0 if either $\overline{X} \geq c_1$ or $\leq c_2$."
   c. Let $\Delta > 0$. For what values of γ (relative to α) will $\beta(\mu_0 + \Delta) < \beta(\mu_0 - \Delta)$?

71. After a period of apprenticeship, an organization gives an exam which must be passed to be eligible for membership. Let $p = P(\text{randomly chosen ap-}$

prentice passes). The organization wishes an exam which most but not all should be able to pass, so decides that $p = .90$ is desirable. For a particular exam the relevant hypotheses are $H_0: p = .90$ versus $H_a: p \neq .90$. Suppose that 10 people take the exam, and let $X =$ the number who pass.

a. Does the lower-tailed region $\{0, 1, \ldots, 5\}$ specify a level .01 test?
b. Show that even though H_a is two-sided, no two-tailed test is a level .01 test.
c. Sketch a graph of $\beta(p')$ as a function of p' for this test. Is this desirable?

Bibliography

See the bibliography at the end of Chapter 7.

Inferences Based on Two Samples

Introduction

Chapters 7 and 8 presented confidence intervals and hypothesis-testing procedures for a single population parameter (μ, p, and σ^2). Here we extend these methods to problem situations involving the means, proportions, and variances of two different population distributions.

9.1   *z* Tests and Confidence Intervals for a Difference between Two Population Means

The inferences discussed in this section concern a difference $\mu_1 - \mu_2$ between the means of two different population distributions. An investigator might, for example, wish to test hypotheses about the difference between true average breaking strengths of two different types of corrugated fiberboard. One such hypothesis would state that $\mu_1 - \mu_2 = 0$; that is, that $\mu_1 = \mu_2$. Alternatively, it may be appropriate to estimate $\mu_1 - \mu_2$ by computing a 95% confidence interval. Such inferences are based on a sample of strength observations for each type of fiberboard.

Basic assumptions

> **1.** $X_1, X_2, \ldots, X_m$ is a random sample from a population with mean μ_1 and variance σ_1^2.
> **2.** $Y_1, Y_2, \ldots, Y_n$ is a random sample from a population with mean μ_2 and variance σ_2^2.
> **3.** The X and Y samples are independent of one another.

The natural estimator of $\mu_1 - \mu_2$ is $\overline{X} - \overline{Y}$, the difference between the corresponding sample means. The test statistic results from standardizing this estimator, so we need expressions for the expected value and standard deviation of $\overline{X} - \overline{Y}$.

Proposition

> The expected value of $\overline{X} - \overline{Y}$ is $\mu_1 - \mu_2$, so $\overline{X} - \overline{Y}$ is an unbiased estimator of $\mu_1 - \mu_2$. The standard deviation of $\overline{X} - \overline{Y}$ is
>
> $$\sigma_{\overline{X} - \overline{Y}} = \sqrt{\frac{\sigma_1^2}{m} + \frac{\sigma_2^2}{n}}$$

Proof.   Both these results depend on the rules of expected value and variance presented in Chapter 5. Since the expected value of a difference is the difference of expected values,

$$E(\overline{X} - \overline{Y}) = E(\overline{X}) - E(\overline{Y}) = \mu_1 - \mu_2$$

Because the X and Y samples are independent, $\overline{X}$ and $\overline{Y}$ are independent quantities; a rule from Section 5.3 then states that the variance of the difference is the *sum* of $V(\overline{X})$ and $V(\overline{Y})$:

$$V(\overline{X} - \overline{Y}) = V(\overline{X}) + V(\overline{Y}) = \frac{\sigma_1^2}{m} + \frac{\sigma_2^2}{n}$$

The standard deviation of $\overline{X} - \overline{Y}$ is the square root of this expression.   ∎

If we think of $\mu_1 - \mu_2$ as a parameter θ, then its estimator is $\hat{\theta} = \overline{X} - \overline{Y}$ with standard deviation $\sigma_{\hat{\theta}}$ given by the above proposition. When σ_1^2 and σ_2^2 both have known values, the test statistic will have the form $(\hat{\theta} - \text{null value})/\sigma_{\hat{\theta}}$; this form of a test statistic was used in several one-sample problems in the previous chapter. When σ_1^2 and σ_2^2 are unknown, the sample variances must be used to estimate $\sigma_{\hat{\theta}}$.

Test Procedures for Normal Populations with Known Variances

In Chapters 7 and 8, the first confidence interval and test procedure for a population mean μ were based on the assumption that the population distribution was normal with the value of the population variance σ^2 known to the investigator. Similarly, we first assume here that *both* population distributions are normal and that the values of *both* σ_1^2 and σ_2^2 are known. Situations in which one or both of these assumptions can be dispensed with will be presented shortly.

Because the population distributions are normal, both $\overline{X}$ and $\overline{Y}$ have normal distributions. This implies that $\overline{X} - \overline{Y}$ is normally distributed, with expected value $\mu_1 - \mu_2$ and standard deviation $\sigma_{\overline{X} - \overline{Y}}$ given in the foregoing proposition. Standardizing $\overline{X} - \overline{Y}$ gives the standard normal variable

$$Z = \frac{\overline{X} - \overline{Y} - (\mu_1 - \mu_2)}{\sqrt{\dfrac{\sigma_1^2}{m} + \dfrac{\sigma_2^2}{n}}} \qquad\qquad (9.1)$$

In a hypothesis-testing problem, the null hypothesis will state that $\mu_1 - \mu_2$ has a specified value. Denoting this null value by Δ_0, the null hypothesis becomes $H_0 : \mu_1 - \mu_2 = \Delta_0$. Often $\Delta_0 = 0$, in which case H_0 says that $\mu_1 = \mu_2$. A test statistic results from replacing $\mu_1 - \mu_2$ in (9.1) by the null value Δ_0. Because the test statistic Z is obtained by standardizing $\overline{X} - \overline{Y}$ under the assumption that H_0 is true, it has a standard normal distribution in this case. Consider the alternative hypothesis $H_a : \mu_1 - \mu_2 > \Delta_0$. A value $\overline{x} - \overline{y}$ that considerably exceeds Δ_0 (the expected value of $\overline{X} - \overline{Y}$ when H_0 is true) provides evidence against H_0 and for H_a. Such a value of $\overline{x} - \overline{y}$ corresponds to a positive and large value of z. Thus H_0 should be rejected in favor of H_a if z is greater than or equal to an appropriately chosen critical value. Because the test statistic Z has a standard normal distribution when H_0 is true, the upper-tailed rejection region $z \geq z_\alpha$ gives a test with significance level (type I error probability) α. Rejection regions for $H_a : \mu_1 - \mu_2 < \Delta_0$ and $H_a : \mu_1 - \mu_2 \neq \Delta_0$ that yield tests with desired significance level α are lower-tailed and two-tailed, respectively.

Null hypothesis: $H_0 : \mu_1 - \mu_2 = \Delta_0$

Test statistic value: $z = \dfrac{\overline{x} - \overline{y} - \Delta_0}{\sqrt{\dfrac{\sigma_1^2}{m} + \dfrac{\sigma_2^2}{n}}}$

Alternative hypothesis	Rejection region for level α test
$H_a : \mu_1 - \mu_2 > \Delta_0$	$z \geq z_\alpha$
$H_a : \mu_1 - \mu_2 < \Delta_0$	$z \leq -z_\alpha$
$H_a : \mu_1 - \mu_2 \neq \Delta_0$	either $z \geq z_{\alpha/2}$ or $z \leq -z_{\alpha/2}$

Because these are z tests, a P-value is computed as it was for the z tests in Chapter 8 (for example, P-value $= 1 - \Phi(z)$ for an upper-tailed test).

Example 9.1   Analysis of a random sample consisting of $m = 20$ specimens of cold-rolled steel to determine yield strengths resulted in a sample average strength of $\overline{x} = 29.8$ ksi. A second random sample of $n = 25$ two-side galvanized steel specimens gave a sample average strength of $\overline{y} = 34.7$ ksi. Assuming that the two yield strength distributions are normal with $\sigma_1 = 4.0$ and $\sigma_2 = 5.0$ (suggested by a graph in the paper "Zinc-Coated Sheet Steel: An Overview," *Automotive Engr.,* Dec. 1984, pp. 39–43), does the data indicate that the corresponding true average yield strengths μ_1 and μ_2 are different? Let's carry out a test at significance level $\alpha = .01$.

1. The parameter of interest is $\mu_1 - \mu_2$, the difference between the true average strengths for the two types of steel.
2. The null hypothesis is $H_0 : \mu_1 - \mu_2 = 0$.
3. The alternative hypothesis is $H_a : \mu_1 - \mu_2 \neq 0$; if H_a is true, then μ_1 and μ_2 are different.

4. With $\Delta_0 = 0$, the test statistic value is

$$z = \frac{\bar{x} - \bar{y}}{\sqrt{\dfrac{\sigma_1^2}{m} + \dfrac{\sigma_2^2}{n}}}$$

5. The inequality in H_a implies that the test is two-tailed. For $\alpha = .01$, $\alpha/2 = .005$ and $z_{\alpha/2} = z_{.005} = 2.58$. H_0 will be rejected if $z \geq 2.58$ or if $z \leq -2.58$.

6. Substituting $m = 20$, $\bar{x} = 29.8$, $\sigma_1^2 = 16.0$, $n = 25$, $\bar{y} = 34.7$, and $\sigma_2^2 = 25.0$ into the formula for z yields

$$z = \frac{29.8 - 34.7}{\sqrt{\dfrac{16.0}{20} + \dfrac{25.0}{25}}} = \frac{-4.90}{1.34} = -3.66$$

That is, the observed value of $\bar{x} - \bar{y}$ is more than three standard deviations below what would be expected were H_0 true.

7. Since $-3.66 < -2.58$, the computed z does fall in the lower tail of the rejection region. H_0 is therefore rejected at level .01 in favor of the conclusion that $\mu_1 \neq \mu_2$. The sample data strongly suggests that the true average yield strength for cold-rolled steel differs from that for galvanized steel. ∎

Example 9.2   A letter in the May 19, 1978, *Journal of the American Medical Association* reported that of 215 male physicians who were Harvard graduates and died between November 1974 and October 1977, the 125 in full-time practice lived an average of 48.9 years beyond graduation, while the 90 with academic affiliations lived an average of 43.2 years beyond graduation. Does the data suggest that the mean lifetime after graduation for doctors in full-time practice exceeds the mean lifetime for those who obtain an academic affiliation (if so, those medical students who say that they are "dying to obtain an academic affiliation" may be closer to the truth than they realize; put another way, is "publish or perish" really "publish and perish")?

Let μ_1 denote the true average number of years lived beyond graduation for physicians in full-time practice, and let μ_2 denote the same quantity for physicians with academic affiliations. Assuming the 125 and 90 physicians to be random samples from populations 1 and 2, respectively (which may not be reasonable if there is reason to believe that Harvard graduates have special characteristics that differentiate them from all other physicians—in this case inferences would be restricted just to the "Harvard populations"), we have $\bar{x} = 48.9$ and $\bar{y} = 43.2$. In order to apply the above test procedure, we must know σ_1^2 and σ_2^2. The letter from which the data was taken gave no information about variances, so for illustration assume that $\sigma_1 = 14.6$ and $\sigma_2 = 14.4$. The hypotheses are $H_0 : \mu_1 - \mu_2 = 0$ versus $H_a : \mu_1 - \mu_2 > 0$, so Δ_0 is zero. The computed value of the test statistic is

$$z = \frac{48.9 - 43.2}{\sqrt{\frac{(14.6)^2}{125} + \frac{(14.4)^2}{90}}} = \frac{5.70}{\sqrt{1.70 + 2.30}} = 2.85$$

The P-value for an upper-tailed test is $1 - \Phi(2.85) = .0022$. At significance level $.01$, H_0 is rejected (because $\alpha > P$-value) in favor of the conclusion that $\mu_1 - \mu_2 > 0$ ($\mu_1 > \mu_2$). This is consistent with the information reported in the letter.   ■

Using a Comparison to Identify Causality

Investigators are often interested in comparing either the effects of two different treatments on a response or the response after treatment with the response after no treatment (treatment versus control). If the individuals or objects to be used in the comparison are not assigned by the investigators to the two different conditions, the study is said to be **observational.** The data in Example 9.2 is the result of a **retrospective** observational study; the investigator did not start out by selecting a sample of doctors and assigning some to the "academic affiliation" treatment and the others to the "full-time practice" treatment, but instead identified members of the two groups by looking backward in time (through obituaries!) to past records.

The difficulty with drawing conclusions based on an observational study is that while statistical analysis may indicate a significant difference in response between the two groups, the difference may be due to some underlying factors which had not been controlled rather than to any difference in treatments. Was the observed difference in Example 9.2 really due to a difference in the type of medical practice after graduation, or is there some other underlying factor (for example, age at graduation) that might also furnish a plausible explanation for the difference? Observational studies have been used to argue for a causal link between smoking and lung cancer. There are many studies that show that the incidence of lung cancer is significantly higher among smokers than among nonsmokers. However, individuals had decided whether or not to become smokers long before investigators arrived on the scene, and factors in making this decision may have played a causal role in the contraction of lung cancer.

A **randomized controlled experiment** results when investigators assign subjects to the two treatments in a random fashion. When statistical significance is observed in such an experiment, the investigator and other interested parties will have more confidence in the conclusion that the difference in response has been caused by a difference in treatments. A very famous example of this type of experiment and conclusion is the Salk polio vaccine experiment described in Section 9.4. These issues are discussed at greater length in the (nonmathematical) books by Moore and by Freedman, Purves, and Pisani listed in the Chapter 1 references.

β and the Choice of Sample Size

The probability of a type II error is easily calculated when both population distributions are normal with known values of σ_1 and σ_2. Consider the case in which the alternative hypothesis is $H_a : \mu_1 - \mu_2 > \Delta_0$, and let Δ' denote a value of $\mu_1 - \mu_2$ that exceeds Δ_0 (a value for which H_0 is false). The upper-tailed rejection region $z \geq z_\alpha$ can be re-expressed in the form $\bar{x} - \bar{y} \geq \Delta_0 + z_\alpha \sigma_{\bar{X} - \bar{Y}}$. Thus the probability of a type II error when $\mu_1 - \mu_2 = \Delta'$ is

$$\beta(\Delta') = P(\text{not rejecting } H_0 \text{ when } \mu_1 - \mu_2 = \Delta')$$
$$= P(\bar{X} - \bar{Y} < \Delta_0 + z_\alpha \sigma_{\bar{X} - \bar{Y}} \text{ when } \mu_1 - \mu_2 = \Delta')$$

When $\mu_1 - \mu_2 = \Delta'$, $\bar{X} - \bar{Y}$ is normally distributed with mean value Δ' and standard deviation $\sigma_{\bar{X} - \bar{Y}}$ (the same standard deviation as when H_0 is true); using these values to standardize the inequality in parentheses gives β.

Alternative hypothesis	$\beta(\Delta') = P(\text{type II error when } \mu_1 - \mu_2 = \Delta')$
$H_a : \mu_1 - \mu_2 > \Delta_0$	$\Phi\left(z_\alpha - \dfrac{\Delta' - \Delta_0}{\sigma} \right)$
$H_a : \mu_1 - \mu_2 < \Delta_0$	$1 - \Phi\left(-z_\alpha - \dfrac{\Delta' - \Delta_0}{\sigma} \right)$
$H_a : \mu_1 - \mu_2 \neq \Delta_0$	$\Phi\left(z_{\alpha/2} - \dfrac{\Delta' - \Delta_0}{\sigma} \right) - \Phi\left(-z_{\alpha/2} - \dfrac{\Delta' - \Delta_0}{\sigma} \right)$

where $\sigma = \sigma_{\bar{X} - \bar{Y}} = \sqrt{\dfrac{\sigma_1^2}{m} + \dfrac{\sigma_2^2}{n}}$ and $\Phi(\cdot)$ is the standard normal c.d.f.

Example 9.3
(Example 9.1 continued)

Suppose that when μ_1 and μ_2 (the true average yield strengths for the two types of steel) differ by as much as 5, the probability of detecting such a departure from H_0 should be .90. Does a level .01 test with sample sizes $m = 20$ and $n = 25$ satisfy this condition? The value of σ for these sample sizes (the denominator of z) was previously calculated as 1.34. The probability of a type II error for the two-tailed level .01 test when $\mu_1 - \mu_2 = \Delta' = 5$ is

$$\beta(5) = \Phi\left(2.58 - \frac{5 - 0}{1.34} \right) - \Phi\left(-2.58 - \frac{5 - 0}{1.34} \right)$$

$$= \Phi(-1.15) - \Phi(-6.31) = .1251$$

It is easy to verify that $\beta(-5) = .1251$ also (because the rejection region is symmetric). Thus the probability of detecting such a departure is $1 - \beta(5) = .8749$. As this is somewhat less than .9, slightly larger sample sizes should be used. ■

As in Chapter 8, sample sizes m and n can be determined that will satisfy both $P(\text{type I error}) = $ a specified α and $P(\text{type II error when } \mu_1 - \mu_2 = \Delta') = $ a specified β. For an upper-tailed test, equating the above expression for $\beta(\Delta')$ to the specified value of β gives

$$\frac{\sigma_1^2}{m} + \frac{\sigma_2^2}{n} = \frac{(\Delta' - \Delta_0)^2}{(z_\alpha + z_\beta)^2}$$

When the two sample sizes are equal, this equation yields

$$m = n = \frac{(\sigma_1^2 + \sigma_2^2)(z_\alpha + z_\beta)^2}{(\Delta' - \Delta_0)^2}$$

These expressions are also correct for a lower-tailed test, whereas α is replaced by $\alpha/2$ for a two-tailed test.

Large-Sample Tests

The assumptions of normal population distributions and known values of σ_1 and σ_2 are unnecessary when both sample sizes are large. In this case, the Central Limit Theorem guarantees that $\overline{X} - \overline{Y}$ has approximately a normal distribution regardless of the underlying population distributions. Furthermore, the two sample variances s_1^2 and s_2^2 will satisfy $s_1^2 \approx \sigma_1^2$ and $s_2^2 \approx \sigma_2^2$ for almost all samples likely to be selected, so the standardized variable

$$Z = \frac{\overline{X} - \overline{Y} - (\mu_1 - \mu_2)}{\sqrt{\dfrac{S_1^2}{m} + \dfrac{S_2^2}{n}}}$$

has approximately a standard normal distribution. A large-sample test statistic results from replacing $\mu_1 - \mu_2$ by Δ_0, the expected value of $\overline{X} - \overline{Y}$ when H_0 is true. This statistic Z then has approximately a standard normal distribution when H_0 is true, so level α tests are obtained by using z critical values exactly as before.

Use of the test statistic value

$$z = \frac{\overline{x} - \overline{y} - \Delta_0}{\sqrt{\dfrac{s_1^2}{m} + \dfrac{s_2^2}{n}}}$$

along with the previously stated upper-, lower-, and two-tailed rejection regions based on z critical values gives large-sample tests whose significance level is approximately α. These tests are usually appropriate if both $m > 30$ and $n > 30$.

Example 9.4   Does proximity of place of residence to heavily traveled roads result in higher blood lead levels? The paper "Blood Lead of Persons Living Near Freeways" (*Arch. Environmental Health,* 1967, pp. 695–702) reported that in a sample of 30 women who did not live near a freeway, the sample average blood lead level was 9.9 with a sample standard deviation of 4.9, while a second sample of 35 females who did live near a freeway resulted in a sample average and sample standard deviation of 16.7 and 7.0, respectively (a histogram of the first sample suggests that the underlying distribution is reasonably close to normal, so we regard the sample size of 30 as sufficiently large to allow use of the large-sample test).

With μ_1 (μ_2) denoting the true average blood lead level for females who do not (do) live close to a freeway, the hypotheses to be tested are $H_0 : \mu_1 - \mu_2 = 0$ versus $H_a : \mu_1 - \mu_2 < 0$. At level .01, the lower-tailed test rejects H_0 if $z \leq -2.33$. The computed value of the test statistic is

$$z = \frac{9.9 - 16.7}{\sqrt{\dfrac{(4.9)^2}{30} + \dfrac{(7.0)^2}{35}}} = \frac{-6.80}{1.483} = -4.58$$

Since -4.58 is ≤ -2.33, H_0 is rejected at level .01. The data suggests that for females there is an *association* between proximity to freeways and blood lead level—but this is far from saying that proximity *causes* increased blood lead. A convincing argument for causality would ultimately rest on much more data from carefully designed experiments.   ■

Example 9.5   In selecting a sulphur concrete for roadway construction in regions that experience heavy frost, it is important that the chosen concrete have a low value of thermal conductivity in order to minimize subsequent damage due to changing temperatures. Suppose that two types of concrete, a graded aggregate and a no-fines aggregate, are being considered for a certain road. Because using the no-fines aggregate involves a higher initial cost, it is decided that the graded aggregate will be used unless experimental data shows conclusively that its thermal conductivity is more than .1 w/mK higher than that of the no-fines concrete. Based on the accompanying summary of experimental data (Table 9.1), is use of the no-fines concrete justified?

Table 9.1

Type	Sample size	Sample average conductivity	Sample SD
Graded	35	.497	.187
No-fines	35	.359	.158

Let μ_1 and μ_2 denote the true average thermal conductivity for the graded aggregate and no-fines aggregate concrete, respectively. The two hypotheses are $H_0 : \mu_1 - \mu_2 = .1$ versus $H_a : \mu_1 - \mu_2 > .1$; only if H_0 can be conclusively rejected in favor of H_a will the no-fines concrete be used. With $\Delta_0 = .1$,

$$z = \frac{(.497 - .359) - .1}{\sqrt{\dfrac{(.187)^2}{35} + \dfrac{(.158)^2}{35}}} = \frac{.038}{.041} = .93$$

The P-value for an upper-tailed z test is $1 - \Phi(z) = 1 - \Phi(.93) = .1762$. Since $.1762 > .10$, H_0 can't be rejected at level .10 or at any smaller significance level. The use of the no-fines concrete does not appear to be justified. ∎

Confidence Intervals for $\mu_1 - \mu_2$

When both population distributions are normal, standardizing $\overline{X} - \overline{Y}$ gives a random variable Z with a standard normal distribution. Since the area under the z curve between $-z_{\alpha/2}$ and $z_{\alpha/2}$ is $1 - \alpha$, it follows that

$$P\left(-z_{\alpha/2} < \frac{\overline{X} - \overline{Y} - (\mu_1 - \mu_2)}{\sqrt{\dfrac{\sigma_1^2}{m} + \dfrac{\sigma_2^2}{n}}} < z_{\alpha/2}\right) = 1 - \alpha$$

Manipulation of the inequalities inside the parentheses to isolate $\mu_1 - \mu_2$ yields the equivalent probability statement

$$P\left(\overline{X} - \overline{Y} - z_{\alpha/2}\sqrt{\frac{\sigma_1^2}{m} + \frac{\sigma_2^2}{n}} < \mu_1 - \mu_2 < \overline{X} - \overline{Y} + z_{\alpha/2}\sqrt{\frac{\sigma_1^2}{m} + \frac{\sigma_2^2}{n}}\right)$$

$$= 1 - \alpha$$

This implies that a $100(1 - \alpha)\%$ confidence interval for $\mu_1 - \mu_2$ has lower limit $\overline{x} - \overline{y} - z_{\alpha/2} \cdot \sigma_{\overline{X} - \overline{Y}}$ and upper limit $\overline{x} - \overline{y} + z_{\alpha/2} \cdot \sigma_{\overline{X} - \overline{Y}}$, where $\sigma_{\overline{X} - \overline{Y}}$ is the square-root expression. This interval is a special case of the general formula $\hat{\theta} \pm z_{\alpha/2} \cdot \sigma_{\hat{\theta}}$.

If both m and n are large, the Central Limit Theorem implies that the above interval is valid even without the assumption of normal populations; in this case, the confidence level is *approximately* $100(1 - \alpha)\%$. Furthermore, use of the sample variances S_1^2 and S_2^2 in the standardized variable Z yields a valid interval in which s_1^2 and s_2^2 replace σ_1^2 and σ_2^2.

Provided that m and n are both large, a confidence interval for $\mu_1 - \mu_2$ with a confidence level of approximately $100(1 - \alpha)\%$ is

$$\overline{x} - \overline{y} \pm z_{\alpha/2}\sqrt{\frac{s_1^2}{m} + \frac{s_2^2}{n}}$$

where $-$ gives the lower limit and $+$ the upper limit of the interval.

The standard rule of thumb for characterizing sample sizes as large is again $m > 30$ and $n > 30$.

Example 9.6   Tensile strength tests were carried out on two different grades of wire rod ("Fluidized Bed Patenting of Wire Rods," *Wire J.*, June 1977, pp. 56–61), resulting in the data in Table 9.2.

Table 9.2

Grade	Sample size	Sample mean (kg/mm²)	Sample SD
AISI 1064	$m = 129$	$\bar{x} = 107.6$	$s_1 = 1.3$
AISI 1078	$n = 129$	$\bar{y} = 123.6$	$s_2 = 2.0$

Let μ_1 and μ_2 denote the true average tensile strengths for grades 1064 and 1078 respectively. Then a 95% confidence interval for $\mu_1 - \mu_2$ is

$$107.6 - 123.6 \pm 1.96 \sqrt{\frac{(1.3)^2}{129} + \frac{(2.0)^2}{129}}$$

$$= -16.0 \pm .41 = (-16.41, -15.59)$$

That is, we can be highly confident that μ_2 exceeds μ_1 by between 15.59 and 16.41. This confidence interval is relatively narrow, suggesting that $\mu_1 - \mu_2$ has been precisely estimated, because both sample sizes are quite large and both sample standard deviations relatively small. ■

If the variances σ_1^2 and σ_2^2 are at least approximately known and the investigator uses equal sample sizes, then the sample size n for each sample that yields a $100(1 - \alpha)\%$ interval of length L is

$$n = \frac{4z_{\alpha/2}^2(\sigma_1^2 + \sigma_2^2)}{L^2}$$

which will generally have to be rounded up to an integer.

Exercises / Section 9.1 (1–16)

1. An article in the November 1983 *Consumer Reports* compared various types of batteries. The average lifetimes of Duracell Alkaline AA batteries and Eveready Energizer Alkaline AA batteries were given as 4.1 h and 4.5 h, respectively. Suppose that these are the population average lifetimes.
   a. Let $\bar{X}$ be the sample average lifetime of 100 Duracell batteries and $\bar{Y}$ be the sample average lifetime of 100 Eveready batteries. What is the mean value of $\bar{X} - \bar{Y}$ (that is, where is the distribution of $\bar{X} - \bar{Y}$ centered)? How does your answer depend on the specified sample sizes?
   b. Suppose that population standard deviations of lifetime are 1.8 h for Duracell batteries and 2.0 h for Eveready batteries. With the sample sizes given in (a), what is the variance of the statistic $\bar{X} - \bar{Y}$, and what is its standard deviation?

   c. For the sample sizes given in (a), draw a picture of the approximate distribution curve of $\bar{X} - \bar{Y}$ (include a measurement scale on the horizontal axis). Would the shape of the curve necessarily be the same for sample sizes of 10 batteries of each type? Explain.

2. Let μ_1 and μ_2 denote true average tread lives for two different brands of size FR78-15 radial tires. Test $H_0 : \mu_1 - \mu_2 = 0$ versus $H_a : \mu_1 - \mu_2 \neq 0$ at level .05 using the following data: $m = 40$, $\bar{x} = 36,500$, $s_1 = 2200$, $n = 40$, $\bar{y} = 33,400$, and $s_2 = 1900$.

3. Let μ_1 denote true average tread life for a certain brand of FR78-15 radial tire, and let μ_2 denote the true average tread life for bias-ply tires of the same brand and size. Test $H_0 : \mu_1 - \mu_2 = 10,000$ versus $H_a : \mu_1 - \mu_2 > 10,000$ at level .01 using the following data: $m = 40$, $\bar{x} = 36,500$, $s_1 = 2200$, $n = 40$, $\bar{y} = 23,800$, and $s_2 = 1500$.

4. Use the data of Exercise 2 to compute a 95% confidence interval for $\mu_1 - \mu_2$. Does the resulting interval suggest that $\mu_1 - \mu_2$ has been precisely estimated?

5. Persons having Reynaud's syndrome are apt to suffer a sudden impairment of blood circulation in fingers and toes. In an experiment to study the extent of this impairment, each subject immersed a forefinger in water and the resulting heat output (cal/cm²/min) was measured. For $m = 10$ subjects with the syndrome, the average heat output was $\bar{x} = .64$, and for $n = 10$ nonsufferers, the average output was 2.05. Let μ_1 and μ_2 denote the true average heat outputs for the two types of subjects. Assume that the two distributions of heat output are normal with $\sigma_1 = .2$ and $\sigma_2 = .4$.

   a. Test $H_0 : \mu_1 - \mu_2 = -1.0$ versus $H_a : \mu_1 - \mu_2 < -1.0$ at level .01 (H_a says that the calorie output for sufferers is more than 1 cal/cm²/min below that for nonsufferers).
   b. Compute the P-value for the value of Z obtained in (a).
   c. What is the probability of a type II error when the actual difference between μ_1 and μ_2 is $\mu_1 - \mu_2 = -1.2$?
   d. Assuming that $m = n$, what sample sizes are required to ensure that $\beta = .1$ when $\mu_1 - \mu_2 = -1.2$?

6. An experiment to compare the tension bond strength of polymer latex modified mortar (portland cement mortar to which polymer latex emulsions have been added during mixing) to that of unmodified mortar resulted in $\bar{x} = 18.12$ kgf/cm² for the modified mortar ($m = 40$) and $\bar{y} = 16.87$ kgf/cm² for the unmodified mortar ($n = 32$). Let μ_1 and μ_2 be the true average tension bond strengths for the modified and unmodified mortars, respectively.

   a. Assuming that $\sigma_1 = 1.6$ and $\sigma_2 = 1.4$, test $H_0 : \mu_1 - \mu_2 = 0$ versus $H_a : \mu_1 - \mu_2 > 0$ at level .01.
   b. Compute the probability of a type II error for the test of (a) when $\mu_1 - \mu_2 = 1$.
   c. Suppose that the investigator decided to use a level .05 test and wished $\beta = .10$ when $\mu_1 - \mu_2 = 1$. If $m = 40$, what value of n is necessary?

   d. How would the analysis and conclusion of (a) change if σ_1 and σ_2 were unknown but $s_1 = 1.6$ and $s_2 = 1.4$?

7. Most educators feel that at the fourth-grade level, children working simple arithmetic problems should show decreasing dependence on techniques such as counting aloud, making tally marks, and the like. A sample of 63 counters (identified by observation) yielded a sample average IQ of 95.7 and a sample standard deviation of 13.0, while a sample of 55 noncounters yielded a sample average IQ of 100.5 with a sample standard deviation of 13.6. Construct a 95% confidence interval for the difference $\mu_1 - \mu_2$ in true average IQ between counters and noncounters at this grade level.

8. A study of reading habits of both college graduates and nongraduates is to be undertaken. An equal number of individuals in the two categories will be selected, and each will be asked to report on the number of books that he or she reads during the next year. On the basis of past studies of reading habits, the investigators feel that values $\sigma_1 = 20$ (grads) and $\sigma_2 = 15$ are conservative (though ultimately s_1 and s_2 will be used). What should the sample sizes be if the 95% confidence interval for the difference in true average number of books read between those with a college degree and those without one is to have a length of at most five?

9. An experiment was performed to compare the fracture toughness of high-purity 18 Ni maraging steel with commercial purity steel of the same type (*Corrosion Science*, 1971, pp. 723–736). For $m = 32$ specimens, the sample average toughness was $\bar{x} = 65.6$ for the high-purity steel, while $\bar{y} = 59.8$ for $n = 38$ specimens of commercial steel. Because the high-purity steel is more expensive, its use for a certain application can be justified only if its fracture toughness exceeds that of commercial purity steel by more than 5.

   a. Assuming that $\sigma_1 = 1.2$ and $\sigma_2 = 1.1$, test the relevant hypotheses using $\alpha = .001$.
   b. Compute β for the test conducted in (a) when $\mu_1 - \mu_2 = 6$.
   c. If $s_1 = 1.2$ and $s_2 = 1.1$ (rather than σ_1 and σ_2), how do the analysis and conclusion of (a) change?

10. In a study carried out to compare the effectiveness of a contract grading method with the traditional method of grading, ninth-grade students were tested for retention of material five weeks after completing a venereal disease unit in a health course.

    Contract: $m = 32, \bar{x} = 30.41, s_1 = 8.00$
    Traditional: $n = 31, \bar{y} = 31.50, s_2 = 8.12$

    (A pretest revealed virtually no difference in knowledge level prior to the unit.) Does the data suggest that there is a difference in true average retention level for the two methods? Compute the *P*-value, and use it to reach a conclusion at level of significance .01 ("Retention of Knowledge: Grade Contract Method Compared to the Traditional Grading Method," *J. Experimental Education,* 1974, pp. 92–96).

11. The accompanying table gives summary data on cube compressive strength (N/mm^2) for concrete specimens made with a pulverized fuel ash mix ("A Study of Twenty-Five Year Old Pulverized Fuel Ash Concrete Used in Foundation Structures," *Proc. Inst. Civ. Engrs.,* March 1985, pp. 149–165).

Age	Sample size	Sample mean	Sample SD
7 days	68	26.99	4.89
28 days	74	35.76	6.43

    Calculate a 99% confidence interval for the difference between true average 7-day strength and true average 28-day strength.

12. A mechanical engineer wishes to compare strength properties of steel beams with similar beams made with a particular alloy. The same number of beams, *n*, of each type will be tested. Each beam will be set in a horizontal position with a support on each end, a force of 2500 lbs will be applied at the center, and the deflection will be measured. From past experience with such beams, the engineer is willing to assume that the true standard deviation of deflection for both types of beam is .05 in. Because the alloy is more expensive, the engineer wishes to test at level .01 whether or not it has smaller average deflection than the steel beam. What value of *n* is appropriate if the desired type II error probability is .05 when the difference in true average deflection favors the alloy by .04 in.?

13. The level of monoamine oxidase activity in blood platelets (nm/mg protein/h) was determined for each individual in a sample of 33 chronic schizophrenics, resulting in $\bar{x} = 2.69$ and $s_1 = 2.30$, as well as for 35 normal subjects, resulting in $\bar{y} = 6.35, s_2 = 4.03$. Does this data strongly suggest that true average MAO activity for normal subjects is more than twice the activity level for schizophrenics? Derive a test procedure and carry out the test using $\alpha = .01$. *Hint: H_0 and H_a* here have a different form from the three standard cases. Let μ_1 and μ_2 refer to true average MAO activity for schizophrenics and normals, respectively, and consider the parameter $\theta = 2\mu_1 - \mu_2$. Write H_0 and H_a in terms of θ, estimate θ, and derive $\hat{\sigma}_{\hat{\theta}}$ ("Reduced Monoamine Oxidase Activity in Blood Platelets from Schizophrenic Patients," *Nature,* July 28, 1972, pp. 225–226).

14. Show for the upper-tailed test with σ_1, σ_2 known that as either *m* or *n* is increased, β decreases when $\mu_1 - \mu_2 > \Delta_0$.

15. For the case of equal sample sizes ($m = n$) and fixed α, what happens to the necessary sample size *n* as β is decreased, where β is the desired type II error probability at a fixed alternative?

16. To decide whether or not two different types of steel have the same true average fracture toughness values, *n* specimens of each type are tested, yielding the following results:

Type	Sample average	Sample SD
1	60.1	1.0
2	59.9	1.0

    Calculate the *P*-value for the appropriate two-sample *z* test assuming that the above data was based on $n = 100$. Then repeat the calculation for $n = 400$. Is the small *P*-value for $n = 400$ indicative of a difference which has practical significance? Would you have been satisfied with just a report of the *P*-value? Comment briefly.

9.2   The Two-Sample t Test and Confidence Interval

In real problems it is virtually always the case that the values of the population variances are unknown. In the previous section we illustrated for large sample sizes the use of a test procedure and confidence interval in which the sample variances were used in place of the population variances. In fact, for large samples, the Central Limit Theorem allows us to use these methods even when the two populations of interest are not normal.

There are many problems, though, in which at least one sample size is small and the population variances have unknown values. We now develop a test and confidence interval appropriate in this situation. However, because sample sizes are small, we will not have the C.L.T. working for us, so to obtain valid procedures we must make stronger assumptions about the underlying populations. The use of these procedures is then restricted to situations in which the assumptions are at least approximately satisfied. The resulting two-sample t test and confidence interval, which are not as broadly applicable as are the two-sample z procedures, are based on the following assumptions.

Assumption 1   Both populations are normal, so that $X_1, X_2, \ldots, X_m$ is a normal random sample and so is $Y_1, \ldots, Y_n$ (with the X's and Y's independent of one another).

Assumption 2   The values of the two population variances σ_1^2 and σ_2^2 are equal, so that their common value can be denoted by σ^2 (which is unknown).

The first assumption is of course analogous to what we assumed in connection with the one-sample t test and t confidence interval, but the second assumption is new. Remember that the mean μ of a normal population specifies the location or center of the bell-shaped curve, while the variance σ^2 controls its spread. Assumption 2 then says that we are not comparing the locations of just any two normal populations, but only populations that are known (or believed) a priori to have the same spread. When Assumption 2 is not reasonable, there are no procedures known to have good properties (for example, a test with controlled α and small β) even when both populations are normal. One possible approach is briefly described at the end of this section. Some authors recommend the use of a formal preliminary test for $\sigma_1^2 = \sigma_2^2$, but there are technical difficulties associated with this. Our approach is simply to "eyeball" the two sample variances; if they are of roughly the same order of magnitude, then one can be comfortable in using this test or confidence interval.

The Pooled Estimator of σ^2

The natural estimator of $\mu_1 - \mu_2$ is still $\overline{X} - \overline{Y}$, but now the variance of this estimator can be expressed as

$$V(\overline{X} - \overline{Y}) = \frac{\sigma^2}{m} + \frac{\sigma^2}{n} = \sigma^2\left(\frac{1}{m} + \frac{1}{n}\right) \tag{9.2}$$

Because the populations are both normal, $\bar{X} - \bar{Y}$ is too, so if we standardize using the square root of (9.2), the result is a standard normal variable. The major result underlying inferential procedures in this section is that if σ in the denominator of the standardized variable is replaced by an appropriate estimator, the resulting variable will have a *t* distribution.

Because σ^2 is the variance of both the X distribution and the Y distribution, the best estimator should depend on both the X_i's and the Y_j's. Furthermore, more weight should be given to the sample corresponding to the larger of the two sample sizes. Both S_1^2 and S_2^2, the two sample variances, are estimators of σ^2; a better estimator than either one individually is the following weighted average of the two.

Definition

> The **pooled estimator** of the common variance σ^2, denoted by S_p^2, is defined by
>
> $$S_p^2 = \frac{(m-1)}{m+n-2} S_1^2 + \frac{(n-1)}{m+n-2} S_2^2$$
>
> $$= \frac{(m-1)S_1^2 + (n-1)S_2^2}{m+n-2} \qquad (9.3)$$
>
> The pooled estimator of σ is S_p.

The middle expression in (9.3) shows that S_p^2 has the form $\lambda S_1^2 + (1 - \lambda)S_2^2$ where $0 < \lambda < 1$, so that S_p^2 is a weighted average. If, for example, m is much larger than n, then much more weight will be placed on S_1^2 than on S_2^2 (if $m = 15$ and $n = 7$, then $\lambda = .7$ and $1 - \lambda = .3$). Only if $m = n$ will $\lambda = 1 - \lambda = .5$, in which case S_p^2 is the ordinary unweighted average of S_1^2 and S_2^2.

Since $(m-1)S_1^2 = \Sigma(X_i - \bar{X})^2$ and $(n-1)S_2^2 = \Sigma(Y_j - \bar{Y})^2$, the numerator of S_p^2 is $\Sigma(X_i - \bar{X})^2 + \Sigma(Y_j - \bar{Y})^2$ (this would not be the case if the weighting factors were m and n, respectively). The X sample contributes $m - 1$ degrees of freedom to the estimator and the Y sample contributes $n - 1$ degrees of freedom, for a total of $m + n - 2$ d.f.

The *t* Statistic and Test

Because of (9.2) the variable

$$Z = \frac{\bar{X} - \bar{Y} - (\mu_1 - \mu_2)}{\sigma\sqrt{\dfrac{1}{m} + \dfrac{1}{n}}}$$

has a standard normal distribution. Substitution of S_p in place of σ yields the following key result on which inferences are based.

Theorem

Under Assumptions 1 and 2 of this section

$$T = \frac{\overline{X} - \overline{Y} - (\mu_1 - \mu_2)}{S_p\sqrt{\dfrac{1}{m} + \dfrac{1}{n}}} \tag{9.4}$$

has a t distribution with $m + n - 2$ degrees of freedom.

The two-sample t statistic for testing $H_0 : \mu_1 - \mu_2 = \Delta_0$ results from replacing $\mu_1 - \mu_2$ in (9.4) by Δ_0. The rejection regions for the various alternatives are similar to those in the one-sample case; each region uses a t critical value based on $m + n - 2$ d.f.

The pooled t test

Null hypothesis: $H_0 : \mu_1 - \mu_2 = \Delta_0$

Test statistic value: $t = \dfrac{\overline{x} - \overline{y} - \Delta_0}{s_p\sqrt{\dfrac{1}{m} + \dfrac{1}{n}}}$

Alternative hypothesis	*Rejection region for level α test*
$H_a : \mu_1 - \mu_2 > \Delta_0$	$t \geq t_{\alpha, m+n-2}$
$H_a : \mu_1 - \mu_2 < \Delta_0$	$t \leq -t_{\alpha, m+n-2}$
$H_a : \mu_1 - \mu_2 \neq \Delta_0$	either $t \geq t_{\alpha/2, m+n-2}$ or $t \leq -t_{\alpha/2, m+n-2}$

The two-sample (pooled) t test can be derived by using the likelihood ratio principle discussed in Chapter 8. It can also be shown that among all reasonable tests for H_0 that have level of significance α, the test has minimum β. Because of these results, statisticians refer to the two-sample t test as a best test (when assumptions 1 and 2 are satisfied).

Example 9.7

The paper "Production of Soluble Organic Nitrogen During Activated Sludge Treatment" (*J. Water Pollution Control Fed.,* 1981, pp. 99–112) reported that for a sample consisting of 14 observations from one type of activated sludge culture, the sample average soluble chemical oxygen demand (SCOD, in mg/l) was 18.1 with a sample standard deviation of 6.0, while for a sample of 16 observations from another type of culture, the sample average SCOD was 15.9 with a sample standard deviation of 5.0. Let μ_1 and μ_2 denote the true average SCOD values for these two types of cultures. Assuming that the sampled distributions are both normal with $\sigma_1 = \sigma_2$, let us use a level .05 test to decide whether sample data strongly suggests that μ_1 and μ_2 differ from one another.

The hypotheses of interest are $H_0 : \mu_1 - \mu_2 = 0$ versus $H_a : \mu_1 - \mu_2 \neq 0$. The test statistic value is

$$t = \frac{\bar{x} - \bar{y}}{s_p \sqrt{\frac{1}{m} + \frac{1}{n}}}$$

and H_0 is rejected if either $t \geq t_{\alpha/2, m+n-2}$ or $t \leq -t_{\alpha/2, m+n-2}$. With $\alpha = .05$, $m = 14$, $n = 16$, $t_{\alpha/2, m+n-2} = t_{.025, 28} = 2.048$; so H_0 will be rejected if $t \geq 2.048$ or $t \leq -2.048$. Since $s_1 = 6.0$ and $s_2 = 5.0$

$$s_p^2 = \frac{(14-1)(6.0)^2 + (16-1)(5.0)^2}{14 + 16 - 2} = 30.11$$

so $s_p = 5.49$ (almost the average of s_1 and s_2, but this will not always be the case). The computed test statistic value is

$$t = \frac{18.1 - 15.9}{5.49\sqrt{\frac{1}{14} + \frac{1}{16}}} = \frac{2.2}{2.01} = 1.10$$

Since 1.10 is neither ≥ 2.048 nor ≤ -2.048, H_0 cannot be rejected at level .05. ■

Example 9.8 A random sample of 15 alumina ceramic insulators doped in a certain manner yielded a sample average holdoff voltage (kV) of 110 and a sample standard deviation of 24, whereas a random sample of 76 plain alumina ceramic insulators resulted in a sample average voltage and sample standard deviation of 101 and 22, respectively (data from "The Effect of Doping on the Voltage Holdoff Performance of Alumina Insulators in Vacuum," *IEEE Transactions on Electrical Insulation,* 1985, pp. 505–509). Assuming that the conditions for use of the pooled t test are met (because one sample size is small, a z test is not appropriate), does the data suggest that true average voltage for doped specimens exceeds that for plain specimens by more than 5 kV?

With μ_1 and μ_2 denoting the true average doped and undoped voltages, respectively, the hypotheses of interest are $H_0 : \mu_1 - \mu_2 = 5$ versus $H_a : \mu_1 - \mu_2 > 5$. At significance level .10, H_0 will be rejected if $t \geq t_{.10, m+n-2} = t_{.10, 89} \approx 1.293$. The computed value of the pooled estimate of σ^2 is $s_p^2 = [14(24)^2 + 75(22)^2]/89 = 498.5$, so $s_p = 22.3$. The test statistic value is then

$$t = \frac{\bar{x} - \bar{y} - 5}{s_p\sqrt{\frac{1}{m} + \frac{1}{n}}} = \frac{110 - 101 - 5}{22.3\sqrt{\frac{1}{15} + \frac{1}{76}}} = \frac{4}{6.3} = .63$$

Because .63 is not in the rejection region $(.63 \not\geq 1.293)$, H_0 is not rejected at significance level .10; that is, P-value $> .10$. The sample data does not indicate that doping increases average holdoff voltage by more than 5 kV. ■

β for the Pooled t Test

In Chapter 8 we showed how β, the probability of a type II error, could be determined for a one-sample t test using the curves in Appendix Table A.13. These same curves can be used to obtain β for the pooled t test when either $\alpha = .05$ or .01. Suppose, for example, that μ_1 is the true average fracture toughness of high purity steel specimens and that μ_2 is the true average fracture toughness of commercial steel specimens. An investigator wishes to test H_0: $\mu_1 - \mu_2 = 0$ versus $H_a : \mu_1 - \mu_2 > 0$ using a level .05 test based on sample sizes $m = 15$ and $n = 15$ (so the probability of rejecting H_0 when H_0 is true is only .05). The value $\mu_1 - \mu_2 = 5$ might be viewed as a substantial departure from H_0, so the investigator would want the probability of not rejecting H_0 when $\mu_1 - \mu_2 = 5$ to be small. This probability is exactly β for the value $\mu_1 - \mu_2 = 5$. The value of β for the one-sample t test depended on σ, the population standard deviation. Similarly, β for the pooled t test depends on σ, the common value of the two population standard deviations. To determine β, first specify the level of significance α, the two sample sizes, and a plausible value of σ. Then select an alternative value Δ' for which β is desired and compute

$$d = \frac{|\Delta' - \Delta_0|}{\sigma} \cdot \sqrt{\frac{mn}{(m + n)(m + n - 1)}}$$

Now locate the appropriate set of curves in Appendix Table A.13 ($\alpha = .05$ or .01 for a one- or two-tailed test), move over to the value of d on the horizontal axis, move up to the $m + n - 2$ d.f. curve, and finally look over to the vertical axis to read the value of β.

Example 9.9   It is important that fabric used in children's clothing be fire resistant. Suppose that an investigator wishes to compare burn times of two different fabrics. A number of fabric specimens will be used in the experiment, and the burn time for each one will be recorded. Assuming that burn times are normally distributed with $\sigma_1 = \sigma_2$, the pooled t test is appropriate for testing $H_0: \mu_1 - \mu_2 = 0$ versus $H_a: \mu_1 - \mu_2 \neq 0$. The chosen sample sizes are $m = 15$ and $n = 15$. If the true standard deviation of each fabric's burn time is $\sigma = 1.50$ and μ_1 actually exceeds μ_2 by 2 ($\Delta' = 2$), then

$$d = \frac{2}{1.5} \sqrt{\frac{(15)(15)}{(30)(29)}} = .68$$

The number of degrees of freedom for the test is $m + n - 2 = 28$. The curve for 28 d.f. does not appear in Appendix Table A.13, but it is very close to the 29 d.f. curve, which does appear. Using a level $\alpha = .05$ two-tailed test gives $\beta \approx .08$, whereas using $\alpha = .01$ gives $\beta \approx .2$. At level .05, if the difference between μ_1 and μ_2 is 2 rather than 0, two samples of 15 in size will result in H_0 being incorrectly not rejected only 8% of the time. The test has good ability to detect this sort of departure from H_0. Of course, if the actual value of σ is greater than 1.5, the value of d would be smaller and β would be larger. The greater the variability in the populations, the more difficult it is to draw the right conclusion.   ■

The t Interval for $\mu_1 - \mu_2$

To obtain a confidence interval for the difference between means of two normal populations when at least one sample size is small, we assume that the two population variances are equal. Then the variable

$$T = \frac{\overline{X} - \overline{Y} - (\mu_1 - \mu_2)}{S_p \sqrt{1/m + 1/n}}$$

has a t distribution with $m + n - 2$ degrees of freedom (where S_p^2 is the pooled estimator of the common variance σ^2), so

$$P(-t_{\alpha/2, m+n-2} \leq T \leq t_{\alpha/2, m+n-2}) = 1 - \alpha \qquad (9.5)$$

Manipulation of the inequalities inside the parentheses in (9.5) in the same manner as before yields the desired interval.

A $100(1 - \alpha)\%$ confidence interval for $\mu_1 - \mu_2$ is

$$\overline{x} - \overline{y} \pm t_{\alpha/2, m+n-2} \cdot s_p \sqrt{\frac{1}{m} + \frac{1}{n}}$$

Example 9.10   Pesticides, particularly the organochlorines, have contributed to the decline of several species of birds. The objective of the research reported in "Aerial Pesticide Applications and Ring-Necked Pheasants" (*J. Wildlife Mgmt.,* vol. 38, pp. 679–685) was to study the effects of several such pesticides on pheasants. In 1970 several applications of these pesticides were made over a small area in Idaho. Subsequently 20 wild pheasants were captured in the area. The average amount of plasma cholinesterase (an enzyme which affects muscle control) for the 20 (in μ moles of acetycholine hydrolized per ml plasma/min) was found to be 1.55 with a sample standard deviation of .39. Nine wild pheasants captured outside the area (the control group) had an average cholinesterase amount of 1.60, with a sample standard deviation of .40. With $m = 20$, $\overline{x} = 1.55$, $s_1 = .39$, $n = 9$, $\overline{y} = 1.60$, and $s_2 = .40$, a 95% interval for the true average difference in plasma cholinesterase is

$$1.55 - 1.60 \pm (2.052)(.39)\sqrt{\frac{1}{20} + \frac{1}{9}} = -.05 \pm .32$$

$$= (-.37, .27) \qquad ■$$

A Test Procedure When $\sigma_1^2 \neq \sigma_2^2$

When Assumptions 1 and 2 are satisfied, there is a general consensus among statisticians that the two-sample t test described above should be used. In fact, it has been shown that if the distributions being sampled are not too nonnormal and/or the two variances are not too different from one another, then the t test works reasonably well in the sense that the actual level of significance is approximately the specified α, and the test continues to have small β. Statisticians customarily summarize these properties by saying that the t test is robust in the presence of mild departures from assumptions.

If, however, σ_1^2 and σ_2^2 are very different from one another, then using the t test, which involves a pooled estimate of a single parameter σ^2, will tend to yield erroneous conclusions. Unfortunately, even when the populations are still assumed normal, there is no test procedure known to have very good properties. The following procedure (called the Smith-Satterthwaite test) is known to be approximately a level α test, but its type II error probabilities have proved difficult to study, so it is not known whether the test is in any sense a best test. The test statistic has approximately a t distribution when H_0 is true, but the number of degrees of freedom ν is estimated from the data (ν will not usually be an integer, so must be rounded to obtain a critical value from the t table).

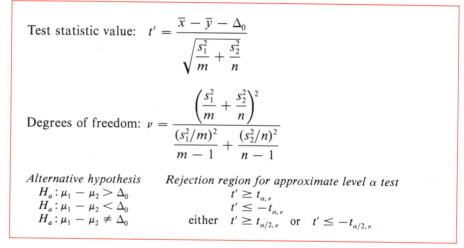

Test statistic value: $t' = \dfrac{\bar{x} - \bar{y} - \Delta_0}{\sqrt{\dfrac{s_1^2}{m} + \dfrac{s_2^2}{n}}}$

Degrees of freedom: $\nu = \dfrac{\left(\dfrac{s_1^2}{m} + \dfrac{s_2^2}{n}\right)^2}{\dfrac{(s_1^2/m)^2}{m-1} + \dfrac{(s_2^2/n)^2}{n-1}}$

Alternative hypothesis	Rejection region for approximate level α test
$H_a : \mu_1 - \mu_2 > \Delta_0$	$t' \geq t_{\alpha, \nu}$
$H_a : \mu_1 - \mu_2 < \Delta_0$	$t' \leq -t_{\alpha, \nu}$
$H_a : \mu_1 - \mu_2 \neq \Delta_0$	either $t' \geq t_{\alpha/2, \nu}$ or $t' \leq -t_{\alpha/2, \nu}$

Example 9.11   A paper in the *Journal of Nervous and Mental Disorders* (1968, vol. 146, pp. 136–146) reported the following data on the amount of dextroamphetamine excreted by a sample of children having organically related disorders and a sample of children with nonorganic disorders (dextroamphetamine is a drug commonly used to treat hyperkinetic children).

Organic: 17.53, 20.60, 17.62, 28.93, 27.10
Nonorganic: 15.59, 14.76, 13.32, 12.45, 12.79

(Observations refer to percentage of recovery of the drug seven hours after its administration.)

The summary values are $\bar{x} = 22.36$, $\bar{y} = 13.78$, $s_1^2 = 28.63$, and $s_2^2 = 1.80$. The data suggests that there is much less variability in percentage of recovery for "nonorganic" children than for "organic" children. To use T' to test

$H_0 : \mu_1 - \mu_2 = 0$ versus $H_a : \mu_1 - \mu_2 \neq 0$, where μ_1 and μ_2 refer to the true average percentages of recovery for the two conditions, we need

$$\nu = \frac{\left(\dfrac{28.63}{5} + \dfrac{1.80}{5}\right)^2}{\dfrac{(28.63/5)^2}{4} + \dfrac{(1.80/5)^2}{4}} = \frac{37.04}{8.20 + .03} = 4.50 \approx 5$$

$$t' = \frac{22.36 - 13.78}{\sqrt{\dfrac{28.63}{5} + \dfrac{1.80}{5}}} = \frac{8.58}{2.47} = 3.47$$

Since $t_{.005, 5} = 4.032$, the data is not significant at level .01. If the pooled two-sample t test had been used, the computed t value would be $t = 4.9$, which would exceed $t_{.005, 8}$. ■

When an investigator can reasonably assume that two populations have the same shape and spread (so $\sigma_1^2 = \sigma_2^2$) but may differ with respect to location and yet the investigator does not wish to impose the assumption of normality, the "distribution-free" procedures presented in Chapter 15 can be used.

Exercises / Section 9.2 (17–29)

17. Let μ_1 and μ_2 denote true average stopping distances for two different types of cars traveling at 50 mph. Assuming normality and $\sigma_1 = \sigma_2$, test $H_0 : \mu_1 - \mu_2 = 0$ versus $H_a : \mu_1 - \mu_2 \neq 0$ at level .05 using the following data: $m = 6$, $\bar{x} = 122.7$, $s_1 = 5.59$, $n = 5$, $\bar{y} = 129.3$, and $s_2 = 5.25$.

18. Suppose that μ_1 and μ_2 are true mean stopping distances at 50 mph for cars of a certain type equipped with disk brakes and with pneumatic brakes, respectively. Use the pooled t test at significance level .01 to test $H_0 : \mu_1 - \mu_2 = -10$ versus $H_a : \mu_1 - \mu_2 < -10$ for the following data: $m = 6$, $\bar{x} = 115.7$, $s_1 = 5.03$, $n = 6$, $\bar{y} = 129.3$, and $s_2 = 5.38$.

19. Use the data of Exercise 18 to calculate a 95% confidence interval for the difference between true average stopping distance for cars equipped with disk brakes and cars equipped with pneumatic brakes. Does the interval suggest that precise information about the value of this difference is available?

20. The article "Mercury in Aquatic Birds at Clay Lake, Western Ontario" (*J. Wildlife Mgmt.*, 1973, pp. 58–61) reported the following data on mercury residues in breast muscles.

Mallard ducks: $m = 16$, $\bar{x} = 6.13$, $s_1 = 2.40$
Blue-winged teals: $n = 17$, $\bar{y} = 6.46$, $s_2 = 1.73$

Compute a 95% confidence interval for the difference between true average mercury residues in these two types of birds in the region of interest. Does the validity of the confidence interval rest on any assumptions? Explain.

21. The paper "The Influence of Corrosion Inhibitor and Surface Abrasion on the Failure of Aluminum-Wired Twist-On Connections" (*IEEE Trans. Components, Hybrids, and Manuf. Tech.*, 1984, pp. 20–25) reported the following summary data on potential drop measurements for one sample of connectors wired with alloy aluminum and another sample wired with EC aluminum.

Type	Sample size	Sample mean	Sample SD
Alloy	20	17.5	.55
EC	20	16.9	.49

Does this data suggest that true average potential drop for alloy connections is higher than for EC connections (as stated in the paper)? Carry out the appropriate test using a significance level of .01. In reaching your conclusion, what type of error might you have committed?

22. The flatwise bending strength (lb/in.) was determined for dowel joints of two different types used in wood furniture frame construction, with the second type having greater rail thickness (1.5 in.) than the first type (1.25 in.).

$$1.25 \text{ in.:} \quad m = 10, \quad \bar{x} = 1215.6, \quad s_1 = 137.4$$
$$1.5 \text{ in.:} \quad n = 10, \quad \bar{y} = 1376.4, \quad s_2 = 155.0$$

a. Does the larger rail thickness of the second type of dowel result in a greater true average bonding strength than for dowels of the first type? Test the relevant hypotheses at level .05.
b. Place upper and lower bounds on the P-value of the data.
c. Suppose that prior to the experiment, the investigator had believed $\sigma = 150$. What would β be for $\mu_1 - \mu_2 = 100$?
d. Assuming $m = n$, what value would be required to yield $\beta = .50$ when $\mu_1 - \mu_2 = 100$?

23. Two agents used in the nozzles of fire-fighting spray systems are compared with respect to the amount of foam expansion. A random sample of five observations with the agent AFFF (aqueous film-forming foam) yields a sample mean of 4.7 and a sample standard deviation of .6. A second random sample of five observations with ATC (alcohol-type concentrates) results in a mean and standard deviation of 6.9 and .8, respectively. Does this data suggest that true average foam expansion for ATC exceeds that for AFFF by more than 1? Stating any assumptions necessary to your analysis, give as much information as you can concerning the P-value. Then use this information to reach a conclusion at significance level .01. (The sample sizes and means are given in "Enhancement of Fire Protection with Directional Cooling Spray Nozzles by Use of AFFF," *Fire Tech.*, 1983, pp. 5–13).

24. In an experiment to study the effects of liming and urea fertilizer application on dimethoate (a pesticide) retention by loamy soil, the following percentages of dimethoate recovery were observed:

Soil treated
  with lime: 28.5,   24.7,   26.2,   23.9,   29.6
Soil treated
  with urea: 38.7,   41.6,   35.9,   41.8,   43.2

Compute a 95% confidence interval for the difference in true average percentage of dimethoate recovery between the two treatments.

25. Serum-prolactin was measured using radioimmunoassay during the menstrual cycle of each of 28 women who had premenstrual syndrome (cyclical recurrence or complaints of tension and headaches only during the premenstrual period, and dramatic and complete relief of symptoms when full menstrual flow begins) and also for 21 women in a control group. Measurements refer to average concentration during the cycle, in ng/ml.

Premenstrual
  syndrome: $m = 28, \quad \Sigma x_i = 986.44,$
          $\Sigma x_i^2 = 36{,}030.31$
Control: $n = 21, \quad \Sigma y_i = 557.55,$
          $\Sigma y_i^2 = 15{,}877.53$

Does this data suggest any difference in the true mean level of serum-prolactin between women with premenstrual syndrome and those without it?

26. Nine observations of surface-soil pH were made at each of two different locations at the Central Soil Salinity Research Institute experimental farm, and the resulting data appeared in the article "Sodium-Calcium Exchange Equilibria in Soils as Affected by Calcium Carbonate and Organic Matter" (*Soil Sci.*, 1984, p. 109). Does the data suggest that the true mean soil pH values differ for the two locations? Test the appropriate hypotheses using a .05 significance level. Be sure to state any assumptions necessary for the validity of your test.

			pH		
Location A	8.53	8.52	8.01	7.99	7.93
	7.89	7.85	7.82	7.80	
Location B	7.85	7.73	7.58	7.40	7.35
	7.30	7.27	7.27	7.23	

27. The paper "Pine Needles as Sensors of Atmospheric Pollution" (*Environ. Monitoring*, 1982, pp. 273–286) reported on the use of neutron activity analysis to determine pollutant concentrations in pine needles. According to the paper's authors, "These observations strongly indicated that for those elements which are determined well by the analytical procedures, the distribution of concentrations is lognormal. Accordingly, in tests of significance the logarithms of concentrations will be used." The given data refers to bromine concentration in needles taken from a site near an oil-fired steam plant and from a relatively clean site. The summary values are means and standard deviations of the log-transformed observations. Let μ_1 be the true average log concentration at the first site and define μ_2 analogously for the second site. Test the hypothesis that the two concentration distribution means are equal against the alternative that they are different by using the pooled t test at level .05 with the log-transformed data.

Site	Sample size	Mean log concentration	Standard deviation of log concentration
Steam plant	8	18	4.9
Clean	9	11	4.6

28. High nitrate intake in food consumption has been shown to have a number of deleterious effects, including lower thyroxin production, increased incidence of cyanosis in newborns, and lower milk production in dairy cows. The following data is the result of an experiment to measure the percentage of weight gain for young laboratory mice given a standard diet and mice given 2000 ppm nitrate in their drinking water.

Nitrate: 12.7, 19.3, 20.5, 10.5, 14.0, 10.8, 16.6, 14.0, 17.2
Control: 18.2, 32.9, 10.0, 14.3, 16.2, 27.6, 15.7

Use the test procedure with statistic T' to decide whether the data indicates at level .01 that a heavy dose of nitrate retards true average percentage weight gain in mice.

29. An experiment is to be performed to assess the effect of companionship at mealtime on food consumption. The amount of food consumed by each of n mice eating alone will be determined. Then each of n similar mice will be allowed to eat with a companion mouse trained to eat continuously. The research hypothesis is that the presence of such a companion will increase true average food consumption, so the investigators plan to use a level .01 upper-tailed test. What value of n is appropriate if σ is taken to be 1.0 and it is desired that the type II error probability should be $\beta = .10$ when the increase in true average food consumption is as much as 1?

Analysis of Paired Data

In Sections 9.1 and 9.2 we considered testing for a difference between two means μ_1 and μ_2. This was done by utilizing the results of a random sample $X_1, X_2, \ldots, X_m$ from the distribution with mean μ_1 and a completely independent (of the X's) sample $Y_1, \ldots, Y_n$ from the distribution with mean μ_2. That is, either m individuals were selected from population one and n different individuals from population two, or m individuals (or experimental objects) were given one treatment and another set of n individuals were given the other treatment. In contrast, there are a number of experimental situations in which there is only one set of n individuals or experimental objects, and two observations are made on each individual or object.

Example 9.12 Trace metals in drinking water affect the flavor, and unusually high concentrations can pose a health hazard. The paper "Trace Metals of South Indian River" (*Envir. Studies,* 1982, pp. 62–66) reports on a study in which six river locations were selected (six experimental objects) and the zinc concentration (mg/L) determined for both surface water and bottom water at each location.

The six pairs of observations are displayed in the accompanying table. Does the data suggest that true average concentration in bottom water exceeds that of surface water?

Location:	1	2	3	4	5	6
Zinc concentration in bottom water (x):	.430	.266	.567	.531	.707	.716
Zinc concentration in surface water (y):	.415	.238	.390	.410	.605	.609
Difference:	.015	.028	.177	.121	.102	.107

Figure 9.1(a) displays a plot of this data. At first glance there appears to be little difference between the x and y samples. From location to location there is a great deal of variability in each sample, and it looks as though any differences between the samples can be attributed to this variability. However, when the observations are identified by location, as in Figure 9.1(b), a different view emerges. At each location, bottom concentration exceeds surface concentration. This is confirmed by the fact that all $x - y$ differences (bottom water concentration — surface water concentration) displayed in the bottom row of the data table are positive. As we shall see, a correct analysis of this data focuses on these differences.

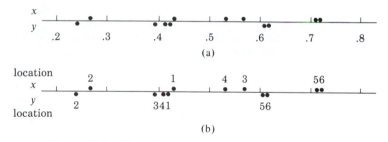

Figure 9.1   Plot of paired data from Example 9.12:
(a) observations not identified by location;
(b) observations identified by location

Basic assumptions

The data consists of n independently selected pairs (X_1, Y_1), (X_2, Y_2), ..., (X_n, Y_n), with $E(X_i) = \mu_1$ and $E(Y_i) = \mu_2$. Let $D_1 = X_1 - Y_1$, $D_2 = X_2 - Y_2, \ldots, D_n = X_n - Y_n$, so the D_i's are the differences within pairs. Then the D_i's are assumed to be normally distributed with variance σ_D^2 (this is usually a consequence of the X_i's and Y_i's themselves being normally distributed).

We are again interested in testing hypotheses about the difference $\mu_1 - \mu_2$. The denominator of the two-sample t test was obtained by first applying the rule $V(\overline{X} - \overline{Y}) = V(\overline{X}) + V(\overline{Y})$. However, with paired data the X and Y observations within each pair are often not independent, so that $\overline{X}$ and $\overline{Y}$ are not independent of one another, and the rule is not valid. We must therefore abandon the two-sample t test and look for an alternative method of analysis.

The Paired t Test

Because different pairs are independent, the D_i's are independent of one another. If we let $D = X - Y$, where X and Y are the first and second observation, respectively, within an arbitrary pair, then the expected difference is

$$\mu_D = E(X - Y) = E(X) - E(Y) = \mu_1 - \mu_2$$

(the rule of expected values used here is valid even when X and Y are dependent). Thus any hypothesis about $\mu_1 - \mu_2$ can be phrased as a hypothesis about the mean difference μ_D. But since the D_i's constitute a normal random sample (of differences) with mean μ_D, hypotheses about μ_D can be tested using a one-sample t test. That is, *to test hypotheses about $\mu_1 - \mu_2$ when data is paired, form the differences $D_1, D_2, \ldots, D_n$ and carry out a one-sample t test (based on $n - 1$ d.f.) on the differences.*

> Null hypothesis: $H_0 : \mu_D = \Delta_0$ (where $D = X - Y$ is the difference between the first and second observations within a pair ($\mu_D = \mu_1 - \mu_2$))
>
> Test statistic value: $t_{\text{paired}} = \dfrac{\overline{d} - \Delta_0}{s_D / \sqrt{n}}$ (where $\overline{d}$ and s_D are the sample mean and standard deviation, respectively, of the d_i's)
>
> | Alternative hypothesis | Rejection region for level α test |
> |---|---|
> | $H_a : \mu_D > \Delta_0$ | $t_{\text{paired}} \geq t_{\alpha, n-1}$ |
> | $H_a : \mu_D < \Delta_0$ | $t_{\text{paired}} \leq -t_{\alpha, n-1}$ |
> | $H_a : \mu_D \neq \Delta_0$ | either $t_{\text{paired}} \geq t_{\alpha/2, n-1}$ or $t_{\text{paired}} \leq -t_{\alpha/2, n-1}$ |

This paired test is valid even if $\sigma_1^2 \neq \sigma_2^2$, since the differences are still normally distributed and s_D^2 estimates $\sigma_D^2 = V(X - Y)$. While a two-sample t test would be based on $2n - 2$ d.f., the paired t test uses only $n - 1$ d.f. To ensure a correct analysis, we give up $n - 1$ degrees of freedom. There are situations in which an experiment can be run in either a paired or an unpaired (independent samples) manner; we will shortly discuss the issues involved in choosing between the two.

Example 9.13
(Example 9.12
continued)

With μ_1 and μ_2 denoting true average zinc concentrations for bottom and surface water, respectively, $\mu_D = \mu_1 - \mu_2$ is the true average difference between bottom and surface concentrations. The hypotheses of interest are $H_0 : \mu_D = 0$ versus $H_a : \mu_D > 0$. The test statistic value is $t_{\text{paired}} = \overline{d}/(s_D/\sqrt{n})$, and H_0 will be rejected if $t_{\text{paired}} \geq t_{\alpha, n-1}$. Since $n - 1 = 5$, a level .01 test rejects H_0 in favor of H_a if $t_{\text{paired}} \geq t_{.01, 5} = 3.365$. Using $\Sigma d_i = .550$ and $\Sigma d_i^2 = .068832$,

$$\bar{d} = \frac{\Sigma d_i}{n} = \frac{.550}{6} = .0917$$

$$s_D^2 = \frac{\Sigma d_i^2 - (\Sigma d_i)^2/n}{n-1} = \frac{.068832 - (.550)^2/6}{5} = .003683$$

and $s_D = \sqrt{.003683} = .0607$. Thus $t_{paired} = .0917/(.0607/\sqrt{6}) = .0917/.0248 = 3.70$. Since $3.70 \geq 3.365$, H_0 is rejected at level .01. The data strongly suggests that true average zinc concentration for bottom water exceeds that for top water. ■

Example 9.14   Scientists and engineers frequently wish to compare two different techniques for measuring or determining the value of a variable. In such situations, interest centers on testing whether or not the mean difference in measurements is zero. The paper "Evaluation of the Deuterium Dilution Technique Against the Test Weighing Procedure for the Determination of Breast Milk Intake" (*Amer. J. Clinical Nutr.*, 1983, pp. 996–1003) reports the accompanying data on amount of milk ingested by each of 14 randomly selected infants.

Infant:	1	2	3	4	5	6	7
Isotopic method:	1509	1418	1561	1556	2169	1760	1098
Test weighing method:	1498	1254	1336	1565	2000	1318	1410
Difference:	11	164	225	−9	169	442	−312

Infant:	8	9	10	11	12	13	14
Isotopic method:	1198	1479	1281	1414	1954	2174	2058
Test weighing method:	1129	1342	1124	1468	1604	1722	1518
Difference:	69	137	157	−54	350	452	540

A normal probability plot of the differences (Figure 9.2) is quite straight, indicating that the assumption of a normal difference distribution is plausible. Use the paired t test at level .05 to see whether or not the true average difference in measured intake for the two methods is zero.

With μ_D denoting the true average difference between intake value measured by the isotopic method and by test weighing, the relevant hypotheses are $H_0: \mu_D = 0$ versus $H_a: \mu_D \neq 0$. The test is therefore two-tailed. From the t table, $t_{.025, 13} = 2.160$, so H_0 will be rejected if $t_{paired} \geq 2.160$ or if $t_{paired} \leq -2.160$. The sample average difference and sample standard deviation of the differences are $\bar{d} = 167.21$ and $s_D = 228.21$, so $t_{paired} = (167.21 - 0)/(228.21/\sqrt{14}) = 167.21/60.99 = 2.74$. Because $2.74 \geq 2.160$, H_0 is rejected in favor of the conclusion that the true average difference is something other than zero.

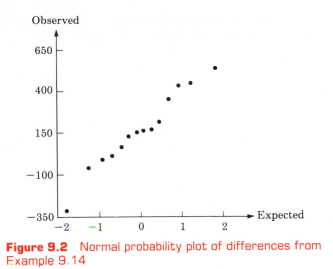

Figure 9.2 Normal probability plot of differences from Example 9.14

When the number of pairs is large, the assumption of a normal difference distribution is not necessary. The Central Limit Theorem validates the resulting z test.

A Confidence Interval for μ_D

In the same way that the t confidence interval for a single population mean μ is based on the t variable $T = (\overline{X} - \mu)/(S/\sqrt{n})$, a t confidence interval for $\mu_D \, (= \mu_1 - \mu_2)$ is based on the fact that

$$T = \frac{\overline{D} - \mu_D}{S_D/\sqrt{n}}$$

has a t distribution with $n - 1$ d.f. Manipulation of this t variable, as in earlier derivations of confidence intervals, yields the following $100(1 - \alpha)\%$ confidence interval:

$$\overline{d} \pm t_{\alpha/2, \, n-1} \cdot s_D/\sqrt{n}$$

When n is small, the validity of this interval requires that the distribution of differences be at least approximately normal. For large n, the Central Limit Theorem ensures that the resulting z interval is valid without any restrictions on the distribution of differences.

Example 9.15 In an experiment reported in the article "Emotional Attributes of Color: A Comparison of Violet and Green" (*Perceptual and Motor Skills,* 1971, pp. 403–406), 14 subjects were exposed to 60-sec intervals of alternating violet and green light for a total of six minutes. Galvanic skin response during the first 12 sec of exposure (calibrations) was measured and averaged over exposure intervals. The results appear in Table 9.3.

Table 9.3

Subject	1	2	3	4	5	6	7	8	9	10	11	12	13	14
Violet	3.0	3.7	4.0	3.2	3.6	3.5	4.2	3.8	3.7	3.4	3.6	3.8	3.4	3.4
Green	2.2	2.7	3.1	2.9	3.3	2.6	2.9	2.8	3.2	2.5	3.5	3.0	2.3	3.5
d_i	.8	1.0	.9	.3	.3	.9	1.3	1.0	.5	.9	.1	.8	1.1	−.1

The computed summary quantities are $\bar{d} = .70$ and $s_D = .41$. With $t_{.025,\,13} = 2.160$, a 95% confidence interval for μ_D is

$$.70 \pm (2.160)(.41)/\sqrt{14} = .70 \pm .24 = (.46, .94)$$ ■

Paired Data and the Two-Sample t Test

To see what happens if the two-sample t test is used (incorrectly) to analyze paired data in which there is dependence within pairs, consider again Example 9.12. The summary quantities are $\bar{x} = .5362$, $\bar{y} = .4445$, $s_1^2 = .0294$, and $s_2^2 = .0201$, $s_p^2 = .0248$, and $s_p = .1573$. The computed value of the two-sample t statistic is

$$t = \frac{\bar{x} - \bar{y} - 0}{s_p\sqrt{\dfrac{1}{n} + \dfrac{1}{n}}} = \frac{.0917}{.1573\sqrt{2/6}} = 1.01$$

At level .01, t is compared to $t_{.01,\,2n-2} = t_{.01,\,10} = 2.764$. Since 1.01 is less than 2.764, H_0 would not be rejected, so the two-sample t test yields a conclusion opposite to that suggested by the correct paired analysis.

Notice that the numerators of t and t_{paired} are identical. This is true in general, since $\bar{d} = \Sigma d_i/n = [\Sigma(x_i - y_i)]/n = (\Sigma x_i)/n - (\Sigma y_i)/n = \bar{x} - \bar{y}$. The difference between the two test statistics is due entirely to the denominators. Each test statistic is obtained by standardizing $\bar{X} - \bar{Y}\,(= \bar{D})$, but in the presence of dependence the standardization in t is incorrect. To see this, recall from Section 5.3 that

$$V(X \pm Y) = V(X) + V(Y) \pm 2\,\text{Cov}(X, Y)$$

Since the correlation between X and Y is $\rho = \text{Corr}(X, Y) = \text{Cov}(X, Y)/\sqrt{V(X)} \cdot \sqrt{V(Y)}$,

$$V(X - Y) = \sigma^2 + \sigma^2 - 2\rho\sigma^2 = 2\sigma^2(1 - \rho)$$

Applying this to $\bar{X} - \bar{Y}$,

$$V(\bar{X} - \bar{Y}) = V(\bar{D}) = V\left(\frac{1}{n}\Sigma D_i\right) = \frac{V(D_i)}{n} = \frac{2\sigma^2(1 - \rho)}{n}$$

The two-sample t test is based on the assumption of independence, in which case $\rho = 0$. But in many paired experiments, there will be a strong *positive* dependence between X and Y (large X associated with large Y), so that ρ will be positive and the variance of $\overline{X} - \overline{Y}$ will be smaller than $2\sigma^2/n$. Thus, *whenever there is positive dependence within pairs, the denominator for t_{paired} should be smaller than for t of the independent samples test.* Often t will be much closer to zero than t_{paired}, considerably understating the significance of the data.

Paired versus Unpaired Experiments

In our examples paired data resulted from two observations on the same individual (Examples 9.14 and 9.15) or experimental object (location in Example 9.12). Even when this cannot be done, paired data with dependence within pairs can be obtained by matching individuals or objects on one or more characteristics thought to influence responses. For example, in a medical experiment to compare the efficacy of two drugs for lowering blood pressure, the experimenter's budget might allow for the treatment of 20 patients. If 10 patients are randomly selected for treatment with the first drug and another 10 independently selected for treatment with the second drug, an independent-samples experiment results.

However, the experimenter, knowing that blood pressure is influenced by age and weight, might decide to pair off patients so that within each of the resulting 10 pairs, age and weight were approximately equal (though there might be sizable differences between pairs). Then each drug would be given to a different patient within each pair for a total of 10 observations on each drug.

Without this matching (or "blocking"), one drug might appear to outperform the other just because patients in one sample were lighter and younger and thus more susceptible to a decrease in blood pressure than the heavier and older patients in the second sample. However, there is a price to be paid for pairing—a smaller number of degrees of freedom for the paired analysis—so we must ask when one type of experiment should be preferred to the other.

There is no straightforward and precise answer to this question, but there are some useful guidelines. The first point is that if we have a choice between two t tests that are both valid (and carried out at the same level of significance α), we should prefer the test that has the larger number of degrees of freedom. The reason for this is that a larger number of d.f. means smaller β for any fixed alternative value of the parameter or parameters. That is, for a fixed type I error probability, the probability of a type II error is decreased by increasing d.f. This is readily verified by examining the β curves in Appendix Table A.13.

However, if the experimental units are quite heterogeneous in their responses (large σ^2), it will be difficult to detect small but significant differences between two treatments. This is essentially what happened in the data set in Example 9.12—for both "treatments" (bottom water and top water) there is great between-location variability, which tends to mask differences in treatments within locations. If there is a high positive correlation within experimental units or subjects, the variance of $\overline{D} = \overline{X} - \overline{Y}$ will be $2\sigma^2(1 - \rho)/n$, which will be much smaller than the unpaired variance $2\sigma^2/n$. Because of this reduced variance due to pairing, it will be easier to detect a difference than if independent samples are used. The pros and cons of pairing can now be summarized as follows:

> **1.** If there is great heterogeneity between subjects (large σ^2) and a large correlation within subjects (large positive ρ), then the loss in degrees of freedom will be compensated for by the increased precision associated with pairing, so a paired experiment is preferable to an independent-samples experiment.
>
> **2.** If the experimental units are relatively homogeneous (small σ^2) and the correlation within pairs is not large, the gain in precision due to pairing will be outweighed by the decrease in degrees of freedom, so an independent-samples experiment should be used.

Of course, the magnitude of σ^2 and ρ will not usually be known very precisely, so an investigator will be required to make a seat-of-the-pants judgment as to whether 1 or 2 obtains. In general, if the number of observations that can be obtained is large, then a loss in degrees of freedom (for example, from 40 to 20) will not be serious, but if the number is small, then the loss (say, from 16 to 8) because of pairing may be serious if not compensated for by increased precision. In making such a decision, the β curves for the t tests should be consulted.

Exercises / Section 9.3 (30–39)

30. The paper "A Supplementary Behavioral Program to Improve Deficient Reading Performance" (*J. Abnormal Child Psychology*, 1973, pp. 390–399) reported the results of an experiment in which seven pairs of children reading below grade level were obtained by matching so that within each pair the two children were equally deficient in reading ability. Then one child from each pair received experimental training, while the other received standard training. Based on the accompanying improvement scores, does the experimental training appear to be superior to the standard training? Use the paired t test to test $H_0: \mu_D = 0$ versus $H_a: \mu_D > 0$ at level .1.

Pair:	1	2	3	4	5	6	7
Experimental (x):	.5	1.0	.6	.1	1.3	.1	1.0
Control (y):	.8	1.1	−.1	.2	.2	1.5	.8

31. Two types of fish attractors, one made from vitrified clay pipes and the other from cement blocks and brush, were used during 16 different time periods spanning four years at Lake Tohopekaliga, Florida ("Two Types of Fish Attractors Compared in Lake Tohopekaliga, Florida," *Trans. Amer. Fisheries Soc.*, 1978, pp. 689–695). The following observations are of fish caught per fishing day.

Period:	1	2	3	4	5	6	7	8
Pipe:	6.64	7.89	1.83	.42	.85	.29	.57	.63
Brush:	9.73	8.21	2.17	.75	1.61	.75	.83	.56

Period:	9	10	11	12	13	14	15	16
Pipe:	.32	.37	.00	.11	4.86	1.80	.23	.58
Brush:	.76	.32	.48	.52	5.38	2.33	.91	.79

Does one attractor appear to be more effective on average than the other?

a. Use the paired t test with $\alpha = .01$ to test $H_0 : \mu_D = 0$ versus $H_a : \mu_D \neq 0$.

b. What happens if the two-sample t test is used ($s_1 = 2.48$ and $s_2 = 2.91$)?

32. The paper "Relative Controllability of Dissimilar Cars" (*Human Factors,* 1962, pp. 375–380) reported results of an experiment to compare handling ability for two cars having quite different lengths, wheelbases, and turning radii. The observations are time in seconds required for subjects to parallel park each car.

Subject:	1	2	3	4	5	6	7
Car A:	37.0	25.8	16.2	24.2	22.0	33.4	23.8
Car B:	17.8	20.2	16.8	41.4	21.4	38.4	16.8

Subject:	8	9	10	11	12	13	14
Car A:	58.2	33.6	24.4	23.4	21.2	36.2	29.8
Car B:	32.2	27.8	23.2	29.6	20.6	32.2	53.8

Does the data suggest that the average person will more easily handle one car than the other? Test the relevant hypotheses using $\alpha = .10$.

33. In an experiment designed to study the effects of illumination level on task performance ("Performance of Complex Tasks Under Different Levels of Illumination," *J. Illuminating Eng.,* 1976, pp. 235–242), subjects were required to insert a fine-tipped probe into the eyeholes of 10 needles in rapid succession both for a low light level with a black background and a higher level with a white background.

Subject:	1	2	3	4	5
Black:	25.85	28.84	32.05	25.74	20.89
White:	18.23	20.84	22.96	19.68	19.50

Subject:	6	7	8	9
Black:	41.05	25.01	24.96	27.47
White:	24.98	16.61	16.07	24.59

Does the data indicate that the higher level of illumination yields a decrease of more than 5 in true average task completion time?

34. The paper "Selection of a Method to Determine Residual Chlorine in Sewage Effluents" (*Water and Sewage Works,* 1971, pp. 360–364) reported the results of an experiment in which two different methods for determining chlorine content were used on samples of Cl_2-demand-free water for various doses and contact times. Observations are in mg/1.

Sample:	1	2	3	4
MSI method:	.39	.84	1.76	3.35
SIB method:	.36	1.35	2.56	3.92

Sample:	5	6	7	8
MSI method:	4.69	7.70	10.52	10.92
SIB method:	5.35	8.33	10.70	10.91

Construct a 99% confidence interval for the difference in true average residual chlorine readings between the two methods.

35. Many researchers have investigated the relationship between stress and reproductive efficiency. One such study is described in the paper "Stress or Acute Adrenocorticotrophin Treatment Suppresses LHRH-Induced LH Release in the Ram" (*J. Repro. and Fertility,* 1984, pp. 385–393). Seven rams were used in the study, and LH (luteinizing hormone) release (ng/min) was recorded before and after treatment with ACTH (adrenocorticotrophin, a drug that results in stimulation of the adrenal gland). Use the accompanying data to determine if there is a significant reduction in mean LH release following treatment with ACTH; first place bounds on the P-value and then carry out a test with $\alpha = .001$.

Ram:	1	2	3	4	5	6	7
Before:	2400	1400	1375	1325	1200	1150	850
After:	2250	1425	1100	800	850	925	700

36. The article "Sex and Race Discrimination in the New-Car Showroom: A Fact or Myth," (*J. Consumer Affairs,* 1977, pp. 107–113) reported the results of an experiment in which individuals of different races and sexes visited car dealerships to request the best possible deal on a certain car. Consider the following representative (hypothetical) data:

Dealership:	1	2	3	4	5
Black female:	4459	4320	4268	4585	4736
White male:	4348	4385	4231	4516	4550

Dealership:	6	7	8	9
Black female:	4262	4440	4398	4823
White male:	4203	4285	4408	4570

Compute a 95% confidence interval for the true average difference between best price offered black females and white males.

37. The urinary fluoride concentration (ppm) was determined for 11 randomly chosen livestock both at the beginning of and in the middle of their grazing period in a region previously exposed to fluoride pollution ("Fluoride Pollution Caused by a Brickworks in the Flemish Countryside of Belgium," *Int. J. Environmental Studies,* 1978, pp. 245–252).

Subject:	1	2	3	4	5	6
Beginning:	24.7	46.1	18.5	29.5	26.3	33.9
Middle:	12.4	14.1	7.6	9.5	19.7	10.6

Subject:	7	8	9	10	11
Beginning:	23.1	20.7	18.0	19.3	23.0
Middle:	9.1	11.5	13.3	8.3	15.0

Does the data suggest that there has been a decrease in the true average urinary fluoride concentration during the period under consideration? *Note:* $s_1 = 8.31$ and $s_2 = 3.53$, suggesting that perhaps $\sigma_1 \neq \sigma_2$, but the paired t test is still valid.

38. Referring back to Exercise 33, compute an interval estimate for the difference between true average task time under the high illumination level and true average time under the low level.

39. Construct a paired data set for which $t_{\text{paired}} = \infty$, so that the data is highly significant when the correct analysis is used, yet t for the two-sample t test is quite near zero, so the incorrect analysis yields an insignificant result.

9.4 Inferences Concerning a Difference between Population Proportions

Having presented methods for comparing the means of two different populations, we now turn to the comparison of two population proportions. The notation for this problem is an extension of the notation used in the corresponding one-population problem. We let p_1 and p_2 denote the proportion of individuals in population one and two, respectively, who possess a particular characteristic. Alternatively, if we use the label S for an individual who possesses the characteristic of interest (does favor a particular proposition, is a member of a particular political party, has read at least one book within the last month, and so on), then p_1 and p_2 represent the probabilities of seeing the label S on a randomly chosen individual from populations one and two, respectively.

We shall assume the availability of a sample of m individuals from the first population and n from the second. The variables X and Y will represent the number of individuals in each sample possessing the characteristic that defines p_1 and p_2. Provided the population sizes are much larger than the sample sizes, the distribution of X can be taken to be binomial with parameters m and p_1, and similarly Y is taken to be a binomial variable with parameters n and p_2. Furthermore, the samples are assumed to be independent of one another, so that X and Y are independent random variables.

The obvious estimator for $p_1 - p_2$, the difference in population proportions, is the corresponding difference in sample proportions $X/m - Y/n$. With $\hat{p}_1 = X/m$ and $\hat{p}_2 = Y/n$, the estimator of $p_1 - p_2$ can be expressed as $\hat{p}_1 - \hat{p}_2$.

Proposition

Let $X \sim \text{Bin}(m, p_1)$ and $Y \sim \text{Bin}(n, p_2)$ with X and Y independent variables. Then

$$E(\hat{p}_1 - \hat{p}_2) = p_1 - p_2$$

so $\hat{p}_1 - \hat{p}_2$ is an unbiased estimator of $p_1 - p_2$, and

$$V(\hat{p}_1 - \hat{p}_2) = \frac{p_1 q_1}{m} + \frac{p_2 q_2}{n} \qquad \text{(where } q_i = 1 - p_i) \qquad (9.6)$$

Proof. Since $E(X) = mp_1$ and $E(Y) = np_2$,

$$E\left(\frac{X}{m} - \frac{Y}{n}\right) = \frac{1}{m} E(X) - \frac{1}{n} E(Y) = \frac{1}{m} mp_1 - \frac{1}{n} np_2 = p_1 - p_2$$

Since $V(X) = mp_1 q_1$, $V(Y) = np_2 q_2$, and X and Y are independent,

$$V\left(\frac{X}{m} - \frac{Y}{n}\right) = V\left(\frac{X}{m}\right) + V\left(\frac{Y}{n}\right) = \frac{1}{m^2} V(X) + \frac{1}{n^2} V(Y) = \frac{p_1 q_1}{m} + \frac{p_2 q_2}{n}$$

■

We shall focus first on situations in which both m and n are large. Then because $\hat{p}_1$ and $\hat{p}_2$ individually have approximately normal distributions, the estimator $\hat{p}_1 - \hat{p}_2$ also has approximately a normal distribution. Using the mean value and variance from the above proposition to standardize $\hat{p}_1 - \hat{p}_2$ yields a variable Z whose distribution is approximately standard normal:

$$Z = \frac{\hat{p}_1 - \hat{p}_2 - (p_1 - p_2)}{\sqrt{\dfrac{p_1 q_1}{m} + \dfrac{p_2 q_2}{n}}}$$

A Large-Sample Test Procedure

Analogously to the hypotheses for $\mu_1 - \mu_2$, the most general null hypothesis an investigator might consider would be of the form $H_0 : p_1 - p_2 = \Delta_0$, where Δ_0 is again a specified number. Although for population means the case $\Delta_0 \neq 0$ presented no difficulties, for population proportions the cases $\Delta_0 = 0$ and $\Delta_0 \neq 0$ must be considered separately. Since the vast majority of actual problems of this sort involve $\Delta_0 = 0$ (that is, the null hypothesis $p_1 = p_2$), we will concentrate on this case. When $H_0 : p_1 - p_2 = 0$ is true, let p denote the common value of p_1 and p_2 (and similarly for q). Then the standardized variable

$$Z = \frac{\hat{p}_1 - \hat{p}_2 - 0}{\sqrt{pq\left(\dfrac{1}{m} + \dfrac{1}{n}\right)}} \qquad (9.7)$$

has approximately a standard normal distribution when H_0 is true. However, this Z cannot serve as a test statistic because the value of p is unknown—H_0 only asserts that there is a common value of p, but doesn't say what that value is. To obtain a test statistic having approximately a standard normal distribution when H_0 is true (so that use of an appropriate z critical value specifies a level α test), p must be estimated from the sample data.

Assuming then that $p_1 = p_2 = p$, instead of separate samples of size m and n from two different populations (two different binomial distributions), we really have a single sample of size $m + n$ from one population with proportion p. Since the total number of individuals in this combined sample having the characteristic of interest is $X + Y$, the estimator of p is

$$\hat{p} = \frac{X + Y}{m + n} = \frac{m}{m + n} \hat{p}_1 + \frac{n}{m + n} \hat{p}_2 \tag{9.8}$$

The far right-hand side of (9.8) shows that p is actually a weighted average of estimators $\hat{p}_1$ and $\hat{p}_2$ obtained from the two samples. If we take (9.8) (with $\hat{q} = 1 - \hat{p}$) and substitute back into (9.7), the resulting statistic has approximately a standard normal distribution when H_0 is true.

Null hypothesis: $H_0 : p_1 - p_2 = 0$

Test statistic value (large samples): $z = \dfrac{\hat{p}_1 - \hat{p}_2}{\sqrt{\hat{p}\hat{q}(1/m + 1/n)}}$

Alternative hypothesis	Rejection region for approximate level α test
$H_a : p_1 - p_2 > 0$	$z \geq z_\alpha$
$H_a : p_1 - p_2 < 0$	$z \leq -z_\alpha$
$H_a : p_1 - p_2 \neq 0$	either $z \geq z_{\alpha/2}$ or $z \leq -z_{\alpha/2}$

Example 9.16   The paper "Evaluation of Telephone Energy Information Centers in Minnesota" (*J. Environmental Systems,* 1980, pp. 229–248) reported that in a random sample of 248 callers to a center operated by the Northern States Power Co., 84 had purchased at least one new appliance rated as energy efficient, while in a control sample of 270 individuals who had not called the center, 59 had purchased such an appliance. Test at level .05 to see if the data indicates that the true proportion of callers who purchased such an appliance exceeds the true proportion of noncallers who made this type of purchase.

With p_1 and p_2 denoting the true proportion of all callers and noncallers, respectively, who purchased an energy-efficient appliance, the hypotheses of interest are $H_0 : p_1 - p_2 = 0$ versus $H_a : p_1 - p_2 > 0$. At level .05, H_0 will be rejected if $z \geq z_{.05} = 1.645$. The necessary sample quantities are $\hat{p}_1 = x/m = 84/248 = .339$, $\hat{p}_2 = y/n = 59/270 = .219$, and $\hat{p} = (x + y)/(m + n) = (84 + 59)/(248 + 270) = 143/518 = .276$, so the computed value of Z is

$$z = \frac{.339 - .219}{\sqrt{(.276)(.724)\left(\frac{1}{248} + \frac{1}{270}\right)}} = \frac{.120}{.0393} = 3.05$$

Since 3.05 is ≥ 1.645, H_0 is rejected in favor of the conclusion that $p_1 > p_2$. ■

Example 9.17 Some defendants in criminal proceedings plead guilty and are sentenced without a trial, whereas others who plead innocent are subsequently found guilty, and then are sentenced. In recent years legal scholars have speculated as to whether sentences of those who plead guilty differ in severity from sentences for those who plead innocent and are subsequently judged guilty. Consider the accompanying data on defendants from San Francisco County accused of robbery, all of whom had previous prison records ("Does It Pay to Plead Guilty? Differential Sentencing and the Functioning of Criminal Courts," *Law and Society Rev.,* 1981–1982, pp. 45–69). Does this data suggest that the proportion of all defendants in these circumstances who plead guilty and are sent to prison differs from the proportion who are sent to prison after pleading innocent and being found guilty?

	Plea	
	Guilty	*Not guilty*
Number judged guilty	$m = 191$	$n = 64$
Number sentenced to prison	$x = 101$	$y = 56$
Sample proportion	$\hat{p}_1 = .529$	$\hat{p}_2 = .875$

Let p_1 and p_2 denote the two population proportions. The hypotheses of interest are $H_0 : p_1 - p_2 = 0$ versus $H_a : p_1 - p_2 \neq 0$. Rather than fixing a level of significance α, stating the corresponding rejection region, and carrying out the test, let's compute a P-value. The combined estimate of the common success proportion is $\hat{p} = (101 + 56)/(191 + 64) = .616$. The value of the test statistic is then

$$z = \frac{.529 - .875}{\sqrt{(.616)(.384)(1/191 + 1/64)}} = \frac{-.346}{.070} = -4.94$$

The P-value for a two-tailed z test is

$$P\text{-value} = 2[1 - \Phi(|z|)] = 2[1 - \Phi(4.94)] < 2[1 - \Phi(3.49)] = .0004$$

A more extensive standard normal table yields P-value $\approx .0000006$. This P-value is so minuscule that at any reasonable level α, H_0 should be rejected. The data very strongly suggests that $p_1 \neq p_2$ and, in particular, that initially pleading guilty may be a good strategy as far as avoiding prison is concerned.

The cited article also reported data on defendants in several other counties. The authors broke down the data by type of crime (burglary or robbery) and by nature of prior record (none, some but no prison, and prison). In every case the conclusion was the same: among defendants judged guilty, those who pleaded that way were less likely to receive prison sentences.   ■

Type II Error Probabilities and Sample Sizes

Here the determination of β is a bit more cumbersome than it was for other large-sample tests. The reason is that the denominator of Z is an estimate of the standard deviation of $\hat{p}_1 - \hat{p}_2$ assuming that $p_1 = p_2 = p$. When H_0 is false, $\hat{p}_1 - \hat{p}_2$ must be restandardized using

$$\sigma_{\hat{p}_1 - \hat{p}_2} = \sqrt{\frac{p_1 q_1}{m} + \frac{p_2 q_2}{n}} \tag{9.9}$$

The form of σ implies that β is not a function of just $p_1 - p_2$, so we denote it by $\beta(p_1, p_2)$.

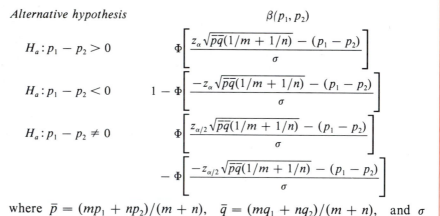

Alternative hypothesis	$\beta(p_1, p_2)$
$H_a : p_1 - p_2 > 0$	$\Phi\left[\dfrac{z_\alpha \sqrt{\bar{p}\bar{q}(1/m + 1/n)} - (p_1 - p_2)}{\sigma}\right]$
$H_a : p_1 - p_2 < 0$	$1 - \Phi\left[\dfrac{-z_\alpha \sqrt{\bar{p}\bar{q}(1/m + 1/n)} - (p_1 - p_2)}{\sigma}\right]$
$H_a : p_1 - p_2 \ne 0$	$\Phi\left[\dfrac{z_{\alpha/2} \sqrt{\bar{p}\bar{q}(1/m + 1/n)} - (p_1 - p_2)}{\sigma}\right]$ $- \Phi\left[\dfrac{-z_{\alpha/2} \sqrt{\bar{p}\bar{q}(1/m + 1/n)} - (p_1 - p_2)}{\sigma}\right]$

where $\bar{p} = (mp_1 + np_2)/(m + n)$, $\quad \bar{q} = (mq_1 + nq_2)/(m + n)$, and σ is given by (9.9).

Proof.   For the upper-tailed test ($H_a : p_1 - p_2 > 0$)

$$\beta(p_1, p_2) = P\left[\hat{p}_1 - \hat{p}_2 < z_\alpha \sqrt{\hat{p}\hat{q}(1/m + 1/n)}\right]$$

$$= P\left[\frac{\hat{p}_1 - \hat{p}_2 - (p_1 - p_2)}{\sigma} < \frac{z_\alpha \sqrt{\hat{p}\hat{q}(1/m + 1/n)} - (p_1 - p_2)}{\sigma}\right]$$

When m and n are both large,

$$\hat{p} = (m\hat{p}_1 + n\hat{p}_2)/(m + n) \approx (mp_1 + np_2)/(m + n) = \bar{p}$$

and $\hat{q} \approx \bar{q}$, which yields the above (approximate) expression for $\beta(p_1, p_2)$.   ■

Alternatively, for specified p_1, p_2 with $p_1 - p_2 = d$, the sample sizes necessary to achieve $\beta(p_1, p_2) = \beta$ can be determined. For example, for the upper-tailed test, we equate $-z_\beta$ to the argument of $\Phi(\cdot)$ (that is, what's inside the parentheses) in the above table. If $m = n$, there is a simple expression for the common value.

For the case $m = n$, the level α test has type II error probability β at the alternative values p_1, p_2 with $p_1 - p_2 = d$ when

$$n = \frac{\left[z_\alpha \sqrt{(p_1 + p_2)(q_1 + q_2)/2} + z_\beta \sqrt{p_1 q_1 + p_2 q_2} \right]^2}{d^2} \tag{9.10}$$

for an upper- or lower-tailed test, with $\alpha/2$ replacing α for a two-tailed test.

Example 9.18   One of the truly impressive applications of statistics occurred in connection with the design of the 1954 Salk polio vaccine experiment and analysis of the resulting data. Part of the experiment focused on the efficacy of the vaccine in combating paralytic polio. Because it was felt that without a control group of children, there would be no sound basis for assessment of the vaccine, it was decided to administer the vaccine to one group and a placebo injection (visually indistinguishable from the vaccine but known to have no effect) to a control group. For ethical reasons and also because it was thought that the knowledge of vaccine administration might have an effect on treatment and diagnosis, the experiment was conducted in a **double blind** manner. That is, neither the individuals receiving injections nor those administering them actually knew who was receiving vaccine and who was receiving the placebo (samples were numerically coded)—remember, at that point it was not at all clear whether the vaccine was beneficial.

Letting p_1 and p_2 be the probabilities of a child getting paralytic polio for the control and treatment condition, respectively, the objective was to test $H_0 : p_1 - p_2 = 0$ versus $H_a : p_1 - p_2 > 0$ (the alternative states that a vaccinated child is less likely to contract polio than an unvaccinated child). Supposing that the true value of p_1 is .0003 (an incidence rate of 30 per 100,000), the vaccine would be a significant improvement if the incidence rate was halved—that is, $p_2 = .00015$. Using a level $\alpha = .05$ test, it would then be reasonable to ask for sample sizes for which $\beta = .1$ when $p_1 = .0003$ and $p_2 = .00015$. Assuming equal sample sizes, the required n is obtained from (9.10) as

$$n = \frac{\left[1.645 \sqrt{(.5)(.00045)(1.99955)} + 1.28 \sqrt{(.00015)(.99985) + (.0003)(.9997)} \right]^2}{(.0003 - .00015)^2}$$

$$= [(.0349 + .0272)/.00015]^2 \approx 171,400$$

The actual data for this experiment appears below. Sample sizes of approximately 200,000 were used. The reader can easily verify that $z = 6.47$, a highly significant value. The vaccine was judged a resounding success! An excellent expository article describing the experiment and results appears in *Statistics: A Guide to the Unknown* edited by Judith Tanur.

placebo: $m = 201{,}229$ $x =$ number of cases of paralytic polio $= 110$
vaccine: $n = 200{,}745$ $y = 33$

■

A Large-Sample Confidence Interval for $p_1 - p_2$

As with means, many two-sample problems involve the objective of comparison through hypothesis testing, but sometimes an interval estimate for $p_1 - p_2$ is appropriate. Both $\hat{p}_1 = X/m$ and $\hat{p}_2 = Y/n$ have approximate normal distributions when m and n are both large. If we identify θ with $p_1 - p_2$, then $\hat{\theta} = \hat{p}_1 - \hat{p}_2$ satisfies the conditions necessary for obtaining a large-sample confidence interval. In particular the estimated standard deviation of $\hat{\theta}$ is $\sqrt{(\hat{p}_1\hat{q}_1/m) + (\hat{p}_2\hat{q}_2/n)}$. The $100(1 - \alpha)\%$ interval $\hat{\theta} \pm z_{\alpha/2} \cdot \hat{\sigma}_{\hat{\theta}}$ then becomes

$$\hat{p}_1 - \hat{p}_2 \pm z_{\alpha/2}\sqrt{\frac{\hat{p}_1\hat{q}_1}{m} + \frac{\hat{p}_2\hat{q}_2}{n}}$$

Notice that the estimated standard deviation of $p_1 - p_2$ (the square root expression) is different here from what it was for hypothesis testing when $\Delta_0 = 0$.

Example 9.19   The paper "The Association of Marijuana Use with Outcome of Pregnancy" (*Amer. J. Public Health*, 1983, pp. 1161–1164) reported the accompanying data on incidence of major malfunctions among newborns both for mothers who were marijuana users and mothers who did not use marijuana.

	User	*Nonuser*
Sample size	1246	11,178
Number of major malfunctions	42	294
$\hat{p}$	.0337	.0263

Let p_1 denote the proportion of births that involve a major malfunction among all mothers who are marijuana users, and define p_2 similarly for nonusers. A 99% confidence interval for $p_1 - p_2$ requires

$$2.58\sqrt{\frac{\hat{p}_1\hat{q}_1}{m} + \frac{\hat{p}_2\hat{q}_2}{n}} = 2.58\sqrt{\frac{(.0337)(.9663)}{1246} + \frac{(.0263)(.9737)}{11{,}178}}$$

$$= (2.58)(.00533) = .0138$$

The confidence interval is then $.0337 - .0263 \pm .0138 = .0074 \pm .0138 = (-.0064, .0212)$. This interval is quite narrow, which suggests that $p_1 - p_2$ has been precisely estimated. Notice that the interval contains zero, so this is one plausible value for $p_1 - p_2$.

■

Small-Sample Inferences

On occasion an inference concerning $p_1 - p_2$ may have to be based on samples for which at least one sample size is small. Appropriate methods for such situations are not as straightforward as those for large samples, and there is more controversy among statisticians as to recommended procedures. One frequently used test, called the Fisher-Irwin test, is based on the hypergeometric distribution. A statistician can be consulted for more information.

Exercises / Section 9.4 (40–48)

40. Is someone who switches brands because of a financial inducement less likely to remain loyal than someone who switches without inducement? Let p_1 and p_2 denote the true proportions of switchers to a certain brand with and without inducement, respectively, who subsequently make a repeat purchase. Test $H_0: p_1 - p_2 = 0$ versus $H_a: p_1 - p_2 < 0$ using $\alpha = .01$ and the following data:

$m = 200$, number of successes $= 30$, $n = 600$, number of successes $= 180$.

(Similar data appears in "Impact of Deals and Deal Retraction on Brand Switching," *J. Marketing*, 1980, pp. 62–70.)

41. A sample of 300 urban adult residents of a particular state revealed 63 who favored increasing the highway speed limit from 55 to 65 mph, while a sample of 180 rural residents yielded 75 who favored the increase. Does this data indicate that the sentiment for increasing the speed limit is different for the two groups of residents?

  a. Test $H_0: p_1 = p_2$ versus $H_a: p_1 \neq p_2$ using $\alpha = .05$, where p_1 refers to the urban population.

  b. If the true proportions favoring the increase are actually $p_1 = .20$ (urban) and $p_2 = .40$ (rural), what is the probability that H_0 will be rejected using a level .05 test with $m = 300$, $n = 180$?

42. A study of the relationship between level of occupational prestige and self-evaluation of competence yielded the following data, where success here refers to an individual who had a high self-competence rating ("The Influence of Educational Attainment on Self-Evaluation of Competence," *Sociology Education,* 1972, pp. 303–311).

Low occupational prestige: $m = 175$, $x = 77$
High occupational prestige: $n = 194$, $y = 107$

Does the data indicate that a person with high occupational prestige is more likely to exhibit high self-esteem than an individual with low occupational prestige? Test the relevant hypotheses using $\alpha = .05$ by first computing the *P*-value.

43. A random sample of 5726 telephone numbers taken in March 1970 yielded 1105 that were unlisted, and one year later a sample of 5384 yielded 980 unlisted numbers.

  a. Test at level .10 to see if there is a difference in true proportions of unlisted numbers between the two years.

  b. If $p_1 = .20$ and $p_2 = .18$, what sample sizes ($m = n$) would be necessary to detect such a difference with probability .90?

44. A recent study (*Lancet,* October 1976) reported the following data on survival rate among patients suffering cardiac arrest both when resuscitation was started by (trained) lay people and when it was delayed until the arrival of an ambulance crew.

Lay: $m = 75$,
              number survived $= x = 27$
Ambulance: $n = 556$,
              number survived $= y = 43$

Does the data suggest that a program to train lay people to perform resuscitation would be beneficial?

45. To study the effect of wording on sentiment for gun registration, two different samples of voters were obtained, and each member of the first (second) sample was asked to respond to question A (B) below.

A: Would you favor or oppose a law that would require a person to obtain a police permit before he could buy a gun?

B: Would you ... a gun, or do you think that such a law would interfere too much with the right of citizens to own guns? (where ... is worded identically to A).

Question A: $m = 615$,
  number who favor $= 463$
Question B: $m = 585$,
  number who favor $= 403$

Compute a 90% confidence interval for the difference $p_1 - p_2$ in true proportions of individuals who would respond, "I favor such a law." ("Attitude Measurement and the Gun Control Paradox," *Public Opinion Q.,* Winter 1977–1978, pp. 427–438.)

46. Sometimes experiments involving success/failure responses are run in a paired or before/after manner. Suppose that before a major policy speech by a political candidate, n individuals are selected and asked whether (S) or not (F) they favor the candidate. Then after the speech the same n people are asked the same question. The responses can be entered in a square table as follows:

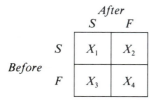

where $X_1 + X_2 + X_3 + X_4 = n$

Let p_1, p_2, p_3, and p_4 denote the four cell probabilities, so that $p_1 = P(S$ before and S after), and so on. We wish to test the hypothesis that the true proportion of supporters (S) after the speech

has not increased against the alternative that it has increased.
  a. State the two hypotheses of interest in terms of p_1, p_2, p_3, and p_4.
  b. Construct an estimator for the after/before difference in success probabilities.
  c. When n is large, it can be shown that the random variable $(X_i - X_j)/n$ has approximately a normal distribution with variance $[p_i + p_j - (p_i - p_j)^2]/n$. Use this to construct a test statistic with approximately a standard normal distribution when H_0 is true (the result is called McNemar's test).
  d. If $x_1 = 350$, $x_2 = 150$, $x_3 = 200$, and $x_4 = 300$, what do you conclude?

47. Two different types of alloy, A and B, have been used to manufacture experimental specimens of a small tension link to be used in a certain engineering application. The ultimate strength (ksi) of each specimen was determined, and the results are summarized in the accompanying frequency distribution.

	A	B
26–under 30	6	4
30–under 34	12	9
34–under 38	15	19
38–under 42	7	10
	$m = 40$	$n = 42$

Compute a 95% confidence interval for the difference between the true proportion of all specimens of alloys A and B that have an ultimate strength of at least 34 ksi.

48. Suppose that a 95% confidence interval for $p_1 - p_2$ is to be constructed based on equal sample sizes from the two populations. For what value of n ($= m$) will the resulting interval have length at most .1 irrespective of the results of the sampling?

9.5   Inferences Concerning Two Population Variances

Methods for comparing two population variances (or standard deviations) are occasionally needed, though such problems arise much less frequently than those involving means or proportions. For the case in which the populations under investigation are normal, the procedures are based on a new family of probability distributions.

The *F* Distribution

The *F* probability distribution has two parameters, denoted by ν_1 and ν_2. The parameter ν_1 is called the *number of numerator degrees of freedom,* and ν_2 is the *number of denominator degrees of freedom;* here ν_1 and ν_2 are positive integers. A random variable that has an *F* distribution cannot assume a negative value. Since the density function is complicated and will not be used explicitly, we omit the formula. There is an important connection between an *F* variable and chi-squared variables. If X_1 and X_2 are independent chi-squared r.v.'s with ν_1 and ν_2 degrees of freedom, respectively, then the r.v.

$$F = \frac{X_1/\nu_1}{X_2/\nu_2} \qquad (9.11)$$

the ratio of the two chi-squared variables divided by their d.f.'s, can be shown to have an *F* distribution.

Figure 9.3 illustrates the graph of a typical *F* density function. Analogous to the notation $t_{\alpha, \nu}$ and $\chi^2_{\alpha, \nu}$, we use F_{α, ν_1, ν_2} for the point on the axis that captures α of the area under the *F* density function with ν_1 and ν_2 d.f. in the upper tail. The density function is not symmetric, so it would seem that both upper- and lower-tail critical values must be tabulated. This is not necessary, though, because of the following property.

$$F_{1-\alpha, \nu_1, \nu_2} = 1/F_{\alpha, \nu_2, \nu_1} \qquad (9.12)$$

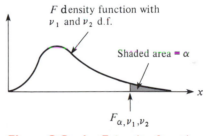

F density function with ν_1 and ν_2 d.f.

Shaded area = α

F_{α, ν_1, ν_2}

Figure 9.3 An *F* density function

Appendix Table A.7 gives F_{α, ν_1, ν_2} for $\alpha = .05$ and $.01$, and various values of ν_1 (in different columns of the table) and ν_2 (in different rows of the table). For example, $F_{.05, 6, 10} = 3.22$ and $F_{.05, 10, 6} = 4.06$. To obtain $F_{.95, 6, 10}$, the number which captures $.95$ of the area to its right (and thus $.05$ to the left) under the *F* curve with $\nu_1 = 6$ and $\nu_2 = 10$, we use (9.12): $F_{.95, 6, 10} = 1/F_{.05, 10, 6} = 1/4.06 = .246$.

Inferential Methods

A test procedure for hypotheses concerning the ratio σ_1^2/σ_2^2 as well as a confidence interval for this ratio are based on the following result.

Theorem

Let $X_1, \ldots, X_m$ be a random sample from a normal distribution with variance σ_1^2, let $Y_1, \ldots, Y_n$ be another random sample (independent of the X_i's) from a normal distribution with variance σ_2^2, and let S_1^2 and S_2^2 denote the two sample variances. Then the random variable

$$F = \frac{S_1^2/\sigma_1^2}{S_2^2/\sigma_2^2} \qquad (9.13)$$

has an F distribution with $\nu_1 = m - 1$ and $\nu_2 = n - 1$.

This theorem results from combining (9.11) with the fact that the variables $(m - 1)\, S_1^2/\sigma_1^2$ and $(n - 1)\, S_2^2/\sigma_2^2$ each have a chi-squared distribution with $m - 1$ and $n - 1$ d.f., respectively (see Section 7.4). Because F involves a ratio rather than a difference, the test statistic is the ratio of sample variances; the claim that $\sigma_1^2 = \sigma_2^2$ is then rejected if the ratio differs by too much from 1.

Null hypothesis: $H_0 : \sigma_1^2 = \sigma_2^2$
Test statistic value: $f = s_1^2/s_2^2$

Alternative hypothesis	Rejection region for a level α test
$H_a : \sigma_1^2 > \sigma_2^2$	$f \geq f_{\alpha,\, m-1,\, n-1}$
$H_a : \sigma_1^2 < \sigma_2^2$	$f \leq f_{1-\alpha,\, m-1,\, n-1}$
$H_a : \sigma_1^2 \neq \sigma_2^2$	either $f \geq f_{\alpha/2,\, m-1,\, n-1}$ or $f \leq f_{1-\alpha/2,\, m-1,\, n-1}$

Since critical values are tabled only for $\alpha = .05$ and $.01$, the two-tailed test can be performed only at levels $.10$ or $.02$. More extensive F tables for testing at other levels are available elsewhere.

Example 9.20

On the basis of data reported in a 1979 article in the *Journal of Gerontology* ("Serum Ferritin in an Elderly Population," pp. 521–524), the authors concluded that the ferritin distribution in the elderly had a smaller variance than in the younger adults. (Serum ferritin is used in diagnosing iron deficiency.) For a sample of 28 elderly males the sample standard deviation of serum ferritin (mg/l) was $s_1 = 52.6$, and for 26 young males the sample standard deviation was $s_2 = 84.2$. Does this data support the above conclusion as applied to males?

Let σ_1^2 and σ_2^2 denote the variance of the serum ferritin distributions for elderly males and young males, respectively. We wish to test $H_0 : \sigma_1^2 = \sigma_2^2$ versus $H_a : \sigma_1^2 < \sigma_2^2$. At level $.01$, H_0 will be rejected if $f \leq F_{.99, 27, 25}$. To obtain the critical value, we need $F_{.01, 25, 27}$. From the F table, $F_{.01, 25, 27} \approx 2.54$ (since $F_{.01, 24, 27} = 2.55$), so $F_{.99, 27, 25} \approx 1/2.54 = .394$. The computed value of F is $(52.6)^2/(84.2)^2 = .390$. Since $.390 \leq .394$, H_0 is rejected at level $.01$ in favor of H_a, so variability does appear to be greater in young males than in elderly males. ∎

The validity of the two-sample t test rested on the assumption that $\sigma_1^2 = \sigma_2^2$. It is tempting to use the F test as a preliminary test to decide whether or not the two-sample t test should be used (use it if $H_0 : \sigma_1^2 = \sigma_2^2$ cannot be rejected in favor of $H_a : \sigma_1^2 \neq \sigma_2^2$), but we do not recommend this. The reason is that the F test may yield a significant value of F not only when H_0 is false but also when the underlying populations are not normal—it is sensitive to departures from normality as well as from H_0. Since the t test is a good test as long as the underlying populations are not too nonnormal, using a preliminary F test might result in not using the t test for the wrong reason. An additional complication is that the compound procedure (F at level α followed by t at level α) will not necessarily have level of significance α.

The confidence interval for σ_1^2/σ_2^2 is based on replacing F in the probability statement

$$P(F_{1-\alpha/2,\, \nu_1,\, \nu_2} < F < F_{\alpha/2,\, \nu_1,\, \nu_2}) = 1 - \alpha$$

by the F variable (9.13) and manipulating the inequalities to isolate σ_1^2/σ_2^2. An interval for σ_1/σ_2 results from taking the square root of each limit. The details are left for an exercise.

Exercises / Section 9.5 (49–54)

49. Obtain or compute the following quantities:
  a. $F_{.05,\,5,\,8}$ **b.** $F_{.05,\,8,\,5}$ **c.** $F_{.95,\,5,\,8}$ **d.** $F_{.95,\,8,\,5}$
  e. the ninety-ninth percentile of the F distribution with $\nu_1 = 10$, $\nu_2 = 12$
  f. the first percentile of the F distribution with $\nu_1 = 10$, $\nu_2 = 12$
  g. $P(F \le 6.16)$ for $\nu_1 = 6$, $\nu_2 = 4$
  h. $P(.177 \le F \le 4.74)$ for $\nu_1 = 10$, $\nu_2 = 5$

50. The sample standard deviation of sodium concentration in whole blood (m-equiv./l) for $m = 20$ marine eels was found to be $s_1 = 40.5$, while the sample standard deviation of concentration for $n = 20$ freshwater eels was $s_2 = 32.1$ ("Ionic Composition of the Plasma and Whole Blood of Marine and Freshwater Eels," *Comp. Biochemistry and Physiology,* 1974, pp. 541–544). Assuming normality of the two concentration distributions, test at level .10 to see if the data suggests any difference between concentration variances for the two types of eels.

51. A general contractor is considering purchasing lumber from one of two different suppliers. A sample of 12 2 in. $\times$ 4 in.'s of a certain length is obtained from each supplier and the length of each board is measured. The sample standard deviation of length for the first supplier's boards is found to be $s_1 = .13$ in., while $s_2 = .17$ in. for the second supplier. Does this data indicate that the lengths of one supplier's 2 in. $\times$ 4 in.'s are subject to more variability than those of the other supplier? Test using $\alpha = .02$, assuming normality.

52. The fat content (%) was determined for $m = 10$ samples of ground chuck and $n = 16$ samples of ground beef purchased at a certain market, yielding $s_1 = 1.4$ and $s_2 = 2.2$. Does this data indicate that there is more variability in the fat content of ground beef than of ground chuck? Assuming normality, test the relevant hypotheses using $\alpha = .05$.

53. In a study of copper deficiency in cattle, the copper values (μg Cu/100 ml blood) were determined both for cattle grazing in an area known to have well-defined molybdenum anomalies (metal values in excess of the normal range of regional variation) and for cattle grazing in a non-anomalous area ("An Investigation into Copper Deficiency in Cattle in the Southern Pennines," *J. Agricultural Soc. Cambridge,* 1972, pp. 157–163), resulting in $s_1 = 21.5$ ($m = 48$) for the anomalous condition and $s_2 = 19.45$ ($n = 45$) for the nonanomalous condition. Test for the equality versus inequality of population variances ($\alpha = .05$).

54. The article "Enhancement of Compressive Properties of Failed Concrete Cylinders with Polymer Impregnation" (*J. Testing and Evaluation*, 1977, pp. 333–337) reported the following data on impregnated compressive modulus (psi $\times 10^6$)

when two different polymers were used to repair cracks in failed concrete.

Epoxy: 1.75, 2.12, 2.05, 1.97
MMA prepolymer: 1.77, 1.59, 1.70, 1.69

Obtain a 90% confidence interval for the ratio of variances by first using the method suggested in the text to obtain a general confidence interval formula.

Supplementary Exercises / Chapter 9 (55–76)

55. The molar percentage of bile salts was determined both for a sample of 12 children suffering from cystic fibrosis who were receiving a pancreatic enzyme treatment and for a second sample of 14 who had ceased taking the treatment for one week. Does the data suggest that cessation of the treatment leads to an increase in the true average molar percentage of bile salts? Test the relevant hypotheses at level .05; be sure to state any assumptions that underlie your analysis.

Treated: 54.8, 70.8, 54.3, 64.5, 50.5, 73.3, 86.4, 67.4, 82.2, 55.1, 60.6, 50.8
Untreated: 61.5, 54.8, 62.4, 74.3, 80.3, 77.7, 78.2, 74.0, 48.3, 61.0, 36.5, 57.0, 40.9, 43.7

("Abnormal Biliary Lipid Composition in Cystic Fibrosis," *New England J. Medicine*, December 15, 1977, pp. 1301–1305.)

56. In Exercise 55, describe how the experiment could be done in a paired fashion using 13 children. What are the advantages and disadvantages of such an approach? Would pairing in this particular situation appear to be a good strategy? Why or why not?

57. A famous paper on the effects of marijuana smoking ("Clinical and Psychological Effects of Marijuana in Man," *Science*, December 1968, pp. 1234–1241) described the results of an experiment in which the change in heartbeat rate was measured for $n = 9$ naive subjects (who had never before used marijuana) both 15 minutes after smoking at a low dose level and 15 minutes after smoking a placebo (untreated) cigarette.

Subject:	1	2	3	4	5	6	7	8	9
Placebo:	16	12	8	20	8	10	4	−8	8
Low dose:	20	24	8	8	4	20	28	20	20

Does the data suggest that marijuana smoking leads to a greater increase in heartbeat rate than does smoking a placebo cigarette? Test using $\alpha = .01$.

58. The accompanying data was obtained in a study to evaluate the liquefaction potential at a proposed nuclear power station ("Cyclic Strengths Compared for Two Sampling Techniques," *J. Geotechnical Division, Am. Soc. Civil Engrs. Proceedings*, 1981, pp. 563–576). Prior to cyclic strength testing, soil samples were gathered using both a pitcher tube method and a block method, resulting in the following observed values of dry density (lbs/ft^3).

Pitcher sampling: 101.1, 111.1, 107.6, 98.1, 99.5, 98.7, 103.3, 108.9, 109.1, 104.1, 110.0, 98.4, 105.1, 104.5, 105.7, 103.3, 100.3, 102.6, 101.7, 105.4, 99.6, 103.3, 102.1, 104.3

Block sampling: 107.1, 105.0, 98.0, 97.9, 103.3, 104.6, 100.1, 98.2, 97.9, 103.2, 96.9.

Calculate a 95% confidence interval for the difference between true average dry densities using the two sampling methods.

59. The accompanying table summarizes data on body weight gain (*g*) for both a sample of control animals and a sample of animals given a 1 mg/pellet dose of a certain soft steroid ("The Soft Drug Approach," *CHEMTECH*, 1984, pp. 28–38).

Treatment	Sample size	Sample mean	Sample SD
Control	10	40.5	2.5
Soft steroid	8	32.8	2.6

Does the data indicate that true average weight gain for control animals differs from that for animals given the soft steroid? Carry out an appropriate test.

60. A sample of both surface soil and subsoil was taken from eight randomly selected agricultural locations in a particular county. The soil samples were analyzed to determine both surface pH and subsoil pH, with the following results:

Location:	1	2	3	4
Surface pH:	6.55	5.98	5.59	6.17
Subsoil pH:	6.78	6.14	5.80	5.91

Location:	5	6	7	8
Surface pH:	5.92	6.18	6.43	5.68
Subsoil pH:	6.10	6.01	6.18	5.88

Compute a 90% confidence interval for the true average difference between surface and subsoil pH for agricultural land in this county. What assumptions have you made about the underlying pH distributions?

61. An experimenter wishes to obtain a confidence interval for the difference between true average breaking strength for cables manufactured by company I and by company II. Suppose that breaking strength is normally distributed for both types of cable with $\sigma_1 = 30$ psi and $\sigma_2 = 20$ psi.
   a. If costs dictate that the sample size for the type I cable should be three times the sample size for the type II cable, how many observations are required if the 99% confidence interval is to be no longer than 20 psi?
   b. Suppose that a total of 400 observations is to be made. How many of the observations should be made on type I cable samples if the length of the resulting interval is to be a minimum?

62. The insulin-binding capacity (pmol/mg protein) was measured for a sample of diabetic rats treated with a low dose of insulin and another sample of diabetic rats treated with a high dose, yielding the following data:

Low dose: $m = 8$, $\bar{x} = 1.98$, $s_1 = .51$
High dose: $n = 12$, $\bar{y} = 1.30$, $s_2 = .35$

(*J. Clinical Investigation*, 1978, pp. 552–560.)
   a. Compute s_p^2, the pooled estimate of σ^2, and s_p.
   b. Does the data indicate that there is any difference in true average insulin-binding capacity due to the dosage level? Use the pooled t test with $\alpha = .001$.
   c. Between what two values does the *P*-value lie?

63. In the article referenced in Exercise 62, there were actually four experimental treatments: control nondiabetic, untreated diabetic, low-dose diabetic, and high-dose diabetic. Denote the sample size for the ith treatment by n_i and the sample variance by S_i^2 ($i = 1, 2, 3, 4$). Assuming that the true variance for each treatment is σ^2, construct a pooled estimator of σ^2 that is unbiased, and verify using rules of expected value that it is indeed unbiased. What is your estimate for the following actual data?

Treatment:	1	2	3	4
Sample size:	16	18	8	12
Sample SD:	.64	.81	.51	.35

64. Suppose that a level .05 test of $H_0 : \mu_1 - \mu_2 = 0$ versus $H_a : \mu_1 - \mu_2 > 0$ is to be performed assuming $\sigma_1 = \sigma_2 = 10$ and normality of both distributions using equal sample sizes ($m = n$). Evaluate the probability of a type II error when $\mu_1 - \mu_2 = 1$ and $n = 25, 100, 2500$, and $10,000$. Can you think of real problems in which the difference $\mu_1 - \mu_2 = 1$ has little practical significance? Would sample sizes of $n = 10,000$ be desirable in such problems?

65. The following data refers to airborne bacteria count (number of colonies/ft³) both for $m = 8$ carpeted hospital rooms and for $n = 8$ uncarpeted rooms ("Microbial Air Sampling in a Carpeted Hospital," *J. Environmental Health*, 1968, p. 405). Does there appear to be a difference in true average bacteria count between carpeted and uncarpeted rooms?

Carpeted: 11.8, 8.2, 7.1, 13.0, 10.8, 10.1, 14.6, 14.0
Uncarpeted: 12.1, 8.3, 3.8, 7.2, 12.0, 11.1, 10.1, 13.7

Suppose that you later learned that all carpeted rooms were in a veterans' hospital while all uncarpeted rooms were in a children's hospital. Would you be able to assess the effect of carpeting? Comment.

66. In a study of the relationship between nightmare frequency and sex, 160 men and 196 women were each asked whether nightmares occurred often (at least one per month) or seldom, with the following results:

Men: $m = 160$, $x = 55$ (often = success)
Women: $n = 192$, $y = 60$

Does the data indicate at level .05 that there is any difference in the true proportions of men and women who often have nightmares? ("Personality Characteristics of Nightmare Sufferers," *J. Nervous and Mental Diseases,* 1971, pp. 29–31).

67. In a sample of 500 male and 500 female high school students enrolled in a college preparatory curriculum, it was found that 230 of the males and only 197 of the females had been encouraged by their male counselors to go on to a four-year college. Does the data suggest that the difference in proportions between males and females encouraged to go on to a four-year college is more than .1? Test using $\alpha = .05$.

68. McNemar's test, developed in Exercise 46, can also be used when individuals are paired off (matched) to yield n pairs and then one member of each pair is given treatment 1 and the other is given treatment 2. Then X_1 is the number of pairs in which both treatments were successful, and similarly for X_2, X_3, and X_4. The test statistic for testing equal efficacy of the two treatments is $(X_2 - X_3)/\sqrt{(X_2 + X_3)}$, which has approximately a standard normal distribution when H_0 is true. Use this to test whether or not the drug ergotamine is effective in the treatment of migraine headaches.

		Ergotamine	
		S	F
Placebo	S	44	34
	F	46	30

The data is fictitious, but the conclusion agrees with that in the paper "Controlled Clinical Trial of Ergotamine Tartrate" (*British Med. J.,* 1970, pp. 325–327).

69. For some time researchers have been unsure as to what actually causes the pharmacological and behavioral effects resulting from smoking marijuana. The two major contenders are Δ^9 tetrahydrocannabinol (Δ^9THC) and one of its metabolites, 11-hydroxy-Δ^9tetrahydrocannabinol (11-OH-Δ^9THC). The article "Intravenous Injection in Man of Δ^9THC and 11-OH-Δ^9THC" (*Science,* 1972, p. 633) reported the results of an experiment in which subjects were given infusions of either Δ^9THC or 11-OH-Δ^9THC and asked to report the moment at which the effect of the drug was first perceived. The (unpaired) data below refers to the necessary dose in micrograms per kilogram of body weight.

Δ^9THC: 19.54, 14.47, 16.00, 24.83, 26.39, 11.49
11-OH-Δ^9THC: 15.95, 25.89, 20.53, 15.52, 14.18, 16.00

Compute a 90% confidence interval for the difference between true average dose-to-perception for Δ^9THC and for 11-OH-Δ^9THC.

70. Tardive dyskinesia denotes a syndrome comprising a variety of abnormal involuntary movements assumed to follow long-term use of antipsychotic drugs. In an experiment to see whether the drug deanol produced a greater improvement from baseline scores than a placebo treatment ("Double Blind Evaluation of Deanol in Tardive Dyskinesia," *J. Amer. Med. Assn.,* 1978, pp. 1997–1998), the two treatments were administered for four weeks each in random order to 14 patients, resulting in the following total severity index (TSI) scores:

Patient:	1	2	3	4	5	6	7
Deanol:	12.4	6.8	12.6	13.2	12.4	7.6	12.1
Placebo:	9.2	10.2	12.2	12.7	12.1	9.0	12.4

Patient:	8	9	10	11	12	13	14
Deanol:	5.9	12.0	1.1	11.5	13.0	5.1	9.6
Placebo:	5.9	8.5	4.8	7.8	9.1	3.5	6.4

(The investigators attributed placebo improvement to frequent attention and the nature of the experiment.) Does the data indicate that on average deanol yields a higher TSI than the placebo treatment? Test using $\alpha = .10$.

71. A random sample of 200 drivers from a particular county who drive a foreign car yielded 115 who use their seatbelts regularly, while another sample of 300 drivers who drive a domestic model yielded 154 who use their seatbelts regularly. Calculate a 99% confidence interval for the difference in proportions of those using seatbelts regularly between drivers of foreign cars and drivers of domestic cars.

72. The paper "Evaluating Variability in Filling Operations" (*Food Tech.*, 1984, pp. 51–55) described two different filling operations used in a ground-beef packing plant. Both filling operations were set to fill packages with 1400 g of ground beef. In a random sample of size 30 taken from each filling operation the resulting means and standard deviations were 1402.24 g and 10.97 g for operation 1 and 1419.63 and 9.96 g for operation 2.
   a. Using a .05 significance level, is there sufficient evidence to indicate that the true mean weight of the packages differs for the two operations?
   b. Does the data from operation 1 suggest that the true mean weight of packages produced by operation 1 is higher than 1400 g? Use a .05 significance level.

73. In Section 9.2 we described a procedure for testing $H_0: \mu_1 - \mu_2 = \Delta_0$ based on a variable T' having approximately a t distribution with d.f. ν estimated from the data.
   a. Use the variable T' to derive an approximate $100(1 - \alpha)\%$ confidence interval for $\mu_1 - \mu_2$ when $\sigma_1 \neq \sigma_2$ (but the populations are still normal).
   b. Compute the 95% interval of (a) for the data in Exercise 54, in which μ_1 and μ_2 refer to the true average compressive modulus for epoxy

polymer and MMA polymer, respectively, used to repair failed concrete.

74. Let $X_1, \ldots, X_m$ be a random sample from a Poisson distribution with parameter λ_1, and let $Y_1, \ldots, Y_n$ be a random sample from another Poisson distribution with parameter λ_2. We wish to test $H_0: \lambda_1 - \lambda_2 = 0$ against one of the three standard alternatives. Since $\mu = \lambda$ for a Poisson distribution, when m and n are large the large-sample z test of Section 9.1 can be used. However, the fact that $V(\overline{X}) = \lambda/n$ suggests that a different denominator should be used in standardizing $\overline{X} - \overline{Y}$. Develop a large-sample test procedure appropriate to this problem, and then apply it to the following data to test whether or not the plant densities for a particular species are equal in two different regions (where each observation is the number of plants found in a randomly located square sampling quadrat having area 1m^2, so for region 1, there were 40 quadrats in which one plant was observed, and so on).

Frequency: 0 1 2 3 4 5 6 7

Region 1: 28 40 28 17 8 2 1 1 $m = 125$

Region 2: 14 25 30 18 49 2 1 1 $n = 140$

75. Referring back to Exercise 74, develop a large-sample confidence interval formula for $\lambda_1 - \lambda_2$, and calculate the interval for the data given there using a confidence level of 95%.

76. Borderline and mildly retarded children attending a hospital developmental evaluation clinic were divided into two groups on the basis of the presence or absence of a probable aetiological factor causing the retardation. Blood-lead concentration was measured for each child, yielding the following data:

Aetiology unknown: 25.5, 23.2, 27.6, 24.3, 26.1, 25.0
Probable aetiology: 21.2, 19.8, 20.3, 21.0, 19.6

Does the data indicate any difference in the extent of variability in blood-lead concentration for the two types of children?

Bibliography

See the bibliography at the end of Chapter 7.

The Analysis of Variance

Introduction

In studying methods for the analysis of quantitative data, we first focused on problems involving a single sample of numbers and then turned to a comparative analysis of two different such samples. In one-sample problems the data consisted of observations on or responses from individuals or experimental objects randomly selected from a single population. In two-sample problems either the two samples were drawn from two different populations and the parameters of interest were the population means, or else two different treatments were applied to experimental units (individuals or objects) selected from a single population; in this latter case the parameters of interest were referred to as true treatment means.

The analysis of variance, or more briefly **ANOVA,** refers broadly to a collection of experimental situations and statistical procedures for the analysis of quantitative responses from experimental units. The simplest analysis of variance problem is referred to variously as a **single-factor, single-classification,** or **one-way** ANOVA and involves the analysis either of data sampled from more than two numerical populations (distributions) or data from experiments in which more than two treatments have been used. The characteristic that differentiates the treatments or populations from one another is called the **factor** under study, and the different treatments or populations are referred to as the **levels** of the factor. Examples of such situations include

1. an experiment to study the effects of five different brands of gasoline on automobile engine operating efficiency (mpg)
2. an experiment to study the effects of the presence of four different sugar solutions (glucose, sucrose, fructose, and a mixture of the three) on bacterial growth
3. an experiment to assess the effects of different amounts of a particular psychedelic drug on manual dexterity

4. an experiment to decide whether varying the tracking weight of a stereo cartridge has any effect on record lifetime

In 1 the factor of interest is gasoline brand, and there are five different levels of the factor. In 2 the factor is sugar, with four levels (or five, if a control solution containing no sugar is used). In both 1 and 2 the factor is qualitative in nature, and the levels correspond to possible categories of the factor. In 3 and 4 the factors are concentration of drug and tracking weight, respectively, and both these factors are quantitative in nature, so that the levels identify different settings of the factor. When the factor of interest is quantitative, statistical techniques from regression analysis (discussed in Chapters 12 and 13) can also be used to analyze the data.

The present chapter focuses on single-factor ANOVA. The question of central interest here is whether or not there are differences in true averages associated with the different treatments or levels of the factor. The null hypothesis states that there are no differences between any of the μ_i's (population means or expected treatment responses), and the alternative hypothesis says that at least two μ_i's differ from one another. The F test for testing H_0 versus H_a is presented in Section 10.1. If H_0 is rejected, further analysis to identify differences is usually appropriate, so multiple comparison methods for detecting such differences are discussed in Section 10.2. The last section covers some other aspects of single-factor ANOVA. The next chapter introduces ANOVA experiments involving more than a single factor.

10.1 Single-Factor Analysis of Variance

Example 10.1, which describes an experiment similar to one reported on in the 1971 volume of *The Research Quarterly,* will be used to illustrate the notation and calculations of single-factor ANOVA.

Example 10.1 A physical education researcher wished to study the effect of four different teaching techniques on swimming proficiency for individuals completing a course in beginning swimming. The techniques used were (a) verbal cues, (b) verbal cues and videotaped feedback, (c) videotaped feedback, and (d) the control treatment (neither verbal cues nor videotaped feedback). A group of 24 nonswimming college women was randomly divided into four groups of six subjects each. Each group of subjects was taught using one of the four techniques, and at the end of the course was timed while swimming 10 yards using the butterfly stroke. The results are shown in Table 10.1.

Table 10.1

Treatment		Time	(in	seconds)			Sample mean	Sample variance
1. Verbal	18.7	21.1	17.9	19.5	22.1	18.3	19.60	2.78
2. Verbal + videotaped	19.9	17.6	18.2	20.0	16.9	17.5	18.35	1.71
3. Videotaped	18.6	20.3	21.7	19.7	20.9	20.8	20.33	1.16
4. Control	19.1	18.9	18.4	18.8	17.7	20.5	18.90	.86

Does the data indicate that the different teaching techniques affect the true average swimming time?   ∎

Notation and Assumptions

Let I denote the number of populations or treatments under investigation ($I = 4$ in Example 10.1). The population means or expected responses to treatments are then $\mu_1, \mu_2, \ldots, \mu_I$. The null hypothesis states that all μ_i's are identical, whereas the alternative hypothesis says that this is not the case:

$$H_0 : \mu_1 = \mu_2 = \cdots = \mu_I$$

versus

$$H_a : \text{at least two of the } \mu_i \text{'s are different}$$

In Example 10.1, H_0 is true only if all four μ_i's are identical. H_a would be true, for example, if $\mu_1 = \mu_2 \neq \mu_3 = \mu_4$, if $\mu_1 = \mu_3 = \mu_4 \neq \mu_2$, or if all four μ_i's differ from one another.

A test of H_0 versus H_a is based on a sample from each of the I populations or treatments. In two-sample problems we used the letters X and Y to differentiate the observations in one sample from those in the other. As this is cumbersome for three or more samples, it is customary to use a single letter with two subscripts. The first subscript identifies the sample number, corresponding to the population or treatment being sampled, and the second subscript denotes the position of the observation within that sample. Let

$X_{i,j}$ = the random variable which denotes the jth measurement taken from the ith population, or the measurement taken on the jth experimental unit which receives the ith treatment, and

$x_{i,j}$ = the observed value of $X_{i,j}$ when the experiment is performed.

The observed data is usually displayed in a rectangular table such as Table 10.1. There samples from the different treatments appear in different rows of the table, and $x_{i,j}$ is the jth number in the ith row. For example, $x_{2,3}$ = the third observation from the second treatment = 18.2, and $x_{4,1} = 19.1$. When there is no ambiguity, we will write x_{ij} rather than $x_{i,j}$ (though if there were 15 observations on each of 12 treatments, x_{112} could mean $x_{1,12}$ or $x_{11,2}$). It is assumed that the X_{ij}'s within any particular sample are independent—a random sample from the ith population or treatment distribution—and that different samples are independent of one another.

In some experiments, different samples contain different numbers of observations. However, the concepts and methods of single-factor ANOVA are most easily developed for the case of equal sample sizes. Restricting ourselves for the moment to this case, let J denote the number of observations in each sample ($J = 6$ in Example 10.1). The data set consists of IJ observations. The mean of the observations in the ith sample is

$$\overline{X}_{i.} = \frac{\sum\limits_{j=1}^{J} X_{ij}}{J} \qquad i = 1, 2, \ldots, I$$

The dot in place of the second subscript signifies that we have added overall values of that subscript while holding the other subscript value fixed, and the horizontal bar indicates division by J to obtain an average. Similarly, the average of all IJ observations, called the **grand mean**, is

$$\overline{X}_{..} = \frac{\sum\limits_{i=1}^{I} \sum\limits_{j=1}^{J} X_{ij}}{IJ}$$

For the swimming data in Table 10.1, $\overline{x}_{1.} = 19.60$, $\overline{x}_{2.} = 18.35$, $\overline{x}_{3.} = 20.33$, $\overline{x}_{4.} = 18.90$, and $\overline{x}_{..} = 19.30$. Additionally, S_i^2 will denote the variance of the observations in the ith sample:

$$S_i^2 = \frac{\sum\limits_{j=1}^{J} (X_{ij} - \overline{X}_{i.})^2}{J - 1} \qquad i = 1, 2, \ldots, I$$

The assumptions that underlie the standard method of analysis in single-factor ANOVA are extensions of those on which the two-sample pooled t test was based.

Assumptions

> The I population or treatment distributions are all normal with the same variance σ^2. That is, each X_{ij} is normally distributed with
>
> $$E(X_{ij}) = \mu_i, \quad V(X_{ij}) = \sigma^2$$

If these assumptions appear to be of doubtful validity in a given problem situation, a knowledgeable statistician should be consulted. Notice that while each of the four treatment distributions in Example 10.1 is assumed to have the same variance, the largest sample variance is roughly three times the smallest one. With four samples of small sample size, this is not surprising. Some authors suggest carrying out a preliminary test of $\sigma_1^2 = \sigma_2^2 = \cdots = \sigma_I^2$, but for technical reasons we do not advocate this.

The Test Statistic

To see whether or not H_0 is plausible, it is natural to examine the I sample means. When H_0 is true, each x_{ij} comes from the same population distribution with mean value μ (the common value of the μ_i's) and variance σ^2, so the $\overline{x}_{i.}$'s should be close to one another and therefore to $\overline{x}_{..}$. If, however, there are substantial differences between the μ_i's, the $\overline{x}_{i.}$'s will tend to be quite discrepant and some will deviate markedly from $\overline{x}_{..}$. A measure of the extent to which the $\overline{x}_{i.}$'s are different is

$$MSTr = \frac{J}{I-1} \sum_{i=1}^{I} (\overline{x}_{i.} - \overline{x}_{..})^2 = J\left\{\frac{\Sigma(\overline{x}_{i.} - \overline{x}_{..})^2}{I-1}\right\} \tag{10.1}$$

The rationale for the acronym *MSTr*, which stands for "mean square for treatments," will be explained shortly. Notice that $MSTr = 0$ when all $\overline{x}_{i.}$'s are equal, and its value increases as the $\overline{x}_{i.}$'s become more discrepant. Thus *MSTr* should tend to be reasonably small when H_0 is true and relatively large when the μ_i's are quite different.

If $\overline{X}_{i.}$ and $\overline{X}_{..}$ replace $\overline{x}_{i.}$ and $\overline{x}_{..}$ in (10.1), *MSTr* becomes a statistic and thus a random variable. The computed or observed value of *MSTr* depends on which samples are selected. Suppose the expected value of the statistic *MSTr*, $E(MSTr)$, can be determined when H_0 is true. Then if the observed value of *MSTr* is much greater than this expected value, rejection of H_0 is appropriate. Each sample mean $\overline{X}_{i.}$, being an average of J observations, has variance $V(\overline{X}_{i.}) = \sigma^2/J$, and $E(\overline{X}_{i.}) = \mu$ when H_0 is true; that is, when H_0 is true, each $\overline{X}_{i.}$ is drawn from a (normal) distribution with mean value μ and variance σ^2/J. This implies that the "sample variance of sample means," $\Sigma(\overline{X}_{i.} - \overline{X}_{..})^2/(I-1)$, is an unbiased estimator of σ^2/J (a sample variance is an unbiased estimator of the variance of the variables on which it is based). Multiplying this sample variance by J gives an unbiased estimator for σ^2 itself, and from (10.1) this estimator is exactly *MSTr*; thus $E(MSTr) = \sigma^2$ when H_0 is true (this is the reason for the factor $J/(I-1)$ in *MSTr*). When H_0 is false, the $\overline{X}_{i.}$'s tend to spread out more than when H_0 is true, and *MSTr* tends to overestimate σ^2.

The statistic

$$MSTr = \frac{J}{I-1} \sum_{i=1}^{I} (\overline{X}_{i.} - \overline{X}_{..})^2$$

is an unbiased estimator of σ^2 $[E(MSTr) = \sigma^2]$ when H_0 is true, but tends to overestimate σ^2 $[E(MSTr) > \sigma^2]$ when H_0 is false.

An *MSTr* value much larger than σ^2 suggests that H_0 is false and should be rejected. Unfortunately, the value of σ^2 is virtually never available for such a comparison, but there is another estimator of σ^2 that is reliable regardless of the status of H_0. Each sample is drawn from a distribution with variance σ^2. Therefore, S_i^2, the sample variance of the ith sample $X_{i1}, X_{i2}, \ldots, X_{iJ}$, is an unbiased estimator of σ^2 whether or not H_0 is true. Because each sample contains the same number of observations, a natural way to combine these I estimators is to form their unweighted average. The test procedure is based on comparing this combined estimator, which we shall denote by *MSE* (for "mean square for error"), to *MSTr* by forming the ratio of these two estimators.

The statistic

$$MSE = \frac{S_1^2 + S_2^2 + \cdots + S_I^2}{I}$$

is an unbiased estimator of the common variance σ^2 whether or not H_0 is true. The test statistic is the ratio $MSTr/MSE$.

When H_0 is true, $MSTr$ and MSE should be comparable in value, resulting in a ratio reasonably close to 1. However, if the μ_i's are quite different, $MSTr$ should considerably exceed MSE and their ratio should be much greater than 1. This suggests a rejection region of the form $MSTr/MSE \geq c$.

F Distributions and the *F* Test

As with other test procedures, the critical value c in the rejection region is selected to control the type I error probability at the desired significance level α. That is, with $F = MSTr/MSE$, c should satisfy $P(F \geq c$ when H_0 is true$) = \alpha$. Once the distribution of F when H_0 is true is known, c is the value that captures upper-tail area α under the associated density curve. In Chapter 9, we introduced a family of probability distributions called F distributions. An F distribution arises in connection with a ratio in which there is one number of degrees of freedom associated with the numerator and another number of degrees of freedom associated with the denominator. Let ν_1 and ν_2 denote the number of numerator and denominator degrees of freedom, respectively, for a variable with an F distribution. Both ν_1 and ν_2 are positive integers. Figure 10.1 pictures an F density curve and the corresponding upper-tail critical value F_{α, ν_1, ν_2}. Appendix Table A.7 gives these critical values for $\alpha = .05$ and for $\alpha = .01$. Values of ν_1 are identified with different columns of the table, and the rows are labeled with various values of ν_2. For example, the F critical value that captures upper-tail area .05 under the F curve with $\nu_1 = 4$ and $\nu_2 = 6$ is $F_{.05, 4, 6} = 4.53$ whereas $F_{.05, 6, 4} = 6.16$. When a test of hypotheses based on an F distribution is carried out using one of the better statistical computer packages, a P-value will be computed. This allows the use of an α other than .05 or .01. The key theoretical result is that the test statistic F has an F distribution when H_0 is true.

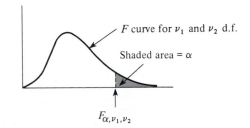

F curve for ν_1 and ν_2 d.f.

Shaded area $= \alpha$

F_{α, ν_1, ν_2}

Figure 10.1   An F curve and critical value F_{α, ν_1, ν_2}

Theorem

Let $F = MSTr/MSE$ be the test statistic in a single-factor ANOVA problem involving I populations or treatments with a random sample of J observations from each one. When H_0 is true and the basic assumptions of this section are satisfied, F has an F distribution with $v_1 = I - 1$ and $v_2 = I(J - 1)$. With f denoting the computed value of F, the rejection region $f \geq F_{\alpha, I-1, I(J-1)}$ then specifies a test with significance level α.*

The rationale for $v_1 = I - 1$ is that while $MSTr$ is based on the I deviations $\overline{X}_{1.} - \overline{X}_{..}, \ldots, \overline{X}_{I.} - \overline{X}_{..}, \Sigma(\overline{X}_{i.} - \overline{X}_{..}) = 0$ so only $I - 1$ of these are freely determined. Because each sample contributes $J - 1$ d.f. to MSE and these samples are independent, $v_2 = (J - 1) + \cdots + (J - 1) = I(J - 1)$.

Example 10.2
(Example 10.1 continued)

The values of I and J for the swimming data are 4 and 6, respectively, so numerator d.f. $= I - 1 = 3$ and denominator d.f. $= I(J - 1) = 20$. At significance level .05, $H_0 : \mu_1 = \mu_2 = \mu_3 = \mu_4$ will be rejected in favor of the conclusion that at least two μ_i's are different if $f \geq F_{.05, 3, 20} = 3.10$. The grand mean is $\overline{x}_{..} = \Sigma\Sigma x_{ij}/IJ = 19.296$, and

$$MSTr = \frac{6}{4 - 1} [(19.60 - 19.296)^2 + (18.35 - 19.296)^2$$

$$+ (20.33 - 19.296)^2 + (18.90 - 19.296)^2]$$

$$= 4.43$$

$$MSE = \frac{1}{4} [2.78 + 1.71 + 1.16 + .86] = 1.63$$

$$f = MSTr/MSE = 4.43/1.63 = 2.72$$

Because $2.72 < 3.10$, H_0 cannot be rejected at level .05. The sample data does not indicate that there are any differences between the true average times for the four techniques. ∎

Computational Formulas for ANOVA

The calculations leading to f can be done efficiently by using formulas similar to the computing formula for a single sample variance s^2. Let $x_{i.}$ represent the *sum* (not average, since there is no bar) of the x_{ij}'s for i fixed (sum of the numbers in the ith row of the table) and $x_{..}$ denote the sum of *all* the x_{ij}'s (the **grand total**).

---

*We have generally distinguished between a statistic and its computed value by using an upper case letter for the former and a lower case letter for the latter, but here we will adhere to tradition and use only upper case letters for sums of squares and mean squares (even when denoting computed values of such quantities). Also, although our development of the F test has been based on heuristics, the test procedure can be derived from the likelihood ratio principle.

Definition

> The **total sum of squares (SST), treatment sum of squares ($SSTr$), and error sum of squares (SSE)** are given by
>
> $$SST = \sum_{i=1}^{I}\sum_{j=1}^{J}(x_{ij} - \overline{x}_{..})^2 = \sum_{i=1}^{I}\sum_{j=1}^{J}x_{ij}^2 - \frac{1}{IJ}x_{..}^2$$
>
> $$SSTr = \sum_{i=1}^{I}\sum_{j=1}^{J}(\overline{x}_{i.} - \overline{x}_{..})^2 = \frac{1}{J}\sum_{i=1}^{I}x_{i.}^2 - \frac{1}{IJ}x_{..}^2$$
>
> $$SSE = \sum_{i=1}^{I}\sum_{j=1}^{J}(x_{ij} - \overline{x}_{i.})^2, \quad \text{where} \quad x_{i.} = \sum_{j=1}^{J}x_{ij}, \quad x_{..} = \sum_{i=1}^{I}\sum_{j=1}^{J}x_{ij}$$

The sum of squares $SSTr$ appears in the numerator of F and SSE appears in the denominator of F; the reason for defining SST will be apparent shortly.

The expressions on the far right-hand side of SST and $SSTr$ are the computing formulas for these sums of squares. Both SST and $SSTr$ involve $x_{..}^2/IJ$ (the square of the grand total divided by IJ), which is usually called the **correction factor for the mean.** After computing the correction factor, SST is obtained by squaring each number in the data table, adding these squares together, and subtracting the correction factor. $SSTr$ results from squaring each row total, summing them, dividing by J, and subtracting the correction factor.

A computing formula for SSE is a consequence of a simple relationship between the three sums of squares.

Fundamental identity

> $$SST = SSTr + SSE \tag{10.2}$$

Thus if any two of the sums of squares are computed, the third can be obtained through (10.2); SST and $SSTr$ are easiest to compute, and then $SSE = SST - SSTr$. The proof follows from squaring both sides of the relationship

$$x_{ij} - \overline{x}_{..} = (x_{ij} - \overline{x}_{i.}) + (\overline{x}_{i.} - \overline{x}_{..}) \tag{10.3}$$

and summing over all i and j. This gives SST on the left, and $SSTr$ and SSE as the two extreme terms on the right. The cross product term is easily seen to be zero.

The interpretation of the fundamental identity is an important aid to an understanding of **ANOVA.** SST is a measure of the total variation in the data—the sum of all squared deviations about the grand mean. The identity says that this total variation can be partitioned into two pieces. SSE measures variation that would be present (within rows) even if H_0 is true, and is thus the part of total variation which is *unexplained* by the truth or falsity of H_0. $SSTr$

is the amount of variation (between rows) that can be *explained* by possible differences in the μ_i's. If explained variation is large relative to unexplained variation, then H_0 is rejected in favor of H_a.

Once $SSTr$ and SSE are computed, F is the ratio of two quantities called **mean squares,** which are simply the two sums of squares divided by their appropriate number of degrees of freedom:

$$MSTr = \frac{SSTr}{I-1}, \quad MSE = \frac{SSE}{I(J-1)}, \quad F = \frac{MSTr}{MSE} \tag{10.4}$$

The computations are often summarized in a tabular format, called an **ANOVA table,** as displayed in Table 10.2.

Table 10.2

Source of variation	d.f.	Sum of squares	Mean square	f
Treatments	$I-1$	$SSTr$	$MSTr = SSTr/(I-1)$	$\dfrac{MSTr}{MSE}$
Error	$I(J-1)$	SSE	$MSE = SSE/I(J-1)$	
Total	$IJ-1$	SST		

Example 10.3

The accompanying data resulted from an experiment comparing the degree of soiling for fabric copolymerized with three different mixtures of methacrylic acid (similar data appeared in the paper "Chemical Factors Affecting Soiling and Soil Release from Cotton DP Fabric," *American Dyestuff Reporter,* 1983, pp. 25–30).

						$x_{i.}$	$\bar{x}_{i.}$
Mixture 1:	.56	1.12	.90	1.07	.94	4.59	.918
Mixture 2:	.72	.69	.87	.78	.91	3.97	.794
Mixture 3:	.62	1.08	1.07	.99	.93	4.69	.938
						$x_{..} = 13.25$	

Let μ_i denote the true average degree of soiling when mixture i is used ($i = 1, 2, 3$). The null hypothesis $H_0: \mu_1 = \mu_2 = \mu_3$ states that the true average degree of soiling is identical for the three mixtures. We shall carry out a test at significance level .01 to see whether H_0 should be rejected in favor of the assertion that true average degree of soiling is not the same for all mixtures. Since $I - 1 = 2$ and $I(J - 1) = 12$, the F critical value for the rejection region is $F_{.01, 2, 12} = 6.93$. Squaring each of the 15 observations and summing gives $\Sigma\Sigma x_{ij}^2 = (.56)^2 + (1.12)^2 + \cdots + (.93)^2 = 12.1351$. The values of the three sums of squares are

$$SST = 12.1351 - (13.25)^2/15 = 12.1351 - 11.7042 = .4309$$

$$SSTr = \frac{1}{5}[(4.59)^2 + (3.97)^2 + (4.69)^2] - 11.7042$$

$$= 11.7650 - 11.7042 = .0608$$

$$SSE = .4309 - .0608 = .3701$$

The remaining computations are summarized in the accompanying ANOVA table. Because $f = .99$ is not at least $F_{.01, 2, 12} = 6.93$, H_0 is not rejected at significance level .01. The mixtures appear to be indistinguishable with respect to degree of soiling.

Source of variation	d.f.	Sum of squares	Mean square	f
Treatments	2	.0608	.0304	.99
Error	12	.3701	.0308	
Total	14	.4309		

When the F test causes H_0 to be rejected, the experimenter will often be interested in further analysis to decide which μ_i's differ from which others. Procedures for doing this are called multiple comparison procedures, and several are described in the next two sections.

Exercises / Section 10.1 (1–9)

1. In an experiment to compare the tensile strengths of $I = 5$ different types of copper wire, $J = 4$ samples of each type were used. The between-samples and within-samples estimates of σ^2 were computed as $MSTr = 2573.3$ and $MSE = 1394.2$, respectively. Use the F test at level .05 to test $H_0: \mu_1 = \mu_2 = \mu_3 = \mu_4 = \mu_5$ versus H_a: at least two μ_i's are unequal.

2. The lumen output was determined for each of $I = 3$ different brands of 60-watt soft-white lightbulbs, with $J = 8$ bulbs of each brand tested. The sums of squares were computed as $SSE = 4773.3$ and $SSTr = 591.2$. State the hypotheses of interest (including word definitions of parameters), and use the F test of ANOVA ($\alpha = .05$) to decide whether or not there are any differences in true average lumen outputs between the three brands for this type of bulb.

3. In a study to assess the effects of malaria infection on mosquito hosts ("Plasmodium Cynomolgi: Effects of Malaria Infection on Laboratory Flight Performance of Anopheles Stephensi Mosquitos," *Experimental Parasitology,* 1977, pp. 397–404), mosquitos were fed on either infective or non-infective rhesus monkeys. Subsequently the distance they flew during a 24-hour period was measured using a flight mill. The mosquitos were divided into four groups of eight mosquitos each: infective rhesus and sporozites present (IRS), infective rhesus and oocysts present (IRD), infective rhesus and no infection developed (IRN), and noninfective (C). The summary data values are $\bar{x}_{1.} = 4.39$ (IRS), $\bar{x}_{2.} = 4.52$ (IRD), $\bar{x}_{3.} = 5.49$ (IRN), $\bar{x}_{4.} = 6.36$ (C), $\bar{x}_{..} = 5.19$, and $\Sigma\Sigma x_{ij}^2 = 911.91$. Use the ANOVA F test at level .05 to decide whether or not there are any differences between true average flight times for the four treatments.

4. Three different laboratory methods for determining the concentration of a contaminant in water are to be compared. A gallon of well water is divided into nine equal parts, and each is placed in a container. The containers are then randomly assigned to the three methods. Summary quantities are:

	J	$\bar{x}_{i.}$	$s_{i.}$
Method 1	3	1.914	.212
Method 2	3	1.949	.095
Method 3	3	2.327	.198

Use this data and a significance level of .01 to test the null hypothesis of no difference in mean contaminant concentration for the three methods.

5. The article "Origin of Precambrian Iron Formations" (*Econ. Geology,* 1964, pp. 1025–1057) reported the following data on total Fe for four types of iron formation (1 = carbonate, 2 = silicate, 3 = magnetite, 4 = hematite).

1:	20.5,	28.1,	27.8,	27.0,	28.0,
	25.2,	25.3,	27.1,	20.5,	31.3
2:	26.3,	24.0,	26.2,	20.2,	23.7,
	34.0,	17.1,	26.8,	23.7,	24.9
3:	29.5,	34.0,	27.5,	29.4,	27.9,
	26.2,	29.9,	29.5,	30.0,	35.6
4:	36.5,	44.2,	34.1,	30.3,	31.4,
	33.1,	34.1,	32.9,	36.3,	25.5

Carry out an analysis of variance F test at significance level .01, and summarize the results in an ANOVA table.

6. In an experiment to investigate the performance of four different brands of spark plugs intended for use on a 125-cc two-stroke motorcycle, five plugs of each brand were tested and the number of miles (at a constant speed) until failure was observed. The partial ANOVA table for the data appears below. Fill in the missing entries, state the relevant hypotheses, and carry out a test.

Source	d.f.	Sum of squares	Mean square	f
Brand				
Error			14,713.69	
Total		310,500.76		

7. A study of the properties of metal plate-connected trusses used for roof support ("Modeling Joints Made with Light-Gauge Metal Connector Plates," *Forest Products J.,* 1979, pp. 39–44) yielded the following observations on axial stiffness index (kips/in) for plate lengths of 4, 6, 8, 10, and 12 in.

4:	309.2,	409.5,	311.0,	326.5,	316.8,
	349.8,	309.7			
6:	402.1,	347.2,	361.0,	404.5,	331.0,
	348.9,	381.7			
8:	392.4,	366.2,	351.0,	357.1,	409.9,
	367.3,	382.0			
10:	346.7,	452.9,	461.4,	433.1,	410.6,
	384.2,	362.6			
12:	407.4,	441.8,	419.9,	410.7,	473.4,
	441.2,	465.8			

Does variation in plate length have any effect on true average axial stiffness? State and test the relevant hypotheses using analysis of variance with $\alpha = .01$. Display your results in an ANOVA table. *Hint:* $\Sigma\Sigma x_{ij}^2 = 5{,}241{,}420.79$.

8. Six samples of each of four types of cereal grain grown in a certain region were analyzed to determine thiamin content, resulting in the following data (micrograms/gram):

Wheat:	5.2,	4.5,	6.0,	6.1,	6.7,	5.8
Barley:	6.5,	8.0,	6.1,	7.5,	5.9,	5.6
Maize:	5.8,	4.7,	6.4,	4.9,	6.0,	5.2
Oats:	8.3,	6.1,	7.8,	7.0,	5.5,	7.2

Does this data suggest that at least two of the grains differ with respect to true average thiamin content? Use a level $\alpha = .05$ test.

9. In single-factor ANOVA with I treatments and J observations per treatment, let $\mu = \frac{1}{I}\Sigma\mu_i$.

a. Express $E(\bar{X}_{..})$ in terms of μ. *Hint:*
$$\bar{X}_{..} = \frac{1}{I}\Sigma\bar{X}_{i.}$$

b. Compute $E(\bar{X}_{i.}^2)$. *Hint:* For any r.v. Y, $E(Y^2) = V(Y) + [E(Y)]^2$.

c. Compute $E(\bar{X}_{..}^2)$.

d. Compute $E(SSTr)$ and then show that
$$E(MSTr) = \sigma^2 + \frac{J}{I-1}\Sigma(\mu_i - \mu)^2$$

e. Using the result of (d), what is $E(MSTr)$ when H_0 is true? When H_0 is false, how does $E(MSTr)$ compare to σ^2?

10.2	**Multiple Comparisons in ANOVA**

When the computed value of the F statistic in single-factor ANOVA is not significant, the analysis is terminated because no differences between the μ_i's have been identified. But when H_0 is rejected, the investigator will usually want to know which of the μ_i's are different from one another. A method for carrying out this further analysis is called a multiple comparisons procedure. There are a number of such procedures in the statistics literature. Here we present two that many statisticians recommend for deciding for each i and j whether $\mu_i = \mu_j$.

Tukey's Procedure (The T Method)

Tukey's procedure involves the use of another probability distribution called the **studentized range distribution.** The distribution depends on two parameters: a numerator d.f. m and a denominator d.f. ν. Let $Q_{\alpha, m, \nu}$ denote the upper-tail α critical value of the studentized range distribution with m numerator d.f. and ν denominator d.f. (analogous to F_{α, ν_1, ν_2}). Values of $Q_{\alpha, m, \nu}$ are given in Appendix Table A.8. Then $Q_{\alpha, I, I(J-1)}$ can be used to obtain simultaneous confidence intervals for all pairwise differences $\mu_i - \mu_j$. Notice that the numerator d.f. for the appropriate Q critical value is I, the number of treatments or populations, and not $I - 1$, as it was for F.

Proposition	With probability $1 - \alpha$, $$\overline{X}_{i.} - \overline{X}_{j.} - Q_{\alpha, I, I(J-1)} \sqrt{MSE/J} \le \mu_i - \mu_j$$ $$\le \overline{X}_{i.} - \overline{X}_{j.} + Q_{\alpha, I, I(J-1)} \sqrt{MSE/J} \qquad (10.5)$$ for every i and j ($i = 1, \ldots, I$ and $j = 1, \ldots, I$) with $i \ne j$.

When the computed $\overline{x}_{i.}$, $\overline{x}_{j.}$, and MSE are substituted into (10.5), the result is a collection of **simultaneous confidence statements** about the true values of all differences $\mu_i - \mu_j$ between true treatment means. *Each interval from (10.5) that does not include zero yields the conclusion that μ_i and μ_j differ significantly at level α.*

The computation of all intervals in (10.5) looks complicated, but a straightforward sequence of steps culminates in a pictorial summary of the conclusions:

The T method for identifying significantly different μ_i's	**1.** Select α and find $Q_{\alpha, I, I(J-1)}$ from Appendix Table A.8. **2.** Determine $w = Q_{\alpha, I, I(J-1)} \cdot \sqrt{MSE/J}$. **3.** List the sample means in increasing order and underline those pairs that differ by less than w. Any pair of sample means not underscored by the same line corresponds to a pair of true treatment means that are judged significantly different.

Example 10.4   It is widely believed that although "house" brands (those sold only through a particular chain of stores) and regionally distributed brands of canned goods tend to be less expensive than their nationally distributed counterparts, they also tend to be less desirable in terms of quantity and quality of contents. To see whether such beliefs have any basis in fact, five different brands of fruit cocktail were chosen: two national brands (1 and 2), a regional brand (3), and a house brand from each of two large supermarket chains (4 and 5). Nine 16-oz cans of each brand were purchased, opened and drained of liquid, and their net contents weighed. The sample average weights for the five brands were $\bar{x}_1. = 14.5$, $\bar{x}_2. = 13.8$, $\bar{x}_3. = 13.3$, $\bar{x}_4. = 14.3$, and $\bar{x}_5. = 13.1$. The ANOVA table summarizing the first part of the analysis appears as Table 10.3.

Table 10.3

Source of variation	d.f.	Sum of squares	Mean square	f
Treatments (brands)	4	13.32	3.33	37.84
Error	40	3.53	.088	
Total	44	16.85		

Since $F_{.05, 4, 40} = 2.61$, H_0 is rejected (decisively) at level .05. We now use Tukey's procedure to look for significant differences among the μ_i's. From Appendix Table A.8, $Q_{.05, 5, 40} = 4.04$ (the second subscript on Q is I and not $I - 1$ as in F), so $w = 4.04\sqrt{.088/9} = .4$. Arranging the five sample means in increasing order, every pair differing by less than .4 is underscored:

$\bar{x}_5.$	$\bar{x}_3.$	$\bar{x}_2.$	$\bar{x}_4.$	$\bar{x}_1.$
13.1	13.3	13.8	14.3	14.5

Thus brands 1 and 4 are not significantly different from one another, but are significantly higher than the other three brands in their true average contents. Brand 2 is significantly better than 3 and 5 but worse than 1 and 4, and brands 3 and 5 do not differ significantly. Note that one house brand (4) is in the top group, while a national brand is in the lowest group. ∎

In Example 10.4 if $\bar{x}_2. = 14.15$ rather than 13.8 with the same computed w, then the configuration of underscored means would be

$\bar{x}_5.$	$\bar{x}_3.$	$\bar{x}_2.$	$\bar{x}_4.$	$\bar{x}_1.$
13.1	13.3	14.15	14.3	14.5

When there are no significant differences among a group of more than two means, it is customary to draw a continuous line underneath the entire group.

Example 10.5   A biologist wished to study the effects of ethanol on sleep time. A sample of 20 rats, matched for age and other characteristics, was selected, and each rat was given an oral injection having a particular concentration of ethanol per body weight. The rapid eye movement (REM) sleep time for each rat was then recorded for a 24-hour period, with the following results:

Treatment (concentration of ethanol)						$x_{i.}$	$\bar{x}_{i.}$
0 (control)	88.6	73.2	91.4	68.0	75.2	396.4	79.28
1 gm/kg	63.0	53.9	69.2	50.1	71.5	307.7	61.54
2 gm/kg	44.9	59.5	40.2	56.3	38.7	239.6	47.92
4 gm/kg	31.0	39.6	45.3	25.2	22.7	163.8	32.76

$$x_{..} = 1107.5, \quad \bar{x}_{..} = 55.375$$

Does the data indicate that the true average REM sleep time depends on the concentration of ethanol? (This example is based on an experiment reported in "Relationship of Ethanol Blood Level to REM and Non-REM Sleep Time and Distribution in the Rat," *Life Sciences*, 1978, pp. 839–846.)

The $\bar{x}_{i.}$'s differ rather substantially from one another, but there is also a great deal of variability within each sample, so to answer the question precisely we must carry out the analysis of variance. With $\Sigma\Sigma x_{ij}^2 = 68,697.61$ and $x_{..}^2/IJ = (1107.5)^2/20 = 61,327.81$, the computing formulas yield

$$SST = 68,697.61 - 61,327.81 = 7369.8$$

$$SSTr = \frac{1}{5}[(396.40)^2 + (307.70)^2 + (239.60)^2 + (163.80)^2] - 61,327.81$$

$$= 67,210.21 - 61,327.81 = 5882.4$$

and

$$SSE = 7369.8 - 5882.4 = 1487.4$$

The ANOVA table appears as Table 10.4.

Table 10.4

Source	d.f.	Sum of squares	Mean square	f
Treatments	3	5882.4	1960.8	21.09
Error	16	1487.4	93.0	
Total	19	7369.8		

Since $F_{.05, 3, 16} = 3.24$, the computed value 21.09 exceeds this critical value, so that the data is significant at level .05. We conclude that true average REM sleep time does indeed depend on the concentration of ethanol in the blood.

There are $I = 4$ treatments and 16 d.f. for error, so $Q_{.05, 4, 16} = 4.05$ and $w = 4.05\sqrt{93.0/5} = 17.47$. Ordering the means and underscoring yields

$\bar{x}_{4.}$	$\bar{x}_{3.}$	$\bar{x}_{2.}$	$\bar{x}_{1.}$
32.76	47.92	61.54	79.28

The interpretation of this underscoring must be done with care, since we seem to have concluded that treatments 2 and 3 do not differ, 3 and 4 do not differ, yet 2 and 4 do differ. The suggested way of expressing this is to say that while evidence allows us to conclude that treatments 2 and 4 differ from one another, neither has been shown to be significantly different from 3. Treatment 1 has a significantly higher true average REM sleep time than any of the other treatments. ∎

The Interpretation of α in Multiple Comparison

Tukey's method involves the simultaneous construction of confidence intervals for all differences $\mu_i - \mu_j$ of pairs of treatment means. In the construction of confidence intervals in the previous chapters, $\alpha = .05$ or 95% confidence referred to the error rate or confidence level for the individual statement about the parameter of interest. If an ANOVA experiment involves comparison of four treatments, then Tukey's procedure obtains simultaneously six different intervals (since there are six different pairs of treatment means). The error rate $\alpha = .05$ no longer refers to a particular interval, but instead refers to the experiment as a whole, so is called an **experimentwise error rate.** If Tukey's method were used on a great many different ANOVA data sets, then in approximately 95% of these experiments no erroneous claim would be made about any of the $\mu_i - \mu_j$'s, while in only 5% would at least one incorrect claim be made. That is, in 95% of all experiments, every confidence interval constructed from (10.5) would include the true value of $\mu_i - \mu_j$, and only 5% of the time would at least one interval fail to cover the true value of a $\mu_i - \mu_j$. Because the confidence level for the entire set of comparisons of means is 95%, the confidence level for any particular comparison is larger than 95% and increases as the number of comparisons (that is, treatment means) increases. This distinction between the experimentwise error rate or confidence level and a "per comparison" error rate or confidence level is important, and should be kept in mind when interpreting the results of any multiple comparison analysis.

Multiple Comparisons and t Intervals

A t interval for a difference $\mu_1 - \mu_2$ was presented in Chapter 9. The interval depended on using the pooled estimator S_p for σ and was based on the t distribution with $m + n - 2$ d.f. In one-way ANOVA, we can replace S_p by $\sqrt{MSE}$ and $m + n - 2$ by $I(J - 1)$ to obtain a $100(1 - \alpha)\%$ confidence interval for a particular difference $\mu_i - \mu_j$:

$$\overline{x}_{i.} - \overline{x}_{j.} \pm t_{\alpha/2, I(J-1)} \sqrt{MSE\left(\frac{1}{J} + \frac{1}{J}\right)} \tag{10.6}$$

If several 95% intervals ($\alpha = .05$) of the form (10.6) are computed for different $\mu_i - \mu_j$'s, the simultaneous confidence level will be less than 95% (for two completely independent intervals, the simultaneous confidence level would be $100(.95)^2 \approx 90\%$, but because MSE appears in each interval computed from (10.6), different intervals are not independent). However, when several intervals are computed, there is a simple lower bound on the simultaneous confidence level.

Proposition

> If an interval of the form (10.6) is computed for each i and j with $i < j$, resulting in $k = \binom{I}{2} = I(I-1)/2$ intervals, the simultaneous confidence that all k intervals cover the true $\mu_i - \mu_j$'s is at least $100(1 - k\alpha)$.*

The multiple comparison procedure resulting from these intervals, called the **Bonferroni** procedure, is to identify μ_i and μ_j as significantly different from one another if the interval for $\mu_i - \mu_j$ does not include zero. While the intervals tend to be somewhat longer than the Tukey intervals (resulting in fewer significantly different μ_i's), the intervals are valid for unequal sample sizes (Section 10.3), whereas the Tukey intervals cannot be used in this case. Each interval (10.6) has the same length (because of the equal sample sizes), so with $w = t_{\alpha/2,\, I(J-1)} \sqrt{2MSE/J}$, the same underscoring method used in the Tukey procedure works here: After the $\bar{x}_{i\cdot}$'s are ordered and those that differ by less than w are underscored, pairs not underscored correspond to significantly different $\mu_i - \mu_j$'s.

Example 10.6

There were four different treatments in the ethanol/sleep-time experiment described in Example 10.5, so $k = 4(3)/2 = 6$. If a 99% interval is computed for each of the six $\mu_i - \mu_j$'s ($\alpha = .01$), the simultaneous confidence level will be at least $100(1 - 6\alpha)\% = 94\%$. With $t_{.005,\, 16} = 2.921$, $MSE = 93.0$, and $J = 5$, $w = 2.921\sqrt{2(93.0)/5} = 17.82$.

$\bar{x}_{4\cdot}$	$\bar{x}_{3\cdot}$	$\bar{x}_{2\cdot}$	$\bar{x}_{1\cdot}$
32.76	47.92	61.54	79.28

Because w is slightly larger than for the T method, the pair corresponding to $i = 1$, $j = 2$ is now underscored, so only the pairs (μ_1, μ_3), (μ_1, μ_4), and (μ_2, μ_4) are judged significantly different. ∎

Confidence Intervals for Other Parametric Functions

In some situations a confidence interval is desired for a function of the μ_i's more complicated than a difference $\mu_i - \mu_j$. Let $\theta = \Sigma c_i \mu_i$, where the c_i's are constants. One such function is $\frac{1}{2}(\mu_1 + \mu_2) - \frac{1}{3}(\mu_3 + \mu_4 + \mu_5)$, which in the context of Example 10.4 measures the difference between fruit cocktail can contents for national and nonnational brands. Because the X_{ij}'s are normally distributed with $E(X_{ij}) = \mu_i$ and $V(X_{ij}) = \sigma^2$, $\hat{\theta} = \Sigma_i c_i \bar{X}_{i\cdot}$ is normally distributed, unbiased for θ, and

---

*To obtain a simultaneous confidence level of at least $100(1 - \alpha)\%$ for k comparisons, values of $t_{\alpha/2k,\, \nu}$ are required. These appear in "Tables of the Bonferroni t Statistic" (*J. Amer. Stat. Assn.*, 1977, pp. 469–478).

$$V(\hat{\theta}) = V\left(\sum_i c_i \overline{X}_{i.}\right) = \sum_i c_i^2 V(\overline{X}_{i.}) = \frac{\sigma^2}{J} \sum_i c_i^2$$

Estimating σ^2 by MSE and forming $\hat{\sigma}_{\hat{\theta}}$ results in a t variable $(\hat{\theta} - \theta)/\hat{\sigma}_{\hat{\theta}}$, which can be manipulated to obtain the following $100(1 - \alpha)\%$ confidence interval for $\Sigma c_i \mu_i$:

$$\Sigma c_i \overline{x}_{i.} \pm t_{\alpha/2,\, I(J-1)} \sqrt{\frac{MSE\, \Sigma c_i^2}{J}} \tag{10.7}$$

Example 10.7
(Example 10.4
continued)

The parametric function for comparing national and nonnational brands is $\theta = \frac{1}{2}(\mu_1 + \mu_2) - \frac{1}{3}(\mu_3 + \mu_4 + \mu_5)$, from which

$$\Sigma c_i^2 = \left(\frac{1}{2}\right)^2 + \left(\frac{1}{2}\right)^2 + \left(-\frac{1}{3}\right)^2 + \left(-\frac{1}{3}\right)^2 + \left(-\frac{1}{3}\right)^2 = \frac{5}{6}$$

With $\hat{\theta} = \frac{1}{2}(\overline{x}_{1.} + \overline{x}_{2.}) - \frac{1}{3}(\overline{x}_{3.} + \overline{x}_{4.} + \overline{x}_{5.}) = .583$ and $MSE = .088$, a 95% interval is

$$.583 \pm 2.201 \sqrt{5(.088)/(6)(9)} = .583 \pm .182 = (.401, .765) \quad \blacksquare$$

If confidence intervals for $k \, (\geq 2)$ parametric functions are desired, the Bonferroni method can be applied using the k intervals computed from (10.7), and the resulting simultaneous confidence level is at least $100(1 - k\alpha)\%$.

Exercises / Section 10.2 (10–22)

10. An experiment to compare the spreading rates of five different brands of yellow interior latex paint available in a particular area used four gallons ($J = 4$) of each paint. The sample average spreading rates (sq ft/gal) for the five brands were $\overline{x}_{1.} = 462.0$, $\overline{x}_{2.} = 512.8$, $\overline{x}_{3.} = 437.5$, $\overline{x}_{4.} = 469.3$, and $\overline{x}_{5.} = 532.1$. The computed value of F was found to be significant at level $\alpha = .05$. With $MSE = 272.8$, use Tukey's procedure to investigate significant differences in the true average spreading rates between brands.

11. In Exercise 10 suppose that $\overline{x}_{3.} = 427.5$. Now which true average spreading rates differ significantly from one another? Be sure to use the method of underscoring to illustrate your conclusions, and also write a paragraph summarizing your results.

12. Repeat Exercise 11 supposing that $\overline{x}_{2.} = 502.8$ in addition to $\overline{x}_{3.} = 427.5$.

13. Use Tukey's procedure on the data in Exercise 3 to identify differences in true average flight times among the four types of mosquitos.

14. Use Tukey's procedure on the data of Exercise 5 to identify differences in true average total Fe between the four types of formations (use $MSE = 15.64$).

15. Use Tukey's procedure to investigate differences in true average axial stiffness for the data in Exercise 7.

16. Repeat Exercise 10 using the Bonferroni procedure in place of Tukey's procedure.

17. Repeat Exercise 14 using the Bonferroni procedure (using $\alpha = .001$ in (10.6)).

18. Refer back to Exercise 4, and compute a 95% t confidence interval for $\theta = \frac{1}{2}(\mu_1 + \mu_2) - \mu_3$.

19. Consider the accompanying data on plant growth after the application of different types of growth hormone.

Hormone					
1	13	17	7	14
2	21	13	20	17
3	18	15	20	17
4	7	11	18	10
5	6	11	15	8

a. Perform an F test at level $\alpha = .05$.

b. What happens when Tukey's procedure is applied?

20. Consider a single-factor ANOVA experiment in which $I = 3$, $J = 5$, $\bar{x}_{1.} = 10$, $\bar{x}_{2.} = 12$, and $\bar{x}_{3.} = 20$. Find a value of SSE for which $F > F_{.05, 2, 12}$, so that $H_0: \mu_1 = \mu_2 = \mu_3$ is rejected, yet when Tukey's procedure is applied none of the μ_i's can be said to differ significantly from one another.

21. Referring to Exercise 20, suppose that $\bar{x}_{1.} = 10$, $\bar{x}_{2.} = 15$, and $\bar{x}_{3.} = 20$. Can you now find a value of SSE that produces such a contradiction between the F test and Tukey's procedure?

22. The article "The Effect of Enzyme Inducing Agents on the Survival Times of Rats Exposed to Lethal Levels of Nitrogen Dioxide" (*Toxicology and Applied Pharmacology*, 1978, pp. 169–174) reported the following data on survival times for rats exposed to nitrogen dioxide (70 ppm) via different injection regimens. There were $J = 14$ rats in each group.

Regimen	$\bar{x}_{i.}$ (min)	s_i
1. Control	166	32
2. 3 Methylcholanthrene	303	53
3. Allylisopropylacetamide	266	54
4. Phenobarbital	212	35
5. Chlorpromazine	202	34
6. p Aminobenzoic Acid	184	31

a. Test the null hypothesis that true average survival time does not depend on injection regimen against the alternative that there is some dependence on injection regimen using $\alpha = .01$.

b. Compute confidence intervals with simultaneous confidence level at least 98% for $\mu_1 - \frac{1}{5}(\mu_2 + \mu_3 + \mu_4 + \mu_5 + \mu_6)$ and $\frac{1}{4}(\mu_2 + \mu_3 + \mu_4 + \mu_5) - \mu_6$.

10.3   More on Single-Factor ANOVA

In this section we briefly consider some issues relating to single-factor ANOVA that were not dealt with in the first two sections. These include an alternative description of the model parameters, β for the F test, the relationship of the test to procedures previously considered, data transformation, a random effects model, and formulas for the case of unequal sample sizes.

An Alternative Description of the ANOVA Model

The assumptions of single-factor ANOVA can be described succinctly by means of the "model equation"

$$X_{ij} = \mu_i + \epsilon_{ij}$$

where ϵ_{ij} represents a random deviation from the population or true treatment mean μ_i. The ϵ_{ij}'s are assumed to be independent normally distributed random variables (implying that the X_{ij}'s are also) with $E(\epsilon_{ij}) = 0$ (so that $E(X_{ij}) = \mu_i$) and $V(\epsilon_{ij}) = \sigma^2$ (from which $V(X_{ij}) = \sigma^2$ for every i and j). An alternative description of single-factor ANOVA will give added insight and suggest appropriate generalizations to models involving more than one factor. Define a parameter μ by

$$\mu = \frac{1}{I} \sum_{i=1}^{I} \mu_i$$

and the parameters $\alpha_1, \ldots, \alpha_I$ by

$$\alpha_i = \mu_i - \mu \quad i = 1, \ldots, I$$

Then the treatment mean μ_i can be written as $\mu + \alpha_i$, where μ represents the true average overall response in the experiment, and α_i is the effect, measured as a departure from μ, due to the ith treatment. Whereas we initially had I parameters, we now have $I + 1$ ($\mu, \alpha_1, \ldots, \alpha_I$). However, because $\Sigma \alpha_i = 0$ (the average departure from the overall mean response is zero), only I of these new parameters are independently determined, so there are as many independent parameters as there were before. In terms of μ and the α_i's, the model becomes

$$X_{ij} = \mu + \alpha_i + \epsilon_{ij} \quad i = 1, \ldots, I \quad j = 1, \ldots, J$$

In Chapter 11 we will develop analogous models for multifactor ANOVA. The claim that the μ_i's are identical is equivalent to the equality of the α_i's, and because $\Sigma \alpha_i = 0$, the null hypothesis becomes

$$H_0: \alpha_1 = \alpha_2 = \cdots = \alpha_I = 0$$

In Section 10.1 it was stated that $MSTr$ is an unbiased estimator of σ^2 when H_0 is true, but otherwise tends to overestimate σ^2. More precisely,

$$E(MSTr) = \sigma^2 + \frac{J}{I-1} \Sigma \alpha_i^2$$

When H_0 is true, $\Sigma \alpha_i^2 = 0$ so $E(MSTr) = \sigma^2$ (MSE is unbiased whether or not H_0 is true). If $\Sigma \alpha_i^2$ is used as a measure of the extent to which H_0 is false, then a larger value of $\Sigma \alpha_i^2$ will result in a greater tendency for $MSTr$ to overestimate σ^2. In the next chapter, formulas for expected mean squares for multifactor models will be used to suggest how to form F ratios to test various hypotheses.

Proof of the formula for *E(MSTr)*. For any random variable Y, $E(Y^2) = V(Y) + [E(Y)]^2$, so

$$E(SSTr) = E\left(\frac{1}{J} \sum_i X_{i\cdot}^2 - \frac{1}{IJ} X_{\cdot\cdot}^2 \right) = \frac{1}{J} \sum_i E(X_{i\cdot}^2) - \frac{1}{IJ} E(X_{\cdot\cdot}^2)$$

$$= \frac{1}{J} \sum_i \{ V(X_{i\cdot}) + [E(X_{i\cdot})]^2 \} - \frac{1}{IJ} \{ V(X_{\cdot\cdot}) + [E(X_{\cdot\cdot})]^2 \}$$

$$= \frac{1}{J} \sum_i \{ J\sigma^2 + [J(\mu + \alpha_i)]^2 \} - \frac{1}{IJ} [IJ\sigma^2 + (IJ\mu)^2]$$

$$= I\sigma^2 + IJ\mu^2 + 2\mu J \sum_i \alpha_i + J \sum_i \alpha_i^2 - \sigma^2 - IJ\mu^2$$

$$= (I - 1)\sigma^2 + J \sum_i \alpha_i^2 \quad \left(\text{since } \sum \alpha_i = 0\right)$$

The result then follows from the relationship $MSTr = SSTr/(I - 1)$.   ■

β for the F Test

Consider a set of parameter values $\alpha_1, \alpha_2, \ldots, \alpha_I$ for which H_0 is not true. The probability of a type II error, β, is the probability that H_0 is not rejected when that set is the set of true values. One might think that β would have to be determined separately for each different configuration of α_i's. Fortunately, since β for the F test depends on the α_i's and σ^2 only through $\Sigma\alpha_i^2/\sigma^2$, it can be simultaneously evaluated for many different alternatives. For example, $\Sigma\alpha_i^2 = 4$ for each of the following sets of α_i's for which H_0 is false, so β is identical for all three alternatives.

1. $\alpha_1 = -1, \quad \alpha_2 = -1, \quad \alpha_3 = 1, \quad \alpha_4 = 1$

2. $\alpha_1 = -\sqrt{2}, \quad \alpha_2 = \sqrt{2}, \quad \alpha_3 = 0 \quad \alpha_4 = 0$

3. $\alpha_1 = -\sqrt{3}, \quad \alpha_2 = \sqrt{1/3}, \quad \alpha_3 = \sqrt{1/3}, \quad \alpha_4 = \sqrt{1/3}$

The quantity $J\Sigma\alpha_i^2/\sigma^2$ is called the **noncentrality parameter** for one-way ANOVA (because when H_0 is false the test statistic has a *noncentral F* distribution with this as one of its parameters), and β is a decreasing function of the value of this parameter. Thus for fixed values of σ^2 and J, the null hypothesis is more likely to be rejected for alternatives far from H_0 (large $\Sigma\alpha_i^2$) than for alternatives close to H_0. For a fixed value of $\Sigma\alpha_i^2$, β decreases as the sample size J on each treatment increases and increases as the variance σ^2 increases (since more underlying variability makes it more difficult to detect any given departure from H_0).

Because hand computation of β and sample size determination for the F test are quite difficult (as in the case of t tests), statisticians have constructed sets of curves from which β can be read. These are called *power curves*. Sets of curves for numerator d.f. $\nu_1 = 3$ and $\nu_1 = 4$ are displayed in Figure 10.2* and Figure 10.3*, respectively. After the values of σ^2 and the α_i's for which β is desired are specified, these are used to compute the value of ϕ, where $\phi^2 = (J/I)\Sigma\alpha_i^2/\sigma^2$. We then enter the appropriate set of curves at the value of ϕ on the horizontal axis, move up to the curve associated with error d.f. ν_2, and move over to the value of power on the vertical axis. Finally, $\beta = 1 - $ power.

---

*Reproduced with permission from E. S. Pearson and H. O. Hartley, "Charts of the Power Function for Analysis of Variance Tests, Derived from the Non-central F Distribution," *Biometrika*, vol. 38, 1951, p. 112.

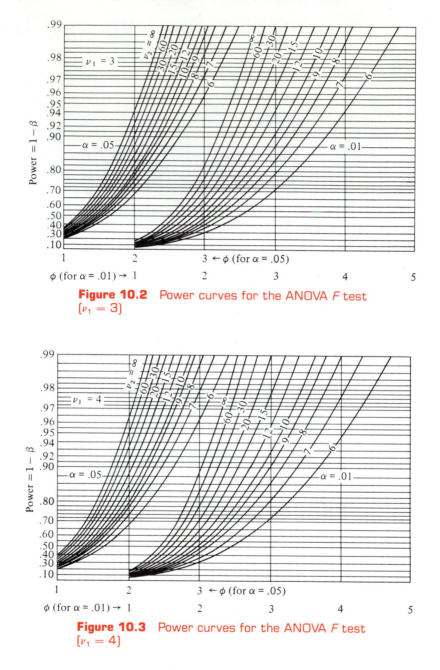

Figure 10.2   Power curves for the ANOVA *F* test $(\nu_1 = 3)$

Figure 10.3   Power curves for the ANOVA *F* test $(\nu_1 = 4)$

Example **10.8**   The effects of four different heat treatments on yield point (tons/in.²) of steel ingots are to be investigated. A total of eight ingots will be cast using each treatment. Suppose that the true standard deviation of yield point for any of the four treatments is $\sigma = 1$. How likely is it that H_0 will not be rejected at level .05 if three of the treatments have the same expected yield point and the other treatment has an expected yield point that is 1 ton/in.^{2} greater than the common value of the other three (that is, the fourth yield is on average one s.d. above those for the first three treatments)?

Supposing that $\mu_1 = \mu_2 = \mu_3$ and $\mu_4 = \mu_1 + 1$, $\mu = (\Sigma\mu_i)/4 = \mu_1 + \frac{1}{4}$. Then $\alpha_1 = \mu_1 - \mu = -\frac{1}{4}$, $\alpha_2 = -\frac{1}{4}$, $\alpha_3 = -\frac{1}{4}$, $\alpha_4 = \frac{3}{4}$ so $\phi^2 = \frac{8}{4}[(-\frac{1}{4})^2 + (-\frac{1}{4})^2 + (-\frac{1}{4})^2 + (\frac{3}{4})^2] = \frac{3}{2}$ and $\phi = 1.22$. The degrees of freedom are $\nu_1 = I - 1 = 3$ and $\nu_2 = I(J - 1) = 28$, so interpolating visually between $\nu_2 = 20$ and $\nu_2 = 30$ gives power $\approx .47$ and $\beta \approx .53$. This β is rather large, so we might decide to increase the value of J. How many ingots of each type would be required to yield $\beta \approx .05$ for the alternative under consideration? By trying different values of J we can verify that $J = 24$ will meet the requirement, but any smaller J will not.   ■

Relationship of the *F* Test to the *t* Test

When the number of treatments or populations is $I = 2$, all formulas and results connected with the F test still make sense, so ANOVA can be used to test $H_0: \mu_1 = \mu_2$ versus $H_a: \mu_1 \neq \mu_2$. In this case a two-tailed two-sample t test can also be used. Which test should be used? Somewhat surprisingly, the two are equivalent. It can be verified from the t and F tables that $t^2_{\alpha/2, \nu} = F_{\alpha, 1, \nu}$, and a bit of algebra shows that the relationship between the two test statistics is also $T^2 = F$ in this case. Thus the same conclusion (reject or accept H_0) will be reached irrespective of which test is used. Of course, if the alternative is one sided, then the one-tailed t test must be used, since the ANOVA alternative is always $H_a: \mu_1 \neq \mu_2$.

Single-Factor ANOVA When Sample Sizes Are Unequal

When the sample sizes from each population or treatment are not equal, let J_1, $J_2, \ldots, J_I$ denote the I sample sizes and let $n = \Sigma_i J_i$ denote the total number of observations. Define sums of squares (along with their computing formulas) by

$$SST = \sum_{i=1}^{I} \sum_{j=1}^{J_i} (X_{ij} - \bar{X}_{..})^2 = \sum_{i=1}^{I} \sum_{j=1}^{J_i} X_{ij}^2 - \frac{1}{n} X_{..}^2$$

$$SSTr = \sum_{i=1}^{I} \sum_{j=1}^{J_i} (\bar{X}_{i.} - \bar{X}_{..})^2 = \sum_{i=1}^{I} \frac{1}{J_i} X_{i.}^2 - \frac{1}{n} X_{..}^2$$

$$SSE = \sum_{i=1}^{I} \sum_{j=1}^{J_i} (X_{ij} - \bar{X}_{i.})^2 = SST - SSTr$$

SSTr and *SSE* have $I - 1$ and $n - I$ $[= \Sigma(J_i - 1)]$ d.f., respectively. With $MSTr = SSTr/(I - 1)$ and $MSE = SSE/(n - I)$, the F statistic is again the ratio of these two mean squares.

Theorem

> In one-way ANOVA with sample sizes $J_1, J_2, \ldots, J_I$ and $n = \Sigma J_i$, when all population or treatment distributions are normal with variance σ^2, a level α test for $H_0 : \mu_1 = \cdots = \mu_I$ versus H_a: at least two μ_i's differ is given by
>
> Test statistic value:
>
> $$f = \frac{MSTr}{MSE} \quad \text{where} \quad MSTr = \frac{SSTr}{I - 1}, \quad MSE = \frac{SSE}{n - I}$$
>
> Rejection region: $f \geq F_{\alpha, I - 1, n - I}$

Example 10.9   The critical flicker frequency (cff) is the highest frequency (in cycles/s) at which a person can detect the flicker in a flickering light source. At frequencies above the cff, the light source appears to be continuous even though it is actually flickering. An investigation carried out to see if true average cff depends on iris color yielded the following data (based on the article "The Effects of Iris Color on Critical Flicker Frequency," *J. General Psych.*, 1973, pp. 91–95).

Iris color		J_i	$x_{i.}$	$\overline{x}_{i.}$
1. Brown	26.8, 27.9, 23.7, 25.0, 26.3, 24.8, 25.7, 24.5	8	204.7	25.59
2. Green	26.4, 24.2, 28.0, 26.9, 29.1	5	134.6	26.92
3. Blue	25.7, 27.2, 29.9, 28.5, 29.4, 28.3	6	169.0	28.17
		$n = 19$	$x_{..} = 508.3$	

Let μ_1, μ_2, and μ_3 denote the true average cff's under the given experimental conditions for individuals with iris color brown, green, and blue, respectively. We shall test $H_0 : \mu_1 = \mu_2 = \mu_3$ against the alternative that at least two of these μ_i's differ. Numerator and denominator d.f. are $I - 1 = 2$ and $n - I = 19 - 3 = 16$, respectively. $F_{.05, 2, 16} = 3.63$, so H_0 will be rejected at level .05 if $f \geq 3.63$. The sum of squares of all 19 observations is $\Sigma \Sigma x_{ij}^2 = 13,659.67$ and the correction factor is $(508.3)^2 / 19 = 13,598.36$. The three sums of squares are

$$SST = 13,659.67 - 13,598.36 = 61.31$$

$$SSTr = \frac{1}{8}(204.7)^2 + \frac{1}{5}(134.6)^2 + \frac{1}{6}(169.0)^2 - 13,598.36 = 23.00$$

$$SSE = 61.31 - 23.00 = 38.31$$

Further calculations are displayed in the accompanying ANOVA table. Because $4.81 \geq 3.63$, H_0 is rejected. True average cff does appear to depend on iris color—a conclusion consistent with that reached in the cited paper.

Source of variation	d.f.	Sum of squares	Mean square	f
Treatments (iris color)	2	23.00	11.50	4.81
Error	16	38.31	2.39	
Total	18	61.31		

Multiple Comparisons When Sample Sizes Are Unequal

There is as yet no suitable modification of the T method for multiple comparisons when the sample sizes are unequal. While the Bonferroni method could be used (with $n - I$ d.f. for error), there is another method, due to Scheffé, which tends to produce shorter intervals when many confidence intervals are to be constructed. If $c_1, \ldots, c_I$ are constants with $\Sigma c_i = 0$, then $\Sigma c_i \mu_i$ is called a **contrast** in the μ_i's. Any pairwise difference, such as $\mu_1 - \mu_2$, is a contrast ($c_1 = 1$, $c_2 = -1$, $c_3 = \cdots = c_I = 0$), and so are $\mu_1 - \frac{1}{3}(\mu_2 + \mu_3 + \mu_4)$ and $\frac{1}{2}(\mu_1 + \mu_2) - \frac{1}{2}(\mu_3 + \mu_4)$. One might examine a particular contrast in order to compare averages of two different sets of μ_i's. Scheffé's method gives simultaneous confidence intervals for *all* possible contrasts in the μ_i's. Each interval is centered at the corresponding estimate $\Sigma c_i \bar{x}_i$.

Proposition

> With simultaneous confidence $100(1 - \alpha)\%$, for every set of constants $c_1, \ldots, c_I$ with $\Sigma c_i = 0$,
>
> $$\sum_{i=1}^{I} c_i \bar{x}_{i.} - \sqrt{\sum_{i=1}^{I} \frac{c_i^2}{J_i} \cdot \sqrt{(I-1)(MSE)F_{\alpha, I-1, n-I}}} \leq \sum_{i=1}^{I} c_i \mu_i$$
>
> $$\leq \sum_{i=1}^{I} c_i \bar{x}_{i.} + \sqrt{\sum_{i=1}^{I} \frac{c_i^2}{J_i} \cdot \sqrt{(I-1)(MSE)F_{\alpha, I-1, n-I}}} \qquad (10.8)$$

Every interval of the form (10.8) that does not include zero is said to correspond to a contrast that differs significantly from zero. Consideration of the intervals for all $\mu_i - \mu_j$'s enables multiple comparisons to be made.

Example 10.10
(Example 10.9 continued)

Confidence intervals with confidence level 95% based on the critical flicker frequency data require $\sqrt{(I-1)(MSE)F_{.05, I-1, n-I}} = \sqrt{2(2.39)(3.63)} = 4.17$. The value of $\Sigma c_i^2 / J_i$ for the contrast $\mu_1 - \mu_2$ is

$$\frac{c_1^2}{J_1} + \frac{c_2^2}{J_2} + \frac{c_3^2}{J_3} = \frac{1^2}{8} + \frac{(-1)^2}{5} + \frac{0^2}{6} = .325$$

The intervals for this contrast and for the contrasts $\mu_1 - \mu_3$ and $\mu_2 - \mu_3$ are

$\mu_1 - \mu_2$: $25.59 - 26.92 \pm \sqrt{.325}(4.17) = -1.33 \pm 2.38 = (-3.71, 1.05)$

$\mu_1 - \mu_3$: $25.59 - 28.17 \pm \sqrt{.292}(4.17) = -2.58 \pm 2.25 = (-4.83, -.33)$

$\mu_2 - \mu_3$: $26.92 - 28.17 \pm \sqrt{.367}(4.17) = -1.25 \pm 2.53 = (-3.78, 1.28)$

Only the second interval does not include zero, so only the true average cff's for those with brown and blue irises are judged significantly different. The contrast $.5\mu_1 + .5\mu_2 - \mu_3$ involves a comparison of those with brown or green irises to those with blue irises. Since

$$\Sigma \frac{c_i^2}{J_i} = \frac{(.5)^2}{8} + \frac{(.5)^2}{5} + \frac{(-1)^2}{6} = .248$$

the confidence interval for this contrast is

$$.5(25.59) + .5(26.92) - 28.17 \pm \sqrt{.248}(4.17) = -1.92 \pm 2.08$$
$$= (-4.00, .16) \blacksquare$$

Scheffé's method can be used when sample sizes are equal, but the resulting intervals for the $\mu_i - \mu_j$'s tend to be longer than those obtained from the T method; of course Scheffé's method also yields intervals for other contrasts.

Data Transformation

The use of ANOVA methods can be invalidated by substantial differences in the variances $\sigma_1^2, \ldots, \sigma_I^2$ (which until now have been assumed equal with common value σ^2). It sometimes happens that $V(X_{ij}) = \sigma_i^2 = g(\mu_i)$, a known function of μ_i (so that when H_0 is false, the variances are not equal). For example, if X_{ij} has a Poisson distribution with parameter λ_i (approximately normal if $\lambda_i \geq 10$), then $\mu_i = \lambda_i$ and $\sigma_i^2 = \lambda_i$ so $g(\mu_i) = \mu_i$ is the known function. In such cases one can often transform the X_{ij}'s to $h(X_{ij})$'s so that they will have approximately equal variances (while leaving the transformed variables approximately normal), and then the F test can be used on the transformed observations. The key idea in choosing a transformation $h(\cdot)$ is that often $V[h(X_{ij})] \approx V(X_{ij}) \cdot [h'(\mu_i)]^2 = g(\mu_i) \cdot [h'(\mu_i)]^2$. We wish the function $h(\cdot)$ for which $g(\mu_i) \cdot [h(\mu_i)]^2 = c$ (a constant) for every i.

Proposition

> If $V(X_{ij}) = g(\mu_i)$, a known function of μ_i, then a transformation $h(X_{ij})$ that "stabilizes the variance" so that $V[h(X_{ij})]$ is approximately the same for each i is given by $h(x) \propto \int [g(x)]^{-1/2} dx$.

In the Poisson case, $g(x) = x$, so $h(x)$ should be proportional to $\int x^{-1/2} dx = 2x^{1/2}$. Thus Poisson data should be transformed to $h(x_{ij}) = \sqrt{x_{ij}}$ before the analysis.

A Random Effects Model

The single-factor problems considered so far have all been assumed to be examples of a **fixed effects** ANOVA model. By this we mean that the chosen levels of the factor under study are the only ones considered relevant by the experimenter. The single-factor fixed effects model is

$$X_{ij} = \mu + \alpha_i + \epsilon_{ij}, \quad \Sigma\alpha_i = 0 \tag{10.9}$$

where the ϵ_{ij}'s are random and both μ and the α_i's are fixed parameters whose values are unknown.

In some single-factor problems the particular levels studied by the experimenter are chosen, either by design or through sampling, from a large population of levels. For example, to study the effects on task performance time of using different operators on a particular machine, a sample of five operators might be chosen from a large pool of operators. Similarly, the effect of soil pH on the yield of maize plants might be studied by using soils with four specific pH values chosen from among the many possible pH levels. When the levels used are selected at random from a larger population of possible levels, the factor is said to be random rather than fixed, and the fixed effects model (10.9) is no longer appropriate. An analogous **random effects** model is obtained by replacing the fixed α_i's in (10.9) by random variables. The resulting model description is

$$X_{ij} = \mu + A_i + \epsilon_{ij} \quad \text{with} \quad E(A_i) = E(\epsilon_{ij}) = 0$$

$$V(\epsilon_{ij}) = \sigma^2, \quad V(A_i) = \sigma_A^2 \qquad (10.10)$$

all A_i's and ϵ_{ij}'s normally distributed and independent of one another.

The condition $E(A_i) = 0$ in (10.10) is similar to the condition $\Sigma\alpha_i = 0$ in (10.11); it states that the expected or average effect of the ith level measured as a departure from μ is zero.

For the random effects model (10.10), the hypothesis of no effects due to different levels is $H_0: \sigma_A^2 = 0$, which says that different levels of the factor contribute nothing to variability of the response. *Although the hypotheses in the single-factor fixed and random effects models are different, they are tested in exactly the same way,* by forming $F = MSTr/MSE$ and rejecting H_0 if $f \geq F_{\alpha, I-1, n-I}$. This can be justified intuitively by noting that $E(MSE) = \sigma^2$ (as for fixed effects), while

$$E(MSTr) = \sigma^2 + \frac{1}{I-1}\left(n - \frac{\Sigma J_i^2}{n}\right)\sigma_A^2 \qquad (10.11)$$

(where $J_1, J_2, \ldots, J_I$ are the sample sizes and $n = \Sigma J_i$).

The factor in parentheses on the right side of (10.11) is nonnegative, so again $E(MSTr) = \sigma^2$ if H_0 is true and $E(MSTr) > \sigma^2$ if H_0 is false.

Example **10.11**   The study of nondestructive forces and stresses in materials furnishes important information for efficient engineering design. The paper "Zero-Force Travel-Time Parameters for Ultrasonic Head-Waves in Railroad Rail" (*Materials Evaluation,* 1985, pp. 854–858) reported on a study of travel time for a certain type of wave that results from longitudinal stress of rails used for railroad track. Three measurements were made on each of six rails randomly selected from a population of rails. The investigators used random effects ANOVA to decide whether some variation in travel time could be attributed to "between-rail variability." The data appears in the accompanying table (each value, in nanoseconds, resulted from subtracting 36.1 μs from the original observation) along with the derived ANOVA table. The value of the F ratio is highly significant, so $H_0: \sigma_A^2 = 0$ is rejected in favor of the conclusion that differences between rails is a source of travel time variability.

				$x_{i.}$
1.	55	53	54	162
2.	26	37	32	95
3.	78	91	85	254
4.	92	100	96	288
5.	49	51	50	150
6.	80	85	83	248

$$x_{..} = 1197$$

Source of variation	d.f.	Sum of squares	Mean square	f
Treatments	5	9310.5	1862.1	115.2
Error	12	194.0	16.17	
Total	17	9504.5		

Exercises / Section 10.3 (23–35)

23. The following data refers to yield of tomatoes (kg/plot) for four different levels of salinity; salinity level here refers to electrical conductivity (EC), where the chosen levels were EC = 1.6, 3.8, 6.0, and 10.2 nmhos/cm.

1.6: 59.5, 53.3, 56.8, 63.1, 58.7
3.8: 55.2, 59.1, 52.8, 54.5
6.0: 51.7, 48.8, 53.9, 49.0
10.2: 44.6, 48.5, 41.0, 47.3, 46.1

Use the F test at level $\alpha = .05$ to test for any differences in true average yield due to the different salinity levels.

24. Apply Scheffé's method to the data in Exercise 23 to identify significant differences among the μ_i's.

25. The following partial ANOVA table is taken from the article "Perception of Spatial Incongruity" (*J. Nervous and Mental Disease*, 1961, p. 222) in which the abilities of three different groups to identify a perceptual incongruity were assessed and compared. All individuals in the experiment had been hospitalized to undergo psychiatric treatment. There were 21 individuals in the depressive group, 32 individuals in the functional "other" group, and 21 individuals in the brain-damaged group. Complete the ANOVA table and carry out the F test at level $\alpha = .01$.

Source	d.f.	Sum of squares	Mean square	f
Groups			76.09	
Error				
Total		1123.14		

26. An article in the *Canadian Entomologist* ("Influence of Natural Diets and Larval Density on Gypsy Moth, Lymantria Dispor, Egg Mass Characteristics," 1977, pp. 1313–1318) reported the following data on egg mass diameters for moths reared on five different diets.

Diet	J_i	$\bar{x}_{i.}$ (mm)	s_i
Red maple '74	13	1.134	.0252
Red oak/red maple	10	1.148	.0253
Red maple '75	20	1.159	.0179
Red oak	16	1.191	.0200
Red oak/white pine	16	1.217	.0160

a. Compute the grand mean $\bar{x}_{..}$, SSTr, and MSTr. Hint: $x_{..} = \Sigma J_i \bar{x}_{i.}$.
b. Compute SSE and MSE. Hint: SSE $= \Sigma(J_i - 1)s_i^2$.
c. Does the data suggest that there are any differences between true average egg diameters for the different diets? Test using $\alpha = .05$.
d. Apply Scheffé's method to look for means which differ significantly from one another.

27. Samples of six different brands of diet/imitation margarine were analyzed to determine the level of physiologically active polyunsaturated fatty acids (PAPFUA, in percentages), resulting in the following data:

Imperial: 14.1, 13.6, 14.4, 14.3
Parkay: 12.8, 12.5, 13.4, 13.0, 12.3
Blue Bonnet: 13.5, 13.4, 14.1, 14.3
Chiffon: 13.2, 12.7, 12.6, 13.9
Mazola: 16.8, 17.2, 16.4, 17.3, 18.0
Fleischmann's: 18.1, 17.2, 18.7, 18.4

(The preceding numbers are fictitious, but the sample means agree with data reported in the January 1975 issue of *Consumer Reports*.)

a. Use ANOVA to test for differences among the true average PAPFUA percentages for the different brands.

b. Use Scheffé's method to compute confidence intervals for all $(\mu_i - \mu_j)$'s.

c. Mazola and Fleischmann's are corn based while the others are soybean based. Compute a Scheffé interval for the contrast

$$\frac{(\mu_1 + \mu_2 + \mu_3 + \mu_4)}{4} - \frac{(\mu_5 + \mu_6)}{2}$$

28. Although tea is the world's most widely consumed beverage after water, little is known about its nutritional value. Folacin is the only B vitamin present in any significant amount in tea, and recent advances in assay methods have made accurate determination of folacin content feasible. Consider the accompanying data on folacin content for randomly selected specimens of the four leading brands of green tea.

Brand	Observations
1	7.9, 6.2, 6.6, 8.6, 8.9, 10.1, 9.6
2	5.7, 7.5, 9.8, 6.1, 8.4
3	6.8, 7.5, 5.0, 7.4, 5.3, 6.1
4	6.4, 7.1, 7.9, 4.5, 5.0, 4.0

(Based on "Folacin Content of Tea," *J. Amer. Dietetic Assoc.*, 1983, pp. 627–632.) Does this data suggest that true average folacin content is the same for all brands? Carry out a test using $\alpha = .05$.

29. For a single-factor ANOVA with sample sizes J_i ($i = 1, 2, \ldots, I$), show that $SSTr = \sum_i J_i(\overline{X}_{i\cdot} - \overline{X}_{\cdot\cdot})^2 = \sum_i J_i \overline{X}_{i\cdot}^2 - n\overline{X}_{\cdot\cdot}^2$, where $n = \sum J_i$.

30. When sample sizes are equal ($J_i = J$), the parameters $\alpha_1, \alpha_2, \ldots, \alpha_I$ of the alternative parameterization are restricted by $\sum \alpha_i = 0$. For unequal sample size, the most natural restriction is $\sum J_i \alpha_i = 0$. Use this to show that

$$E(MSTr) = \sigma^2 + \frac{1}{I-1} \sum J_i \alpha_i^2$$

What is $E(MSTr)$ when H_0 is true? [This expectation is correct if $\sum J_i \alpha_i = 0$ is replaced by the restriction $\sum \alpha_i = 0$ (or any other single linear restriction on the α_i's used to reduce the model to I independent parameters), but $\sum J_i \alpha_i = 0$ simplifies the algebra and yields natural estimates for the model parameters (in particular, $\hat{\alpha}_i = \overline{x}_{i\cdot} - \overline{x}_{\cdot\cdot}$).]

31. Reconsider Example 10.8 involving an investigation of the effects of different heat treatments on the yield point of steel ingots.

a. If $J = 8$ and $\sigma = 1$, what is β for a level .05 F test when $\mu_1 = \mu_2, \mu_3 = \mu_1 - 1, \mu_4 = \mu_1 + 1$?

b. For the alternative of (a), what value of J is necessary to obtain $\beta = .05$?

c. If there are $I = 5$ heat treatments, $J = 10$, and $\sigma = 1$, what is β for the level .05 F test when four of the μ_i's are equal and the fifth differs by 1 from the other four?

32. When sample sizes are not equal, the noncentrality parameter is $\sum J_i \alpha_i^2 / 2\sigma^2$ and $\phi^2 = (1/I) \sum J_i \alpha_i^2 / \sigma^2$. Referring to Exercise 10.23 above, what is the power of the test when $\mu_2 = \mu_3, \mu_1 = \mu_2 - \sigma, \mu_4 = \mu_2 + \sigma$?

33. In an experiment to compare the quality of four different brands of reel-to-reel recording tape, five 2400-ft reels of each brand were selected and the number of flaws in each reel was determined.

	A	10, 5, 12, 14, 8
Brand	B	14, 12, 17, 9, 8
	C	13, 18, 10, 15, 18
	D	17, 16, 12, 22, 14

It is believed that the number of flaws has approximately a Poisson distribution for each brand. Analyze the data at level .01 to see whether the expected number of flaws per reel is the same for each brand.

34. Suppose that X_{ij} is a binomial variable with parameters n and p_i (so approximately normal when $np_i \geq 5$ and $nq_i \geq 5$). Then since $\mu_i = np_i$, $V(X_{ij}) = \sigma_i^2 = np_i(1 - p_i) = \mu_i(1 - \mu_i/n)$. How should the X_{ij}'s be transformed so as to stabilize the variance? *Hint:* $g(\mu_i) = \mu_i(1 - \mu_i/n)$.

35. Simplify $E(MSTr)$ for the random effects model when $J_1 = J_2 = \cdots = J_I = J$.

Supplementary Exercises / Chapter 10 (36–45)

36. The paper "Computer-Assisted Instruction Augmented with Planned Teacher/Student Contacts" (*J. Exp. Educ.*, Winter, 1980–1981, pp. 120–126) compared five different methods for teaching descriptive statistics. The five methods were traditional lecture and discussion (L/D), programmed textbook instruction (R), programmed text with lectures (R/L), computer instruction (C), and computer instruction with lectures (C/L). Forty-five students were randomly assigned, nine to each method. After completing the course, the students took a 1-hour exam. In addition, a 10-minute retention test was administered 6 weeks later. Summary quantities are given.

	Exam		Retention test	
Method	$\bar{x}_{i.}$	s_i	$\bar{x}_{i.}$	s_i
L/D	29.3	4.99	30.20	3.82
R	28.0	5.33	28.80	5.26
R/L	30.2	3.33	26.20	4.66
C	32.4	2.94	31.10	4.91
C/L	34.2	2.74	30.20	3.53

The grand mean for the exam was 30.82 and the grand mean for the retention test was 29.30.

a. Does the data suggest that there is a difference between the five teaching methods with respect to true mean exam score? Use $\alpha = .05$.

b. Using a .05 significance level, test the null hypothesis of no difference between the true mean retention test scores for the five different teaching methods.

37. The article "Major Appliance Prices in the Chicago Area" (*J. Business*, 1977, pp. 231–235) reported the following data on prices for a certain size of refrigerator at a sample of stores in the Chicago area.

Brand	Sample size	$\bar{x}_{i.}$	s_i
Frigidaire	18	423.21	18.71
GE	28	418.13	21.15
Whirlpool	19	421.27	17.87

Does this data suggest any differences in the true average prices charged for the three brands?

38. An article in the British scientific journal *Nature* ("Sucrose Induction of Hepatic Hyperplasis in the Rat," August 25, 1972, p. 461) reported on an experiment in which each of five groups consisting of six rats was put on a diet with a different carbohydrate. At the conclusion of the experiment the DNA content of the liver of each rat was determined (mg/g liver), with the following results:

Carbohydrate	$\bar{x}_{i.}$
Starch	2.58
Sucrose	2.63
Fructose	2.13
Glucose	2.41
Maltose	2.49

Assuming also that $\Sigma\Sigma x_{ij}^2 = 183.4$, does the data indicate that true average DNA content is affected by the type of carbohydrate in the diet? Construct an ANOVA table and use a .05 level of significance.

39. Referring to Exercise 38, construct a t confidence interval for

$$\theta = \mu_1 - (\mu_2 + \mu_3 + \mu_4 + \mu_5)/4$$

which measures the difference between the average DNA content for the starch diet and the combined average for the four other diets. Does the resulting interval include zero?

40. Referring back to Exercise 38, what is β for the test when true average DNA content is identical for three of the starches and falls below this common value by one standard deviation (σ) for the other two starches?

41. Four laboratories are randomly selected from a large population and each is asked to make three determinations of the percentage of methyl alcohol in specimens of a compound taken from a single batch. Based on the accompanying data, are differences between laboratories a source of variation in the percentage of methyl alcohol? State and test the relevant hypotheses using significance level .05.

	1	85.06	85.25	84.87
Laboratory	2	84.99	84.28	84.88
	3	84.48	84.72	85.10
	4	84.10	84.55	84.05

42. Four types of mortars—ordinary cement mortar (OCM), polymer impregnated mortar (PIM), resin mortar (RM), and polymer cement mortar (PCM)—were subjected to a compression test to measure strength (MPa). Three strength observations for each mortar type appeared in the paper "Polymer Mortar Composite Matrices for Maintenance-Free Highly Durable Ferrocement" (*J. Ferrocement,* 1984, pp. 337–345) and are reproduced below. Construct an ANOVA table. Using a .05 significance level, determine whether the data suggests that the true mean strength is not the same for all four mortar types. If you determine that the true mean strengths are not all equal, use Tukey's method to identify the significant differences.

	OCM	32.15	35.53	34.20
Type	PIM	126.32	126.80	134.79
	RM	117.91	115.02	114.58
	PCM	29.09	30.87	29.80

43. Suppose that the x_{ij}'s are "coded" by $y_{ij} = cx_{ij} + d$. How does the value of the F statistic computed from the y_{ij}'s compare to the value computed from the x_{ij}'s? Justify your assertion.

44. In Example 10.11, subtract $\bar{x}_{i\cdot}$ from each observation in the ith sample $(i = 1, \ldots, 6)$ to obtain a set of 18 residuals. Then construct a normal probability plot and comment on the plausibility of the normality assumption.

45. Consider a single-factor ANOVA in the case $I = 2$. Show that
   a. $MSE = S_p^2$, the pooled estimator of σ^2 for the pooled t test.
   b. $T^2 = F$, where T is the pooled t statistic for testing $H_0 : \mu_1 - \mu_2 = 0$.

Bibliography

Montgomery, Douglas, *Design and Analysis of Experiments* (2nd ed.), Wiley, New York, 1984. A very up-to-date presentation of ANOVA models and methodology.

Neter, John, Wasserman, William, and Kutner, Michael, *Applied Linear Statistical Models* (2nd ed.), Irwin, Homewood, Ill., 1985. The second half of this book contains a very well-presented survey of ANOVA; the level is comparable to that of the present text, but the discussion is more comprehensive, making the book an excellent reference.

Ott, Lyman, *An Introduction to Statistical Methods and Data Analysis* (2nd ed.), Duxbury Press, Boston, 1985. Includes several chapters on ANOVA methodology that can profitably be read by students desiring a very nonmathematical exposition; there is a good chapter on various multiple comparison methods.

Multifactor Analysis of Variance

Introduction

In the previous chapter we used the analysis of variance to test for equality of either I different population means or the true average responses associated with I different levels of a single factor (alternatively referred to as I different treatments). In many experimental situations there are two or more factors that are of simultaneous interest. This chapter extends the methods of Chapter 10 to investigate such multifactor situations.

In the first two sections we concentrate on the case of two factors of interest. We shall use I to denote the number of levels of the first factor (A) and J to denote the number of levels of the second factor (B). Then there are IJ possible combinations consisting of one level of factor A and one of factor B; each such combination is called a treatment, so there are IJ different treatments. The number of observations made on treatment (i, j) will be denoted by K_{ij}. In Section 11.1 we present the model and analysis when $K_{ij} = 1$. An important special case of this type is a randomized block design, in which a single factor A is of primary interest but another factor, "blocks," is created in order to control for extraneous variability in experimental units or subjects. In Section 11.2 we focus on the case $K_{ij} = K > 1$, and mention briefly the difficulties associated with unequal K_{ij}'s.

Section 11.3 discusses experiments involving more than two factors, including a Latin square design, which controls for the effects of two extraneous factors thought to influence the response variable. When the number of factors is large, an experiment consisting of at least one observation for each treatment would be quite expensive and time consuming. An important special case, which we discuss in the last section, is that in which there are p factors, each of which has two levels. There are then 2^p different treatments, and we consider both the case in which observations are made on all these treatments (a complete design) and the case in which observations are made for only a selected subset of treatments (an incomplete design).

11.1 Two-Factor ANOVA with $K_{ij} = 1$

When factor A consists of I levels and factor B consists of J levels, there are IJ different combinations (pairs) of levels of the two factors, each called a treatment. With K_{ij} = the number of observations on the treatment consisting of factor A at level i and factor B at level j, we focus in this section on the case $K_{ij} = 1$, so that the data consists of IJ observations. We shall first discuss the fixed-effects model, in which the only levels of interest for the two factors are those actually represented in the experiment. The case in which one or both factors are random is discussed briefly at the end of the section.

Example 11.1

In a study on automobile traffic and air pollution reported in the *International Journal of Environmental Studies* ("Automobile Traffic and Air Pollution in a Developing Country," 1977, pp. 197–203), air samples taken at four different times and at five different locations were analyzed to obtain the amount of particulate matter present in the air (mg/m^3).

		Factor B: *Location*					Row total
		1	*2*	*3*	*4*	*5*	
Factor A:	Oct. 1975	76	67	81	56	51	331
Time	Jan. 1976	82	69	96	59	70	376
	May 1976	68	59	67	54	42	290
	Sept. 1976	63	56	64	58	37	278
	Column total	289	251	308	227	200	1275

Is there any difference in true average amount of particulate matter present in the air due to either different sampling times or different locations? ■

The Notation

As in single-factor ANOVA, double subscripts are used to identify random variables and observed values. Let

X_{ij} = the random variable denoting the measurement when factor A is held at level i and factor B is held at level j

x_{ij} = the observed value of X_{ij}

The x_{ij}'s are usually presented in a two-way table in which the ith row contains the observed values on factor A held at level i and the jth column contains the observed values on factor B held at level j. In the air pollution experiment of Example 11.1, the number of levels of factor A is $I = 4$ and the number of levels of factor B is $J = 5$.

Whereas in single-factor ANOVA we were interested only in row means and the grand mean, here we are interested also in column means. Let

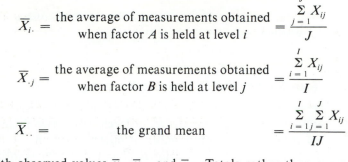

$$\overline{X}_{i\cdot} = \begin{array}{c}\text{the average of measurements obtained}\\\text{when factor } A \text{ is held at level } i\end{array} = \frac{\sum\limits_{j=1}^{J} X_{ij}}{J}$$

$$\overline{X}_{\cdot j} = \begin{array}{c}\text{the average of measurements obtained}\\\text{when factor } B \text{ is held at level } j\end{array} = \frac{\sum\limits_{i=1}^{I} X_{ij}}{I}$$

$$\overline{X}_{\cdot\cdot} = \text{the grand mean} = \frac{\sum\limits_{i=1}^{I}\sum\limits_{j=1}^{J} X_{ij}}{IJ}$$

with observed values $\overline{x}_{i\cdot}$, $\overline{x}_{\cdot j}$, and $\overline{x}_{\cdot\cdot}$. Totals rather than averages are denoted by omitting the horizontal bar (so $x_{\cdot j} = \sum_i x_{ij}$, and so on). Intuitively, to see whether there is any effect due to the levels of factor A, we should compare the observed $\overline{x}_{i\cdot}$'s with one another, while information about the different levels of factor B should come from the $\overline{x}_{\cdot j}$'s.

The Model

Proceeding by analogy to single-factor ANOVA, one's first inclination in specifying a model is to let $\mu_{ij} =$ the true average response when factor A is at level i and factor B at level j, giving IJ mean parameters. Then let

$$X_{ij} = \mu_{ij} + \epsilon_{ij}$$

where ϵ_{ij} is the random amount by which the observed value differs from its expectation and the ϵ_{ij}'s are assumed normal and independent with common variance σ^2. Unfortunately there is no valid test procedure for this choice of parameters. The reason is that under the alternative hypothesis of interest, the μ_{ij}'s are free to take on any values whatsoever, while σ^2 can be any value greater than zero, so that there are $IJ + 1$ freely varying parameters. But there are only IJ observations, so after using each x_{ij} as an estimate of μ_{ij}, there is no way to estimate σ^2.

To rectify this problem of a model having more parameters than observed values, we must specify a model that is realistic yet involves relatively few parameters. Suppose that we assume the existence of I parameters $\alpha_1, \alpha_2, \ldots, \alpha_I$ and J parameters $\beta_1, \beta_2, \ldots, \beta_J$ such that

$$X_{ij} = \alpha_i + \beta_j + \epsilon_{ij} \quad i = 1, \ldots, I; \quad j = 1, \ldots, J \tag{11.1}$$

That is, we now assume that each μ_{ij} can be written in the form

$$\mu_{ij} = \alpha_i + \beta_j \tag{11.2}$$

Including σ^2, there are now $I + J + 1$ model parameters, so if $I \geq 3$ and $J \geq 3$, then there will be fewer parameters than observations (in fact, we will shortly modify (11.2) so that even $I = 2$ and/or $J = 2$ will be accommodated).

The model specified in (11.1) and (11.2) is called an **additive model,** because each mean response μ_{ij} is the sum of an effect due to factor A at level i (α_i) and an effect due to factor B at level j (β_j). The difference between mean responses for factor A at level i and level i' when B is held at level j is $\mu_{ij} - \mu_{i'j}$. When the model is additive,

$$\mu_{ij} - \mu_{i'j} = (\alpha_i + \beta_j) - (\alpha_{i'} + \beta_j) = \alpha_i - \alpha_{i'}$$

which is independent of the level j of the second factor. A similar result holds for $\mu_{ij} - \mu_{ij'}$. Thus additivity means that the difference in mean responses for two levels of one of the factors is the same for all levels of the other factor. Figure 11.1a shows a set of mean responses that satisfy the condition of additivity, while Figure 11.1b shows a nonadditive configuration of mean responses.

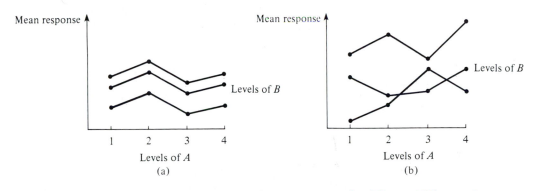

Figure 11.1   Mean responses for (a) an additive, and (b) a nonadditive model

Example 11.2
(Example 11.1 continued)

If we plot the observed x_{ij}'s in a manner analogous to that of Figure 11.1, the result is shown in Figure 11.2. While there is some "crossing over" in the observed x_{ij}'s, the configuration is reasonably representative of what would be expected under additivity with just one observation per treatment.

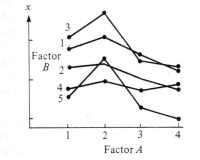

Figure 11.2   Plot of data from Example 11.2   ■

Expression (11.2) is not quite the final model description, because the α_i's and β_j's are not uniquely determined. Pictured below are two different configurations of the α_i's and β_j's that yield the same additive μ_{ij}'s.

	$\beta_1 = 1$	$\beta_2 = 4$			$\beta_1 = 2$	$\beta_2 = 5$
$\alpha_1 = 1$	$\mu_{11} = 2$	$\mu_{12} = 5$		$\alpha_1 = 0$	$\mu_{11} = 2$	$\mu_{12} = 5$
$\alpha_2 = 2$	$\mu_{21} = 3$	$\mu_{22} = 6$		$\alpha_2 = 1$	$\mu_{21} = 3$	$\mu_{22} = 6$

By subtracting any constant c from all α_i's and adding c to all β_j's, other configurations corresponding to the same additive model are obtained. This non-uniqueness is eliminated by use of the model

$$X_{ij} = \mu + \alpha_i + \beta_j + \epsilon_{ij}, \quad \text{where} \quad \sum_{i=1}^{I} \alpha_i = 0 \quad \text{and} \quad \sum_{j=1}^{J} \beta_j = 0 \qquad (11.3)$$

which is analogous to the alternative choice of parameters for single-factor ANOVA discussed in Section 10.3. It is not difficult to verify that (11.3) is an additive model in which the parameters are uniquely determined (for example, for the μ_{ij}'s mentioned above, $\mu = 4$, $\alpha_1 = -.5$, $\alpha_2 = .5$, $\beta_1 = -1.5$, and $\beta_2 = 1.5$). Notice that there are only $I - 1$ independently determined α_i's and $J - 1$ independently determined β_j's, so (including μ) (11.3) specifies $I + J - 1$ mean parameters.

The interpretation of the parameters of (11.3) is straightforward: μ is the true grand mean (mean response averaged over all levels of both factors), α_i is the effect of factor A at level i (measured as a deviation from μ), and β_j is the effect of factor B at level j. Unbiased estimators for these parameters are

$$\hat{\mu} = \overline{X}_{..}, \quad \hat{\alpha}_i = \overline{X}_{i.} - \overline{X}_{..}, \quad \text{and} \quad \hat{\beta}_j = \overline{X}_{.j} - \overline{X}_{..}$$

These are also maximum likelihood estimators when the ϵ_{ij}'s are normal.

The Hypotheses

There are two different hypotheses of interest in a two-factor experiment with $K_{ij} = 1$. The first, denoted by H_{0A}, states that the different levels of factor A have no effect on true average response, while the second, denoted by H_{0B}, states that there is no factor B effect:

$$H_{0A} : \alpha_1 = \alpha_2 = \cdots = \alpha_I = 0 \quad \text{versus} \quad H_{aA} : \text{at least one } \alpha_i \neq 0$$
$$H_{0B} : \beta_1 = \beta_2 = \cdots = \beta_J = 0 \quad \text{versus} \quad H_{aB} : \text{at least one } \beta_j \neq 0 \tag{11.4}$$

(No factor A effect implies all α_i's are equal, so they must all be 0 since they sum to 0, and similarly for the β_j's.)

The Test Procedures

The description and analysis now follow closely that for single-factor ANOVA. The relevant sums of squares and their computing forms are given by

$$SST = \sum_{i=1}^{I} \sum_{j=1}^{J} (X_{ij} - X_{..})^2 = \sum_{i=1}^{I} \sum_{j=1}^{J} X_{ij}^2 - \frac{1}{IJ} X_{..}^2$$

$$SSA = \sum_{i=1}^{I} \sum_{j=1}^{J} (X_{i.} - X_{..})^2 = \frac{1}{J} \sum_{i=1}^{I} X_{i.}^2 - \frac{1}{IJ} X_{..}^2 \tag{11.5}$$

$$SSB = \sum_{i=1}^{I} \sum_{j=1}^{J} (X_{.j} - X_{..})^2 = \frac{1}{I} \sum_{j=1}^{J} X_{.j}^2 - \frac{1}{IJ} X_{..}^2$$

and

$$SSE = \sum_{i=1}^{I} \sum_{j=1}^{J} (X_{ij} - X_{i.} - X_{.j} + X_{..})^2$$

(SSE comes from substituting parameter estimators into $\Sigma\Sigma[X_{ij} - (\mu + \alpha_i + \beta_j)]^2$.) The fundamental identity for two-factor ANOVA with one observation per treatment is

$$SST = SSA + SSB + SSE \tag{11.6}$$

Thus SSE can be obtained by subtraction once the other three sums of squares have been computed. As in single-factor ANOVA, total variation is split into a part (SSE) that is not explained by either the truth or the falsity of H_{0A} or H_{0B} and two parts that can be explained by possible falsity of the two null hypotheses.

The numbers of degrees of freedom for SSA and SSB are $I - 1$ and $J - 1$ respectively. There are IJ observations, but $(I - 1) + (J - 1) + 1$ mean parameters are independently estimated, leaving $IJ - [1 + (I - 1) + (J - 1)] = (I - 1)(J - 1)$ d.f. for error. There are then three mean squares, with $MS = SS/\text{d.f.}$ The test procedures for the two pairs of hypotheses are

Hypotheses: H_{0A} versus H_{aA}

Test statistic value: $f_A = \dfrac{MSA}{MSE}$

Rejection region: $f_A \geq F_{\alpha, I-1, (I-1)(J-1)}$

Hypotheses: H_{0B} versus H_{aB}

Test statistic value: $f_B = \dfrac{MSB}{MSE}$

Rejection region: $f_B \geq F_{\alpha, J-1, (I-1)(J-1)}$

Example 11.3
(Example 11.2 continued)

The $x_{i\cdot}$'s (row totals) and $x_{\cdot j}$'s (column totals) for the particulate matter data are displayed along the right and bottom margins of the data table given earlier. In addition, $\Sigma\Sigma x_{ij}^2 = 84{,}853$ and the correction factor is $x_{\cdot\cdot}^2/IJ = (1275)^2/20 = 81{,}281.25$. The sums of squares are then

$$SST = 84{,}853 - 81{,}281.25 = 3571.75$$

$$SSA = \frac{1}{5}\left[(331)^2 + (376)^2 + (290)^2 + (278)^2\right] - 81{,}281.25 = 1182.95$$

$$SSB = \frac{1}{4}\left[(289)^2 + (251)^2 + (308)^2 + (227)^2 + (200)^2\right] - 81{,}281.25$$

$$= 1947.50$$

$$SSE = 3571.75 - (1182.95 + 1947.50) = 441.30$$

The accompanying ANOVA table (Table 11.1) summarizes further calculations.

Table 11.1

Source of variation	d.f.	Sum of squares	Mean square	f
Factor A (time)	$I - 1 = 3$	$SSA = 1182.95$	$MSA = 394.32$	$f_A = 10.72$
Factor B (location)	$J - 1 = 4$	$SSB = 1947.50$	$MSB = 486.88$	$f_B = 13.24$
Error	$(I-1)(J-1) = 12$	$SSE = 441.30$	$MSE = 36.78$	
Total	$IJ - 1 = 19$	$SST = 3571.75$		

The critical value for testing H_{0A} at level of significance .05 is $F_{.05, 3, 12} = 3.49$. Since $10.72 \geq 3.49$, H_{0A} is rejected in favor of the claim that the average amount of particulate matter does vary with time of sampling. Similarly, $F_{.05, 4, 12} = 3.26$ and $13.24 \geq 3.26$, so H_{0B} is rejected at significance level .05 in favor of the assertion that pollution varies with sampling location. ∎

Expected Mean Squares

The plausibility of using the F tests just described is demonstrated by computing the expected mean squares. After some tedious algebra,

$$E(MSE) = \sigma^2 \quad \text{(when the model is additive)}$$

$$E(MSA) = \sigma^2 + \frac{J}{I-1}\sum_{i=1}^{I}\alpha_i^2, \quad E(MSB) = \sigma^2 + \frac{I}{J-1}\sum_{j=1}^{J}\beta_j^2$$

When H_{0A} is true, MSA is an unbiased estimator of σ^2, so F is a ratio of two unbiased estimators of σ^2. When H_{0A} is false, MSA tends to overestimate σ^2, so H_{0A} should be rejected when the ratio F_A is too large. Similar comments apply to MSB and H_{0B}.

Multiple Comparisons in Two-Factor ANOVA

When either H_{0A} or H_{0B} has been rejected, Tukey's procedure can be used to identify significant differences between the levels of the factor under investigation. The steps in the analysis are identical to those for a single-factor ANOVA:

1. For comparing levels of factor A, obtain $Q_{\alpha, I, (I-1)(J-1)}$.
   For comparing levels of factor B, obtain $Q_{\alpha, J, (I-1)(J-1)}$.
2. Compute

   $$w = Q \cdot \text{(estimated standard deviation of the sample means being compared)}$$

   $$= Q_{\alpha, I, (I-1)(J-1)} \cdot \sqrt{MSE/J} \quad \text{for factor } A \text{ comparisons}$$

   $$\text{or } Q_{\alpha, J, (I-1)(J-1)} \cdot \sqrt{MSE/I} \quad \text{for factor } B \text{ comparisons}$$

   (because, for example, the standard deviation of $\overline{X}_{i\cdot}$ is $\sigma/\sqrt{J}$).
3. Arrange the sample means in increasing order, underscore those pairs differing by less than w, and identify pairs not underscored by the same line as corresponding to significantly different levels of the given factor.

Example 11.4
(Example 11.3 continued)

Both the hypothesis of no sampling time effects and that of no location effects were rejected at level .05. To identify significant differences among the four sampling times (factor A), we need $Q_{.05, 4, 12} = 4.20$ and $w = 4.20\sqrt{36.78/5} = 11.39$. The four factor A sample means (row averages) are now listed in increasing order and any pair differing by less than 11.39 is underscored by a line segment:

$\overline{x}_{4\cdot}$	$\overline{x}_{3\cdot}$	$\overline{x}_{1\cdot}$	$\overline{x}_{2\cdot}$
55.60	58.00	66.20	75.20

We conclude that time 2 differs significantly from both time 3 and time 4, but no other significant differences are identified. For factor B (location), $w = 13.68$ and the underscoring is as follows:

$\overline{x}_{.5}$	$\overline{x}_{.4}$	$\overline{x}_{.2}$	$\overline{x}_{.1}$	$\overline{x}_{.3}$
50.00	56.75	62.75	72.75	77.00

The location pairs identified as significantly different from one another are 1 and 5, 3 and 5, 1 and 4, 3 and 4, and finally 2 and 3. Here the underscoring pattern makes interpretation difficult. ∎

Randomized Block Experiments

In using single-factor ANOVA to test for the presence of effects due to the I different treatments under study, once the IJ subjects or experimental units have been chosen, treatments should be allocated in a completely random fashion. That is, J subjects should be chosen at random for the first treatment, then another sample of J chosen at random from the remaining $IJ - J$ subjects for the second treatment, and so on.

It frequently happens, though, that subjects or experimental units exhibit heterogeneity with respect to other variables that may affect the observed responses. When this is the case, the presence or absence of a significant F value may be due to this extraneous variation rather than to the presence or absence of factor effects. This was the reason for introducing a paired experiment in Chapter 9. The analogy to a paired experiment when $I > 2$ is called a **randomized block** experiment. An extraneous factor, "blocks," is constructed by dividing the IJ units into J groups with I units in each group. This grouping or blocking is done in such a way that within each block, the I units are homogeneous with respect to other factors thought to affect the responses. Then within each homogeneous block, the I treatments are randomly assigned to the I units or subjects in the block.

Example 11.5 A consumer product-testing organization wished to compare the annual power consumption for five different brands of dehumidifier. Because power consumption depends on the prevailing humidity level, it was decided to monitor each brand at four different levels ranging from moderate to heavy humidity (thus blocking on humidity level). Within each level, brands were randomly assigned to the five selected locations. The resulting amount of power consumption (annual kwh) appears in Table 11.2.

Table 11.2

Treatments (brands)	Blocks (humidity level) 1	2	3	4	$x_{i.}$	$\overline{x}_{i.}$
1	685	792	838	875	3190	797.50
2	722	806	893	953	3374	843.50
3	733	802	880	941	3356	839.00
4	811	888	952	1005	3656	914.00
5	828	920	978	1023	3749	937.25
$x_{.j}$	3779	4208	4541	4797	17,325	

Since $\Sigma\Sigma x_{ij}^2 = 15,178,901.00$ and $x_{..}^2/IJ = 15,007,781.25$,

$$SST = 15,178,901.00 - 15,007,781.25 = 171,119.75$$

$$SSA = \frac{1}{4}[60,244,049] - 15,007,781.25 = 53,231.00$$

$$SSB = \frac{1}{5}[75,619,995] - 15,007,781.25 = 116,217.75$$

and

$$SSE = 171,119.75 - 53,231.00 - 116,217.75 = 1671.00$$

Table 11.3

Source of variation	d.f.	Sum of squares	Mean square	f
Treatments (brands)	4	53,231.00	13,307.75	$f_A = 95.57$
Blocks	3	116,217.75	38,739.25	$f_B = 278.20$
Error	12	1671.00	139.25	
Total	19	171,119.75		

Since $F_{.05, 4, 12} = 3.26$ and $f_A = 95.57 \geq 3.26$, H_0 is rejected in favor of H_a, and we conclude that power consumption does depend on the brand of humidifier. To identify significantly different brands, we use Tukey's procedure. $Q_{.05, 5, 12} = 4.51$ and $w = 4.51\sqrt{139.25/4} = 26.6$.

$\bar{x}_{1.}$	$\bar{x}_{3.}$	$\bar{x}_{2.}$	$\bar{x}_{4.}$	$\bar{x}_{5.}$
797.50	839.00	843.50	914.00	937.25

The underscoring indicates that the brands can be divided into three groups with respect to power consumption.

Because the block factor is of secondary interest, $F_{.05, 3, 12}$ is not needed, though the computed value of F_B is clearly highly significant.   ∎

In many experimental situations in which treatments are to be applied to subjects, a single subject can receive all I of the treatments. Blocking is then often done on the subjects themselves to control for variability between subjects; each subject is then said to act as its own control. Social scientists sometimes refer to such experiments as repeated-measures designs. The "units" within a block are then the different "instances" of treatment application. Similarly, blocks are often taken as different time periods, locations, or observers.

Example 11.6   The data in Table 11.4 appeared in the paper "Compounding of Discriminative Stimuli from the Same and Different Sensory Modalities" (*J. Experimental Analysis Behavior*, 1971, pp. 337–342). Rat response was maintained by fixed interval schedules of reinforcement in the presence of a tone or two separate lights. The lights were either of moderate ($L1$) or low intensity ($L2$). Observations are given as the mean number of responses emitted by each subject during single and compound stimuli presentations over a four-day period.

Table 11.4

| Stimulus | Subject | | | | $x_i.$ | $\bar{x}_i.$ |
	1	2	3	4		
L1	8.0	17.3	52.0	22.0	99.3	24.8
L2	6.9	19.3	63.7	21.6	111.5	27.9
Tone (T)	9.3	18.8	60.0	28.3	116.4	29.1
L1 + L2	9.2	24.9	82.4	44.9	161.4	40.3
L1 + T	12.0	31.7	83.8	37.4	164.9	41.2
L2 + T	9.4	33.6	96.6	40.6	180.2	45.1
$x._j$	54.8	145.6	438.5	194.8	833.7	

With $\Sigma\Sigma x_{ij}^2 = 44,614.21$, $SST = 15,653.56$, $SSA = 1428.28$, $SSB = 13,444.63$, and $SSE = 780.65$.

Table 11.5

Source of variation	d.f.	Sum of squares	Mean square	f
Stimuli (A)	5	1428.28	285.60	$f_A = 5.49$
Subjects (B)	3	13,444.63	4481.54	$f_B = 86.12$
Error	15	780.65	52.04	
Total	23	15,653.56		

Since $F_{.05, 5, 15} = 2.90$ and $5.49 \geq 2.90$, we conclude that there are differences in the true average responses associated with the different stimuli. For Tukey's procedure, $w = 4.59\sqrt{52.04/4} = 16.56$.

$\bar{x}_1.$	$\bar{x}_2.$	$\bar{x}_3.$	$\bar{x}_4.$	$\bar{x}_5.$	$\bar{x}_6.$
24.8	27.9	29.1	40.3	41.2	45.1

Thus both $L1$ and $L2$ are significantly different from $L2 + T$, and there are no other significant differences between the stimuli. ∎

In most randomized block experiments in which subjects serve as blocks, the subjects actually participating in the experiment are selected from a large population. The subjects then contribute random rather than fixed effects. This does not affect the procedure for comparing treatments when $K_{ij} = 1$ (one observation per "cell," as in this section), but the procedure is altered if $K_{ij} = K > 1$. We shall shortly consider two-factor models in which effects are random.

More on Blocking

When $I = 2$, either the F test or the paired differences t test can be used to analyze the data. The resulting conclusion will not depend on which procedure is used, since $T^2 = F$ and $t_{\alpha/2, \nu}^2 = F_{\alpha, 1, \nu}$.

Just as with pairing, blocking entails both a potential gain and a potential loss in precision. If there is a great deal of heterogeneity in experimental units, the value of the variance parameter σ^2 in the one-way model will be large. The effect of blocking is to filter out the variation σ^2 in the two-way model appropriate for a randomized block experiment. Other things being equal, a smaller value of σ^2 results in a test that is more likely to detect departures from H_0 (that is, a test with greater power).

However, other things are not equal here, since the single-factor F test is based on $I(J - 1)$ degrees of freedom for error, while the two-factor F test is based on $(I - 1)(J - 1)$ degrees of freedom for error. Fewer degrees of freedom for error results in a decrease in power, essentially because the denominator estimator of σ^2 is not as precise. This loss in degrees of freedom can be especially serious if the experimenter can afford only a small number of observations. Nevertheless, if it appears that blocking will significantly reduce variability, it is probably worth the loss in degrees of freedom.

Models for Random Effects

In many experiments the actual levels of a factor used in the experiment, rather than being the only ones of interest to the experimenter, have been selected from a much larger population of possible levels of the factor. In a two-factor situation, when this is the case for both factors, a **random effects** model is appropriate. The case in which the levels of one factor are the only ones of interest and the levels of the other factor are selected from a population of levels leads to a **mixed effects** model. The two-factor random effects model when $K_{ij} = 1$ is

$$X_{ij} = \mu + A_i + B_j + \epsilon_{ij} \quad i = 1, \ldots, I; j = 1, \ldots, J$$

where the A_i's, B_j's, and ϵ_{ij}'s are all independent, normally distributed random variables with mean zero and variances σ_A^2, σ_B^2, and σ^2, respectively. The hypotheses of interest are then $H_{0A} : \sigma_A^2 = 0$ (expected response does not depend on which level of factor A is selected) versus $H_{aA} : \sigma_A^2 > 0$ and $H_{0B} : \sigma_B^2 = 0$ versus $H_{aB} : \sigma_B^2 > 0$. While $E(MSE) = \sigma^2$ as before, the expected mean squares for factors A and B are now

$$E(MSA) = \sigma^2 + J\sigma_A^2, \qquad E(MSB) = \sigma^2 + I\sigma_B^2$$

Thus when H_{0A} (H_{0B}) is true, F_A (F_B) is still a ratio of two unbiased estimators of σ^2. It can be shown that a level α test for H_{0A} versus H_{aA} still rejects H_{0A} if $f_A \geq F_{\alpha, I-1, (I-1)(J-1)}$, and similarly the same procedure as before is used to decide between H_{0B} and H_{aB}.

For the case in which factor A is fixed and factor B is random, the mixed model is

$$X_{ij} = \mu + \alpha_i + B_j + \epsilon_{ij} \quad i = 1, \ldots, I; j = 1, \ldots, J$$

where $\Sigma\alpha_i = 0$ and the B_j's and ϵ_{ij}'s are normally distributed with mean zero and variances σ_B^2 and σ^2, respectively. Now the two null hypotheses are

$$H_{0A} : \alpha_1 = \cdots = \alpha_I = 0 \quad \text{and} \quad H_{0B} : \sigma_B^2 = 0$$

with expected mean squares

$$E(MSE) = \sigma^2, \quad E(MSA) = \sigma^2 + \frac{J}{I-1}\Sigma\alpha_i^2, \quad E(MSB) = \sigma^2 + I\sigma_B^2$$

The test procedures for H_{0A} versus H_{aA} and H_{0B} versus H_{aB} are exactly as before. For example, in the analysis of the air pollution data in Example 11.1, if the five locations were randomly selected, then because $f_B = 13.24$ and $F_{.05, 4, 12} = 3.26$, $H_{0B} : \sigma_B^2 = 0$ is rejected in favor of $H_{aB} : \sigma_B^2 > 0$. An estimate of the "variance component" σ_B^2 is then given by $(MSB - MSE)/I = 112.53$.

Summarizing, when $K_{ij} = 1$, although the hypotheses and expected mean squares differ from the case of both effects fixed, the test procedures are identical.

Exercises / Section 11.1 [1–12]

1. The number of miles of useful tread wear (in thousands) was determined for tires of each of five different makes of subcompact car (factor A, with $I = 5$) in combination with each of four different brands of radial tires (factor B, with $J = 4$), resulting in $IJ = 20$ observations. The values $SSA = 30.6$, $SSB = 44.1$, and $SSE = 59.2$ were then computed. Assume that an additive model is appropriate.

   a. Test $H_0 : \alpha_1 = \alpha_2 = \alpha_3 = \alpha_4 = \alpha_5 = 0$ (no differences in true average tire lifetime due to makes of cars) versus H_a : at least one $\alpha_i \neq 0$ using a level .05 test.

   b. Test $H_0 : \beta_1 = \beta_2 = \beta_3 = \beta_4 = 0$ (no differences in true average tire lifetime due to brands of tires) versus H_a : at least one $\beta_j \neq 0$ using a level .05 test.

2. Four different coatings are being considered for corrosion protection of metal pipe. The pipe will be buried in three different types of soil. To investigate whether the amount of corrosion depends either on the coating or on the type of soil, 12 pieces of pipe are selected. Each piece is coated with one of the four coatings and buried in one of the three types of soil for a fixed time, after which the amount of corrosion (depth of maximum pits, in .0001 in.) is determined. The data appears in the accompanying table.

		Soil type (B)		
		1	2	3
Coating (A)	1	64	49	50
	2	53	51	48
	3	47	45	50
	4	51	43	52

   a. Assuming the validity of the additive model, carry out the ANOVA analysis using an ANOVA table to see whether the amount of corrosion depends on either the type of coating used or the type of soil. Use $\alpha = .05$.

   b. Compute $\hat{\mu}$, $\hat{\alpha}_1$, $\hat{\alpha}_2$, $\hat{\alpha}_3$, $\hat{\alpha}_4$, $\hat{\beta}_1$, $\hat{\beta}_2$, and $\hat{\beta}_3$.

3. The paper "Adiabatic Humidification of Air with Water in a Packed Tower" (*Chem. Eng. Prog.*, 1952, pp. 362–370) reported data on gas film heat transfer coefficient (Btu/hr ft^2 on °F) as a function of gas rate (factor A) and liquid rate (factor B).

			B		
		1(190)	2(250)	3(300)	4(400)
	1(200)	200	226	240	261
A	2(400)	278	312	330	381
	3(700)	369	416	462	517
	4(1100)	500	575	645	733

   a. After constructing an ANOVA table, test at level .01 both the hypothesis of no gas rate effect against the appropriate alternative and the hypothesis of no liquid rate effect against the appropriate alternative.

b. Use Tukey's procedure to investigate differences in expected heat transfer coefficient due to different gas rates.

c. Repeat (b) for liquid rates.

4. In an experiment to see whether the amount of coverage of light blue interior latex paint depended either on the brand of paint or on the brand of roller used, one gallon of each of four brands of paint was applied using each of three brands of roller, resulting in the following data (number of square feet covered).

		Roller brand		
		1	2	3
Paint	1	454	446	451
brand	2	446	444	447
	3	439	442	444
	4	444	437	443

a. Construct the ANOVA table. *Hint:* The computations can be expedited by subtracting 400 (or any other convenient number) from each observation. This does not affect the final results.

b. State and test hypotheses appropriate for deciding whether paint brand has any effect on coverage. Use $\alpha = .05$.

c. Repeat (b) for brand of roller.

d. Use Tukey's method to identify significant differences among brands. Is there one brand that seems clearly preferable to the others?

5. In an experiment to assess the effect of the angle of pull on the force required to cause separation in electrical connectors, four different angles (factor A) were used and each of a sample of five connectors (factor B) was pulled once at each angle ("A Mixed Model Factorial Experiment in Testing Electrical Connectors," *Industrial Quality Control,* 1960, pp. 12–16). The data appears in the accompanying table.

			B		
	1	2	3	4	5
$0°$	45.3	42.2	39.6	36.8	45.8
$2°$	44.1	44.1	38.4	38.0	47.2
A $4°$	42.7	42.7	42.6	42.2	48.9
$6°$	43.5	45.8	47.9	37.9	56.4

Does the data suggest that true average separation force is affected by the angle of pull? State

and test the appropriate hypotheses at level .01 by first constructing an ANOVA table.

6. A particular county employs three assessors who are responsible for determining the value of residential property in the county. To see whether or not these assessors differ systematically in their assessments, five houses are selected and each assessor is asked to determine the market value of each house. With factor A denoting assessors ($I = 3$) and factor B denoting houses ($J = 5$), suppose that $SSA = 11.7$, $SSB = 113.5$, and $SSE = 25.6$.

a. Test $H_0: \alpha_1 = \alpha_2 = \alpha_3 = 0$ at level .05 (H_0 states that there are no systematic differences between assessors).

b. Explain why a randomized block experiment with only five houses was used rather than a one-way ANOVA experiment involving a total of 15 different houses with each assessor asked to assess five different houses (a different group of five for each assessor).

7. The article "Rate of Stuttering Adaptation Under Two Electro-Shock Conditions" (*Behavior Research Therapy,* 1967, pp. 49–54) gave adaptation scores for three different treatments: (1) no shock, (2) shock following each stuttered word, and (3) shock during each moment of stuttering. These treatments were used on each of 18 stutterers.

a. Summary statistics include $x_{1\cdot} = 905$, $x_{2\cdot} = 913$, $x_{3\cdot} = 936$, $x_{\cdot\cdot} = 2754$, $\sum x_{\cdot j}^2 = 430{,}295$, and $\sum\sum x_{ij}^2 = 143{,}930$. Construct the ANOVA table and test at level .05 to see whether true average adaptation score depends on the treatment given.

b. Judging from the F ratio for subjects (factor B), do you think that blocking on subjects was effective in this experiment? Explain.

8. The accompanying table gives plasma epinephrine concentration for 10 experimental subjects during (1) isoflurane, (2) halothane, and (3) cyclopropane anesthesia ("Sympathoadrenal and Hemodynamic Effects of Isoflurane, Halothane, and Cyclopropane in Dogs," *Anesthesiology,* 1974, pp. 465–470).

a. Does the choice of anesthetic affect true average concentration? Test $H_0: \alpha_1 = \alpha_2 = \alpha_3 = 0$ at level .05 after constructing the ANOVA table.

	Subject (B)				
	1	2	3	4	5
Anesthetic (A) 1	.28	.51	1.00	.39	.29
2	.30	.39	.63	.38	.21
3	1.07	1.35	.69	.28	1.24

	6	7	8	9	10
Anesthetic (A) 1	.36	.32	.69	.17	.33
2	.88	.39	.51	.32	.42
3	1.53	.49	.56	1.02	.30

$$\Sigma\Sigma x_{ij}^2 = 13.7980$$

b. Use Tukey's procedure to investigate significant differences among the anesthetics.

9. Suppose that in the experiment described in Exercise 6 the five houses had actually been selected at random from among those of a certain age and size, so that factor B is random rather than fixed. Test $H_0 : \sigma_B^2 = 0$ versus $H_a : \sigma_B^2 > 0$ using a level .01 test.

10. **a.** Show that a constant d can be added to (or subtracted from) each x_{ij} without affecting any of the ANOVA sums of squares.
    b. Suppose that each x_{ij} is multiplied by a nonzero constant c. How does this affect the ANOVA sums of squares? How does this affect the values of the F statistics F_A and F_B? What effect does "coding" the observations by $y_{ij} = cx_{ij} + d$ have on the conclusions resulting from the ANOVA procedures?

11. Use the fact that $E(X_{ij}) = \mu + \alpha_i + \beta_j$ with $\Sigma\alpha_i = \Sigma\beta_j = 0$ to show that $E(\overline{X}_{i.} - \overline{X}_{..}) = \alpha_i$, so that $\hat{\alpha}_i = \overline{X}_{i.} - \overline{X}_{..}$ is an unbiased estimator for α_i.

12. The power curves of Figures 10.2 and 10.3 can be used to obtain $\beta = P(\text{type II error})$ for the F test in two-factor ANOVA. For fixed values of $\alpha_1, \alpha_2, \ldots, \alpha_I$, the quantity $\phi^2 = (J/I)\Sigma\alpha_i^2/\sigma^2$ is computed. Then the figure corresponding to $\nu_1 = I - 1$ is entered on the horizontal axis at the value ϕ, the power is read on the vertical axis from the curve labeled $\nu_2 = (I - 1)(J - 1)$, and $\beta = 1 - \text{power}$.
    a. For the corrosion experiment described in Exercise 2, find β when $\alpha_1 = 4$, $\alpha_2 = 0$, $\alpha_3 = \alpha_4 = -2$, and $\sigma = 4$. Repeat if $\alpha_1 = 6$, $\alpha_2 = 0$, $\alpha_3 = \alpha_4 = -3$, and $\sigma = 4$.
    b. By symmetry, what is β for the test of H_{0B} versus H_{aB} in Example 11.1 when $\beta_1 = 4$, $\beta_2 = \beta_3 = \beta_4 = \beta_5 = -1$, and $\sigma = 4$?

11.2   Two-Factor ANOVA with $K_{ij} > 1$

In Section 11.1 we analyzed data from a two-factor experiment in which there was one observation for each of the IJ combinations of levels of the two factors. To obtain valid test procedures, the μ_{ij}'s were assumed to have an additive structure with $\mu_{ij} = \mu + \alpha_i + \beta_j$, $\Sigma\alpha_i = \Sigma\beta_j = 0$. Additivity means that the difference in true average responses for any two levels of the factors is the same for each level of the other factor. For example, $\mu_{ij} - \mu_{i'j} = (\mu + \alpha_i + \beta_j) - (\mu + \alpha_{i'} + \beta_j) = \alpha_i - \alpha_{i'}$ independent of the level j of the second factor. This is shown in Figure 11.1a, in which the lines connecting true average responses are parallel.

Figure 11.1b depicts a set of true average responses that does not have additive structure. The lines connecting these μ_{ij}'s are not parallel, which means that the difference in true average responses for different levels of one factor does depend on the level of the other factor. When additivity does not hold, we say that there is **interaction** between the different levels of the factors. The assumption of additivity allowed us in Section 11.1 to obtain an estimator of the random error variance σ^2 (*MSE*) that was unbiased whether or not either null hypothesis of interest was true. When $K_{ij} > 1$ for at least one (i, j) pair, a valid estimator of σ^2 can be obtained without assuming additivity. In specifying the appropriate model and deriving test procedures, we will focus on the case $K_{ij} = K > 1$, so the number of observations per "cell" (for each combination of levels) is constant.

Parameters for the Fixed Effects Model with Interaction

Rather than use the μ_{ij}'s themselves as model parameters, it is usual to use an equivalent set that reveals more clearly the role of interaction. Let

$$\mu = \frac{1}{IJ} \sum_i \sum_j \mu_{ij}, \quad \mu_{i\cdot} = \frac{1}{J} \sum_j \mu_{ij}, \quad \text{and} \quad \mu_{\cdot j} = \frac{1}{I} \sum_i \mu_{ij} \qquad (11.7)$$

Thus μ is the expected response averaged over all levels of both factors (the true grand mean), $\mu_{i\cdot}$ is the expected response averaged over levels of the second factor when the first factor A is held at level i, and similarly for $\mu_{\cdot j}$. Now define

$$\alpha_i = \mu_{i\cdot} - \mu = \text{the effect of factor } A \text{ at level } i$$
$$\beta_j = \mu_{\cdot j} - \mu = \text{the effect of factor } B \text{ at level } j \qquad (11.8)$$

and

$$\gamma_{ij} = \mu_{ij} - (\mu + \alpha_i + \beta_j)$$

Then

$$\mu_{ij} = \mu + \alpha_i + \beta_j + \gamma_{ij} \qquad (11.9)$$

and the model is additive if and only if all γ_{ij}'s $= 0$. The γ_{ij}'s are referred to as the **interaction parameters**. The α_i's are called the **main effects for factor A,** while the β_j's are the **main effects for factor B.** Although there are I α_i's, J β_j's, and IJ γ_{ij}'s in addition to μ, the conditions $\Sigma \alpha_i = 0$, $\Sigma \beta_j = 0$, $\Sigma_j \gamma_{ij} = 0$ for any i, and $\Sigma_i \gamma_{ij} = 0$ for any j [all by virtue of (11.7) and (11.8)] imply that only IJ of these new parameters are independently determined: μ, $I - 1$ of the α_i's, $J - 1$ of the β_j's, and $(I - 1)(J - 1)$ of the γ_{ij}'s.

There are now three sets of hypotheses which will be considered:

$$H_{0AB} : \gamma_{ij} = 0 \quad \text{for all } i, j \quad \text{versus} \quad H_{aAB} : \text{at least one } \gamma_{ij} \neq 0$$
$$H_{0A} : \alpha_1 = \cdots = \alpha_I = 0 \quad \text{versus} \quad H_{aA} : \text{at least one } \alpha_i \neq 0$$
$$H_{0B} : \beta_1 = \cdots = \beta_J = 0 \quad \text{versus} \quad H_{aB} : \text{at least one } \beta_j \neq 0$$

The no-interaction hypothesis H_{0AB} is usually tested first. If H_{0AB} is not rejected, then the other two hypotheses can be tested to see whether or not the main effects are significant. If H_{0AB} is rejected and H_{0A} is then tested and accepted, the resulting model $\mu_{ij} = \mu + \beta_j + \gamma_{ij}$ does not lend itself to straightforward interpretation. In such a case it is best to construct a picture similar to that of Figure 11.1*b* to try to visualize the way in which the factors interact.

Notation, Model, and Analysis

We now use triple subscripts for both random variables and observed values, with X_{ijk} and x_{ijk} referring to the kth observation (replication) when factor A is at level i and factor B is at level j. The model is then

$$X_{ijk} = \mu + \alpha_i + \beta_j + \gamma_{ij} + \epsilon_{ijk}$$
$$i = 1, \ldots, I; j = 1, \ldots, J; k = 1, \ldots, K \qquad (11.10)$$

where the ϵ_{ij}'s are independent and normally distributed, each with mean zero and variance σ^2.

Again a dot in place of a subscript means that we have summed over all values of that subscript, while a horizontal bar denotes averaging. Thus $X_{ij.}$ is the total of all K observations made for factor A at level i and factor B at level j [all observations in the (i, j)th cell] and $\overline{X}_{ij.}$ is the average of these K observations.

Example 11.7   Three different varieties of tomato (Harvester, Pusa Early Dwarf, and Ife No. 1) and four different plant densities (10, 20, 30, and 40 thousand plants per hectare) are being considered for planting in a particular region. To see whether either variety or plant density affects yield, each combination of variety and plant density is used in three different plots, resulting in the data on yields in Table 11.6 (based on the article "Effects of Plant Density on Tomato Yields in Western Nigeria," *Experimental Agriculture*, 1976, pp. 43–47):

Table 11.6

Variety	10,000	Planting density 20,000	30,000	40,000	$x_{i..}$	$\overline{x}_{i..}$
H	10.5, 9.2, 7.9	12.8, 11.2, 13.3	12.1, 12.6, 14.0	10.8, 9.1, 12.5	136.0	11.33
Ife	8.1, 8.6, 10.1	12.7, 13.7, 11.5	14.4, 15.4, 13.7	11.3, 12.5, 14.5	146.5	12.21
P	16.1, 15.3, 17.5	16.6, 19.2, 18.5	20.8, 18.0, 21.0	18.4, 18.9, 17.2	217.5	18.13
$x_{.j.}$	103.3	129.5	142.0	125.2	500.00	
$\overline{x}_{.j.}$	11.48	14.39	15.78	13.91		13.89

Here $I = 3$, $J = 4$, and $K = 3$, for a total of $IJK = 36$ observations.   ■

To test the hypotheses of interest, we again define sums of squares and present computing formulas:

$$SST = \sum_i \sum_j \sum_k (X_{ijk} - \overline{X}_{...})^2 = \sum_i \sum_j \sum_k X_{ijk}^2 - \frac{1}{IJK} X_{...}^2$$

$$SSE = \sum_i \sum_j \sum_k (X_{ijk} - \overline{X}_{ij.})^2 = \sum_i \sum_j \sum_k X_{ijk}^2 - \frac{1}{K} \sum_i \sum_j X_{ij.}^2$$

$$SSA = \sum_i \sum_j \sum_k (\overline{X}_{i..} - \overline{X}_{...})^2 = \frac{1}{JK} \sum_i X_{i..}^2 - \frac{1}{IJK} X_{...}^2$$

$$SSB = \sum_i \sum_j \sum_k (\overline{X}_{.j.} - \overline{X}_{...})^2 = \frac{1}{IK} \sum_j X_{.j.}^2 - \frac{1}{IJK} X_{...}^2$$

$$SSAB = \sum_i \sum_j \sum_k (\overline{X}_{ij.} - \overline{X}_{i..} - \overline{X}_{.j.} + \overline{X}_{...})^2$$

$SSAB$ is called the **interaction sum of squares;** there is no computing formula for it because the fundamental ANOVA identity enables it to be obtained by subtraction once the other SS's have been computed:

$$SST = SSA + SSB + SSAB + SSE$$

Total variation is thus partitioned into four pieces: unexplained (SSE—which would be present whether or not any of the three null hypotheses was true) and three pieces that may be explained by the truth or falsity of the three H_0's.

Each SS has associated with it a number of degrees of freedom: $IJK - 1$ for SST, $IJ(K - 1)$ for SSE, $I - 1$ for SSA, $J - 1$ for SSB, and $(I - 1) \cdot (J - 1)$ for $SSAB$. Each of four mean squares is defined by $MS = SS/\text{d.f.}$ The expected mean squares suggest that each set of hypotheses should be tested using the appropriate ratio of mean squares with MSE in the denominator:

$$E(MSE) = \sigma^2$$

$$E(MSA) = \sigma^2 + \frac{JK}{I - 1} \sum_{i=1}^{I} \alpha_i^2$$

$$E(MSB) = \sigma^2 + \frac{IK}{J - 1} \sum_{j=1}^{J} \beta_j^2$$

$$E(MSAB) = \sigma^2 + \frac{K}{(I - 1)(J - 1)} \sum_{i=1}^{I} \sum_{j=1}^{J} \gamma_{ij}^2$$

Each of the three mean square ratios can be shown to have an F distribution when the associated H_0 is true, which yields the following level α test procedures:

Hypotheses	Test statistic value	Rejection region
H_{0A} versus H_{aA}	$f_A = \dfrac{MSA}{MSE}$	$f_A \geq F_{\alpha,\, I-1,\, IJ(K-1)}$
H_{0B} versus H_{aB}	$f_B = \dfrac{MSB}{MSE}$	$f_B \geq F_{\alpha,\, J-1,\, IJ(K-1)}$
H_{0AB} versus H_{aAB}	$f_{AB} = \dfrac{MSAB}{MSE}$	$f_{AB} \geq F_{\alpha,\, (I-1)(J-1),\, IJ(K-1)}$

As before, the results of the analysis are summarized in an ANOVA table.

Example 11.8
(Example 11.7 continued)

From the given data, $x_{\cdot\cdot\cdot}^2 = (500)^2 = 250{,}000$,

$$\sum_i \sum_j \sum_k x_{ijk}^2 = (10.5)^2 + (9.2)^2 + \cdots + (18.9)^2 + (17.2)^2 = 7404.80$$

$$\sum_i x_{i\cdot\cdot}^2 = (136.0)^2 + (146.5)^2 + (217.5)^2 = 87{,}264.50$$

and

$$\sum_j x_{\cdot j\cdot}^2 = 63{,}280.18$$

The cell totals ($x_{ij\cdot}$'s) are

	10,000	20,000	30,000	40,000
H	27.6	37.3	38.7	32.4
Ife	26.8	37.9	43.5	38.3
P	48.9	54.3	59.8	54.5

from which $\sum_i \sum_j x_{ij\cdot}^2 = (27.6)^2 + \cdots + (54.5)^2 = 22{,}100.28$. Then

$$SST = 7404.80 - \frac{1}{36}(250{,}000) = 7404.80 - 6944.44 = 460.36$$

$$SSA = \frac{1}{12}(87{,}264.50) - 6944.44 = 327.60$$

$$SSB = \frac{1}{9}(63{,}280.18) - 6944.44 = 86.69$$

$$SSE = 7404.80 - \frac{1}{3}(22{,}100.28) = 38.04$$

and

$$SSAB = 460.36 - 327.60 - 86.69 - 38.04 = 8.03$$

Table 11.7 now summarizes the computations:

Table 11.7

Source of variation	d.f.	Sum of squares	Mean square	f
Varieties	2	327.60	163.8	$f_A = 103.02$
Density	3	86.69	28.9	$f_B = 18.18$
Interaction	6	8.03	1.34	$f_{AB} = .84$
Error	24	38.04	1.59	
Total	35	460.36		

Since $F_{.01, 6, 24} = 3.63$ and $f_{AB} = .84$ is not ≥ 3.63, H_{0AB} cannot be rejected at level .01, so we conclude that the interaction effects are not significant. Now the presence or absence of main effects can be investigated. Since $F_{.01, 2, 24} = 5.61$ and $f_A = 103.2 \geq 5.61$, H_{0A} is rejected at level .01 in favor of the conclusion that different varieties do affect the true average yields. Similarly, $f_B = 18.18 \geq 4.24 = F_{.01, 3, 24}$, so we conclude that true average yield also depends on plant density. ■

Multiple Comparisons

When the no-interaction hypothesis H_{0AB} is not rejected and at least one of the two main effect null hypotheses is rejected, Tukey's method can be used to identify significant differences in levels. For identifying differences among the α_i's when H_{0A} is rejected,

1. Obtain $Q_{\alpha, I, IJ(K-1)}$, where the second subscript I identifies the number of levels being compared and the third subscript refers to the number of degrees of freedom for error.
2. Compute $w = Q\sqrt{MSE/JK}$, where JK is the number of observations averaged to obtain each of the $\bar{x}_{i..}$'s compared in step 3.
3. Order the $\bar{x}_{i..}$'s from smallest to largest and, as before, underscore all pairs that differ by less than w. Pairs not underscored correspond to significantly different levels of factor A.

To identify different levels of factor B when H_{0B} is rejected, replace the second subscript in Q by J, replace JK by IK in w, and replace $\bar{x}_{i..}$ by $\bar{x}_{.j.}$.

Example 11.9
(Example 11.8 continued)

For factor A (varieties), $I = 3$, so with $\alpha = .01$ and $IJ(K-1) = 24$, $Q_{.01, 3, 24} = 4.55$. Then $w = 4.55\sqrt{1.59/12} = 1.66$, so ordering and underscoring gives

$\bar{x}_{1..}$	$\bar{x}_{2..}$	$\bar{x}_{3..}$
11.33	12.21	18.13

The Harvester and Ife varieties do not appear to differ significantly from one another in effect on true average yield, but both differ from the Pusa variety.
    For factor B (density), $J = 4$ so $Q_{.01, 4, 24} = 4.91$ and $w = 4.91\sqrt{1.59/9} = 2.06$.

$\overline{x}_{.1.}$	$\overline{x}_{.4.}$	$\overline{x}_{.2.}$	$\overline{x}_{.3.}$
11.48	13.91	14.39	15.78

Thus with experimentwise error rate .01, which is quite conservative, only the lowest density appears to differ significantly from all others. Even with $\alpha = .05$ (so that $w = 1.64$), densities 2 and 3 cannot be judged significantly different from one another in their effect on yield.   ■

Models with Mixed and Random Effects

In some problems the levels of either factor may have been chosen from a large population of possible levels, so that the effects contributed by the factor are random rather than fixed. As in Section 11.1, if both factors contribute random effects, the model is referred to as a random effects model, while if one factor is fixed and the other is random, a mixed effects model results. We shall consider here the analysis for a mixed effects model in which factor A (rows) is the fixed factor and factor B (columns) is the random factor. The case in which both factors are random is dealt with in the exercises.

The mixed effects model in this situation is

$$X_{ijk} = \mu + \alpha_i + B_j + G_{ij} + \epsilon_{ijk}$$
$$i = 1, \ldots, I; j = 1, \ldots, J; k = 1, \ldots, K$$

where μ and α_i's are constants with $\Sigma\alpha_i = 0$, and the B_j's, G_{ij}'s, and ϵ_{ijk}'s are normally distributed random variables with expected value 0 and variances σ_B^2, σ_G^2, and σ^2, respectively. Because the I levels of the fixed factor A are the only ones under consideration, we also assume that $\Sigma_i G_{ij} = 0$ (this implies that the G_{ij}'s for fixed j are not independent of one another but are negatively correlated). The three hypotheses of interest are then

$$H_{0A}: \alpha_1 = \alpha_2 = \cdots = \alpha_I = 0 \quad \text{versus} \quad H_{aA}: \text{at least one } \alpha_i \neq 0$$
$$H_{0B}: \sigma_B^2 = 0 \qquad\qquad\qquad \text{versus} \quad H_{aB}: \sigma_B^2 > 0$$
$$H_{0G}: \sigma_G^2 = 0 \qquad\qquad\qquad \text{versus} \quad H_{aG}: \sigma_G^2 > 0$$

It is customary to test H_{0A} and H_{0B} only if the no-interaction hypothesis H_{0G} cannot be rejected.

The relevant sums of squares and mean squares needed for the test procedures are defined and computed exactly as in the fixed effects case. The expected mean squares for the mixed model are

$$E(MSE) = \sigma^2$$

$$E(MSA) = \sigma^2 + \frac{IK}{I-1}\sigma_G^2 + \frac{JK}{I-1}\Sigma\alpha_i^2$$

$$E(MSB) = \sigma^2 + IK\sigma_B^2$$

and

$$E(MSAB) = \sigma^2 + \frac{IK}{I-1}\sigma_G^2$$

Thus to test the no-interaction hypothesis, the ratio $f_{AB} = MSAB/MSE$ is again appropriate, with H_{0G} rejected if $f_{AB} \geq F_{\alpha, (I-1)(J-1), IJ(K-1)}$. Similarly, the F ratio and test procedure for testing H_{0B} are exactly as they were in the fixed effects case. However, for testing H_{0A} versus H_{aA}, the expected mean squares suggest that while the numerator of the F ratio should still be MSA, the denominator should be $MSAB$ rather than MSE, and this is indeed the case:

> For testing H_{0A} versus H_{aA} (factors A fixed, B random), the test statistic value is $f_A = MSA/MSAB$, and the rejection region is $f_A \geq F_{\alpha, I-1, (I-1)(J-1)}$.

Example 11.10 A study of lifetimes of stereo cartridges carried out by staff members of a magazine devoted to high fidelity equipment focused on four different brands of cartridges, each designed to track at two grams. To see whether lifetime was affected by choice of record label, three different labels were randomly selected from the population of all labels, and two lifetime observations (hours of playing time) were obtained for each combination of brand of cartridge and record label (see Table 11.8).

Table 11.8

Brand of cartridge	Record label 1	2	3	$x_{i..}$
1	697, 658	718, 688	640, 679	4080
2	635, 684	700, 736	696, 665	4116
3	670, 696	693, 659	675, 703	4096
4	651, 678	668, 709	715, 687	4108
$x_{.j.}$	5369	5571	5460	16,400

Additional summary quantities are $\sum\sum\sum_{i\,j\,k} x_{ijk}^2 = 11,220,964$ and $\sum\sum_{i\,j} x_{ij.}^2 = 22,427,518$.

Factor A, cartridges, is fixed here, since there are only four brands of interest to the experimenters and they are all represented in the experiment. Factor B, however, is random, since it is not the three specific labels actually chosen that are of interest; instead the question is whether or not there is variability in lifetime associated with the population of all labels. The three null hypotheses are then $H_{0A}: \alpha_1 = \alpha_2 = \alpha_3 = \alpha_4 = 0$, $H_{0B}: \sigma_B^2 = 0$, and $H_{0G}: \sigma_G^2 = 0$. The ANOVA table is Table 11.9; remember that $f_A = MSA/MSAB$, not MSA/MSE as with fixed effects.

Table 11.9

Source of variation	d.f.	Sum of squares	Mean square	f
Cartridges (A)	3	122.66	40.89	$f_A = .055$
Labels (B)	2	2558.58	1279.29	$f_B = 2.13$
Interaction	6	4411.09	735.18	$f_{AB} = 1.22$
Error	12	7205	600.42	
Total	23	14,297.33		

None of the three computed F's exceeds $F_{.05}$ for appropriate numerator and denominator degrees of freedom, so none of the three null hypotheses can be rejected. In particular the data suggests no differences in lifetime between the four brands of cartridge and no variability in lifetime due to different record labels. ∎

When at least two of the K_{ij}'s are unequal, the ANOVA computations are much more complex than for the case $K_{ij} = K$, and there are no nice formulas for the appropriate test statistics. One of the chapter references can be consulted for more information.

Exercises / Section 11.2 (13–21)

13. In an experiment to assess the effects of curing time (factor A) and type of mix (factor B) on the compressive strength of hardened cement cubes, three different curing times were used in combination with four different mixes, with three observations obtained for each of the 12 curing time/mix combinations. The resulting sums of squares were computed to be $SSA = 30,763.0$, $SSB = 34,185.6$, $SSE = 97,436.8$, and $SST = 205,966.6$.

a. Construct an ANOVA table.

b. Test at level .05 the null hypothesis H_{0AB}: all γ_{ij}'s $= 0$ (no interaction of factors) against H_{aAB}: at least one $\gamma_{ij} \neq 0$.

c. Test at level .05 the null hypothesis H_{0A}: $\alpha_1 = \alpha_2 = \alpha_3 = 0$ (factor A main effects are absent) against H_{aA}: at least one $\alpha_i \neq 0$.

d. Test H_{0B}: $\beta_1 = \beta_2 = \beta_3 = \beta_4 = 0$ versus H_{aB}: at least one $\beta_j \neq 0$ using a level .05 test.

e. The values of the $\bar{x}_{i..}$'s were $\bar{x}_{1..} = 4010.88$, $\bar{x}_{2..} = 4029.10$, and $\bar{x}_{3..} = 3960.02$. Use Tukey's procedure to investigate significant differences among the three curing times.

14. The accompanying data table gives observations on total acidity of coal samples of three different types, with determinations made using three different concentrations of ethanolic NaOH

("Chemistry of Brown Coals," *Australian J. Applied Science,* 1958, pp. 375–379).

		Type of coal		
		Morwell	Yallourn	Maddingley
NaOH conc.	.404N	8.27, 8.17	8.66, 8.61	8.14, 7.96
	.626N	8.03, 8.21	8.42, 8.58	8.02, 7.89
	.786N	8.60, 8.20	8.61, 8.76	8.13, 8.07

Additionally, $\sum\sum\sum_{ijk} x_{ijk}^2 = 1240.1525$ and $\sum\sum_{ij} x_{ij.}^2 = 2479.9991$.

a. Assuming both effects to be fixed, construct an ANOVA table, test for the presence of interaction, and then test for the presence of main effects for each factor (all using level .01).

b. Use Tukey's procedure to identify significant differences among the types of coal.

15. The current (in microamperes) necessary to produce a certain level of brightness of a television tube was measured for two different types of glass and three different types of phosphor, resulting in the accompanying data ("Fundamentals of Analysis of Variance," *Industrial Quality Control,* 1956, pp. 5–8).

		Phosphor type		
		1	2	3
Glass type	1	280, 290, 285	300, 310, 295	270, 285, 290
	2	230, 235, 240	260, 240, 235	220, 225, 230

Assuming that both factors are fixed, test H_{0AB} versus H_{aAB} at level .01. Then if H_{0AB} cannot be rejected, test the two sets of main effect hypotheses.

16. In an experiment to investigate the effect of "cement factor" (number of sacks of cement per cubic yard) on flexural strength of the resulting concrete ("Studies of Flexural Strength of Concrete. Part 3: Effects of Variation in Testing Procedure," *Proceedings ASTM*, 1957, pp. 1127–1139), $I = 3$ different factor values were used, $J = 5$ different batches of cement were selected, and $K = 2$ beams were cast from each cement factor/batch combination. Summary values include $\Sigma\Sigma\Sigma x_{ijk}^2 = 12,280,103$, $\Sigma\Sigma x_{ij.}^2 = 24,529,699$, $\Sigma x_{i..}^2 = 122,380,901$, $\Sigma x_{.j.}^2 = 73,427,483$, and $x_{...} = 19,143$.
   a. Construct the ANOVA table.
   b. Assuming a mixed model with cement factor (A) fixed and batches (B) random, test the three pairs of hypotheses of interest at level .05.

17. The accompanying data was obtained in an experiment to investigate whether compressive strength of concrete cylinders depended on the type of capping material used or variability in different batches ("The Effect of Type of Capping Material on the Compressive Strength of Concrete Cylinders," *Proceedings ASTM*, 1958, pp. 1166–1186). Each number is a cell total ($x_{ij.}$) based on $K = 3$ observations.

		Batch			
	1	2	3	4	5
Capping 1	1847	1942	1935	1891	1795
material 2	1779	1850	1795	1785	1626
3	1806	1892	1889	1891	1756

In addition, $\Sigma\Sigma\Sigma x_{ijk}^2 = 16,815,853$ and $\Sigma\Sigma x_{ij.}^2 = 50,433,409$. Obtain the ANOVA table, and then test at level .01 the hypotheses H_{0G} versus H_{aG}, H_{0A} versus H_{aA}, and H_{0B} versus H_{aB} assuming that capping is a fixed effect and batches is a random effect.

18. Show that $E(\overline{X}_{i..} - \overline{X}_{...}) = \alpha_i$, so that $\overline{X}_{i..} - \overline{X}_{...}$ is an unbiased estimator for α_i (in the fixed effects model).

19. With $\hat{\gamma}_{ij} = \overline{X}_{ij.} - \overline{X}_{i..} - \overline{X}_{.j.} + \overline{X}_{...}$, show that $\hat{\gamma}_{ij}$ is an unbiased estimator for γ_{ij} (in the fixed effects model).

20. Show how a $100(1 - \alpha)\%$ t confidence interval for $\alpha_i - \alpha_{i'}$ can be obtained. Then compute a 95% interval for $\alpha_2 - \alpha_3$ using the data from Exercise 14. *Hint:* With $\theta = \alpha_2 - \alpha_3$, the result of Exercise 18 indicates how to obtain $\hat{\theta}$. Then compute $V(\hat{\theta})$ and $\sigma_{\hat{\theta}}$, and obtain an estimate of $\sigma_{\hat{\theta}}$ by using $\sqrt{MSE}$ to estimate σ (which identifies the appropriate number of d.f.).

21. When both factors are random in a two-way ANOVA experiment with K replications per combination of factor levels, the expected mean squares are $E(MSE) = \sigma^2$, $E(MSA) = \sigma^2 + K\sigma_G^2 + JK\sigma_A^2$, $E(MSB) = \sigma^2 + K\sigma_G^2 + IK\sigma_B^2$, and $E(MSAB) = \sigma^2 + K\sigma_G^2$.
   a. What F ratio is appropriate for testing $H_{0G}: \sigma_G^2 = 0$ versus $H_{aG}: \sigma_G^2 > 0$?
   b. Answer (a) for testing $H_{0A}: \sigma_A^2 = 0$ versus $H_{aA}: \sigma_A^2 > 0$ and $H_{0B}: \sigma_B^2 = 0$ versus $H_{aB}: \sigma_B^2 > 0$.

11.3 Three-Factor ANOVA

To indicate the nature of models and analyses when ANOVA experiments involve more than two factors, we shall focus here on the case of three fixed factors—A, B, and C. The numbers of levels of the three factors will be denoted by I, J, and K, respectively, and $L_{ijk} = $ the number of observations made with factor A at level i, factor B at level j, and factor C at level k. As with two-factor ANOVA, the analysis is quite complicated when the L_{ijk}'s are not all equal, so we further specialize to $L_{ijk} = L$. X_{ijkl} and x_{ijkl} then denote the observed value, before and after the experiment is performed, of the lth replication ($l = 1, 2, \ldots, L$) when the three factors are fixed at levels i, j, and k.

To understand the parameters that will appear in the three-factor ANOVA model, first recall that in two-factor ANOVA with replications, $E(X_{ijk}) = $

$\mu_{ij} = \mu + \alpha_i + \beta_j + \gamma_{ij}$, where the restrictions $\Sigma_i \alpha_i = \Sigma_j \beta_j = 0$, $\Sigma_i \gamma_{ij} = 0$ for every j, and $\Sigma_j \gamma_{ij} = 0$ for every i were necessary to obtain a unique set of parameters. If we use dot subscripts on the μ_{ij}'s to denote averaging (rather than summation), then

$$\mu_{i.} - \mu_{..} = \frac{1}{J} \sum_j \mu_{ij} - \frac{1}{IJ} \sum_i \sum_j \mu_{ij} = \alpha_i$$

is the effect of factor A at level i averaged over levels of factor B, while

$$\mu_{ij} - \mu_{.j} = \mu_{ij} - \frac{1}{I} \sum_i \mu_{ij} = \alpha_i + \gamma_{ij}$$

is the effect of factor A at level i specific to factor B at level j. If the effect of A at level i depends on the level of B, then there is interaction between the factors, and the γ_{ij}'s are not all zero. In particular

$$\mu_{ij} - \mu_{.j} - \mu_{i.} + \mu_{..} = \gamma_{ij} \qquad (11.11)$$

The Three-Factor Fixed Effects Model

The model for three-factor ANOVA with $L_{ijk} = L$ is

$$X_{ijkl} = \mu_{ijk} + \epsilon_{ijkl} \qquad \begin{array}{l} i = 1, \ldots, I;\ j = 1, \ldots, J; \\ k = 1, \ldots, K;\ l = 1, \ldots, L \end{array} \qquad (11.12)$$

where the ϵ_{ijkl}'s are independent and normally distributed with mean zero and variance σ^2, and

$$\mu_{ijk} = \mu + \alpha_i + \beta_j + \delta_k + \gamma_{ij}^{AB} + \gamma_{ik}^{AC} + \gamma_{jk}^{BC} + \gamma_{ijk} \qquad (11.13)$$

The restrictions necessary to obtain uniquely defined parameters are that the sum over any subscript of any parameter on the right-hand side of (11.13) equal zero.

The parameters γ_{ij}^{AB}, γ_{ik}^{AC}, and γ_{jk}^{BC} are called two-factor interactions, while γ_{ijk} is called a three-factor interaction; the α_i's, β_j's, and δ_k's are the main effects parameters. For any fixed level k of the third factor, analogous to (11.11),

$$\mu_{ijk} - \mu_{i.k} - \mu_{.jk} + \mu_{..k} = \gamma_{ij}^{AB} + \gamma_{ijk}$$

is the interaction of the ith level of A with the jth level of B specific to the kth level of C, while

$$\mu_{ij.} - \mu_{i..} - \mu_{.j.} + \mu_{...} = \gamma_{ij}^{AB}$$

is the interaction between A at level i and B at level j averaged over levels of C. If the interaction of A at level i and B at level j does not depend on k, then all

γ_{ijk}'s equal 0. Thus nonzero γ_{ijk}'s represent nonadditivity of the two-factor γ_{ij}^{AB}'s over the various levels of the third factor C. If the experiment included more than three factors, there would be corresponding higher-order interaction terms with analogous interpretations. Note that in the above argument, if we had considered fixing the level of either A or B (rather than C, as was done) and examining the γ_{ijk}'s, their interpretation would be the same—if any of the interactions of two factors depend on the level of the third factor, then there are nonzero γ_{ijk}'s.

The Analysis of a Three-Factor Experiment

When $L > 1$, there is a sum of squares for each main effect, each two-factor interaction, and the three-factor interaction. To write these in a way that indicates how sums of squares are defined when there are more than three factors, note that any of the model parameters in (11.13) can be estimated unbiasedly by averaging X_{ijkl} over appropriate subscripts and taking differences. Thus

$$\hat{\mu} = \overline{X}_{\ldots}, \quad \hat{\alpha}_i = \overline{X}_{i\ldots} - \overline{X}_{\ldots}, \quad \hat{\gamma}_{ij}^{AB} = \overline{X}_{ij\ldots} - \overline{X}_{i\ldots} - \overline{X}_{\cdot j \cdot \cdot} + \overline{X}_{\ldots},$$

$$\hat{\gamma}_{ijk} = \overline{X}_{ijk\cdot} - \overline{X}_{ij\cdot\cdot} - \overline{X}_{i\cdot k\cdot} - \overline{X}_{\cdot jk\cdot} + \overline{X}_{i\ldots} + \overline{X}_{\cdot j\cdot\cdot} + \overline{X}_{\cdot\cdot k\cdot} - \overline{X}_{\ldots}$$

with other main effects and interaction estimators obtained by symmetry. Then sums of squares are

$$SST = \sum_i \sum_j \sum_k \sum_l (X_{ijkl} - \overline{X}_{\ldots})^2 = \sum_i \sum_j \sum_k \sum_l X_{ijkl}^2 - \frac{X_{\ldots}^2}{IJKL}$$

$$SSA = \sum_i \sum_j \sum_k \sum_l \hat{\alpha}_i^2 = JKL \sum_i (\overline{X}_{i\ldots} - \overline{X}_{\ldots})^2$$

$$= \frac{1}{JKL} \sum_i X_{i\ldots}^2 - \frac{X_{\ldots}^2}{IJKL}$$

$$SSAB = \sum_i \sum_j \sum_k \sum_l (\hat{\gamma}_{ij}^{AB})^2$$

$$= \frac{1}{KL} \sum_i \sum_j X_{ij\cdot\cdot}^2 - \frac{1}{JKL} \sum_i X_{i\ldots}^2 - \frac{1}{IKL} \sum_j X_{\cdot j\cdot\cdot}^2 + \frac{X_{\ldots}^2}{IJKL}$$

$$SSABC = \sum_i \sum_j \sum_k \sum_l \hat{\gamma}_{ijk}^2, \quad SSE = \sum_i \sum_j \sum_k \sum_l (X_{ijkl} - \overline{X}_{ijk\cdot})^2$$

$$= \sum_i \sum_j \sum_k \sum_l X_{ijkl}^2 - \frac{1}{L} \sum_i \sum_j \sum_k X_{ijk\cdot}^2$$

with the other main effect and two factor interaction SS's obtained by symmetry.

The fundamental ANOVA identity here is

$$
\begin{aligned}
SST = SSA + SSB + SSC + SSAB + SSAC + SSBC \\
+ SSABC + SSE
\end{aligned}
\tag{11.14}
$$

so that $SSABC$ can be obtained by subtraction. Each main effect and interaction SS has associated with it a number of degrees of freedom equal to the number of independently determined parameters which define the SS. Thus SSA has $I - 1$ d.f., $SSAB$ has $(I - 1)(J - 1)$ d.f., $SSABC$ has $(I - 1)(J - 1)(K - 1)$ d.f., and so on, while SST has $IJKL - 1$ d.f. and SSE has $IJK(L - 1)$ d.f.

Each SS (excepting SST) when divided by its d.f. gives a mean square, with

$$E(MSE) = \sigma^2$$

$$E(MSA) = \sigma^2 + \frac{JKL}{I - 1} \sum_i \alpha_i^2$$

$$E(MSAB) = \sigma^2 + \frac{KL}{(I - 1)(J - 1)} \sum_i \sum_j (\gamma_{ij}^{AB})^2$$

$$E(MSABC) = \sigma^2 + \frac{L}{(I - 1)(J - 1)(K - 1)} \sum_i \sum_j \sum_k (\gamma_{ijk})^2$$

and similar expressions for the other expected mean squares. Main effect and interaction hypotheses are tested by forming F ratios with MSE in each denominator:

Null hypothesis	Test statistic value	Rejection region
H_{0A}: all α_i's $= 0$	$f_A = \dfrac{MSA}{MSE}$	$f_A \geq F_{\alpha, I - 1, IJK(L - 1)}$
H_{0AB}: all γ_{ij}^{AB}'s $= 0$	$f_{AB} = \dfrac{MSAB}{MSE}$	$f_{AB} \geq F_{\alpha, (I - 1)(J - 1), IJK(L - 1)}$
H_{0ABC}: all γ_{ijk}'s $= 0$	$f_{ABC} = \dfrac{MSABC}{MSE}$	$f_{ABC} \geq F_{\alpha, (I - 1)(J - 1)(K - 1), IJK(L - 1)}$

Usually the main effect hypotheses are tested only if all interactions are judged not significant.

The above analysis assumes that $L_{ijk} = L > 1$. If $L = 1$, then as in the two-factor case, the highest-order interactions must be assumed to equal zero in order to obtain an MSE that estimates σ^2. Setting $L = 1$ and disregarding the fourth subscript summation over l, the above formulas for SS's are still valid, and $SSE = \sum_i \sum_j \sum_k \hat{\gamma}_{ijk}^2$ with $\overline{X}_{ijk.} = X_{ijk}$ in the expression for $\hat{\gamma}_{ijk}$.

Example **11.11** The following observations (body temperature $-$ 100 °F) were reported in an experiment to study heat tolerance of cattle ("The Significance of the Coat in Heat Tolerance of Cattle," *Australian J. Agriculture Research*, 1959, pp. 744–748). Measurements were made at four different periods (factor A, with $I = 4$) on two different strains of cattle (factor B, with $J = 2$) having four different types of coat (factor C, with $K = 4$); $L = 3$ observations were made for each of the $4 \times 2 \times 4 = 32$ combinations of levels of the three factors.

	\multicolumn{4}{c}{*B1*}	\multicolumn{4}{c}{*B2*}
	C1	*C2*	*C3*	*C4*	*C1*	*C2*	*C3*	*C4*
A1	3.6	3.4	2.9	2.5	4.2	4.4	3.6	3.0
	3.8	3.7	2.8	2.4	4.0	3.9	3.7	2.8
	3.9	3.9	2.7	2.2	3.9	4.2	3.4	2.9
A2	3.8	3.8	2.9	2.4	4.4	4.2	3.8	2.0
	3.6	3.9	2.9	2.2	4.4	4.3	3.7	2.9
	4.0	3.9	2.8	2.2	4.6	4.7	3.4	2.8
A3	3.7	3.8	2.9	2.1	4.2	4.0	4.0	2.0
	3.9	4.0	2.7	2.0	4.4	4.6	3.8	2.4
	4.2	3.9	2.8	1.8	4.5	4.5	3.3	2.0
A4	3.6	3.6	2.6	2.0	4.0	4.0	3.8	2.0
	3.5	3.7	2.9	2.0	4.1	4.4	3.7	2.2
	3.8	3.9	2.9	1.9	4.2	4.2	3.5	2.3

The table of cell totals ($x_{ijk.}$'s) for all combinations of the three factors is

$x_{ijk.}$	\multicolumn{4}{c}{*B1*}	\multicolumn{4}{c}{*B2*}
	C1	*C2*	*C3*	*C4*	*C1*	*C2*	*C3*	*C4*
A1	11.3	11.0	8.4	7.1	12.1	12.5	10.7	8.7
A2	11.4	11.6	8.6	6.8	13.4	13.2	10.9	7.7
A3	11.8	11.7	8.4	5.9	13.1	13.1	11.1	6.4
A4	10.9	11.2	8.4	5.9	12.3	12.6	11.0	6.5

Figure 11.3 displays plots of the corresponding cell means. There are three tables for cell totals of different pairs of factors:

$x_{ij..}$	*B1*	*B2*	$x_{i...}$
A1	37.8	44.0	81.8
A2	38.4	45.2	83.6
A3	37.8	43.7	81.5
A4	36.4	42.4	78.8
$x_{.j..}$	150.4	175.3	325.7

$x_{i.k.}$	*C1*	*C2*	*C3*	*C4*	$x_{i...}$
A1	23.4	23.5	19.1	15.8	81.8
A2	24.8	24.8	19.5	14.5	83.6
A3	24.9	24.8	19.5	12.3	81.5
A4	23.2	23.8	19.4	12.4	78.8
$x_{..k.}$	96.3	96.9	77.5	55.0	325.7

$x_{.jk.}$	$C1$	$C2$	$C3$	$C4$	$x_{.j..}$
$B1$	45.4	45.5	33.8	25.7	150.4
$B2$	50.9	51.4	43.7	29.3	175.3
$x_{..k.}$	96.3	96.9	77.5	55.0	325.7

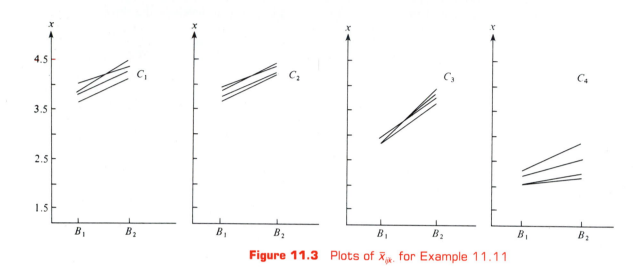

Figure 11.3 Plots of $\bar{x}_{ijk.}$ for Example 11.11

The correction factor $x^2_{....}/IJKL$ is $(325.7)^2/96 = 1105.01$ and $\sum_i\sum_j\sum_k\sum_l x^2_{ijkl} = 1166.17$, so $SST = 1166.17 - 1105.01 = 61.16$. The other sums of squares are

$$SSA = \frac{(81.8)^2 + (83.6)^2 + (81.5)^2 + (78.8)^2}{24} - 1105.01$$

$$= 1105.50 - 1105.01 = .49$$

$$SSB = 1111.46 - 1105.01 = 6.45$$

$$SSC = 1153.94 - 1105.01 = 48.93$$

$$SSAB = 1111.97 - 1105.50 - 1111.46 + 1105.01 = .02$$

$$SSAC = 1.61, \quad SSBC = .88, \quad SSE = 1166.17 - 1163.64 = 2.53$$

and

$$SSABC = 61.16 - [.49 + 6.45 + 48.93 + .02 + 1.61 + .88 + 2.53]$$

$$= .25$$

These calculations are of course typically done using a computer.

Table 11.10

Source	d.f.	Sum of squares	Mean square	f
A	$I - 1 = 3$	.49	.163	4.13
B	$J - 1 = 1$	6.45	6.45	163.29
C	$K - 1 = 3$	48.93	16.31	412.91
AB	$(I - 1)(J - 1) = 3$	.02	.0067	.170
AC	$(I - 1)(K - 1) = 9$	1.61	.179	4.53
BC	$(J - 1)(K - 1) = 3$	.88	.293	7.42
ABC	$(I - 1)(J - 1)(K - 1) = 9$	.25	.0278	.704
Error	$IJK(L - 1) = 64$	2.53	.0395	
Total	$IJKL - 1 = 95$	61.16		

Since $F_{.01, 9, 64} \approx 2.70$ and $f_{ABC} = MSABC/MSE = .704$ does not exceed 2.70, we conclude that three-factor interactions are not significant. However, although the AB interactions are also not significant, both AC and BC interactions as well as all main effects seem to be necessary in the model. When there are no ABC or AB interactions, a plot of the $\bar{x}_{ijk \cdot}$'s $(= \hat{\mu}_{ijk})$ separately for each level of C should reveal no substantial interactions (if only the ABC interactions are zero, plots are more difficult to interpret; see the article "Two-Dimensional Plots for Interpreting Interactions in the Three-Factor Analysis of Variance Model," *Amer. Statistician,* May 1979, pp. 63–69). ■

Tukey's procedure can be used in three-factor (or more) ANOVA. The second subscript on Q is the number of sample means being compared, while the third is degrees of freedom for error.

Models with random and mixed effects can also be analyzed. Sums of squares and degrees of freedom are identical to the fixed effects case, but expected mean squares are of course different for the random main effects or interactions. A good reference is the book by Montgomery listed in the chapter references.

Latin Square Designs

When several factors are to be studied simultaneously, an experiment in which there is at least one observation for every possible combination of levels is referred to as a **complete layout.** If the factors are A, B, and C with I, J, and K levels, respectively, a complete layout requires at least IJK observations. Frequently an experiment of this size is either impracticable, because of cost, time, or space constraints, or literally impossible. For example, if the response variable is sales of a certain product and the factors are different display configurations, different stores, and different time periods, then only one display configuration can realistically be used in a given store during a given time period.

A three-factor experiment in which fewer than IJK observations are made is called an incomplete layout. There are some incomplete layouts in which the pattern of combinations of factors is such that the analysis is straightforward.

One such three-factor design is called a **Latin square.** It is appropriate when $I = J = K$ (for example, four display configurations, four stores, and four time periods) and all two- and three-factor interaction effects are assumed absent. If the levels of factor A are identified with the rows of a two-way table and the levels of B with the columns of the table, then the defining characteristic of a Latin square design is that *every level of factor C appears exactly once in each row and exactly once in each column.* Pictured in Figure 11.4 are examples of a 3×3, 4×4, and 5×5 Latin square. There are 12 different 3×3 Latin squares, and the number of different $N \times N$ Latin squares increases rapidly with N (for example, every permutation of rows of a given Latin square yields a Latin square, and similarly for column permutations). It is recommended that the square actually used in a particular experiment be chosen at random from the set of all possible squares of the desired dimension; for further details, one of the chapter references can be consulted.

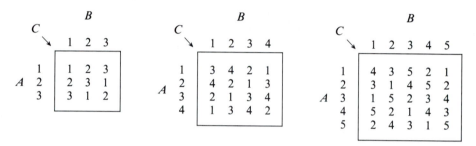

Figure 11.4   Examples of Latin squares

The Model and Analysis for Latin Squares

The letter N will be used to denote the common value of I, J, and K. Then a complete layout with one observation per combination would require N^3 observations, whereas a Latin square requires only N^2 observations. Once a particular square has been chosen, the value of k (the level of factor C) is completely determined by the values of i and j. To emphasize this, we use $x_{ij(k)}$ to denote the observed value when the three factors are at levels i, j, and k, respectively, with k taking on only one value for each i, j pair. The model is then

$$X_{ij(k)} = \mu + \alpha_i + \beta_j + \delta_k + \epsilon_{ij(k)} \qquad i = 1, \ldots, N;\ j = 1, \ldots, N; \\ k = 1, \ldots, N$$

where the $\epsilon_{ij(k)}$'s are independent and normally distributed with mean zero and variance σ^2, and $\Sigma\alpha_i = \Sigma\beta_j = \Sigma\delta_k = 0$.

We employ the following notation for totals and averages:

$$X_{i..} = \sum_j X_{ij(k)}, \quad X_{.j.} = \sum_i X_{ij(k)}, \quad X_{..k} = \sum_{i,j} X_{ij(k)}, \quad X_{...} = \sum_i \sum_j X_{ij(k)},$$

$$\overline{X}_{i..} = \frac{X_{i..}}{N}, \quad \overline{X}_{.j.} = \frac{X_{.j.}}{N}, \quad \overline{X}_{..k} = \frac{X_{..k}}{N}, \quad \overline{X}_{...} = \frac{X_{...}}{N^2}$$

Note that although $X_{i..}$ previously suggested a double summation, now it corresponds to a single sum over all j (and the associated values of k). The sums of squares used in the analysis are then

$$SST = \sum_i \sum_j (X_{ij(k)} - \overline{X}_{...})^2 = \sum_i \sum_j X_{ij(k)}^2 - \frac{X_{...}^2}{N^2}$$

$$SSA = \sum_i \sum_j (\overline{X}_{i..} - \overline{X}_{...})^2 = \frac{1}{N} \sum_i X_{i..}^2 - \frac{X_{...}^2}{N^2}$$

$$SSB = \sum_i \sum_j (\overline{X}_{.j.} - \overline{X}_{...})^2 = \frac{1}{N} \sum_j X_{.j.}^2 - \frac{X_{...}^2}{N^2}$$

$$SSC = \sum_i \sum_j (\overline{X}_{..k} - \overline{X}_{...})^2 = \frac{1}{N} \sum_k X_{..k}^2 - \frac{X_{...}^2}{N^2}$$

and

$$SSE = \sum_i \sum_j (X_{ij(k)} - \hat{\mu} - \hat{\alpha}_i - \hat{\beta}_j - \hat{\delta}_k)^2$$

$$= \sum_i \sum_j (X_{ij(k)} - \overline{X}_{i..} - \overline{X}_{.j.} - \overline{X}_{..k} + 2\overline{X}_{...})^2$$

where $\hat{\mu}$, $\hat{\alpha}_i$, $\hat{\beta}_j$, and $\hat{\delta}_k$ are the usual unbiased estimators for the respective parameters.

An ANOVA identity is still valid here:

$$SST = SSA + SSB + SSC + SSE$$

so that SSE is computed by subtraction after the other four SS's have been computed. There are $N^2 - 1$ degrees of freedom for SST and $N - 1$ d.f. for each of the main effect SS's, so error d.f. $= N^2 - 1 - (N - 1) - (N - 1) - (N - 1) = (N - 1) \cdot (N - 2)$. Mean squares are then

$$MSA = \frac{SSA}{N - 1}, \quad MSB = \frac{SSB}{N - 1}, \quad MSC = \frac{SSC}{N - 1}$$

and

$$MSE = \frac{SSE}{(N-1)(N-2)}$$

For testing $H_{0C}: \delta_1 = \delta_2 = \cdots = \delta_N = 0$, the test statistic value is $f_C = MSC/MSE$, with H_{0C} rejected if $f_C \geq F_{\alpha, N-1, (N-1)(N-2)}$. The other two main-effect null hypotheses are also rejected if the corresponding F ratio exceeds $F_{\alpha, N-1, (N-1)(N-2)}$.

If any of the null hypotheses is rejected, significant differences can be identified by using Tukey's procedure. After computing $w = Q_{\alpha, N, (N-1)(N-2)} \cdot \sqrt{MSE/N}$, pairs of sample means (the $\bar{x}_{i..}$'s, $\bar{x}_{.j.}$'s, or $\bar{x}_{..k}$'s) differing by more than w correspond to significant differences between associated factor effects (the α_i's, β_j's, or δ_k's).

The hypothesis H_{0C} is frequently the one of central interest, and a Latin square design is used to control for extraneous variation in the A and B factors, as was done by a randomized block design for the case of a single extraneous factor. Thus in the product sales example mentioned earlier, variation due to both stores and time periods is controlled by a Latin square design, enabling an investigator to test for the presence of effects due to different product display configurations.

Example 11.12 In an experiment to investigate the effect of relative humidity on abrasion resistance of leather cut from a rectangular pattern ("The Abrasion of Leather," *J. Inter. Soc. Leather Trades' Chemists,* 1946, p. 287), a 6×6 Latin square was used to control for possible variability due to row and column position in the pattern. The six levels of relative humidity studied were $1 = 25\%$, $2 = 37\%$, $3 = 50\%$, $4 = 62\%$, $5 = 75\%$, and $6 = 87\%$, with the following results.

				B (columns)				
		1	2	3	4	5	6	$x_{i..}$
A (rows)	1	[3] 7.38	[4] 5.39	[6] 5.03	[2] 5.50	[5] 5.01	[1] 6.79	35.10
	2	[2] 7.15	[1] 8.16	[5] 4.96	[4] 5.78	[3] 6.24	[6] 5.06	37.35
	3	[4] 6.75	[6] 5.64	[3] 6.34	[5] 5.31	[1] 7.81	[2] 8.05	39.90
	4	[1] 8.05	[3] 6.45	[2] 6.31	[6] 5.46	[4] 6.05	[5] 5.51	37.83
	5	[6] 5.65	[5] 5.44	[1] 7.27	[3] 6.54	[2] 7.03	[4] 5.96	37.89
	6	[5] 6.00	[2] 6.55	[4] 5.93	[1] 8.02	[6] 5.80	[3] 6.61	38.91
	$x_{.j.}$	40.98	37.63	35.84	36.61	37.94	37.98	

Also, $x_{..1} = 46.10$, $x_{..2} = 40.59$, $x_{..3} = 39.56$, $x_{..4} = 35.86$, $x_{..5} = 32.23$, $x_{..6} = 32.64$, $x_{...} = 226.98$, and $\sum_i \sum_j x^2_{ij(k)} = 1462.89$. Further computations are summarized in Table 11.11.

Table 11.11

Source of variation	d.f.	Sum of squares	Mean square	f
A (rows)	5	2.19	.438	2.50
B (columns)	5	2.57	.514	2.94
C (treatments)	5	23.53	4.706	26.89
Error	20	3.49	.175	
Total	35	31.78		

Since $F_{.05, 5, 20} = 2.71$ and $26.89 \geq 2.71$, H_{0C} is rejected in favor of the hypothesis that relative humidity does on average affect abrasion resistance.

To apply Tukey's procedure, $w = Q_{.05, 6, 20} \cdot \sqrt{MSE/6} = 4.45 \sqrt{.175/6} = .76$. Ordering the $\bar{x}_{..k}$'s and underscoring yields

75%	87%	62%	50%	37%	25%
5.37	5.44	5.98	6.59	6.77	7.68

In particular, the lowest relative humidity appears to result in a true average abrasion resistance significantly higher than for any other relative humidity studied. ■

Exercises / Section 11.3 [22–30]

22. The output of a continuous extruding machine that coats steel pipe with plastic was studied as a function of the thermostat temperature profile (A, at three levels), type of plastic (B, at three levels), and the speed of the rotating screw that forces the plastic through a tube-forming die (C, at three levels). There were two replications ($L = 2$) at each combination of levels of the factor, yielding a total of 54 observations on output. The summary statistics include $x_{....} = 29,184$, $\Sigma x_{i...}^2 = 284,156,552$, $\Sigma x_{.j..}^2 = 284,001,155$, $\Sigma x_{..k.}^2 = 288,306,487$, $\Sigma\Sigma x_{ij..}^2 = 94,758,336$, $\Sigma\Sigma x_{i.k.}^2 = 96,187,405$, $\Sigma\Sigma x_{.jk.}^2 = 96,137,220$, $\Sigma\Sigma\Sigma x_{ijk.}^2 = 32,078,455$, and $\Sigma\Sigma\Sigma\Sigma x_{ijkl}^2 = 16,042,355$.

   a. Construct the ANOVA table.
   b. Use appropriate F tests to show that none of the F ratios for two- or three-factor interactions is significant at level .05.
   c. Which main effects appear significant?
   d. With $x_{..1.} = 8242$, $x_{..2.} = 9732$, and $x_{..3.} = 11,210$, use Tukey's procedure to identify significant differences among the levels of factor C.

23. To see whether thrust force in drilling is affected by drilling speed (A), feed rate (B), or material used (C), an experiment using four speeds, three rates, and two materials was performed, with two samples ($L = 2$) drilled at each combination of levels of the three factors. Summary statistics include $x_{....} = 17,709$, $\Sigma x_{i...}^2 = 78,631,967$, $\Sigma x_{.j..}^2 = 145,960,989$, $\Sigma x_{..k.}^2 = 160,582,841$, $\Sigma\Sigma x_{ij..}^2 = 36,779,799$, $\Sigma\Sigma x_{i.k.}^2 = 40,314,811$, $\Sigma\Sigma x_{.jk.}^2 = 74,975,035$, $\Sigma\Sigma\Sigma x_{ijk.}^2 = 18,919,719$, and $\Sigma\Sigma\Sigma\Sigma x_{ijkl}^2 = 9,516,679$. Construct the ANOVA table and identify significant interactions using $\alpha = .01$. Is there any single factor that appears to have no effect on thrust force (does any factor appear nonsignificant in every effect in which it appears)?

24. The paper "An Analysis of Variance Applied to Screw Machines" (*Industrial Quality Control*, 1956, pp. 8–9) described an experiment to investigate how the length of steel bars was affected by time of day (A), heat treatment applied (B), and screw machine used (C). The three times were 8:00 A.M., 11:00 A.M., and 3:00 P.M., and there were two treatments and four machines (a $3 \times 2 \times 4$ factorial experiment), resulting in the accompanying data [coded as 1000(length − 4.380), which doesn't affect the analysis].

B_1

	C_1	C_2	C_3	C_4
A_1	6, 9, 1, 3	7, 9, 5, 5	1, 2, 0, 4	6, 6, 7, 3
A_2	6, 3, 1, −1	8, 7, 4, 8	3, 2, 1, 0	7, 9, 11, 6
A_3	5, 4, 9, 6	10, 11, 6, 4	−1, 2, 6, 1	10, 5, 4, 8

B_2

	C_1	C_2	C_3	C_4
A_1	4, 6, 0, 1	6, 5, 3, 4	−1, 0, 0, 1	4, 5, 5, 4
A_2	3, 1, 1, −2	6, 4, 1, 3	2, 0, −1, 1	9, 4, 6, 3
A_3	6, 0, 3, 7	8, 7, 10, 0	0, −2, 4, −4	4, 3, 7, 0

a. Construct the ANOVA table for this data.

b. Test to see whether any of the interaction effects are significant at level .05.

c. Test to see whether any of the main effects are significant at level .05 (that is, H_{0A} versus H_{aA}, and so on).

d. Use Tukey's procedure to investigate significant differences between the four machines.

25. The following summary quantities were computed from an experiment involving four levels of nitrogen (A), two times of planting (B), and two levels of potassium (C) ("Use and Misuse of Multiple Comparison Procedures," *Agronomy J.*, 1977, pp. 205–208). Only one observation (N content, in percentages, of corn grain) was made for each of the 16 combinations of levels. Computed summary quantities are $x_{\ldots} = 20.96$ (since $L = 1$, a fourth subscript is unnecessary), $\Sigma x_{i\ldots}^2 = 110.7354$, $\Sigma x_{\cdot j\cdot}^2 = 219.6610$, $\Sigma x_{\cdot\cdot k}^2 = 219.6896$, $\Sigma\Sigma x_{ij\cdot}^2 = 55.3764$, $\Sigma\Sigma x_{i\cdot k}^2 = 55.3762$, $\Sigma\Sigma x_{\cdot jk}^2 = 109.8474$, and $\Sigma\Sigma\Sigma x_{ijk}^2 = 27.6960$.

a. Construct the ANOVA table.

b. Assume that there are no three-way interaction effects, so that $MSABC$ is a valid esti-

mate of σ^2, and test at level .05 for interaction and main effects.

c. The nitrogen averages are $\bar{x}_{1\ldots} = 1.1200$, $\bar{x}_{2\ldots} = 1.3025$, $\bar{x}_{3\ldots} = 1.3875$, and $\bar{x}_{4\ldots} = 1.4300$. Use Tukey's method to examine differences in percentage N among the nitrogen levels ($Q_{.05, 3, 3} = 5.91$).

26. The paper "Kolbe-Schmitt Carbonation of 2-Napthol" (*Industrial and Eng. Chemistry: Process and Design Development*, 1969, pp. 165–173) presented the accompanying data on percentage yield of BON acid as a function of reaction time (1, 2, and 3 hours), temperature (30, 70, and 100 °C), and pressure (30, 70, and 100 psi). Assume that there is no three-factor interaction, so that $SSE = SSABC$ provides an estimate of σ^2. Construct the ANOVA table, and carry out all appropriate tests.

B_1

	C_1	C_2	C_3
A_1	68.5	73.0	68.7
A_2	74.5	75.0	74.6
A_3	70.5	72.5	74.7

B_2

	C_1	C_2	C_3
A_1	72.8	80.1	72.0
A_2	72.0	81.5	76.0
A_3	69.5	84.5	76.0

B_3

	C_1	C_2	C_3
A_1	72.5	72.5	73.1
A_2	75.5	70.0	76.0
A_3	65.0	66.5	70.5

27. When factors A and B are fixed but factor C is random, the expected mean squares are

$E(MSE) = \sigma^2,$

$E(MSA) = \sigma^2 + JL\sigma_{AC}^2 + \dfrac{JKL}{I-1}\Sigma\alpha_i^2,$

$E(MSB) = \sigma^2 + IL\sigma_{BC}^2 + \dfrac{IKL}{J-1}\Sigma\beta_j^2,$

$E(MSC) = \sigma^2 + IJL\sigma_C^2,$

$E(MSAB) = \sigma^2 + L\sigma_{ABC}^2$

$\qquad + \dfrac{KL}{(I-1)(J-1)}\sum_i\sum_j(\gamma_{ij}^{AB})^2,$

$E(MSAC) = \sigma^2 + JL\sigma_{AC}^2,$
$E(MSBC) = \sigma^2 + IL\sigma_{BC}^2,$
$E(MSABC) = \sigma^2 + L\sigma_{ABC}^2.$

a. Based on these expected mean squares, what F ratios would you use to test $H_0: \sigma_{ABC}^2 = 0$; $H_0: \sigma_C^2 = 0$; $H_0: \gamma_{ij}^{AB} = 0$ for all i, j; and $H_0: \alpha_1 = \cdots = \alpha_I = 0$?

b. In an experiment to assess the effects of age, type of soil, and day of production on compressive strength of cement/soil mixtures, two ages (A), four types of soil (B), and three days (C, assumed random) were used, with $L = 2$ observations made for each combination of factor levels. The resulting sums of squares were $SSA = 14{,}318.24$, $SSB = 9656.40$, $SSC = 2270.22$, $SSAB = 3408.93$, $SSAC = 1442.58$, $SSBC = 3096.21$, $SSABC = 2832.72$, and $SSE = 8655.60$. Obtain the ANOVA table and carry out all tests using level .01.

28. Because of potential variability in aging due to different castings and segments on the castings, a Latin square design with $N = 7$ was used to investigate the effect of heat treatment on aging. With A = castings, B = segments, C = heat treatments, summary statistics include $x_{\cdots} = 3815.8$, $\Sigma x_{i\cdots}^2 = 297{,}216.90$, $\Sigma x_{\cdot j\cdot}^2 = 297{,}200.64$, $\Sigma x_{\cdot\cdot k}^2 = 297{,}155.01$, and $\Sigma\Sigma x_{ij(k)}^2 = 297{,}317.65$. Compute the ANOVA table and test at level .05 the hypothesis that heat treatment has no effect on aging.

29. The article "The Responsiveness of Food Sales to Shelf Space Requirements" (*J. Marketing Research*, 1964, pp. 63–67) reported the use of a Latin square design to investigate the effect of shelf space on food sales. The experiment was carried out over a six-week period using six dif-

ferent stores, resulting in the following data on sales of powdered coffee cream (with shelf space index in parentheses).

		Weeks	
	1	2	3
1	27 (5)	14 (4)	18 (3)
2	34 (6)	31 (5)	34 (4)
3	39 (2)	67 (6)	31 (5)
Stores 4	40 (3)	57 (1)	39 (2)
5	15 (4)	15 (3)	11 (1)
6	16 (1)	15 (2)	14 (6)

		Weeks	
	4	5	6
1	35 (1)	28 (6)	22 (2)
2	46 (3)	37 (2)	23 (1)
3	49 (4)	38 (1)	48 (3)
Stores 4	70 (6)	37 (4)	50 (5)
5	9 (2)	18 (5)	17 (6)
6	12 (5)	19 (3)	22 (4)

Construct the ANOVA table, and state and test at level .01 the hypothesis that shelf space does not affect sales against the appropriate alternative.

30. The article "Variation in Moisture and Ascorbic Acid Content from Leaf to Leaf and Plant to Plant in Turnip Greens" (*Southern Cooperative Services Bull.*, 1951, pp. 13–17) used a Latin square design in which factor A is plant, factor B is leaf size (smallest to largest), factor C is time of weighing, and the response variable is moisture content.

Time (C, in parentheses)	Leaf size (B)		
	1	2	3
Plant (A) 1	6.67 (5)	7.15 (4)	8.29 (1)
2	5.40 (2)	4.77 (5)	5.40 (4)
3	7.32 (3)	8.53 (2)	8.50 (5)
4	4.92 (1)	5.00 (3)	7.29 (2)
5	4.88 (4)	6.16 (1)	7.83 (3)

Time (C, in parentheses)	Leaf size (B)	
	4	5
Plant (A) 1	8.95 (3)	9.62 (2)
2	7.54 (1)	6.93 (3)
3	9.99 (4)	9.68 (1)
4	7.85 (5)	7.08 (4)
5	5.83 (2)	8.51 (5)

When all three factors are random, the expected mean squares are $E(MSA) = \sigma^2 + N\sigma_A^2$, $E(MSB) = \sigma^2 + N\sigma_B^2$, $E(MSC) = \sigma^2 + N\sigma_C^2$, and $E(MSE) = \sigma^2$. This implies that the F ratios for testing $H_{0A} : \sigma_A^2 = 0$, $H_{0B} : \sigma_B^2 = 0$, and $H_{0C} : \sigma_C^2 = 0$ are identical to those for fixed effects. Obtain the ANOVA table and test at level .05 to see if there is any variation in moisture content due to the factors.

11.4 2^p Factorial Experiments

If an experimenter wishes to study simultaneously the effect of p different factors on a response variable and the factors have $I_1, I_2, \ldots, I_p$ levels, respectively, then a complete experiment requires at least $I_1 \cdot I_2 \cdots I_p$ observations. In such situations the experimenter can often perform a "screening experiment" with each factor at only two levels to obtain preliminary information about factor effects. An experiment in which there are p factors, each at two levels, is referred to as a **2^p factorial experiment.** The analysis of data from such an experiment is computationally simpler than for more general factorial experiments. In addition, a 2^p experiment provides a simple setting for introducing the important concepts of confounding and fractional replications.

2^3 Experiments

As in Section 11.3 we let X_{ijkl} and x_{ijkl} refer to the observation from the lth replication with factors A, B, and C at levels i, j, and k, respectively. The model for this situation is

$$X_{ijkl} = \mu + \alpha_i + \beta_j + \delta_k + \gamma_{ij}^{AB} + \gamma_{ik}^{AC} + \gamma_{jk}^{BC} + \gamma_{ijk} + \epsilon_{ijkl} \qquad (11.15)$$

for $i = 1, 2$; $j = 1, 2$; $k = 1, 2$; $l = 1, \ldots, n$. The ϵ_{ijkl}'s are assumed independent, normally distributed, with mean zero and variance σ^2. Because there are only two levels of each factor, the side conditions on the parameters of (11.15) that uniquely specify the model are simply stated: $\alpha_1 + \alpha_2 = 0, \ldots,$ $\gamma_{11}^{AB} + \gamma_{21}^{AB} = 0, \gamma_{12}^{AB} + \gamma_{22}^{AB} = 0, \gamma_{11}^{AB} + \gamma_{12}^{AB} = 0, \gamma_{21}^{AB} + \gamma_{22}^{AB} = 0$, and the like. These conditions imply that there is only one functionally independent parameter of each type (for each main effect and interaction). For example, $\alpha_2 = -\alpha_1$, while $\gamma_{21}^{AB} = -\gamma_{11}^{AB}, \gamma_{12}^{AB} = -\gamma_{11}^{AB}$, and $\gamma_{22}^{AB} = \gamma_{11}^{AB}$. Because of this, each sum of squares in the analysis will have a single degree of freedom.

The parameters of the model can be estimated by taking averages over various subscripts of the X_{ijkl}'s and then forming appropriate linear combinations of the averages. For example,

$$\hat{\alpha}_1 = \overline{X}_{1\cdots} - \overline{X}_{\cdots}$$
$$= \frac{(X_{111\cdot} + X_{121\cdot} + X_{112\cdot} + X_{122\cdot} - X_{211\cdot} - X_{212\cdot} - X_{221\cdot} - X_{222\cdot})}{8n}$$

and

$$\hat{\gamma}_{11}^{AB} = \overline{X}_{11..} - \overline{X}_{1...} - \overline{X}_{.1..} + \overline{X}_{....}$$
$$= \frac{(X_{111.} - X_{121.} - X_{211.} + X_{221.} + X_{112.} - X_{122.} - X_{212.} + X_{222.})}{8n}$$

Each estimator is, except for the factor $1/8n$, a linear function of the cell totals ($X_{ijk.}$'s) in which each coefficient is $+1$ or -1, with an equal number of each; such functions are called **contrasts** in the $X_{ijk.}$'s. Furthermore, the estimators satisfy the same side conditions satisfied by the parameters themselves. For example

$$\hat{\alpha}_1 + \hat{\alpha}_2 = \overline{X}_{1...} - \overline{X}_{....} + \overline{X}_{2...} - \overline{X}_{....} = \overline{X}_{1...} + \overline{X}_{2...} - 2\overline{X}_{....}$$
$$= \frac{1}{4n} X_{1...} + \frac{1}{4n} X_{2...} - \frac{2}{8n} X_{....} = \frac{1}{4n} X_{....} - \frac{1}{4n} X_{....} = 0$$

Example 11.13 In an experiment to investigate the compressive strength properties of cement/soil mixtures, two different aging periods were used in combination with two different aging temperatures and two different soils. Two replications were made for each combination of levels of the three factors, resulting in the following data.

Age	Temperature	Soil 1	Soil 2
1	1	471, 413	385, 434
	2	485, 552	530, 593
2	1	712, 637	770, 705
	2	712, 789	741, 806

The computed cell totals are $x_{111.} = 884$, $x_{211.} = 1349$, $x_{121.} = 1037$, $x_{221.} = 1501$, $x_{112.} = 819$, $x_{212.} = 1475$, $x_{122.} = 1123$, and $x_{222.} = 1547$, so $x_{....} = 9735$. Then

$$\hat{\alpha}_1 = (884 - 1349 + 1037 - 1501 + 819 - 1475 + 1123 - 1547)/16$$
$$= -125.5625 = -\hat{\alpha}_2$$
$$\hat{\gamma}_{11}^{AB} = (884 - 1349 - 1037 + 1501 + 819 - 1475 - 1123 + 1547)/16$$
$$= -14.5625 = -\hat{\gamma}_{12}^{AB} = -\hat{\gamma}_{21}^{AB} = \hat{\gamma}_{22}^{AB}$$

The other parameter estimates can be computed in the same manner. ∎

Sums of Squares and Analysis for a 2^3 Experiment

The reason for computing parameter estimates is that sums of squares for the various effects are easily obtained from the estimates. For example

$$SSA = \sum_i \sum_j \sum_k \sum_l \hat{\alpha}_i^2 = 4n \sum_{i=1}^{2} \hat{\alpha}_i^2 = 4n[\hat{\alpha}_1^2 + (-\hat{\alpha}_1)^2] = 8n\hat{\alpha}_1^2$$

and

$$SSAB = \sum_i \sum_j \sum_k \sum_l (\hat{\gamma}_{ij}^{AB})^2$$

$$= 2n \sum_{i=1}^{2} \sum_{j=1}^{2} (\hat{\gamma}_{ij}^{AB})^2 = 2n[(\hat{\gamma}_{11}^{AB})^2 + (-\hat{\gamma}_{11}^{AB})^2 + (-\hat{\gamma}_{11}^{AB})^2 + (\hat{\gamma}_{11}^{AB})^2]$$

$$= 8n(\hat{\gamma}_{11}^{AB})^2$$

Since each estimate is a contrast in the cell totals multiplied by $1/8n$, each sum of squares has the form $(\text{contrast})^2/8n$. Thus to compute the various sums of squares, we need to know the coefficients ($+1$ or -1) of the appropriate contrasts. The signs ($+$ or $-$) on each $x_{ijk.}$ in each effect contrast are most conveniently displayed in a table. We shall use the notation (1) for the experimental condition $i = 1, j = 1, k = 1$, a for $i = 2, j = 1, k = 1$, ab for $i = 2, j = 2, k = 1$, and so on. If level 1 is thought of as "low" and level 2 as "high," any letter that appears denotes a high level of the associated factor. In Table 11.12 each column gives the signs for a particular effect contrast in the $x_{ijk.}$'s associated with the different experimental conditions.

Table 11.12   Signs for computing effect contrasts

Experimental condition	Cell total	*A*	*B*	*C*	*AB*	*AC*	*BC*	*ABC*
(1)	$x_{111.}$	$-$	$-$	$-$	$+$	$+$	$+$	$-$
a	$x_{211.}$	$+$	$-$	$-$	$-$	$-$	$+$	$+$
b	$x_{121.}$	$-$	$+$	$-$	$-$	$+$	$-$	$+$
ab	$x_{221.}$	$+$	$+$	$-$	$+$	$-$	$-$	$-$
c	$x_{112.}$	$-$	$-$	$+$	$+$	$-$	$-$	$+$
ac	$x_{212.}$	$+$	$-$	$+$	$-$	$+$	$-$	$-$
bc	$x_{122.}$	$-$	$+$	$+$	$-$	$-$	$+$	$-$
abc	$x_{222.}$	$+$	$+$	$+$	$+$	$+$	$+$	$+$

In each of the first three columns, the sign is $+$ if the corresponding factor is at the high level and $-$ if it is at the low level. Every sign in the AB column is then the "product" of the signs in the A and B columns, with $(+)(+) = (-)(-) = +$ and $(+)(-) = (-)(+) = -$, and similarly for the AC and BC columns. Finally, the signs in the ABC column are the products of AB with C (or B with AC or A with BC). Thus, for example,

$$AC \text{ contrast} = +x_{111.} - x_{211.} + x_{121.} - x_{221.} - x_{112.} + x_{212.} - x_{122.} + x_{222.}$$

Once the seven effect contrasts are computed,

$$SS(\text{effect}) = \frac{(\text{effect contrast})^2}{8n}$$

Even with a table of signs, calculation of the contrasts is tedious. An efficient computational technique, due to Yates, is as follows. Write in a column the eight cell totals in the **standard order** as given in the table of signs and establish three further columns. In each of these three columns, the first four entries are the sums of entries 1 and 2, 3 and 4, 5 and 6, 7 and 8 of the previous columns. The last four entries are the differences between entries 2 and 1, 4 and 3, 6 and 5, and 8 and 7 of the previous column. The last column then contains $x_{....}$ and the seven effect contrasts in standard order. Squaring each contrast and dividing by $8n$ then gives the seven sums of squares.

Example 11.14
(Example 11.13 continued)

Since $n = 2$, $8n = 16$.

Table 11.13

Treatment condition	$x_{ijk.}$	1	2	Effect contrast	$SS = (contrast)^2/16$
$(1) = x_{111.}$	884	2233	4771	9735	
$a = x_{211.}$	1349	2538	4964	2009	252,255.06
$b = x_{121.}$	1037	2294	929	681	28,985.06
$ab = x_{221.}$	1501	2670	1080	−233	3,393.06
$c = x_{112.}$	819	465	305	193	2,328.06
$ac = x_{212.}$	1475	464	376	151	1,425.06
$bc = x_{122.}$	1123	656	−1	71	315.06
$abc = x_{222.}$	1547	424	−232	−231	3,335.06
					292,036.42

From the original data, $\sum_i \sum_j \sum_k \sum_l x_{ijkl}^2 = 6,232,289$, and

$$\frac{x_{....}^2}{16} = 5,923,139.06$$

so

$$SST = 6,232,289 - 5,923,139.06 = 309,149.94$$
$$SSE = SST - [SSA + \cdots + SSABC] = 309,149.94 - 292,036.42$$
$$= 17,113.52$$

Table 11.14

Source of variation	d.f.	Sum of squares	Mean square	f
A	1	252,255.06	252,255.06	117.92
B	1	28,985.06	28,985.06	13.55
C	1	3,393.06	3,393.06	1.59
AB	1	2,328.06	2,328.06	1.09
AC	1	1,425.06	1,425.06	.67
BC	1	315.06	315.06	.15
ABC	1	3,335.06	3,335.06	1.56
Error	8	17,113.52	2,139.19	
Total	15	309,149.94		

2^p Experiments for $p > 3$

Although the computations when $p > 3$ are quite tedious, the analysis parallels that of the three-factor case. For example, if there are four factors A, B, C, and D, there are 16 different experimental conditions. The first eight in standard order are exactly those already listed for a three-factor experiment, while the second eight are obtained by placing the letter d beside each condition in the first group. Yates' method is then initiated by computing totals across replications, listing these totals in standard order, and proceeding as before; with p factors, the pth column to the right of the treatment totals will give the effect contrasts.

For $p > 3$ there will often be no replications of the experiment (so only one complete replicate is available). To obtain an error sum of squares for testing various hypotheses, the higher-order interactions are usually assumed absent and their SS's added to obtain an SSE.

Confounding

It is often not possible to carry out all 2^p experimental conditions of a 2^p factorial experiment in a homogeneous experimental environment. In such situations it may be possible to separate the experimental conditions into 2^r homogeneous blocks ($r < p$), so that there are 2^{p-r} experimental conditions in each block. The blocks may, for example, correspond to different laboratories, different time periods, or different operators or work crews. In the simplest case, $p = 3$ and $r = 1$, so that there are two blocks with each block consisting of four of the eight experimental conditions.

As always, blocking is effective in reducing variation associated with extraneous sources. However, when the 2^p experimental conditions are placed in 2^r blocks, the price paid for this blocking is that $2^r - 1$ of the factor effects cannot be estimated. This is because $2^r - 1$ factor effects (main effects and/or interactions) are mixed up or **confounded** with the block effects. The allocation of experimental conditions to blocks is then usually done so that only higher-level interactions are confounded, while main effects and low-order interactions remain estimable and can be tested for.

To see how allocation to blocks is accomplished, consider first a 2^3 experiment with two blocks ($r = 1$) and four treatments per block. Suppose that we select ABC as the effect to be confounded with blocks. Then any experimental condition having an odd number of letters in common with ABC, such as b (one letter) or abc (three letters), is placed in one block, while any condition having an even number of letters in common with ABC (where zero is even) goes in the other block. Figure 11.5 shows this allocation of treatments to the two blocks.

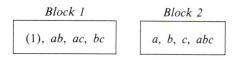

Block 1

(1), ab, ac, bc

Block 2

a, b, c, abc

Figure 11.5   Confounding ABC in a 2^3 experiment

In the absence of replications, the data from such an experiment would usually be analyzed by assuming that there were no two-factor interactions (additivity) and using $SSE = SSAB + SSAC + SSBC$ with three degrees of freedom to test for the presence of main effects. Most frequently, though, there are replications when just three factors are being studied. Suppose that there are u replicates, resulting in a total of $2^r \cdot u$ blocks in the experiment. Then after subtracting from SST all sums of squares associated with effects not confounded with blocks (computed using Yates' method), the block sum of squares is computed using the $2^r \cdot u$ block totals and then subtracted to yield SSE (so there are $2^r \cdot u - 1$ d.f. for blocks).

Example 11.15 The article "Factorial Experiments in Pilot Plant Studies" (*Industrial and Eng. Chemistry*, 1951, pp. 1300–1306) reports the results of an experiment to assess the effects of reactor temperature (A), gas throughput (B), and concentration of active constituent (C) on strength of the product solution (measured in arbitrary units) in a recirculation unit. Two blocks were used, with the ABC effect confounded with blocks, and there were two replications, resulting in the data in Figure 11.6. The four block $\times$ replication totals are 288, 212, 88, and 220, with a grand total of 808, so

$$SSBl = \frac{(288)^2 + (212)^2 + (88)^2 + (220)^2}{4} - \frac{(808)^2}{16} = 5204.00$$

	Replication 1				Replication 2		
Block 1		Block 2		Block 1		Block 2	
(1)	99	a	18	(1)	46	a	18
ab	52	b	51	ab	-47	b	62
ac	42	c	108	ac	22	c	104
bc	95	abc	35	bc	67	abc	36

Figure 11.6

The other sums of squares are computed by Yates' method using the eight experimental condition totals, resulting in the following ANOVA table:

Table 11.15

Source of variation	d.f.	Sum of squares	Mean square	f
A	1	12,996	12,996	39.82
B	1	702.25	702.25	2.15
C	1	2,756.25	2,756.25	8.45
AB	1	210.25	210.25	.64
AC	1	30.25	30.25	.093
BC	1	25	25	.077
Blocks	3	5,204	1,734.67	5.32
Error	6	1,958	326.33	
Total	15	23,857		

By comparison with $F_{.05, 1, 6} = 5.99$, we conclude that only the main effects for A and C differ significantly from zero. ∎

Confounding Using More Than Two Blocks

In the case $r = 2$ (four blocks), three effects are confounded with blocks. The experimenter first chooses two defining effects to be confounded. For example, in a five-factor experiment (A, B, C, D, and E), the two three-factor interactions BCD and CDE might be chosen for confounding. The third effect confounded is then the **generalized interaction** of the two, obtained by writing the two chosen effects side by side and then cancelling any letters common to both: $(BCD)(CDE) = BE$. Notice that if ABC and CDE are chosen for confounding, their generalized interaction is $(ABC)(CDE) = ABDE$, so that no main-effects or two-factor interactions are confounded.

Once the two defining effects have been selected for confounding, one block consists of all treatment conditions having an even number of letters in common with both defining effects, the second block consists of all conditions having an even number of letters in common with the first defining contrast and an odd number of letters in common with the second contrast, while the third and fourth blocks consist of the "odd/even" and "odd/odd" contrasts. In a five-factor experiment with defining effects ABC and CDE, this results in the following allocation to blocks (with the number of letters in common with each defining contrast appearing beside each experimental condition):

Block 1		Block 2		Block 3		Block 4	
(1)	(0, 0)	d	(0, 1)	a	(1, 0)	c	(1, 1)
ab	(2, 0)	e	(0, 1)	b	(1, 0)	ad	(1, 1)
de	(0, 2)	ac	(2, 1)	cd	(1, 2)	ae	(1, 1)
acd	(2, 2)	bc	(2, 1)	ce	(1, 2)	bd	(1, 1)
ace	(2, 2)	abd	(2, 1)	ade	(1, 2)	be	(1, 1)
bcd	(2, 2)	abe	(2, 1)	bde	(1, 2)	abc	(3, 1)
bce	(2, 2)	$acde$	(2, 3)	$abcd$	(3, 2)	cde	(1, 3)
$abde$	(2, 2)	$bcde$	(2, 3)	$abce$	(3, 2)	$abcde$	(3, 3)

Figure 11.7 Four blocks in a 2^5 factorial experiment with defining effects ABC and CDE

The block containing (1) is called the **principal block.** Once it has been constructed, a second block can be obtained by selecting any experimental condition not in the principal block and obtaining its generalized interaction with every condition in the principal block. The other blocks are then constructed in the same way by first selecting a condition not in a block already constructed and finding generalized interactions with the principal block.

For experimental situations with $p > 3$, there is often no replication, so sums of squares associated with nonconfounded higher-order interactions are

usually pooled to obtain an error sum of squares that can be used in the denominators of the various F statistics. All computations can again be carried out
using Yates' technique, with $SSBl$ being the sum of SS's associated with confounded effects.

When $r > 2$, one first selects r defining effects to be confounded with
blocks, making sure that no one of the effects chosen is the generalized interaction of any other two selected. The additional $2^r - r - 1$ effects confounded
with the blocks are then the generalized interactions of all effects in the defining set (including not only generalized interactions of pairs of effects, but also
of sets of three, four, and so on). For more details the book by Johnson and
Leone can be consulted.

Fractional Replication

When the number of factors p is large, even a single replicate of a 2^p experiment
can be expensive and time consuming. For example, one replicate of a 2^6 factorial experiment involves an observation for each of the 64 different experimental conditions. An appealing strategy in such situations is to make observations
for only a fraction of the 2^p conditions. Provided that care is exercised in the
choice of conditions to be observed, much information about factor effects can
still be obtained.

Suppose that we decide to include only 2^{p-1} (half) of the 2^p possible conditions in our experiment; this is usually called a **half-replicate.** The price paid for
this economy is twofold. First, information about a single effect (determined by
the 2^{p-1} conditions selected for observation) is completely lost to the experimenter in the sense that no reasonable estimate of the effect is possible. Second,
the remaining $2^p - 2$ main effects and interactions are paired up so that any
one effect in a particular pair is confounded with the other effect in the same
pair. For example, one such pair may be $\{A, BCD\}$, so that separate estimates of
the A main effect and BCD interaction are not possible. It is desirable, then, to
select a half-replicate for which main effects and low-order interactions are
paired off (confounded) only with higher-order interactions rather than with
one another.

The first step in selecting a half-replicate is to select a defining effect as
the nonestimable effect. Suppose that in a five-factor experiment, $ABCDE$ is
chosen as the defining effect. Now the $2^5 = 32$ possible treatment conditions
are divided into two groups with 16 conditions each, one group consisting of all
conditions having an odd number of letters in common with $ABCDE$ and the
other containing an even number of letters in common with the defining contrast. Then either group of 16 conditions is used as the half-replicate. The "odd"
group is

a, b, c, d, e, abc, abd, abe, acd, ace, ade, bcd, bce, bde, cde, abcde

Each main effect and interaction other than $ABCDE$ is then confounded with (**aliased** with) its generalized interaction with $ABCDE$. Thus

$(AB)(ABCDE) = CDE$, so the AB interaction and CDE interaction are confounded with each other. The resulting **alias pairs** are

$$\{A, BCDE\}, \quad \{B, ACDE\}, \quad \{C, ABDE\}, \quad \{D, ABCE\}, \quad \{E, ABCD\}, \quad \{AB, CDE\},$$
$$\{AC, BDE\}, \quad \{AD, BCE\}, \quad \{AE, BCD\}, \quad \{BC, ADE\}, \quad \{BD, ACE\}, \quad \{BE, ACD\},$$
$$\{CD, ABE\}, \quad \{CE, ABD\}, \quad \{DE, ABC\}$$

Note in particular that every main effect is aliased with a four-factor interaction. Assuming these interactions to be negligible allows for testing for the presence of main effects.

To select a quarter-replicate of a 2^p factorial experiment (2^{p-2} of the 2^p possible treatment conditions), two defining effects must be selected. These two and their generalized interaction become the nonestimable effects. Instead of alias pairs as in the half-replicate, each remaining effect is now confounded with three other effects, each being its generalized interaction with one of the three nonestimable effects.

Example 11.16   The article "More on Planning Experiments to Increase Research Efficiency" (*Industrial and Eng. Chemistry*, 1970, pp. 60–65) reported on the results of a quarter-replicate of a 2^5 experiment in which the five factors were A = condensation temperature, B = amount of material B, C = solvent volume, D = condensation time, and E = amount of material E. The response variable was the yield of the chemical process. The chosen defining contrasts were ACE and BDE, with generalized interaction $(ACE)(BDE) = ABCD$. The remaining 28 main effects and interactions can now be partitioned into seven groups of four effects each such that the effects within a group cannot be assessed separately. For example, the generalized interactions of A with the nonestimable effects are $(A)(ACE) = CE$, $(A)(BDE) = ABDE$, and $(A)(ABCD) = BCD$, so one alias group is $\{A, CE, ABDE, BCD\}$. The complete set of alias groups is

$$\{A, CE, ABDE, BCD\}, \quad \{B, ABCE, DE, ACD\}, \quad \{C, AE, BCDE, ABD\},$$
$$\{D, ACDE, BE, ABC\}, \quad \{E, AC, BD, ABCDE\}, \quad \{AB, BCE, ADE, CD\},$$
$$\{AD, CDE, ABE, BC\}$$

■

Analysis of a Fractional Replicate

Once the defining contrasts have been chosen for a quarter-replicate, they are used as in the discussion of confounding to divide the 2^p treatment conditions into four groups of 2^{p-2} conditions each. Then any one of the four groups is selected as the set of conditions for which data will be collected. Similar comments apply to a $1/2^r$ replicate of a 2^p factorial experiment.

Having made observations for the selected treatment combinations, a table of signs similar to Table 11.2 is constructed. The table contains a row only for each of the treatment combinations actually observed rather than the full 2^p rows, and there is a single column for each alias group (since each effect in the group would have the same set of signs for the treatment conditions selected for observation). The signs in each column indicate as usual how contrasts for the various sums of squares are computed. Yates' method can also be used, but the rule for arranging observed conditions in standard order must be modified.

The difficult part of a fractional replication analysis typically involves deciding what to use for error sum of squares. Since there will usually be no replication (though one could observe, for example, two replicates of a quarter-replicate), some effect SS's must be pooled to obtain an error sum of squares. In a half-replicate of a 2^8 experiment, for example, an alias structure can be chosen so that the eight main effects and 28 two-factor interactions are each confounded only with higher-order interactions and that there are an additional 27 alias groups involving only higher-order interactions. Assuming the absence of higher-order interaction effects, the resulting 27 SS's can then be added to yield an error sum of squares, allowing one-degree-of-freedom tests for all main effects and two-factor interactions. However, in many cases tests for main effects can be obtained only by pooling some or all of the SS's associated with alias groups involving two factor interactions, while the corresponding two-factor interactions cannot be investigated.

Example 11.17
(Example 11.16 continued)

The set of treatment conditions chosen and resulting yields for the quarter-replicate of the 2^5 experiment were

e	ab	ad	bc	cd	ace	bde	$abcde$
23.2	15.5	16.9	16.2	23.8	23.4	16.8	18.1

The abbreviated table of signs is displayed in Figure 11.8.

	A	B	C	D	E	AB	AD
e	−	−	−	−	+	+	+
ab	+	+	−	−	−	+	−
ad	+	−	−	+	−	−	+
bc	−	+	+	−	−	−	+
cd	−	−	+	+	−	+	−
ace	+	−	+	−	+	−	−
bde	−	+	−	+	+	−	−
$abcde$	+	+	+	+	+	+	+

Figure 11.8

With SSA denoting the sum of squares for effects in the alias group $\{A, CE, ABDE, BCD\}$,

$$SSA = \frac{(-23.2 + 15.5 + 16.9 - 16.2 - 23.8 + 23.4 - 16.8 + 18.1)^2}{8}$$

$$= 4.65$$

Similarly, $SSB = 53.56$, $SSC = 10.35$, $SSD = .91$, $SSE' = 10.35$ (the ' differentiates this quantity from error sum of squares SSE), $SSAB = 6.66$, and $SSAD = 3.25$, giving $SST = 4.65 + 53.56 + \cdots + 3.25 = 89.73$. To test for main effects, we use $SSE = SSAB + SSAD = 9.91$ with two degrees of freedom.

Table 11.16

Source	d.f.	Sum of squares	Mean square	f
A	1	4.65	4.65	.94
B	1	53.56	53.56	10.80
C	1	10.35	10.35	2.09
D	1	.91	.91	.18
E	1	10.35	10.35	2.09
Error	2	9.91	4.96	
Total	7	89.73		

Since $F_{.05, 1, 2} = 18.51$, none of the five main effects can be judged significant. Of course, with only two degrees of freedom for error, the test is not very powerful (that is, it is quite likely to fail to detect the presence of effects). The article from *Industrial and Engineering Chemistry* from which the data came actually had an independent estimate of the standard error of the treatment effects based on prior experience, so used a somewhat different analysis. Our analysis was done here only for illustrative purposes, since one would ordinarily want many more than two degrees of freedom for error. ■

The subjects of factorial experimentation, confounding, and fractional replication encompass many models and techniques we have not discussed. For more information, the chapter references should be consulted.

Exercises / Section 11.4 [31–41]

31. The accompanying data resulted from a 2^3 experiment with three replications per combination of treatments designed to study the effects of concentration of detergent (*A*), concentration of sodium carbonate (*B*), and concentration of sodium carboxy-methyl cellulose (*C*) on cleaning ability of a solution in washing tests (a large number indicates better cleaning ability than a small number).

Factor levels				
A	B	C	Condition	Observations
1	1	1	(1)	106, 93, 116
2	1	1	a	198, 200, 214
1	2	1	b	197, 202, 185
2	2	1	ab	329, 331, 307
1	1	2	c	149, 169, 135
2	1	2	ac	243, 247, 220
1	2	2	bc	255, 230, 252
2	2	2	abc	383, 360, 364

a. After obtaining cell totals $x_{ijk.}$, compute estimates of β_1, γ_{11}^{AC}, and γ_{21}^{AC}.

b. Use the cell totals along with Yates' method to compute the effect contrasts and sums of squares. Then construct an ANOVA table and test all appropriate hypotheses using $\alpha = .05$.

32. In a study of processes used to remove impurities from cellulose goods ("Optimization of Rope-Range Bleaching of Cellulosic Fabrics," *Textile Research J.*, 1976, pp. 493–496), the following data resulted from a 2^4 experiment involving the desizing process. The four factors were enzyme concentration (*A*), pH (*B*), temperature (*C*), and time (*D*).

a. Use Yates' algorithm to obtain sums of squares and the ANOVA table.

b. Do there appear to be any second-, third-, or fourth-order interaction effects present? Explain your reasoning. Which main effects appear to be significant?

Treat- ment	Enzyme, g/l	pH	Temp, °C	Time, h	Starch % by weight 1st repl.	2nd repl.
(1)	.50	6.0	60.0	6	9.72	13.50
a	.75	6.0	60.0	6	9.80	14.04
b	.50	7.0	60.0	6	10.13	11.27
ab	.75	7.0	60.0	6	11.80	11.30
c	.50	6.0	70.0	6	12.70	11.37
ac	.75	6.0	70.0	6	11.96	12.05
bc	.50	7.0	70.0	6	11.38	9.92
abc	.75	7.0	70.0	6	11.80	11.10
d	.50	6.0	60.0	8	13.15	13.00
ad	.75	6.0	60.0	8	10.60	12.37
bd	.50	7.0	60.0	8	10.37	12.00
abd	.75	7.0	60.0	8	11.30	11.64
cd	.50	6.0	70.0	8	13.05	14.55
acd	.75	6.0	70.0	8	11.15	15.00
bcd	.50	7.0	70.0	8	12.70	14.10
abcd	.75	7.0	70.0	8	13.20	16.12

33. In Exercise 31, suppose that a low water temperature has been used to obtain the data. The entire experiment is then repeated with a higher water temperature to obtain the following data. Use Yates' algorithm on the entire set of 48 observations to obtain the sums of squares and ANOVA table, and then test appropriate hypotheses at level .05.

Condition	Observations
d	144, 154, 158
ad	239, 227, 244
bd	232, 242, 246
abd	364, 362, 346
cd	194, 162, 203
acd	284, 295, 291
bcd	291, 287, 297
abcd	411, 406, 395

34. The following data on power consumption in electric furnace heats (kilowatts consumed per ton of melted product) resulted from a 2^4 factorial experiment with three replicates ("Studies on a 10-cwt Arc Furnace," *J. Iron and Steel Institute,* 1956, p. 22). The factors were nature of roof *A* (low, high), power setting *B* (low, high), scrap used *C* (tube, plate), and charge *D* (700 lb, 1000 lb).

Treatment	x_{ijklm}	Treatment	x_{ijklm}
(1)	866, 862, 800	d	988, 808, 650
a	946, 800, 840	ad	966, 976, 876
b	774, 834, 746	bd	702, 658, 650
ab	709, 789, 646	abd	784, 700, 596
c	1017, 990, 954	cd	922, 808, 868
ac	1028, 906, 977	acd	1056, 870, 908
bc	817, 783, 771	bcd	798, 726, 700
abc	829, 806, 691	abcd	752, 714, 714

Construct the ANOVA table and test all hypotheses of interest using $\alpha = .01$.

35. The article "Statistical Design and Analysis of Qualification Test Program for a Small Rocket Engine" (*Industrial Quality Control,* 1964, pp. 14–18) presented data from an experiment to assess the effects of vibration (*A*), temperature cycling (*B*), altitude cycling (*C*), and temperature for altitude cycling and firing (*D*) on thrust duration. A subset of the data appears below (in the paper there were four levels of *D* rather than just two). Use the Yates method to compute sums of squares and the ANOVA table. Then assume that three- and four-factor interactions are absent, pool the corresponding sums of squares to obtain an estimate of σ^2, and test all appropriate hypotheses at level .05.

		D_1		D_2	
		C_1	C_2	C_1	C_2
A_1	B_1	21.60	21.60	11.54	11.50
	B_2	21.09	22.17	11.14	11.32
A_2	B_1	21.60	21.86	11.75	9.82
	B_2	19.57	21.85	11.69	11.18

36. a. In a 2^4 experiment, suppose that two blocks are to be used, and it is decided to confound the *ABCD* interaction with the block effect. Which treatments should be carried out in the first block [the one containing the treatment (1)], and which treatments are allocated to the second block?

b. In an experiment to investigate niacin retention in vegetables as a function of cooking temperature (*A*), sieve size (*B*), type of processing (*C*), and cooking time (*D*), each factor was held at two levels. Two blocks were used, with the allocation of blocks as given in (a) in order to confound only the *ABCD* interaction with blocks. Use Yates' procedure to obtain the ANOVA table for the accompanying data.

Treatment	x_{ijkl}	Treatment	x_{ijkl}
(1)	91	d	72
a	85	ad	78
b	92	bd	68
ab	94	abd	79
c	86	cd	69
ac	83	acd	75
bc	85	bcd	72
abc	90	abcd	71

  c. Assume that all three-way interaction effects are absent, so that the associated sums of squares can be combined to yield an estimate of σ^2, and carry out all appropriate tests at level .05.

37. a. An experiment was carried out to investigate the effects on audio sensitivity of varying resistance (A), two capacitances (B, C), and inductance of a coil (D) in part of a television circuit. If four blocks were used with four treatments per block, and the defining effects for confounding were AB and CD, which treatments appeared in each block?

  b. Suppose that two replications of the experiment described in (a) were performed, resulting in the accompanying data. Obtain the ANOVA table and test all relevant hypotheses at level .01.

Treat-ment	x_{ijkl1}	x_{ijkl2}	Treat-ment	x_{ijkl1}	x_{ijkl2}
(1)	618	598	d	598	585
a	583	560	ad	587	541
b	477	525	bd	480	508
ab	421	462	abd	462	449
c	601	595	cd	603	577
ac	550	589	acd	571	552
bc	505	484	bcd	502	508
abc	452	451	abcd	449	455

38. In an experiment involving four factors A, B, C, and D and four blocks, show that at least one main-effect or two-factor interaction effect must be confounded with the block effect.

39. a. In a seven-factor experiment ($A, \ldots, G$), suppose that a quarter-replicate is actually car-

ried out. If the defining effects are $ABCDE$ and $CDEFG$, what is the third nonestimable effect and what treatments are in the group containing (1)? What are the alias groups of the seven main effects?

  b. If the quarter-replicate is to be carried out using four blocks (with eight treatments per block), what are the blocks if the chosen confounding effects are ACF and BDG?

40. Suppose that in the rocket thrust problem of Exercise 35, enough resources had been available for only a half-replicate of the 2^4 experiment.

  a. If the effect $ABCD$ is chosen as the defining effect for the replicate and the group of eight treatments for which data is obtained includes treatment (1), what other treatments are in the observed group and what are the alias pairs?

  b. Suppose that the results of carrying out the experiment as described in (a) are as recorded below (given in standard order after deleting the half not observed). Assuming that two- and three-factor interactions are negligible, test at level .05 for the presence of main effects.

19.09, 20.11, 21.66, 20.44, 13.72, 11.26, 11.72, 12.29

41. A half-replicate of a 2^5 experiment to investigate the effects of heating time (A), quenching time (B), drawing time (C), position of heating coils (D), and measurement position (E) on hardness of steel castings resulted in the accompanying data. Construct the ANOVA table and (assuming second- and higher-order interactions to be negligible) test at level .01 for the presence of main effects.

Treat-ment	Observation	Treat-ment	Observation
a	70.4	acd	66.6
b	72.1	ace	67.5
c	70.4	ade	64.0
d	67.4	bcd	66.8
e	68.0	bce	70.3
abc	73.8	bde	67.9
abd	67.0	cde	65.9
abe	67.8	abcde	68.0

Supplementary Exercises / Chapter 11 (42–48)

42. The results of a study on the effectiveness of line drying on the smoothness of fabric was summarized in the paper "Line-Dried vs. Machine-Dried Fabrics: Comparison of Appearance, Hand, and Consumer Acceptance" (*Home Econ. Research J.*, 1984, pp. 27–35). Smoothness scores were given for nine different types of fabric and five different drying methods: (1) machine dry, (2) line dry, (3) line dry followed by 15-min tumble, (4) line dry with softener, and (5) line dry with air movement. Regarding the different types of fabric as blocks, construct an ANOVA table. Using a .05 significance level, test to see if there is a difference in the true mean smoothness score for the drying methods.

		Drying method				
		1	2	3	4	5
	Crepe	3.3	2.5	2.8	2.5	1.9
	Double knit	3.6	2.0	3.6	2.4	2.3
	Twill	4.2	3.4	3.8	3.1	3.1
	Twill mix	3.4	2.4	2.9	1.6	1.7
Fabric	Terry	3.8	1.3	2.8	2.0	1.6
	Broadcloth	2.2	1.5	2.7	1.5	1.9
	Sheeting	3.5	2.1	2.8	2.1	2.2
	Corduroy	3.6	1.3	2.8	1.7	1.8
	Denim	2.6	1.4	2.4	1.3	1.6

43. The water absorption of two types of mortar used to repair damaged cement was discussed in the paper "Polymer Mortar Composite Matrices for Maintenance-Free, Highly Durable Ferrocement" (*J. Ferrocement,* 1984, pp. 337–345). Specimens of ordinary cement mortar (OCM) and polymer cement mortar (PCM) were submerged for varying lengths of time (5, 9, 24, or 48 h) and water absorption (% by weight) was recorded. With mortar type as factor A (with two levels) and submersion period as factor B (with four levels), three observations were made for each factor-level combination. Data included in the paper was used to compute the sums of squares, which were $SSA = 322.667$, $SSB = 35.623$, $SSAB = 8.557$, and $SST = 372.113$. Use this information to construct an ANOVA table, and then use a .05 significance level to test the appropriate hypotheses.

44. Four plots were available for an experiment to compare clover accumulation for four different sowing rates ("Performance of Overdrilled Red Clover with Different Sowing Rates and Initial Grazing Managements," *N. Zeal. J. Exp. Ag.,* 1984, pp. 71–81). Since the four plots had been grazed differently prior to the experiment and it was thought that this might affect clover accumulation, a randomized block experiment was used with all four sowing rates tried on a section of each plot. Use the given data to test the null hypothesis of no difference in true mean clover accumulation (kg DM/ha) for the different sowing rates.

		Sowing rate (kg/ha)			
		3.6	6.6	10.2	13.5
	1	1155	2255	3505	4632
Plot	2	123	406	564	416
	3	68	416	662	379
	4	62	75	362	564

45. A chemical engineer has carried out an experiment to study the effects of the fixed factors vat pressure (A), cooking time of pulp (B), and hardwood concentration (C) on the strength of paper. The experiment involved two pressures, four cooking times, three concentrations, and two observations at each combination of these levels. Calculated sums of squares are $SSA = 6.94$, $SSB = 5.61$, $SSC = 12.33$, $SSAB = 4.05$, $SSAC = 7.32$, $SSBC = 15.80$, $SSE = 14.40$, and $SST = 70.82$. Construct the ANOVA table and carry out appropriate tests at significance level .05.

46. The paper "Food Consumption and Energy Requirements of Captive Bald Eagles" (*J. Wildlife Mgmt.,* 1982, pp. 646–654) investigated mean gross daily energy intake (the response variable) for different diet types (factor A, with three levels) and temperature (factor B, with three levels). Summary quantities given in the paper were used to generate data, resulting in $SSA = 18,138$, $SSB = 5182$, $SSAB = 1737$, $SST = 36,348$, and error d.f. = 36. Construct an ANOVA table and test the relevant hypotheses.

47. Analogous to a Latin square, a Graeco-Latin square design can be used when it is suspected that three extraneous factors may affect the response variable and all four factors (the three ex-

traneous ones and the one of interest) have the same number of levels. In a Latin square, each level of the factor of interest (C) appears once in each row (with each level of A) and once in each column (with each level of B). In a Graeco-Latin square, each level of factor D appears once in each row, in each column, and also with each level of the third extraneous factor C. Alternatively, the design can be used when the four factors are all of equal interest, the number of levels of each is N, and resources are available for only N^2 observations. A 5×5 square is pictured in (a), with (k, l) in each cell denoting the kth level of C and lth level of D. In (b) we present data on weight loss in silicon bars used for semiconductor material as a function of volume of etch (A), color of nitric acid in the etch solution (B), size of bars (C), and time in the etch solution (D) (from "Applications of Analytic Techniques to the Semiconductor Industry," Fourteenth Midwest Quality Control Conference, 1959).

Let $X_{ij(kl)}$ denote the observed weight loss when factor A is at level i, B is at level j, C is at level k, and D is at level l. Assuming no interaction between factors, total sum of squares SST (with $N^2 - 1$ d.f.) can be partitioned into SSA, SSB, SSC, SSD, and SSE. Give expressions for these sums of squares, including computing formulas, obtain the ANOVA table for the above data, and test each of the four main effect hypotheses using $\alpha = .05$.

48. In Exercise 3, after calculating $\hat{\mu}$, $\hat{\alpha}_i$, and $\hat{\beta}_j$, compute the residuals $x_{ij} - (\hat{\mu} + \hat{\alpha}_i + \hat{\beta}_j)$. Then construct a normal probability plot and comment.

			B		
(C, D)	1	2	3	4	5
1	(1, 1)	(2, 3)	(3, 5)	(4, 2)	(5, 4)
2	(2, 2)	(3, 4)	(4, 1)	(5, 3)	(1, 5)
A 3	(3, 3)	(4, 5)	(5, 2)	(1, 4)	(2, 1)
4	(4, 4)	(5, 1)	(1, 3)	(2, 5)	(3, 2)
5	(5, 5)	(1, 2)	(2, 4)	(3, 1)	(4, 3)

(a)

65	82	108	101	126
84	109	73	97	83
105	129	89	89	52
119	72	76	117	84
97	59	94	78	106

(b)

Bibliography

Box, George, Hunter, William, and Hunter, Stuart, *Statistics for Experimenters,* Wiley, New York, 1978. Contains a wealth of suggestions and insights on data analysis based on the authors' extensive consulting experience.

Hocking, Ronald, *The Analysis of Linear Models,* Brooks/Cole, Monterey, Ca., 1985. A very general treatment of analysis of variance written by one of the foremost authorities in this field.

Johnson, Norman, and Leone, Frederick, *Statistics and Experimental Design in Engineering and the Physical Sciences,* vol. II (2nd ed.), Wiley, New York, 1977. Somewhat difficult to read because of awkward notation, but contains much information on various experimental designs.

Kleinbaum, David, and Kupper, Lawrence, *Applied Regression Analysis and Other Multivariable Methods,* Duxbury Press, Boston, 1978. Contains an especially good discussion of problems associated with analysis of "unbalanced data"—that is, unequal K_{ij}'s.

Montgomery, Douglas, *Design and Analysis of Experiments* (2nd ed.), Wiley, New York, 1984. See the Chapter 10 bibliography.

Neter, John, Wasserman, William, and Kutner, Michael, *Applied Linear Statistical Models,* Irwin, Homewood, Ill., 1985. See the Chapter 10 bibliography.

Steel, Robert, and Torrie, James, *Principles and Procedures of Statistics* (2nd ed.), McGraw-Hill, New York, 1979. A good reference book for those interested only in methods.

Simple Linear Regression and Correlation

In the two-sample problems discussed in Chapter 9, we were interested in comparing values of parameters for the x distribution and the y distribution. Even when observations were paired, we did not try to use information about one of the variables in studying the other variable. This is precisely the objective of regression analysis: to exploit the relationship between two (or more) variables so that we can gain information about one of them through knowing values of the other(s).

Much of mathematics is devoted to studying variables that are *deterministically* related. Saying that x and y are related in this manner means that once we are told the value of x, the value of y is completely specified. For example, suppose we decide to rent a car for a weekend and that the rental cost is $15.00 plus $.10 per mile driven. If we let x = the number of miles driven and y = the amount we will pay the rental agency, then $y = 15 + .1x$. If we drive the car 100 miles ($x = 100$), then $y = 15 + .1(100) = 25$. As another example, if the initial velocity of a particle is v_o and it undergoes constant acceleration a, then distance traveled $= y = v_o x + \frac{1}{2}ax^2$ where x = time.

There are many variables x and y that would appear to be related to one another, but not in a deterministic fashion. A familiar example to many students is given by variables x = high school grade point average and y = college grade point average. The value of y cannot be determined just from knowledge of x, and two different students could have the same x value but have very different y values. Yet there is a tendency for those students who have high (low) high school g.p.a.'s to also have high (low) college g.p.a.'s. Knowledge of a student's high school g.p.a. should be quite helpful in enabling us to predict how that person will do in college.

Other examples of variables related in a nondeterministic fashion include $x =$ age of a child and $y =$ size of that child's vocabulary, $x =$ size of an engine in cubic centimeters and $y =$ miles per gallon for an automobile equipped with that engine, and $x =$ applied tensile force and $y =$ amount of elongation in a metal strip. Many other examples will undoubtedly occur to the reader in connection with his or her own discipline.

Regression analysis is the part of statistics that deals with investigation of the relationship between two or more variables related in a nondeterministic fashion. In this chapter we generalize the linear relation $y = \beta_0 + \beta_1 x$ to a linear probabilistic relationship, develop procedures for making inferences about the parameters of the model, and obtain a quantitative measure (the correlation coefficient) of the extent to which the two variables are related. In the next chapter we consider techniques for validating a particular model and investigate nonlinear relationships and relationships involving more than two variables.

12.1   The Simple Linear Regression Model

The simplest deterministic mathematical relationship between two variables x and y is a linear relationship $y = \beta_0 + \beta_1 x$. The set of pairs (x, y) for which $y = \beta_0 + \beta_1 x$ determines a straight line with slope β_1 and y intercept β_0.* The objective of this section is to develop a linear probabilistic model.

If the two variables are not deterministically related, then for a fixed value of x the value of the second variable is random. For example, if we are investigating the relationship between age of child and size of vocabulary and decide to select a child of age $x = 5.0$ years, then before the selection is made, vocabulary size is a random variable Y. After a particular five-year-old child has been selected and tested, a vocabulary of 2000 words may result. We would then say that the observed value of Y associated with fixing $x = 5.0$ was $y = 2000$.

More generally the variable whose value is fixed by the experimenter will be denoted by x and will be called the **independent variable.** For fixed x the second variable will be random; we denote this random variable and its observed value by Y and y, respectively, and refer to it as the **dependent variable.**

Usually observations will be made for a number of settings of the independent variable. Let $x_1, x_2, \ldots, x_n$ denote values of the independent variable for which observations are made, and let Y_i and y_i respectively denote the random variable and observed value associated with x_i. The available data then consists of the n pairs $(x_1, y_1), (x_2, y_2), \ldots, (x_n, y_n)$.

Example 12.1   The paper "A Study of Stainless Steel Stress-Corrosion Cracking by Potential Measurements" (*Corrosion*, 1962, pp. 425–432) reported on the relationship

---

*The slope of a line is the change in y for a one-unit increase in x. For example, if $y = -3x + 10$, then y decreases by 3 when x increases by 1, so the slope is -3. The y intercept is the height at which the line crosses the vertical axis, and is obtained by setting $x = 0$ in the equation.

$$\text{slope} = \frac{y - y_1}{x - x_1}$$

between applied stress (the independent variable x, in kg/mm^2) and time to fracture (the dependent variable y, in hours) for 18-8 stainless steel under uniaxial tensile stress in a 40% CaCl$_2$ solution at 100°C. Ten different settings of applied stress were used, and the resulting data values (as read from a graph which appeared in the paper) were

i	1	2	3	4	5	6	7	8	9	10
x_i	2.5	5	10	15	17.5	20	25	30	35	40
y_i	63	58	55	61	62	37	38	45	46	19

■

In Example 12.1, the value of x_i increased as i increased, but this need not be the case. Furthermore, the values of the x_i's need not all be distinct. The investigators might have made several independent determinations of fracture time for a stress of 15 kg/mm^2.

A first step in regression analysis involving two variables is to construct a scatter plot of the observed data. In such a plot, each (x_i, y_i) is represented as a point plotted on a two-dimensional coordinate system. Figure 12.1 pictures a scatter plot of the data from Example 12.1. It is clear that no simple curve will pass through all the points in the plot. There is, however, a strong tendency for a large stress value to be associated with a small value of time-to-fracture, which suggests a relationship of some sort.

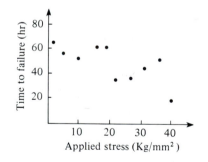

Figure 12.1   Scatter plot for data of Example 12.1

A Linear Probabilistic Model

For the deterministic model $y = \beta_0 + \beta_1 x$, the actual observed value of y is a linear function of x. The appropriate generalization of this to a probabilistic model assumes that *the expected value of Y is a linear function of x*, but that for fixed x, the variable Y differs from its expected value by a random amount.

The simple linear regres- sion model

> There exist parameters β_0, β_1, and σ^2 such that for any fixed value of the independent variable x, the dependent variable is related to x through the model equation
>
> $$Y = \beta_0 + \beta_1 x + \epsilon \qquad (12.1)$$
>
> The quantity ϵ in the model equation is a random variable, assumed to be normally distributed with $E(\epsilon) = 0$ and $V(\epsilon) = \sigma^2$.

The variable ϵ is usually referred to as the **random deviation** or **random error term** in the model. Without ϵ, any observed pair (x, y) would correspond to a point falling exactly on the line $y = \beta_0 + \beta_1 x$, called the **true regression line.** The inclusion of the random error term allows (x, y) to fall either above the true regression line (when $\epsilon > 0$) or below the line (when $\epsilon < 0$). The points $(x_1, y_1), \ldots, (x_n, y_n)$ resulting from n independent observations will then be scattered about the true regression line. On occasion, the appropriateness of the simple linear regression model may be suggested by theoretical considerations (for example, there is an exact linear relationship between the two variables, with ϵ representing measurement error). Much more frequently, though, the reasonableness of the model is indicated by a scatter plot exhibiting a substantial linear pattern (as in Figure 12.1). In Chapter 13, we will look at methods for checking the adequacy of any particular regression model.

An understanding of the model's properties is facilitated by the introduction of some notation. Let $\mu_{Y \cdot x}$ (or alternatively $E(Y|x)$) denote the expected value of Y for a fixed value of x. For example, in the context of Example 12.1, $\mu_{Y \cdot 20}$ would denote the expected value of time-to-fracture when applied stress is 20 kg/mm^2. Similarly, let $\sigma^2_{Y \cdot x}$ denote the variance of Y for a fixed x value. If we think of an entire population of (x, y) pairs, then $\mu_{Y \cdot x}$ is the mean of all y values for which the independent variable has value x, and $\sigma^2_{Y \cdot x}$ is a measure of how much these values of y spread out about the mean value. If, for example, $x =$ age of a child and $y =$ vocabulary size, then $\mu_{Y \cdot 5}$ is the average vocabulary size for all five-year-old children in the population and $\sigma^2_{Y \cdot 5}$ describes the amount of variability in vocabulary size for this part of the population. Once x is fixed, the only randomness on the right-hand side of the model equation (12.1) is in the random error ϵ, and its mean value and variance are 0 and σ^2, respectively, whatever the value of x. This implies that

$$\mu_{Y \cdot x} = E(\beta_0 + \beta_1 x + \epsilon) = \beta_0 + \beta_1 x + E(\epsilon) = \beta_0 + \beta_1 x, \text{ and}$$

$$\sigma^2_{Y \cdot x} = V(\beta_0 + \beta_1 x + \epsilon) = V(\beta_0 + \beta_1 x) + V(\epsilon) = 0 + \sigma^2 = \sigma^2$$

The relation $\mu_{Y \cdot x} = \beta_0 + \beta_1 x$ says that the mean value of Y, rather than Y itself, is a linear function of x. The true regression line $y = \beta_0 + \beta_1 x$ is thus the *line of mean values;* its height above any particular x value is the expected value of Y for that value of x. The slope β_1 of the true regression line is interpreted as the *expected* change in Y associated with a one-unit increase in the value of x. The second relation states that the amount of variability in the distribution of Y values is the same at each different value of x (homogeneity of variance). In the example involving age of a child and vocabulary size, the model implies that average vocabulary size changes linearly with age (hopefully β_1 is positive) and that the amount of variability in vocabulary size at any particular age is the same as at any other age. Finally, for fixed x, Y is the sum of a constant $\beta_0 + \beta_1 x$ and a normally distributed random variable ϵ, so itself has a normal distribution. These properties are illustrated in Figure 12.2. The variance parameter σ^2 determines the extent to which each normal curve spreads out about its mean value (the height of the line). When σ^2 is small, an observed point (x, y) will almost always fall quite close to the true regression line, whereas observations may deviate considerably from their expected values (corresponding to points far from the line) when σ^2 is large.

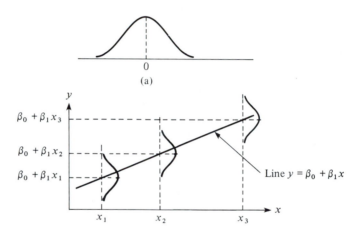

Figure 12.2 (*a*) Distribution of ϵ; (*b*) distribution of Y for different values of x

Example 12.2
(Example 12.1 continued)

Suppose that the relationship between applied stress x and time-to-failure y is described by the simple linear regression model with true regression line $y = 65 - 1.2x$ and $\sigma = 8$. Then for any fixed value of stress, time-to-failure has a normal distribution with mean value $65 - 1.2x$ and standard deviation 8. Roughly speaking, a typical (x, y) point will deviate (vertically) from the true regression line by the amount 8. For $x = 20$, Y has mean value $\mu_{Y \cdot 20} = 65 - 1.2(20) = 41$, so

$$P(Y > 50 \text{ when } x = 20) = P\left(Z > \frac{50 - 41}{8}\right) = 1 - \Phi(1.13) = .1292$$

The probability that time-to-failure exceeds 50 when applied stress is 25 is, since $\mu_{Y \cdot 25} = 35$,

$$P(Y > 50 \text{ when } x = 25) = P\left(Z > \frac{50 - 35}{8}\right) = 1 - \Phi(1.88) = .0301$$

These probabilities are illustrated as the shaded areas in Figure 12.3.

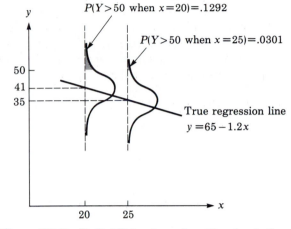

Figure 12.3   Probabilities based on the simple linear regression model

Suppose that Y_1 denotes an observation on time-to-failure made with $x = 25$, and Y_2 denotes an independent observation made with $x = 24$. Then $Y_1 - Y_2$ is normally distributed with mean value $E(Y_1 - Y_2) = \beta_1 = -1.2$, variance $V(Y_1 - Y_2) = \sigma^2 + \sigma^2 = 128$, and standard deviation $\sqrt{128} = 11.314$. The probability that Y_1 exceeds Y_2 is

$$P(Y_1 - Y_2 > 0) = P\left(Z > \frac{0 - (-1.2)}{11.314}\right) = P(Z > .11) = .4562$$

That is, even though we expect Y to decrease when x increases by one unit, it is not unlikely that the observed Y at $x + 1$ will be larger than the observed Y at x. ∎

Exercises / Section 12.1 (1–5)

1. The article "Some Field Experience in the Use of an Accelerated Method in Estimating 28-Day Strength of Concrete" (*J. Amer. Concrete Institute,* 1969, p. 895) considered regressing y = 28-day standard-cured strength (psi) against x = accelerated strength (psi). Suppose that the equation of the true regression line is $y = 1800 + 1.3x$.

a. What is the expected value of 28-day strength when accelerated strength = 2500?

b. By how much can we expect 28-day strength to change when accelerated strength increases by 1 psi?

c. Answer (b) for an increase of 100 psi.

d. Answer (b) for a decrease of 100 psi.

2. Referring back to Exercise 1, suppose that the standard deviation of the random deviation ϵ is 350 psi.

a. What is the probability that the observed value of 28-day strength will exceed 5000 psi when the value of accelerated strength is 2000?

b. Repeat (a) with 2500 in place of 2000.

c. Consider making two independent observations on 28-day strength, the first for an accelerated strength of 2000 and the second for $x = 2500$. What is the probability that the second observation will exceed the first by more than 1000 psi?

d. Let Y_1 and Y_2 denote observations on 28-day strength when $x = x_1$ and $x = x_2$, respectively. By how much would x_2 have to exceed x_1 in order that $P(Y_2 > Y_1) = .95$?

3. The flow rate y (m^3/min) in a device used for air quality measurement depends on the pressure drop x (in. of water) across the device's filter. Suppose that for x values between 5 and 20, the two variables are related according to the simple linear regression model with true regression line $y = -.12 + .095x$.

a. What is the expected change in flow rate associated with a 1-inch increase in pressure drop? Explain.

b. What change in flow rate can be expected when pressure drop decreases by 5 inches?

c. What is the expected flow rate for a pressure drop of 10 inches? A drop of 15?

d. Suppose that $\sigma = .025$, and consider a pressure drop of 10. What is the probability that the observed value of flow rate will exceed .835? That observed flow rate will exceed .840?

e. What is the probability that an observation on flow rate when pressure drop is 10 will exceed an observation on flow rate made when pressure drop is 11?

4. Suppose that the expected cost of a production run is related to the size of the run by the equation $y = 4000 + 10x$. Let Y denote an observation on the cost of a run. If the variables *size* and *cost* are related according to the simple linear regression model, could it be the case that $P(Y > 5500$ when $x = 1000) = .05$ and $P(Y > 6500$ when $x = 2000) = .10$? Explain.

5. Suppose that in a certain chemical process the reaction time y (hr) is related to the temperature ($^\circ$F) in the chamber in which the reaction takes place according to the simple linear regression model with equation $y = 5.00 - .01x$ and $\sigma = .075$.

a. What is the expected change in reaction time for a 1° F increase in temperature? For a 10° F increase in temperature?

b. What is the expected reaction time when temperature is 200° F? When temperature is 250° F?

c. Suppose that five observations are made independently on reaction time, each one for a temperature of 250° F. What is the probability that all five times are between 2.4 and 2.6 hours?

d. What is the probability that two independently observed reaction times for temperatures one degree apart are such that the time at the higher temperature exceeds the time at the lower temperature?

12.2   **Estimating Model Parameters**

We shall assume in this and the next several sections that the variables x and y are related according to the simple linear regression model. The values of β_0, β_1, and σ^2 will almost never be known to an investigator. Instead, sample data consisting of n observed pairs $(x_1, y_1), \ldots, (x_n, y_n)$ will be available, from which the model parameters and the true regression line itself can be estimated. These observations are assumed to have been obtained independently of one another. That is, y_i is the observed value of a random variable Y_i, where $Y_i =$

$\beta_0 + \beta_1 x_i + \epsilon_i$ and the n deviations $\epsilon_1, \epsilon_2, \ldots, \epsilon_n$ are independent random variables. Independence of $Y_1, Y_2, \ldots, Y_n$ follows from the independence of the ϵ_i's.

According to the model, the observed points will be distributed about the true regression line in a random manner. Figure 12.4 shows a typical plot of observed pairs along with two candidates for the estimated regression line, $y = a_0 + a_1 x$ and $y = b_0 + b_1 x$. Intuitively the line $y = a_0 + a_1 x$ is not a reasonable estimate of the true line $y = \beta_0 + \beta_1 x$, since if $y = a_0 + a_1 x$ were the true line, the observed points would almost surely have been closer to this line. The line $y = b_0 + b_1 x$ is a more plausible estimate, since the observed points are scattered rather closely about this line.

Figure 12.4 and the foregoing discussion suggest that our estimate of $y = \beta_0 + \beta_1 x$ should be a line that provides in some sense a best fit to the observed data points. This is what motivates the principle of least squares, which can be traced back to the German mathematician Gauss (1777–1855). According to this principle, a line provides a good fit to the data if the vertical distances (deviations) from the observed points to the line are small (see Figure 12.5). The measure of the goodness of fit is the sum of the squares of these deviations. The best-fit line is then the one having the smallest possible sum of squared deviations.

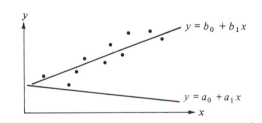

Figure 12.4 Two different estimates of the true regression line

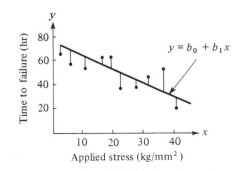

Figure 12.5 Deviations of Example 12.1 data from line $y = b_0 + b_1 x$

The sum of squared vertical deviations from the points $(x_1, y_1), \ldots,$ (x_n, y_n) to the line with equation $y = b_0 + b_1 x$ is

$$f(b_0, b_1) = \sum_{i=1}^{n} [y_i - (b_0 + b_1 x_i)]^2$$

The point estimates of β_0 and β_1, denoted by $\hat{\beta}_0$ and   and called the **least squares estimates,** are those values that minimize $f(b_0, b_1)$. That is, $\hat{\beta}_0$ and $\hat{\beta}_1$ are such that $f(\hat{\beta}_0, \hat{\beta}_1) \le f(b_0, b_1)$ for any b_0 and b_1. The **estimated regression line** or **least squares line** is then the line whose equation is $y = \hat{\beta}_0 + \hat{\beta}_1 x$.

The minimizing values of b_0 and b_1 are found by taking partial derivatives of $f(b_0, b_1)$ with respect to both b_0 and b_1, equating them both to zero (analogously to $f'(b) = 0$ in univariate calculus), and solving the equations:

$$\frac{\partial f(b_0, b_1)}{\partial b_0} = \Sigma 2(y_i - b_0 - b_1 x_i)(-1) = 0$$

$$\frac{\partial f(b_0, b_1)}{\partial b_1} = \Sigma 2(y_i - b_0 - b_1 x_i)(-x_i) = 0$$

Cancellation of the -2 factor and rearrangement gives the following system of equations, called the **normal equations:**

$$nb_0 + (\Sigma x_i) b_1 = \Sigma y_i$$
$$(\Sigma x_i) b_0 + (\Sigma x_i^2) b_1 = \Sigma x_i y_i$$

The normal equations are linear in the two unknowns b_0 and b_1. Provided that at least two of the x_i's are different, the least squares estimates are the unique solution to this system.

The least squares estimates of the coefficients β_0 and β_1 of the true regression line are

$$b_1 = \hat{\beta}_1 = \frac{\Sigma(x_i - \bar{x})(y_i - \bar{y})}{\Sigma(x_i - \bar{x})^2} = \frac{n\Sigma x_i y_i - (\Sigma x_i)(\Sigma y_i)}{n\Sigma x_i^2 - (\Sigma x_i)^2}$$

$$b_0 = \hat{\beta}_0 = \frac{\Sigma y_i - \hat{\beta}_1 \Sigma x_i}{n} = \bar{y} - \hat{\beta}_1 \bar{x} \qquad (12.2)$$

The expression for $\hat{\beta}_1$ on the right in (12.2) should be used for computation; it requires only the summary statistics Σx_i, Σy_i, Σx_i^2, $\Sigma x_i y_i$ (Σy_i^2 will be needed shortly), and minimizes the effects of roundoff. In computing $\hat{\beta}_0$, use extra digits in $\hat{\beta}_1$, since if Σx_i is large in magnitude rounding will affect the final answer. We emphasize that *before using (12.2), the scatter plot should be examined to see if a linear probabilistic model is plausible*. If the points do not tend to cluster about a straight line with roughly the same degree of spread for all x, other models should be investigated. In practice, plots and regression calculations are usually done by using a statistical computer package.

Example 12.3   Plasma etching is essential to the fine-line pattern transfer in current semiconductor processes. The paper "Ion Beam-Assisted Etching of Aluminum with Chlorine" (*J. Electrochem. Soc.*, 1985, pp. 2010–2012) gave the accompanying data (read from a graph) on chlorine flow (x, in SCCM) through a nozzle used in the etching mechanism and etch rate (y, in 100A/min.). The scatter plot (Figure 12.6) shows a substantial linear pattern, suggesting that the simple linear regression model is a reasonable way to relate x and y. The necessary summary quantities for the analysis are most conveniently obtained via a tabular format with columns for x, y, x^2, xy, and y^2.

x:	1.5	1.5	2.0	2.5	2.5	3.0	3.5	3.5	4.0
y:	23.0	24.5	25.0	30.0	33.5	40.0	40.5	47.0	49.0

Observation	x	y	x^2	xy	y^2
1	1.5	23.0	2.25	34.50	529.00
2	1.5	24.5	2.25	36.75	600.25
3	2.0	25.0	4.00	50.00	625.00
4	2.5	30.0	6.25	75.00	900.00
5	2.5	33.5	6.25	83.75	1122.25
6	3.0	40.0	9.00	120.00	1600.00
7	3.5	40.5	12.25	141.75	1640.25
8	3.5	47.0	12.25	164.50	2209.00
9	4.0	49.0	16.00	196.00	2401.00
	24.0	312.5	70.50	902.25	11626.75
	$\uparrow$	$\uparrow$	$\uparrow$	$\uparrow$	$\uparrow$
	Σx_i	Σy_i	Σx_i^2	$\Sigma x_i y_i$	Σy_i^2

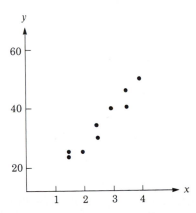

Figure 12.6   Scatter plot of the data from Example 12.3

The least squares estimates are

$$\hat{\beta}_1 = \frac{9(902.25) - (24.0)(312.5)}{9(70.50) - (24.0)^2} = \frac{620.25}{58.50} = 10.602564$$

$$\hat{\beta}_0 = \frac{312.5 - (10.602564)(24.0)}{9} = \frac{58.038464}{9} = 6.448718$$

The equation of the estimated regression line is $y = 6.448718 + 10.602564x \approx 6.45 + 10.60x$. ∎

The estimated regression line can immediately be used for two different purposes. For a fixed x value, $\hat{\beta}_0 + \hat{\beta}_1 x$ (the height of the line above the given x value) gives either (a) a point estimate of the expected value of Y at that x value, or (b) a prediction of the Y value that will result from a single new observation made at the specified x value.

Example 12.4
(Example 12.1
continued)

From the data on applied stress (x) and time-to-failure (y) given in the previous section, it is easily verified that

$$n = 10, \ \Sigma x_i = 200, \ \Sigma x_i^2 = 5412.5, \ \Sigma y_i = 484, \text{ and } \Sigma x_i y_i = 8407.5, \text{ so}$$

$$\hat{\beta}_1 = \frac{(10)(8407.5) - (200)(484)}{(10)(5412.5) - (200)^2} = \frac{-12725}{14125} = -.900885$$

and

$$\hat{\beta}_0 = \frac{484 - (-.900885)(200)}{10} = 66.417699$$

The equation of the estimated regression line is thus $y = 66.42 - .901x$. A point estimate for expected time-to-failure when applied stress is 22.5 results from substituting this x value into the estimated equation:

$$\hat{\mu}_{Y \cdot 22.5} = \hat{\beta}_0 + \hat{\beta}_1(22.5) = 66.42 - .901(22.5) = 46.1$$

If the investigator were to make one more observation on time-to-failure when applied stress is 30, a prediction of the resulting failure time is

$$\hat{y} = (\text{predicted value of } y \text{ when } x = 30) = 66.42 - .901(30) = 39.4$$

The data set contained an observation made for $x = 30$; the corresponding y value was 45, which is larger than $\hat{y}$ because the point $(30, 45)$ lies above the estimated regression line. ∎

Estimating σ^2

The parameter σ^2 is a measure of the amount of variability inherent in the regression model. A large value of σ^2 will lead to observed (x_i, y_i)'s that are quite spread out about the true regression line, whereas when σ^2 is small the observed points will tend to fall very close to the true line (see Figure 12.7). An estimate of σ^2 will be used in confidence interval formulas and hypothesis testing procedures presented in the next two sections. Because the equation of the true line is unknown, the estimate is based on the extent to which the sample observations deviate from the estimated line. Many large deviations (residuals) suggest a large value of σ^2, whereas deviations all of which are small in magnitude suggest that σ^2 is small.

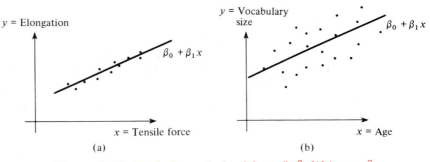

Figure 12.7   Typical sample for (a) small σ^2, (b) large σ^2

<table>
<tr><td>Definition</td><td>The ith **fitted** (or **predicted**) **value**, denoted by $\hat{y}_i$, is $\hat{y}_i = \hat{\beta}_0 + \hat{\beta}_1 x_i$ ($i = 1, \ldots, n$), and the ith **residual** is $y_i - \hat{y}_i$.</td></tr>
</table>

In words, the predicted value $\hat{y}_i$ is the value of y that we would predict or expect when using the estimated regression line with $x = x_i$; $\hat{y}_i$ is the height of the estimated regression line above the value x_i for which the ith observation was made. The residual $y_i - \hat{y}_i$ is the difference between the observed y_i and the predicted $\hat{y}_i$. If the residuals are all small in magnitude, then much of the variability in observed y values appears to be due to the linear relationship between x and y, while many large residuals suggest quite a bit of inherent variability in y relative to the amount due to the linear relation. Assuming that the line in Figure 12.5 is the least squares line, the residuals are identified by the vertical line segments from the observed points to the line.

Example 12.5   An investigation of the relationship between traffic flow x (1000's of cars per 24 hours) and lead content y of bark on trees near the highway ($\mu g/g$ dry wt) yielded the data in the x_i and y_i columns below:

i	x_i	y_i	$\hat{y}_i$	$y_i - \hat{y}_i$
1	8.3	227	287.48	−60.48
2	8.3	312	287.48	24.52
3	12.1	362	424.98	−62.98
4	12.1	521	424.98	96.02
5	17.0	640	602.28	37.72
6	17.0	539	602.28	−63.28
7	17.0	728	602.28	125.72
8	24.3	945	866.43	78.57
9	24.3	738	866.43	−128.43
10	24.3	759	866.43	−107.43
11	33.6	1263	1202.94	60.06

The summary statistics are $\Sigma x_i = 198.3$, $\Sigma x_i^2 = 4198.03$, $\Sigma y_i = 7034$, $\Sigma y_i^2 = 5,390,382$, and $\Sigma x_i y_i = 149,354.4$, so

$$\hat{\beta}_1 = \frac{11(149,354.4) - (198.3)(7034)}{11(4198.03) - (198.3)^2} = 36.18385, \hat{\beta}_0 = -12.84159$$

The estimated regression line is $y = -12.84 + 36.18x$. For numerical accuracy, the fitted values $\hat{y}_i$ are calculated from $\hat{y}_i = -12.84159 + 36.18385x_i$. The residuals should sum to zero (a consequence of the first normal equation), but the sum here is .01 because of rounding. A positive residual results from a point lying above the estimated regression line ($y_i > \hat{y}_i$) and a negative residual from a point lying below the line.   ■

In much the same way that the deviations from the mean in a one-sample situation were combined to obtain the estimate $s^2 = \Sigma(x_i - \bar{x})^2/(n - 1)$, the estimate of σ^2 in regression analysis is based on squaring and summing the residuals. We shall continue to use the symbol s^2 for this estimated variance, so please don't confuse it with our earlier s^2.

Definition

> The **error sum of squares**, denoted by SSE, is
>
> $$SSE = \Sigma(y_i - \hat{y}_i)^2 = \Sigma[y_i - (\hat{\beta}_0 + \hat{\beta}_1 x_i)]^2$$
>
> and the estimate of σ^2 is
>
> $$\hat{\sigma}^2 = s^2 = \frac{SSE}{n - 2} = \frac{\Sigma(y_i - \hat{y}_i)^2}{n - 2}$$

The divisor $n - 2$ in s^2 is the number of d.f. associated with the estimate (or equivalently with error sum of squares). This is because to obtain s^2, the two parameters β_0 and β_1 must first be estimated, which results in a loss of 2 d.f. (just as μ had to be estimated in one-sample problems, resulting in an estimated variance based on $n - 1$ d.f.). Replacing each y_i in the formula for s^2 by the random variable Y_i gives the estimator S^2. It can be shown that S^2 is an unbiased estimator for σ^2 (though the estimator S is not unbiased for σ).

Example 12.6
(Example 12.5 continued)

The residuals for the traffic flow/lead content data were calculated earlier. The corresponding error sum of squares is

$$SSE = (-60.48)^2 + (24.52)^2 + \cdots + (60.06)^2 = 76,493.98$$

The estimate of σ^2 is then $\hat{\sigma}^2 = s^2 = 76,493.98/(11 - 2) = 8499.33$ and the estimated standard deviation is $\hat{\sigma} = s = \sqrt{8499.33} = 92.19$. Roughly speaking, 92.19 is the magnitude of a typical deviation from the estimated regression line.   ■

Computation of *SSE* from the defining formula involves much tedious arithmetic, since both the predicted values and residuals must first be calculated. Use of the following computational formula does not require these quantities.

$$SSE = \Sigma y_i^2 - \hat{\beta}_0 \Sigma y_i - \hat{\beta}_1 \Sigma x_i y_i$$

This expression results from substituting $\hat{y}_i = \hat{\beta}_0 + \hat{\beta}_1 x_i$ into $\Sigma(y_i - \hat{y}_i)^2$, squaring the summand, carrying through the sum to the resulting three terms, and simplifying. This computational formula is especially sensitive to the effects of rounding in $\hat{\beta}_0$ and $\hat{\beta}_1$, so carrying as many digits as possible in intermediate computations will protect against roundoff error.

Example 12.7   The paper "Promising Quantitative Nondestructive Evaluation Techniques for Composite Materials" (*Materials Evaluation*, 1985, pp. 561–565) reported on a study to investigate how the propagation of an ultrasonic stress wave through a substance depends on the properties of the substance. The accompanying data on fracture strength (*x*, as a percentage of ultimate tensile strength) and attenuation (*y*, in neper/cm, the decrease in amplitude of the stress wave) in fiber-glass-reinforced polyester composites was read from a graph that appeared in the paper. The simple linear regression model is suggested by the substantial linear pattern in the scatter plot.

x:	12	30	36	40	45	57	62	67	71	78	93	94	100	105
y:	3.3	3.2	3.4	3.0	2.8	2.9	2.7	2.6	2.5	2.6	2.2	2.0	2.3	2.1

The necessary summary quantities are $n = 14$, $\Sigma x_i = 890$, $\Sigma x_i^2 = 67{,}182$, $\Sigma y_i = 37.6$, $\Sigma y_i^2 = 103.54$, and $\Sigma x_i y_i = 2234.30$, from which

$$\hat{\beta}_1 = \frac{14(2234.30) - (890)(37.6)}{14(67{,}182) - (890)^2} = \frac{-2183.80}{148{,}448} = -.0147109$$

and $\hat{\beta}_0 = 3.6209072$. The computational formula for *SSE* gives

$$SSE = 103.54 - (3.6209072)(37.6) - (-.0147109)(2234.30)$$

$$= .2624532$$

so $s^2 = .2624532/12 = .0218711$ and $s = .1479$. When $\hat{\beta}_0$ and $\hat{\beta}_1$ are rounded to three decimal places in the computational formula for *SSE*, the result is

$$SSE = 103.54 - (3.621)(37.6) - (-.015)(2234.30) = .905$$

which is more than three times the correct value.   ∎

The Coefficient of Determination
Figure 12.8 shows three different scatter plots of bivariate data. In all three plots the heights of the different points vary substantially, indicating that there

is much variability in observed y values. The points in the first plot all fall exactly on a straight line. In this case all (100%) of the sample variation in y can be attributed to the fact that x and y are linearly related in combination with variation in x. The points in Figure 12.8b don't fall exactly on a line, but compared to overall y variability, the deviations from the least squares line are small. It is reasonable to conclude in this case that much of the observed y variation can be attributed to the approximate linear relationship between the variables postulated by the simple linear regression model. When the scatter plot looks like that of Figure 12.8c, there is substantial variation about the least squares line relative to overall y variation, so the simple linear regression model fails to explain variation in y by relating y to x.

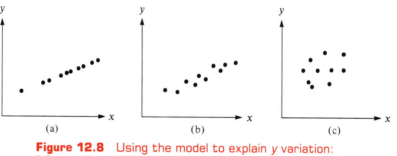

Figure 12.8 Using the model to explain y variation: (a) data for which all variation is explained; (b) data for which most variation is explained; (c) data for which little variation is explained

The error sum of squares SSE can be interpreted as a measure of how much variation in y is left unexplained by the model—that is, how much cannot be attributed to a linear relationship. In Figure 12.8a, $SSE = 0$ and there is no unexplained variation, whereas unexplained variation is small for the data of Figure 12.8b and much larger in Figure 12.8c. A quantitative measure of the total amount of variation in observed y values is given by the **total sum of squares** $SST = \Sigma(y_i - \bar{y})^2 = \Sigma y_i^2 - (\Sigma y_i)^2/n$. Total sum of squares is the sum of squared deviations about the sample mean of the observed y values. Thus the same number $\bar{y}$ is subtracted from each y_i in SST, whereas SSE involves subtracting each different predicted value $\hat{y}_i$ from the corresponding observed y_i. Just as SSE is the sum of squared deviations about the least squares line $y = \hat{\beta}_0 + \hat{\beta}_1 x$, SST is the sum of squared deviations about the horizontal line at height $\bar{y}$ (since then vertical deviations are $y_i - \bar{y}$) as pictured in Figure 12.9. Furthermore, because the sum of squared deviations about the least squares line is smaller than the sum of squared deviations about *any* other line, $SSE < SST$ unless the horizontal line itself is the least squares line. The ratio SSE/SST is the proportion of total variation that cannot be "explained by" the simple linear regression model, and $1 - SSE/SST$ (a number between 0 and 1) is the proportion of observed y variation explained by the model.

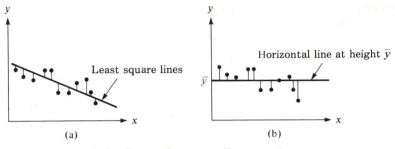

Figure 12.9 Sums of squares illustrated:
(a) SSE = sum of squared deviations about the least
squares line; (b) SST = sum of squared deviations
about the horizontal line

Definition

> The **coefficient of determination,** denoted by r^2, is given by
>
> $$r^2 = 1 - \frac{SSE}{SST}$$
>
> It is interpreted as the proportion of observed y variation that can be
> explained by the simple linear regression model (attributed to an approxi-
> mate linear relationship between y and x).

The higher the value of r^2, the more successful is the simple linear regression model in explaining y variation. When regression analysis is done by a
statistical computer package, either r^2 or $100r^2$ (the percentage of variation
explained by the model) is a prominent part of the output. If r^2 is small, an
analyst will usually want to search for an alternative model (either a nonlinear
model or a multiple regression model that involves more than a single independent variable) that can more effectively explain y variation.

Example 12.8
(Example 12.7
continued)

The value of SSE for the fracture strength/attenuation data was previously
calculated as .2625. The value of total sum of squares is

$$SST = \Sigma y_i^2 - (\Sigma y_i)^2/n = 103.54 - (37.6)^2/14 = 2.5571$$

The value of r^2 is thus $1 - .2625/2.5571 = 1 - .103 = .897$, so 89.7% of the
observed variation in attenuation can be attributed to the approximate linear
relationship between attenuation and fracture strength. This very high r^2 value
confirms the initial impression gleaned from the scatter plot that the simple
linear regression model will give a good fit to the data.　■

The coefficient of determination can be written in a slightly different way
by introducing a third sum of squares, **regression sum of squares** SSR, given
by $SSR = SST - SSE$. Regression sum of squares is interpreted as the

amount of total variation that *is* explained by the model. Then $r^2 = 1 - SSE/SST = (SST - SSE)/SST = SSR/SST$, the ratio of explained variation to total variation.

Example **12.9**
(Example 12.4
continued)

The sums of squares for the applied stress and time-to-failure data given earlier are

$$SST = \Sigma y_i^2 - (\Sigma y_i)^2/n = 25{,}238 - (484)^2/10 = 1812.40$$
$$SSE = \Sigma y_i^2 - \hat{\beta}_0 \Sigma y_i - \hat{\beta}_1 \Sigma x_i y_i$$
$$= 25{,}238 - (66.417699)(484) - (-.900885)(8407.5) = 666.0243,$$

and $SSR = 1812.40 - 666.02 = 1146.38$. Thus $r^2 = 1146.38/1812.40 = .633$. In many social science studies, a model that explains 63.3% of observed y variation would be considered highly satisfactory, but in the hard sciences and engineering such is usually not the case. Because the scatter plot reveals no obvious curved pattern, the investigator would probably want to relate time-to-failure to attributes other than applied stress via a multiple regression model.

■

Further Comments: Terminology and Scope of Regression Analysis

The phrase "regression analysis" was first used by Francis Galton in the late nineteenth century in connection with his work on the relationship between father's height x and son's height y. After collecting a number of pairs (x_i, y_i), Galton used the principle of least squares to obtain the equation of the estimated regression line with the objective of using it to predict son's height from father's height. In using the derived line, Galton found that if a father was above average in height, the son would also be expected to be above average in height, *but not by as much as the father was*. Similarly, the son of a shorter than average father would also be expected to be shorter than average, but not by as much as the father. Thus the predicted height of a son was "pulled back in" toward the mean; since "regression" means a coming or going back, Galton adopted the terminology "regression line." This phenomenon of being pulled back in toward the mean has been observed in many other situations (for example, batting averages from year to year in baseball), and is called **the regression effect.**

Our discussion thus far has presumed that the independent variable is under the control of the investigator, so that only the dependent variable Y is random. This was not, however, the case with Galton's experiment; fathers' heights were not preselected, but instead both X and Y were random. Methods and conclusions of regression analysis can be applied both when the values of the independent variable are fixed in advance and when they are random, but because the derivations and interpretations are more straightforward in the former case, we shall continue to work explicitly with it. For more commentary, see the excellent book by Neter et al. listed in the chapter bibliography.

Finally, when a scatter plot suggests the appropriateness of the simple linear regression model, the model should not be assumed valid much outside the range of x observations in the sample. In Example 12.4, use of the estimated regression line to predict failure time when stress is 75 yields $\hat{y} = -1.16$, a ridiculous prediction. This is an extreme example of the **danger of extrapolation,** using the estimated relationship for x values well outside the range of observation. Even when the prediction is not ridiculous, it may be very much in error because the form of the model changes considerably for x values much above the largest value in the sample or much below the smallest value.

Exercises / Section 12.2 (6–20)

6. A study reported in the article "The Effects of Water Vapor Concentration on the Rate of Combustion of an Artificial Graphite in Humid Air Flow" (*Combustion and Flame,* 1983, pp. 107–118) gave data on x = temperature of a nitrogen-oxygen mixture (1000° F) under specified conditions and y = oxygen diffusivity. Summary quantities are $n = 9$, $\Sigma x_i = 12.6$, $\Sigma y_i = 27.68$, $\Sigma x_i^2 = 18.24$, $\Sigma x_i y_i = 40.968$, and $\Sigma y_i^2 = 93.3448$. Assume that the two variables are related according to the simple linear regression model.

   a. Calculate the least squares estimates of the slope and y intercept of the population regression line.

   b. Use the equation of the estimated regression line to predict what value of oxygen diffusivity would be observed when $x = 1.5$.

   c. Give a point estimate of the true average diffusivity for a temperature of 1.75.

   d. The observed value of diffusivity for a temperature of 1.5 was 3.39. Calculate the value of the corresponding residual.

7. The March 29, 1975, issue of *Lancet* reported on the relationship between ages of a number of children who had high levels of lead absorption and a measure of wrist flexor and extensor muscle function for these children ("Neuropsychological Dysfunction in Children with Chronic Low-Level Lead Absorption"). The measure involved the number of taps with a stylus on a single metal plate during a 10-sec period. Representative data follows, with x = age in months and y = taps/10 sec.

   x: 73 84 98 112 116 132 150 160 164 180
   y: 35 40 50 42 46 41 52 52 51 66

   a. Write the system of normal equations for this data.

   b. Compute $\hat{\beta}_0$ and $\hat{\beta}_1$, and verify that they solve the normal equations of (a).

   c. What would be your estimate of expected change in the measure of muscle function for a one-month increase in age?

   d. Answer (c) for a one-year increase in age.

   e. Estimate the true average measure of muscle function for 10-year-old children in the population sampled.

8. The following summary statistics were obtained from a study that used regression analysis to investigate the relationship between pavement deflection and surface temperature of the pavement at various locations on a state highway. Here x = temperature (°F) and y = deflection adjustment factor.

   $n = 15$, $\Sigma x_i = 1425$, $\Sigma y_i = 10.68$,
   $\Sigma x_i^2 = 139{,}037.25$, $\Sigma x_i y_i = 987.645$,
   $\Sigma y_i^2 = 7.8518$

   (Many more than 15 observations were made in the study; the reference is "Flexible Pavement Evaluation and Rehabilitation," *Transportation Eng. J.,* 1977, pp. 75–85).

   a. Compute $\hat{\beta}_1$, $\hat{\beta}_0$, and the equation of the estimated regression line. Graph the estimated line.

   b. What is the estimate of expected change in the deflection adjustment factor when temperature is increased by 1 °F?

   c. Suppose that temperature was measured in °C rather than °F. What would be the estimated regression line? Answer (b) for an increase of 1 °C. *Hint:* °F $= \frac{9}{5}$°C $+ 32$; now substitute for the "old x" in terms of the "new x."

d. Supposing that a 200 °F surface temperature was within the realm of possibility, would you use the estimated line of (a) to predict deflection factor for this temperature? Why or why not?

9. The following data is representative of that reported in the article "An Experimental Correlation of Oxides of Nitrogen Emissions from Power Boilers Based on Field Data" (*J. Eng. for Power*, July 1973, pp. 165–170), with $x =$ burner area liberation rate (MBtu/hr-ft^2) and $y = NO_x$ emission rate (ppm).

x:	100	125	125	150	150	200	200
y:	150	140	180	210	190	320	280

x:	250	250	300	300	350	400	400
y:	400	430	440	390	600	610	670

a. Assuming that the simple linear regression model is valid, obtain the least squares estimate of the true regression line.

b. What is the estimate of expected NO_x emission rate when burner area liberation rate equals 225?

c. Estimate the amount by which you expect NO_x emission rate to change when burner area liberation rate is decreased by 50?

d. For each x_i used in the experiment, compute the predicted value $\hat{y}_i = \hat{\beta}_0 + \hat{\beta}_1 x_i$ ($i = 1, \ldots, 14$). Then graph $\hat{y}$ versus y (predicted versus observed). If the linear relationship were actually deterministic (so predictions were perfect), what would this plot look like? Does the plot indicate that the prediction relationship is effective for the observed data?

10. The paper "Effects of Bike Lanes on Driver and Bicyclist Behavior" (*ASCE Transportation Eng. J.*, 1977, pp. 243–256) reported the results of a regression analysis with $x =$ available travel space in feet (a convenient measure of roadway width, defined as the distance between a cyclist and the roadway center line) and separation distance y between a bike and a passing car (determined by photography). The data, for 10 streets with bike lanes, appears below.

x:	12.8	12.9	12.9	13.6	14.5	14.6	15.1	17.5	19.5	20.8
y:	5.5	6.2	6.3	7.0	7.8	8.3	7.1	10.0	10.8	11.0

a. Verify that $\Sigma x_i = 154.20$, $\Sigma y_i = 80$, $\Sigma x_i^2 = 2452.18$, $\Sigma x_i y_i = 1282.74$, and $\Sigma y_i^2 = 675.16$.

b. Derive the equation of the estimated regression line.

c. What separation distance would you predict for another street that has 15.0 as its available travel space value?

d. What would be the estimate of expected separation distance for all streets having available travel space value 15?

e. Letting Y_{new} denote the actual separation distance for a new road with travel space value 15, what is the error of prediction in part (c)?

f. Write an expression for the error of estimation in (d), and compare with the error of prediction from (e). Which entails more uncertainty, estimation or prediction? Why?

11. Refer back to the summary quantities given in Exercise 6 for the regression of $y =$ oxygen diffusivity on $x =$ temperature. Calculate

a. the value of SSE.

b. estimates of σ^2 and σ.

c. and interpret the value of r^2.

12. a. Use the summary quantities from Exercise 8 to estimate the standard deviation of the random deviation ϵ in the simple linear regression model.

b. Based on the summary quantities from Exercise 8, what proportion of variation in deflection adjustment factor can be explained by the simple linear regression relationship between adjustment factor and temperature?

13. a. Obtain SSE for the data in Exercise 9 from the defining formula ($SSE = \Sigma(y_i - \hat{y}_i)^2$), and compare to the value calculated from the computational formula.

b. Calculate the value of total sum of squares. Does the simple linear regression model appear to do an effective job of explaining variation in emission rate? Justify your assertion.

14. The accompanying data resulted from a study carried out to examine the relationship between a measure of the corrosion of iron (y) and the concentration of $NaPO_4$ (x, in ppm) ("Mechanism of Action of Acidified Sodium Phosphate Solution as a Corrosion Inhibitor of Iron in Tap Water," *Brit. Corrosion J.,* 1979, pp. 176–178).

x:	2.50	5.03	7.60	11.60	13.00	19.60
y:	7.68	6.95	6.30	5.75	5.01	1.43

x:	26.20	33.00	40.00	50.00	55.00
y:	.93	.72	.68	.65	.56

a. Construct a scatter plot of the data. Does the simple linear regression model appear to be plausible?

b. Calculate the equation of the estimated regression line, use it to predict the value of the corrosion rate that would be observed for a concentration of 33 ppm, and calculate the corresponding residual.

c. Estimate the standard deviation of observations about the true regression line.

d. What percentage of sample variation in corrosion can be attributed to the model relationship?

15. The accompanying data was read from a graph that appeared in the paper "Reactions on Painted Steel Under the Influence of Sodium Chloride, and Combinations Thereof" (*Ind. Engr. Chem. Prod. Res. Dev.,* 1985, pp. 375–378). The independent variable is SO_2 deposition rate $(mg/m^2/day)$ and the dependent variable is steel weight loss (gm/m^2).

x:	14	18	40	43	45	112
y:	280	350	470	500	560	1200

a. Construct a scatter plot. Does the simple linear regression model appear to be reasonable in this situation?

b. Calculate the equation of the estimated regression line.

c. What percentage of observed variation in steel weight loss can be attributed to the model relationship in combination with variation in deposition rate?

d. Because the largest x value in the sample greatly exceeds the others, this observation may have been very influential in determining

the equation of the estimated line. Delete this observation and recalculate the equation. Does the new equation appear to differ substantially from the original one (you might consider predicted values)?

16. Show that b_1 and b_0 of Expression (12.2) satisfy the normal equations.

17. Show that the "point of averages" $(\bar{x}, \bar{y})$ lies on the estimated regression line.

18. Suppose that an investigator has data on the amount of shelf space x devoted to display of a particular product and sales revenue y for that product. Then the investigator may wish to fit a model for which the true regression line passes through $(0, 0)$. The appropriate model is $Y = \beta_1 x + \epsilon$. Assume that $(x_1, y_1), \ldots, (x_n, y_n)$ are observed pairs generated from this model, and derive the least squares estimator of β_1. *Hint:* Write the sum of squared deviations as a function of b_1, a trial value, and use calculus to find the minimizing value of b_1.

19. a. Consider the data in Exercise 7. Suppose that instead of the least squares line passing through the points $(x_1, y_1), \ldots, (x_n, y_n)$ we wish the least squares line passing through $(x_1 - \bar{x}, y_1), \ldots, (x_n - \bar{x}, y_n)$. Construct a scatter plot of the (x_i, y_i) points and then of the $(x_i - \bar{x}, y_i)$ points. Use the plots to explain intuitively how the two least squares lines are related to one another.

b. Suppose that instead of the model $Y_i = \beta_0 + \beta_1 x_i + \epsilon_i$ $(i = 1, \ldots, n)$, we wish to fit a model of the form $Y_i = \beta_0^* + \beta_1^* (x_i - \bar{x}) + \epsilon_i$ $(i = 1, \ldots, n)$. What are the least squares estimators of β_0^* and β_1^*, and how do they relate to $\hat{\beta}_0$ and $\hat{\beta}_1$?

20. Consider the following three data sets, in which the variables of interest are x = commuting distance and y = commuting time. Based on a scatter plot and the values of s and r^2, in which situation would simple linear regression be most (least) effective, and why?

Data set		1		2		3
	x	y	x	y	x	y
	15	42	5	16	5	8
	16	35	10	32	10	16
	17	45	15	44	15	22
	18	42	20	45	20	23
	19	49	25	63	25	31
	20	46	50	115	50	60
$\Sigma(x_i - \bar{x})^2$		17.50		1270.8333		1270.8333
$\Sigma(x_i - \bar{x})(y_i - \bar{y})$		29.50		2722.5		1431.6667
$\hat{\beta}_1$		1.685714		2.142295		1.126557
$\hat{\beta}_0$		13.666672		7.868852		3.196729
SST		114.83		5897.5		1627.33
SSE		65.10		65.10		14.48

12.3   Inferences about the Slope Parameter β_1

In Section 12.2 we used the least squares criteria to obtain point estimates for the two parameters β_0 and β_1 of the true regression line. The values of the x_i's are assumed to be chosen before the experiment is performed, so only the Y_i's are random. The estimators (statistics, and thus random variables) for β_0 and β_1 are obtained by replacing y_i by Y_i in (12.2):

$$\hat{\beta}_1 = \frac{\Sigma(x_i - \bar{x})(Y_i - \bar{Y})}{\Sigma(x_i - \bar{x})^2}$$

$$\hat{\beta}_0 = \frac{\Sigma Y_i - \hat{\beta}_1 \Sigma x_i}{n}$$

Similarly, the estimator for σ^2 results from replacing each y_i in the formula for s^2 by the random variable Y_i:

$$\hat{\sigma}^2 = S^2 = \frac{\Sigma Y_i^2 - \hat{\beta}_0 \Sigma Y_i - \hat{\beta}_1 \Sigma x_i Y_i}{n - 2}$$

In earlier chapters confidence intervals and hypothesis test procedures for any particular parameter were based on properties of an estimator for the parameter. The development of inferential procedures for β_1, the expected change in Y associated with a one-unit increase in x, rests on certain facts about the distribution of the estimator $\hat{\beta}_1$ (the intercept parameter β_0 is usually of less interest than is β_1, so we concentrate on the latter parameter). The denominator of $\hat{\beta}_1$, $\Sigma(x_i - \bar{x})^2$, depends only on the x_i's and not on the Y_i's, so it is a constant c. Then because $\Sigma(x_i - \bar{x})\bar{Y} = \bar{Y}\Sigma(x_i - \bar{x}) = \bar{Y} \cdot 0 = 0$, the slope estimator can be written as

$$\hat{\beta}_1 = \frac{\Sigma(x_i - \bar{x})Y_i}{c} = \Sigma c_i Y_i \quad \text{where } c_i = (x_i - \bar{x})/c$$

That is, $\hat{\beta}_1$ is a linear function of the independent random variables $Y_1, Y_2, \ldots, Y_n$, each of which is normally distributed. Invoking properties of a linear function of random variables discussed in Chapter 5 leads to the following results.

1. The mean value of $\hat{\beta}_1$ is $E(\hat{\beta}_1) = \mu_{\hat{\beta}_1} = \beta_1$, so $\hat{\beta}_1$ is an unbiased estimator of β_1 (the distribution of $\hat{\beta}_1$ is always centered at the value of β_1).
2. The variance of $\hat{\beta}_1$ is

$$V(\hat{\beta}_1) = \sigma_{\hat{\beta}_1}^2 = \frac{\sigma^2}{\Sigma(x_i - \overline{x})^2} = \frac{\sigma^2}{\Sigma x_i^2 - (\Sigma x_i)^2/n} \qquad (12.3)$$

Replacing σ^2 by its estimate s^2 gives an estimate for $\sigma_{\hat{\beta}_1}^2$ (the estimated variance of β_1):

$$s_{\hat{\beta}_1}^2 = \frac{s^2}{\Sigma(x_i - \overline{x})^2}$$

(this estimated variance can also be denoted by $\hat{\sigma}_{\hat{\beta}_1}^2$). The estimated standard deviation of $\hat{\beta}_1$ is $s_{\hat{\beta}_1} = \sqrt{s_{\hat{\beta}_1}^2}$.

3. The estimator $\hat{\beta}_1$ has a normal distribution (because it is a linear function of independent normal random variables).

According to (12.3), the variance of $\hat{\beta}_1$ equals the variance σ^2 of the random error term—or equivalently of any Y_i—divided by $\Sigma(x_i - \overline{x})^2$. Because $\Sigma(x_i - \overline{x})^2$ is a measure of how spread out the x_i's are about $\overline{x}$, we conclude that making observations at x_i values that are quite spread out results in a more precise estimator of the slope parameter (smaller variance of $\hat{\beta}_1$) whereas values of x_i all close to one another imply a highly variable estimator. Of course, if the x_i's are spread out too far, a linear model may not be appropriate throughout the range of observation.

Many inferential procedures discussed earlier were based on standardizing an estimator by first subtracting its mean value and then dividing by its estimated standard deviation. In particular, test procedures and a confidence interval for the mean μ of a normal population utilized the fact that the standardized variable $(\overline{X} - \mu)/(S/\sqrt{n})$—that is, $(\overline{X} - \mu)/S_{\hat{\mu}}$—had a t distribution with $n - 1$ degrees of freedom. A similar result here provides the key to further inferences concerning β_1.

Theorem

The assumptions of the simple linear regression model imply that the standardized variable

$$T = \frac{\hat{\beta}_1 - \beta_1}{S/\sqrt{\Sigma x_i^2 - (\Sigma x_i)^2/n}} = \frac{\hat{\beta}_1 - \beta_1}{S_{\hat{\beta}_1}}$$

has a t distribution with $n - 2$ d.f.

A Confidence Interval for β_1

As in the derivation of earlier confidence intervals, we begin with a probability statement:

$$P\left(-t_{\alpha/2, n-2} < \frac{\hat{\beta}_1 - \beta_1}{S_{\hat{\beta}_1}} < t_{\alpha/2, n-2}\right) = 1 - \alpha$$

Manipulation of the inequalities inside the parentheses to isolate β_1 and substitution of estimates in place of the estimators gives the confidence interval formula.

> A $100(1 - \alpha)\%$ confidence interval for the slope β_1 of the true regression line is
>
> $$\hat{\beta}_1 \pm t_{\alpha/2, n-2} \cdot s_{\hat{\beta}_1}$$

This interval has the same general form as did many of our previous intervals. It is centered at the point estimate of the parameter, and the amount it extends out to either side of the estimate depends on the desired confidence level (through the t critical value) and on the amount of variability in the estimator $\hat{\beta}_1$ (through $s_{\hat{\beta}_1}$, which will tend to be small when there is little variability in the distribution of $\hat{\beta}_1$ and large otherwise).

Example 12.10 Silane coupling agents have been used in the rubber industry to improve the performance of fillers in rubber compounds. The accompanying data on $y = $ tensile modulus (in MPa, a measure of silane coupling effectiveness) and $x = $ bound rubber content (%) appeared in the paper "The Effect of the Structure of Sulfur-Containing Silane Coupling Agents on Their Activity in Silica-Filled SBR" (*Rubber Chem. and Tech.*, 1984, pp. 675–685).

x	16.1	31.5	21.5	22.4	20.5	28.4	30.3	25.6	32.7	29.2	34.7
y	4.41	6.81	5.26	5.99	5.92	6.14	6.84	5.87	7.03	6.89	7.87

A scatter plot of the data exhibits a strong linear pattern, so it is reasonable to assume that bound rubber content and tensile modulus are related according to the simple linear regression model. In this context, β_1 is the expected change in tensile modulus associated with a 1% increase in bound rubber content. The necessary summary quantities are $\Sigma x_i = 292.90$, $\Sigma y_i = 69.03$, $\Sigma x_i^2 = 8141.75$, $\Sigma x_i y_i = 1890.200$, and $\Sigma y_i^2 = 442.1903$, from which

$$\hat{\beta}_1 = \frac{11(1890.200) - (292.90)(69.03)}{11(8141.75) - (292.90)^2} = \frac{573.313}{3768.84} = .152119$$

$$\hat{\beta}_0 = \frac{69.03 - (.152119)(292.90)}{11} = 2.224940$$

$$SSE = 442.1903 - (2.224940)(69.03) - (.152119)(1890.20)$$
$$= 1.067358$$

$$SST = 442.1903 - (69.03)^2/11 = 8.995673$$
$$r^2 = 1 - SSE/SST = 1 - 1.067/8.996 = .881$$

This high r^2 value provides further support for our choice of the simple linear regression model. A confidence interval for β_1 requires $s^2 = SSE/(n-2) = 1.067358/9 = .118595$, $s = \sqrt{.118595} = .344376 \approx .3444$, and

$$s_{\hat{\beta}_1} = \frac{s}{\sqrt{\Sigma x_i^2 - (\Sigma x_i)^2/n}} = \frac{.3444}{\sqrt{8141.75 - (292.90)^2/11}} = \frac{.3444}{18.5100}$$
$$= .0186$$

The t critical value for a confidence level of 95% is $t_{.025, 9} = 2.262$, and the interval is

$$.152 \pm (2.262)(.0186) = .152 \pm .042 = (.110, .194)$$

That is, we can be highly confident that when bound rubber content is increased by 1%, the associated expected change in tensile modulus is between .110 and .194. Even with a high confidence level and small sample size, the narrowness of this interval indicates that β_1 has been rather precisely estimated. ■

Hypothesis Testing Procedures

As before, the null hypothesis in a test about β_1 will be an equality statement. The null value (value of β_1 claimed true by the null hypothesis) will be denoted by β_{10} (read "beta one nought," not "beta ten"). The test statistic results from replacing β_1 in the standardized variable T by the null value β_{10}—that is, from standardizing the estimator of β_1 under the assumption that H_0 is true. The test statistic thus has a t distribution with $n - 2$ d.f. when H_0 is true, so the type I error probability is controlled at the desired level α by using an appropriate t critical value.

Null hypothesis: $H_0 : \beta_1 = \beta_{10}$

Test statistic value: $t = \dfrac{\hat{\beta}_1 - \beta_{10}}{s_{\hat{\beta}_1}}$

Alternative hypothesis	Rejection region for level α test
$H_a : \beta_1 > \beta_{10}$	$t \geq t_{\alpha, n-2}$
$H_a : \beta_1 < \beta_{10}$	$t \leq -t_{\alpha, n-2}$
$H_a : \beta_1 \neq \beta_{10}$	either $t \geq t_{\alpha/2, n-2}$ or $t \leq -t_{\alpha/2, n-2}$

Example 12.11 In anthropological studies a characteristic of fossils that is of central importance is cranial capacity. Frequently skulls are at least partially decomposed, so it is necessary to use other characteristics to obtain information about capacity. One such measurement that has been used is the length of the lambda-opisthion chord (lambda is the point at which the occipital and the left and right parietal bones meet, while the opisthion is the most posterior point on the edge of the hole in the base of the skull through which the spinal cord passes). A paper that appeared in the 1971 *American Journal of Physical Anthropology* reported the following data for $n = 7$ *Homo erectus* fossils.

x (chord length in mm):	78	75	78	81	84	86	87
y (cranial capacity in cm³):	850	775	750	975	915	1015	1030

The summary statistics are $\Sigma x_i = 569$, $\Sigma x_i^2 = 46{,}375$, $\Sigma y_i = 6310$, $\Sigma y_i^2 = 5{,}764{,}600$, and $\Sigma x_i y_i = 515{,}660$.

Suppose that from previous evidence, anthropologists had believed that for each 1-mm increase in chord length, cranial capacity would be expected to increase by 20 cm³. Does this new experimental data strongly contradict prior belief? That is, should $H_0 : \beta_1 = 20$ be rejected in favor of $H_a : \beta_1 \neq 20$?

We calculate

$$\hat{\beta}_1 = \frac{7(515{,}660) - (569)(6310)}{7(46{,}375) - (569)^2} = 22.25694 \quad \hat{\beta}_0 = -907.74306$$

$$SSE = 5{,}764{,}600 - (-907.74306)(6310) - (22.25694)(515{,}660)$$

$$= 15{,}445.03$$

$$s^2 = 3089.01, \quad s = 55.58, \quad \text{and} \quad s_{\hat{\beta}_1} = 55.58/\sqrt{123.43} = 5.00$$

This gives

$$t = \frac{22.26 - 20}{5.00} = .45$$

Since for $\alpha = .05$, $t_{\alpha/2, n-2} = t_{.025, 5} = 2.571$ and $.45$ is neither ≥ 2.571 nor ≤ -2.571, H_0 cannot be rejected at level .05. The data does not suggest that β_1 differs from 20. ■

The most commonly encountered pair of hypotheses about β_1 is $H_0 : \beta_1 = 0$ versus $H_a : \beta_1 \neq 0$. When this null hypothesis is true, $\mu_{Y \cdot x} = \beta_0$ independent of x, so knowledge of x gives no information about the value of the dependent variable. A test of these two hypotheses is often referred to as the **model utility test** in simple linear regression. Unless n is quite small, H_0 will be rejected and the utility of the model confirmed precisely when r^2 is large.

Example 12.12
(Example 12.10 continued)

The question of whether or not the simple linear regression model prescribes a useful relationship between bound rubber content (x) and tensile modulus (y) can be answered by carrying out the model utility test. The hypotheses are $H_0 : \beta_1 = 0$ versus $H_a : \beta_1 \neq 0$. Since the null value is zero, the test statistic value is the "t ratio" $t = \hat{\beta}_1/s_{\hat{\beta}_1}$. A level .05 two-tailed test based on $n - 2 = 9$ d.f. requires $t_{.025, 9} = 2.262$. H_0 will be rejected and the utility of the model confirmed if either $t \geq 2.262$ or $t \leq -2.262$. From our earlier calculations, the t ratio has value $t = .152/.0186 = 8.17$. This computed value is well into the upper tail of the rejection region, so H_0 is rejected and the model is judged useful. This conclusion is consistent with the pattern in the scatter plot and the high value of r^2. ■

In Example 12.12 the fact that $\hat{\beta}_1$ was rather small in magnitude might have suggested that the t test would judge β_1 not significantly different from zero, while exactly the reverse happened. A value of $\hat{\beta}_1$ near zero does not necessarily imply a weak relationship between x and y. In fact, $\hat{\beta}_1$ can be made very near zero by multiplying each y by a small number c, which amounts to changing the units of measurement on y (say, from inches to miles). The new $\hat{\beta}_1$ will be c times the old one, but the new s will also be c times the old one, so the t statistic will have the same value.

While the test procedures for hypotheses about β_1 were derived intuitively by standardizing $\hat{\beta}_1$, it can be shown that these procedures are all likelihood ratio tests. Exercise 32 discusses the determination of type II error probabilities for these t tests.

Regression and ANOVA

The splitting of the total sum of squares $\Sigma(y_i - \bar{y})^2$ into a part SSE, which measures unexplained variation, and a part SSR, which measures variation explained by the linear relationship, is strongly reminiscent of one-way ANOVA. In fact, the null hypothesis $H_0: \beta_1 = 0$ can be tested against $H_a: \beta_1 \neq 0$ by constructing an ANOVA table (Table 12.1) and rejecting H_0 if $f \geq F_{\alpha, 1, n-2}$.

Table 12.1

Source of variation	d.f.	Sum of squares	Mean square	f
Regression	1	SSR	SSR	$\dfrac{SSR}{SSE/(n-2)}$
Error	$n-2$	SSE	$s^2 = \dfrac{SSE}{(n-2)}$	
Total	$n-1$	SST		

The F test gives exactly the same result as the model utility t test because $t^2 = f$ and $t^2_{\alpha/2, n-2} = F_{\alpha, 1, n-2}$. Virtually all computer packages that have regression options include such an ANOVA table in the printout.

Example 12.13
(Example 12.10 continued)

The ANOVA table for the regression of tensile modulus on bound rubber content appears in Table 12.2.

Table 12.2

Source of variation	d.f.	Sum of squares	Mean square	f
Regression	1	7.929	7.929	66.63
Error	9	1.067	.119	
Total	10	8.996		

Since $F_{.05, 1, 9} = 5.12$ and $66.63 \geq 5.12$, $H_0: \beta_1 = 0$ is rejected in favor of the alternative $H_a: \beta_1 \neq 0$. This is the same conclusion as that reached in Example 12.12 via the t test. ■

Exercises / Section 12.3 (21–32)

21. Reconsider the situation described in Exercise 1, in which $x =$ accelerated strength of concrete and $y =$ 28-day cured strength. Suppose that the simple linear regression model is valid for x between 1000 and 4000 and that $\beta_1 = 1.25$ and $\sigma = 350$. Consider an experiment in which $n = 7$ and the x values at which observations are made are $x_1 = 1000$, $x_2 = 1500$, $x_3 = 2000$, $x_4 = 2500$, $x_5 = 3000$, $x_6 = 3500$, and $x_7 = 4000$.

  a. Calculate $\sigma_{\hat{\beta}_1}$, the standard deviation of $\hat{\beta}_1$.

  b. What is the probability that the estimated slope based on such observations will be between 1.00 and 1.50?

  c. Suppose that it is also possible to make a single observation at each of the $n = 11$ values $x_1 = 2000$, $x_2 = 2100, \ldots, x_{11} = 3000$. If a major objective is to estimate β_1 as accurately as possible, would the experiment with $n = 11$ be preferable to the one with $n = 7$?

22. Reconsider the summary quantities given in Exercise 8 for the regression of $y =$ deflection factor on $x =$ temperature.

  a. Compute the estimated standard deviation $s_{\hat{\beta}_1}$.

  b. Calculate a 95% confidence interval for β_1, the expected change in deflection factor associated with a one-degree increase in temperature.

23. In a study of a reactive sputtering technique for the deposit of silicon nitride films on substrate material, the following measurements were obtained on P-etch rate (y) as a function of sputtering voltage (x) ("Preparation and Properties of Reactively Sputtered Silicon Nitride," *J. Vacuum Science and Technology,* 1967, pp. 37–40).

x:	400	600	800	800	1000
y:	44.0	39.9	35.0	33.8	29.1

Assuming that the variables are related according to the simple linear regression model, obtain a 99% confidence interval for the expected change in P-etch rate when sputtering voltage is increased by one unit. Does the interval suggest that precise information about the value of this expected change is available?

24. An article in a recent volume of *J. Public Health Eng.* reported the results of a regression analysis based on $n = 15$ observations in which $x =$ filter application temperature (°C) and $y = \%$ efficiency of BOD removal. Calculated quantities include $\Sigma x_i = 402$, $\Sigma x_i^2 = 11{,}098$, $s = 3.725$, and $\hat{\beta}_1 = 1.7035$.

  a. Test at level .01 $H_0 : \beta_1 = 1$, which states that the expected increase in % BOD removal is 1 when filter application temperature increases by 1° C, against the alternative $H_a : \beta_1 > 1$.

  b. Compute a 99% confidence interval for β_1, the expected increase in % BOD removal for a 1° C increase in filter application temperature.

25. The article "Hydrogen, Oxygen, and Nitrogen in Cobalt Metal" (*Metallurgia,* 1969, pp. 121–127) contained a plot of the following data pairs, where $x =$ pressure of extracted gas (microns) and $y =$ extraction time (min).

x: 40 130 155 160 260 275 325 370 420 480
y: 2.5 3.0 3.1 3.3 3.7 4.1 4.3 4.8 5.0 5.4

  a. Estimate σ and the standard deviation of $\hat{\beta}_1$.

  b. Suppose that the investigators had believed prior to the experiment that $\beta_1 = .0060$. Does the data contradict this prior belief? Test using $\alpha = .10$.

26. Using the data from Exercise 14 on $x =$ NaPO$_4$ concentration and $y =$ corrosion rate, can it be concluded that the simple linear regression model specifies a useful relationship between the two variables? State and test the appropriate hypotheses at significance level .05.

27. Refer to the data on $x =$ liberation rate and $y = $ NO$_x$ emission rate given in Exercise 9.

  a. Does the simple linear regression model specify a useful relationship between the two rates? Use the appropriate test procedure to obtain bounds on the P-value, and then reach a conclusion at significance level .01.

  b. Compute a 95% confidence interval for the expected change in emission rate associated with a 10 MBtu/hr-ft^2 increase in liberation rate.

28. Carry out the model utility test using ANOVA approach for the traffic flow/lead content data of

Example 12.5. Verify that it gives a result equivalent to that of the t test.

29. Use the rules of expected value to show that $\hat{\beta}_0$ is an unbiased estimator for β_0.

30. **a.** Verify that $E(\hat{\beta}_1) = \beta_1$ by using the rules of expected value from Chapter 5.
    b. Use the rules of variance from Chapter 5 to verify the expression for $V(\hat{\beta}_1)$ given in this section.

31. Verify that if each x_i is multiplied by a positive constant c, and each y_i is multiplied by another positive constant d, the t statistic for testing $H_0: \beta_1 = 0$ versus $H_a: \beta_1 \neq 0$ is unchanged in value.

32. The probability of a type II error for the t test for $H_0: \beta_1 = \beta_{10}$ can be computed in the same manner as it was computed for the t tests of Chapter 8. If the alternative value of β_1 is denoted by β_1', the value of

$$d = \frac{|\beta_{10} - \beta_1'|}{\sigma \sqrt{\dfrac{n-1}{\Sigma x_i^2 - (\Sigma x_i)^2/n}}}$$

is first calculated, then the appropriate set of curves in Table A.13 is entered on the horizontal axis at the value of d, and β is read from the curve for $n - 2$ d.f. Use this to compute P(type II error) for the test of Exercise 24 when $\beta_1' = 2$ and $\sigma = 4$.

12.4   Inferences Concerning $\mu_{Y \cdot x}$ and the Prediction of Future Y Values

Let x^* denote a specified value of the independent variable x. Then once the estimates $\hat{\beta}_0$ and $\hat{\beta}_1$ have been calculated, $\hat{\beta}_0 + \hat{\beta}_1 x^*$ can be regarded either as a point estimate of $\mu_{Y \cdot x^*}$ (the expected or true average value of Y when $x = x^*$) or as a prediction of the Y value that will result from a single observation made when $x = x^*$. The point estimate or prediction by itself gives no information concerning how precisely $\mu_{Y \cdot x^*}$ has been estimated or Y has been predicted. This can be remedied by developing a confidence interval for $\mu_{Y \cdot x^*}$ and a prediction interval for a single Y value.

If we think of $\hat{\beta}_0$ and $\hat{\beta}_1$ as random variables (before observations on the dependent variable have been made), then $\hat{\beta}_0 + \hat{\beta}_1 x^*$ is also a random variable. The confidence interval and prediction interval, as well as test procedures for hypotheses about $\mu_{Y \cdot x^*}$, are obtained by exploiting properties of the distribution of $\hat{\beta}_0 + \hat{\beta}_1 x^*$. Substitution of the expressions for $\hat{\beta}_0$ and $\hat{\beta}_1$ into $\hat{\beta}_0 + \hat{\beta}_1 x^*$ followed by some algebraic manipulation leads to the representation of $\hat{\beta}_0 + \hat{\beta}_1 x^*$ as a linear function of the Y_i's:

$$\hat{\beta}_0 + \hat{\beta}_1 x^* = \sum_{i=1}^{n} \left[\frac{1}{n} + \frac{(x^* - \bar{x})(x_i - \bar{x})}{\Sigma(x_i - \bar{x})^2} \right] Y_i = \sum_{i=1}^{n} d_i Y_i$$

The coefficients $d_1, d_2, \ldots, d_n$ in this linear function involve the x_i's and x^*, all of which are fixed. Application of the rules of Chapter 5 to this linear function gives the following properties.

1. The mean value of $\hat{\beta}_0 + \hat{\beta}_1 x^*$ is

$$E(\hat{\beta}_0 + \hat{\beta}_1 x^*) = \mu_{\hat{\beta}_0 + \hat{\beta}_1 x^*} = \beta_0 + \beta_1 x^*$$

Thus $\hat{\beta}_0 + \hat{\beta}_1 x^*$ is an unbiased estimator for $\beta_0 + \beta_1 x^*$ (that is, for $\mu_{Y \cdot x^*}$).

2. The variance of $\hat{\beta}_0 + \hat{\beta}_1 x^*$ is

$$V(\hat{\beta}_0 + \hat{\beta}_1 x^*) = \sigma^2_{\hat{\beta}_0 + \hat{\beta}_1 x^*} = \sigma^2 \left[\frac{1}{n} + \frac{n(x^* - \bar{x})^2}{n\Sigma x_i^2 - (\Sigma x_i)^2} \right]$$

and the standard deviation $\sigma_{\hat{\beta}_0 + \hat{\beta}_1 x^*}$ is the square root of this expression. The estimated standard deviation of $\hat{\beta}_0 + \hat{\beta}_1 x^*$, denoted by $s_{\hat{\beta}_0 + \hat{\beta}_1 x^*}$, results from replacing σ by its estimate s:

$$s_{\hat{\beta}_0 + \hat{\beta}_1 x^*} = s \sqrt{\left[\frac{1}{n} + \frac{n(x^* - \bar{x})^2}{n\Sigma x_i^2 - (\Sigma x_i)^2} \right]}$$

3. $\hat{\beta}_0 + \hat{\beta}_1 x^*$ has a normal distribution.

The variance of $\hat{\beta}_0 + \hat{\beta}_1 x^*$ is smallest when $x^* = \bar{x}$ and increases as x^* moves away from $\bar{x}$ in either direction. Thus the estimator of $\mu_{Y \cdot x^*}$ is more precise when x^* is near the center of the x_i's than when it is far from the x values at which observations have been made. This will imply that both the confidence interval and prediction interval are narrower for an x^* near $\bar{x}$ than for an x^* far from $\bar{x}$.

Inferences Concerning $\mu_{Y \cdot x^*}$

Just as inferential procedures for β_1 were based on the t variable obtained by standardizing $\hat{\beta}_1$, a t variable obtained by standardizing $\hat{\beta}_0 + \hat{\beta}_1 x^*$ leads to a confidence interval and test procedures here.

Theorem

The variable

$$T = \frac{\hat{\beta}_0 + \hat{\beta}_1 x^* - (\beta_0 + \beta_1 x^*)}{S_{\hat{\beta}_0 + \hat{\beta}_1 x^*}}$$

has a t distribution with $n - 2$ d.f.

As for β_1 in the previous section, a probability statement involving this standardized variable can be manipulated to yield a confidence interval for $\mu_{Y \cdot x^*}$.

> A $100(1 - \alpha)\%$ confidence interval for $\mu_{Y \cdot x^*}$, the expected value of Y when $x = x^*$, is
>
> $$\hat{\beta}_0 + \hat{\beta}_1 x^* \pm t_{\alpha/2, n-2} \cdot s_{\hat{\beta}_0 + \hat{\beta}_1 x^*} \qquad (12.4)$$

This confidence interval is centered at the point estimate for $\mu_{Y \cdot x^*}$, and extends out to either side by an amount that depends on the confidence level and on the extent of variability in the estimator on which the point estimate is based.

Example 12.14   For a high-performance tissue machine used in processing of paper by paper mills, the following data was collected on machine speed x (m/min) and temperature in the drying hood y (°C) ("Gas Turbines for Process Improvement of Industrial Thermal Power Plants," *Combustion,* April 1976, pp. 35–41).

x:	1000	1100	1200	1250	1300	1400	1450
y:	220	280	350	375	450	470	500

To obtain a 99% confidence interval for true average hood temperature when machine speed equals 1200, we first calculate $\Sigma x_i = 8700$, $\Sigma x_i^2 = 10{,}965{,}000$, $\Sigma y_i = 2645$, $\Sigma y_i^2 = 1{,}063{,}325$, $\Sigma x_i y_i = 3{,}384{,}750$, $\hat{\beta}_1 = .640$, $\hat{\beta}_0 = -417.70$, $s = 17.62$, and $\bar{x} = 1242.86$. The estimated standard deviation of $\hat{\beta}_0 + \hat{\beta}_1(1200)$ is

$$s_{\hat{\beta}_0 + \hat{\beta}_1(1200)} = 17.62 \sqrt{\frac{1}{7} + \frac{7(1200 - 1242.86)^2}{7(10{,}965{,}000) - (8700)^2}}$$

$$= 17.62 \sqrt{.1429 + .0121} = 6.937$$

Using the t critical value $t_{.005, 5} = 4.032$ gives the interval

$$-417.70 + (.640)(1200) \pm (4.032)(6.937) = 350.30 \pm 27.97$$

$$= (322.33, 378.27)$$

We can be highly confident that the value of $\mu_{Y \cdot 1200}$ is between 322.33 and 378.27. The interval is rather wide because the sample size is small, the confidence level is quite high, and the amount of variability in $\hat{\beta}_0 + \hat{\beta}_1(1200)$ is not insubstantial. A 99% confidence interval for $\mu_{Y \cdot 1300}$ would be somewhat wider than the above interval, since 1300 is further from $\bar{x}$ than is 1200.   ■

In some situations a confidence interval is desired not just for a single x value but for two or more x values. Suppose that an investigator wishes a confidence interval for both $\mu_{Y \cdot v}$ and for $\mu_{Y \cdot w}$, where v and w are two different values of the independent variable. It is tempting to compute the interval (12.4) first for $x = v$ and then for $x = w$. Suppose that we use $\alpha = .05$ in each computation to get two 95% intervals. Then if the variables involved in computing the two intervals were independent of one another, the joint confidence coefficient would be $(.95) \cdot (.95) \approx .90$.

However, the intervals are not independent, because the same $\hat{\beta}_0$, $\hat{\beta}_1$, and S are used in each. We therefore cannot assert that the joint confidence coefficient for the two intervals is exactly 90%. It can be shown, though, that if the $100(1 - \alpha)\%$ confidence interval (12.4) is computed both for $x = v$ and for $x = w$ to obtain joint confidence intervals for $\mu_{Y \cdot v}$ and $\mu_{Y \cdot w}$, then *the joint confidence coefficient on the resulting pair of intervals is at least $100(1 - 2\alpha)\%$*. In particular using $\alpha = .05$ results in a joint confidence coefficient of *at least* 90%, while using $\alpha = .01$ results in at least 98% confidence.

Example 12.15
(Example 12.14 continued)

The 99% confidence interval for true average machine speed when hood temperature is 1200 was computed as (322.33, 378.27). The 99% confidence interval for a hood temperature of 1350 is (413.11, 479.49). Thus we can be at least 98% confident that both $\mu_{Y \cdot 1200}$ lies in (322.33, 378.27) and $\mu_{Y \cdot 1350}$ lies in (413.11, 479.49). ■

The validity of these joint or simultaneous confidence intervals rests on a probability result called the Bonferroni inequality, so the joint confidence intervals are referred to as Bonferroni intervals (the same type of intervals used in the second multiple comparisons method presented in Chapter 10). The method is easily generalized to yield joint intervals for k different $\mu_{Y \cdot x}$'s. *Using the interval (12.4) separately first for $x = x_1^*$, then for $x = x_2^*, \ldots$, and finally for $x = x_k^*$ yields a set of k confidence intervals for which the joint or simultaneous confidence level is at least $100(1 - k\alpha)\%$.*

Testing Hypotheses Concerning $\mu_{Y \cdot x}$

Replacing $\beta_0 + \beta_1 x$ in the standardized variable T by its hypothesized value under H_0 produces a statistic for testing various hypotheses about $\mu_{Y \cdot x}$ for a fixed value of x.

Null hypothesis: $H_0 : \beta_0 + \beta_1 x^* = \mu_0$

Test statistic value: $t = \dfrac{\hat{\beta}_0 + \hat{\beta}_1 x^* - \mu_0}{s_{\hat{\beta}_0 + \hat{\beta}_1 x^*}}$

Alternative hypothesis	Rejection region for level α test
$H_a : \beta_0 + \beta_1 x^* > \mu_0$	$t \geq t_{\alpha, n-2}$
$H_a : \beta_0 + \beta_1 x^* < \mu_0$	$t \leq -t_{\alpha, n-2}$
$H_a : \beta_0 + \beta_1 x^* \neq \mu_0$	either $t \geq t_{\alpha/2, n-2}$ or $t \leq -t_{\alpha/2, n-2}$

Example 12.16

An article "Performance Test Conducted for a Gas Air-Conditioning System" (*Amer. Soc. Heating, Refrigerating, and Air Conditioning Eng.*, October 1969, p. 54) reported the following data on maximum outdoor temperature (x) and hours of chiller operation per day (y) for a 3-ton residential gas air-conditioning system.

x:	72	78	80	86	88	92
y:	4.8	7.2	9.5	14.5	15.7	17.9

Suppose that this system is actually a prototype model, which the manufacturer does not wish to produce unless the data strongly indicates that when maximum outdoor temperature is 82, the true average number of hours of chiller operation is less than 12. The appropriate hypotheses are then $H_0 : \beta_0 + 82\beta_1 = 12$ versus $H_a : \beta_0 + 82\beta_1 < 12$. The summary statistics are $\Sigma x_i = 496$, $\Sigma y_i = 69.60$, $\Sigma x_i^2 = 41,272$, $\Sigma y_i^2 = 942.28$, and $\Sigma x_i y_i = 5942.60$, from which we calculate $\hat{\beta}_1 = .702$, $\hat{\beta}_0 = -46.42$, and $s = .755$. The test statistic value is

$$t = \frac{-46.42 + (.702)(82) - 12}{.755\sqrt{\dfrac{1}{6} + \dfrac{6(82 - 82.67)^2}{1616}}} = \frac{-.856}{.310} = -2.76$$

For $\alpha = .01$, H_0 is rejected if $t \leq -t_{.01, 4} = -3.747$. Since -2.76 is not ≤ -3.747, H_0 is not rejected at level .01; experimental evidence does not strongly indicate that when the maximum outdoor temperature is 82, true average daily chiller operation time is less than 12. ∎

A Prediction Interval for a Future Value of Y

Analogous to the confidence interval (12.4) for $\mu_{Y \cdot x*}$, one frequently wishes to obtain an interval of plausible values for the value of Y associated with some future experiment when the independent variable has value $x*$. For instance, in the example in which vocabulary size y is related to the age x of a child, for $x = 6$ years (12.4) would be a confidence interval for the true average vocabulary size of all six-year-old children. Alternatively, we might wish an interval of plausible values for the vocabulary size of a particular six-year-old child.

A confidence interval refers to a parameter, or population characteristic, whose value is fixed but unknown to us. In contrast, a future value of Y is not a parameter but instead a random variable; for this reason we refer to an interval of plausible values for a future Y as a **prediction interval** rather than a confidence interval. The error of estimation is $\beta_0 + \beta_1 x* - (\hat{\beta}_0 + \hat{\beta}_1 x*)$, a difference between a fixed (but unknown) quantity and a random variable. The error of prediction is $Y - (\hat{\beta}_0 + \hat{\beta}_1 x*)$, a difference between two random variables. There is thus more uncertainty in prediction than in estimation, so a prediction interval will be wider than a confidence interval. Because the future value Y is independent of the observed Y_i's,

$$\begin{aligned}
V[Y - (\hat{\beta}_0 + \hat{\beta}_1 x*)] &= \text{variance of prediction error} \\
&= V(Y) + V(\hat{\beta}_0 + \hat{\beta}_1 x*) \\
&= \sigma^2 + \sigma^2 \left[\frac{1}{n} + \frac{n(x* - \overline{x})^2}{n\Sigma x_i^2 - (\Sigma x_i)^2} \right] \\
&= \sigma^2 \left[1 + \frac{1}{n} + \frac{n(x* - \overline{x})^2}{n\Sigma x_i^2 - (\Sigma x_i)^2} \right]
\end{aligned}$$

Furthermore, because $E(Y) = \beta_0 + \beta_1 x^*$ and $E(\hat{\beta}_0 + \hat{\beta}_1 x^*) = \beta_0 + \beta_1 x^*$, the expected value of the prediction error is $E(Y - (\hat{\beta}_0 + \hat{\beta}_1 x^*)) = 0$. It can then be shown that the standardized variable

$$T = \frac{Y - (\hat{\beta}_0 + \hat{\beta}_1 x^*)}{S\sqrt{1 + \dfrac{1}{n} + \dfrac{n(x^* - \bar{x})^2}{n\Sigma x_i^2 - (\Sigma x_i)^2}}}$$

has a t distribution with $n - 2$ d.f. Substituting this T into the probability statement $P(-t_{\alpha/2, n-2} < T < t_{\alpha/2, n-2}) = 1 - \alpha$ and manipulating to isolate Y between the two inequalities yields the following interval.

A $100(1 - \alpha)\%$ prediction interval for a future Y observation made when $x = x^*$ is

$$\hat{\beta}_0 + \hat{\beta}_1 x^* \pm t_{\alpha/2, n-2} \cdot s \sqrt{1 + \frac{1}{n} + \frac{n(x^* - \bar{x})^2}{n\Sigma x_i^2 - (\Sigma x_i)^2}} \qquad (12.5)$$

The interpretation of the prediction level $100(1 - \alpha)\%$ is identical to that of earlier confidence levels—if (12.5) is used repeatedly, in the long run the resulting intervals will actually contain the observed y values $100(1 - \alpha)\%$ of the time. Notice that the one underneath the square root symbol makes the prediction interval (12.5) longer than the confidence interval (12.4), though the intervals are both centered at $\hat{\beta}_0 + \hat{\beta}_1 x^*$. Also, as $n \to \infty$ the length of the confidence interval approaches zero, while the length of the prediction interval does not (because even with perfect knowledge of β_0 and β_1, there will still be uncertainty in prediction).

Example 12.17   The article "The Incorporation of Uranium and Silver by Hydrothermally Synthesized Galena" (*Econ. Geology,* 1964, pp. 1003–1024) reported on the determination of silver content of galena crystals grown in a closed hydrothermal system over a range of temperature. With $x =$ crystallization temperature in °C and $y = Ag_2S$ in mol %, the data follows:

x: 398 292 352 575 568 450 550 408 484 350 503 600 600
y: .15   .05   .23   .43   .23   .40   .44   .44   .45   .09   .59   .63   .60

From the summary values $\Sigma x_i = 6130$, $\Sigma x_i^2 = 3{,}022{,}050$, $\Sigma y_i = 4.73$, $\Sigma y_i^2 = 2.1785$, and $\Sigma x_i y_i = 2418.74$, the regression estimates are computed to be $\hat{\beta}_1 = .00143$, $\hat{\beta}_0 = -.311$, and $s = .131$. If the next crystallization temperature chosen is $500°$ C, a 95% prediction interval for the resulting silver content is

$$-.311 + (.00143)(500) \pm (2.201)(.131)\sqrt{1 + \frac{1}{13} + \frac{13(500 - 471.54)^2}{1{,}709{,}750}}$$

$$= .40 \pm .30 = (.10, .70), \quad \text{a very wide interval.}$$

Exercises / Section 12.4 (33–42)

33. Reconsider the filter application temperature – % BOD removal experiment described in Exercise 24. In addition to information given there, $\hat{\beta}_0 = 8.2141$.

a. Compute a 90% confidence interval for $\beta_0 + 25\beta_1$, the expected % BOD removal when filter application temperature is $25°$ C.

b. Test at level .01 the hypotheses $H_0 : \beta_0 + 25\beta_1 = 50$ versus $H_a : \beta_0 + 25\beta_1 > 50$ (the alternative hypothesis states that expected % BOD removal exceeds 50 when filter application temperature is 25).

34. Reconsider the data of Example 12.3, in which $x =$ chlorine flow and $y =$ etch rate.

a. Calculate a 95% confidence interval for $\mu_{Y \cdot 3.0}$, the true average etch rate when flow $= 3$. Has this average been precisely estimated?

b. Calculate a 95% prediction interval for a single future observation on etch rate to be made when flow $= 3.0$. Is the prediction likely to be accurate?

c. Would the 95% confidence and prediction intervals when flow $= 2.5$ be wider or narrower than the corresponding intervals of (a) and (b)? Answer without actually computing the intervals.

d. Would you recommend calculating a 95% prediction interval for a flow of 6.0? Explain.

35. Does the data of Example 12.3 suggest that $\mu_{Y \cdot 2.5}$, the expected etch rate when flow $= 2.5$, is something other than 30? Test $H_0 : \mu_{Y \cdot 2.5} = 30$ against the appropriate alternative using a significance level of .01.

36. An experiment to measure the macroscopic magnetic relaxation time in crystals (μ sec) as a function of the strength of the external biasing magnetic field (KG) yielded the following data ("An Optical Faraday Rotation Technique for the Determination of Magnetic Relaxation Times," *IEEE Trans. Magnetics,* June 1968, pp. 175–178, with data read from a graph that appeared in the paper).

x:	11.0	12.5	15.2	17.2	19.0	20.8
y:	187	225	305	318	367	365

x:	22.0	24.2	25.3	27.0	29.0
y:	400	435	450	506	558

The summary statistics are $\Sigma x_i = 223.2$, $\Sigma y_i = 4116$, $\Sigma x_i^2 = 4877.50$, $\Sigma x_i y_i = 90{,}096.1$, and $\Sigma y_i^2 = 1{,}666{,}782$. Compute

a. a 95% confidence interval for expected relaxation time when field strength equals 18.

b. a 95% prediction interval for future relaxation time when field strength equals 18.

c. simultaneous confidence intervals for expected relaxation time when field strength equals both 15 and 20; your joint confidence coefficient should be at least 90%.

37. Use the data in Example 12.16 to compute a 99% confidence interval for expected daily hours of chiller operation when maximum outdoor temperature equals $85°$. Repeat for $95°$. Why is this second interval much wider?

38. Using the data in Example 12.14, obtain simultaneous confidence intervals for true average hood temperature when machine speed equals 1200, 1250, and 1300; your joint confidence coefficient should be at least 90%.

39. Exercise 25 presented data on $x =$ pressure of extracted gas and $y =$ extraction time.

a. Does the data strongly contradict prior belief that when extraction pressure is 300, expected extraction time is at most 4 min? Use a level .01 test. What can you say about the P-value (upper and/or lower bound)?

b. Extraction time is to be observed once for a pressure of 200 and once for a pressure of 300. Obtain a pair of prediction intervals for the future observed values such that the joint prediction coefficient is at least 90%.

40. Use the data given in Exercise 14 to predict the amount of corrosion resulting from a single experimental run in which $NaPO_4$ concentration is 15.0. Your calculations should give information about the accuracy of such a prediction by indicating a range of plausible values.

41. Based on the data in Example 12.17, does the expected silver content when temperature equals $400°$ C appear to differ significantly from .250?

Test the appropriate hypotheses by first obtaining information about the *P*-value and then drawing a conclusion at significance level .01.

42. Verify that $V(\hat{\beta}_0 + \hat{\beta}_1 x)$ is indeed given by the expression in the text. *Hint:* $V(\Sigma d_i Y_i) = \Sigma d_i^2 \cdot V(Y_i)$.

12.5 Correlation

There are many situations in which the objective in studying the joint behavior of two variables is to see whether or not they are related, rather than to use one to predict the value of the other. In this section we first develop the sample correlation coefficient *r* as a measure of how strongly related two variables *x* and *y* are in a sample, and then relate *r* to the correlation coefficient ρ defined in Chapter 5.

The Sample Correlation Coefficient *r*

Given *n* pairs of observations $(x_1, y_1), (x_2, y_2), \ldots, (x_n, y_n)$, it is natural to speak of *x* and *y* having a positive relationship if large *x*'s are paired with large *y*'s and small *x*'s with small *y*'s. Similarly, if large *x*'s are paired with small *y*'s and small *x*'s with large *y*'s, then a negative relationship between the variables is implied. Consider the quantity

$$s_{xy} = \sum_{i=1}^{n} (x_i - \bar{x})(y_i - \bar{y})$$

Then if the relationship is strongly positive, an x_i above the mean $\bar{x}$ will tend to be paired with a y_i above the mean $\bar{y}$, so that $(x_i - \bar{x})(y_i - \bar{y}) > 0$, and this product will also be positive whenever both x_i and y_i are below their respective means. Thus a positive relationship implies that s_{xy} will be positive. An analogous argument shows that when the relationship is negative, s_{xy} will be negative, since most of the products $(x_i - \bar{x})(y_i - \bar{y})$ will be negative. This is illustrated in Figure 12.10.

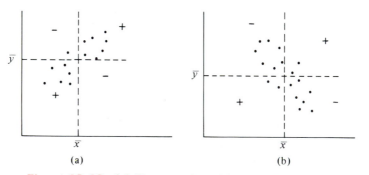

(a) (b)

Figure 12.10 (*a*) Scatter plot with s_{xy} positive; (*b*) scatter plot with s_{xy} negative [+ means $(x_i - \bar{x})(y_i - \bar{y}) > 0$, and − means $(x_i - \bar{x})(y_i - \bar{y}) < 0$]

While s_{xy} seems a plausible measure of the strength of a relationship, we do not yet have any idea of how positive or negative it can be. Unfortunately, s_{xy} has a serious defect: by changing the units of measurement of either x or y, s_{xy} can be made either arbitrarily large in magnitude or arbitrarily close to zero. For example, if $s_{xy} = 25$ when x is measured in meters, then $s_{xy} = 25,000$ when x is measured in millimeters and .025 when x is expressed in kilometers. A reasonable condition to impose on any measure of how strongly x and y are related is that the calculated measure should not depend on the particular units used to measure them. This condition is achieved by modifying s_{xy} to obtain the sample correlation coefficient.

Definition

> The **sample correlation coefficient** for the n pairs $(x_1, y_1), \ldots, (x_n, y_n)$ is
>
> $$r = \frac{s_{xy}}{\sqrt{\Sigma(x_i - \bar{x})^2} \sqrt{\Sigma(y_i - \bar{y})^2}}$$
>
> $$= \frac{n\Sigma x_i y_i - (\Sigma x_i)(\Sigma y_i)}{\sqrt{n\Sigma x_i^2 - (\Sigma x_i)^2} \sqrt{n\Sigma y_i^2 - (\Sigma y_i)^2}} \qquad (12.6)$$

The second expression in (12.6) is more convenient for hand computation.

Example 12.18 The April 2, 1977, issue of the medical journal *Lancet* reported results on water/lead concentration in the maternal home during pregnancy and blood/lead concentration (both in μmol/liter) for a group of mentally retarded children. Representative data appears below.

x (blood):	1.98	1.44	2.02	1.20	1.57	1.82	1.45	1.80
y (water):	5.6	7.7	8.8	5.1	6.8	3.9	4.5	5.8

According to (12.6) r can be calculated from the five quantities Σx_i, Σx_i^2, Σy_i, Σy_i^2, and $\Sigma x_i y_i$; these are 13.28, 22.63, 48.2, 309.44, and 80.81, respectively. Thus

$$r = \frac{8(80.81) - (13.28)(48.2)}{\sqrt{8(22.63) - (13.28)^2} \sqrt{8(309.44) - (48.2)^2}} = \frac{6.38}{(2.16)(12.34)} = .239$$

∎

Properties of r

The most important properties of r are as follows.

1. The value of r does not depend on which of the two variables under study is labelled x and which is labelled y.
2. The value of r is independent of the units in which x and y are measured.
3. $-1 \leq r \leq 1$
4. $r = 1$ if and only if all (x_i, y_i) pairs lie on a straight line with positive slope, and $r = -1$ if and only if all (x_i, y_i) pairs lie on a straight line with negative slope.

5. The square of the sample correlation coefficient gives the value of the coefficient of determination that would result from fitting the simple linear regression model—in symbols, $(r)^2 = r^2$.

Property 1 stands in marked contrast to what happens in regression analysis, where virtually all quantities of interest (the estimated slope, estimated y intercept, s^2, and so on) depend on which of the two variables is treated as the dependent variable. However, property 5 shows that the proportion of variation in the dependent variable explained by fitting the simple linear regression model does not depend on which variable plays this role.

Property 2 is equivalent to saying that r is unchanged if each x_i is replaced by cx_i and if each y_i is replaced by dy_i (a change in the scale of measurement), as well as if each x_i is replaced by $x_i - a$ and y_i by $y_i - b$ (which changes the location of zero on the measurement axis). This implies, for example, that r is the same whether temperature is measured in $°F$ or $°C$.

The third property tells us that the maximum value of r, corresponding to the largest possible degree of positive relationship, is $r = 1$, while the most negative relationship is identified with $r = -1$. According to property 4 the largest positive and largest negative correlations are achieved only when all points lie along a straight line. Any other configuration of points, even if the configuration suggests a deterministic relationship between variables, will yield an r value less than one in absolute magnitude. Thus, *r measures the degree of linear relationship* among variables. A value of r near zero is not evidence of a lack of strong relationship, but only the absence of a linear relation, so that such a value of r must be interpreted with caution. Figure 12.11 illustrates several configurations of points associated with different values of r.

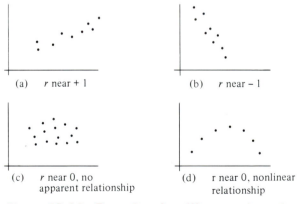

(a) r near $+ 1$

(b) r near $- 1$

(c) r near 0, no
       apparent relationship

(d)   r near 0, nonlinear
       relationship

Figure 12.11   Data plots for different values of r

A frequently asked question is, "When can it be said that there is a strong correlation between the variables, and when is the correlation weak?" A reasonable rule of thumb is to say that the correlation is weak if $0 \le |r| \le .5$, strong if

$.8 \leq |r| \leq 1$, and moderate otherwise. It may surprise you that $r = .5$ is considered weak, but $r^2 = .25$ implies that in a regression of y on x, only 25% of observed y variation would be explained by the model. In Example 12.18, the correlation between blood/lead concentration and water/lead concentration would be described as weak.

The Population Correlation Coefficient ρ and Inferences about Correlation

The correlation coefficient r is a measure of how related x and y are in the observed sample. We can think of the pairs (x_i, y_i) as having been drawn from a bivariate population of pairs, with (X_i, Y_i) having joint probability distribution $f(x, y)$. In Chapter 5 we defined the correlation coefficient $\rho(X, Y)$ by

$$\rho = \rho(X, Y) = \frac{\text{Cov}(X, Y)}{\sigma_X \cdot \sigma_Y}, \quad \text{where}$$

$$\text{Cov}(X, Y) = \begin{cases} \sum_x \sum_y (x - \mu_X)(y - \mu_Y) f(x, y) & (X, Y) \text{ discrete} \\ \int_{-\infty}^{\infty} \int_{-\infty}^{\infty} (x - \mu_X)(y - \mu_Y) f(x, y) \, dx \, dy & (X, Y) \text{ continuous} \end{cases}$$

If we think of $f(x, y)$ as describing the distribution of pairs of values within the entire population, ρ becomes a measure of how strongly related x and y are in that population. In Chapter 5, we listed properties of ρ analogous to those for r.

The population correlation coefficient ρ is a parameter or population characteristic, just as μ_X, μ_Y, σ_X, and σ_Y are, so we can use the sample correlation coefficient to make various inferences about ρ. In particular, r is a point estimate for ρ, and the corresponding estimator is

$$\hat{\rho} = R = \frac{\Sigma(X_i - \overline{X})(Y_i - \overline{Y})}{\sqrt{\Sigma(X_i - \overline{X})^2} \sqrt{\Sigma(Y_i - \overline{Y})^2}}$$

Example 12.19 In some locations there is a strong association between concentrations of two different pollutants. The paper "The Carbon Component of the Los Angeles Aerosol: Source Apportionment and Contributions to the Visibility Budget" (*J. Air Pollution Control Fed.*, 1984, pp. 643–650) reported the accompanying data on ozone concentration x (ppm) and secondary carbon concentration y ($\mu g/m^3$).

x:	.066	.088	.120	.050	.162	.186	.057	.100
y:	4.6	11.6	9.5	6.3	13.8	15.4	2.5	11.8

x:	.112	.055	.154	.074	.111	.140	.071	.110
y:	8.0	7.0	20.6	16.6	9.2	17.9	2.8	13.0

The summary quantities are $n = 16$, $\Sigma x_i = 1.656$, $\Sigma y_i = 170.6$, $\Sigma x_i^2 = .196912$, $\Sigma x_i y_i = 20.0397$, and $\Sigma y_i^2 = 2253.56$, from which

$$r = \frac{16(20.0397) - (1.656)(170.6)}{\sqrt{16(.196912) - (1.656)^2}\,\sqrt{16(2253.56) - (170.6)^2}}$$

$$= \frac{38.1216}{(.6389)(83.3823)} = .716$$

The point estimate of the population correlation coefficient ρ between ozone concentration and secondary carbon concentration is $\hat{\rho} = r = .716$. ∎

In our previous work concerning inference, we typically did not have to make strong assumptions about the population distribution in order to obtain point estimators and estimates of parameters. However, in order to construct confidence intervals and test procedures, we have usually assumed that the sample comes from a normal population. To test hypotheses about ρ, we must make an analogous assumption about the distribution of pairs of (x, y) values in the population. We are now assuming that *both X and Y are random*, whereas much of our regression work focused on x fixed by the experimenter.

Assumption

The joint probability distribution of (X, Y) is specified by

$$f(x, y) = \frac{1}{2\pi \cdot \sigma_1 \sigma_2 \sqrt{1 - \rho^2}} e^{-\left[\left(\frac{x - \mu_1}{\sigma_1}\right)^2 - 2\rho\left(\frac{x - \mu_1}{\sigma_1}\right)\left(\frac{y - \mu_2}{\sigma_2}\right) + \left(\frac{y - \mu_2}{\sigma_2}\right)^2\right]/2(1 - \rho^2)}$$

$$-\infty < x < \infty$$
$$-\infty < y < \infty \qquad\qquad (12.7)$$

where μ_1 and σ_1 are the mean and standard deviation of X, and μ_2 and σ_2 are the mean and standard deviation of Y; $f(x, y)$ is called the **bivariate normal probability distribution.**

The bivariate normal distribution is obviously rather complicated, but for our purposes we only need a passing acquaintance with several of its properties. The surface determined by $f(x, y)$ lies entirely above the x-y plane $[f(x, y) \geq 0]$ and has a three-dimensional bell- or mound-shaped appearance, as illustrated in Figure 12.12. If we slice through the surface with any plane perpendicular to the (x, y) plane and look at the shape of the curve sketched out on the "slicing plane," the result is a normal curve. More precisely, if X is fixed at value x, it can be shown that the (conditional) distribution of Y is normal with mean $\mu_{Y \cdot x} = \mu_2 - \rho\mu_1\sigma_2/\sigma_1 + \rho\sigma_2 x/\sigma_1$ and variance $(1 - \rho^2)\sigma_2^2$. This is exactly the model used in simple linear regression with $\beta_0 = \mu_2 - \rho\mu_1\sigma_2/\sigma_1$, $\beta_1 = \rho\sigma_2/\sigma_1$, and $\sigma^2 = (1 - \rho^2)\sigma_2^2$ independent of x. The implication is that *if the observed pairs (x_i, y_i) are actually drawn from a bivariate normal distribution, then the simple linear regression model is an appropriate way of studying the behavior of Y for fixed x.* If $\rho = 0$, then $\mu_{Y \cdot x} = \mu_2$ independent of x; in fact, when $\rho = 0$ the joint p.d.f. $f(x, y)$ of (12.7) can be factored into a part involving x only and a part involving y only, which implies that X and Y are independent variables.

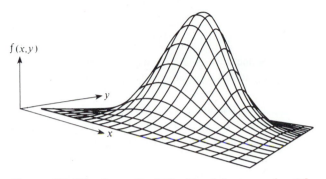

$f(x, y)$

y

x

Figure 12.12 A graph of the bivariate normal p.d.f.

Assuming that the pairs are drawn from a bivariate normal distribution allows us to test hypotheses about ρ and construct a confidence interval. There is no completely satisfactory way to check the plausibility of the bivariate normality assumption. A partial check involves constructing two separate normal probability plots, one for the sample x_i's and another for the sample y_i's, since bivariate normality implies that the marginal distributions of both X and Y are normal. If either plot deviates substantially from a straight line pattern, the following inferential procedures should not be used when the sample size n is small.

Procedure for Testing $H_0 : \rho = 0$

Test statistic: $T = \dfrac{R \sqrt{n - 2}}{\sqrt{1 - R^2}}$

When H_0 is true, T has a t distribution with $n - 2$ degrees of freedom.

Alternative hypothesis	*Rejection region for level α test*
$H_a : \rho > 0$	$t \geq t_{\alpha, n-2}$
$H_a : \rho < 0$	$t \leq -t_{\alpha, n-2}$
$H_a : \rho \neq 0$	either $\quad t \geq t_{\alpha/2, n-2}$ or $\quad t \leq -t_{\alpha/2, n-2}$

Example 12.20
(Example 12.19 continued)

Separate normal probability plots of the 16 ozone concentration measurements and 16 secondary carbon concentration measurements are quite straight, so we assume that the 16 pairs were drawn from a bivariate normal population distribution and use the above t test to decide whether the data suggests a positive relationship between the two concentrations in the population consisting of all concentration pairs in the sampled locale. The hypotheses of interest are $H_0 : \rho = 0$ versus $H_a : \rho > 0$. The necessary t critical value for a level .01 test is $t_{.01, 14} = 2.624$. We previously computed $r = .716$, so

$$t = \frac{.716 \sqrt{14}}{\sqrt{1 - (.716)^2}} = 3.84$$

Since $3.84 \geq 2.624$, we reject H_0 and conclude that the two concentrations are positively related in the population.   ■

Because ρ measures the extent to which there is a linear relationship between the two variables in the population, the null hypothesis $H_0 : \rho = 0$ states that there is no such population relationship. In Section 12.3, we used the t ratio $\hat{\beta}_1 / s_{\hat{\beta}_1}$ to test for a linear relationship between the two variables in the context of regression analysis. It turns out that the two test procedures are completely equivalent because $r \sqrt{n - 2} / \sqrt{1 - r^2} = \hat{\beta}_1 / s_{\hat{\beta}_1}$. When interest lies only in assessing the strength of any linear relationship rather than in fitting a model and using it to estimate or predict, the test statistic formula just presented requires fewer computations than does the t ratio.

Other Inferences Concerning ρ

The procedure for testing $H_0 : \rho = \rho_0$ when $\rho_0 \neq 0$ is not equivalent to any procedure from regression analysis. The test statistic is based on a transformation of R called the Fisher transformation.

Proposition

> When $(X_1, Y_1), \ldots, (X_n, Y_n)$ is a sample from a bivariate normal distribution, the random variable
>
> $$V = \frac{1}{2} \ln \left(\frac{1 + R}{1 - R} \right) \tag{12.8}$$
>
> has approximately a normal distribution with mean and variance
>
> $$\mu_V = \frac{1}{2} \ln \left(\frac{1 + \rho}{1 - \rho} \right), \quad \sigma_V^2 = \frac{1}{n - 3}$$

The rationale for the transformation is to obtain a function of R that has a variance independent of ρ; this would not be the case with R itself. Also, the transformation should not be used if n is quite small, since the approximation will not be valid.

> Test statistic: $Z = \dfrac{V - \dfrac{1}{2} \ln \left[(1 + \rho_0) / (1 - \rho_0) \right]}{1 / \sqrt{n - 3}}$

where V is given by (12.8). When H_0 is true, Z has approximately a standard normal distribution.

Alternative hypothesis	Rejection region for level α test
$H_a: \rho > \rho_0$	$z \geq z_\alpha$
$H_a: \rho < \rho_0$	$z \leq -z_\alpha$
$H_a: \rho \neq \rho_0$	either $z \geq z_{\alpha/2}$ or $z \leq -z_{\alpha/2}$

Example 12.21   The paper "A Study of a Partial Nutrient Removal System for Wastewater Treatment Plants" (*Water Research*, 1972, pp. 1389–1397) reported on a method of nitrogen removal that involves the treatment of the supernatant from an aerobic digester. Both the influent total nitrogen x (mg/l) and the percentage of nitrogen removed were recorded for 20 days, with resulting summary statistics $\Sigma x_i = 285.9$, $\Sigma x_i^2 = 4409.55$, $\Sigma y_i = 690.30$, $\Sigma y_i^2 = 29{,}040.29$, and $\Sigma x_i y_i = 10{,}869.71$. Does the data indicate that influent total nitrogen and percentage of nitrogen removed are at least moderately positively correlated?

Our earlier interpretation of moderate positive correlation was $.5 < \rho < .8$, so we wish to test $H_0: \rho = .5$ versus $H_a: \rho > .5$. The computed value of r is .733, so

$$\frac{1}{2} \ln\left(\frac{1 + .733}{1 - .733}\right) = .935, \quad \frac{1}{2} \ln\left(\frac{1 + .5}{1 - .5}\right) = .549$$

This gives $z = (.935 - .549)\sqrt{17} = 1.59$. Since $1.59 < 1.645$, at level .05, we cannot conclude that $\rho > .5$, so the relationship has not been proved to be even moderately strong (a somewhat surprising conclusion since $r = .73$, but when n is small a large r may result even when ρ is small). ■

To obtain a confidence interval for ρ, we first derive an interval for $\mu_V = \frac{1}{2} \ln\left[(1 + \rho)/(1 - \rho)\right]$. Standardizing V, writing a probability statement, and manipulating the resulting inequalities yields

$$\left(v - \frac{z_{\alpha/2}}{\sqrt{n - 3}}, \; v + \frac{z_{\alpha/2}}{\sqrt{n - 3}}\right) \tag{12.9}$$

as a $100(1 - \alpha)\%$ interval for μ_V, where $v = \frac{1}{2} \ln\left[(1 + r)/(1 - r)\right]$. This interval can then be manipulated to yield a confidence interval for ρ:

The interval

$$\left(\frac{e^{2c_1} - 1}{e^{2c_1} + 1}, \; \frac{e^{2c_2} - 1}{e^{2c_2} + 1}\right)$$

is a $100(1 - \alpha)\%$ confidence interval for ρ, where c_1 and c_2 are the left and right endpoints, respectively, of the interval (12.9).

Example 12.22
(Example 12.21 continued)

The sample correlation coefficient between influent nitrogen and percentage nitrogen removed was $r = .733$, giving $v = .935$. With $n = 20$, a 95% interval for μ_V is $(.935 - 1.96/\sqrt{17}, .935 + 1.96/\sqrt{17}) = (.460, 1.410) = (c_1, c_2)$. The 95% interval for ρ is

$$\left[\frac{e^{2(.46)} - 1}{e^{2(.46)} + 1}, \frac{e^{2(1.41)} - 1}{e^{2(1.41)} + 1}\right] = (.43, .89)$$

■

In Chapter 5 we cautioned that a large value of the correlation coefficient (near 1 or -1) implies only association and not causation. This applies both to ρ and to r.

Exercises / Section 12.5 (43–52)

43. Observations were made on $x =$ June–August precipitation (cm) and $y =$ perennial grass production (kg/ha) for certain pasture areas during each year of a 10-year period, with the following results:

x:	22.05	35.74	30.48	11.89	27.28
y:	291	629	823	307	660
x:	9.63	17.63	22.20	17.27	19.63
y:	263	375	366	563	558

("Influence of Precipitation on Perennial Grass Production in the Semidesert Southwest," *Ecology,* 1975, pp. 981–986).

a. Compute the sample correlation coefficient for this data.

b. If you were to do a regression analysis, what would the value of the coefficient of determination be?

44. Sixteen different air samples were obtained at Herald Square in New York City, and both the carbon monoxide concentration x (ppm) and benzo(a) pyrene concentration y $(\mu g/10^3 m^3)$ measured for each sample ("Carcinogenic Air Pollutants in Relation to Automobile Traffic in New York City," *Environmental Science and Technology,* 1971, pp. 145–150).

x:	2.8	15.5	19.0	6.8	5.5	5.6	9.6	13.3
y:	.5	.1	.8	.9	1.0	1.1	3.9	4.0
x:	5.5	12.0	5.6	19.5	11.0	12.8	5.5	10.5
y:	1.3	5.7	1.5	6.0	7.3	8.1	2.2	9.5

a. Compute the sample correlation coefficient for this data.

b. Test the hypothesis $H_0: \rho = 0$ against $H_a: \rho \neq 0$ at level .01.

45. Data on per capita disposable personal income x (in the United States) and per capita food expenditure y was obtained for a 24-year period. The summary statistics are $\Sigma x_i = 55,661$, $\Sigma y_i = 10,276$, $\Sigma x_i^2 = 144,051,189$, $\Sigma x_i y_i = 25,479,315$, and $\Sigma y_i^2 = 4,582,506$.

a. Compute r for this data.

b. If you decided to obtain an estimated regression line for this data, what proportion of variation in food expenditure would be explained by variation in disposable income?

46. Physical properties of six flame retardant fabric samples were investigated in the paper "Sensory and Physical Properties of Inherently Flame-Retardant Fabrics" (*Textile Research,* 1984, pp. 61–68). Use the accompanying data and a .05 significance level to determine if a linear relationship exists between stiffness x (mg-cm) and thickness y (mm).

x:	7.98	24.52	12.47	6.92	24.11	35.71
y:	.28	.65	.32	.27	.81	.57

47. The article "Pedunculate Oak Woodland in a Severe Environment" (*J. Ecology,* 1978, pp. 707–740) reports the following data on $x =$ age (yrs) and $y =$ annual trunk diameter growth increment (mm) for a sample of trees in a certain region.

x:	17	23	30	37.5	40
y:	2.20	1.25	.85	1.30	1.70

x:	46.5	50	54	55	93
y:	.75	.75	.50	1.00	.70

a. Does the data strongly indicate that the true correlation coefficient ρ (for the population of all such trees in this area) differs from zero? Test using $\alpha = .05$. Place an upper and/or lower bound on the P-value.

b. Compute a 95% confidence interval for ρ.

48. An investigation of the relationship between water temperature x and calling rate y for a particular type of hybrid toad ("The Mating Call of Hybrids of the Fire-Bellied Toad and Yellow-Bellied Toad," *Oecologia*, 1974, pp. 61–71) yielded the following summary statistics: $n = 17$, $\Sigma x_i = 376.20$, $\Sigma y_i = 752$, $\Sigma x_i^2 = 8563.70$, $\Sigma x_i y_i = 17,140.40$, and $\Sigma y_i^2 = 34,496.00$. Does this data indicate that there is a positive correlation between water temperature and calling rate? Test the appropriate hypotheses using $\alpha = .05$.

49. The article "Increases in Steroid Binding Globulins Induced by Tamofixen in Patients with Carcinoma of the Breast" (*J. Endocrinology*, 1978, pp. 219–226) reported data on the effects of the drug tamofixen on change in the level of cortisol binding globulin of patients during treatment. With age $= x$ and $\Delta CBG = y$, summary values are $n = 26$, $\Sigma x_i = 1613$, $\Sigma (x_i - \bar{x})^2 = 3756.96$, $\Sigma y_i = 281.9$, $\Sigma (y_i - \bar{y})^2 = 465.34$, and $\Sigma x_i y_i = 16,731$.

a. Compute a 90% confidence interval for the true correlation coefficient ρ.

b. Test $H_0 : \rho = -.5$ versus $H_a : \rho < -.5$ at level .05.

c. In a regression analysis of y on x, what proportion of variation in change of cortisol binding globulin level could be explained by variation in patient age within the sample?

d. If you decided to perform a regression analysis with age as the dependent variable, what proportion of variation in age is explainable by variation in ΔCBG?

50. The paper "Chronological Trend in Blood Lead Levels" (*N. Eng. J. Med.*, 1983, pp. 1373–1377) gave the following data on $y =$ average blood lead level of white children age 6 months to 5 years and $x =$ amount of lead used in gasoline production (in 1000 tons) for ten 6-month periods.

x:	48	59	79	80	95
y:	9.3	11.0	12.8	14.1	13.6
x:	95	97	102	102	107
y:	13.8	14.6	14.6	16.0	18.2

a. Construct separate normal probability plots for x and y. Do you think that it is reasonable to assume that the (x, y) pairs are from a bivariate normal population?

b. Does the data provide sufficient evidence to conclude that there is a linear relationship between blood lead level and the amount of lead used in gasoline production? Use $\alpha = .01$.

51. A sample of $n = 500$ (x, y) pairs was collected and a test of $H_0 : \rho = 0$ versus $H_a : \rho \neq 0$ was carried out. The resulting P-value was computed to be .00032.

a. What conclusion would be appropriate at level of significance .001?

b. Does this small P-value indicate that there is a very strong linear relationship between x and y (a value of ρ that differs considerably from zero)? Explain.

52. A sample of $n = 10,000$ (x, y) pairs resulted in $r = .022$. Test $H_0 : \rho = 0$ versus $H_a : \rho \neq 0$ at level .05. Is the result statistically significant? Comment on the practical significance of your analysis.

Supplementary Exercises / Chapter 12 (53–64)

53. The paper "Refuse-Derived Fuel Evaluation in an Industrial Spreader-Stoker Boiler" (*J. Engr. for Gas Turbines and Power*, 1984, pp. 782–788) reported the accompanying data on $x = \%$ refuse derived fuel (RDF) heat input and $y = \%$ efficiency for a certain boiler.

x:	37	30	48	29	27	16	0	20
y:	78.0	77.2	74.4	77.7	76.9	79.0	82.1	76.5

a. Obtain the equation of the estimated regression line.

b. Does the simple linear regression model specify a useful relationship between % RDF heat input and % efficiency? State and test the appropriate hypotheses.

c. To obtain an accurate estimate of β_1, would it have been preferable to make four observations at $x = 0$ and four observations at $x = 50$ (assuming that the model is valid for x between 0 and 50)? What about three observations at $x = 0$ and three at $x = 50$? Explain.

d. Estimate true average % efficiency when % RDF heat input is 25 using a 95% confidence interval. Does it appear that true average % efficiency has been precisely estimated? Explain.

54. The accompanying data on x = diesel oil consumption rate measured by the drain-weigh method and y = rate measured by the Cl-trace method, both in g/h, was read from a graph in the paper "A New Measurement Method of Diesel Engine Oil Consumption Rate" (*J. Society Auto Engr.*, 1985, pp. 28–33).

x: 4 5 8 11 12 16 17 20 22 28 30 31 39
y: 5 7 10 10 14 15 13 25 20 24 31 28 39

a. Assuming that x and y are related by the simple linear regression model, test $H_0 : \beta_1 = 1$ versus $H_a : \beta_1 \neq 1$ using a significance level of .05.

b. Calculate the value of the sample correlation coefficient for this data.

55. The paper "Root Regeneration and Early Growth of Red Oak Seedlings: Influence of Soil Temperature" (*Forest Science*, 1970, pp. 442–446) reported the results of a regression analysis in which the independent variable x was daily degree hours of soil heat and the dependent variable y was shoot elongation per seedling (cm). The accompanying data values were read from a graph that appeared in the paper.

x:	300	350	400	400
y:	5.8	4.5	5.9	6.2
x:	450	450	480	480
y:	6.0	7.5	6.1	8.6
x:	530	530	580	580
y:	8.9	8.2	14.2	11.9
x:	620	620	670	700
y:	11.1	11.5	14.5	14.8

a. What proportion of variation in shoot elongation is explained by variation in soil heat?

b. Does there appear to be a useful linear relationship between the two variables? Test the appropriate hypotheses.

c. Can it be concluded that expected amount of shoot elongation differs significantly from 10 when daily degree hours of soil heat equals 500? Use a level .05 test.

56. Eye weight (g) and cornea thickness (μm) were recorded for nine randomly selected calves, and the resulting data from "The Collagens of the Developing Bovine Cornea" (*Exper. Eye Research*, 1984, pp. 639–652) is given. Use this data and a .05 significance level to test the null hypothesis of no correlation between eye weight and cornea thickness versus the alternative hypothesis of a positive correlation between eye weight and cornea thickness.

Eye weight:	.2	1.4	2.2	2.7	4.9
Thickness:	416	673	733	801	957

Eye weight:	5.3	8.0	8.8	9.6
Thickness:	1035	883	736	567

57. The accompanying data is a subset of the data that appeared in the paper "Radial Tension Strength of Pipe and Other Curved Flexural Members" (*J. Amer. Concrete Inst.*, 1980, pp. 33–39). The variables are age of a pipe specimen (x, in days) and load necessary to obtain a first crack (y, in 1000 lb/ft).

x:	20	20	20	25	25
y:	11.45	10.42	11.14	10.84	11.17
x:	25	31	31	31	
y:	10.54	9.47	9.19	9.54	

a. Calculate the equation of the estimated regression line.

b. Suppose that a theoretical model suggested that the expected decrease in load associated with a one-day increase in age is at most .10. Does the data contradict this assertion? State and test the appropriate hypotheses at significance level .05.

c. For purposes of estimating the slope of the true regression line as accurately as possible, would it have been preferable to make a single observation at each of the ages 20, 21, 22, . . . , 30, and 31? Explain.

d. Calculate an estimate of true average load to first crack when age is 28 days. Your estimate should convey information regarding precision of estimation.

58. The paper "Photocharge Effects in Dye Sensitized Ag[Br,I] Emulsions at Millisecond Range Exposures" (*Photographic Sci. and Engr.,* 1981, pp. 138–144) gave the accompanying data on $x = \%$ light absorption at 5800A and $y =$ peak photovoltage.

x:	4.0	8.7	12.7	19.1	21.4
y:	.12	.28	.55	.68	.85

x:	24.6	28.9	29.8	30.5
y:	1.02	1.15	1.34	1.29

a. Construct a scatter plot of this data. What does it suggest?

b. Assuming that the simple linear regression model is appropriate, obtain the equation of the estimated regression line.

c. How much of the observed variation in peak photovoltage can be explained by the model relationship?

d. Predict peak photovoltage when % absorption is 19.1, and compute the value of the corresponding residual.

e. The paper's authors claimed that there is a useful linear relationship between % absorption and peak photovoltage. Do you agree? Carry out a formal test.

f. Give an estimate of the change in expected peak photovoltage associated with a 1% increase in light absorption. Your estimate should convey information about the precision of estimation.

g. Repeat (f) for the expected value of peak photovoltage when % light absorption is 20.

59. In Section 12.4 we presented a formula for $V(\hat{\beta}_0 + \hat{\beta}_1 x^*)$ and a confidence interval for $\beta_0 + \beta_1 x^*$. Taking $x^* = 0$ gives $\sigma_{\hat{\beta}_0}^2$ and a confidence interval for β_0. Use the data of Example 12.7 to calculate the estimated standard deviation of $\hat{\beta}_0$ and a 95% confidence interval for the y intercept of the true regression line.

60. Let s_x and s_y denote the sample standard deviations of the observed x's and y's, respectively [so $s_x^2 = \Sigma(x_i - \bar{x})^2/(n - 1)$ and similarly for s_y^2].

a. Show that an alternative expression for the estimated regression line $y = \hat{\beta}_0 + \hat{\beta}_1 x$ is

$$y = \bar{y} + r \cdot \frac{s_y}{s_x}(x - \bar{x})$$

b. This expression for the regression line can be interpreted as follows. Suppose that $r = .5$. What then is the predicted y for an x that lies one standard deviation (s_x units) above the mean of the x_i's? If r were 1, the prediction would be for y to lie one standard deviation above its mean $\bar{y}$, but since $r = .5$, we predict a y that is only .5 standard deviations (s_y units) above $\bar{y}$. Using the data in Exercise 49 for a patient whose age is one standard deviation below the average age in the sample, by how many standard deviations is the patient's predicted ΔCBG above or below the average ΔCBG for the sample?

61. Verify that the t statistic for testing $H_0: \beta_1 = 0$ in Section 12.3 is identical to the t statistic in Section 12.5 for testing $H_0: \rho = 0$.

62. Use the formula for computing SSE to verify that $r^2 = 1 - SSE/SST$.

63. The paper "Increased Oxygen Consumption During the Uptake of Water by the Eversible Vesicles of Petrobius Brevistylis" (*J. Insect Physiology,* 1977, pp. 1285–1294) presented the results of a regression of $y =$ increased oxygen uptake (in μl) above the mean resting rate on $x =$ weight increase (mg) when dehydrated insects were allowed access to distilled water. A sample size of $n = 20$ was used, and the computed summary statistics were (approximately, based on numbers read from a graph) $\Sigma x_i = 63.5$, $\Sigma y_i = 17.26$, $\Sigma x_i^2 = 311.74$, $\Sigma x_i y_i = 71.51$, and $\Sigma y_i^2 = 19.9625$.

a. Compute the equation of the estimated regression line.

b. There was only one observation made for an x value larger than 7: for $x_{20} = 9.8$, $y_{20} = 1.9$. The investigator would like to know whether the exclusion of this point greatly alters the estimated regression relationship. Compute the estimated regression line based just on the 19 pairs with (9.8, 1.9) deleted from the sample. What y would you predict using this new line when $x = 9.8$? *Hint:* First recompute the summary statistics; for example, new $\Sigma x_i = $ old $\Sigma x_i - 9.8$.

64. Reconsider the situation of Exercise 53, in which $x = \%$ RDF heat input and $y = \%$ efficiency for a particular boiler were related via the simple linear regression model $Y = \beta_0 + \beta_1 x + \epsilon$. Suppose that for a second boiler, these variables are also related via the simple linear regression model $Y = \gamma_0 + \gamma_1 x + \epsilon$ and that $V(\epsilon) = \sigma^2$ for both boilers. If the data set consists of n_1 observations on the first boiler and n_2 on the second, and SSE_1 and SSE_2 denote the two error sums of squares, then a pooled estimation of σ^2 is $\hat{\sigma}^2 = (SSE_1 + SSE_2)/(n_1 + n_2 - 4)$. Let ssx_1 and ssx_2 denote $\Sigma(x_i - \overline{x})^2$ for the data on the first and second boilers, respectively. A test of $H_0 : \beta_1 - \gamma_1 = 0$ (equal slopes) is based on the statistic

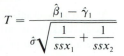

$$T = \frac{\hat{\beta}_1 - \hat{\gamma}_1}{\hat{\sigma}\sqrt{\dfrac{1}{ssx_1} + \dfrac{1}{ssx_2}}}$$

When H_0 is true, T has a t distribution with $n_1 + n_2 - 4$ d.f. Suppose that the six observations on the second boiler are $(0, 81.3)$, $(10, 78.4)$, $(20, 78.2)$, $(25, 79.1)$, $(30, 77.6)$, and $(40, 77.4)$. Using this along with the data of Exercise 53, carry out a test at level .05 to see whether or not expected change in % efficiency associated with a 1% increase in RDF heat input is identical for the two boilers.

Bibliography

Draper, Norman, and Smith, Harry, *Applied Regression Analysis* (2nd ed.), Wiley, New York, 1981. The most comprehensive and authoritative book on regression analysis currently in print.

Neter, John, Wasserman, William, and Kutner, Michael, *Applied Linear Statistical Models* (2nd ed.), Irwin, Homewood, Ill., 1985. The first 15 chapters constitute an extremely readable and informative survey of regression analysis.

Younger, Mary Sue, *A Handbook for Linear Regression* (2nd ed.), Duxbury Press, Boston, 1985. A readable presentation of regression models and methodology; its discussion of the capabilities of the BMD, SAS, and SPSS "canned" statistical computer packages is especially enlightening.

Nonlinear and Multiple Regression

Introduction

The probabilistic model studied in Chapter 12 specified that the observed value of the dependent variable Y deviated from the linear regression function $\mu_{Y \cdot x} = \beta_0 + \beta_1 x$ by a random amount. Here we consider two ways of generalizing the simple linear regression model. The first way is to replace $\beta_0 + \beta_1 x$ by a nonlinear function of x, while the second is to use a regression function involving more than a single independent variable. After fitting a regression function of the chosen form to the given data, it is of course important to have methods available for making inferences about the parameters of the chosen model. Before these methods are used, though, the data analyst should first assess the validity of the chosen model. In Section 13.1 we discuss methods, based primarily on a graphical analysis of the residuals (observed minus predicted y's), for checking the aptness of a fitted model.

In Section 13.2 we consider nonlinear regression functions of a single independent variable x that are "intrinsically linear." By this we mean that it is possible to transform one or both of the variables so that the relationship between the new variables is linear. Another class of nonlinear relations is obtained by using polynomial regression functions of the form $\mu_{Y \cdot x} = \beta_0 + \beta_1 x + \beta_2 x^2 + \cdots + \beta_k x^k$; these polynomial models are the subject of Section 13.3. Multiple regression analysis involves building models for relating Y to $k (\geq 2)$ independent variables $x_1, \ldots, x_k$. The focus in Section 13.4 is on interpretation of the parameters of various multiple regression models and on understanding and using the regression output of widely available statistical computer packages. The last section of the chapter surveys some extensions and pitfalls of multiple regression modeling.

13.1 Aptness of the Model and Model Checking

A plot of the observed pairs (x_i, y_i) is a necessary first step in deciding on the form of a mathematical relationship between x and y. It is possible to fit many functions other than a linear one $y = b_0 + b_1 x$ to the data, using either the principle of least squares or another fitting method. Once a function of the chosen form has been fitted, it is important to check the fit of the model to see whether it is in fact appropriate. One way to study the fit is to superimpose a graph of the best-fit function on the scatter plot of the data. However, any tilt or curvature of the best-fit function may obscure some aspects of the fit that should be investigated. Furthermore, the scale on the vertical axis may make it difficult to assess the extent to which observed values deviate from the best-fit functions.

Residuals and Fitted Values

A more effective approach to assessment of model adequacy is to compute the fitted or predicted values $\hat{y}_i$ and the residuals $e_i = y_i - \hat{y}_i$, and then plot various functions of these computed quantities. We then examine the plots either to confirm our choice of model or for indications that the model is not appropriate. Suppose that the simple linear regression model is correct, and let $y = \hat{\beta}_0 + \hat{\beta}_1 x$ be the equation of the estimated regression line. Then the ith residual is $e_i = y_i - (\hat{\beta}_0 + \hat{\beta}_1 x)$. To derive properties of the residuals, let $e_i = Y_i - \hat{Y}_i$ represent the ith residual as a random variable (before observations are actually made). Then

$$E(Y_i - \hat{Y}_i) = E(Y_i) - E(\hat{\beta}_0 + \hat{\beta}_1 X_i) = \beta_0 + \beta_1 x_i - (\beta_0 + \beta_1 x_i) = 0$$

(13.1)

so each residual has expected value zero. Because $\hat{Y}_i(=\hat{\beta}_0 + \hat{\beta}_1 x_i)$ is a linear function of the Y_j's, so is $Y_i - \hat{Y}_i$ (where the coefficients depend on the x_j's). Thus the normality of the Y_j's implies that each residual is normally distributed. It can also be shown that

$$V(Y_i - \hat{Y}_i) = \sigma^2 \cdot \left[1 - \frac{1}{n} - \frac{(x_i - \bar{x})^2}{\Sigma(x_j - \bar{x})^2} \right]$$

(13.2)

Since we shall want to know whether any given residual is unusually large, the **standardized residuals**

$$e_i^* = \frac{y_i - \hat{y}_i}{s\sqrt{1 - \dfrac{1}{n} - \dfrac{(x_i - \bar{x})^2}{\Sigma(x_j - \bar{x})^2}}}$$

(13.3)

will be useful. Notice that the variances of the residuals differ from one another. If n is reasonably large, though, the bracketed term in (13.2) will be approximately 1, so some authors use e_i/s as the standardized residual. Computation of the e_i^*'s can be tedious, but several of the most widely used statistical computer packages (such as MINITAB) automatically provide them and (upon request) can construct various plots involving them.

Example 13.1 Exercise 9 in Chapter 12 presented data on $x =$ burner area liberation rate and $y = NO_x$ emissions. Here we reproduce the data and give the fitted values, residuals, and standardized residuals. The estimated regression line is $y = -45.55 + 1.71x$, and $r^2 = .961$. Notice that the standardized residuals are not a constant multiple of the residuals (that is, $e_i^* \neq e_i/s$).

x_i	y_i	$\hat{y}_i$	e_i	e_i^*
100	150	125.6	24.4	.75
125	140	168.4	−28.4	−.84
125	180	168.4	11.6	.35
150	210	211.1	−1.1	−.03
150	190	211.1	−21.1	−.62
200	320	296.7	23.3	.66
200	280	296.7	−16.7	−.47
250	400	382.3	17.7	.50
250	430	382.3	47.7	1.35
300	440	467.9	−27.9	−.80
300	390	467.9	−77.9	−2.24
350	600	553.4	46.6	1.39
400	610	639.0	−29.0	−.92
400	670	639.0	31.0	.99

■

Diagnostic Plots

The basic plots that many statisticians recommend for an assessment of model validity and usefulness are

1. e_i^* (or e_i) on the vertical axis versus x_i on the horizontal axis,
2. e_i^* (or e_i) on the vertical axis versus $\hat{y}_i$ on the horizontal axis,
3. $\hat{y}_i$ on the vertical axis versus y_i on the horizontal axis, and
4. a normal probability plot of the standardized residuals.

Plots (1) and (2) are called **residual plots** (against the independent variable and fitted values, respectively), while (3) is a plot of fitted against observed values.

If plot (3) yields points close to the 45° line [slope +1 through (0, 0)], then the estimated regression function gives accurate predictions of the values actually observed. Thus (3) provides a visual assessment of model effectiveness in making predictions. Provided that the model is correct, both residual plots should exhibit no distinct patterns. The residuals should be randomly distributed about zero according to a normal distribution, so all but a very few standardized residuals should lie between −2 and +2 (that is, all but a few residuals within two standard deviations of their expected value zero). The plot of standardized residuals versus $\hat{y}$ is really a combination of the two other plots, showing implicitly both how residuals vary with x and how fitted values compare with observed values. This latter plot is the single one most often recommended for multiple regression analysis. The fourth plot allows the analyst to assess the plausibility of the assumption that ϵ has a normal distribution.

Example 13.2
(Example 13.1
continued)

Figure 13.1 presents, in addition to the scatter plot of the data, the four plots recommended in (1)–(4) above. The plot of $\hat{y}$ versus y confirms the impression given by r^2 that x is effective in predicting y, and also indicates that there is no observed y for which the predicted value is terribly far off the mark. Both

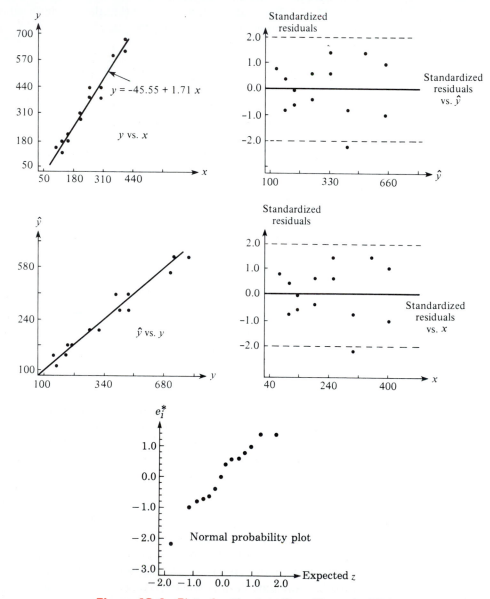

Figure 13.1   Plots for the data from Example 13.1

residual plots show no unusual pattern or discrepant values. There is one standardized residual slightly outside the interval $(-2, 2)$, but this is not surprising in a sample of size 14. The normal probability plot of the standardized residuals is reasonably straight (in Chapter 14 we present a formal test for normality based on the points in a normal probability plot; when applied to the data here, it appears quite plausible that the ϵ distribution is normal). In summary, the plots leave us with no qualms about either the appropriateness of a simple linear relationship or the fit to the given data.   ■

Difficulties and Remedies

While we hope that our analysis will yield plots like those of Figure 13.1, quite frequently the plots will suggest one or more of the following difficulties:

1. A nonlinear probabilistic relationship between x and y is appropriate.
2. The variance of ϵ (and of Y) is not a constant σ^2, but depends on x.
3. The selected model fits the data well except for a very few discrepant or outlying data values, which may have greatly influenced the choice of the best-fit function.
4. The error term ϵ does not have a normal distribution.
5. When the subscript i indicates the time order of the observations, the ϵ_i's exhibit dependence over time.
6. One or more relevant independent variables have been omitted from the model.

Figure 13.2 presents residual plots corresponding to (1)–(3), (5), and (6). In Chapter 4 we discussed patterns in normal probability plots that cast doubt on the assumption of an underlying normal distribution. Notice that the residuals from the data in Figure 13.2d with the circled point included would not by themselves necessarily suggest further analysis, yet when a new line is fit with that point deleted, the new line differs considerably from the original line. This type of behavior is more difficult to identify in multiple regression. It is most likely to arise when there is a single (or very few) data point(s) with independent variable value(s) far removed from the remainder of the data.

We now indicate briefly what remedies are available for the types of difficulties. For a more comprehensive discussion, one or more of the references on regression analysis should be consulted. If the residual plot looks something like that of Figure 13.2a, exhibiting a curved pattern, then a nonlinear function of x may be fit; in Section 13.2 we show how transformations can be used to this end, while in Section 13.3 we discuss polynomial models.

The residual plot of Figure 13.2b suggests that while a straight-line relationship may be reasonable, the assumption that $V(Y_i) = \sigma^2$ for each i is of doubtful validity. When the assumptions of Chapter 12 are valid, it can be shown that among all unbiased estimators of β_0 and β_1, the ordinary least squares estimators have minimum variance. These estimators give equal weight to each (x_i, Y_i). If the variance of Y increases with x, then Y_i's for large x_i

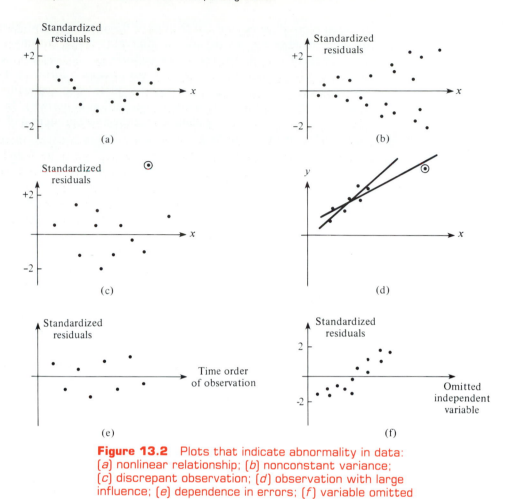

Figure 13.2   Plots that indicate abnormality in data:
(*a*) nonlinear relationship; (*b*) nonconstant variance;
(*c*) discrepant observation; (*d*) observation with large
influence; (*e*) dependence in errors; (*f*) variable omitted

should be given less weight than those with small x_i. This suggests that β_0 and β_1 should be estimated by minimizing

$$f_w(b_0, b_1) = \Sigma w_i[y_i - (b_0 + b_1 x_i)]^2 \tag{13.4}$$

where the w_i's are weights that decrease with increasing x_i. Minimization of (13.4) yields **weighted least squares** estimates. For example, if the standard deviation of Y is proportional to x (for $x > 0$)—that is, $V(Y) = kx^2$—then it can be shown that the weights $w_i = 1/x_i^2$ yield best estimators of β_0 and β_1. The books by Neter et al. and by Chatterjee and Price contain more detail. Weighted least squares is used quite frequently by econometricians (economists who use statistical methods) to estimate parameters.

When plots or other evidence suggest that the data set contains outliers or points having large influence on the resulting fit, one possible approach is to omit these outlying points and recompute the estimated regression equation.

This would certainly be correct if it was found that the outliers resulted from errors in recording data values or experimental errors. If no assignable cause can be found for the outliers, it is still desirable to report the estimated equation both with and without outliers omitted. Yet another approach is to retain possible outliers but to use an estimation principle that puts relatively less weight on outlying values than does the principle of least squares. One such principle is MAD (minimize absolute deviations), which selects $\hat{\beta}_0$ and $\hat{\beta}_1$ to minimize $\Sigma|y_i - (b_0 + b_1 x_i)|$. Unlike the estimates of least squares, there are no nice formulas for the MAD estimates: their values must be found by using an iterative computational procedure. More information about alternative fitting procedures can be found in Chapter 14 of *Data Analysis and Regression* by Mosteller and Tukey. Such procedures are also used when it is suspected that the ϵ_i's have a distribution that is not normal but instead has "heavy tails" (making it much more likely than for the normal distribution that discrepant values will enter the sample); robust regression procedures are those that produce reliable estimates for a wide variety of underlying error distributions. Least squares estimators are not robust in the same way that the sample mean $\overline{X}$ is not a robust estimator for μ.

When a plot suggests time dependence in the error terms, an appropriate analysis may involve a transformation of the y's or else a model explicitly including a time variable. Lastly, a plot such as that of Figure 13.2f, which shows a pattern in the residuals when plotted against an omitted variable, suggests that a multiple regression model that includes the previously omitted variable should be considered. We discuss such models in the last two sections of the chapter.

Exercises / Section 13.1 (1–12)

1. Suppose that the variables $x =$ commuting distance and $y =$ commuting time are related according to the simple linear regression model with $\sigma = 10$.

  a. If $n = 5$ observations are made at the x values $x_1 = 5$, $x_2 = 10$, $x_3 = 15$, $x_4 = 20$, and $x_5 = 25$, calculate the standard deviations of the five corresponding residuals.

  b. Repeat (a) for $x_1 = 5$, $x_2 = 10$, $x_3 = 15$, $x_4 = 20$, and $x_5 = 50$.

  c. What do the results of (a) and (b) imply about the deviation of the estimated line from the observation made at the largest sampled x value?

2. The x values and standardized residuals for the chlorine flow-etch rate data of Example 12.3 are displayed in the accompanying table. Construct a standardized residual plot and comment on its appearance.

x :	1.50	1.50	2.00	2.50	2.50
e^* :	.31	1.02	-1.15	-1.23	.23

x :	3.00	3.50	3.50	4.00
e^* :	.73	-1.36	1.53	.07

3. Example 12.5 presented the residuals from a simple linear regression of bark lead content y on traffic flow x.

  a. Plot the residuals against x. Does the resulting plot suggest that a straight-line regression function is a reasonable choice of model? Explain your reasoning.

  b. Using $s = 92.19$, compute the values of the standardized residuals. Is $e_i^* \approx e_i/s$ for $i = 1, \ldots, n$, or are the e_i^*'s not close to being proportional to the e_i's?

  c. Plot the standardized residuals against x. Does the plot differ significantly in general appearance from the plot of (a)?

4. Exercise 36 in Chapter 12 presented data on y = magnetic relaxation time and x = strength of the external magnetic field. A simple linear regression yields the estimated regression function $y = -8.78 + 18.87x$, with $s = 16.61$.

   a. Compute the values of the residuals and plot the residuals against x. Does the plot suggest that a linear regression function is inappropriate?

   b. Compute the values of the standardized residuals and plot them against x. Are there any unusually large (positive or negative) standardized residuals? Does this plot give the same message as the plot of (a) as far as the appropriateness of a linear regression function?

   c. Plot the predicted against the observed values. Does the plot indicate that the model gives reliable predictions?

   d. Construct a normal probability plot of the standardized residuals, and comment on its appearance.

5. The article "Effects of Gamma Radiation on Juvenile and Mature Cuttings of Quaking Aspen" (*Forest Science,* 1967, pp. 240–245) reported the following data on exposure time to radiation (x, in kr/16 hr) and dry weight of roots (y, in mg $\times 10^{-1}$).

x	0	2	4	6	8
y	110	123	119	86	62

   a. Construct a scatter plot. Does the plot suggest that a linear probabilistic relationship is appropriate?

   b. A linear regression results in the least squares line $y = 127 - 6.65x$, with $s = 16.94$. Compute the residuals and standardized residuals, and then construct residual plots. What do these plots suggest? What type of function should provide a better fit to the data than a straight line does?

6. Exercise 47 in Chapter 12 reported data on x = age of tree and y = annual trunk diameter growth increment. The estimated regression line is $y = 1.78 - .0154x$, with $s = .434$. Does a scatter plot suggest that a particular observation has excessively influenced the choice of a best fit line? Is this impression confirmed by the standardized residuals?

7. Consider the following four (x, y) data sets; the first three have the same x values, so these values are listed only once (from Frank Anscombe, "Graphs in Statistical Analysis," *Amer. Statistician,* 1973, pp. 17–21).

Data set	1–3	1	2	3	4	4
Variable	x	y	y	y	x	y
	10.0	8.04	9.14	7.46	8.0	6.58
	8.0	6.95	8.14	6.77	8.0	5.76
	13.0	7.58	8.74	12.74	8.0	7.71
	9.0	8.81	8.77	7.11	8.0	8.84
	11.0	8.33	9.26	7.81	8.0	8.47
	14.0	9.96	8.10	8.84	8.0	7.04
	6.0	7.24	6.13	6.08	8.0	5.25
	4.0	4.26	3.10	5.39	19.0	12.50
	12.0	10.84	9.13	8.15	8.0	5.56
	7.0	4.82	7.26	6.42	8.0	7.91
	5.0	5.68	4.74	5.73	8.0	6.89

For each of these four data sets, the values of the summary statistics Σx_i, Σx_i^2, Σy_i, Σy_i^2, and $\Sigma x_i y_i$ are identical, so all quantities computed from these five will be identical for the four sets—the least squares line ($y = 3 + .5x$), SSE, s^2, r^2, t intervals, t statistics, and so on. The summary statistics provide no way of distinguishing among the four data sets. Based on a scatter plot and a residual plot for each set, comment on the appropriateness or inappropriateness of fitting a straight-line model; include in your comments any specific suggestions for how a "straight-line analysis" might be modified or qualified.

8. a. Show that $\sum\limits_{i=1}^{n} e_i = 0$ when the e_i's are the residuals from a simple linear regression.

   b. Are the residuals from a simple linear regression independent of one another, positively correlated, or negatively correlated? Explain.

   c. Show that $\sum\limits_{i=1}^{n} x_i e_i = 0$ for the residuals from a simple linear regression. [This result along with (a) shows that there are two linear restrictions on the e_i's, resulting in a loss of 2 d.f. when the squared residuals are used to estimate σ^2.]

   d. Is it true that $\sum\limits_{i=1}^{n} e_i^* = 0$? Give a proof or counterexample.

9. a. Express the ith residual $Y_i - \hat{Y}_i$ (where $\hat{Y}_i = \hat{\beta}_0 + \hat{\beta}_1 x_i$) in the form $\Sigma c_j Y_j$, a linear function of the Y_j's. Then use rules of variance to verify that $V(Y_i - \hat{Y}_i)$ is given by Expression (13.2).

b. It can be shown that $\hat{Y}_i$ and $Y_i - \hat{Y}_i$ (the ith predicted value and residual) are independent of one another. Use this fact, the relation $Y_i = \hat{Y}_i + (Y_i - \hat{Y}_i)$, and the expression for $V(\hat{Y})$ from Section 12.4 to again verify (13.2).

c. As x_i moves further away from $\bar{x}$, what happens to $V(\hat{Y}_i)$ and to $V(Y_i - \hat{Y}_i)$?

10. a. Could a linear regression result in residuals 23, -27, 5, 17, -8, 9, and 15? Why or why not?

b. Could a linear regression result in residuals 23, -27, 5, 17, -8, -12, and 2 corresponding to x values 3, -4, 8, 12, -14, -20, and 25? Why or why not? *Hint:* See Exercise 8.

11. Recall that $\hat{\beta}_0 + \hat{\beta}_1 x$ has a normal distribution with expected value $\beta_0 + \beta_1 x$ and variance

$$\sigma^2 \left\{ \frac{1}{n} + \frac{(x - \bar{x})^2}{\Sigma(x_i - \bar{x})^2} \right\}$$

so that

$$Z = \frac{\hat{\beta}_0 + \hat{\beta}_1 x - (\beta_0 + \beta_1 x)}{\sigma \left(\frac{1}{n} + \frac{(x - \bar{x})^2}{\Sigma(x_i - \bar{x})^2} \right)^{1/2}}$$

has a standard normal distribution. If $S = \sqrt{SSE/(n-2)}$ is substituted for σ, the resulting variable has a t distribution with $n - 2$ degrees of freedom. By analogy, what is the distribution of any particular standardized residual? If $n = 25$, what is the probability that a particular standardized residual falls outside the interval $(-2.50, 2.50)$?

12. If there is at least one x value at which more than one Y has been observed, there is a formal test procedure for testing

$H_0 : \mu_{Y \cdot x} = \beta_0 + \beta_1 x$ for some values β_0, β_1 (the true regression function is linear)

versus

$H_a : H_0$ is not true (the true regression function is not linear)

Suppose that observations are made at $x_1, x_2, \ldots, x_c$, and let $Y_{11}, Y_{12}, \ldots, Y_{1n_1}$ denote the n_1 observations when $x = x_1; \ldots; Y_{c1}, Y_{c2}, \ldots, Y_{cn_c}$ denote the n_c observations when $x = x_c$. With $n = \Sigma n_i$ (the total number of observations), SSE has $n - 2$ d.f. We break SSE into two pieces, $SSPE$ (pure error) and $SSLF$ (lack of fit), as follows:

$$SSPE = \sum_i \sum_j (Y_{ij} - \bar{Y}_{i \cdot})^2$$

$$= \Sigma \Sigma Y_{ij}^2 - \Sigma n_i \bar{Y}_{i \cdot}^2.$$

$$SSLF = SSE - SSPE$$

The n_i observations at x_i contribute $n_i - 1$ d.f. to $SSPE$, so the number of d.f. for $SSPE$ is $\sum_i (n_i - 1) = n - c$ and the d.f. for $SSLF$ is $n - 2 - (n - c) = c - 2$. Let $MSPE = SSPE/(n - c)$ and $MSLF = SSLF/(c - 2)$. Then it can be shown that while $E(MSPE) = \sigma^2$ whether or not H_0 is true, $E(MSLF) = \sigma^2$ if H_0 is true and $E(MSLF) > \sigma^2$ if H_0 is false.

Test statistic: $F = \dfrac{MSLF}{MSPE}$

Rejection region: $f \geq F_{\alpha, c - 2, n - c}$

The following data comes from the paper "Changes in Growth Hormone Status Related to Body Weight of Growing Cattle" (*Growth*, 1977, pp. 241–247), with $x =$ body weight and $y =$ metabolic clearance rate/body weight.

x	110	110	110	230	230	230	360
y	235	198	173	174	149	124	115

x	360	360	360	505	505	505	505
y	130	102	95	122	112	98	96

(So $c = 4$, $n_1 = n_2 = 3$, $n_3 = n_4 = 4$.)

a. Test H_0 versus H_a at level .05 using the above lack of fit test.

b. Does a scatter plot of the data suggest that the relationship between x and y is linear? How does this compare with the result of (a)? (A nonlinear regression function was used in the paper.)

13.2   Regression with Transformed Variables

The necessity for an alternative model to the linear probabilistic model $Y = \beta_0 + \beta_1 x + \epsilon$ may be suggested either by a theoretical argument or else by examining diagnostic plots from a linear regression analysis. In either case it is desirable to settle on a model whose parameters can be easily estimated. An important class of such models is specified by means of functions that are "intrinsically linear."

Definition

> A function relating y to x is **intrinsically linear** if by means of a transformation on x and/or y, the function can be expressed as $y' = \beta_0 + \beta_1 x'$, where $x' = $ the transformed independent variable and $y' = $ the transformed dependent variable.

Four of the most useful intrinsically linear functions are given in Table 13.1 below. In each case the appropriate transformation is either a log transformation—either base 10 or natural logarithm (base e)—or a reciprocal transformation. Representative graphs of the four functions appear in Figure 13.3.

Table 13.1   Useful Intrinsically Linear Functions*

Function	Transformation(s) to linearize	Linear form
(a) Exponential: $y = \alpha e^{\beta x}$	$y' = \ln(y)$	$y' = \ln(\alpha) + \beta x$
(b) Power: $y = \alpha x^{\beta}$	$y' = \log(y),\ x' = \log(x)$	$y' = \log(\alpha) + \beta x'$
(c) $y = \alpha + \beta \cdot \log(x)$	$x' = \log(x)$	$y = \alpha + \beta x'$
(d) Reciprocal: $y = \alpha + \beta \cdot \dfrac{1}{x}$	$x' = \dfrac{1}{x}$	$y = \alpha + \beta x'$

*When $\log\,(\cdot)$ appears, either a base 10 or base e logarithm can be used.

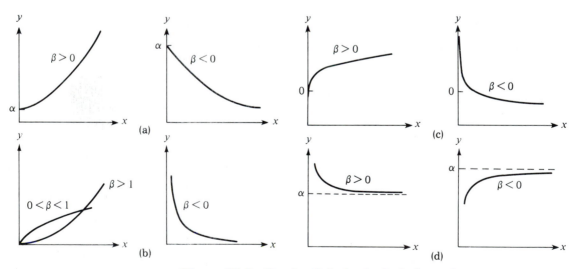

Figure 13.3   Graphs of the intrinsically linear functions given in Table 13.1

Thus for an exponential function relationship, only y is transformed to achieve linearity, while for a power function relationship, both x and y are transformed. Because the variable x is in the exponent in an exponential relationship, y increases (if $\beta > 0$) or decreases (if $\beta < 0$) much more rapidly as x increases than is the case for the power function, though over a short interval of x values it can be difficult to differentiate between the two functions. Examples of functions that are not intrinsically linear are $y = \alpha + \gamma e^{\beta x}$ and $y = \alpha + \gamma x^{\beta}$.

Intrinsically linear functions lead directly to probabilistic models which, though not linear in x as a function, have parameters whose values are easily estimated using ordinary least squares.

Definition

> A probabilistic model relating Y to x is **intrinsically linear** if by means of a transformation on Y and/or x, it can be reduced to a linear probabilistic model $Y' = \beta_0 + \beta_1 x' + \epsilon'$.

Corresponding to the four functions of Table 13.1, the intrinsically linear probabilistic models are

a. $Y = \alpha e^{\beta x} \cdot \epsilon$, a multiplicative exponential model, so that $\ln(Y) = Y' = \beta_0 + \beta_1 x' + \epsilon'$ with $x' = x$, $\beta_0 = \ln(\alpha)$, $\beta_1 = \beta$, and $\epsilon' = \ln(\epsilon)$;

b. $Y = \alpha x^{\beta} \cdot \epsilon$, a multiplicative power model, so that $\log(Y) = Y' = \beta_0 + \beta_1 x' + \epsilon'$ with $x' = \log(x)$, $\beta_0 = \log(\alpha)$, $\beta_1 = \beta$, and $\epsilon' = \log(\epsilon)$;

c. $Y = \alpha + \beta \log(x) + \epsilon$, so that $x' = \log(x)$ immediately linearizes the model;

d. $Y = \alpha + \beta \cdot 1/x + \epsilon$, so that $x' = 1/x$ yields a linear model.

The additive exponential and power models, $Y = \alpha e^{\beta x} + \epsilon$ and $Y = \alpha x^{\beta} + \epsilon$, are not intrinsically linear. Notice that both (a) and (b) require a transformation on Y and, as a result, a transformation on the error variable ϵ. In fact, if ϵ has a lognormal distribution (see Chapter 4) with $E(\epsilon) = e^{\sigma^2/2}$ and $V(\epsilon) = \tau^2$ independent of x, then the transformed models for both (a) and (b) will satisfy all the assumptions of Chapter 12 regarding the linear probabilistic model; this in turn implies that all inferences for the parameters of the transformed model based on these assumptions will be valid. If σ^2 is small, $\mu_{Y \cdot x} \approx \alpha e^{\beta x}$ in (a) or αx^{β} in (b).

The major advantage of an intrinsically linear model is that the parameters β_0 and β_1 of the transformed model can be immediately estimated using the principle of least squares simply by substituting x' and y' into the estimating formulas:

$$\hat{\beta}_1 = \frac{n \Sigma x_i' y_i' - \Sigma x_i' \Sigma y_i'}{n \Sigma (x_i')^2 - (\Sigma x_i')^2}$$

$$\hat{\beta}_0 = \frac{\Sigma y_i' - \hat{\beta}_1 \Sigma x_i'}{n}$$

(13.5)

Parameters of the original nonlinear model can then be estimated by transforming back $\hat{\beta}_0$ and/or $\hat{\beta}_1$ if necessary. In cases (a) and (b), when σ^2 is small an approximate confidence interval for $\mu_{Y \cdot x}$ results from taking antilogs in the interval for $\beta_0 + \beta_1 x$.*

Example 13.3

Taylor's equation for tool life y as a function of cutting time x states that $xy^c = k$, or equivalently that $y = \alpha x^{\beta}$. The paper "The Effect of Experimental Error on the Determination of Optimum Metal Cutting Conditions" (*J. Eng. for Industry*, 1967, pp. 315–322) observes that the relationship is not exact (deterministic), and that the parameters α and β must be estimated from data. Thus an appropriate model is the multiplicative power model $Y = \alpha \cdot x^{\beta} \cdot \epsilon$, which the author fit to the accompanying data consisting of 12 carbide tool life observations. In addition to the x, y, x', and y' values, the predicted transformed values ($\hat{y}'$) and the predicted values on the original scale ($\hat{y}$, after transforming back) are given.

Table 13.2

	x	y	$x' = \ln(x)$	$y' = \ln(y)$	$\hat{y}'$	$\hat{y} = e^{\hat{y}'}$
1	600.	2.3500	6.39693	0.85442	1.12754	3.0881
2	600.	2.6500	6.39693	0.97456	1.12754	3.0881
3	600.	3.0000	6.39693	1.09861	1.12754	3.0881
4	600.	3.6000	6.39693	1.28093	1.12754	3.0881
5	500.	6.4000	6.21461	1.85630	2.11203	8.2650
6	500.	7.8000	6.21461	2.05412	2.11203	8.2650
7	500.	9.8000	6.21461	2.28238	2.11203	8.2650
8	500.	16.5000	6.21461	2.80336	2.11203	8.2650
9	400.	21.5000	5.99146	3.06805	3.31694	27.5760
10	400.	24.5000	5.99146	3.19867	3.31694	27.5760
11	400.	26.0000	5.99146	3.25810	3.31694	27.5760
12	400.	33.0000	5.99146	3.49651	3.31694	27.5760

The summary statistics for fitting a straight line to the transformed data are $\Sigma x_i' = 74.41200$, $\Sigma y_i' = 26.22601$, $\Sigma x_i'^2 = 461.75874$, $\Sigma y_i'^2 = 67.74609$, and $\Sigma x_i' y_i' = 160.84601$, so

$$\hat{\beta}_1 = \frac{12(160.84601) - (74.41200)(26.22601)}{12(461.75874) - (74.41200)^2} = -5.3996$$

$$\hat{\beta}_0 = \frac{26.22601 - (-5.3996)(74.41200)}{12} = 35.6684$$

The estimated values of α and β, the parameters of the power function model, are $\hat{\beta} = \hat{\beta}_1 = -5.3996$ and $\hat{\alpha} = e^{\hat{\beta}_0} = 3.094491530 \cdot 10^{15}$. Thus the estimated regression function is $\hat{\mu}_{Y \cdot x} = 3.094491530 \cdot 10^{15} \cdot x^{-5.3996}$. To recapture

---

*Strictly speaking, taking antilogs gives a confidence interval for the median of the Y distribution for the fixed x value—that is, for $\tilde{\mu}_{Y \cdot x}$. Because the lognormal distribution is positively skewed, $\mu_{Y \cdot x} > \tilde{\mu}_{Y \cdot x}$; the two are approximately equal if σ^2 is close to zero.

Taylor's (estimated) equation, set $y = 3.094491530 \cdot 10^{15} \cdot x^{-5.3996}$, whence $xy^{.185} = 740$.

Figure 13.4a gives a plot of the standardized residuals from the linear regression using transformed variables (for which $r^2 = .922$); there is no apparent pattern in the plot, though one standardized residual is a bit large, and the residuals look as they should for a simple linear regression. Figure 13.4b pictures a plot of $\hat{y}$ versus y, which indicates satisfactory predictions on the original scale.

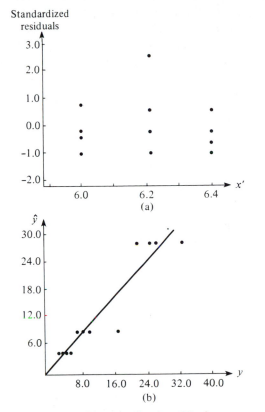

Figure 13.4 (a) Standardized residuals versus x' from Example 13.3; (b) $\hat{y}$ versus y from Example 13.3

To obtain a confidence interval for median tool life when cutting time is 500, we transform $x = 500$ to $x' = 6.21461$. Then $\hat{\beta}_0 + \hat{\beta}_1 x' = 2.1120$, and a 95% confidence interval for $\beta_0 + \beta_1(6.21461)$ is (from Section 12.4) $2.1120 \pm (2.228)(.0824) = (1.928, 2.296)$. The 95% confidence interval for $\tilde{\mu}_{Y \cdot 500}$ is then obtained by taking antilogs: $(e^{1.928}, e^{2.296}) = (6.876, 9.930)$. It is easily checked that for the transformed data $s^2 = \hat{\sigma}^2 \approx .081$. Because this is quite small, $(6.876, 9.930)$ is an approximate interval for $\mu_{Y \cdot 500}$. ■

Example **13.4**   In the article "Ethylene Synthesis in Lettuce Seeds: Its Physiological Significance" (*Plant Physiology,* 1972, pp. 719–722), ethylene content of lettuce seeds
(y, in nl/g dry wt) was studied as a function of exposure time (x, in min) to an
ethylene absorbant. Figure 13.5 presents both a scatter plot of the data and a
plot of the residuals generated from a linear regression of y on x. Both plots
show a strong curved pattern, suggesting that a transformation to achieve linearity is appropriate. In addition, a linear regression gives negative predictions
for $x = 90$ and $x = 100$.

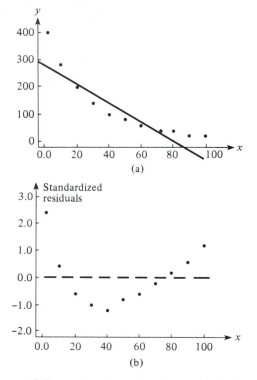

Figure 13.5   (*a*) Scatter plot; (*b*) residual plot
from linear regression for the data in Example 13.4

The author did not give any argument for a theoretical model, but his plot
of $y' = \ln(y)$ versus x shows a strong linear relationship, suggesting that an
exponential function will provide a good fit to the data. Below we record the
data values and other information from a linear regression of y' on x. The
estimates of parameters of the linear model are $\hat{\beta}_1 = -.0323$ and $\hat{\beta}_0 = 5.941$,
with $r^2 = .995$. The estimated regression function for the exponential model is
$\hat{\mu}_{Y \cdot x} \approx e^{\hat{\beta}_0} \cdot e^{\hat{\beta}_1 x} = 380.32 e^{-.0323x}$. The predicted values $\hat{y}_i$ can then be obtained
by substitution of $x_i (i = 1, \ldots, n)$ into $\hat{\mu}_{Y \cdot x}$ or else by computing $\hat{y}_i = e^{\hat{y}'_i}$
where the $\hat{y}'_i$'s are the predictions from the transformed straight-line model.
Figure 13.6 presents both a plot of $e'_i{}^*$ versus x (the standardized residuals from

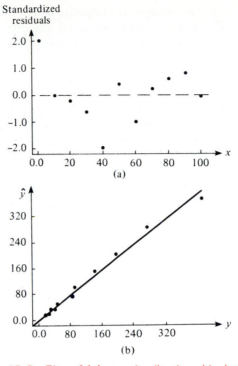

Figure 13.6 Plot of (a) standardized residuals (after transforming) versus x; (b) ŷ versus y for data in Example 13.4

a linear regression) and a plot of $\hat{y}$ versus y. These plots support the choice of an exponential model.

Table 13.3

x	y	$y' = \ln(y)$	$\hat{y}'$	$\hat{y} = e^{\hat{y}'}$
2	408	6.01	5.876	353.32
10	274	5.61	5.617	275.12
20	196	5.28	5.294	199.12
30	137	4.92	4.971	144.18
40	90	4.50	4.647	104.31
50	78	4.36	4.324	75.50
60	51	3.93	4.001	54.64
70	40	3.69	3.677	39.55
80	30	3.40	3.354	28.62
90	22	3.09	3.031	20.72
100	15	2.71	2.708	15.00

In analyzing transformed data, one should keep in mind the following points:

1. Estimating β_1 and β_0 as in (13.5) above and then transforming back to obtain estimates of the original parameters is not equivalent to using the

principle of least squares directly on the original model. Thus for the exponential model, we could estimate α and β by minimizing $\Sigma(y_i - \alpha e^{\beta x_i})^2$. The resulting estimates would not be equal: $\hat{\alpha} \neq e^{\hat{\beta}_0}$ and $\hat{\beta} \neq \hat{\beta}_1$.

2. If the chosen model is not intrinsically linear, the approach summarized in (13.5) cannot be used. Instead, least squares (or some other fitting procedure) would have to be applied to the untransformed model. Thus for the additive exponential model $Y = \alpha e^{\beta x} + \epsilon$, least squares would involve minimizing $\Sigma(y_i - \alpha e^{\beta x_i})^2$. Taking partial derivatives with respect to α and β results in two nonlinear normal equations in α and β; these equations must then be solved using an iterative procedure. The book by Draper and Smith gives more details.

3. When the transformed linear model satisfies all the assumptions listed in Chapter 12, least squares yields best estimates of the transformed parameters. However, estimates of the original parameters may not be best in any sense, though they will be reasonable. For example, in the exponential model the estimator $\hat{\alpha} = e^{\hat{\beta}_0}$ will not be unbiased, though it will be the maximum likelihood estimator of α if the error variable ϵ' is normally distributed. It is conceivable that using least squares directly (without transforming) could yield better estimates, though the computations would be quite burdensome.

4. If a transformation on Y has been made and one wishes to use the standard formulas to test hypotheses or construct confidence intervals, ϵ' should be at least approximately normally distributed. To check this, the residuals from the transformed regression should be examined.

5. When Y is transformed, the r^2 value from the resulting regression refers to variation in the y_i''s explained by the transformed regression model. While a high value of r^2 here indicates a good fit of the estimated original nonlinear model to the observed y_i's, r^2 does not refer to these original observations. Perhaps the best way to assess the quality of the fit is to compute the predicted values $\hat{y}_i'$ using the transformed model, transform them back to the original y scale to obtain $\hat{y}_i$, and then plot $\hat{y}$ versus y. A good fit is then evidenced by points close to the 45° line. One could compute $SSE = \Sigma(y_i - \hat{y}_i)^2$ as a numerical measure of the goodness of fit. When the model was linear, we compared this to $SST = \Sigma(y_i - \bar{y})^2$, the total variation about the horizontal line at height $\bar{y}$; this led to r^2. In the nonlinear case, though, it is not necessarily informative to measure total variation in this way, so an r^2 value is not as useful as in the linear case.

Exercises / Section 13.2 (13–21)

13. The accompanying data resulted from an experiment to investigate the relationship between applied stress x and time to rupture y for notched cold-rolled brass specimens exposed to a cracking solution ("On Initiation and Growth of Stress Corrosion Cracks in Tarnished Brass," *J. Electrochemical Soc.*, 1965, pp. 131–138).

x	22.5	25.0	28.0	30.5	38.0	40.5
y	44.0	42.0	33.5	28.0	18.0	13.6

x	42.5	48.0	54.5	55.0	70.0
y	15.0	10.3	9.0	6.3	4.0

A plot on log–log paper shows a pronounced linear pattern, suggesting a power model.

a. Show that fitting $Y' = \beta_0 + \beta_1 x' + \epsilon'$, where $Y' = \ln(Y)$ and $x' = \ln(x)$, results in $\hat{\beta}_1 = -2.16$ and $\hat{\beta}_0 = 10.67$. What are the estimates of the parameters of the power model?

b. Does a residual plot based on the transformed data and analysis confirm the choice of a power model?

c. Does a plot of $\hat{y}$ versus y confirm the efficacy of the power model for prediction of y?

d. Assuming that the power model is correct, test $H_0: \tilde{\mu}_{Y \cdot x} = \alpha x^{-2}$. $SSE = .128$ for the transformed data.

e. Obtain a 95% confidence interval for the exponent β in the power model.

f. Obtain a 95% confidence interval for the median time to rupture when applied stress $= 40$.

14. The paper "The Luminosity–Spectral Index Relationship for Radio Galaxies" (*Nature*, 1972, pp. 88–89) suggested that for class S galaxies, $\ln(L_{178})$ is linearly related to spectral index, where L_{178} denotes luminosity at 178 MHz. Representative data appears below.

Spectral index	.59	.67	.72	.80	.85	.90
$\ln(L_{178})$	23.6	25.6	26.4	25.7	26.8	26.7

Spectral index	.94	.66	1.00	.86	1.03	.70
$\ln(L_{178})$	27.0	24.9	27.1	27.2	26.9	25.2

a. Estimate the parameters of the exponential model implied by the linear relationship between $\ln(L_{178})$ and spectral index.

b. What value of L_{178} would you predict for a spectral index of .75?

c. Compute a 95% prediction interval for luminosity when the spectral index is .95.

15. The following data on mass rate of burning x and flame length y is representative of that which appeared in the article "Some Burning Characteristics of Filter Paper" (*Combustion Science and Technology*, 1971, pp. 103–120).

x	1.7	2.2	2.3	2.6	2.7	3.0	3.2
y	1.3	1.8	1.6	2.0	2.1	2.2	3.0

x	3.3	4.1	4.3	4.6	5.7	6.1
y	2.6	4.1	3.7	5.0	5.8	5.3

a. Estimate the parameters of a power function model.

b. Construct diagnostic plots to check whether a power function is an appropriate model choice.

c. Test $H_0: \beta = \frac{4}{3}$ versus $H_a: \beta < \frac{4}{3}$, using a level .05 test.

d. Test the null hypothesis that states that the median flame length when burning rate is 5.0 is twice the median flame length when burning rate is 2.5 against the alternative that this is not the case.

16. An investigation of the influence of sodium benzoate concentration on the critical minimum pH necessary for the inhibition of Fe ("Mechanism of the Corrosion Inhibition of Fe by Sodium Benzoate," *Corrosion Science*, 1971, pp. 675–682) yielded the accompanying data, which suggests that expected critical minimum pH is linearly related to the natural logarithm of concentration.

Concentration	.01	.025	.1	.95
pH	5.1	5.5	6.1	7.3

a. What is the implied probabilistic model, and what are the estimates of the model parameters?

b. What critical minimum pH would you predict for a concentration of 1.0? Obtain a 95% prediction interval for critical minimum pH when concentration is 1.0?

17. Thermal endurance tests were performed to study the relationship between temperature and lifetime of polyester enameled wire ("Thermal Endurance of Polyester Enameled Wires Using Twisted Wire Specimens," *IEEE Trans. Insulation*, 1965, pp. 38–44).

Temp.:	200	200	200	200	200	200
Lifetime:	5933	5404	4947	4963	3358	3878

Temp.:	220	220	220	220	220	220
Lifetime:	1561	1494	747	768	609	777

Temp.:	240	240	240	240	240	240
Lifetime:	258	299	209	144	180	184

a. Does a scatter plot of the data suggest a linear probabilistic relationship between lifetime and temperature?

b. What model is implied by a linear relationship between expected ln(lifetime) and 1/temperature? Does a scatter plot of the transformed data appear consistent with this relationship?

c. Estimate the parameters of the model suggested in (b). What lifetime would you predict for a temperature of 220?

d. Because there are multiple observations at each x value, the method in Exercise 12 can be used to test the null hypothesis which states that the model suggested in (b) is correct. Carry out the test at level .01.

18. Exercise 12 presented data on body weight x and MCR/body weight y. Consider the following intrinsically linear functions for specifying the relationship between the two variables: (a) $\ln(y)$ versus x, (b) $\ln(y)$ versus $\ln(x)$, (c) y versus $\ln(x)$, (d) y versus $1/x$, and (e) $\ln(y)$ versus $1/x$. Use any appropriate diagnostic plots and analyses to decide which of these functions you would select to specify a probabilistic model. Explain your reasoning.

19. A plot appearing in the article "Thermal Conductivity of Polyethylene: The Effects of Crystal Size, Density, and Orientation on the Thermal Conductivity" (*Polymer Eng. and Science*, 1972,

pp. 204–208) suggests that the expected value of thermal conductivity y is a linear function of $10^4 \cdot 1/x$ where x is lamellar thickness.

x	240	410	460	490	520	590	745	8300
y	12	14.7	14.7	15.2	15.2	15.6	16.0	18.1

a. Estimate the parameters of the regression function and the regression function itself.

b. Predict the value of thermal conductivity when lamellar thickness is 500 angstroms.

20. In each of the following cases, decide whether the given function is intrinsically linear. If so, identify x' and y', and then explain how a random error term ϵ can be introduced so as to yield an intrinsically linear probabilistic model.

a. $y = 1/(\alpha + \beta x)$ **b.** $y = 1/(1 + e^{\alpha + \beta x})$
c. $y = e^{e^{\alpha + \beta x}}$ (a Gompertz curve)
d. $y = \alpha + \beta e^{\lambda x}$

21. Suppose that x and y are related according to a probabilistic exponential model $Y = \alpha e^{\beta x} \cdot \epsilon$ with $V(\epsilon)$ a constant independent of x (as was the case in the simple linear model $Y = \beta_0 + \beta_1 x + \epsilon$). Is $V(Y)$ a constant independent of x [as was the case for $Y = \beta_0 + \beta_1 x + \epsilon$, where $V(Y) = \sigma^2$]? Explain your reasoning. Draw a picture of a prototype scatter plot resulting from this model. Answer the same questions for the power model $Y = \alpha x^\beta \cdot \epsilon$.

13.3   Polynomial Regression

The nonlinear yet intrinsically linear models of Section 13.2 involved functions of the independent variable x that were either strictly increasing or strictly decreasing. In many situations either theoretical reasoning or else a scatter plot of the data suggests that the true regression function $\mu_{Y \cdot x}$ has one or more peaks or valleys—that is, at least one relative minimum or maximum. In such cases a polynomial function $y = \beta_0 + \beta_1 x + \cdots + \beta_k x^k$ may provide a satisfactory approximation to the true regression function. The probabilistic model that we shall study in this section is

$$Y = \beta_0 + \beta_1 x + \beta_2 x^2 + \cdots + \beta_k x^k + \epsilon \tag{13.6}$$

where ϵ is a random error variable with

$$\mu_\epsilon = 0, \qquad \sigma_\epsilon^2 = \sigma^2 \tag{13.7}$$

From (13.6) and (13.7) it follows immediately that

$$\mu_{Y \cdot x} = \beta_0 + \beta_1 x + \cdots + \beta_k x^k, \quad \sigma^2_{Y \cdot x} = \sigma^2 \qquad (13.8)$$

In words, the expected value of Y is a kth-degree polynomial function of x, while the variance of Y, which controls the spread of observed values about the regression function, is the same for each value of x. The observed pairs $(x_1, y_1), \ldots, (x_n, y_n)$ are assumed to have been generated independently from the model (13.6). In addition, to obtain confidence intervals for and test hypotheses about the parameters of the model, we shall assume a normal distribution for ϵ, so that the Y_i's will be independent normal variables with mean and variance given by (13.8). Figure 13.7 illustrates both a quadratic and cubic model.

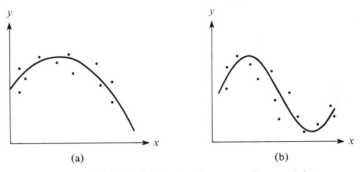

(a)   (b)

Figure 13.7 (a) Quadratic regression model; (b) cubic regression model

Estimation of Parameters Using Least Squares

In order to estimate $\beta_0, \beta_1, \ldots, \beta_k$, consider a trial regression function $y = b_0 + b_1 x + \cdots + b_k x^k$. Then the "goodness of fit" of this function to the observed data can be measured by computing the sum of squared deviations

$$f(b_0, b_1, \ldots, b_k) = \sum_{i=1}^{n} [y_i - (b_0 + b_1 x_i + b_2 x_i^2 + \cdots + b_k x_i^k)]^2 \quad (13.9)$$

According to the principle of least squares, the estimates $\hat{\beta}_0, \hat{\beta}_1, \ldots, \hat{\beta}_k$ are those values of $b_0, b_1, \ldots, b_k$ which minimize (13.9). It should be noted that when $x_1, x_2, \ldots, x_k$ are all different, there is a polynomial of degree $n - 1$ that fits the data perfectly, so that the minimizing value of (13.9) is zero when $k = n - 1$. However, in virtually all applications the polynomial model (13.6) with k large is quite unrealistic, and in most applications $k = 2$ (quadratic) or $k = 3$ (cubic) is appropriate.

To find the minimizing values in (13.9), we take the $k + 1$ partial derivatives $\partial f / \partial b_0, \partial f / \partial b_1, \ldots, \partial f / \partial b_k$ and equate them to zero, resulting in the system of normal equations for the estimates. Because the trial function $b_0 + b_1 x + \cdots + b_k x^k$ is linear in $b_0, \ldots, b_k$ (though not in x), the $k + 1$ normal equations are linear in these unknowns. The system is

$$
\begin{aligned}
b_0 n + b_1 \Sigma x_i + b_2 \Sigma x_i^2 + \cdots + b_k \Sigma x_i^k &= \Sigma y_i \\
b_0 \Sigma x_i + b_1 \Sigma x_i^2 + b_2 \Sigma x_i^3 + \cdots + b_k \Sigma x_i^{k+1} &= \Sigma x_i y_i \\
\vdots \qquad\qquad \vdots \qquad\qquad \vdots \qquad\qquad& \\
b_0 \Sigma x_i^k + b_1 \Sigma x_i^{k+1} + \cdots + b_k \Sigma x_i^{2k} &= \Sigma x_i^k y_i
\end{aligned}
\tag{13.10}
$$

All standard statistical computer packages will automatically solve the system (13.10) and print out the estimates. To solve by hand, the values of the summary statistics $\Sigma x_i, \Sigma x_i^2, \ldots, \Sigma x_i^{2k}, \Sigma y_i, \Sigma x_i y_i, \ldots,$ and $\Sigma x_i^k y_i$ must first be computed. Then after substitution into (13.10), the system can be solved by Gaussian elimination.

For the special case of quadratic regression, explicit formulas for the estimates [the solution of (13.10)] are (with $\overline{x^2} = \Sigma x_i^2 / n$)*

$$
\hat{\beta}_1 = \frac{s_{1y} s_{22} - s_{2y} s_{12}}{s_{11} s_{22} - s_{12}^2}, \quad \hat{\beta}_2 = \frac{s_{2y} s_{11} - s_{1y} s_{12}}{s_{11} s_{22} - s_{12}^2}, \quad \hat{\beta}_0 = \overline{y} - \hat{\beta}_1 \overline{x} - \hat{\beta}_2 \overline{x^2}
\tag{13.11}
$$

where

$$
\begin{aligned}
s_{1y} &= \Sigma x_i y_i - n \overline{x}\,\overline{y}, \; s_{2y} = \Sigma x_i^2 y_i - n \overline{x^2}\,\overline{y} \\
s_{11} &= \Sigma x_i^2 - n \overline{x}^2, \; s_{12} = \Sigma x_i^3 - n \overline{x}\,\overline{x^2}
\end{aligned}
\tag{13.12}
$$

and

$$
s_{22} = \Sigma x_i^4 - n (\overline{x^2})^2
$$

Example **13.5** The article "Determination of Biological Maturity and Effect of Harvesting and Drying Conditions on Milling Quality of Paddy" (*J. Agricultural Eng. Research*, 1975, pp. 353–361) reported the following data on date of harvesting x (number of days after flowering) and yield y (kg/ha) of paddy, a grain farmed in India.

x:	16	18	20	22	24	26	28	30
y:	2508	2518	3304	3423	3057	3190	3500	3883

x:	32	34	36	38	40	42	44	46
y:	3823	3646	3708	3333	3517	3241	3103	2776

---

*If all s_{ij}'s are multiplied by n, formulas like those of simple linear regression, which involve Σx_i, Σx_i^2, and the like rather than $\overline{x}$, $\overline{x^2}$, and so on will result, but then the numbers in the numerator and denominator of $\hat{\beta}_1$ and $\hat{\beta}_2$ may be extremely large.

Both a scatter plot and a residual plot from a linear regression suggest that the authors' choice of a quadratic model is reasonable. The summary quantities are $\Sigma x_i = 496$, $\Sigma x_i^2 = 16{,}736$, $\Sigma x_i^3 = 603{,}136$, $\Sigma x_i^4 = 22{,}825{,}094$, $\Sigma y_i = 52{,}530$, $\Sigma x_i y_i = 1{,}645{,}108$, and $\Sigma x_i^2 y_i = 55{,}565{,}884$, so that $\bar{x} = 31.0000$, $\overline{x^2} = 1046.0000$, and $\bar{y} = 3283.1250$. Substituting into (13.11) and (13.12) gives

$$s_{1y} = 1{,}645{,}108 - 16(31)(3283.125) = 16{,}678$$

$$s_{2y} = 55{,}565{,}884 - 16(1046)(3283.125) = 619{,}504$$

$$s_{11} = 1360, \; s_{12} = 84{,}320, \text{ and } s_{22} = 5{,}319{,}238, \text{ yielding}$$

$$\hat{\beta}_1 = \frac{(16{,}678)(5{,}319{,}238) - (619{,}504)(84{,}320)}{(1360)(5{,}319{,}238) - (84{,}320)^2} = 293.4618$$

$$\hat{\beta}_2 = \frac{(619{,}504)(1360) - (16{,}678)(84{,}320)}{(1360)(5{,}319{,}238) - (84{,}320)^2} = -4.5355$$

$$\hat{\beta}_0 = 3283.125 - (293.4618)(31) - (-4.5355)(1046) = -1070.0578$$

The estimated regression function is then $y = -1070.05 + 293.46x - 4.54x^2$, from which predicted values and residuals can be computed. A residual plot indicates that a quadratic regression function is appropriate, though there is somewhat more spread in the residuals for small x than for large x. Weighted least squares can be used to obtain a more satisfactory fit, but the authors did not consider this. ∎

$\hat{\sigma}^2$ and R^2

To make further inferences about the parameters of the regression function, the error variance σ^2 must be estimated. With $\hat{y}_i = \hat{\beta}_0 + \hat{\beta}_1 x_i + \cdots + \hat{\beta}_k x_i^k$, the ith residual is $y_i - \hat{y}_i$ and the sum of squared residuals (error sum of squares) is $SSE = \Sigma (y_i - \hat{y}_i)^2$. A computational formula for SSE is

$$SSE = \Sigma y_i^2 - \hat{\beta}_0 \Sigma y_i - \hat{\beta}_1 \Sigma x_i y_i - \cdots - \hat{\beta}_k \Sigma x_i^k y_i \tag{13.13}$$

which can be verified by squaring the residuals and summing the resulting terms. The estimate for σ^2 is then

$$\hat{\sigma}^2 = s^2 = \frac{SSE}{n - (k+1)} = MSE \tag{13.14}$$

where the denominator $n - (k+1)$ is used because $k+1$ degrees of freedom are lost in estimating $\beta_0, \beta_1, \ldots, \beta_k$.

Example 13.6
(Example 13.5 continued)

After computing $\Sigma y_i^2 = 175{,}087{,}792$,

$$SSE = 175{,}087{,}792 - (-1070.0578)(52{,}530) - (293.4618)(1{,}645{,}108)$$
$$- (-4.5355)(55{,}565{,}884) = 540{,}640.2$$

so

$$s^2 = \frac{540{,}640.2}{16 - 3} = 41{,}587.71 \text{ and } \hat{\sigma} = s = 203.93 \qquad ∎$$

If we again let $SST = \Sigma(y_i - \bar{y})^2$, then SSE/SST is the proportion of the total variation in the observed y_i's not explained by the polynomial model. The quantity $1 - SSE/SST$, the proportion of variation explained by the model, is called the **coefficient of multiple determination** and is denoted by R^2. For the data of Example 13.6, $SST = 2{,}625{,}235.7$, so the coefficient of multiple determination is $R^2 = 1 - (540{,}640.2)/(2{,}625{,}235.7) = .794$ and the quadratic model explains 79.4% of the observed variation in y.

Suppose that we consider fitting a cubic model to the data in Example 13.5. Because the cubic model includes the quadratic as a special case, the fit to a cubic will be at least as good as the fit to a quadratic. More generally, with SSE_k = the error sum of squares from a kth degree polynomial, $SSE_{k'} \leq SSE_k$ and $R^2_{k'} \geq R^2_k$ whenever $k' > k$. Because the objective of regression analysis is to find a model that is both simple (relatively few parameters) and provides a good fit to the data, a higher-degree polynomial may not specify a better model than a lower-degree model in spite of its higher R^2 value. To balance the cost of using more parameters against the gain in R^2, many statisticians use the **adjusted coefficient of multiple determination**

$$\text{adjusted } R^2 = 1 - \frac{n-1}{n-(k+1)} \cdot \frac{SSE}{SST} = \frac{(n-1)R^2 - k}{n-1-k} \qquad (13.15)$$

Adjusted R^2 adjusts the proportion of unexplained variation upward (since $(n-1)/(n-k-1) > 1$), which results in adjusted $R^2 < R^2$. Thus if $R^2_2 = .66$, $R^2_3 = .70$, and $n = 10$, then

$$\text{adjusted } R^2_2 = \frac{9(.66) - 2}{10 - 3} = .563, \qquad \text{adjusted } R^2_3 = \frac{9(.70) - 3}{10 - 4} = .550$$

so the small gain in R^2 in going from a quadratic to a cubic model is not enough to offset the cost of adding an extra parameter to the model.

In addition to computing R^2 and adjusted R^2, one should examine the usual diagnostic plots to determine whether model assumptions are valid or whether modification may be appropriate.

Confidence Intervals and Test Procedures

Because the y_i's appear in the normal equations (13.10) only on the right-hand side and in a linear fashion, the resulting estimates $\hat{\beta}_0, \ldots, \hat{\beta}_k$ are themselves linear functions of the y_i's. Thus the estimators are linear functions of the Y_i's, so each $\hat{\beta}_i$ has a normal distribution. It can also be shown that each $\hat{\beta}_i$ is an unbiased estimator of β_i.

Let $\sigma_{\hat{\beta}_i}$ denote the standard deviation of the estimator $\hat{\beta}_i$. This standard deviation has the form

$$\sigma_{\hat{\beta}_i} = \sigma \cdot \left\{ \begin{array}{l} \text{a complicated expression involving } \textit{all} \\ x_j\text{'s, } x_j^2\text{'s, } \ldots, \text{ and } x_j^k\text{'s} \end{array} \right\}$$

Fortunately, the expression in braces has been programmed into all of the most frequently used statistical computer packages—MINITAB, BMD, SAS, and

SPSS. The estimated standard deviation of $\hat{\beta}_i$, $s_{\hat{\beta}_i}$, results from substituting s in place of σ in the above expression for $\sigma_{\hat{\beta}_i}$. These estimated standard deviations $s_{\hat{\beta}_0}, s_{\hat{\beta}_1}, \ldots,$ and $s_{\hat{\beta}_k}$ appear on output from all the aforementioned statistical packages. Let $S_{\hat{\beta}_i}$ denote the estimator of $\sigma_{\hat{\beta}_i}$—that is, the random variable whose observed value is $s_{\hat{\beta}_i}$. Then it can be shown that the standardized variable

$$T = \frac{\hat{\beta}_i - \beta_i}{S_{\hat{\beta}_i}} \tag{13.16}$$

has a t distribution based on $n - (k + 1)$ d.f. This leads to the following inferential procedures.

A $100(1 - \alpha)\%$ confidence interval for β_i, the coefficient of x^i in the polynomial regression function, is

$$\hat{\beta}_i \pm t_{\alpha/2,\, n - (k + 1)} \cdot s_{\hat{\beta}_i}$$

A test of $H_0 : \beta_i = \beta_{i0}$ is based on the t statistic value

$$t = \frac{\hat{\beta}_i - \beta_{i0}}{s_{\hat{\beta}_i}}$$

The test utilizes a t critical value with $n - (k + 1)$ d.f., and is upper, lower, or two-tailed according to whether the inequality in H_a is $>$, $<$, or $\neq$.

A point estimate of $\mu_{Y \cdot x}$—that is, of $\beta_0 + \beta_1 x + \cdots + \beta_k x^k$, is $\hat{\mu}_{Y \cdot x} = \hat{\beta}_0 + \hat{\beta}_1 x + \cdots + \hat{\beta}_k x^k$. The estimated standard deviation of the corresponding estimator is rather complicated. Several computer packages will give this estimated standard deviation for *any* x value when requested to do so by a user, and MINITAB automatically prints it for each x value in the sample ($x = x_1$, $x = x_2, \ldots, x = x_n$). This along with an appropriate standardized t variable, can be used to justify the following procedures.

Let x^* denote a specified value of the independent variable x. A $100(1 - \alpha)\%$ confidence interval for $\mu_{Y \cdot x^*}$ is

$$\hat{\mu}_{Y \cdot x^*} \pm t_{\alpha/2,\, n - (k + 1)} \cdot \left\{ \begin{array}{c} \text{estimated S.D. of} \\ \hat{\mu}_{Y \cdot x^*} \end{array} \right\}$$

A $100(1 - \alpha)\%$ prediction interval for a future y value to be observed when $x = x^*$ is

$$\hat{\mu}_{Y \cdot x^*} \pm t_{\alpha/2,\, n - (k + 1)} \cdot \left\{ s^2 + \left(\begin{array}{c} \text{estimated S.D.} \\ \text{of } \hat{\mu}_{Y \cdot x^*} \end{array} \right)^2 \right\}^{1/2}$$

Example **13.7** A computer analysis yielded the following (partial) results for the date of harvesting/paddy yield data first presented in Example 13.5.

Table 13.4

Coefficient	Estimate $\hat{\beta}_i$	Estimated S.D. $s_{\hat{\beta}_i}$
β_0	-1070.05	617.25
β_1	293.46	42.18
β_2	-4.54	.674

Table 13.5

x_i	y_i	$\hat{y}_i = \hat{\mu}_{Y \cdot x_i}$	Estimated S.D. of $\hat{\mu}_{Y \cdot x_i}$
16	2508	2464	136
18	2518	2743	105
.	.	.	.
28	3500	3591	74
30	3883	3652	76
32	3823	3676	76
.	.	.	.
46	2776	2832	136

a. For testing $H_0: \beta_2 = 0$ versus $H_a: \beta_2 \neq 0$ (H_0 here says that the quadratic term in the model is unnecessary), the test statistic is $T = \hat{\beta}_2 / S_{\hat{\beta}_2}$, with computed value $-4.54/.674 = -6.74$. The test is based on $n - (k + 1) = 16 - 3 = 13$ d.f.; with $t_{.025, 13} = 2.179$, H_0 is rejected at level .05 if either $t \geq 2.179$ or $t \leq -2.179$. Since $-6.74 \leq -2.179$, H_0 is rejected at level .05, validating the inclusion of the quadratic term.

b. A 95% confidence interval for $\mu_{Y \cdot 30}$ $[= \beta_0 + \beta_1(30) + \beta_2(30)^2]$ is $\hat{\mu}_{Y \cdot 30} \pm t_{.025, 13} \cdot$ (estimated S.D. of $\hat{\mu}_{Y \cdot 30}$) $= 3652 \pm (2.179)(76) = (3486.4, 3817.6)$.

c. A 95% prediction interval for y when $x = 28$ is (since $s^2 = 41,587.71$)

$$3591 \pm (2.179)\sqrt{41,587.71 + (76)^2} = 3591 \pm (2.179)(217.6)$$
$$= (3116.8, 4065.2)$$

This interval is quite wide because $s = 203.93$ is relatively large. ∎

Centering x Values

For the quadratic model with regression function $\mu_{Y \cdot x} = \beta_0 + \beta_1 x + \beta_2 x^2$, the parameters β_0, β_1, and β_2 characterize the behavior of the function near $x = 0$. For example, β_0 is the height at which the regression function crosses the vertical axis $x = 0$, while β_1 is the first derivative of the function at $x = 0$ (instantaneous rate of change of $\mu_{Y \cdot x}$ at $x = 0$). If the x_i's all lie far from zero, we may not have precise information about the values of these parameters. Let $\bar{x} =$ the average of the x_i's for which observations are to be taken, and consider the model

$$Y = \beta_0^* + \beta_1^*(x - \bar{x}) + \beta_2^*(x - \bar{x})^2 + \epsilon \qquad (13.17)$$

In the model (13.17) $\mu_{Y \cdot x} = \beta_0^* + \beta_1^*(x - \bar{x}) + \beta_2^*(x - \bar{x})^2$ and the parameters now describe the behavior of the regression function near the center $\bar{x}$ of the data.

To estimate the parameters of (13.18), we simply subtract $\bar{x}$ off each x_i to obtain $x_i' = x_i - \bar{x}$, and then use the x_i''s in place of the x_i's. An important benefit of this is that the coefficients of $b_0, \ldots, b_k$ in the normal equations (13.10) will be of much smaller magnitude than would be the case were the original x_i's used. When the system is solved by computer, this centering protects against any roundoff error that may result.

Example 13.8

The paper "A Method for Improving the Accuracy of Polynomial Regression Analysis" (*J. Quality Technology*, 1971, pp. 149–155) reported the following data on $x =$ cure temperature (°F) and $y =$ ultimate shear strength of a rubber compound (psi), with $\bar{x} = 297.13$.

x	280	284	292	295	298	305	308	315
x'	−17.13	−13.13	−5.13	−2.13	.87	7.87	10.87	17.87
y	770	800	840	810	735	640	590	560

A computer analysis yielded the following results:

Table 13.6

Parameter	Estimate	Estimated S.D.	Parameter	Estimate	Estimated S.D.
β_0	−26,219.64	11,912.78	β_0^*	759.36	23.20
β_1	189.21	80.25	β_1^*	−7.61	1.43
β_2	−.3312	.1350	β_2^*	−.3312	.1350

The estimated regression function using the original model is $y = -26,219.64 + 189.21x - .3312x^2$, while for the centered model the function is $y = 759.36 - 7.61(x - 297.13) - .3312(x - 297.13)^2$. These estimated functions are identical; the only difference is that different parameters have been estimated for the two models. The estimated standard deviations indicate clearly that β_0^* and β_1^* have been more accurately estimated than β_0 and β_1. The quadratic parameters are identical ($\beta_2 = \beta_2^*$), as can be seen by comparing the x^2 term in (13.17) with the original model. We emphasize again that a major benefit of centering is the gain in computational accuracy, not only in quadratic but also in higher-degree models. ∎

The book by Neter, Wasserman, and Kutner is again a good source for more information about polynomial regression.

Exercises / Section 13.3 (22–30)

22. The following data on $y =$ glucose concentration (g/L) and $x =$ fermentation time (days) for a particular blend of malt liquor was read from a scatter plot appearing in the paper "Improving

Fermentation Productivity with Reverse Osmosis" (*Food Tech.*, 1984, pp. 92–96).

x	1	2	3	4	5	6	7	8
y	74	54	52	51	52	53	58	71

a. Verify that a scatter plot of the data is consistent with the choice of a quadratic regression model.

b. The estimated quadratic regression equation is $y = 84.482 - 15.875x + 1.7679x^2$. Predict the value of glucose concentration for a fermentation time of 6 days, and compute the corresponding residual.

c. Using $SSE = 61.77$, what proportion of observed variation can be attributed to the quadratic regression relationship?

d. The $n = 8$ standardized residuals based on the quadratic model are 1.91, -1.95, $-.25$, .58, .90, .04, $-.66$, and .20. Construct a plot of the standardized residuals versus x and a normal probability plot. Do the plots exhibit any troublesome features?

e. The estimated standard deviation of $\hat{\mu}_{Y \cdot 6}$— that is, $\hat{\beta}_0 + \hat{\beta}_1(6) + \hat{\beta}_2(36)$—is 1.69. Compute a 95% confidence interval for $\mu_{Y \cdot 6}$.

f. Compute a 95% prediction interval for a glucose concentration observation made after 6 days of fermentation time.

23. The viscosity of an oil (y) was measured by a cone and plate viscometer at six different cone speeds (x). It was assumed that a quadratic regression model was appropriate, and the estimated regression function resulting from the $n = 6$ observations was

$$y = -113.0937 + 3.3684x - .01780x^2$$

a. Estimate $\mu_{Y \cdot 75}$, the expected viscosity when speed is 75 rpm.

b. What viscosity would you predict for a cone speed of 60 rpm?

c. If $\Sigma y_i^2 = 8386.43$, $\Sigma y_i = 210.70$, $\Sigma x_i y_i = 17,002.00$, and $\Sigma x_i^2 y_i = 1,419,780$, compute SSE, s^2, and s.

d. From (c), $SST = 8386.43 - (210.70)^2/6 = 987.35$. Using SSE computed in (c), what is the computed value of the coefficient of multiple determination R^2?

e. If the estimated standard deviation of $\hat{\beta}_2$ is $s_{\hat{\beta}_2} = .00226$, test $H_0 : \beta_2 = 0$ versus $H_a : \beta_2 \neq 0$ at level .01.

24. Exercise 5 presented the following data on exposure time to radiation x and dry weight of roots y.

x	0	2	4	6	8
y	110	123	119	86	62

a. Show that the estimated quadratic regression function is $y = 111.89 + 8.06x - 1.84x^2$.

b. Compute the predicted values and residuals. Then compute SSE and s^2 using the residuals.

c. Compute SSE using the computational formula, and then compute s^2.

d. Compute the coefficient of multiple determination R^2.

e. The estimated standard deviation of $\hat{\beta}_2$, the estimator of the quadratic coefficient β_2, is $s_{\hat{\beta}_2} = .480$. Does the quadratic term belong in the model? State and test the appropriate hypotheses at level .05.

f. The estimated standard deviation of $\hat{\beta}_1$ is $s_{\hat{\beta}_1} = 4.01$. Use this and the information in (e) to obtain joint confidence intervals for β_1 and β_2 with joint confidence level (at least) 95%.

g. The estimated standard deviation of $\hat{\mu}_{Y \cdot 4}$ $(= \hat{\beta}_0 + 4\hat{\beta}_1 + 16\hat{\beta}_2)$ is 5.01. Compute a 90% confidence interval for $\mu_{Y \cdot 4}$.

h. Estimate the exposure time that maximizes expected dry weight of roots.

25. The article "A Simulation-Based Evaluation of Three Cropping Systems on Cracking-Clay Soils in a Summer Rainfall Environment" (*Agricultural Meteorology*, 1976, pp. 211–229) proposed a quadratic model for the relationship between water supply index (x) and farm wheat yield (y). Representative data and the resulting summary quantities appear below.

x	1.2	1.3	1.5	1.8	2.1	2.3	2.5
y	790	950	740	1230	1000	1465	1370

x	2.9	3.1	3.2	3.3	3.9	4.0	4.3
y	1420	1625	1600	1720	1500	1550	1560

$\Sigma x_i = 37.40$, $\Sigma x_i^2 = 113.42$, $\Sigma x_i^3 = 375.8961$, $\Sigma x_i^4 = 1322.7388$, $\Sigma y_i = 18,520$, $\Sigma y_i^2 = 25,871,756$, $\Sigma x_i y_i = 53,111$, $\Sigma x_i^2 y_i = 168,248.3$

a. Estimate β_0, β_1, β_2, and the quadratic regression function $\beta_0 + \beta_1 x + \beta_2 x^2$.

b. Compute SSE, estimate σ^2, and compute the coefficient of multiple determination.

c. The estimated standard deviation of $\hat{\beta}_2$ is $s_{\hat{\beta}_2} = 41.97$. Obtain a 95% confidence interval for β_2.

d. The estimated standard deviation of $\hat{\beta}_0 + \hat{\beta}_1 x + \hat{\beta}_2 x^2$ when $x = 2.5$ is 53.5. Test $H_0 : \mu_{Y \cdot 2.5} = 1500$ versus $H_a : \mu_{Y \cdot 2.5} < 1500$ using $\alpha = .01$.

e. Obtain a 95% prediction interval for wheat yield when the water supply index is 2.5 by using the information given in (d).

26. The accompanying data was obtained from a study of a certain method for preparing pure alcohol from refinery streams ("Direct Hydration of Olefins," *Industrial and Eng. Chemistry,* 1961, pp. 209–211). The independent variable x is volume hourly space velocity, and the dependent variable y is the amount of conversion of Isobutylene.

x	1	1	2	4	4	4	6
y	23.0	24.5	28.0	30.9	32.0	33.6	20.0

a. Assuming that a quadratic probabilistic model is appropriate, estimate the regression function.

b. Compute the predicted values and residuals, and construct a residual plot. Does the plot look roughly as expected when the quadratic model is correct? Does the plot indicate that any observation has had a great influence on the fit? Does a scatter plot identify a point having large influence? If so, which point?

c. Compute s^2 and R^2. Does the quadratic model provide a good fit to the data?

d. In Exercise 9, it was noted that the predicted value $\hat{Y}_j$ and the residual $Y_j - \hat{Y}_j$ are independent of one another, so that $\sigma^2 = V(Y_j) = V(\hat{Y}_j) + V(Y_j - \hat{Y}_j)$. A computer printout gives the estimated standard deviations of the predicted values as .955, .955, .712, .777, .777, .777, and 1.407. Use these values along with s^2 to compute the estimated standard deviation of each residual. Then compute the standardized residuals and plot them against x. Does the plot look much like the plot of (b)? Suppose that you had standardized the residuals using just s in the denominator. Would the resulting values be much different than the correct values?

e. Using information given in (d), compute a 90% prediction interval for Isobutylene conversion when volume hourly space velocity is 4.

27. The following data is a subset of data obtained in an experiment to study the relationship between soil pH x and $y = $ Al. Concentration/EC ("Root Responses of Three Gramineae Species to Soil Acidity in an Oxisol and an Ultisol," *Soil Science,* 1973, pp. 295–302).

x	4.01	4.07	4.08	4.10	4.18
y	1.20	.78	.83	.98	.65

x	4.20	4.23	4.27	4.30	4.41
y	.76	.40	.45	.39	.30

x	4.45	4.50	4.58	4.68	4.70	4.77
y	.20	.24	.10	.13	.07	.04

A cubic model was proposed in the paper, but the version of MINITAB used by the author of the present text refused to include the x^3 term in the model, stating that "x^3 is highly correlated with other predictor variables." To remedy this, $\bar{x} = 4.3456$ was subtracted off each x value to yield $x' = x - \bar{x}$. A cubic regression was then requested to fit the model having regression function

$$y = \beta_0^* + \beta_1^* x' + \beta_2^* (x')^2 + \beta_3^* (x')^3$$

The following computer output resulted:

Parameter	Estimate	Estimated S.D.
β_0^*	.3463	.0366
β_1^*	−1.2933	.2535
β_2^*	2.3964	.5699
β_3^*	−2.3968	2.4590

a. What is the estimated regression function for the "centered" model?

b. What is the estimated value of the coefficient β_3 in the "uncentered" model with regression function $y = \beta_0 + \beta_1 x + \beta_2 x^2 + \beta_3 x^3$? What is the estimate of β_2?

c. Using the cubic model, what value of y would you predict when soil pH is 4.5?

d. Carry out a test to decide whether or not the cubic term should be retained in the model.

28. In many polynomial regression problems, rather than fitting a "centered" regression function using

$x' = x - \bar{x}$, computational accuracy can be improved by using a function of the standardized independent variable $x' = (x - \bar{x})/s_x$, where s_x is the standard deviation of the x_i's. Consider fitting the cubic regression function $y = \beta_0^* + \beta_1^* x' + \beta_2^* (x')^2 + \beta_3^* (x')^3$ to the following data resulting from a study of the relation between thrust efficiency y of supersonic propelling rockets and the half-divergence angle x of the rocket nozzle ("More on Correlating Data," *CHEMTECH*, 1976, pp. 266–270).

x	5	10	15	20	25	30	35
y	.985	.996	.988	.962	.940	.915	.878

Parameter	Estimate	Estimated S.D.
β_0^*	.9671	.0026
β_1^*	−.0502	.0051
β_2^*	−.0176	.0023
β_3^*	.0062	.0031

a. What value of y would you predict when the half-divergence angle is 20? When $x = 25$?

b. What is the estimated regression function $\hat{\beta}_0 + \hat{\beta}_1 x + \hat{\beta}_2 x^2 + \hat{\beta}_3 x^3$ for the "unstandardized" model?

c. Use a level .05 test to decide whether or not the cubic term should be deleted from the model.

d. What can you say about the relationship between SSE's and R^2's for the standardized and unstandardized models? Explain.

e. SSE for the cubic model is .00006300, while for a quadratic model SSE is .00014367. Compute R^2 for each model. Does the difference between the two suggest that the cubic term can be deleted?

29. The following data resulted from an experiment to assess the potential of unburnt colliery spoil as a medium for plant growth. The variables are x = acid extractable cations and y = exchangeable acidity/total cation exchange capacity ("Ex-

changeable Acidity in Unburnt Colliery Spoil," *Nature*, 1969, p. 161).

x	−23	−5	16	26	30	38
y	1.50	1.46	1.32	1.17	.96	.78

x	52	58	67	81	96	100	113
y	.77	.91	.78	.69	.52	.48	.55

Standardizing the independent variable x to obtain $x' = (x - \bar{x})/s_x$ and fitting the regression function $y = \beta_0^* + \beta_1^* x' + \beta_2^* (x')^2$ yielded the accompanying computer output.

Parameter	Estimate	Estimated S.D.
β_0^*	.8733	.0421
β_1^*	−.3255	.0316
β_2^*	.0448	.0319

a. Estimate $\mu_{Y \cdot 50}$.

b. Compute the value of the coefficient of multiple determination.

c. What is the estimated regression function $\hat{\beta}_0 + \hat{\beta}_1 x + \hat{\beta}_2 x^2$ using the unstandardized variable x?

d. What is the estimated standard deviation of $\hat{\beta}_2$ computed in (c)?

e. Carry out a test using the standardized estimates to decide whether or not the quadratic term should be retained in the model. Repeat using the unstandardized estimates. Do your conclusions differ?

30. The paper "The Respiration in Air and in Water of the Limpets Patella Caerulea and Patella Lusitanica" (*Comp. Biochemistry and Physiology*, 1975, pp. 407–411) proposed a simple power model for the relationship between respiration rate y and temperature x for P. caerulea in air. However, a plot of $\ln(y)$ versus x exhibits a curved pattern. Fit the quadratic power model $Y = \alpha e^{\beta x + \gamma x^2} \cdot \epsilon$ to the accompanying data.

x	10	15	20	25	30
y	37.1	70.1	109.7	177.2	222.6

13.4 Multiple Regression Analysis

In multiple regression the objective is to build a probabilistic model that relates a dependent variable Y to more than one independent or predictor variable. The general form of the model to be studied here is

$$Y = \beta_0 + \beta_1 x_1 + \beta_2 x_2 + \cdots + \beta_k x_k + \epsilon \qquad (13.18)$$

where $E(\epsilon) = 0$ and $V(\epsilon) = \sigma^2$. This in turn implies that $\sigma^2_{Y \cdot x_1, x_2, \ldots, x_k}$, the variance of Y for given values of $x_1, \ldots, x_k$, equals σ^2 independently of the given values, and that the regression function (expected Y as a function of the predictors) is

$$\mu_{Y \cdot x_1, \ldots, x_k} = \beta_0 + \beta_1 x_1 + \cdots + \beta_k x_k \qquad (13.19)$$

To construct confidence and prediction intervals and to test hypotheses about the model parameters, we shall also assume that ϵ has a normal distribution.

Each x_i in (13.18) will be called a **carrier,** since it carries information about Y in the model. Thus the model (13.18) contains k carriers. The reason for not using "variable" in place of "carrier" is that we may wish to build a model involving two independent variables x_1 and x_2, with several of the carriers being themselves functions of x_1 and x_2. For example, with $x_3 = x_1^2$ and $x_4 = x_1 x_2$, the model

$$Y = \beta_0 + \beta_1 x_1 + \beta_2 x_2 + \beta_3 x_3 + \beta_4 x_4 + \epsilon$$

has the general form of (13.18) with four carriers, but x_1, x_2, x_3, and x_4 are not functionally independent variables. Thus even a model built from just two independent variables may have a number of carriers.

While the regression function (13.19) will not always be a linear function of the independent variables used to build the model, further analysis depends critically on the fact that it is *a linear function of the unknown parameters* $\beta_0, \beta_1, \ldots, \beta_k$. This property makes it easy to apply the principle of least squares to obtain estimators that can be used for inference and prediction. Before discussing this, we first consider interpretations of different multiple regression models.

Model Interpretation

For the case of two independent variables x_1 and x_2, four useful multiple regression models are

1. the first-order model, with $Y = \beta_0 + \beta_1 x_1 + \beta_2 x_2 + \epsilon$
2. the second-order no-interaction model, with $Y = \beta_0 + \beta_1 x_1 + \beta_2 x_2 + \beta_3 x_1^2 + \beta_4 x_2^2 + \epsilon$
3. the first-order interaction model, with $Y = \beta_0 + \beta_1 x_1 + \beta_2 x_2 + \beta_3 x_1 x_2 + \epsilon$
4. the second-order model with interaction, specified by $Y = \beta_0 + \beta_1 x_1 + \beta_2 x_2 + \beta_3 x_1^2 + \beta_4 x_2^2 + \beta_5 x_1 x_2 + \epsilon$

Understanding the differences between these models is an important first step in building realistic regression models from the independent variables under study.

The first-order model is the most straightforward generalization of simple linear regression. It states that for a fixed value of either variable, the expected value of Y is a linear function of the other variable, and that the expected change in Y for a unit increase in x_1 (x_2) is β_1 (β_2) independent of the level of x_2 (x_1). Thus if we graph the regression function as a function of x_1 for several different values of x_2, we obtain as contours of the regression function a collection of parallel lines as pictured in Figure 13.8a. The function $y = \beta_0 + \beta_1 x_1 + \beta_2 x_2$ specifies a plane in three-dimensional space; the first-order model says that each observed value of the dependent variable deviates from this plane by a random amount ϵ.

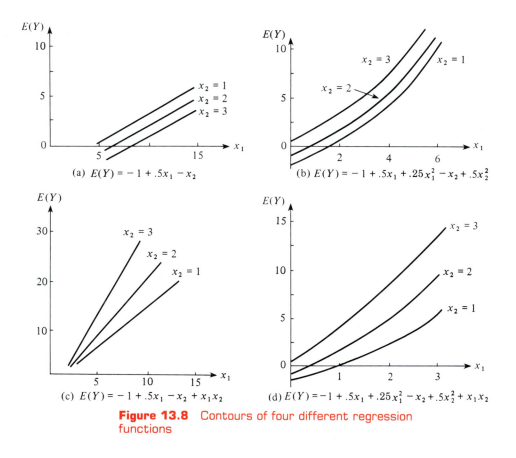

Figure 13.8 Contours of four different regression functions

According to the second-order no-interaction model, if x_2 is fixed, the expected change in Y for a one unit increase in x_1 is

$$\beta_0 + \beta_1(x_1 + 1) + \beta_2 x_2 + \beta_3(x_1 + 1)^2 + \beta_4 x_2^2 -$$
$$(\beta_0 + \beta_1 x_1 + \beta_2 x_2 + \beta_3 x_1^2 + \beta_4 x_2^2) = \beta_1 + 2\beta_3 x_1$$

Because this expected change does not depend on x_2, the contours of the regression function for different values of x_2 are still parallel to one another. However, the dependence of the expected change on the value of x_1 means that the contours are now curves rather than straight lines. This is pictured in Figure 13.8b. In this case the regression surface is no longer a plane in three-space but is instead a curved surface.

The contours of the regression function for the first-order interaction model are nonparallel straight lines. This is because the expected change in Y when x_1 is increased by 1 is

$$\beta_0 + \beta_1(x_1 + 1) + \beta_2 x_2 + \beta_3(x_1 + 1)x_2 -$$
$$(\beta_0 + \beta_1 x_1 + \beta_2 x_2 + \beta_3 x_1 x_2) = \beta_0 + \beta_1 + \beta_3 x_2$$

This expected change depends on the value of x_2, so each contour line must have a different slope as in Figure 13.8c. The word "interaction" reflects the fact that an expected change in Y when one variable increases depends on the value of the other variable.

Finally, for the second-order interaction model, the expected change in Y when x_2 is held fixed while x_1 is increased by one unit is $\beta_1 + 2\beta_3 x_1 + \beta_5 x_2$, which is a function of both x_1 and x_2. This implies that the contours of the regression function are both curved and not parallel to one another, as illustrated in Figure 13.8d.

Similar considerations apply to models constructed from more than two independent variables. In general, the presence of interaction terms in the model implies that the expected change in Y depends not only on the variable being increased or decreased but also on the values of some of the fixed variables. As in ANOVA, it is possible to have higher-way interaction terms (for example, $x_1 x_2 x_3$), making model interpretation more difficult.

Implicit in the discussion thus far is the assumption that x_1 and x_2 are quantitative variables. It is also possible to build models involving one or more qualitative variables. Suppose that x_1 is a quantitative variable and the other variable of interest is qualitative with three different levels (say, manufacturers 1, 2, and 3). Then two possible models are

$$Y = \beta_0 + \beta_1 x_1 + \beta_2 x_2 + \beta_3 x_3 + \epsilon$$

and

$$Y = \beta_0 + \beta_1 x_1 + \beta_2 x_2 + \beta_3 x_3 + \beta_4 x_1 x_2 + \beta_5 x_1 x_3 + \epsilon$$

The carriers x_2 and x_3 here both refer to the qualitative variable, with $x_2 = 0$, $x_3 = 0$ corresponding to level 1 of the variable, $x_2 = 1$ and $x_3 = 0$ to level 2, and $x_2 = 0$, $x_3 = 1$ to level 3. These carriers are often called *dummy* or *indicator variables*. The contour of the regression function for each of the three levels of the qualitative variable is pictured in Figure 13.9 for both these models.

Chapter 9 of the book by Neter et al. contains an excellent discussion of models containing qualitative variables.

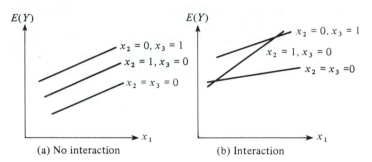

Figure 13.9   Regression function contours for a model with one quantitative variable (x_1) and one qualitative variable having three levels

Estimating Parameters

Instead of n pairs, the data now consists of n $k + 1$-tuples $(x_{11}, x_{21}, \ldots, x_{k1}, y_1)$, $(x_{12}, x_{22}, \ldots, x_{k2}, y_2), \ldots, (x_{1n}, x_{2n}, \ldots, x_{kn}, y_n)$, where x_{ij} is the value of the ith carrier x_i associated with the observed value y_j. The y_j's are assumed to have been observed independently of one another according to the model (13.18). To estimate the parameters $\beta_0, \beta_1, \ldots, \beta_k$ using the principle of least squares, form the sum of squared deviations of the observed y_j's from a trial function $y = b_0 + b_1 x_1 + \cdots + b_k x_k$:

$$f(b_0, b_1, \ldots, b_k) = \sum_j [y_j - (b_0 + b_1 x_{1j} + b_2 x_{2j} + \cdots + b_k x_{kj})]^2 \quad (13.20)$$

The least squares estimates are those values of $b_0, b_1, \ldots, b_k$ that minimize $f(b_0, \ldots, b_k)$. Upon taking the partial derivative of f with respect to each b_i ($i = 0, 1, \ldots, k$) and equating all partials to zero, the following system of **normal equations** is obtained:

$$b_0 n + b_1 \Sigma x_{1j} + b_2 \Sigma x_{2j} + \cdots + b_k \Sigma x_{kj} = \Sigma y_j$$
$$b_0 \Sigma x_{1j} + b_1 \Sigma x_{1j}^2 + b_2 \Sigma x_{1j} x_{2j} + \cdots + b_k \Sigma x_{1j} x_{kj} = \Sigma x_{1j} y_j \quad (13.21)$$
$$\vdots \qquad\qquad \vdots \qquad\qquad\qquad \vdots$$
$$b_0 \Sigma x_{kj} + b_1 \Sigma x_{1j} x_{kj} + \cdots + b_{k-1} \Sigma x_{k-1, j} x_{kj} + b_k \Sigma x_{kj}^2 = \Sigma x_{kj} y_j$$

That these equations are linear in the unknowns $b_0, b_1, \ldots, b_k$ is a consequence of the regression function being linear in the parameters. Solving (13.21) yields the least squares estimates $\hat\beta_0, \hat\beta_1, \ldots, \hat\beta_k$. In general the system (13.21) can be solved by hand by first computing all coefficients of the b_i's and then using a technique such as Gaussian elimination. This is quite tedious, but fortunately any of the standard statistical regression packages will automatically solve for and print out $\hat{\beta}_0, \ldots, \hat{\beta}_k$.

For the case $k = 2$ (2 carriers) the solution to (13.21) can be written down as was done for quadratic regression. Let

$$s_{1y} = \Sigma x_{1j}y_j - n\bar{x}_1\bar{y}, \quad s_{2y} = \Sigma x_{2j}y_j - n\bar{x}_2\bar{y}, \quad s_{11} = \Sigma x_{1j}^2 - n\bar{x}_1^2,$$
$$s_{12} = \Sigma x_{1j}x_{2j} - n\bar{x}_1\bar{x}_2, \quad s_{22} = \Sigma x_{2j}^2 - n\bar{x}_2^2$$

Then

$$\hat{\beta}_1 = \frac{s_{1y}s_{22} - s_{2y}s_{12}}{s_{11}s_{22} - s_{12}^2}, \quad \hat{\beta}_2 = \frac{s_{2y}s_{11} - s_{1y}s_{12}}{s_{11}s_{22} - s_{12}^2}$$

and   (13.22)

$$\hat{\beta}_0 = \bar{y} - \hat{\beta}_1\bar{x}_1 - \hat{\beta}_2\bar{x}_2$$

Example 13.9   In the paper "An Ultracentrifuge Flour Absorption Method" (*Cereal Chemistry*, 1978, pp. 96–101), the authors studied the relationship between water absorption for wheat flour and various characteristics of the flour. In particular the authors used a first-order multiple linear regression model to relate absorption y (%) to flour protein x_1 (%) and starch damage x_2 (Farrand units). The data is shown in Table 13.7.

Table 13.7

x_1	x_2	y	x_1	x_2	y
8.5	2	30.9	12.9	24	47.0
8.9	3	32.7	12.0	25	46.8
10.6	3	36.7	12.9	28	45.9
10.2	20	41.9	13.1	28	48.8
9.8	22	40.9	11.4	32	46.2
10.8	20	42.9	13.2	28	47.8
11.6	31	46.3	11.6	35	49.2
12.0	32	47.6	12.1	34	48.3
12.5	31	47.2	11.3	35	48.6
10.9	28	44.0	11.1	40	50.2
12.2	36	47.7	11.5	45	49.6
11.9	28	43.9	11.6	50	53.2
11.3	30	46.8	11.7	55	54.3
13.0	27	46.2	11.7	57	55.8

The necessary summary quantities are $n = 28$, $\Sigma x_{1j} = 322.3$, $\Sigma x_{2j} = 829.0$, $\Sigma x_{1j}^2 = 3746.4$, $\Sigma x_{2j}^2 = 29{,}327.0$, $\Sigma x_{1j}x_{2j} = 9746.6$, $\Sigma y_j = 1287.4$, $\Sigma x_{1j}y_j = 14{,}940.06$, and $\Sigma x_{2j}y_j = 40{,}016.0$. This gives $s_{11} = 36.49679$, $s_{12} = 204.21786$, $s_{22} = 4782.67857$, $s_{1y} = 121.16643$, and $s_{2y} = 1899.76429$, which in turn yields $\hat{\beta}_1 = 1.44176$, $\hat{\beta}_2 = .33566$, and $\hat{\beta}_0 = 19.44495$. The estimated regression function is $\hat{\mu}_{Y \cdot x_1, x_2} = 19.45 + 1.44x_1 + .34x_2$. ∎

$\hat{\sigma}^2$ and R^2

As for both simple linear regression and polynomial regression, the estimate of σ^2 is based on the sum of squared residuals

$$SSE = \Sigma[y_j - (\hat{\beta}_0 + \hat{\beta}_1 x_{1j} + \cdots + \hat{\beta}_k x_{kj})]^2$$

By squaring out the residuals and carrying the summation through to each term, it can be verified that

$$SSE = \Sigma y_j^2 - \hat{\beta}_0 \Sigma y_j - \hat{\beta}_1 \Sigma x_{1j} y_j - \cdots - \hat{\beta}_k \Sigma x_{kj} y_j$$

Because $k + 1$ parameters $\beta_0, \beta_1, \ldots, \beta_k$ have been estimated, $k + 1$ degrees of freedom are lost, so $n - (k + 1)$ degrees of freedom is associated with SSE, and

$$\hat{\sigma}^2 = s^2 = \frac{SSE}{n - (k + 1)} = MSE$$

With $SST = \Sigma(y_i - \bar{y})^2$, the proportion of total variation explained by the multiple regression model is $R^2 = 1 - SSE/SST$, the **coefficient of multiple determination.** As in polynomial regression, R^2 is often adjusted for the number of parameters in the model by the formula adjusted $R^2 = [(n - 1)R^2 - k]/[n - (k + 1)]$. The positive square root of the coefficient of multiple determination is called the multiple correlation coefficient R. It can be shown that R is the sample correlation coefficient r between the observed y_j's and the predicted $\hat{y}_j$'s (that is, using $x_j = \hat{y}_j$ in the formula for r results in $r = R$).*

Example **13.10** (Example 13.9 continued)

With $\Sigma y_j = 1287.4$ and $\Sigma y_j^2 = 60,035.12$,

$$SSE = 60,035.12 - (19.44495)(1287.4) - (1.44176)(14,940.6)$$
$$- (.33566)(40,016) = 29.496$$

so $\hat{\sigma}^2 = 29.496/[28 - (2 + 1)] = 1.18$ and $\hat{\sigma} = s = 1.09$. Since $SST = 60,035.12 - (1287.4)^2/28 = 842.31$, $R^2 = 1 - (29.496)/(842.31) = .965$, so 96.5% of the variation in the observed y's can be explained by the multiple regression model. Adjusted $R^2 = .962$. ∎

Inferences about Model Parameters

Before testing hypotheses, constructing confidence intervals, and making predictions, one should first examine diagnostic plots to see whether the model needs modification or whether there are outliers in the data. The recommended

---

*Because each $\hat{\beta}_i$ is a linear function of the observed y_j's, $\hat{y} = \hat{\beta}_0 + \hat{\beta}_1 x_1 + \cdots + \hat{\beta}_k x_k$ is also a linear function of the y_j's. Consider any other such linear function $y' = \Sigma c_j y_j$. Then it can be shown that the correlation between the y_j's and $\hat{y}_j$'s is at least as large as between the y_j's and the y'_j's. Put another way, the estimated regression function yields predicted values that have maximum correlation with observed values.

plots are (standardized) residuals versus each independent variable, residuals versus $\hat{y}$, $\hat{y}$ versus y, and a normal probability plot of the standardized residuals. Potential problems are suggested by the same patterns discussed in Section 13.1. Of particular importance is the identification of points that have a large influence on the fit. In the next section we describe several diagnostic tools suitable for this task.

Because each $\hat{\beta}_i$ is a linear function of the y_i's, the standard deviation of each $\hat{\beta}_i$ is the product of σ and a function of the x_{ij}'s, so an estimate $s_{\hat{\beta}_i}$ is obtained by substituting s for σ. Unfortunately the function of the x_{ij}'s is quite complicated, but all standard regression computer packages compute and print the $s_{\hat{\beta}_i}$'s. Inferences concerning a single β_i are based on the standardized variable

$$T = \frac{\hat{\beta}_i - \beta_i}{S_{\hat{\beta}_i}}$$

which has a t distribution with $n - (k + 1)$ d.f.

Let $\mu_{Y \cdot x_1^*, \ldots, x_k^*}$ denote the expected value of Y when $x_1 = x_1^*, \ldots, x_k = x_k^*$. The point estimate of this expected value is $\hat{\mu}_{Y \cdot x_1^*, \ldots, x_k^*} = \hat{\beta}_0 + \hat{\beta}_1 x_1^* + \cdots + \hat{\beta}_k x_k^*$. The estimated standard deviation of the corresponding estimator is again a complicated expression involving the sample x_{ij}'s. However, the better statistical computer packages will calculate it on request. Inferences about $\mu_{Y \cdot x_1^*, \ldots, x_k^*}$ are based on standardizing its estimator to obtain a t variable having $n - (k + 1)$ d.f.

Inferences based on the model $Y = \beta_0 + \beta_1 x_1 + \cdots + \beta_k x_k + \epsilon$

a. A $100(1 - \alpha)\%$ confidence interval for β_i, the coefficient of x_i in the regression function, is

$$\hat{\beta}_i \pm t_{\alpha/2, n - (k + 1)} \cdot s_{\hat{\beta}_i}$$

Simultaneous confidence intervals for several β_i's for which the simultaneous confidence level is controlled can be obtained by applying the Bonferroni technique.

b. A test for $H_0 : \beta_i = \beta_{i0}$ uses the t statistic value $t = (\hat{\beta}_i - \beta_{i0})/s_{\hat{\beta}_i}$ along with a t critical value with $n - (k + 1)$ d.f. The test is upper-, lower-, or two-tailed according to whether H_a contains the inequality $>$, $<$, or $\neq$.

c. A $100(1 - \alpha)\%$ confidence interval for $\mu_{Y \cdot x_1^*, \ldots, x_k^*}$ is

$$\hat{\mu}_{Y \cdot x_1^*, \ldots, x_k^*} \pm t_{\alpha/2, n - (k + 1)} \cdot \{\text{estimated S.D. of } \hat{\mu}_{Y \cdot x_1^*, \ldots, x_k^*}\}$$

d. A $100(1 - \alpha)\%$ prediction interval for a future y value is

$$\hat{\mu}_{Y \cdot x_1^*, \ldots, x_k^*} \pm t_{\alpha/2, n - (k + 1)} \cdot \{s^2 + (\text{estimated S.D. of } \hat{\mu}_{Y \cdot x_1^*, \ldots, x_k^*})^2\}^{1/2}$$

Example 13.11 Soil and sediment adsorption, the extent to which chemicals collect in a condensed form on the surface, is an important characteristic influencing the effectiveness of pesticides and various agricultural chemicals. The paper "Adsorption of Phosphate, Arsenate, Methanearsonate, and Cacodylate by Lake and Stream Sediments: Comparisons with Soils" (*J. of Environ. Qual.,* 1984, pp. 499–504) gave the accompanying data on y = phosphate adsorption index, x_1 = amount of extractable iron, and x_2 = amount of extractable aluminum.

Table 13.8

Observation	$x_1 = Extractable$ iron	$x_2 = Extractable$ aluminum	$y = Adsorption$ index
1	61	13	4
2	175	21	18
3	111	24	14
4	124	23	18
5	130	64	26
6	173	38	26
7	169	33	21
8	169	61	30
9	160	39	28
10	244	71	36
11	257	112	65
12	333	88	62
13	199	54	40

The paper proposed the model

$$Y = \beta_0 + \beta_1 x_1 + \beta_2 x_2 + \epsilon$$

A computer analysis yielded the following information:

Parameter β_i	Estimate $\hat{\beta}_i$	Estimated S.D. $s_{\hat{\beta}_i}$
β_0	-7.351	3.485
β_1	.11273	.02969
β_2	.34900	.07131

$R^2 = .948$, adjusted $R^2 = .938$, $s = 4.379$

$\hat{\mu}_{Y \cdot 160, 39} = -7.351 + (.11273)(160) + (.34900)(39) = 24.30$

estimated S.D. of $\hat{\mu}_{Y \cdot 160, 39} = 1.30$

A 99% confidence interval for β_1, the change in expected adsorption associated with a one-unit increase in extractable iron while extractable aluminum is held fixed, requires $t_{.005, 13 - (2 + 1)} = t_{.005, 10} = 3.169$. The confidence interval is

$$.11273 \pm (3.169)(.02969) = .11273 \pm .09409 \approx (.019, .207)$$

Similarly, a 99% interval for β_2 is

$$.34900 \pm (3.169)(.07131) = .34900 \pm .22598 \approx (.123, .575)$$

The Bonferroni technique implies that the simultaneous confidence level for both intervals is at least 98%.

A 95% confidence interval for $\mu_{Y \cdot 160, 39}$, expected adsorption when extractable iron $= 160$ and extractable aluminum $= 39$, is

$$24.30 \pm (2.228)(1.30) = 24.30 \pm 2.90 = (21.40, 27.20)$$

A 95% prediction interval for a future value of adsorption to be observed when $x_1 = 160$ and $x_2 = 39$ is

$$24.30 \pm (2.228)\{(4.379)^2 + (1.30)^2\}^{1/2} = 24.30 \pm 10.18 = (14.12, 34.48)$$

■

Conclusions resulting from individually testing hypotheses about the regression parameters can be misleading. To decide whether both $\beta_1 = 0$ and $\beta_2 = 0$ for a model with two carriers x_1 and x_2, it might be tempting to examine the t ratios $\hat{\beta}_1/s_{\hat{\beta}_1}$ and $\hat{\beta}_2/s_{\hat{\beta}_2}$. The difficulty is that the t ratio $\hat{\beta}_1/s_{\hat{\beta}_1}$ for testing $H_0 : \beta_1 = 0$ is computed assuming that x_2 belongs in the model, and similarly for testing $H_0 : \beta_2 = 0$ using $t = \hat{\beta}_2/s_{\hat{\beta}_2}$. It can happen that both $\hat{\beta}_1/s_{\hat{\beta}_1}$ and $\hat{\beta}_2/s_{\hat{\beta}_2}$ have nonsignificant values, so that neither x_i should be in the model when the other one is, yet this does not imply that both x_i's should be deleted. This is especially likely to occur when the sample values of the x_{1j}'s and x_{2j}'s are highly correlated, so that either carrier can serve as a "proxy" for the other.

The coefficient of multiple determination can be used to test simultaneously whether the parameters associated with all carriers equal zero; intuitively this tells us whether there exists a useful linear relationship between any of the carriers and the dependent variable.

Test of model utility

Null hypothesis: $H_0 : \beta_1 = \beta_2 = \cdots = \beta_k = 0$

Alternative hypothesis: H_a : at least one $\beta_i \neq 0$ $(i = 1, \ldots, k)$

Test statistic value:
$$f = \frac{R^2/k}{(1 - R^2)/[n - (k + 1)]} \tag{13.23}$$

Rejection region for a level α test: $f \geq F_{\alpha, k, n - (k + 1)}$

Except for a constant multiple, the test statistic here is $R^2/(1 - R^2)$, the ratio of explained to unexplained variation. If the proportion of explained variation is high relative to unexplained, we would naturally want to reject H_0 and confirm the utility of the model.

Example 13.12
(Example 13.10 continued)

Referring back to the flour absorption data, we previously computed $R^2 = .965$ for the model with $k = 2$ carriers x_1 and x_2. With $n = 28$ and $\alpha = .01$, $F_{.01, 2, 25} = 5.57$. To test $H_0 : \beta_1 = \beta_2 = 0$, we compute

$$f = \frac{.965/2}{(1 - .965)/25} = 344.64$$

Since $344.64 \geq 5.57$, H_0 is rejected in favor of the conclusion that Y is linearly related to at least one of the x_i's. ■

The above F test was appropriate for testing whether or not all k β_i's associated with carriers were zero. In many problems one first builds a model involving k carriers and then wishes to know whether a particular subset of l carriers provides almost as good a fit as the "full" k-carrier model. Label the parameters associated with carriers as $\beta_1, \ldots, \beta_l, \beta_{l+1}, \ldots, \beta_k$ so that the first l are associated with the carriers of the "reduced" model. We then wish to test

H_0: model is $Y = \beta_0 + \beta_1 x_1 + \cdots + \beta_l x_l + \epsilon$ (reduced model)

versus

H_a: model is $Y = \beta_0 + \beta_1 x_1 + \cdots + \beta_l x_l + \cdots + \beta_k x_k + \epsilon$

(full model)

Since the full model contains not only the parameters of the reduced model but also some extra parameters, it should fit the data at least as well as the reduced model. That is, if we let SSE_k be the sum of squared residuals for the full model and SSE_l be the corresponding sum for the reduced model, then $SSE_k \leq SSE_l$.* Intuitively, if SSE_k is a great deal smaller than SSE_l, the full model provides a much better fit than the reduced model; the appropriate test statistic should then depend on the reduction $SSE_l - SSE_k$ in unexplained variation. The formal procedure is

SSE_k = unexplained variation for the full model

SSE_l = unexplained variation for the reduced model

Test statistic value: $f = \dfrac{(SSE_l - SSE_k)/(k - l)}{SSE_k/[n - (k + 1)]}$ (13.24)

Rejection region: $f \geq F_{\alpha, k - l, n - (k + 1)}$

Example 13.13   The accompanying data was taken from the paper "Applying Stepwise Multiple Regression Analysis to the Reaction of Formaldehyde with Cotton Cellulose" (*Textile Research J.*, 1984, pp. 157–165). The dependent variable y is durable press rating, a quantitative measure of wrinkle resistance. The four independent variables used in the model building process are $x_1 = $ HCHO (formaldehyde)

---

*The estimates $\hat{\beta}_0, \hat{\beta}_1, \ldots, \hat{\beta}_l$ will in general be different for the full and reduced models, so in general two different multiple regressions must be run to obtain SSE_l and SSE_k. If the variables are listed in the above order, though, most computer packages provide an ANOVA table for the full model that can be used to avoid fitting the reduced model.

concentration, x_2 = catalyst ratio, x_3 = curing temperature, and x_4 = curing time.

Table 13.9

Observation	x_1	x_2	x_3	x_4	y	Observation	x_1	x_2	x_3	x_4	y
1	8	4	100	1	1.4	16	4	10	160	5	4.6
2	2	4	180	7	2.2	17	4	13	100	7	4.3
3	7	4	180	1	4.6	18	10	10	120	7	4.9
4	10	7	120	5	4.9	19	5	4	100	1	1.7
5	7	4	180	5	4.6	20	8	13	140	1	4.6
6	7	7	180	1	4.7	21	10	1	180	1	2.6
7	7	13	140	1	4.6	22	2	13	140	1	3.1
8	5	4	160	7	4.5	23	6	13	180	7	4.7
9	4	7	140	3	4.8	24	7	1	120	7	2.5
10	5	1	100	7	1.4	25	5	13	140	1	4.5
11	8	10	140	3	4.7	26	8	1	160	7	2.1
12	2	4	100	3	1.6	27	4	1	180	7	1.8
13	4	10	180	3	4.5	28	6	1	160	1	1.5
14	6	7	120	7	4.7	29	4	1	100	1	1.3
15	10	13	180	3	4.8	30	7	10	100	7	4.6

Consider the full model consisting of $k = 14$ carriers: x_1, x_2, x_3, x_4, $x_5 = x_1^2, \ldots, x_8 = x_4^2$, $x_9 = x_1 x_2, \ldots, x_{14} = x_3 x_4$ (all first- and second-order predictors). Is the inclusion of the second-order predictors justified? That is, should the reduced model consisting of just the carriers x_1, x_2, x_3, and x_4 ($l = 4$) be used? Output resulting from fitting the two models follows.

Parameter	Estimate for reduced model	Estimate for full model
β_0	−.9122	−8.807
β_1	.16073	.1768
β_2	.21978	.7580
β_3	.011226	.10400
β_4	.10197	.5052
β_5	—	−.04393
β_6	—	−.035887
β_7	—	−.00003271
β_8	—	−.01646
β_9	—	.00588
β_{10}	—	.002702
β_{11}	—	.01178
β_{12}	—	−.0006547
β_{13}	—	.00242
β_{14}	—	.002526
R^2	.692	.921
SSE	17.4951	4.4782

With $k = 14$ and $l = 4$, the F critical value for a test with $\alpha = .01$ is $F_{.01, 10, 15} = 3.80$. The test statistic value is

$$f = \frac{(17.4951 - 4.4782)/10}{4.4782/15} = \frac{1.3017}{.2985} = 4.36$$

Since $4.36 \geq 3.80$, H_0 is rejected. We conclude that the appropriate model should include at least one of the second-order predictors (Example 13.17 contains more information about selecting a model for this situation).   ■

Exercises / Section 13.4 [31–41]

31. The paper "The Value of Information for Selected Appliances" (*J. of Marketing Research,* 1980, pp. 14–25) suggested the plausibility of the general multiple regression model for relating the dependent variable y = the price of an air conditioner to $k = 3$ independent variables: x_1 = Btu-per-hour rating, x_2 = energy efficiency ratio, and x_3 = number of settings. Suppose that the model equation is

$$Y = -70 + .025x_1 + 20x_2 + 7.5x_3 + \epsilon$$

and that $\sigma = 20$.

a. What is the expected value of price when Btu rating = 6000, energy efficiency ratio = 8.0, and number of settings = 5?

b. What is the expected change in price associated with a one-unit increase in energy efficiency ratio when the values of x_1 and x_3 are held fixed?

c. Assuming that ϵ has a normal distribution, what is the probability that a single observation on price is between 240 and 300 when x_1, x_2, and x_3 have the values specified in (a)?

32. A chemist is studying how the yield of a product (y) from a certain chemical reaction depends on reaction temperature (x_1) and pressure (x_2). The following two models have been proposed for temperatures between 80 and 110 in combination with pressure values ranging between 50 and 70.

(i) $Y = 1200 + 15x_1 - 35x_2 + \epsilon$

(ii) $Y = -4500 + 75x_1 + 60x_2 - x_1 x_2 + \epsilon$

a. Give the expected value of yield for each model when the temperature is 90. Repeat for a temperature of 95 and again for a temperature of 100. Graph the three mean values for each of the two models.

b. Referring back to (a), what is the expected change in yield associated with a one-unit increase in pressure for a temperature of 90? Answer this for a temperature of 95 and again for a temperature of 100. How do these expected changes differ for the two models?

33. The paper "Readability of Liquid Crystal Displays: A Response Surface" (*Human Factors,* 1983, pp. 185–190) used a multiple regression model with four independent variables to study accuracy in reading liquid crystal displays. The variables were

y = error percentage for subjects reading a four-digit liquid crystal display

x_1 = level of backlight (ranging from 0 to 122 cd/m²)

x_2 = character subtense (ranging from .025° to 1.34°)

x_3 = viewing angle (ranging from 0° to 60°)

x_4 = level of ambient light (ranging from 20 to 1500 lx)

The model fit to data was $Y = \beta_0 + \beta_1 x_1 + \beta_2 x_2 + \beta_3 x_3 + \beta_4 x_4 + \epsilon$. The resulting estimated coefficients were $\hat{\beta}_0 = 1.52$, $\hat{\beta}_1 = .02$, $\hat{\beta}_2 = -1.40$, $\hat{\beta}_3 = .02$, and $\hat{\beta}_4 = -.0006$.

a. Calculate an estimate of expected error percentage when $x_1 = 10$, $x_2 = .5$, $x_3 = 50$, and $x_4 = 100$.

b. Estimate the mean error percentage associated with a backlight level of 20, character subtense of .5, viewing angle of 10, and ambient light level of 30.

c. What is the estimated expected change in error percentage when the level of ambient light is increased by one unit while all other variables are fixed at the values given in (a)? Answer for a 100-unit increase in ambient light level.

d. Explain why the answers in (c) don't depend on the fixed values of x_1, x_2, and x_3. Under what conditions would there be such a dependence?

34. The estimated model discussed in Exercise 33 was based on $n = 30$ observations. The paper also reported the values $SST = 39.2$ and $SSE = 20.0$. Calculate the coefficient of multiple determination, and then carry out the model utility test using $\alpha = .05$.

35. The paper "Development of a Model for Use in Maize Replant Decisions" (*Agronomy J.*, 1980, pp. 459–464) reported a summary of regression analyses using as variables $y = \%$ expected maize yield, $x_1 =$ planting date (days after April 20), and $x_2 =$ plant density [.0004047 × (plants/ha)]. The five planting dates were April 20, May 6, May 22, May 31, and June 10, the four plant densities were 30,890, 41,180, 51,480, and 61,780 plants/ha, and n was 180. The model

$$Y = \beta_0 + \beta_1 x_1 + \beta_2 x_2 + \beta_3 x_1^2 + \beta_4 x_2^2 + \epsilon$$

with $R^2 = .820$ gave a good fit to the data. Parameter estimates and estimated SD's were

Parameter	Estimate	Estimated S.D.
β_0	21.09	—
β_1	.653	.14
β_2	5.488	1.407
β_3	−.02059	.0027
β_4	−.10155	.0373

a. Graph the contours of the regression function for each of the given values of x_1 (estimated expected Y versus x_2 for each of the five x_1 values).

b. Assuming that the true model is given as shown above, what is the expected change in yield resulting from an increase of one in x_2? Estimate this expected change when plant density is 41,180 plants/ha.

36. Refer back to the maize yield regression discussed in Exercise 35.

a. Test the hypothesis $H_0: \beta_1 = \beta_2 = \beta_3 = \beta_4 = 0$ (no linear relationship between Y and any of the carriers) against the alternative $H_a: \beta_i \neq 0$ for at least one i.

b. Compute joint confidence intervals for β_3 and β_4 such that the joint confidence level is at least 95%.

c. Suppose that $SSE = 1275.75$ (not given in the paper). Compute $\hat{\sigma}$ and also SST.

d. The authors also fit the regression function $y = \beta_0 + \beta_1 x_1 + \beta_2 x_2 + \beta_3 x_1^2 + \beta_4 x_2^2 + \beta_5 x_1^3 + \beta_6 x_2^3$. Supposing that the resulting SSE was 1247.30 and using SSE of (c) for the model in Exercise 35, test $H_0: \beta_5 = \beta_6 = 0$ at level .05.

37. The paper "The Influence of Temperature and Sunshine on the Alpha-Acid Contents of Hops (*Agricultural Meteorology*, 1974, pp. 375–382) reported the following data on yield (y), mean temperature over the period between date of coming into hops and date of picking (x_1), and mean percentage of sunshine during the same period (x_2), for the fuggle variety of hop.

x_1:	16.7	17.4	18.4	16.8	18.9	17.1
x_2:	30	42	47	47	43	41
y:	210	110	103	103	91	76

x_1:	17.3	18.2	21.3	21.2	20.7	18.5
x_2:	48	44	43	50	56	60
y:	73	70	68	53	45	31

The summary quantities are $n = 12$, $\Sigma x_{1j} = 222.5$, $\Sigma x_{2j} = 551$, $\Sigma y_j = 1033$, $\Sigma x_{1j}^2 = 4156.47$, $\Sigma x_{2j}^2 = 25{,}937$, $\Sigma x_{1j} x_{2j} = 10{,}276.3$, $\Sigma x_{1j} y_j = 18{,}680$, $\Sigma x_{2j} y_j = 44{,}169$, and $\Sigma y_j^2 = 112{,}123$.

a. Assuming that $\mu_{Y \cdot x_1, x_2} = \beta_0 + \beta_1 x_1 + \beta_2 x_2$ (as was done in the paper), use the summary quantities to verify that the estimated regression function is $\hat{\mu}_{Y \cdot x_1, x_2} = 415.1131 - 6.5928 x_1 - 4.5036 x_2$.

b. What yield would you predict for a mean temperature of 20 and mean percentage sunshine of 40? What is $\hat{\mu}_{Y \cdot 18.9, 43}$ and what is the residual for these values of x_1 and x_2?

c. Compute SSE, $\hat{\sigma}$, and R^2.

d. Test $H_0: \beta_1 = \beta_2 = 0$ versus $H_a:$ either β_1 or $\beta_2 \neq 0$ at level .05.

e. The estimated standard deviation of $\hat{\beta}_0 + \hat{\beta}_1 x_1 + \hat{\beta}_2 x_2$ when $x_1 = 18.9$ and $x_2 = 43$ is 8.20. Use this to obtain a 95% confidence interval for $\mu_{Y \cdot 18.9, 43}$.

f. Use the information in (e) to obtain a 95% prediction interval for yield in a future experiment when $x_1 = 18.9$ and $x_2 = 43$.

g. The estimated S.D. of $\hat{\beta}_1$ is $s_{\hat{\beta}_1} = 4.86$. Test to see whether x_1 belongs in the model.

h. When the model $Y = \beta_0 + \beta_2 x_2 + \epsilon$ is fit, the resulting value of R^2 is .721. Verify that the F statistic for testing $H_0: Y = \beta_0 + \beta_2 x_2 + \epsilon$ versus $H_a: Y = \beta_0 + \beta_1 x_1 + \beta_2 x_2 + \epsilon$ satisfies $t^2 = f$, where t is the value of the t statistic computed in part (g).

38. a. When the model $Y = \beta_0 + \beta_1 x_1 + \beta_2 x_2 + \beta_3 x_1^2 + \beta_4 x_2^2 + \beta_5 x_1 x_2 + \epsilon$ is fit to the hops data of Exercise 37, the estimate of β_5 is $\hat{\beta}_5 = .557$ with estimated standard deviation $s_{\hat{\beta}_5} = .94$. Test $H_0: \beta_5 = 0$ versus $H_a: \beta_5 \neq 0$.

b. Each t ratio $\hat{\beta}_i/s_{\hat{\beta}_i}$ ($i = 1, 2, 3, 4, 5$) for the model of (a) is less than 2 in absolute value, yet $R^2 = .861$ for this model. Would it be correct to drop each term from the model because of its small t ratio? Explain.

c. Using $R^2 = .861$ for the model of (a), test $H_0: \beta_3 = \beta_4 = \beta_5 = 0$ (which says that all second-order terms can be deleted).

39. The article "The Undrained Strength of Some Thawed Permafrost Soils" (*Canadian Geotechnical J.*, 1979, pp. 420–427) contained the following data on undrained shear strength of sandy soil (y, in kPa), depth (x_1, in m), and water content (x_2, in %). The predicted values and residuals were computed by fitting a full quadratic model, which resulted in the estimated regression function

$$y = -151.36 - 16.22x_1 + 13.48x_2 + .094x_1^2 - .253x_2^2 + .492x_1x_2$$

	y	x_1	x_2	$\hat{y}$	$y - \hat{y}$	Standardized residual e^*
1	14.7000	8.9000	31.5000	23.35	−8.65	−1.50
2	48.0000	36.6000	27.0000	46.38	1.62	0.54
3	25.6000	36.8000	25.9000	27.13	−1.53	−0.53
4	10.0000	6.1000	39.1000	10.99	−0.99	−0.17
5	16.0000	6.9000	39.2000	14.10	1.90	0.33
6	16.8000	6.9000	38.3000	16.54	0.26	0.04
7	20.7000	7.3000	33.9000	23.34	−2.64	−0.42
8	38.8000	8.4000	33.8000	25.43	13.37	2.17
9	16.9000	6.5000	27.9000	15.63	1.27	0.23
10	27.0000	8.0000	33.1000	24.29	2.71	0.44
11	16.0000	4.5000	26.3000	15.36	0.64	0.20
12	24.9000	9.9000	37.8000	29.61	−4.71	−0.91
13	7.3000	2.9000	34.6000	15.38	−8.08	−1.53
14	12.8000	2.0000	36.4000	7.96	4.84	1.02

a. Do plots of e^* versus x_1, e^* versus x_2, and e^* versus $\hat{y}$ suggest that the full quadratic model should be modified? Explain your answer.

b. The value of R^2 for the full quadratic model is .759. Test at level .05 the null hypothesis stating that there is no linear relationship between the dependent variable and any of the five carriers.

c. It can be shown that $V(Y) = \sigma^2 = V(\hat{Y}) + V(Y - \hat{Y})$. The estimate of σ is $\hat{\sigma} = s = 6.99$ (from the full quadratic model). First obtain the estimated S.D. of $Y - \hat{Y}$, and then esti-

mate the standard deviation of $\hat{Y}$ (that is, of $\hat{\beta}_0 + \hat{\beta}_1x_1 + \hat{\beta}_2x_2 + \hat{\beta}_3x_1^2 + \hat{\beta}_4x_2^2 + \hat{\beta}_5x_1x_2$) when $x_1 = 8.0$ and $x_2 = 33.1$. Finally, compute a 95% confidence interval for $\mu_{Y \cdot 8.0, 33.1}$. *Hint:* What is $(y - \hat{y})/e^*$?

d. Fitting the first-order model with regression function $\mu_{Y \cdot x_1, x_2} = \beta_0 + \beta_1x_1 + \beta_2x_2$ results in $SSE = 894.95$. Test at level .05 the null hypothesis that states that all quadratic terms can be deleted from the model.

40. In an experiment to study factors influencing the wood specific gravity of slash pines ("Anatomical Factors Influencing Wood Specific Gravity of Slash Pines and the Implications for the Development of a High-Quality Pulpwood," *TAPPI*, 1964, pp. 401–404), a sample of 20 mature wood samples was obtained, and measurements were taken on number of fibers/mm² in springwood (x_1), number of fibers/mm² in summerwood (x_2), % springwood (x_3), light absorption in springwood (x_4), and light absorption in summerwood (x_5).

a. Fitting the regression function
$$\mu_{Y \cdot x_1, x_2, x_3, x_4, x_5} = \beta_0 + \beta_1x_1 + \cdots + \beta_5x_5$$
resulted in $R^2 = .769$. Does the data indicate that there is a linear relationship between specific gravity and at least one of the carriers? Test using $\alpha = .01$.

b. When x_2 is dropped from the model, the value of R^2 remains at .769. Compute adjusted R^2 for both the full model and the model with x_2 deleted.

c. When x_1, x_2, and x_4 are all deleted, the resulting value of R^2 is .654. The total sum of squares is $SST = .0196610$. Does the data suggest that all of x_1, x_2, and x_4 have zero coefficients in the true regression model? Test the relevant hypotheses at level .05.

41. The accompanying data resulted from a study of the relationship between brightness of finished paper (y) and the variables $H_2O_2\%$ by weight (x_1), NaOH% by weight (x_2), silicate % by weight (x_3), and process temperature (x_4) ("Advantages of CEHDP Bleaching for High Brightness Kraft Pulp Production," *TAPPI*, 1964, pp.

Test no.	H_2O_2 conc. (x_1)	NaOH conc. (x_2)	Silicate conc. (x_3)	Temp. (x_4)	Bright. y
1	−1	−1	−1	−1	83.9
2	+1	−1	−1	−1	84.9
3	−1	+1	−1	−1	83.4
4	+1	+1	−1	−1	84.2
5	−1	−1	+1	−1	83.8
6	+1	−1	+1	−1	84.7
7	−1	+1	+1	−1	84.0
8	+1	+1	+1	−1	84.8
9	−1	−1	−1	+1	84.5
10	+1	−1	−1	+1	86.0
11	−1	+1	−1	+1	82.6
12	+1	+1	−1	+1	85.1
13	−1	−1	+1	+1	84.5
14	+1	−1	+1	+1	86.0
15	−1	+1	+1	+1	84.0
16	+1	+1	+1	+1	85.4
17	−2	0	0	0	82.9
18	+2	0	0	0	85.5
19	0	−2	0	0	85.2
20	0	+2	0	0	84.5
21	0	0	−2	0	84.7
22	0	0	+2	0	85.0
23	0	0	0	−2	84.9
24	0	0	0	+2	84.0
25	0	0	0	0	84.5
26	0	0	0	0	84.7
27	0	0	0	0	84.6
28	0	0	0	0	84.9
29	0	0	0	0	84.9
30	0	0	0	0	84.5
31	0	0	0	0	84.6

Variables		−2	−1	0	+1	+2
x_1	Hydrogen peroxide (100%), %wt	0.1	0.2	0.3	0.4	0.5
x_2	NaOH, %wt	0.1	0.2	0.3	0.4	0.5
x_3	Silicate (41°Bé), %wt	0.5	1.5	2.5	3.5	4.5
x_4	Process temp., °F	130	145	160	175	190

170A–173A). Each independent variable was allowed to assume five different values, and these values were coded for regression analysis as −2, −1, 0, 1, and 2.

a. When a (coded) model involving all linear terms, all quadratic terms, and all cross-product terms was fit, the estimated regression function was

$$y = 84.67 + .650x_1 − .258x_2 + .133x_3$$
$$+ .108x_4 − .135x_1^2 + .028x_2^2 + .028x_3^2$$
$$− .072x_4^2 + .038x_1x_2 − .075x_1x_3$$
$$+ .213x_1x_4 + .200x_2x_3 − .188x_2x_4$$
$$+ .050x_3x_4$$

Use this estimated model to predict brightness when H_2O_2 is .4%, NaOH is .4%, silicate is 3.5%, and temperature is 175. What are the values of the residuals for these values of the variables?

b. Express the estimated regression function in uncoded form.

c. $SST = 17.2567$, and R^2 for the model of (a) is .885. When a model that includes only the four linear terms is fit, the resulting value of R^2 is .721. State and test at level .05 the null hypothesis that specifies that the coefficients of all quadratic and cross-product terms in the regression function are zero.

d. The estimated (coded) regression function when only linear terms are included is
$\hat{\mu}_{Y \cdot x_1, x_2, x_3, x_4} = 85.5548 + .6500x_1 − .2583x_2 + .1333x_3 + .1083x_4$. When $x_1 = x_2 = x_3 = x_4 = 0$, the estimated S.D. of $\hat{\mu}_{Y \cdot 0,0,0,0}$ is .0772. Suppose that it had been believed that expected brightness for these values of the x_i's was at least 85.0. Does the given information contradict this belief? State and test the appropriate hypotheses.

13.5   Other Issues in Multiple Regression

This section touches upon a number of issues that may arise when a multiple regression analysis is carried out. Consult the chapter references for a more extensive treatment of any particular topic.

Transformations in Multiple Regression

Often theoretical considerations suggest a nonlinear relation between an independent variable and two or more dependent variables, while on other occasions diagnostic plots indicate that some type of nonlinear function should be used. Frequently a transformation will linearize the model.

Example 13.14   An article in *Lubrication Eng.* ("Accelerated Testing of Solid Film Lubricants," 1972, pp. 365–372) reported on an investigation of wear life for solid film lubricant. Three sets of journal bearing tests were run on a Mil-L-8937 type film at each combination of three loads (3000, 6000, and 10,000 psi) and three speeds (20, 60, and 100 rpm), and the wear life (hours) was recorded for each run.

Table 13.10

s	$l(1000\text{'s})$	w	s	$l(1000\text{'s})$	w
20	3	300.2	60	6	65.9
20	3	310.8	60	10	10.7
20	3	333.0	60	10	34.1
20	6	99.6	60	10	39.1
20	6	136.2	100	3	26.5
20	6	142.4	100	3	22.3
20	10	20.2	100	3	34.8
20	10	28.2	100	6	32.8
20	10	102.7	100	6	25.6
60	3	67.3	100	6	32.7
60	3	77.9	100	10	2.3
60	3	93.9	100	10	4.4
60	6	43.0	100	10	5.8
60	6	44.5			

The article contains the comment that a lognormal distribution is appropriate for W, since $\ln(W)$ is known to follow a normal law (recall from Chapter 4 that this is what defines a lognormal distribution). The model that appears is $W = (c/s^a l^b) \cdot \epsilon$, from which $\ln(W) = \ln(c) - a\ln(s) - b\ln(l) + \ln(\epsilon)$; so with $Y = \ln(W)$, $x_1 = \ln(s)$, $x_2 = \ln(l)$, $\beta_0 = \ln(c)$, $\beta_1 = -a$, and $\beta_2 = -b$, we have a multiple linear regression model. After computing $\ln(w_i)$, $\ln(s_i)$, and $\ln(l_i)$ for the above data, a first-order model in the transformed variables yielded the results shown in Table 13.11.

Table 13.11

Parameter β_i	Estimate $\hat{\beta}_i$	Estimated S.D. $s_{\hat{\beta}_i}$	$t = \hat{\beta}_i / s_{\hat{\beta}_i}$
β_0	10.8719	.7871	13.81
β_1	−1.2054	.1710	−7.05
β_2	−1.3979	.2327	−6.01

The multiple coefficient of determination (for the transformed observations) has value $R^2 = .781$. The estimated regression function for the transformed variables is

$$\ln(w) = 10.87 - 1.21 \ln(s) - 1.40 \ln(l)$$

so that the original regression function is estimated as

$$w = e^{10.87} \cdot s^{-1.21} \cdot l^{-1.40}$$

The Bonferroni approach can be used to obtain simultaneous confidence intervals for β_1 and β_2, and because $\beta_1 = -a$ and $\beta_2 = -b$, intervals for a and b are then immediately available.   ■

Standardizing Variables

In Section 13.3 we considered transforming x to $x' = x - \bar{x}$ before fitting a polynomial. For multiple regression, especially when values of variables are large in magnitude, it is advantageous to carry this coding one step further. Let $\bar{x}_i$ and s_i be the sample average and sample standard deviation of the x_{ij}'s ($j = 1, \ldots, n$). We now code each variable x_i by $x'_i = (x_i - \bar{x}_i)/s_i$. The coded variable x'_i simply reexpresses any x_i value in units of standard deviation above or below the mean. Thus if $\bar{x}_i = 100$ and $s_i = 20$, $x_i = 130$ becomes $x'_i = 1.5$ because 130 is 1.5 standard deviations above the mean of the values of x_i. For example, the coded full second-order model with two independent variables has regression function

$$E(Y) = \beta_0 + \beta_1 \left(\frac{x_1 - \bar{x}_1}{s_1}\right) + \beta_2 \left(\frac{x_2 - \bar{x}_2}{s_2}\right) + \beta_3 \left(\frac{x_1 - \bar{x}_1}{s_1}\right)^2$$

$$+ \beta_4 \left(\frac{x_2 - \bar{x}_2}{s_2}\right)^2 + \beta_5 \left(\frac{x_1 - \bar{x}_1}{s_1}\right)\left(\frac{x_2 - \bar{x}_2}{s_2}\right)$$

$$= \beta_0 + \beta_1 x'_1 + \beta_2 x'_2 + \beta_3 x'_3 + \beta_4 x'_4 + \beta_5 x'_5$$

The benefits of coding are (a) increased numerical accuracy in all computations (through less computer roundoff error), and (b) more accurate estimation than for the parameters of the uncoded model because the individual parameters of the coded model characterize the behavior of the regression function near the center of the data rather than near the origin.

Example 13.15   The paper "The Value and the Limitations of High-Speed Turbo-Exhausters for the Removal of Tar-Fog from Carburetted Water-Gas" (*Soc. Chemical Industry J.,* 1946, pp. 166–168) presents the accompanying data on $y = $ tar content (grains/100 ft³) of a gas stream as a function of $x_1 = $ rotor speed (rpm) and $x_2 = $ gas inlet temperature (°F). The data is also considered in the paper "Some Aspects of Nonorthogonal Data Analysis" (*J. Quality Technology,* 1973, pp. 67–79), which suggests using the coded model described above. The means and standard deviations are $\bar{x}_1 = 2991.13$, $s_1 = 387.81$, $\bar{x}_2 = 58.468$, and $s_2 = 6.944$, so $x'_1 = (x_1 - 2991.13)/387.81$ and $x'_2 = (x_2 - 58.468)/6.944$. With $x'_3 = (x'_1)^2$, $x'_4 = (x'_2)^2$, and $x'_5 = x'_1 \cdot x'_2$, fitting the full second-order model requires solving the system of six normal equations in six unknowns. A computer analysis yielded $\hat{\beta}_0 = 40.2660$, $\hat{\beta}_1 = -13.4041$, $\hat{\beta}_2 = 10.2553$, $\hat{\beta}_3 = 2.3313$, $\hat{\beta}_4 = -2.3405$, and $\hat{\beta}_5 = 2.5978$. The estimated regression equation is then

$$\hat{y} = 40.27 - 13.40x_1' + 10.26x_2' + 2.33x_3' - 2.34x_4' + 2.60x_5'$$

Thus if $x_1 = 3200$ and $x_2 = 57.0$, $x_1' = .539$, $x_2' = -.211$, $x_3' = (.539)^2 = .2901$, $x_4' = (-.211)^2 = .0447$, and $x_5' = (.539)(-.211) = -.1139$, so

$$\hat{y} = 40.27 - (13.40)(.539) + (10.26)(-.211) + (2.33)(.2901)$$
$$- (2.34)(.0447) + (2.60)(-.1139) = 31.16$$

Table 13.12

Run	y	x_1	x_2	x_1'	x_2'
1	60.0000	2400.00	54.5000	-1.52428	-0.57145
2	61.0000	2450.00	56.0000	-1.39535	-0.35543
3	65.0000	2450.00	58.5000	-1.39535	0.00461
4	30.5000	2500.00	43.0000	-1.26642	-2.22763
5	63.5000	2500.00	58.0000	-1.26642	-0.06740
6	65.0000	2500.00	59.0000	-1.26642	0.07662
7	44.0000	2700.00	52.5000	-0.75070	-0.85948
8	52.0000	2700.00	65.5000	-0.75070	1.01272
9	54.5000	2700.00	68.0000	-0.75070	1.37276
10	30.0000	2750.00	45.0000	-0.62177	-1.93960
11	26.0000	2775.00	45.5000	-0.55731	-1.86759
12	23.0000	2800.00	48.0000	-0.49284	-1.50755
13	54.0000	2800.00	63.0000	-0.49284	0.65268
14	36.0000	2900.00	58.5000	-0.23499	0.00461
15	53.5000	2900.00	64.5000	-0.23499	0.86870
16	57.0000	3000.00	66.0000	0.02287	1.08472
17	33.5000	3075.00	57.0000	0.21627	-0.21141
18	34.0000	3100.00	57.5000	0.28073	-0.13941
19	44.0000	3150.00	64.0000	0.40966	0.79669
20	33.0000	3200.00	57.0000	0.53859	-0.21141
21	39.0000	3200.00	64.0000	0.53859	0.79669
22	53.0000	3200.00	69.0000	0.53859	1.51677
23	38.5000	3225.00	68.0000	0.60305	1.37276
24	39.5000	3250.00	62.0000	0.66752	0.50866
25	36.0000	3250.00	64.5000	0.66752	0.86870
26	8.5000	3250.00	48.0000	0.66752	-1.50755
27	30.0000	3500.00	60.0000	1.31216	0.22063
28	29.0000	3500.00	59.0000	1.31216	0.07662
29	26.5000	3500.00	58.0000	1.31216	-0.06740
30	24.5000	3600.00	58.0000	1.57002	-0.06740
31	26.5000	3900.00	61.0000	2.34360	0.36465

■

Variable Selection

Often an experimenter will have data on a large number of carriers and then wish to build a regression model involving a subset of the carriers. The use of the subset will make the resulting model more manageable, especially if more data is to be subsequently collected, and also result in a model that is easier to interpret and understand than one with many more carriers. Two fundamental questions in connection with variable selection are:

1. If we can examine regressions involving all possible subsets of the carriers for which data is available, what criteria should be used to select a model?
2. If the number of carriers is too large to permit all regressions to be examined, is there a way of examining a reduced number of subsets among which a good model (or models) will be found?

To address (1) first, if the number of carriers is small (≤ 5, say), then it would not be too tedious to examine all possible regressions using any one of the readily available statistical computer packages (MINITAB, SAS, BMD, SPSS). If data on at least six carriers is available, all possible regressions involve at least 64 ($= 2^6$) different models. There are efficient computer codes available for examining all regressions for up to 12 carriers; a brief discussion of such codes appears in the review article by Hocking listed in the chapter bibliography. Even when it is not possible to examine all regressions, there is a program called SELECT (discussed in the Hocking article), which for each fixed number of possible carriers will identify the subset with the smallest SSE.* These SSE's (or functions of them) can then be compared according to any of the criteria discussed below to decide on a model.

Criteria for Variable Selection

As before we use a subscript k to denote a quantity (say, SSE_k) computed from a model with k carriers (and thus $k + 1$ β_i's, because β_0 will always be included). For a fixed value of k, it is reasonable to identify the best model as the one having minimum SSE_k. The more difficult issue concerns comparison of SSE_k's for different values of k. Three different criteria, each one a simple function of SSE_k, are widely used.

1. R_k^2, the coefficient of multiple determination for a k carrier model. Because R_k^2 will virtually always increase as k does (and can never decrease), it is not the k which maximizes R_k^2 that interests us. Instead we wish to identify a small k for which R_k^2 is nearly as large as R^2 for all carriers in the model.
2. $MSE_k = SSE/(n - k - 1)$, the mean square error for a k carrier model. This is often used in place of R_k^2, since while R_k^2 never decreases with increasing k, a small decrease in SSE_k obtained with one extra carrier can be more than offset by a decrease of one in the denominator of MSE_k. The objective is then to identify the model having minimum MSE_k. Since adjusted $R_k^2 = 1 - MSE_k/MST$ where $MST = SST/(n - 1)$ is constant in k, examination of adjusted R_k^2 is equivalent to consideration of MSE_k.

---

*If the number of carriers is at most 27, the BMDP package will obtain for any specified m between 1 and 10 the best m subsets consisting of 1 carrier, of 2 carriers, of 3 carriers, and so on.

3. The rationale for the third criterion, C_k, is more difficult to understand, but the criterion is gaining increasing acceptance among data analysts. Suppose that the true regression model is specified by m carriers—that is,

$$Y = \beta_0 + \beta_1 x_1 + \cdots + \beta_m x_m + \epsilon, \, V(\epsilon) = \sigma^2$$

so that

$$E(Y) = \beta_0 + \beta_1 x_1 + \cdots + \beta_m x_m$$

Consider fitting a model by using a subset of k of these m carriers; for simplicity of notation, suppose that we use $x_1, x_2, \ldots, x_k$. Then by solving the system of normal equations, estimates $\hat{\beta}_0, \hat{\beta}_1, \ldots, \hat{\beta}_k$ are obtained (but not, of course, estimates of any β's corresponding to carriers not in the fitted model). The true expected value $E(Y)$ can then be estimated by $\hat{Y} = \hat{\beta}_0 + \hat{\beta}_1 x_1 + \cdots + \hat{\beta}_k x_k$. Now consider the **normalized expected total error of estimation**

$$\Gamma_k = \frac{E\left(\sum_{i=1}^{n} [\hat{Y}_i - E(Y_i)]^2\right)}{\sigma^2} = \frac{E(SSE_k)}{\sigma^2} + 2(k+1) - n \qquad (13.25)$$

The second equality in (13.25) must be taken on faith, because it requires a tricky expected value argument. A particular subset is then appealing if its Γ_k value is small. Unfortunately, though, $E(SSE_k)$ and σ^2 are not known. To remedy this, let s^2 denote the estimate of σ^2 based on the model that includes all carriers for which data is available, and define

$$C_k = \frac{SSE_k}{s^2} + 2(k+1) - n$$

A desirable model is then specified by carriers for which C_k is small.

Example 13.16   The article by Hocking reports on an analysis of data taken from the 1974 issues of *Motor Trend* magazine. The dependent variable y was gas mileage, there were $n = 32$ observations, and the carriers for which data was obtained were x_1 = engine shape (1 = straight and 0 = V), x_2 = number of cylinders, x_3 = transmission type (1 = manual and 0 = auto), x_4 = number of transmission speeds, x_5 = engine size, x_6 = horsepower, x_7 = number of carburetor barrels, x_8 = final drive ratio, x_9 = weight, and x_{10} = quarter-mile time. In Table 13.13 we present summary information from the analysis. The table describes for each k the subset having minimum SSE_k; reading down the variables column indicates which variable is added in going from k to $k+1$ (in going from $k = 2$ to $k = 3$, both x_3 and x_{10} are added, and x_2 is deleted). Figure 13.10 contains plots of R_k^2, adjusted R_k^2, and C_k against k; these plots are an important visual aid in selecting a subset. The estimate of σ^2 is $s^2 = 6.24$, which is MSE_{10}. A simple model that rates highly according to all criteria is the one containing carriers x_3, x_9, and x_{10}.

Table 13.13 Best subsets for gas mileage data of Example 13.16

k = number of carriers	Variables	SSE_k	R_k^2	Adjusted R_k^2	C_k
1	9	247.2	.756	.748	11.6
2	2	169.7	.833	.821	1.2
3	3, 10, −2	150.4	.852	.836	.1
4	6	142.3	.860	.839	.8
5	5	136.2	.866	.840	1.8
6	8	133.3	.869	.837	3.4
7	4	132.0	.870	.832	5.2
8	7	131.3	.871	.826	7.1
9	1	131.1	.871	.818	9.0
10	2	131.0	.871	.809	11.0

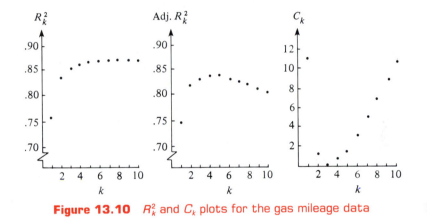

Figure 13.10 R_k^2 and C_k plots for the gas mileage data ■

Generally speaking, when a subset of k carriers ($k < m$) is used to fit a model, the estimators $\hat{\beta}_0, \hat{\beta}_1, \ldots, \hat{\beta}_k$ will be biased for $\beta_0, \beta_1, \ldots, \beta_k$ and $\hat{Y}$ will also be a biased estimator for the true $E(Y)$ (all this because $m - k$ carriers are missing from the fitted model). However, as measured by the total normalized expected error Γ_k, estimates based on a subset can actually provide more precision than would be obtained using all possible carriers; essentially this greater precision is obtained at the price of introducing a bias in the estimators. A value of k for which $C_k \approx k + 1$ indicates that the bias associated with this k carrier model would be small.

Example 13.17
(Example 13.13 continued)

Reconsider the durable press rating data introduced at the end of Section 13.4. The data set consisted of observations on four independent variables, from which we created a model containing 14 carriers: the four x_i's, the four x_i^2's, and the six $x_i x_j$'s. This "full" model was judged superior to the "reduced" model that involved only the carriers x_1, x_2, x_3, and x_4. The R^2 value for the full model was .921. However, computer output shows that several t ratios $\hat{\beta}_i / s_{\hat{\beta}_i}$ are quite small in magnitude, suggesting that there may be several among the 14 carriers that can be deleted.

Figure 13.11 presents selected output resulting from use of the BMDP package to identify the best three models of each given size. Only information on the best three six-carrier and the best three seven-carrier models is given. The best six-carrier model seemed superior to models including fewer carriers, whereas very little appeared to be gained by considering models with more than seven carriers. The choice between competing models is not exactly clear-cut. It would be nice if there were a good model that did not include any carriers based on one or more of the primary variables (for example, the model with only x_2, x_4, x_2^2, x_4^2, and x_2x_4). Unfortunately this does not happen (though if we had only looked at the best single model of any given size, such a model might have been overlooked). The best six-carrier model seems a reasonable choice.

```
                                         SUBSETS WITH     6 VARIABLES

                 ADJUSTED
R-SQUARED        R-SQUARED        CK

   .884952         .854939       5.92   VARIABLE        COEFFICIENT     T-STATISTIC
                                        X2                 .959914         6.66
                                        X1SQ             -.0372760        -3.82
                                        X2SQ             -.0389469        -5.82
                                        X1X3              .00368402        4.91
                                        X1X4              .0192505         3.23
                                        X2X3             -.00128271       -1.91
                                        INTERCEPT        -1.21835

   .881662         .850791       6.54   VARIABLE        COEFFICIENT     T-STATISTIC
                                        X2                 .779668         7.93
                                        X4                 .408872         2.98
                                        X1SQ             -.0313804        -3.40
                                        X2SQ             -.0394407        -5.81
                                        X1X3              .00356717        5.12
                                        X3X4             -.00208494       -2.27
                                        INTERCEPT        -1.34754

   .880382         .849177       6.79   VARIABLE        COEFFICIENT     T-STATISTIC
                                        X2                 .760472         7.71
                                        X1SQ             -.0380752        -3.46
                                        X2SQ             -.0383770        -5.64
                                        X1X3              .00319522        5.13
                                        X1X4              .0384039         2.92
                                        X3X4             -.000908222      -1.62
                                        INTERCEPT        -.632344

                                         SUBSETS WITH     7 VARIABLES

                 ADJUSTED
R-SQUARED        R-SQUARED        CK

   .899208         .867138       5.20   VARIABLE        COEFFICIENT     T-STATISTIC
                                        X2                 .800294         8.40
                                        X3                 .0966644        2.41
                                        X4                 .129477         3.49
                                        X1SQ             -.0392306        -3.13
                                        X2SQ             -.0416758        -6.35
                                        X3SQ             -.000388272      -2.70
                                        X1X3              .00435437        4.08
                                        INTERCEPT        -7.27515
```

.898832	.866643	5.27	VARIABLE	COEFFICIENT	T-STATISTIC
			X2	.825845	8.60
			X3	.0974167	2.42
			X1SQ	−.0392516	−3.12
			X2SQ	−.0433512	−6.57
			X3SQ	−.000390110	−2.71
			X4SQ	.0158501	3.47
			X1X3	.00436212	4.08
			INTERCEPT	−7.25606	

.896670	.863792	5.68	VARIABLE	COEFFICIENT	T-STATISTIC
			X2	.786050	8.15
			X3	.0820370	2.02
			X1SQ	−.0413732	−3.21
			X2SQ	−.0407510	−6.13
			X3SQ	−.000325208	−2.24
			X1X3	.00401885	3.79
			X1X4	.0195025	3.36
			INTERCEPT	−6.06342	

Figure 13.11   Partial BMDP output for Example 13.17   ■

Stepwise Regression

Because algorithms such as SELECT are not yet readily available, when the number of carriers is too large to allow examination of all possible subsets, there are several alternative selection procedures that generally will identify good models. The simplest such procedure is the backward elimination (BE) method. This method starts with the model in which all carriers under consideration are used. Let the set of all such carriers be $x_1, \ldots, x_m$. Then each t ratio $\hat{\beta}_i / s_{\hat{\beta}_i}$ $(i = 1, \ldots, m)$ appropriate for testing $H_0 : \beta_i = 0$ versus $H_a : \beta_i \neq 0$ is examined. If the t ratio with the smallest absolute value is less than a prespecified constant t_{out}, that is, if

$$\min_{i=1,\ldots,m} \left| \frac{\hat{\beta}_i}{s_{\hat{\beta}_i}} \right| < t_{out}$$

then the carrier corresponding to the smallest ratio is eliminated from the model. The reduced model is now fit, the $m - 1$ t ratios are again examined, and another carrier is eliminated if it corresponds to the smallest absolute t ratio smaller than t_{out}. In this way the algorithm continues until at some stage, all absolute t ratios are at least t_{out}. The model used is the one containing all carriers that were not eliminated. The value $t_{out} = 2$ is often recommended since most $t_{.05}$ values are near 2.

Example 13.18
(Example 13.15 continued)
For the coded full quadratic model in which $y =$ tar content, the five potential carriers are x_1', x_2', $x_3' = x_1'^2$, $x_4' = x_2'^2$, and $x_5' = x_1' x_2'$ (so $m = 5$). Without specifying t_{out}, the carrier with the smallest absolute t ratio (asterisked) was eliminated at each stage, resulting in the sequence of models shown in Table 13.14.

Table 13.14

Step	Carriers	$|t\ ratio|$ 1	2	3	4	5
1	1, 2, 3, 4, 5	16.0	10.8	2.9	2.8	1.8*
2	1, 2, 3, 4	15.4	10.2	3.7	2.0*	—
3	1, 2, 3	14.5	12.2	4.3*	—	—
4	1, 2	10.9	9.1*	—	—	—
5	1	4.4*	—	—	—	—

Using $t_{out} = 2$, the resulting model would be based on x_1', x_2', and x_3', since at step 3 no carrier could be eliminated. It can be verified that each subset above is actually the best subset of its size, though this is by no means always the case. ∎

An alternative to the BE procedure is forward selection (FS). FS starts with no carriers in the model and considers fitting in turn the model with only x_1, only x_2, ..., and finally only x_m. The variable which, when fit, yields the largest absolute t ratio enters the model provided that the ratio exceeds the specified constant t_{in}. Suppose that x_1 enters the model. Then models with (x_1, x_2), (x_1, x_3), ..., (x_1, x_m) are considered in turn. The largest $|\hat{\beta}_j/s_{\hat{\beta}_j}|$ $(j = 2, \ldots, m)$ then specifies the entering carrier provided that this maximum also exceeds t_{in}. This continues until at some step no absolute t ratios exceed t_{in}. The entered carriers then specify the model. The value $t_{in} = 2$ is often used for the same reason that $t_{out} = 2$ is used in BE. For the tar-content data, FS resulted in the sequence of models given at steps 5, 4, ..., 1 above, so agreed with BE. This will not always be the case.

The stepwise procedure most widely used is a combination of FS and BE, denoted by FB. This procedure starts off as does forward selection, by adding variables to the model, but after each addition examines those variables previously entered to see if any is a candidate for elimination. For example, if there are eight carriers under consideration and the current set consists of x_2, x_3, x_5, and x_6 with x_5 having just been added, the t ratios $\hat{\beta}_2/s_{\hat{\beta}_2}$, $\hat{\beta}_3/s_{\hat{\beta}_3}$, and $\hat{\beta}_6/s_{\hat{\beta}_6}$ are examined. If the smallest absolute ratio is less than t_{out}, then the corresponding variable is eliminated from the model. The idea behind FB is that with forward selection, a single variable may be more strongly related to y than either of two or more other variables individually, but the combination of these variables may make the single variable subsequently redundant. This actually happened with the gas mileage data of Example 13.16, with x_2 entering and subsequently leaving the model.

The FB procedure is part of several standard computer packages. The BMDP package specifies $t_{in} = 2$ and $t_{out} = \sqrt{3.9}$ (most packages actually use $f = t^2$ rather than t itself).

While in most situations these automatic selection procedures will identify a good model, there is no guarantee that the best or even a nearly best model will result. Close scrutiny should be given to data sets for which there appear to be strong relationships between some of the potential carriers; we shall say more about this shortly.

Identification of Influential Observations

In simple linear regression it is easy to spot an observation whose x value is much larger or much smaller than other x values in the sample. Such an observation may have a great impact on the estimated regression equation (whether or not it actually does depends on how consistent the corresponding y value is with the remainder of the data). In multiple regression it is also desirable to know whether the values of the carriers for a particular observation are such that it has the potential for exerting great influence on the estimated equation. One method for identifying potentially influential observations relies on the fact that because each $\hat{\beta}_i$ is a linear function of $y_1, y_2, \ldots, y_n$, each predicted y value of the form $\hat{y} = \hat{\beta}_0 + \hat{\beta}_1 x_1 + \cdots + \hat{\beta}_k x_k$ is also a linear function of the y_j's. In particular, the predicted values corresponding to sample observations can be written as follows:

$$\hat{y}_1 = h_{11} y_1 + h_{12} y_2 + \cdots + h_{1n} y_n$$
$$\hat{y}_2 = h_{21} y_1 + h_{22} y_2 + \cdots + h_{2n} y_n$$
$$\vdots \qquad \vdots \qquad \vdots \qquad \qquad \vdots$$
$$\hat{y}_n = h_{n1} y_1 + h_{n2} y_2 + \cdots + h_{nn} y_n$$

Each coefficient h_{ij} is a function only of the x_{ij}'s in the sample and not of the y_j's. It can be shown that $h_{ij} = h_{ji}$ and that $0 \le h_{ij} \le 1$.

Let's focus on the "diagonal" coefficients $h_{11}, h_{22}, \ldots, h_{nn}$. The coefficient h_{jj} is the weight given to y_j in computing the corresponding predicted value $\hat{y}_j$. This quantity can also be expressed as a measure of the distance between the point $(x_{1j}, \ldots, x_{kj})$ in k-dimensional space and the center of the data $(\bar{x}_{1.}, \ldots, \bar{x}_{k.})$. It is therefore natural to characterize an observation whose h_{jj} is relatively large as one with potentially large influence. Unless there is a perfect linear relationship between the k carriers, $\sum_{j=1}^{n} h_{jj} = k + 1$, so the average of the h_{jj}'s is $(k + 1)/n$. Some statisticians suggest that if $h_{jj} > 2(k + 1)/n$, the jth observation be cited as having potentially large influence; others use $3(k + 1)/n$ as the dividing line.

Example 13.19   The accompanying data appeared in the paper "Testing for the Inclusion of Variables in Linear Regression by a Randomization Technique" (*Techno-*

Beam number	Specific gravity (x_1)	Moisture content (x_2)	Strength (y)
1	0.499	11.1	11.14
2	0.558	8.9	12.74
3	0.604	8.8	13.13
4	0.441	8.9	11.51
5	0.550	8.8	12.38
6	0.528	9.9	12.60
7	0.418	10.7	11.13
8	0.480	10.5	11.70
9	0.406	10.5	11.02
10	0.467	10.7	11.41

metrics, 1966, pp. 695–699), and was reanalyzed in Hoaglin and Welsch, "The Hat Matrix in Regression and ANOVA" (*Amer. Statistician,* 1978, pp. 17–22): The h_{ij}'s (with elements below the diagonal omitted by symmetry) are

	1	*2*	*3*	*4*	*5*	*6*	*7*	*8*	*9*	*10*
1	.418	−.002	.079	−.274	−.046	.181	.128	.222	.050	.242
2		.242	.292	.136	.243	.128	−.041	.033	−.035	.004
3			.417	−.019	.273	.187	−.126	.044	−.153	.004
4				.604	.197	−.038	.168	−.022	.275	−.028
5					.252	.111	−.030	.019	−.010	−.010
6						.148	.042	.117	.012	.111
7							.262	.145	.277	.174
8								.154	.120	.168
9									.315	.148
10										.187

Here $k = 2$ so $(k + 1)/n = \frac{3}{10} = .3$; since $h_{44} = .604 > 2(.3)$, the fourth data point is identified as potentially influential. ∎

Another technique for assessing the influence of the jth observation that takes into account y_j as well as the carrier values involves deleting the jth observation from the data set and performing a regression based on the remaining observations. If the estimated coefficients from the "deleted observation" regression differ greatly from the estimates based on the full data, the jth observation has clearly had a substantial impact on the fit. One way to judge whether estimated coefficients change greatly is to express each change relative to the estimated standard deviation of the coefficient:

$$\frac{(\hat{\beta}_i \text{ before deletion}) - (\hat{\beta}_i \text{ after deletion})}{s_{\hat{\beta}_i}} = \frac{\text{change in } \hat{\beta}_i}{s_{\hat{\beta}_i}}$$

There exist efficient computational formulas that allow all this information to be obtained from the "no deletion" regression, so that the additional n regressions are unnecessary.

Example 13.20
(Example 13.19 continued)

Consider separately deleting observations 1 and 6, whose residuals are the largest, and observation 4, where h_{jj} is large.

Table 13.15

Parameter	No deletions estimates	Estimated S.D.	Change when point j is deleted		
			j = 1	*j = 4*	*j = 6*
β_0	10.302	1.896	2.710	−2.109	−.642
β_1	8.495	1.784	−1.772	1.695	.748
β_2	.2663	.1273	−.1932	.1242	.0329
e_j:			−3.25	−.96	2.20
h_{jj}:			.418	.604	.148

For deletion of both point 1 and point 4, the change in each estimate is in the range 1–1.5 standard deviation, which is reasonably substantial (this does not tell us what would happen if both points were simultaneously omitted). For point 6, however, the change is roughly .25 standard deviation. Thus points 1 and 4, but not 6, might well be omitted in calculating a regression equation. ■

Multicollinearity

In many multiple regression data sets the carriers $x_1, x_2, \ldots, x_k$ are highly interdependent. Suppose that we consider the usual model

$$Y = \beta_0 + \beta_1 x_1 + \cdots + \beta_k x_k + \epsilon$$

with data $(x_{1j}, \ldots, x_{kj}, y_j)$ $(j = 1, \ldots, n)$ available for fitting. If we use the principle of least squares to regress x_i on the other carriers $x_1, \ldots, x_{i-1}, x_{i+1}, \ldots, x_k$, obtaining

$$\hat{x}_i = a_0 + a_1 x_1 + \cdots + a_{i-1} x_{i-1} + a_{i+1} x_{i+1} + \cdots + a_k x_k$$

it can be shown that

$$V(\hat{\beta}_i) = \frac{\sigma^2}{\displaystyle\sum_{j=1}^{n} (x_{ij} - \hat{x}_{ij})^2} \tag{13.26}$$

When the sample x_i values can be predicted very well from the other carrier values, the denominator of (13.26) will be small, so $V(\hat{\beta}_i)$ will be quite large. If this is the case for at least one carrier, the data is said to exhibit **multicollinearity**. Multicollinearity is often suggested by a regression computer output in which R^2 is large but some of the t ratios $\hat{\beta}_i/s_{\hat{\beta}_i}$ are small for predictors that, based on prior information and intuition, seem important. Another clue to the presence of multicollinearity lies in a $\hat{\beta}_i$ value that has the opposite sign from that which intuition would suggest, indicating that another carrier or collection of carriers is serving as a "proxy" for x_i.

An assessment of the extent of multicollinearity can be obtained by regressing each carrier in turn on the remaining $k - 1$ carriers. Let R_i^2 denote the value of R^2 in the regression with dependent variable x_i and carriers $x_1, \ldots, x_{i-1}, x_{i+1}, \ldots, x_k$. It has been suggested that severe multicollinearity is present if $R_i^2 > .9$ for any i. MINITAB will refuse to include a carrier in the model when its R_i^2 value is close to 1.

There is unfortunately no consensus among statisticians as to what remedies are appropriate when severe multicollinearity is present. One possibility involves continuing to use a model that includes all the carriers but estimating parameters by using something other than least squares. Please consult a chapter reference for more details.

Exercises / Section 13.5 (42–47)

42. The paper "Bank Full Discharge of Rivers" (*Water Resources J.*, 1978, pp. 1141–1154) reported data on discharge amount (q, in m³/sec), flow area (a, in m²), and slope of the water surface (b, in m/m) obtained at a number of floodplain stations. A subset of the data appears below. The paper proposed a multiplicative power model $Q = \alpha a^\beta b^\gamma \epsilon$.

q	17.6	23.8	5.7	3.0	7.5
a	8.4	31.6	5.7	1.0	3.3
b	.0048	.0073	.0037	.0412	.0416

q	89.2	60.9	27.5	13.2	12.2
a	41.1	26.2	16.4	6.7	9.7
b	.0063	.0061	.0036	.0039	.0025

a. Use an appropriate transformation to make the model linear, and then estimate the regression parameters for the transformed model. Finally, estimate α, β, and γ (the parameters of the original model). What would be your prediction of discharge amount when flow area is 10 and slope is .01?

b. Without actually doing any analysis, how would you fit a multiplicative exponential model $Q = \alpha e^{\beta a} e^{\gamma b} \epsilon$?

c. After the transformation to linearity in (a), a 95% confidence interval for the value of the transformed regression function when $a = 3.3$ and $b = .0046$ was obtained from computer output as (.217, 1.755). Obtain a 95% confidence interval for $\alpha a^\beta b^\gamma$ when $a = 3.3$, $b = .0046$.

43. a. Referring back to Exercise 40, the mean and standard deviation of x_3 were 52.540 and 5.4447, respectively, while those of x_5 were 89.195 and 3.6660, respectively. When the model involving these two standardized variables as fit, the estimated regression equation was $y = .5255 - .0236x_3' + .0097x_5'$. What value of specific gravity would you predict for a wood sample with % springwood $= 50$ and % light absorption in summerwood $= 90$?

b. The estimated standard deviation of the estimated coefficient $\hat{\beta}_3$ of x_3' (that is, $s_{\hat{\beta}_3}$ for $\hat{\beta}_3$ of the standardized model) was .0046. Obtain a 95% confidence interval for β_3.

c. Using the information in (a) and (b), what is the estimated coefficient of x_3 in the unstandardized model (using only carriers x_3 and x_5), and what is the estimated standard deviation of the coefficient estimator (that is, $s_{\hat{\beta}_3}$ for $\hat{\beta}_3$ in the unstandardized model)?

d. The estimate of σ for the two-carrier model is $s = .02001$, while the estimated standard deviation of $\hat{\beta}_0 + \hat{\beta}_3 x_3' + \hat{\beta}_5 x_5'$ when $x_3' =$

$-.3747$ and $x_5' = -.2769$ (that is, when $x_3 = 50.5$ and $x_5 = 88.9$) is .00482. Compute a 95% prediction interval for specific gravity when % springwood $= 50.5$ and % light absorption in summerwood $= 88.9$.

44. In the accompanying table we give the smallest SSE for each number of carriers k ($k = 1, 2, 3, 4$) for a regression problem in which $y =$ cumulative heat of hardening in cement, $x_1 = \%$ tricalcium aluminate, $x_2 = \%$ tricalcium silicate, $x_3 = \%$ aluminum ferrate, and $x_4 = \%$ dicalcium silicate.

Number of carriers k	Carrier(s)	SSE
1	x_4	880.85
2	x_1, x_2	58.01
3	x_1, x_2, x_3	49.20
4	x_1, x_2, x_3, x_4	47.86

In addition $n = 13$ and $SST = 2715.76$.

a. Use the criteria discussed in the text to recommend the use of a particular regression model.

b. Would forward selection result in the best two carrier models? Explain.

45. The paper "Creep and Fatigue Characteristics of Ferrocement Slabs" (*J. Ferrocement*, 1984, pp. 309–322) reported data on $y =$ tensile strength (MPa), $x_1 =$ slab thickness (cm), $x_2 =$ load (kg), $x_3 =$ age at loading (days), and $x_4 =$ time under test (days) resulting from stress tests of $n = 9$ reinforced concrete slabs. The results of applying the backward elimination method of variable selection are summarized in the accompanying tabular format. Explain what occurred at each step of the procedure.

Step	1	2	3
Constant	8.496	12.670	12.989
x_1	-0.29	-0.42	-0.49
T-RATIO	-1.33	-2.89	-3.14
x_2	0.0104	0.0110	0.0116
T-RATIO	6.30	7.40	7.33
x_3	0.0059		
T-RATIO	0.83		
x_4	-0.023	-0.023	
T-RATIO	-1.48	-1.53	
S	0.533	0.516	0.570
R-SQ	95.81	95.10	92.82

46. An analysis of the hops yield data of Exercise 37 yields the accompanying h_{ii}'s for the model $Y = \beta_0 + \beta_1 x_1 + \beta_2 x_2 + \epsilon$.

i	1	2	3	4	5	6	7
h_{ii}	.486	.131	.087	.219	.112	.159	.172

i	8	9	10	11	12
h_{ii}	.090	.460	.314	.301	.468

According to the rule of thumb given in the text, do any of the data points have a potentially large influence on the fit?

47. Refer back to the water discharge data given in Exercise 42, and let $y = \ln(q)$, $x_1 = \ln(a)$, and $x_2 = \ln(b)$. Consider fitting the model $Y = \beta_0 + \beta_1 x_1 + \beta_2 x_2 + \epsilon$.

a. The resulting h_{ii}'s are .138, .302, .266, .604, .464, .360, .215, .153, .214, and .284. Does any observation appear to be influential?

b. The estimated coefficients are $\hat{\beta}_0 = 1.5652$, $\hat{\beta}_1 = .9450$, $\hat{\beta}_2 = .1815$, and the corresponding estimated standard deviations are $s_{\hat{\beta}_0} = .7328$, $s_{\hat{\beta}_1} = .1528$, and $s_{\hat{\beta}_2} = .1752$. The second standardized residual is $e_2^* = 2.19$. When the second observation is omitted from the data set, the resulting estimated coefficients are $\hat{\beta}_0 = 1.8982$, $\hat{\beta}_1 = 1.025$, and $\hat{\beta}_2 = .3085$. Do any of these changes indicate that the second observation is influential?

c. Deletion of the fourth observation (why?) yields $\hat{\beta}_0 = 1.4592$, $\hat{\beta}_1 = .9850$, and $\hat{\beta}_2 = .1515$. Is this observation influential?

Supplementary Exercises / Chapter 13 (48–54)

48. The accompanying data on $x = $ frequency (MHz) and $y = $ output power (W) for a certain laser configuration was read from a graph in the paper "Frequency Dependence in RF Discharge Excited Waveguide CO_2 Lasers" (*IEEE J. Quantum Electronics*, 1984, pp. 509–514).

x:	60	63	77	100	125	157	186	222
y:	16	17	19	21	22	20	15	5

A computer analysis yielded the following information for a quadratic regression model: $\hat{\beta}_0 = -1.5127$, $\hat{\beta}_1 = .391901$, $\hat{\beta}_2 = -.00163141$, $s_{\hat{\beta}_2} = .00003391$, $SSE = .29$, $SST = 202.88$, and (estimated standard deviation of $\hat{\mu}_{y \cdot 100}$) = .1141.

a. Does the quadratic model appear to be suitable for explaining observed variation in output power by relating it to frequency?

b. Would the simple linear regression model be nearly as satisfactory as the quadratic model?

c. Do you think it would be worth considering a cubic model?

d. Compute a 95% confidence interval for expected power output when frequency is 100.

e. Use a 95% prediction interval to predict the power from a single experimental run when frequency is 100.

49. Conductivity is one important characteristic of glass. The paper "Structure and Properties of Rapidly Quenched Li_2O-Al_2O-Nb_2O_5 Glasses" (*J. Amer. Ceramic Soc.*, 1983, pp. 890–892) reported the accompanying data on $x = Li_2O$ content of a certain type of glass and $y = $ conductivity at 500° K.

x:	19	20	24	27	29	30
y:	$10^{-8.0}$	$10^{-7.1}$	$10^{-7.2}$	$10^{-6.7}$	$10^{-6.2}$	$10^{-6.8}$

x:	31	39	40	43	45	50
y:	$10^{-5.8}$	$10^{-5.3}$	$10^{-6.0}$	$10^{-4.7}$	$10^{-5.4}$	$10^{-5.1}$

(This is a subset of the data that appeared in the paper.) Propose a suitable model for relating y to x, estimate the model parameters, and predict conductivity when Li_2O content is 35.

50. The effect of manganese (Mn) on wheat growth is examined in the article "Manganese Deficiency and Toxicity Effects on Growth, Development and Nutrient Composition in Wheat" (*Agronomy J.*, 1984, pp. 213–217). A quadratic regression model was used to relate $y = $ plant height (cm) to $x = \log_{10}(\text{added Mn})$, with μM as the units for added Mn. The accompanying data was read from a scatter diagram appearing in the paper.

x:	−1.0	−.4	0	.2	1.0
y:	32	37	44	45	46

x:	2.0	2.8	3.2	3.4	4.0
x:	42	42	40	37	30

In addition, $\hat{\beta}_0 = 41.7422$, $\hat{\beta}_1 = 6.581$, $\hat{\beta}_2 = -2.3621$, $s_{\hat{\beta}_0} = .8522$, $s_{\hat{\beta}_1} = 1.002$, $s_{\hat{\beta}_2} = .3073$, and $SSE = 26.98$.

a. Is the quadratic model useful for describing the relationship between x and y? *Hint:* Quadratic regression is a special case of multiple regression with $k = 2$, $x_1 = x$, and $x_2 = x^2$. Apply an appropriate test procedure.

b. Should the quadratic predictor be eliminated?

c. Estimate expected height for wheat treated with 10 μM of Mn using a 90% confidence interval. *Hint:* The estimated standard deviation of $\hat{\beta}_0 + \hat{\beta}_1 + \hat{\beta}_2$ is 1.031.

51. Reconsider the situation of Exercise 45, and test the utility of the model identified by the backward elimination procedure.

52. A sample of $n = 20$ companies was selected, and the values of y = stock price and $k = 15$ predictor variables (such as quarterly dividend, previous year's earnings, debt ratio, and so on) were determined. When the multiple regression model using these 15 predictors was fit to the data, $R^2 = .90$ resulted.

a. Does the model appear to specify a useful relationship between y and the predictor variables? Carry out a test using significance level .05. (*Hint:* The F critical value for 15 numerator and 4 denominator d.f. is 5.86.)

b. Based on the result of (a), does a high R^2 value by itself imply that a model is useful? Under what circumstances might you be suspicious of a model with a high R^2 value?

c. With n and k as given above, how large would R^2 have to be for the model to be judged useful at the .05 level of significance?

53. Does exposure to air pollution result in decreased life expectancy? This question was examined in the paper "Does Air Pollution Shorten Lives?" (*Statistics and Public Policy*, Reading, MA, Addison-Wesley, 1977). Data on

y = total mortality rate (deaths per 10,000)
x_1 = mean suspended particle reading (μg/m^3)
x_2 = smallest sulfate reading ([μg/m^3] $\times$ 10)
x_3 = population density (people/mi^2)
x_4 = (percent nonwhite) $\times$ 10
x_5 = (percent over 65) $\times$ 10

for the year 1960 was recorded for $n = 117$ randomly selected standard metropolitan statistical areas. The estimated regression equation was

$$y = 19.607 + .041x_1 + .071x_2 + .001x_3 + .041x_4 + .687x_5$$

a. For this model, $R^2 = .827$. Using a .05 significance level, perform a model utility test.

b. The estimated standard deviation of $\hat{\beta}_1$ was .016. Calculate and interpret a 90% confidence interval for β_1.

c. Given that the estimated standard deviation of $\hat{\beta}_4$ is .007, determine if percent nonwhite is an important variable in the model. Use a .01 significance level.

d. In 1960, the values of x_1, x_2, x_3, x_4, and x_5 for Pittsburgh were 166, 60, 788, 68, and 95, respectively. Use the given regression equation to predict Pittsburgh's mortality rate. How does your prediction compare with the actual 1960 value of 103 deaths per 10,000?

54. Given that $R^2 = .723$ for the model containing predictors x_1, x_4, x_5, and x_8 and $R^2 = .689$ for the model with predictors x_1, x_3, x_5, and x_6, what can you say about R^2 for the model containing predictors

a. x_1, x_3, x_4, x_5, x_6, and x_8? Explain.

b. x_1 and x_4? Explain.

Bibliography

Chatterjee, S., and Price, Bertram, *Regression Analysis by Example,* Wiley, New York, 1977. A brief but informative discussion of selected topics, especially multicollinearity and the use of biased estimation methods.

Daniel, Cuthbert, and Wood, Fred, *Fitting Equations to Data* (2nd ed.), Wiley, New York, 1980. Contains many insights and methods that evolved from the authors' extensive consulting experience.

Draper, Norman, and Smith, Harry, *Applied Regression Analysis* (2nd ed.), Wiley, New York, 1981. See the Chapter 12 bibliography.

Hoaglin, David, and Welsch, Roy, "The Hat Matrix in Regression and ANOVA," *American Statistician,* 1978, pp. 17–23. Describes methods for detecting influential observations in a regression data set.

Hocking, Ron, "The Analysis and Selection of Variables in Linear Regression," *Biometrics,* 1976, pp. 1–49. An excellent survey of some recent developments.

Mosteller, Frederick, and Tukey, John, *Data Analysis and Regression,* Addison-Wesley, Reading, Mass., 1977. Contains many interesting ideas exposited by two pioneers of the methods of exploratory data analysis.

Neter, John, Wasserman, William, and Kutner, Michael, *Applied Linear Statistical Models* (2nd ed.), Irwin, Homewood, Ill., 1985. See the Chapter 12 bibliography.

CHAPTER 14

The Analysis of Categorical Data

Introduction

In the simplest type of situation considered in this chapter, each observation in a sample is classified as belonging to one of a finite number of categories (for example, blood type could be one of the four categories O, A, B, or AB). With p_i denoting the probability that any particular observation belongs in category i (or the proportion of the population belonging to category i), we wish to test a null hypothesis that completely specifies the values of all the p_i's (such as $H_0 : p_1 = .45$, $p_2 = .35$, $p_3 = .15$, $p_4 = .05$, when there are four categories). The test statistic will be a measure of the discrepancy between the observed numbers in the categories and the expected numbers when H_0 is true. Because a decision will be reached by comparing the computed value of the test statistic to a critical value of the chi-squared distribution, the procedure is called a chi-squared goodness-of-fit test.

Sometimes the null hypothesis specifies that the p_i's depend on some smaller number of parameters without specifying the values of these parameters. For example, with three categories the null hypothesis might state that $p_1 = \theta^2$, $p_2 = 2\theta(1 - \theta)$, and $p_3 = (1 - \theta)^2$. For a chi-squared test to be performed, the values of any unspecified parameters must be estimated from the sample data. These problems are discussed in Section 14.2. The methods are then applied to test a null hypothesis that states that the sample comes from a particular family of distributions, such as the Poisson family (with λ estimated from the sample) or the normal family (with μ and σ estimated).

Chi-squared tests for two different situations are presented in Section 14.3. In the first, the null hypothesis states that the p_i's are the same for several different populations. The second type of situation involves taking a sample from a single population and classifying each individual with respect to two different categorical factors (such as religious preference and political party registration). The null hypothesis in this situation is that the two factors are independent within the population.

14.1   Goodness-of-Fit Tests When Category Probabilities Are Completely Specified

A binomial experiment consists of a sequence of independent trials in which each trial can result in one of two possible outcomes (the same two possibilities for each trial). The two possibilities are labeled S (for success) and F (for failure). The probability of success, denoted by p, is assumed to be constant from trial to trial, and the number of trials n is fixed at the outset of the experiment. In Chapter 8 we presented a large-sample z test for testing $H_0: p = p_0$. Notice that this null hypothesis specifies both $P(S)$ and $P(F)$, since if $P(S) = p_0$, then $P(F) = 1 - p_0$. Denoting $P(F)$ by q and $1 - p_0$ by q_0, the null hypothesis can be written as $H_0: p = p_0$, $q = q_0$. The z test is two-tailed when the alternative of interest is $p \neq p_0$.

A **multinomial experiment** generalizes a binomial experiment by allowing each trial to result in one of k possible outcomes, where k is an integer greater than 2. As an example, suppose that a store accepts three different types of credit cards. A multinomial experiment would result from observing the type of credit card used—type 1, type 2, or type 3—by each of the next n customers who pay with a credit card. In general, we shall refer to the k possible outcomes on any given trial as categories, and p_i will denote the probability that a trial results in category i. If the experiment consists of selecting n individuals or objects from a population and categorizing each one, then p_i is the proportion of the population falling in the ith category (such an experiment will be approximately multinomial provided that n is much smaller than the population size).

The null hypothesis of interest will specify the value of each p_i. For example, in the case $k = 3$, we might have $H_0: p_1 = .5$, $p_2 = .3$, $p_3 = .2$. The alternative hypothesis will state that H_0 is not true—that is, that at least one of the p_i's has a value different from that asserted by H_0 (in which case at least two must be different, since they sum to one). The symbol p_{i0} will represent the value of p_i claimed by the null hypothesis. In the example just given, $p_{10} = .5$, $p_{20} = .3$, and $p_{30} = .2$.

Before the multinomial experiment is performed, the number of trials that will result in category i ($i = 1, 2, \ldots,$ or k) is a random variable—just as the number of successes and the number of failures in a binomial experiment are random variables. This random variable will be denoted by N_i and its observed value by n_i. Since each trial results in exactly one of the k categories, $\Sigma N_i = n$, and the same is true of the n_i's. As an example, an experiment with $n = 100$ and $k = 3$ might yield $N_1 = 46$, $N_2 = 35$, and $N_3 = 19$.

The expected number of successes and expected number of failures in a binomial experiment are np and nq, respectively. When $H_0: p = p_0$, $q = q_0$ is true, the expected numbers of successes and failures are np_0 and nq_0, respectively. Similarly, in a multinomial experiment the expected number of trials resulting in category i is $E(N_i) = np_i$ $(i = 1, \ldots, k)$. When $H_0 : p_1 = p_{10}, \ldots,$ $p_k = p_{k0}$ is true, these expected values become $E(N_1) = np_{10}$, $E(N_2) = np_{20}$, $\ldots, E(N_k) = np_{k0}$. For the case $k = 3$, $H_0 : p_1 = .5$, $p_2 = .3$, $p_3 = .2$; and $n = 100$, $E(N_1) = 100(.5) = 50$, $E(N_2) = 30$, and $E(N_3) = 20$ when H_0 is true. The n_i's are often displayed in a tabular format consisting of a row of k cells, one for each category, as illustrated in Figure 14.1. The expected values when H_0 is true are displayed just below the observed values. The N_i's and n_i's are usually referred to as *observed cell counts* (or *observed cell frequencies*), and $np_{10}, np_{20}, \ldots, np_{k0}$ are the corresponding *expected cell counts* under H_0.

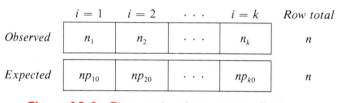

Figure 14.1 Observed and expected cell counts

The n_i's should all be reasonably close to the corresponding np_{i0}'s when H_0 is true. On the other hand, several of the observed counts should differ substantially from these expected counts when the actual values of the p_i's differ markedly from what the null hypothesis asserts. The test procedure involves measuring the discrepancy between the n_i's and the np_{i0}'s, with H_0 being rejected when the measured discrepancy is sufficiently large. It is natural to base a measure of discrepancy on the squared deviations $(n_1 - np_{10})^2$, $(n_2 - np_{20})^2$, $\ldots, (n_k - np_{k0})^2$. An obvious way to combine these into an overall measure is to add them together to obtain $\Sigma(n_i - np_{i0})^2$. However, suppose that $np_{10} = 100$ and $np_{20} = 10$. Then if $n_1 = 95$ and $n_2 = 5$, the two categories contribute the same squared deviations to the proposed measure. Yet n_1 is only 5% less than what would be expected when H_0 is true, whereas n_2 is 50% less.

To take relative magnitudes of the deviations into account, we shall divide each squared deviation by the corresponding expected count and then combine. Before giving a more detailed description, we must discuss a type of probability distribution called the *chi-squared distribution*. This distribution was first introduced in Section 4.4, and was used in Chapter 7 to obtain a confidence interval for the variance σ^2 of a normal population. The chi-squared distribution has a single parameter ν, called the number of degrees of freedom (d.f.) of the distribution, with possible values $1, 2, 3, \ldots$. Analogous to the critical value $t_{\alpha, \nu}$ for the t distribution, $\chi^2_{\alpha, \nu}$ is the value such that α of the area under the χ^2 curve in Figure 14.2 with ν degrees of freedom lies to the right of $\chi^2_{\alpha, \nu}$. Selected values of $\chi^2_{\alpha, \nu}$ are given in Appendix Table A.6.

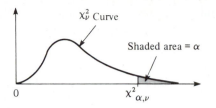

Figure 14.2 A critical value for a chi-squared distribution

Theorem

Provided that $np_i \geq 5$ for every i ($i = 1, 2, \ldots, k$), the random variable

$$\chi^2 = \sum_{i=1}^{k} \frac{(N_i - np_i)^2}{np_i} = \sum_{\text{all cells}} \frac{(\text{observed} - \text{expected})^2}{\text{expected}}$$

has approximately a chi-squared distribution with $k - 1$ d.f.

The fact that d.f. $= k - 1$ is a consequence of the restriction $\Sigma N_i = n$. Although there are k observed cell counts, once any $k - 1$ are known, the remaining one is uniquely determined. That is, there are only $k - 1$ "freely determined" cell counts, and thus $k - 1$ d.f.

If np_{i0} is substituted for np_i in χ^2, the resulting test statistic has a chi-squared distribution when H_0 is true. Rejection of H_0 is appropriate when $\chi^2 \geq c$ (because large deviations lead to a large value of χ^2), and the choice $c = \chi^2_{\alpha, k-1}$ yields a test with significance level α.

Null hypothesis: $H_0 : p_1 = p_{10}, p_2 = p_{20}, \ldots, p_k = p_{k0}$

Alternative hypothesis: $H_a :$ at least one p_i does not equal p_{i0}

Test statistic value: $\chi^2 = \sum_{\text{all cells}} \frac{(\text{observed} - \text{expected})^2}{\text{expected}} = \sum_{i=1}^{k} \frac{(n_i - np_{i0})^2}{np_{i0}}$

Rejection region: $\chi^2 \geq \chi^2_{\alpha, k-1}$

Example 14.1

If we focus on two different characteristics of an organism, each controlled by a single gene, and cross a pure strain having genotype AABB with a pure strain having genotype aabb (capital letters denoting dominant alleles and small letters recessive alleles), the resulting genotype will be AaBb. If these first-generation organisms are then crossed among themselves (a dihybrid cross), there will be four phenotypes depending on whether or not a dominant allele of either type is present. Mendel's laws of inheritance imply that these four phenotypes should have probabilities $\frac{9}{16}$, $\frac{3}{16}$, $\frac{3}{16}$, and $\frac{1}{16}$ of arising in any given dihybrid cross.

The paper "Linkage Studies of the Tomato" (*Trans. Royal Canadian Institute*, 1931, pp. 1–19) reported the following data on phenotypes from a dihybrid cross of tall cut-leaf tomatoes with dwarf potato-leaf tomatoes. There are $k = 4$ categories corresponding to the four possible phenotypes, with the null hypothesis being

$$H_0: p_1 = \frac{9}{16}, \ p_2 = \frac{3}{16}, \ p_3 = \frac{3}{16}, \ p_4 = \frac{1}{16}$$

The expected cell counts are $9n/16$, $3n/16$, $3n/16$, and $n/16$, and χ^2 is based on $k - 1 = 3$ d.f. The total sample size was $n = 1611$.

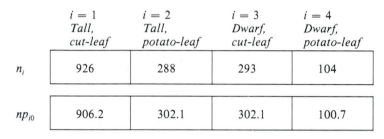

	$i = 1$ Tall, cut-leaf	$i = 2$ Tall, potato-leaf	$i = 3$ Dwarf, cut-leaf	$i = 4$ Dwarf, potato-leaf
n_i	926	288	293	104
np_{i0}	906.2	302.1	302.1	100.7

Figure 14.3   Observed and expected cell counts for Example 14.1

The contribution to χ^2 from the first cell is

$$\frac{(n_1 - np_{10})^2}{np_{10}} = \frac{(926 - 906.2)^2}{906.2} = .433$$

Cells 2, 3, and 4 contribute .658, .274, and .108, respectively, so $\chi^2 = .433 + .658 + .274 + .108 = 1.473$. A test with significance level .10 requires $\chi^2_{.10, 3}$, the number in the 3 d.f. row and .10 column of Appendix Table A.6. This critical value is 6.251. Since 1.473 is not at least 6.251, H_0 cannot be rejected even at this rather large level of significance. The data is quite consistent with Mendel's laws. ∎

Although we have developed the chi-squared test for situations in which $k > 2$, it can also be used when $k = 2$. The null hypothesis in this case can be stated as $H_0: p_1 = p_{10}$, since the relations $p_2 = 1 - p_1$ and $p_{20} = 1 - p_{10}$ make the inclusion of $p_2 = p_{20}$ in H_0 redundant. The alternative hypothesis is $H_a: p_1 \neq p_{10}$. These hypotheses can also be tested using a two-tailed z test with test statistic

$$Z = \frac{(N_1/n) - p_{10}}{\sqrt{\dfrac{p_{10}(1 - p_{10})}{n}}} = \frac{\hat{p}_1 - p_{10}}{\sqrt{\dfrac{p_{10}p_{20}}{n}}}$$

Surprisingly, the two test procedures are completely equivalent. This is because it can be shown that $Z^2 = \chi^2$ and $(z_{\alpha/2})^2 = \chi^2_{1,\alpha}$, so that $\chi^2 \geq \chi^2_{1,\alpha}$ if and only if $|Z| \geq z_{\alpha/2}$.*

Example 14.2

The developers of an English proficiency exam to be used in a large university system believe that 60% of all incoming freshmen will be able to pass the exam. In a random sample of 200 incoming freshmen, 105 pass the exam. Does this contradict the claim of the developers?

We wish to test $H_0: p_1 = .6$, $p_2 = .4$ versus $H_a: H_0$ is not true. The expected cell counts are $np_{10} = 200(.6) = 120$ and $np_{20} = 80$, so the computed χ^2 is

$$\chi^2 = \frac{(105 - 120)^2}{120} + \frac{(95 - 80)^2}{80} = 1.88 + 2.81 = 4.69$$

Using $\alpha = .05$, $\chi^2_{.05, 1} = 3.843$. Since $4.69 \geq 3.843$, H_0 is rejected at level .05. Notice that $\chi^2_{.025, 1} = 5.025$, so the P-value satisfies $.025 < P < .05$. With $\hat{p}_1 = 105/200 = .525$, the value of the z statistic is

$$z = \frac{.525 - .6}{\sqrt{\frac{(.6)(.4)}{200}}} = -2.165$$

and the z critical value for a level .05 test is $z_{.025} = 1.96$. Thus $z^2 = (-2.165)^2 = 4.69 = \chi^2$ and $(z_{\alpha/2})^2 = (1.96)^2 = 3.84 = \chi^2_{1, .05}$, as must be the case. ∎

If the alternative hypothesis is either $H_a: p_1 > p_{10}$ or $H_a: p_1 < p_{10}$, the chi-squared test cannot be used. One must then revert to an upper- or lower-tailed z test.

As is the case with all test procedures, one must be careful not to confuse statistical significance with practical significance. A computed χ^2 that exceeds $\chi^2_{\alpha, k-1}$ may be a result of a very large sample size rather than any practical differences between the hypothesized p_{i0}'s and true p_i's. Thus if $p_{10} = p_{20} = p_{30} = \frac{1}{3}$, but the true p_i's have values .330, .340, and .330, a large value of χ^2 is sure to arise with a sufficiently large n. Before rejecting H_0, the $\hat{p}_i$'s should be examined to see if they suggest a model different from that of H_0 from a practical point of view.

χ^2 When the p_i's Are Functions of Other Parameters

Frequently the p_i's are hypothesized to depend on a smaller number of parameters $\theta_1, \ldots, \theta_m$ $(m < k)$. Then a specific hypothesis involving the θ_i's yields specific p_{i0}'s, which are then used in the χ^2 test.

---

*The fact is that $(z_{\alpha/2})^2 = \chi^2_{1,\alpha}$ is a consequence of the relationship between the standard normal distribution and the chi-squared distribution with 1 d.f.; if $Z \sim N(0, 1)$, then Z^2 has a chi-squared distribution with $\nu = 1$.

Example **14.3** In a well-known genetics paper ("The Progeny in Generations F_{12} to F_{17} of a Cross Between a Yellow-Wrinkled and a Green-Round Seeded Pea," *J. Genetics*, 1923, pp. 255–331), the early statistician G. U. Yule analyzed data resulting from crossing garden peas. The dominant alleles in the experiment were Y = yellow color and R = round shape, resulting in the double dominant YR. Yule examined 269 four-seed pods resulting from a dihybrid cross and counted the number of YR seeds in each pod. Letting X denote the number of YR's in a randomly selected pod, possible X values are 0, 1, 2, 3, 4, which we identify with cells 1, 2, 3, 4, and 5 of a rectangular table (so, for example, a pod with $X = 4$ yields an observed count in cell 5).

The hypothesis that the Mendelian laws are operative and that genotypes of individual seeds within a pod are independent of one another implies that X has a binomial distribution with $n = 4$ and $\theta = \frac{9}{16}$. We thus wish to test $H_0 : p_1 = p_{10}, \ldots, p_5 = p_{50}$ where

$$p_{i0} = P(i - 1 \text{ YR's among 4 seeds when } H_0 \text{ is true})$$
$$= \binom{4}{i-1} \theta^{i-1}(1 - \theta)^{4-(i-1)} \quad i = 1, 2, 3, 4, 5; \; \theta = \frac{9}{16}$$

Yule's data and the computations appear in Figure 14.4 with expected cell counts $np_{i0} = 269p_{i0}$.

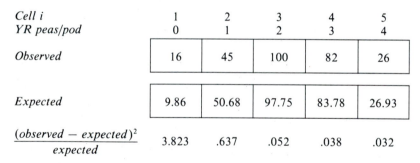

Cell i	1	2	3	4	5
YR peas/pod	0	1	2	3	4
Observed	16	45	100	82	26
Expected	9.86	50.68	97.75	83.78	26.93
$\dfrac{(observed - expected)^2}{expected}$	3.823	.637	.052	.038	.032

Figure 14.4 Observed and expected cell counts for Example 14.3

Thus $\chi^2 = 3.823 + \cdots + .032 = 4.582$. Since $\chi^2_{.01, k-1} = \chi^2_{.01, 4} = 13.277$, H_0 is not rejected at level .01. ∎

In Section 14.2 we will discuss goodness-of-fit tests when the values of parameters $\theta_1, \ldots, \theta_m$ are unspecified by H_0, so that we must estimate expected cell counts by first estimating the θ_i's.

χ^2 When the Underlying Distribution Is Continuous

We have so far assumed that the k categories are naturally defined in the context of the experiment under consideration. The χ^2 test can also be used to test whether a sample comes from a specific underlying continuous distribution.

Let X denote the variable being sampled, and suppose that the hypothesized p.d.f. of X is $f(x)$. As in the construction of a frequency distribution in Chapter 1, subdivide the measurement scale of X into k intervals $[a_0, a_1), [a_1, a_2), \ldots,$ $[a_{k-1}, a_k)$, where the interval $[a_{i-1}, a_i)$ includes the value a_{i-1} but not a_i. The cell probabilities specified by H_0 are then

$$p_{i0} = P(a_{i-1} \le X < a_i) = \int_{a_{i-1}}^{a_i} f(x)\, dx$$

The cells should be chosen so that $np_{i0} \ge 5$ for $i = 1, \ldots, k$. Often they are selected so that the np_{i0}'s are equal.

Example 14.4

To see whether the time of onset of labor among expectant mothers is uniformly distributed throughout a 24-hour day, we can divide a day into k periods, each of length $24/k$. The null hypothesis states that $f(x)$ is the uniform p.d.f. on the interval $[0, 24]$, so that $p_{i0} = 1/k$. The article "The Hour of Birth" (*British J. Preventive and Social Medicine,* 1953, pp. 43–59) reported on 1186 onset times, which were categorized into $k = 24$ one-hour intervals beginning at midnight, resulting in cell counts of 52, 73, 89, 88, 68, 47, 58, 47, 48, 53, 47, 34, 21, 31, 40, 24, 37, 31, 47, 34, 36, 44, 78, and 59. Each expected cell count is $1186 \cdot \frac{1}{24} = 49.42$, and the resulting value of χ^2 is 162.77. Since $\chi^2_{.01, 23} = 41.64$, the computed value is highly significant and the null hypothesis is resoundingly rejected. Generally speaking, it appears that labor is much more likely to commence very late at night than during normal waking hours. ∎

For testing whether or not a sample comes from a specific normal distribution, the fundamental parameters are $\theta_1 = \mu$ and $\theta_2 = \sigma$, and each p_{i0} will be a function of these parameters.

Example 14.5

At a certain university final exams are supposed to last two hours. The psychology department constructed a departmental final for an elementary course that was believed to satisfy the following criteria: (a) actual time taken to complete the exam is normally distributed, (b) $\mu = 100$ min, and (c) exactly 90% of all students will finish within the two-hour period. To see whether this is actually the case, 120 students were randomly selected and their completion times recorded. It was decided that $k = 8$ intervals should be used. The criteria imply that the ninetieth percentile of the completion time distribution is $\mu + 1.28\sigma = 120$. Since $\mu = 100$, this implies that $\sigma = 15.63$.

The eight intervals that divide the standard normal scale into eight equally likely segments are $[0, .32), [.32, .675), [.675, 1.15), [1.15, \infty)$, and their four counterparts on the other side of 0. For $\mu = 100$ and $\sigma = 15.63$, these intervals become $[100, 105), [105, 110.55), [110.55, 117.97)$, and $[117.97, \infty)$. Thus $p_{i0} = \frac{1}{8} = .125$ ($i = 1, \ldots, 8$), so each expected cell count is $np_{i0} = 120(.125) = 15$. The observed cell counts were 21, 17, 12, 16, 10, 15, 19, and 10, resulting in a χ^2 of 7.73. Since $\chi^2_{.10, 7} = 12.017$ and 7.73 is not ≥ 12.017, there is no evidence for concluding that the criteria have not been met. ∎

Exercises / Section 14.1 (1–7)

1. Student consultants at a particular computer center face questions about programs written in FORTRAN (1), BASIC (2), PASCAL (3), and PL1 (4). The consultants have been hired based on the assumption that 40% of all questions concern FORTRAN programs, 25% concern BASIC programs, 25% concern PASCAL programs, and 10% concern PL1 programs. Let p_i denote the probability that a randomly selected question concerns a program written in language i ($i = 1, 2, 3, 4$). Use the accompanying data to test

$$H_0 : p_1 = .4, \ p_2 = .25, \ p_3 = .25, \ p_4 = .10$$

versus

$$H_a : H_0 \text{ is not true}$$

using a level .05 chi-squared test.

Cell	1	2	3	4
Frequency	52	38	21	9

2. It is hypothesized that when homing pigeons are disoriented in a certain manner, they will exhibit no preference for any direction of flight after take-off (so that the direction X should be uniformly distributed on the interval from $0°$ to $360°$). To test this, 120 pigeons are disoriented, let loose, and the direction of flight of each is recorded; the resulting data appears below. Use the chi-squared test at level .10 to see if the data supports the hypothesis.

Direction	$0- < 45°$	$45- < 90°$	$90- < 135°$
Frequency	12	16	17

Direction	$135- < 180°$	$180- < 225°$	$225- < 270°$
Frequency	15	13	20

Direction	$270- < 315°$	$315- < 360°$
Frequency	17	10

3. Information has been loaded onto a single-disk storage device with 10 concentric tracks in such a way that the probability of the access arm next being required to access track i is given by $p_i = (5.5 - |i - 5.5|)/30$ for $i = 1, \ldots, 10$ (that is, the p_i's are $\frac{1}{30}, \frac{2}{30}, \frac{3}{30}, \frac{4}{30}, \frac{5}{30}, \frac{5}{30}, \frac{4}{30}, \frac{3}{30}, \frac{2}{30}, \frac{1}{30}$). A sample of 200 track numbers accessed resulted in the following data. Use the chi-squared test at level .10 to decide whether the data is consistent with the above access probabilities.

Track number	1	2	3	4	5	6	7	8	9	10
Frequency	4	15	23	25	38	31	32	14	10	8

4. Sorghum is an important cereal crop whose quality and appearance could be affected by the presence of pigments in the pericarp (the walls of the plant ovary). The paper "A Genetic and Biochemical Study on Pericarp Pigments in a Cross Between Two Cultivars of Grain Sorghum, Sorghum Bicolor" (*Heredity,* 1976, pp. 413–416) reported on an experiment that involved an initial cross between CK60 sorghum (an American variety with white seeds) and Abu Taima (an Ethiopian variety with yellow seeds) to produce plants with red seeds, and then a self-cross of the red-seeded plants. According to genetic theory, this F_2 cross should produce plants with red, yellow, or white seeds in the ratio $9 : 3 : 4$. The data from the experiment appears below; does the data confirm or contradict the genetic theory? Test at level .05.

Seed color	Red	Yellow	White
Observed frequency	195	73	100

5. The response time of a computer system to a request for a certain type of information is hypothesized to have an exponential distribution with parameter $\lambda = 1$ sec (so if $X =$ response time, the p.d.f. of X under H_0 is $f(x) = e^{-x}$ for $x \geq 0$).

a. If you had observed $X_1, X_2, \ldots, X_n$ and wanted to use the chi-squared test with five class intervals having equal probability under H_0, what would the resulting class intervals be?

b. Carry out the chi-squared test using the following data resulting from a random sample of 40 response times:

.10, .99, 1.14, 1.26, 3.24, .12, .26, .80, .79, 1.16, 1.76, .41, .59, .27, 2.22, .66, .71, 2.21, .68, .43, .11, .46, .69, .38, .91, .55, .81, 2.51, 2.77, .16, 1.11, .02, 2.13, .19, 1.21, 1.13, 2.93, 2.14, .34, .44

6. a. Show that another expression for the chi-squared statistic is

$$\chi^2 = \sum_{i=1}^{k} \frac{N_i^2}{np_{i0}} - n$$

Why is it more efficient to compute χ^2 using this formula?

b. When the null hypothesis is $H_0 : p_1 = p_2 = \cdots = p_k = 1/k$ (that is, $p_{i0} = 1/k$ for all i), how does the formula of (a) simplify? Use the simplified expression to calculate χ^2 for the pigeon/direction data in Exercise 2.

7. a. Having obtained a random sample from a population, you wish to use a chi-squared test to decide whether or not the population distribution is standard normal. If you base the test on six class intervals having equal probability under H_0, what should the class intervals be?

b. If you wish to use a chi-squared test to test H_0: the population distribution is normal with $\mu = .5$, $\sigma = .002$, and the test is to be based on six equiprobable (under H_0) class intervals, what should these intervals be?

c. Use the chi-squared test with the intervals of (b) to decide, based on the following 45 bolt diameters, whether bolt diameter is a normally distributed variable with $\mu = .5$ in., $\sigma = .002$ in.

.4974, .4976, .4991, .5014, .5008, .4993, .4994, .5010, .4997, .4993, .5013, .5000, .5017, .4984, .4967, .5028, .4975, .5013, .4972, .5047, .5069, .4977, .4961, .4987, .4990, .4974, .5008, .5000, .4967, .4977, .4992, .5007, .4975, .4998, .5000, .5008, .5021, .4959, .5015, .5012, .5056, .4991, .5006, .4987, .4968

14.2 Goodness of Fit for Composite Hypotheses

In the previous section we presented a goodness-of-fit test based on a χ^2 statistic for deciding between $H_0 : p_1 = p_{10}, \ldots, p_k = p_{k0}$ and the alternative H_a stating that H_0 is not true. The null hypothesis was a **simple hypothesis** in the sense that each p_{i0} was a specified number, so that the expected cell counts when H_0 was true were uniquely determined numbers.

In many situations there are k naturally occurring categories, but H_0 states only that the p_i's are functions of other parameters $\theta_1, \ldots, \theta_m$ without specifying the values of these θ's. For example, a population may be in equilibrium with respect to proportions of the three genotypes AA, Aa, and aa. With $p_1, p_2,$ and p_3 denoting these proportions (probabilities), one may wish to test

$$H_0 : p_1 = \theta^2, \ p_2 = 2\theta(1 - \theta), \ p_3 = (1 - \theta)^2 \tag{14.1}$$

where θ represents the proportion of gene A in the population. This hypothesis is **composite** because knowing that H_0 is true does not uniquely determine the cell probabilities and expected cell counts, but only their general form. To carry out a χ^2 test, the unknown θ_i's must first be estimated.

Similarly, we may be interested in testing to see whether a sample came from a particular family of distributions without specifying any particular member of the family. To use the χ^2 test to see whether the distribution is Poisson, for example, the parameter λ must be estimated. In addition, because there are actually an infinite number of possible values of a Poisson variable, these values must be grouped so that there are a finite number of cells. If H_0 states that the underlying distribution is normal, use of a χ^2 test must be preceded by a choice of cells and estimation of μ and σ.

χ^2 When Parameters Are Estimated

As before, k will denote the number of categories or cells and p_i will denote the probability of an observation falling in the ith cell. The null hypothesis now states that each p_i is a function of a small number of parameters $\theta_1, \ldots, \theta_m$ with the θ_i's otherwise unspecified:

$$H_0: p_1 = \pi_1(\boldsymbol{\theta}), \ldots, p_k = \pi_k(\boldsymbol{\theta}) \quad \text{where} \quad \boldsymbol{\theta} = (\theta_1, \ldots, \theta_m)$$
$$H_a: \text{the hypothesis } H_0 \text{ is not true}$$

(14.2)

For example, for H_0 of (14.1), $m = 1$ (there is only one θ), $\pi_1(\theta) = \theta^2$, $\pi_2(\theta) = 2\theta(1 - \theta)$, and $\pi_3(\theta) = (1 - \theta)^2$.

In the case $k = 2$ there is really only a single random variable N_1 (since $N_1 + N_2 = n$), which has a binomial distribution. The joint probability that $N_1 = n_1$ and $N_2 = n_2$ is then

$$P(N_1 = n_1, N_2 = n_2) = \binom{n}{n_1} p_1^{n_1} \cdot p_2^{n_2} \propto p_1^{n_1} \cdot p_2^{n_2}$$

where $p_1 + p_2 = 1$ and $n_1 + n_2 = n$. For general k, the joint distribution of $N_1, \ldots, N_k$ is the multinomial distribution (Section 5.1) with

$$P(N_1 = n_1, \ldots, N_k = n_k) \propto p_1^{n_1} \cdot p_2^{n_2} \ldots p_k^{n_k}$$

(14.3)

When H_0 is true, (14.3) becomes

$$P(N_1 = n_1, \ldots, N_k = n_k) \propto [\pi_1(\boldsymbol{\theta})]^{n_1} \ldots [\pi_k(\boldsymbol{\theta})]^{n_k}$$

(14.4)

To apply a chi-squared test, $\boldsymbol{\theta} = (\theta_1, \ldots, \theta_m)$ must be estimated.

Method of estimation

Let $n_1, n_2, \ldots, n_k$ denote the observed values of $N_1, \ldots, N_k$. Then $\hat{\theta}_1, \ldots, \hat{\theta}_m$ are those values of the θ_i's which maximize (14.4).

The resulting estimators $\hat{\theta}_1, \ldots, \hat{\theta}_m$ are the **maximum likelihood estimators** of $\theta_1, \ldots, \theta_m$; this principle of estimation was discussed in Section 6.2.

Example 14.6

In humans there is a blood group, the MN group, which is composed of individuals having one of the three blood types M, MN, and N. Type is determined by two alleles and there is no dominance, so the three possible genotypes give rise to three phenotypes. A population consisting of individuals in the MN group is in equilibrium if

$$P(\text{M}) = p_1 = \theta^2$$
$$P(\text{MN}) = p_2 = 2\theta(1 - \theta)$$
$$P(\text{N}) = p_3 = (1 - \theta)^2$$

for some θ. Suppose that a sample from such a population yielded the following results:

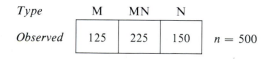

Type	M	MN	N	
Observed	125	225	150	$n = 500$

Figure 14.5 Observed counts for Example 14.6

Then

$$[\pi_1(\theta)]^{n_1}[\pi_2(\theta)]^{n_2}[\pi_3(\theta)]^{n_3} = [(\theta^2)]^{n_1}[2\theta(1-\theta)]^{n_2}[(1-\theta)^2]^{n_3}$$
$$= 2^{n_2} \cdot \theta^{2n_1 + n_2} \cdot (1-\theta)^{n_2 + 2n_3}$$

Maximizing this with respect to θ (or, equivalently, maximizing the natural logarithm of this quantity, which is easier to differentiate) yields

$$\hat{\theta} = \frac{2n_1 + n_2}{[(2n_1 + n_2) + (n_2 + 2n_3)]} = \frac{2n_1 + n_2}{2n}$$

With $n_1 = 125$ and $n_2 = 225$, $\hat{\theta} = 475/1000 = .475$. ∎

Once $\boldsymbol{\theta} = (\theta_1, \ldots, \theta_m)$ has been estimated by $\hat{\boldsymbol{\theta}} = (\hat{\theta}_1, \ldots, \hat{\theta}_m)$, the estimated expected cell counts are the $n\pi_i(\hat{\boldsymbol{\theta}})$'s. These are now used in place of the np_{i0}'s of Section 14.1 to specify a χ^2 statistic.

Theorem

> Under general "regularity" conditions on $\theta_1, \ldots, \theta_m$ and the $\pi_i(\boldsymbol{\theta})$'s, if $\theta_1, \ldots, \theta_m$ are estimated by the method of maximum likelihood as described above and n is large,
>
> $$\chi^2 = \sum_{\text{all cells}} \frac{(\text{observed} - \text{estimated expected})^2}{\text{estimated expected}} = \sum_{i=1}^{k} \frac{[N_i - n\pi_i(\hat{\boldsymbol{\theta}})]^2}{n\pi_i(\hat{\boldsymbol{\theta}})}$$
>
> has approximately a chi-squared distribution with $k - 1 - m$ degrees of freedom when H_0 of (14.2) is true. An approximately level α test of H_0 versus H_a is then to reject H_0 if $\chi^2 \geq \chi^2_{\alpha, k-1-m}$. In practice, the test can be used if $n\pi_i(\hat{\boldsymbol{\theta}}) \geq 5$ for every i.

Notice that *the number of degrees of freedom is reduced by the number of θ_i's estimated.*

Example 14.7
(Example 14.6 continued)

With $\hat{\theta} = .475$ and $n = 500$, the estimated expected cell counts are $n\pi_1(\hat{\theta}) = 500(\hat{\theta})^2 = 112.81$, $n\pi_2(\hat{\theta}) = (500)(2)(.475)(1 - .475) = 249.98$, and $n\pi_3(\hat{\theta}) = 500 - 112.81 - 249.98 = 137.82$. Then

$$\chi^2 = \frac{(125 - 112.81)^2}{112.81} + \frac{(225 - 249.98)^2}{249.98} + \frac{(150 - 137.82)^2}{137.82} = 4.89$$

Since $\chi^2_{.05, k-1-m} = \chi^2_{.05, 3-1-1} = \chi^2_{.05, 1} = 3.843$ and $4.89 \geq 3.843$, H_0 is rejected. ∎

Example 14.8

Consider a series of games between two teams, I and II, which terminates as soon as one team has won four games (with no possibility of a tie). A simple probability model for such a series assumes that outcomes of successive games are independent and that the probability of team I winning any particular game is a constant θ. We arbitrarily designate I the better team, so that $\theta \geq .5$. Any particular series can then terminate after 4, 5, 6, or 7 games. Let $\pi_1(\theta)$, $\pi_2(\theta)$, $\pi_3(\theta)$, $\pi_4(\theta)$ denote the probability of termination in 4, 5, 6, and 7 games, respectively. Then

$$\pi_1(\theta) = P(\text{I wins in 4 games}) + P(\text{II wins in 4 games})$$
$$= \theta^4 + (1 - \theta)^4$$
$$\pi_2(\theta) = P(\text{I wins 3 of the first 4 and the fifth})$$
$$\quad + P(\text{I loses 3 of the first 4 and the fifth})$$
$$= \binom{4}{3}\theta^3(1 - \theta) \cdot \theta + \binom{4}{1}\theta(1 - \theta)^3 \cdot (1 - \theta)$$
$$= 4\theta(1 - \theta)[\theta^3 + (1 - \theta)^3]$$
$$\pi_3(\theta) = 10\theta^2(1 - \theta)^2[\theta^2 + (1 - \theta)^2]$$
$$\pi_4(\theta) = 20\theta^3(1 - \theta)^3$$

The paper "Seven Game Series in Sports" by Groeneveld and Meeden (*Mathematics Magazine*, 1975, pp. 187–192) tested the fit of this model to results of National Hockey League playoffs during the period 1943–1967 (when league membership was stable). The data appears in Figure 14.6.

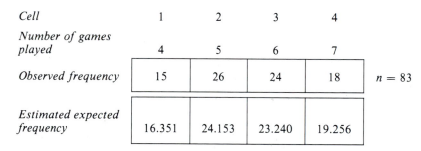

Cell	1	2	3	4	
Number of games played	4	5	6	7	
Observed frequency	15	26	24	18	$n = 83$
Estimated expected frequency	16.351	24.153	23.240	19.256	

Figure 14.6   Observed and expected counts for the simple model

The estimated expected cell counts are $83\pi_i(\hat{\theta})$, where $\hat{\theta}$ is the value of θ that maximizes

$$\{\theta^4 + (1 - \theta)^4\}^{15} \cdot \{4\theta(1 - \theta)[\theta^3 + (1 - \theta)^3]\}^{26}$$
$$\cdot \{10\theta^2(1 - \theta)^2[\theta^2 + (1 - \theta)^2]\}^{24} \cdot \{20\theta^3(1 - \theta)^3\}^{18} \qquad (14.5)$$

Standard calculus methods fail to yield a nice formula for the maximizing value $\hat{\theta}$, so it must be computed using numerical methods. The result is $\hat{\theta} = .654$, from which $\pi_i(\hat{\theta})$ and the estimated expected cell counts are computed. The computed value of χ^2 is .360, while (since $k - 1 - m = 4 - 1 - 1 = 2$) $\chi^2_{.10, 2} = 4.605$. There is thus no reason to reject the simple model as applied to NHL playoff series.

The cited paper also considered World Series data for the period 1903–1973. For the simple model, $\chi^2 = 5.97$, so the model does not seem appropriate. The suggested reason for this is that for the simple model

$$P(\text{series lasts six games} \mid \text{series lasts at least six games}) \geq .5 \qquad (14.6)$$

while of the 38 series that actually lasted at least six games, only 13 lasted exactly six. The following alternative model is then introduced:

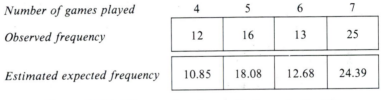

The first two π_i's are identical to the simple model, while θ_2 is the conditional probability of (14.6) (which can now be any number between zero and one). The values of $\hat{\theta}_1$ and $\hat{\theta}_2$ that maximize the expression analogous to (14.5) are determined numerically as $\hat{\theta}_1 = .614, \hat{\theta}_2 = .342$. A summary appears in Figure 14.7 with $\chi^2 = .384$. Since two parameters are estimated, d.f. $= k - 1 - m = 1$ with $\chi^2_{.10,1} = 2.706$, indicating a good fit of the data to this new model.

Number of games played	4	5	6	7
Observed frequency	12	16	13	25
Estimated expected frequency	10.85	18.08	12.68	24.39

Figure 14.7 Observed and expected counts for the more complex model

One of the regularity conditions on the θ_i's in the theorem is that they be functionally independent of one another. That is, no single θ_i can be determined from the values of other θ_i's, so that m is the number of functionally independent parameters estimated. A general rule of thumb for d.f. in a chi-squared test is

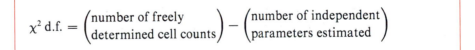

This rule will be used in connection with several different chi-squared tests in the next section.

Goodness of Fit for Discrete Distributions

Many experiments involve observing a random sample $X_1, X_2, \ldots, X_n$ from some discrete distribution. One may then wish to investigate whether the underlying distribution is a member of a particular family, such as the Poisson or negative binomial family. In the case of both a Poisson and a negative binomial distribution, the set of possible values is infinite, so the values must be grouped into k subsets before a chi-squared test can be used. The groupings should be done so that the expected frequency in each cell (group) is at least five. The last cell will then correspond to X values of $c, c + 1, c + 2, \ldots$ for some value c.

This grouping can considerably complicate the computation of the $\hat{\theta}_i$'s and estimated expected cell counts. This is because the theorem requires that the $\hat{\theta}_i$'s be obtained from the cell counts $N_1, \ldots, N_k$ rather than the sample values $X_1, \ldots, X_n$.

Example 14.9   The article "Some Sampling Characteristics of Plants and Arthropods of the Arizona Desert" (*Ecology*, 1962, pp. 567–571) reported the following count data on the number of Larrea divaricata plants found in each of 48 sampling quadrats.

Cell	1	2	3	4	5
Number of plants	0	1	2	3	≥ 4
Frequency	9	9	10	14	6

Figure 14.8   Observed counts for Example 14.9

The author fit a Poisson distribution to the data. Let λ denote the Poisson parameter, and suppose for the moment that the six counts in cell 5 were actually 4, 4, 5, 5, 6, 6. Then denoting sample values by $x_1, \ldots, x_{48}$, nine of the x_i's were 0, nine were 1, and so on. The likelihood of the observed sample is

$$\frac{e^{-\lambda}\lambda^{x_1}}{x_1!} \cdots \frac{e^{-\lambda}\lambda^{x_{48}}}{x_{48}!} = \frac{e^{-48\lambda}\lambda^{\Sigma x_i}}{x_1! \ldots x_{48}!} = \frac{e^{-48\lambda}\lambda^{101}}{x_1! \ldots x_{48}!}$$

The value of λ for which this is maximized is $\hat{\lambda} = \Sigma x_i/n = 101/48 = 2.10$ (the value reported in the paper).

However, the $\hat{\lambda}$ required for χ^2 is obtained by maximizing expression (14.4) rather than the likelihood of the full sample. The cell probabilities are

$$\pi_i(\lambda) = \frac{e^{-\lambda}\lambda^{i-1}}{(i-1)!} \quad i = 1, 2, 3, 4; \quad \pi_5(\lambda) = 1 - \sum_{i=0}^{3} \frac{e^{-\lambda}\lambda^i}{i!}$$

so the right-hand side of (14.4) becomes

$$\left[\frac{e^{-\lambda}\lambda^0}{0!}\right]^9 \left[\frac{e^{-\lambda}\lambda^1}{1!}\right]^9 \left[\frac{e^{-\lambda}\lambda^2}{2!}\right]^{10} \left[\frac{e^{-\lambda}\lambda^3}{3!}\right]^{14} \left[1 - \sum_{i=0}^{3} \frac{e^{-\lambda}\lambda^i}{i!}\right]^6$$

There is no nice formula for $\hat{\lambda}$, the maximizing value of λ, in this latter expression, so it must be obtained numerically. ■

Because the parameter estimates are usually much more difficult to compute from the grouped data than from the full sample, they are virtually always computed using this latter method. When these "full" estimators are used in the chi-squared statistic, the distribution of the statistic is altered and a level α test is no longer specified by the critical value $\chi^2_{\alpha, k-1-m}$.

Theorem

Let $\hat{\theta}_1, \ldots, \hat{\theta}_m$ be the maximum likelihood estimators of $\theta_1, \ldots, \theta_m$ based on the full sample $X_1, \ldots, X_n$, and let χ^2 denote the statistic based on these estimators. Then the critical value c_α that specifies a level α upper-tailed test satisfies

$$\chi^2_{\alpha, k-1-m} \leq c_\alpha \leq \chi^2_{\alpha, k-1} \tag{14.7}$$

The test procedure implied by this theorem is:

$$
\begin{aligned}
&\text{if } \quad \chi^2 \geq \chi^2_{\alpha, k-1}, \quad \text{reject } H_0 \\
&\text{if } \quad \chi^2 \leq \chi^2_{\alpha, k-1-m}, \quad \text{don't reject } H_0 \\
&\text{if } \quad \chi^2_{\alpha, k-1-m} < \chi^2 < \chi^2_{\alpha, k-1}, \quad \text{withhold judgment}
\end{aligned}
\tag{14.8}
$$

Example 14.10
(Example 14.9 continued)

Using $\hat{\lambda} = 2.10$, the estimated expected cell counts are computed from $n\pi_i(\hat{\lambda})$ where $n = 48$. For example

$$n\pi_1(\hat{\lambda}) = 48 \cdot \frac{e^{-2.1}(2.1)^0}{0!} = (48)(e^{-2.1}) = 5.88$$

Similarly, $n\pi_2(\hat{\lambda}) = 12.34$, $n\pi_3(\hat{\lambda}) = 12.96$, $n\pi_4(\hat{\lambda}) = 9.07$, and $n\pi_5(\hat{\lambda}) = 48 - 5.88 - \cdots - 9.07 = 7.75$. Then

$$\chi^2 = \frac{(9 - 5.88)^2}{5.88} + \cdots + \frac{(6 - 7.75)^2}{7.75} = 6.31$$

Since $m = 1$ and $k = 5$, at level .05 we need $\chi^2_{.05, 3} = 7.815$ and $\chi^2_{.05, 4} = 9.488$. Because $6.31 \leq 7.815$, we do not reject H_0; at the 5% level, the Poisson distribution provides a reasonable fit to the data. Notice that $\chi^2_{.10, 3} = 6.251$ and $\chi^2_{.10, 4} = 7.779$, so at level .10 we would have to withhold judgment on whether or not the Poisson distribution was appropriate. ∎

Sometimes even the maximum likelihood estimates based on the full sample are quite difficult to compute. This is the case, for example, for the two-parameter (generalized) negative binomial distribution. In such situations method-of-moments estimates are often used and the resulting χ^2 compared to $\chi^2_{\alpha, k-1-m}$, though it is not known to what extent the use of moments estimators affects the true critical value.

Goodness of Fit for Continuous Distributions

The chi-squared test can also be used to test whether or not the sample comes from a specified family of continuous distributions, such as the exponential family or the normal family. The choice of cells (class intervals) is even more arbitrary in the continuous case than in the discrete case. To ensure that the chi-squared test is valid, the cells should be chosen independently of the sample observations. Once the cells are chosen, it is almost always quite difficult to estimate unspecified parameters (such as μ and σ in the normal case) from the observed cell counts, so instead maximum likelihood estimates based on the full sample are computed. The critical value c_α again satisfies (14.7), and the test procedure is given by (14.8).

Example 14.11   The Institute of Nutrition of Central America and Panama (INCAP) has carried out extensive dietary studies and research projects in Central America. In one study reported in the November 1964 issue of the *American Journal of Clinical Nutrition* ("The Blood Viscosity of Various Socioeconomic Groups in Guatemala"), serum total cholesterol measurements for a sample of 49 low-income rural Indians were reported as follows (in mg/L):

> 204, 108, 140, 152, 158, 129, 175, 146, 157, 174, 192, 194, 144, 152, 135, 223, 145,
> 231, 115, 131, 129, 142, 114, 173, 226, 155, 166, 220, 180, 172, 143, 148, 171, 143,
> 124, 158, 144, 108, 189, 136, 136, 197, 131, 95, 139, 181, 165, 142, 162

Is it plausible that serum cholesterol level is normally distributed for this population? Suppose that prior to sampling, it was felt that plausible values for μ and σ were 150 and 30, respectively. The seven equiprobable class intervals for the standard normal distribution are $(-\infty, -1.07)$, $(-1.07, -.57)$, $(-.57, -.18)$, $(-.18, .18)$, $(.18, .57)$, $(.57, 1.07)$, and $(1.07, \infty)$, with each endpoint also giving the distance in standard deviations from the mean for any other normal distribution. For $\mu = 150$ and $\sigma = 30$, these intervals become $(-\infty, 117.9)$, $(117.9, 132.9)$, $(132.9, 144.6)$, $(144.6, 155.4)$, $(155.4, 167.1)$, $(167.1, 182.1)$, and $(182.1, \infty)$.

To obtain the estimated cell probabilities $\pi_1(\hat{\mu}, \hat{\sigma}), \ldots, \pi_7(\hat{\mu}, \hat{\sigma})$, we first need the maximum likelihood estimates $\hat{\mu}$ and $\hat{\sigma}$. In Chapter 6 the maximum likelihood estimate of σ was shown to be $[\Sigma(x_i - \bar{x})^2/n]^{1/2}$ (rather than s), so with $s = 31.75$

$$\hat{\mu} = \bar{x} = 157.02, \quad \hat{\sigma} = \left[\frac{\Sigma(x_i - \bar{x})^2}{n}\right]^{1/2} = \left[\frac{(n-1)s^2}{n}\right]^{1/2} = 31.42$$

Each $\pi_i(\hat{\mu}, \hat{\sigma})$ is then the probability that a normal random variable X with mean 157.02 and standard deviation 31.42 falls in the ith class interval. For example,

$$\pi_2(\hat{\mu}, \hat{\sigma}) = P(117.9 \le X \le 132.9) = P(-1.25 \le Z \le -.77) = .1150$$

so $n\pi_2(\hat{\mu}, \hat{\sigma}) = 49(.1150) = 5.64$. Observed and estimated expected cell counts are shown in Figure 14.9.

Cell	$(-\infty, 117.9)$	$(117.9, 132.9)$	$(132.9, 144.6)$	$(144.6, 155.4)$
Observed	5	5	11	6
Estimated expected	5.17	5.64	6.08	6.64

Cell	$(155.4, 167.1)$	$(167.1, 182.1)$	$(182.1, \infty)$
Observed	6	7	9
Estimated expected	7.12	7.98	10.38

Figure 14.9 Observed and expected counts for Example 14.11

The computed χ^2 is 4.60. With $k = 7$ cells and $m = 2$ parameters estimated, $\chi^2_{.05, k-1} = \chi^2_{.05, 6} = 12.592$ and $\chi^2_{.05, k-1-m} = \chi^2_{.05, 4} = 9.488$. Since $4.60 \leq 9.488$, a normal distribution provides quite a good fit to the data. ■

Example 14.12   The paper "Some Studies on Tuft Weight Distribution in the Opening Room" (*Textile Research J.*, 1976, pp. 567–573) reported the accompanying data on the distribution of output tuft weight X (mg) of cotton fibers for the input weight $x_0 = 70$.

Interval	0–8	8–16	16–24	24–32	32–40	40–48	48–56	56–64	64–70
Observed frequency	20	8	7	1	2	1	0	1	0
Expected frequency	18.0	9.9	5.5	3.0	1.8	.9	.5	.3	.1

The authors postulated a truncated exponential distribution:

$$H_0 : f(x) = \frac{e^{-\lambda x}}{1 - e^{-\lambda x_0}} \quad 0 \leq x \leq x_0$$

The mean of this distribution is

$$\mu = \int_0^{x_0} x f(x)\, dx = \frac{1}{\lambda} - \frac{x_0 e^{-\lambda x_0}}{1 - e^{-\lambda x_0}}$$

The parameter λ was estimated by replacing μ by $\bar{x} = 13.086$ and solving the resulting equation to obtain $\hat{\lambda} = .0742$ (so $\hat{\lambda}$ is a method-of-moments estimate and not a maximum likelihood estimate). Then with $\hat{\lambda}$ replacing λ in $f(x)$, the estimated expected cell frequencies as displayed above are computed as

$$40\pi_i(\hat{\lambda}) = 40P(a_{i-1} \leq X < a_i) = 40\int_{a_{i-1}}^{a_i} f(x)\, dx = \frac{40(e^{-\hat{\lambda}a_{i-1}} - e^{-\hat{\lambda}a_i})}{1 - e^{-\hat{\lambda}x_0}}$$

where $[a_{i-1}, a_i)$ is the ith class interval. To obtain expected cell counts of at least 5, the last six cells are combined to yield observed counts of 20, 8, 7, 5 and expected counts of 18.0, 9.9, 5.5, 6.6. The computed value of chi-squared is then $\chi^2 = 1.34$. Because $\chi_{.05,2}^2 = 5.992$, H_0 is not rejected, so the truncated exponential model provides a good fit.   ■

A Special Test for Normality

Probability plots were introduced in Section 4.6 as an informal method for assessing the plausibility of any specified population distribution as the one from which the given sample was selected. The straighter the probability plot, the more plausible is the distribution on which the plot is based. A normal probability plot is used for checking whether *any* member of the normal distribution family is plausible. Let's denote the sample x_i's when ordered from smallest to largest by $x_{(1)}, x_{(2)}, \ldots, x_{(n)}$. Then the plot suggested for checking normality was a plot of the points $(x_{(i)}, y_i)$, where $y_i = \Phi^{-1}((i - .5)/n)$.

A quantitative measure of the extent to which points cluster about a straight line is the sample correlation coefficient r introduced in Chapter 12. Consider calculating r for the n pairs $(x_{(1)}, y_1), \ldots, (x_{(n)}, y_n)$. The y_i's here are not observed values in a random sample from a y population, so properties of this r are quite different from those described in Section 12.5. However, it is true that the more the r deviates from 1, the less the probability plot resembles a straight line (remember that a probability plot must slope upward). This idea can be extended to yield a formal test procedure: reject the hypothesis of population normality if $r \leq c_\alpha$, where c_α is a critical value chosen to yield the desired significance level α. That is, the critical value is chosen so that when the population distribution is actually normal, the probability of obtaining an r value that is at most c_α (and thus incorrectly rejecting H_0) is the desired α. The developers of the MINITAB statistical computer package give critical values for $\alpha = .10$, .05, and .01 in combination with different sample sizes. These critical values are based on a slightly different definition of the y_i's than that given earlier.

MINITAB will also construct a normal probability plot based on these y_i's. The plot will be almost identical in appearance to that based on the earlier y_i's. When there are several tied $x_{(i)}$'s, MINITAB computes r by using the average of the corresponding y_i's as the second number in each pair.

Let $y_i = \Phi^{-1}\left(\dfrac{i - .375}{n + .25}\right)$, and compute the sample correlation coefficient r for the n pairs $(x_{(1)}, y_1), \ldots, (x_{(n)}, y_n)$. A test of

H_0: the population distribution is normal

versus

H_a: the population distribution is not normal

consists of rejecting H_0 when $r \le c_\alpha$. Critical values c_α are given in Appendix Table A.14 for various significance levels α and sample sizes n.

Example 14.13   Consider the accompanying random sample of $n = 20$ observations, each the value of the width-to-length ratio of a beaded rectangle used by Shoshoni Indians to decorate leather handicrafts (*Lowie's Selected Papers in Anthropology*, 1960, pp. 137–142). Let's carry out a test at level .01 to see whether it is plausible that the population distribution of width-to-length ratios is normal. The critical value for $n = 20$ is .929, and the calculated value of r is .905. Since $.905 \le .929$, H_0 is rejected. A normal probability plot exhibits curvature of the sort associated with a positively skewed distribution having a long upper tail.

	$x_{(i)}$	y_i	$x_{(i)}^2$	y_i^2	$x_{(i)} y_i$
1	0.553	-1.87129	0.305809	3.50172	-1.03482
2	0.570	-1.40377	0.324900	1.97057	-0.80015
3	0.576	-1.12690	0.331776	1.26990	-0.64909
4	0.601	-0.91718	0.361201	0.84121	-0.55122
5	0.606	-0.66267	0.367236	0.43913	-0.40158
6	0.606	-0.66267	0.367236	0.43913	-0.40158
7	0.609	-0.44602	0.370881	0.19893	-0.27162
8	0.611	-0.31325	0.373321	0.09812	-0.19139
9	0.615	-0.18593	0.378225	0.03457	-0.11435
10	0.628	-0.06165	0.394384	0.00380	-0.03872
11	0.654	0.06165	0.427716	0.00380	0.04032
12	0.662	0.18593	0.438244	0.03457	0.12309
13	0.668	0.31325	0.446224	0.09812	0.20925
14	0.670	0.44602	0.448900	0.19893	0.29883
15	0.672	0.58740	0.451584	0.34504	0.39473
16	0.690	0.74198	0.476100	0.55054	0.51197
17	0.693	0.91718	0.480249	0.84121	0.63560
18	0.749	1.12690	0.561001	1.26990	0.84405
19	0.844	1.40377	0.712336	1.97057	1.18478
20	0.933	1.87129	0.870489	3.50172	1.74591
Sum:	13.210	.00404	8.887812	17.61148	1.53401

■

Exercises / Section 14.2 (8–18)

8. Consider a large population of families in which each family has exactly three children. If the sexes of the three children in any family are independent of one another, the number of male children in a randomly selected family will have a binomial distribution based on three trials.

a. Suppose that a random sample of 160 families yields the following results. Test the relevant hypotheses by proceeding as in Example 14.6.

Number of male children:	0	1	2	3
Frequency:	14	66	64	16

b. Suppose that a random sample of families in a non-human population resulted in observed frequencies of 15, 20, 12, and 3, respectively. Would the chi-squared test be based on the same number of d.f. as the test in (a)? Explain.

9. A study of sterility in the fruit fly ("Hybrid Dysgenesis in Drosophila Melanogaster: The Biology of Female and Male Sterility," *Genetics*, 1979, pp. 161–174) reported the following data on the number of ovaries developed for each female fly in a sample of size 1388. One model for unilateral sterility states that each ovary develops with some probability p independently of the other ovary. Test the fit of this model using χ^2.

x = number of ovaries developed	0	1	2
Observed count	1212	118	58

10. The article "Feeding Ecology of the Red-Eyed Vireo and Associated Foliage-Gleaning Birds" (*Ecological Monographs*, 1971, pp. 129–152) presented the accompanying data on the variable $X =$ the number of hops before the first flight and preceded by a flight. The author then proposed and fit a geometric probability distribution $[p(x) = P(X = x) = p^{x-1} \cdot q$ for $x = 1, 2, \ldots$ where $q = 1 - p]$ to the data. The total sample size was $n = 130$.

x	1	2	3	4	5	6	7	8	9	10	11	12
Number of times x observed	48	31	20	9	6	5	4	2	1	1	2	1

a. The likelihood is $(p^{x_1 - 1} \cdot q) \ldots (p^{x_n - 1} \cdot q) = p^{\Sigma x_i - n} \cdot q^n$. Show that the maximum likelihood estimate of p is $\hat{p} = (\Sigma x_i - n)/\Sigma x_i$, and compute $\hat{p}$ for the given data.

b. Estimate the expected cell counts using $\hat{p}$ of (a) [expected cell counts = $n \cdot (\hat{p})^{x-1} \cdot \hat{q}$ for $x = 1, 2, \ldots$] and test the fit of the model using a χ^2 test by combining the counts for $x = 7, 8, \ldots$, and 12 into one cell ($x \geq 7$).

11. A certain type of flashlight is sold with the four batteries included. A random sample of 150 flashlights is obtained and the number of defective batteries in each is determined, resulting in the following data:

Number defective	0	1	2	3	4
Frequency	26	51	47	16	10

Let X be the number of defective batteries in a randomly selected flashlight. Test the null hypothesis that the distribution of X is Bin(4, θ). That is, with $p_i = P(i$ defectives), test

$$H_0 : p_i = \binom{4}{i} \theta^i (1 - \theta)^{4 - i} \quad i = 0, 1, 2, 3, 4$$

Hint: To obtain the maximum likelihood estimate of θ, write the likelihood (the function to be maximized) as $\theta^u (1 - \theta)^v$ where the exponents u and v are linear functions of the cell counts. Then take the natural log, differentiate with respect to θ, equate the result to zero, and solve for $\hat{\theta}$.

12. In a genetic experiment, investigators looked at 300 chromosomes of a particular type and counted the number of sister-chromatid exchanges on each ("On the Nature of Sister-Chromatid Exchanges in 5-Bromodeoxyuridine-Substituted Chromosomes," *Genetics*, 1979, pp. 1251–1264). A Poisson model was hypothesized for the distribution of the number of exchanges. Test the fit of a Poisson distribution to the data by first estimating λ and then combining the counts for $x = 8$ and $x = 9$ into one cell.

X = number of exchanges	0	1	2	3	4	5	6	7	8	9
Observed counts	6	24	42	59	62	44	41	14	6	2

13. A paper in *Annals of Mathematical Statistics* reported the following data on the number of borers in each of 120 groups of borers. Does the Poisson p.m.f. provide a plausible model for the distribution of the number of borers in a group? *Hint:* Add the frequencies for 7, 8, ..., 12 to establish a single category "≥ 7."

Number of borers	0	1	2	3	4	5	6	7	8	9	10	11	12
Frequency	24	16	16	18	15	9	6	5	3	4	3	0	1

14. The article "A Probabilistic Analysis of Dissolved Oxygen–Biochemical Oxygen Demand

Relationship in Streams" (*J. Water Resources Control Fed.*, 1969, pp. 73–90) reported data on the rate of oxygenation in streams at 20° C in a certain region. The sample mean and standard deviation were computed as $\bar{x} = .173$ and $s = .066$. Based on the accompanying frequency distribution, can it be concluded that oxygenation rate is a normally distributed variable? Use the chi-squared test with $\alpha = .05$.

Rate (per day)	Frequency
below .100	12
.100–below .150	20
.150–below .200	23
.200–below .250	15
.250 or more	13

15. Each headlight on an automobile undergoing an annual vehicle inspection can be focused either too high (H), too low (L), or properly (N). Checking the two headlights simultaneously (and not distinguishing between left and right) results in the six possible outcomes HH, LL, NN, HL, HN, and LN. If the probabilities (population proportions) for the single headlight focus direction are $P(H) = \theta_1$, $P(L) = \theta_2$, and $P(N) = 1 - \theta_1 - \theta_2$, and the two headlights are focused independently of one another, the probabilities of the six outcomes for a randomly selected car are

$$p_1 = \theta_1^2, \quad p_2 = \theta_2^2, \quad p_3 = (1 - \theta_1 - \theta_2)^2$$
$$p_4 = 2\theta_1\theta_2, \quad p_5 = 2\theta_1(1 - \theta_1 - \theta_2),$$
$$p_6 = 2\theta_2(1 - \theta_1 - \theta_2)$$

Use the data given below to test the null hypothesis

$$H_0: p_1 = \pi_1(\theta_1, \theta_2), \ldots, p_6 = \pi_6(\theta_1, \theta_2)$$

where the $\pi_i(\theta_1, \theta_2)$'s are given above.

Outcome	HH	LL	NN	HL	HN	LN
Frequency	49	26	14	20	53	38

Hint: Write the likelihood as a function of θ_1 and θ_2, take the natural log, then compute $\partial/\partial\theta_1$ and $\partial/\partial\theta_2$, equate them to zero, and solve for $\hat{\theta}_1, \hat{\theta}_2$.

16. The following data set consists of 26 observations on fracture toughness of base plate of 18% nickel miraging steel (from "Fracture Testing of Weldments," *ASTM Special Publ. No. 381*, 1965, pp. 328–356). Use the normal probability plot correlation coefficient test to decide whether a normal distribution provides a plausible model for fracture toughness.

69.5, 71.9, 72.6, 73.1, 73.3, 73.5, 74.1, 74.2, 75.3, 75.5, 75.7, 75.8, 76.1, 76.2, 76.9, 77.0, 77.9, 78.1, 79.6, 79.7, 79.9, 80.1, 82.2, 83.7, 93.7

17. The paper "Nonbloated Burned Clay Aggregate Concrete" (*J. Materials*, 1972, pp. 555–563) reported the following data on seven-day flexural strength of nonbloated burned clay aggregate concrete samples (psi):

257, 327, 317, 300, 340, 340, 343, 374, 377, 386, 383, 393, 407, 407, 434, 427, 440, 407, 450, 440, 456, 460, 456, 476, 480, 490, 497, 526, 546, 700

Test at level .10 to decide whether or not flexural strength is a normally distributed variable.

18. Use the normal probability plot correlation coefficient test on the standardized residuals of Example 13.1 to assess the plausibility of assuming that the random deviation in the regression model is normally distributed.

14.3 Two-Way Contingency Tables

In the previous two sections we discussed inferential problems in which the count data was displayed in a rectangular table of cells. Each table consisted of one row and a specified number of columns, where the columns corresponded to categories into which the population had been divided. We now study problems in which the data also consists of counts or frequencies, but the data table will now have I rows ($I \geq 2$) and J columns, so IJ cells. There are two commonly encountered situations in which such data arises:

1. There are I populations of interest, each corresponding to a different row of the table, and each population is divided into the same J categories. A sample is taken from the ith population $(i = 1, \ldots, I)$ and the counts are entered in the cells in the ith row of the table.

2. There is a single population of interest, with each individual in the population categorized with respect to two different factors. There are I categories associated with the first factor and J categories associated with the second factor. A single sample is taken, and the number of individuals belonging in category i of factor 1 and category j of factor 2 is entered in the cell in row i, column j $(i = 1, \ldots, I; j = 1, \ldots, J)$.

Let n_{ij} denote the number of individuals in the sample(s) falling in the (i, j)th cell (row i, column j) of the table—that is, the (i, j)th cell count. The table displaying the n_{ij}'s is called a **two-way contingency table.**

Figure 14.10   A two-way contingency table

In situations of type 1, we want to investigate whether or not the proportions in the different categories are the same for all populations. The null hypothesis states that the populations are **homogeneous** with respect to these categories. In type-2 situations, we investigate whether or not the categories of the two factors occur independently of one another in the population.

In Chapter 8 we developed a large-sample z test for testing the null hypothesis that the proportion of successes in one population is identical to the proportion in a second population. Instead of using S and F to label the two sides of the dichotomy, let's label them as category 1 and category 2. The symbols p_{11} and p_{12} will represent the proportions of population 1 belonging to categories 1 and 2, respectively (in place of p_1 and q_1); p_{21} and p_{22} will denote the analogous proportions for the second population. In our use of doubly subscripted quantities, the first subscript will refer to the population or sample from that population, and the second subscript will identify the category. The null hypothesis that the two dichotomous populations are homogeneous with respect to category proportions, previously stated as $H_0 : p_1 = p_2$, now becomes $H_0 : p_{11} = p_{21}$ (in which case $p_{12} = p_{22}$ also).

Homogeneity for I Dichotomous Populations

Suppose now that there are I different dichotomous populations ($I > 2$), each consisting of the same two categories. Let p_{i1} (p_{i2}) denote the proportion of population i falling in category 1 (2). For example, we might focus on $I = 4$ different brands of television sets, with category 1 consisting of all sets that need no warranty repair work and category 2 consisting of all sets that do need such work. Then p_{11}, p_{21}, p_{31}, and p_{41} denote the proportions of the four brands falling in the first category. The null hypothesis of homogeneity states that $p_{11} = p_{21} = p_{31} = p_{41}$ (implying that $p_{12} = p_{22} = p_{32} = p_{42}$). More generally, we wish to test $H_0 : p_{11} = p_{21} = \cdots = p_{I1}$ against the alternative that these category 1 proportions are not all the same.

To do so, we need sample data. Consider taking a random sample of size n_i from the ith population ($i = 1, \ldots, I$). The I samples are assumed to have been selected independently of one another. The total sample size is $n = \Sigma n_i$. Let n_{i1} = the number among the n_i sampled from the ith population that fall in category 1, and $n_{i2} = n_i - n_{i1}$. Before the samples are selected, the numbers of individuals falling into categories 1 and 2 are random variables—N_{11}, $N_{21}, \ldots, N_{I1}$ for category 1 and $N_{12}, N_{22}, \ldots, N_{I2}$ for category 2. The n_{ij}'s are the observed values of the N_{ij}'s. These are customarily displayed in a tabular format as illustrated in Figure 14.11, and are called *observed cell counts*. $N_{.1}$ and $n_{.1}$ refer to the total number among the n sampled that fall in category 1—the total of the observed counts from the first column; $N_{.2}$ and $n_{.2}$ refer to the total for category 2. Notice that the row totals in the table are fixed in advance of the experiment, whereas the column totals are not.

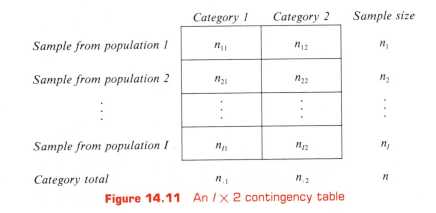

	Category 1	Category 2	Sample size
Sample from population 1	n_{11}	n_{12}	n_1
Sample from population 2	n_{21}	n_{22}	n_2
$\vdots$	$\vdots$	$\vdots$	$\vdots$
Sample from population I	n_{I1}	n_{I2}	n_I
Category total	$n_{.1}$	$n_{.2}$	n

Figure 14.11   An $I \times 2$ contingency table

The expected number of observations in the sample from population i that fall in the first category is $E(N_{i1}) = $ (sample size)(category proportion) $= n_i p_{i1}$ ($i = 1, \ldots, I$). When the null hypothesis is true, let p_1 denote the common value of $p_{11}, p_{21}, \ldots,$ and p_{I1}. In this case the expected counts for category 1 are $n_1 p_1, n_2 p_1, \ldots,$ and $n_I p_1$, and they are $n_1(1 - p_1), \ldots,$ and $n_I(1 - p_1)$ for the second category. However, whereas H_0 asserts that there is a common proportion p_1, it doesn't specify what that value is. Construction of a test procedure requires that p_1 be estimated from the data. Since p_1 represents the proportion in category 1 for every population, the natural estimate is the fraction among all those sampled that fall in category 1:

$$\hat{p}_1 = \frac{n_{11} + n_{21} + \cdots + n_{I1}}{n_1 + n_2 + \cdots + n_I} = \frac{n_{.1}}{n}$$

Substituting this estimate into the expressions for expected counts when H_0 is true yields *estimated expected cell counts*:

Let $\hat{e}_{ij}$ denote the point estimate of $E(N_{ij})$, the expected number from the ith sample falling in category j, computed under the assumption that H_0 is true. Then

$$\hat{e}_{ij} = \begin{cases} n_i \hat{p}_1 = \dfrac{n_i \cdot n_{.1}}{n} & j = 1 \\[2mm] n_i(1 - \hat{p}_1) = \dfrac{n_i \cdot n_{.2}}{n} & j = 2 \end{cases}$$

That is, the estimated expected cell counts are given by

$$\hat{e}_{ij} = \frac{(i\text{th row total})(j\text{th column total})}{n} \tag{14.9}$$

Example 14.14   Three different sites off the coast of California are being considered as possible locations for a liquid natural gas terminal. To get information on public opinion regarding the LNG project, a separate random sample of individuals in each region is obtained. Each individual is then asked whether he/she favors building an LNG terminal at the proposed location in the region in which the individual resides. The observed cell counts appear below.

The hypothesis of interest is $H_0: p_{11} = p_{21} = p_{31}$, where p_{i1} is the true proportion of individuals in region i who favor an LNG facility in that region. To compute the estimated expected cell counts, the column totals are computed (the row totals n_1, n_2, and n_3 are fixed in advance, since three different regions are sampled). The row and column totals are then entered on the margins of a table having the same dimensions as the observed count table, and formula (14.9) is employed. Calculations for this example are displayed in Figure 14.12.

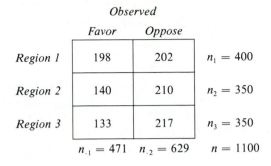

Observed

	Favor	Oppose	
Region 1	198	202	$n_1 = 400$
Region 2	140	210	$n_2 = 350$
Region 3	133	217	$n_3 = 350$
	$n_{.1} = 471$	$n_{.2} = 629$	$n = 1100$

Estimated expected

	Favor	Oppose	
Region 1	$\dfrac{(400)(471)}{1100} = 171.27$	$\dfrac{(400)(629)}{1100} = 228.73$	400
Region 2	$\dfrac{(350)(471)}{1100} = 149.86$	$\dfrac{(350)(629)}{1100} = 200.14$	350
Region 3	$\dfrac{(350)(471)}{1100} = 149.86$	$\dfrac{(350)(629)}{1100} = 200.14$	350
	471	629	

Figure 14.12 Observed and estimated expected cell counts for Example 14.14 ∎

Replacing $n_{.1}$ by $N_{.1}$ and $n_{.2}$ by $N_{.2}$ in $\hat{e}_{ij}$ gives the estimators (random variables $\hat{E}_{ij}$ $(i = 1, \ldots, I; j = 1, 2)$. These estimators can be used to specify a chi-squared goodness-of-fit statistic that has the same general form as those in Sections 14.1 and 14.2:

$$\chi^2 = \sum_{\text{all cells}} \frac{(\text{observed} - \text{estimated expected})^2}{\text{estimated expected}} = \sum_{i=1}^{I} \sum_{j=1}^{2} \frac{(N_{ij} - \hat{E}_{ij})^2}{\hat{E}_{ij}}$$

A theorem from mathematical statistics states that when $H_0 : p_{11} = p_{21} = \ldots = p_{I1}$ is true and all I sample sizes are large, χ^2 has approximately a chi-squared distribution with $I - 1$ d.f. This leads to the following test for homogeneity in I dichotomous populations.

Null hypothesis: $H_0: p_{11} = p_{21} = \ldots = p_{I1}$

Alternative hypothesis: H_a: at least two of the p_{i1}'s are unequal

Test statistic value:

$$\chi^2 = \sum_{\text{all cells}} \frac{(\text{observed} - \text{estimated expected})^2}{\text{estimated expected}} = \sum_{i=1}^{I} \sum_{j=1}^{2} \frac{(n_{ij} - \hat{e}_{ij})^2}{\hat{e}_{ij}}$$

Rejection region: $\chi^2 \geq \chi^2_{\alpha, I-1}$

In practice, the test can safely be applied as long as $\hat{e}_{ij} \geq 5$ for every i, j (all cells).

Example 14.15   The accompanying data resulted from an experiment in which seeds of five different types were planted and the number that germinated within five weeks of planting was recorded for each seed type ("Nondestructive Optical Methods of Food Quality Evaluation," *Food Science and Nutr.*, 1984, pp. 232–279).

		Observed				Estimated		
		Germinated	Failed to germinate			Germinated	Failed to germinate	
	1	31	7	$n_1 = 38$		22.5	15.5	38
	2	57	33	$n_2 = 90$		53.3	36.7	90
Seed type	3	87	60	$n_3 = 147$		87.1	59.9	147
	4	52	44	$n_4 = 96$		56.9	39.1	96
	5	10	19	$n_5 = 29$		17.2	11.8	29
		$n_{.1} = 237$	$n_{.2} = 163$	$n = 400$		237	163	400

Figure 14.13   Observed and estimated expected counts for Example 14.15

Let $p_{i1} =$ the true proportion of seeds of type i that germinate within five weeks of planting. The hypotheses are

$H_0: p_{11} = p_{21} = p_{31} = p_{41} = p_{51}$

H_a: at least two p_{i1}'s are unequal

The test is based on $I - 1 = 4$ d.f., and since $\chi^2_{.01, 4} = 13.277$, H_0 will be rejected at significance level .01 if $\chi^2 \geq 13.277$. The computed value of the test statistic is

$$\chi^2 = \frac{(31 - 22.5)^2}{22.5} + \frac{(57 - 53.3)^2}{53.3} + \cdots + \frac{(19 - 11.8)^2}{11.8}$$

$$= 3.2 + .25 + \cdots + 4.37 = 16.86$$

Because $16.86 \geq 13.277$, H_0 is rejected at significance level .01. The true proportion of seeds that germinate appears to depend on seed type. ∎

The number of d.f. for the test comes from our general rule of thumb: there are I freely determined cell counts (one in each row, since row totals are fixed), and the single parameter p_1 is estimated, giving d.f. $= I - 1$. The test procedure can be used even when $I = 2$ (a comparison of two different population proportions). In this case, d.f. $= 1$ and the chi-squared test is equivalent to the two-tailed z test of Section 9.4 because $z^2 = \chi^2$ and $(z_{\alpha/2})^2 = \chi^2_{\alpha, 1}$. If H_0 is rejected in favor of the conclusion that the I populations are not homogeneous, there are multiple comparison methods available for identifying significant differences among the p_{i1}'s.

Homogeneity for *I* Populations, *J* Categories

We now assume that each individual in every one of the I populations belongs in exactly one of J categories. A sample of n_i individuals is taken from the ith population; let $n = \Sigma n_i$ and

$n_{ij} =$ the number of individuals in the ith sample who fall into category j

$$n_{\cdot j} = \sum_{i=1}^{I} n_{ij} = \quad \text{the total number of individuals among the } n \text{ sampled who fall into category } j$$

The n_{ij}'s are recorded in a two-way contingency table with I rows and J columns. The sum of the n_{ij}'s in the ith row is n_i, while the sum of entries in the jth column is $n_{\cdot j}$.

With $J > 2$, the notation necessary to state the hypotheses is more cumbersome than in the case $J = 2$ just considered. Let

$$p_{ij} = \quad \text{the proportion of the individuals in population } i \text{ who fall into category } j$$

Thus for population 1 the J proportions are $p_{11}, p_{12}, \ldots, p_{1J}$ (which sum to 1), and similarly for the other populations. The **null hypothesis of homogeneity** states that the proportion of individuals in category j is the same for each population, and that this is true for every category; that is, for every j, $p_{1j} = p_{2j} = \ldots = p_{Ij}$.

When H_0 is true, we can use $p_1, p_2, \ldots, p_J$ to denote the population proportions in the J different categories; these proportions are common to all I populations. The expected number of individuals in the ith sample who fall in the jth category when H_0 is true is then $E(N_{ij}) = n_i \cdot p_j$. To estimate $E(N_{ij})$, we must first estimate p_j, the proportion in category j. Among the total sample of n individuals, $N_{.j}$ fall into category j, so we use $\hat{p}_j = N_{.j}/n$ as the estimator (this can be shown to be the maximum likelihood estimator of p_j). Substitution of the estimate $\hat{p}_j$ for p_j in $n_i p_j$ yields the same formula for estimated expected counts under H_0 as in the dichotomous case:

$$\hat{e}_{ij} = \text{estimated expected count in cell } (i, j) = n_i \cdot \frac{n_{.j}}{n}$$

$$= \frac{(i\text{th row total})(j\text{th column total})}{n} \tag{14.10}$$

The test statistic also has the same form as in earlier problem situations. The number of degrees of freedom comes from the general rule of thumb. In each row of the table there are $J - 1$ freely determined cell counts (the sample size n_i is fixed), so there are a total of $I(J - 1)$ freely determined cells. Parameters $p_1, \ldots, p_J$ are estimated, but because $\Sigma p_i = 1$, only $J - 1$ of these are independent. Therefore d.f. $= I(J - 1) - (J - 1) = (J - 1)(I - 1)$.

Null hypothesis: $H_0: p_{1j} = p_{2j} = \cdots = p_{Ij}$ $j = 1, 2, \ldots, J$

Alternative hypothesis: $H_a: H_0$ is not true

Test statistic value:

$$\chi^2 = \sum_{\text{all cells}} \frac{(\text{observed} - \text{estimated expected})^2}{\text{estimated expected}} = \sum_{i=1}^{I} \sum_{j=1}^{J} \frac{(n_{ij} - \hat{e}_{ij})^2}{\hat{e}_{ij}}$$

Rejection region: $\chi^2 \geq \chi^2_{\alpha, (I-1)(J-1)}$

The test can safely be applied as long as $\hat{e}_{ij} \geq 5$ for all cells.

Example 14.16   In a study to investigate the extent to which individuals are aware of industrial odors in a certain region ("Annoyance and Health Reactions to Odor from Refineries and Other Industries in Carson, California," *Environmental Research,* 1978, pp. 119–132), a sample of individuals was obtained from each of three different areas near industrial facilities. Each individual was asked whether he/she noticed odors (1) every day, (2) at least once/week, (3) at least once/month, (4) less often than once/month, or (5) not at all, resulting in the data given in Figure 14.14, with the estimated expected cell counts computed from (14.10).

Category

Observed	1	2	3	4	5	n_i
Area I	20	28	23	14	12	97
Area II	14	34	21	14	12	95
Area III	4	12	10	20	53	99
$n_{\cdot j}$	38	74	54	48	77	291

Expected	1	2	3	4	5	
Area I	12.67	24.67	18.00	16.00	25.67	97
Area II	12.41	24.16	17.63	15.67	25.14	95
Area III	12.93	25.18	18.37	16.33	26.20	99
	38	74	54	48	77	

Figure 14.14 Observed and estimated expected counts for Example 14.16

Then

$$\chi^2 = \frac{(20 - 12.67)^2}{12.67} + \cdots + \frac{(53 - 26.20)^2}{26.20} = 71.36$$

With $(I - 1)(J - 1) = (2)(4) = 8$, $\chi^2_{.005, 8} = 21.954$. Because $71.36 \geq 21.954$, the hypothesis of homogeneity is rejected at level .005 in favor of the conclusion that there are differences in perception of odors among the three areas. ▪

One way to extend the chi-squared analysis described here is to partition the overall χ^2 statistic into independent components. In Example 14.16 χ^2 can be broken down into χ^2 for area I versus area II with d.f. = 4 and areas I and II versus area III with d.f. = 4. The conclusion is that areas I and II don't differ significantly from one another while in combination they differ significantly from area III. This type of partitioning should be planned before the data is actually obtained. For unplanned comparisons, there are methods of multiple comparisons that can be employed. The book by Everitt contains a good exposition of these topics.

Independence of Factors in a Single Population

We focus now on the relationship between two different factors in a single population. The number of categories of the first factor will be denoted by I and the number of categories of the second factor by J. Each individual in the population is assumed to belong in exactly one of the I categories associated with the first factor and exactly one of the J categories associated with the second factor. For example, the population of interest might consist of all individuals who regularly watch the national news on television, with the first factor being preferred network (ABC, CBS, NBC, or PBS, so $I = 4$) and the second factor political philosophy (liberal, moderate, or conservative, giving $J = 3$).

For a sample of n individuals taken from the population, let n_{ij} denote the number among the n who fall both in category i of the first factor and category j of the second factor. The n_{ij}'s can be displayed in a two-way contingency table with I rows and J columns. In the case of homogeneity for I populations, the row totals were fixed in advance and only the J column totals were random. Now only the total sample size is fixed, and both the $n_{i.}$'s and $n_{.j}$'s are observed values of random variables. To state the hypotheses of interest, let

p_{ij} = the proportion of individuals in the population who belong in category i of factor 1 and category j of factor 2

$\quad = P$(a randomly selected individual falls in both category i of factor 1 and category j of factor 2)

Then

$$p_{i.} = \sum_j p_{ij} = P(\text{a randomly selected individual falls in category } i \text{ of factor 1})$$

$$p_{.j} = \sum_i p_{ij} = P(\text{a randomly selected individual falls in category } j \text{ of factor 2})$$

Recall from probability that two events A and B are independent if $P(A \cap B) = P(A) \cdot P(B)$. The *null hypothesis here says that an individual's category with respect to factor 1 is independent of the category with respect to factor 2.* In symbols, this becomes $p_{ij} = p_{i.} \cdot p_{.j}$ for every pair (i, j).

The expected count in cell (i, j) is $n \cdot p_{ij}$, so when H_0 is true, $E(N_{ij}) = n \cdot p_{i.} \cdot p_{.j}$. To obtain a chi-squared statistic, we must therefore estimate the $p_{i.}$'s $(i = 1, \ldots, I)$ and $p_{.j}$'s $(j = 1, \ldots, J)$. The (maximum likelihood) estimates are

$$\hat{p}_{i.} = \frac{n_{i.}}{n} = \text{sample proportion for category } i \text{ of factor 1}$$

and

$$\hat{p}_{.j} = \frac{n_{.j}}{n} = \text{sample proportion for category } j \text{ of factor 2}$$

This gives estimated expected cell counts identical to those in the case of homogeneity.

$$\hat{e}_{ij} = n \cdot \hat{p}_{i\cdot} \cdot \hat{p}_{\cdot j} = n \cdot \frac{n_{i\cdot}}{n} \cdot \frac{n_{\cdot j}}{n} = \frac{n_{i\cdot} \cdot n_{\cdot j}}{n}$$
$$= \frac{(i\text{th row total})(j\text{th column total})}{n}$$

The test statistic is also identical to that used in testing for homogeneity, as is the number of d.f. This is because the number of freely determined cell counts is $IJ - 1$, since only the total n is fixed in advance. There are I $p_{i\cdot}$'s estimated, but only $I - 1$ are independently estimated since $\Sigma p_{i\cdot} = 1$, and similarly $J - 1$ $p_{\cdot j}$'s are independently estimated, so $I + J - 2$ parameters are independently estimated. The rule of thumb now yields d.f. $= IJ - 1 - (I + J - 2) = IJ - I - J + 1 = (I - 1) \cdot (J - 1)$.

Null hypothesis: $H_0: p_{ij} = p_{i\cdot} \cdot p_{\cdot j} \quad i = 1, \ldots, I \quad \text{and} \quad j = 1, \ldots, J$

Alternative hypothesis: $H_a: H_0$ is not true

Test statistic value:

$$\chi^2 = \sum_{\text{all cells}} \frac{(\text{observed} - \text{estimated expected})^2}{\text{estimated expected}} = \sum_{i=1}^{I} \sum_{j=1}^{J} \frac{(n_{ij} - \hat{e}_{ij})^2}{\hat{e}_{ij}}$$

Rejection region: $\chi^2 \geq \chi^2_{\alpha, (I-1)(J-1)}$

The test can safely be used provided that $\hat{e}_{ij} \geq 5$ for every cell.

Example 14.17 A study of the relationship between facility conditions at gasoline stations and aggressiveness in the pricing of gasoline ("An Analysis of Price Aggressiveness in Gasoline Marketing," *J. Marketing Research,* 1970, pp. 36–42) reported the accompanying data based on a sample of $n = 441$ stations. At level .01, does the data suggest that facility conditions and pricing policy are independent of one another?

Observed Pricing Policy

		Aggressive	*Neutral*	*Nonaggressive*	$n_{i\cdot}$
	Substandard	24	15	17	56
Condition	*Standard*	52	73	80	205
	Modern	58	86	36	180
	$n_{\cdot j}$	134	174	133	441

Expected Pricing Policy

17.02	22.10	16.89	56
62.29	80.88	61.83	205
54.69	71.02	54.29	180
134	174	133	441

Figure 14.15 Observed and estimated expected counts for Example 14.17

Thus

$$\chi^2 = \frac{(24 - 17.02)^2}{17.02} + \cdots + \frac{(36 - 54.29)^2}{54.29} = 22.47$$

and because $\chi^2_{.01, 4} = 13.277$, the hypothesis of independence is rejected.

We conclude that knowledge of a station's pricing policy does give information about the condition of facilities at the station. In particular, stations with an aggressive pricing policy appear more likely to have substandard facilities than stations with a neutral or nonaggressive policy. ■

Models and methods for analyzing data in which each individual is categorized with respect to three or more factors (multidimensional contingency tables) are discussed in several of the chapter references.

Exercises / Section 14.3 (19–30)

19. The accompanying data refers to leaf marks found on white clover samples selected from both long-grass areas and short-grass areas ("The Biology of the Leaf Mark Polymorphism in Trifolium Repens L.," *Heredity*, 1976, pp. 306–325). Use a χ^2 test to decide whether the true proportions of different marks are identical for the two types of regions.

	Type of mark					*Sample size*
	L	*LL*	*Y + YL*	*O*	*Others*	
Long-grass areas	409	11	22	7	277	726
Short-grass areas	512	4	14	11	220	761

20. The following data resulted from an experiment to study the effects of leaf removal on the ability of fruit of a certain type to mature ("Fruit Set, Herbivory, Fruit Reproduction, and the Fruiting Strategy of Catalpa Speciosa," *Ecology*, 1980, pp. 57–64).

Treatment	*Number of fruits matured*	*Number of fruits aborted*
Control	141	206
Two leaves removed	28	69
Four leaves removed	25	73
Six leaves removed	24	78
Eight leaves removed	20	82

Does the data suggest that the chance of a fruit maturing is affected by the number of leaves removed? State and test the appropriate hypotheses at level .01.

21. The article "Human Lateralization from Head to Foot: Sex-Related Factors" (*Science*, 1978, pp. 1291–1292) reported for both a sample of right-handed males and a sample of right-handed females the number of individuals whose feet were the same size, had a bigger left than right foot (a difference of half a shoe size or more), or had a bigger right than left foot.

	L > R	*L = R*	*L < R*	*Sample size*
Males	2	10	28	40
Females	55	18	14	87

Does the data indicate that sex has a strong effect on the development of foot asymmetry? State the appropriate null and alternative hypotheses, compute the value of χ^2, and place a bound or bounds on the *P*-value.

22. The paper "Susceptibility of Mice to Audiogenic Seizure Is Increased by Handling Their Dams During Gestation" (*Science*, 1976, pp. 427–428) reported on research into the effect of different injection treatments on the frequencies of audiogenic seizures.

Treatment	No response	Wild running	Clonic seizure	Tonic seizure
Thienylalanine	21	7	24	44
Solvent	15	14	20	54
Sham	23	10	23	48
Unhandled	47	13	28	32

Does the data suggest that the true percentages in the different response categories depend on the nature of the injection treatment? State and test the appropriate hypotheses using $\alpha = .005$.

23. The accompanying data on sex combinations of two recombinants resulting from six different male genotypes appeared in the article "A New Method for Distinguishing Between Meiotic and Premeiotic Recombinational Events in Drosophila Melonogaster" (*Genetics*, 1979, pp. 543–554). Does the data support the hypothesis that the frequency distribution among the three sex combinations is homogeneous with respect to the different genotypes? Define the parameters of interest, state the appropriate H_0 and H_a, and perform the analysis.

		Sex combination		
		M/M	M/F	F/F
	1.	35	80	39
	2.	41	84	45
Male	3.	33	87	31
genotype	4.	8	26	8
	5.	5	11	6
	6.	30	65	20

24. Each of 325 individuals participating in a certain drug program was categorized both with respect to the presence or absence of hypoglycemia and with respect to mean daily dosage of insulin ("Relation of Body Weight and Insulin Dose to the Frequency of Hypoglycemia," *J. Amer. Medical Assn.*, 1974, pp. 192–194). Does the accompanying data support the claim that the presence/absence of hypoglycemia is independent of insulin dosage? Test using $\alpha = .05$.

		Mean daily insulin dose				
		<.25	.25– .49	.50– .74	.75– .99	≥1.0
Hypo-glycemia condition	Present	4	21	28	15	12
	Absent	40	74	59	26	46

25. A random sample of individuals who drive alone to work in a large metropolitan area was obtained, and each individual was categorized with respect to both size of car and commuting distance. Does the accompanying data suggest that commuting distance and size of car are related in the population sampled? State the appropriate hypotheses and use a level .05 chi-squared test.

		Commuting distance		
		0–<10	10–<20	≥20
Size of car	Subcompact	6	27	19
	Compact	8	36	17
	Midsize	21	45	33
	Full-size	14	18	6

26. Each individual in a random sample of high school and college students was cross classified with respect to both political views and marijuana usage, resulting in the data displayed in the accompanying two-way table ("Attitudes About Marijuana and Political Views," *Psychological Reports*, 1973, pp. 1051–1054). Does the data support the hypothesis that political views and marijuana usage level are independent within the population? Test the appropriate hypotheses using level of significance .01.

		Usage level		
		Never	Rarely	Frequently
Political views	Liberal	479	173	119
	Conservative	214	47	15
	Other	172	45	85

27. Show that the chi-squared statistic for the test of independence can be written in the form

$$\chi^2 = \sum_{i=1}^{I} \sum_{j=1}^{J} \left(\frac{N_{ij}^2}{\hat{E}_{ij}} \right) - n$$

Why is this formula more efficient computationally than the defining formula for χ^2?

28. Suppose that in Exercise 26 each student had been categorized with respect to political views, marijuana usage, and religious preference, with the categories of this latter factor being Protestant, Catholic, and other. The data could be displayed in three different two-way tables, one corresponding to each category of the third factor. With $p_{ijk} = P$(political category i, marijuana category j, and religious category k), the null hypothesis of independence of all three factors states that $p_{ijk} = p_{i..} \cdot p_{.j.} \cdot p_{..k}$. Let n_{ijk} denote the observed frequency in cell (i, j, k). Show how to estimate the expected cell counts assuming that H_0 is true ($\hat{e}_{ijk} = n\hat{p}_{ijk}$, so the $\hat{p}_{ijk}$'s must be determined). Then use the general rule of thumb to determine the number of degrees of freedom for the chi-squared statistic.

29. Suppose that in a particular state consisting of four distinct regions, a random sample of n_k voters is obtained from the kth region for $k = 1, 2, 3, 4$. Each voter is then classified according to which of candidates 1, 2, or 3 he/she prefers and according to voter registration (1 = Dem.,

2 = Rep., 3 = Indep.). Let p_{ijk} denote the proportion of voters in region k who belong in candidate category i and registration category j. The null hypothesis of homogeneous regions is $H_0 : p_{ij1} = p_{ij2} = p_{ij3} = p_{ij4}$ for all i, j (that is, the proportion within each candidate-registration combination is the same for all four regions). Assuming that H_0 is true, determine $\hat{p}_{ijk}$ and $\hat{e}_{ijk}$ as functions of the observed n_{ijk}'s, and use the general rule of thumb to obtain the number of degrees of freedom for the chi-squared test.

30. Consider the accompanying 2×3 table displaying the sample proportions that fell in the various combinations of categories (for example, 13% of those in the sample were in the first category of both factors).

	1	2	3
1	.13	.19	.28
2	.07	.11	.22

a. Suppose that the sample consisted of $n = 100$ people. Use the chi-squared test for independence with significance level .10.

b. Repeat (a) assuming that the sample size was $n = 1000$.

c. What is the smallest sample size n for which these observed proportions would result in rejection of the independence hypothesis?

Supplementary Exercises / Chapter 14 (31–37)

31. The paper "Birth Order and Political Success" (*Psych. Reports,* 1971, pp. 1239–1242) reported that among 31 randomly selected candidates for political office who came from families with four children, 12 were firstborn, 11 were middleborn, and 8 were lastborn. Use this data to test the null hypothesis that a political candidate from such a family is equally likely to be in any one of the four ordinal positions.

32. The results of an experiment to assess the effect of crude oil on fish parasites were described in the paper "Effects of Crude Oils on the Gastrointestinal Parasites of Two Species of Marine Fish" (*J. Wildlife Diseases,* 1983, pp. 253–258).

Three treatments (corresponding to populations in the procedure described) were compared: (1) no contamination, (2) contamination by 1-year-old weathered oil, and (3) contamination by new oil. For each treatment condition, a sample of fish was taken and then each fish was classified as either parasitized or not parasitized. Data compatible with that in the paper is given. Does the data indicate that the three treatments differ with respect to the true proportion of parasitized and nonparasitized fish? Test using $\alpha = .01$.

Treatment	Parasitized	Nonparasitized
Control	30	3
Old oil	16	8
New oil	16	16

33. The accompanying two-way frequency table appeared in the paper "Marijuana Use in College" (*Youth and Society*, 1979, pp. 323–334). Four hundred and forty-five college students were classified according to both frequency of marijuana use and parental use of alcohol and psychoactive drugs. Does the data suggest that parental usage and student usage are independent in the population from which the sample was drawn? Use the *P*-value method to reach a conclusion.

Student level of marijuana use

		Never	Occasional	Regular
Parental use of alcohol and drugs	*Neither*	141	54	40
	One	68	44	51
	Both	17	11	19

34. In a study of 2989 cancer deaths, the location of death (home, acute-care hospital, or chronic-care facility) and age at death were recorded, resulting in the given two-way frequency table ("Where Cancer Patients Die," *Public Health Reports*, 1983, p. 173). Using a .01 significance level, test the null hypothesis that age at death and location of death are independent.

Location

		Home	Acute care	Chronic care
Age	15–54	94	418	23
	55–64	116	524	34
	65–74	156	581	109
	Over 74	138	558	238

35. Refer back to the data on fatigue crack propagation life given in Exercise 20 in Chapter 1. Carry out a test for normality based on the correlation coefficient between observed values and expected standard normal scores. Use a significance level of .01.

36. Many shoppers have expressed unhappiness over plans by grocery stores to stop putting prices on individual grocery items. The paper "The Impact of Item Price Removal on Grocery Shopping Behavior" (*J. Marketing*, 1980, pp. 73–93) reported on a study in which each shopper in a sample was classified by age and by whether or not he or she felt the need for item pricing. Based on the accompanying data, does the need for item pricing appear to be independent of age?

	Age				
	<30	30–39	40–49	50–59	≥60
Number in sample	150	141	82	63	49
Number who want item pricing	127	118	77	61	41

37. Let p_1 denote the proportion of successes in a particular population. The test statistic value in Chapter 8 for testing $H_0: p_1 = p_{10}$ was $z = (\hat{p}_1 - p_{10})/\sqrt{p_{10}p_{20}/n}$, where $p_{20} = 1 - p_{10}$. Show that for the case $k = 2$, the chi-squared test statistic value of Section 14.1 satisfies $\chi^2 = z^2$. *Hint:* First show that $(n_1 - np_{10})^2 = (n_2 - np_{20})^2$.

Bibliography

Everitt, B. S., *The Analysis of Contingency Tables,* Halsted Press, New York, 1977. A compact but informative survey of methods for analyzing categorical data, exposited with a minimum of mathematics.

Fienberg, Stephen, *The Analysis of Cross-Classified Categorical Data,* MIT Press, Cambridge, Mass., 1977. A good introduction to the analysis of multidimensional contingency tables.

Mosteller, Frederick, and Rourke, Richard, *Sturdy Statistics,* Addison-Wesley, Reading, Mass., 1973. Contains several very readable chapters on the varied uses of chi-square.

Distribution-Free Procedures

Introduction

When the underlying population or populations are nonnormal, the t and F tests and t confidence intervals of Chapters 7–13 will in general have actual levels of significance or confidence levels that differ from the nominal levels (those prescribed by the experimenter through the choice of, say, $t_{.025}$, $F_{.01}$, and so on) α and $100(1 - \alpha)\%$, although the difference between actual and nominal levels may not be large when the departure from normality is not too severe. Because the t and F procedures require the distributional assumption of normality, they are not "distribution-free" procedures—alternatively, because they are based on a particular parametric family of distributions (normal), they are not "nonparametric" procedures.

In this chapter we describe procedures which are valid [actual level α or confidence level $100(1 - \alpha)\%$] simultaneously for many different types of underlying distributions. Such procedures are called **distribution-free** or **nonparametric.** Sections 15.1 and 15.2 discuss two test procedures for analyzing a single sample of data; Section 15.3 presents a test procedure for use in two-sample problems. In Section 15.4 we develop distribution-free confidence intervals for μ and $\mu_1 - \mu_2$, while Section 15.5 describes distribution-free ANOVA procedures. These procedures are all competitors of the parametric (t and F) procedures described in earlier chapters, so it is important to compare the performance of the two types of procedures under both normal and nonnormal population models. Generally speaking, the distribution-free procedures perform almost as well as their t and F counterparts on the "homeground" of the normal distribution and will often yield a considerable improvement under nonnormal conditions.

15.1   The Sign Test

The sign test is a procedure for testing hypotheses about the median of a continuous distribution. Recall from Chapter 4 that the median $\tilde{\mu}$ of such a distribution is a number such that half the area under the density curve lies to the left of $\tilde{\mu}$ and half lies to the right of $\tilde{\mu}$. If X denotes the random variable whose distribution is under investigation, then $P(X \le \tilde{\mu}) = P(X \ge \tilde{\mu}) = .5$. The general null hypothesis will have the form $H_0 : \tilde{\mu} = \tilde{\mu}_0$. To test H_0 against one of the three commonly encountered alternative hypotheses, we will have available a random sample $X_1, \ldots, X_n$ from the distribution of interest.

In Chapter 8 the t test was used to test hypotheses about the mean of a normal population. Because any normal distribution is symmetric, $\tilde{\mu} = \mu$, so the sign test can also be used to test hypotheses about a normal mean. We shall comment shortly on the relative merits of the two test procedures. For now, we note that the sign test is valid whenever the distribution is continuous, whereas the t test was designed specifically for the normal distribution. Because the sign test can be used for a wide variety of underlying distributions, it is often referred to as a distribution-free test.

Testing $H_0 : \tilde{\mu} = 0$

When $\tilde{\mu} = 0$, any X_i is equally likely to be positive or negative. If, however, the true value of $\tilde{\mu}$ is much larger than zero, X_i is much more likely to be positive than negative (see Figure 15.1), so we would expect most of the observed X_i's to be positive in this case. When most x_i's are positive, we would suspect that $\tilde{\mu} > 0$ rather than $\tilde{\mu} = 0$.

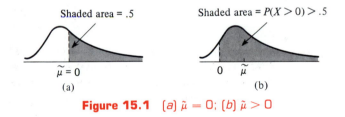

Shaded area = .5   Shaded area = $P(X > 0) > .5$

$\tilde{\mu} = 0$ 0 $\tilde{\mu}$

(a)   (b)

Figure 15.1   (a) $\tilde{\mu} = 0$; (b) $\tilde{\mu} > 0$

Define a test statistic Y by

Y = the number of X_i's for which $X_i > 0$

For testing $H_0 : \tilde{\mu} = 0$ versus $H_a : \tilde{\mu} > 0$, the sign test rejects H_0 when $y \ge c$ (when a large number of the x_i's are positive). The constant c should now be chosen to control the probability of a type I error (rejecting H_0 when H_0 is true). Regard the observation of each X_i as constituting a trial, so that the experiment consists of n identical trials. If we now identify a positive X_i with a success and a nonpositive X_i with a failure, then when H_0 is true,

$$p = P(\text{success}) = P(X_i > 0) = P(X_i > \tilde{\mu}) = .5$$

Since the trials are independent, we have

Proposition

> When $H_0 : \tilde{\mu} = 0$ is true, the statistic Y has a binomial distribution with parameters n and $p = .5$.

According to this proposition, a critical value c that ensures that $P(\text{reject } H_0 \text{ when } H_0 \text{ is true}) = \alpha$ for a specified α can be found from the binomial distribution with appropriate n and $p = .5$.

Example 15.1

The use of a particular cloud-seeding technique for inducing rain in a series of 15 independent experiments performed under similar climatic conditions yielded the following observations on the net gain in inches of rain after the seeding: $-.30$, 1.50, $.41$, $-.11$, 2.03, $.28$, $.76$, $-.29$, -1.14, $.09$, $-.92$, $.71$, $.36$, $.25$, $-.50$. Does this data indicate that the cloud-seeding technique is successful in its objective under the given conditions?

Let X represent the net gain in inches resulting from one particular seeding experiment, and let $\tilde{\mu}$ denote the median of the distribution of X (assumed continuous). Then we shall say that seeding is successful if $\tilde{\mu} > 0$, so that more than 50% of all seeding experiments result in a positive gain in rainfall. Because seeding is expensive, we will be convinced of its validity only if the data strongly indicates that $\tilde{\mu} > 0$, so we test $H_0 : \tilde{\mu} = 0$ versus $H_a : \tilde{\mu} > 0$.

The test statistic Y has a binomial distribution with $n = 15$ and $p = .5$ when H_0 is true. From the binomial tables, $P(Y \geq 12) = 1 - B(11; 15, .5) = .018$ while $P(Y \geq 13) = .004$. Thus a test with level of significance approximately .02 rejects H_0 if $y \geq 12$. Since only nine of the 15 x_i's in the sample are positive, the observed value of Y is $y = 9$, which is not in the rejection region. At the chosen level of significance, H_0 cannot be rejected, so we conclude that seeding is not successful. The P-value is $P(Y \geq 9) = 1 - B(8; 15, .5) = .304$. ∎

The test statistic Y has a discrete distribution (binomial), so for any given level α there will not in general be a critical value that yields a test with exactly the desired level. In Example 15.1, there was no value of c yielding exactly level .01, so we used a test with a slightly higher level of significance. When the distribution of a test statistic is discrete, as is usually the case with a distribution-free procedure, the rejection region is chosen to yield an α as close to the desired level as possible.

The Sign Test for $H_0 : \tilde{\mu} = \tilde{\mu}_0$

A minor modification of the test procedure for $\tilde{\mu}_0 = 0$ yields a procedure appropriate for any other null value. The new test statistic is the number of positive $(X_i - \tilde{\mu}_0)$'s, or equivalently the number of sample observations that exceed $\tilde{\mu}_0$.

Null hypothesis: $H_0: \tilde{\mu} = \tilde{\mu}_0$

Test statistic value: $y =$ the number of x_i's that exceed $\tilde{\mu}_0$

Alternative hypothesis	Rejection region
$H_a: \tilde{\mu} > \tilde{\mu}_0$	$y \geq c_1$ (large y)
$H_a: \tilde{\mu} < \tilde{\mu}_0$	$y \leq c_2$ (small y)
$H_a: \tilde{\mu} \neq \tilde{\mu}_0$	either $y \geq c$ or $y \leq n - c$ (small or large y)

where the critical values c_1, c_2, and c are obtained from the binomial distribution Bin$(n, .5)$ to yield the desired α.

Example 15.2 An electronics company will shortly begin to market a kit for a CB radio. The company wishes to claim that the kit can be assembled quite easily even by someone with no previous electronics experience. Let X be the assembly time for a randomly selected person of this type, with $f(x)$ denoting the distribution of X and $\tilde{\mu}$ the median of this distribution. It is decided that the claim will be validated only if experimental evidence strongly suggests that at least half of all such individuals can assemble the kit in less than 20 hours. The null hypothesis is then $H_0: \tilde{\mu} = 20$, while the appropriate alternative is $H_a: \tilde{\mu} < 20$; this choice of H_a ensures that the burden of proof lies with the claim of easy assembly.

With a sample size of $n = 15$, the test that rejects H_0 when Y (the number among the 15 assembly times that exceed 20) is 0, 1, 2, or 3 has $\alpha = .018$. For the sample $x_1 = 18.5$, $x_2 = 19.1$, $x_3 = 22.2$, $x_4 = 16.9$, $x_5 = 19.6$, $x_6 = 18.8$, $x_7 = 19.8$, $x_8 = 19.0$, $x_9 = 18.3$, $x_{10} = 19.2$, $x_{11} = 21.6$, $x_{12} = 17.5$, $x_{13} = 18.9$, $x_{14} = 18.0$, and $x_{15} = 19.4$, the test statistic has value $y = 2$. Since 2 falls in the rejection region for a level .018 test, H_0 is rejected at this level in favor of the conclusion that $\tilde{\mu} < 20$. ∎

The Normal Approximation

When $p = .5$, the binomial distribution can be approximated by a normal distribution for n as small as 10. Since $\mu_Y = np = .5n$ and $\sigma_Y^2 = np(1 - p) = .25n$, the test statistic

$$Z = \frac{Y - .5n}{.5\sqrt{n}}$$

has approximately a standard normal distribution when H_0 is true and $n \geq 10$. The upper-tailed test then rejects H_0 in favor of $H_a: \tilde{\mu} > \tilde{\mu}_0$ when $z \geq z_\alpha$, and rejection regions for the other two standard alternatives are identical to those of previous lower- and two-tailed z tests.

Example 15.3 Each of 20 randomly selected homeowners in a particular city was asked to reveal the amount by which his or her property tax bill had increased as a result of the most recent reassessment. The sample (dollar) amounts were

342, 176, 517, 296, 312, 143, 279, 412, 228, 209, 195, 241, 211, 285, 329, 137, 188, 260, 233, 357

The county assessor has stated that the median increase for all county homeowners was \$300. Does the sample data strongly indicate that the true median increase for residents of this city was less than that for the county as a whole? Use the sign test with $\alpha = .01$.

The null hypothesis here is $H_0 : \tilde{\mu} = 300$ and the alternative is $H_a : \tilde{\mu} < 300$, where $\tilde{\mu}$ is the median increase for city homeowners. The number of sample values that exceed 300 is $y = 6$, so

$$z = \frac{y - .5n}{.5\sqrt{n}} = \frac{6 - (.5)(20)}{.5\sqrt{20}} = -1.79$$

The P-value for a lower-tailed z test is $\Phi(z) = \Phi(-1.79) = .0367$. Since $.01 < .0367$, H_0 cannot be rejected at significance level .01. ∎

β for the Sign Test

For testing $H_0 : \tilde{\mu} = \tilde{\mu}_0$, as long as the underlying distribution is continuous the sign test will control the probability of a type I error at the desired level α. To assess the ability of the test procedure to draw correct conclusions, we should also examine probabilities of type II errors. The test should be able to detect departures from the null hypothesis as evidenced by small β for alternatives of interest.

The difficulty with β calculations for the sign test is that we must assume not only a particular alternative value of $\tilde{\mu}$, but also a specific shape for the underlying density function. For example, if $\tilde{\mu} = \tilde{\mu}_0 + 1$ rather than $\tilde{\mu}_0$, β will also depend on whether the distribution assumed for the calculation is normal, uniform, exponential, or whatever. However, *for any particular underlying distribution, the test statistic Y is a binomial random variable,* but if $\tilde{\mu} \neq \tilde{\mu}_0$ then p will be something other than .5. Thus all β calculations reduce to binomial probability computations.

Example 15.4
(Example 15.1
continued)

Suppose that H_0 is false because $\tilde{\mu} = 1$, and consider the two possible distributions of Figure 15.2.

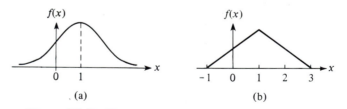

(a)   (b)

Figure 15.2   Two distributions for which $H_0 : \tilde{\mu} = 0$ is false: (a) normal, $\tilde{\mu} = 1$, $\sigma = 1$; (b) symmetric triangular, $\tilde{\mu} = 1$

For the normal distribution, the probability of a positive net change is

$$p = P(X_i > 0) = P(Z > -1) = 1 - \Phi(-1) = .8413$$

Thus Y has a binomial distribution with $n = 15$ and $p = .8413$. Using the rejection region $\{y \geq 12\} = \{12, 13, 14, 15\}$, the probability of rejecting H_0 is

$$\beta(1) = P(Y \leq 11 \quad \text{when} \quad Y \sim \text{Bin}(15, .8413))$$

$$= \sum_{y=0}^{11} \binom{15}{y}(.8413)^y(.1587)^{15-y} = .2050$$

If the underlying distribution is triangular as in Figure 15.2b, then a geometrical argument yields $p = .875$, so that

$$\beta(1) = P(Y \leq 11 \quad \text{when} \quad Y \sim \text{Bin}(15, .875)) = .1079 \qquad \blacksquare$$

The above calculations show that β for the sign test depends not just on the alternative value of $\tilde{\mu}$ but on the area under the density curve to the right of $\tilde{\mu}_0$. This area in turn depends on the particular shape of the density function.

Comparison of the Sign and t Tests

If it is assumed that the underlying distribution is normal, then either the sign test or the t test can be used to test $H_0 : \tilde{\mu} = \tilde{\mu}_0$ against one of the usual alternative hypotheses. In this case the t test is known to have minimum type II error probabilities among all level α tests, so is certainly a better test than the sign test. When the population distribution is symmetric but not normal and μ is finite (so $\tilde{\mu} = \mu$), β for the sign test will be larger than for the t test unless the distribution has very heavy tails compared to the normal distribution. The test procedure to be presented in the next section is better than the sign test and compares very favorably to the t test in symmetric situations. The sign test is thus better regarded as a procedure for drawing conclusions about a population median than as a serious competitor to the t test.

Exercises / Section 15.1 (1–7)

1. Determine the median of
   a. a uniform distribution on the interval $[a, b]$.
   b. an exponential distribution with parameter λ.

2. The July 3, 1978, *Los Angeles Times* reported that under conditions of the desegregation plan to be implemented, any students who travel more than 45 minutes one way on a bus are required to have only three years of integrated schooling, while those traveling at most 45 minutes face a five-year requirement. Let $\tilde{\mu}$ be the median of the distribution of travel time between two particular locations, and suppose that a random sample of 20 travel times consists of

   47.3, 44.1, 44.5, 43.7, 46.2, 45.3, 47.0, 42.5, 43.9, 45.6, 46.0, 44.8, 45.7, 47.8, 46.6, 44.2, 47.2, 46.5, 43.6, 45.4

   Use the sign test at level .10, with $Y =$ the number of times that exceed 45 minutes, to test $H_0 : \tilde{\mu} = 45$ versus $H_a : \tilde{\mu} > 45$.

3. Ten soil samples taken from a particular region were subjected to chemical analysis to determine the pH of each sample. The sample pH's were found to be

   5.93, 6.08, 5.86, 5.91, 6.12, 5.90, 5.95, 5.89, 5.98, 5.96

   It had previously been believed that the median soil pH in this region was 6.0. Does this sample data strongly indicate that the true median pH is something other than 6.0? Let $Y =$ the number of pH's in the sample that exceed 6.0, and use the sign test at level .05 to test the appropriate hypotheses.

4. Suppose that the distribution of times between successive alarms at a fire station is exponential, and we wish to test $H_0: \tilde{\mu} \leq 3$ hrs versus $H_a: \tilde{\mu} > 3$, where $\tilde{\mu}$ is the median time between successive alarms. Suppose that a sample of 10 such times is obtained, let $Y =$ the number of times that exceed 3 hours, and consider using the sign test with rejection region $\{8, 9, 10\}$. What is the probability of
   a. a type I error when $\lambda = .231$ (λ is the parameter of the exponential distribution, which is related to $\tilde{\mu}$ by the result of Exercise 1(b))?
   b. a type II error if $\lambda = .1189$?

5. In the manufacture of a certain type of cylindrical rod, a lathe is periodically reset to the desired diameter of the rod. A production supervisor has recently become concerned over a possible tendency for the lathes to produce rods with larger and larger diameters immediately subsequent to setting (an upward trend in observed diameters). A sample of 12 rods is obtained and the measured diameters are

   $x_1 = 4.935$, $x_2 = 4.967$, $x_3 = 4.989$, $x_4 = 5.012$, $x_5 = 5.020$, $x_6 = 4.997$, $x_7 = 5.025$, $x_8 = 5.041$, $x_9 = 5.008$, $x_{10} = 5.035$, $x_{11} = 5.016$, and $x_{12} = 5.022$

   a. Let $p =$ the probability that the $(i + 6)$th diameter X_{i+6} exceeds the ith diameter X_i. If there is actually no upward or downward trend in diameters, so that X_{i+6} is no more or less likely to exceed X_i than it is to fall below X_i, what is the value of p?
   b. If $Y =$ the number of values of i for which $X_{i+6} > X_i$, and there is no upward or downward trend in diameters, what probability distribution does Y have?
   c. Use the result of (b) to test H_0: there is no trend versus H_a: there is an upward trend. Your test statistic should be Y, and you should use a level .1 (approximately) test. This simple

test for trends, a modification of the sign test, is due to Cox and Stuart.

6. Suppose that the underlying population is normal with $\sigma = 1$. For testing $H_0: \mu = 0$ versus $H_a: \mu > 0$ based on a sample of size $n = 15$, consider the z test that rejects H_0 if $z \geq 2.10$ (where $Z = (\overline{X} - \mu_0)/(\sigma/\sqrt{n}) = \sqrt{n}\,\overline{X}$).
   a. What is α for this test?
   b. When $\mu = 1$, how does β for this test compare to β for the sign test that rejects H_0 if $y \geq 12$?

7. A process for producing stainless steel bars was originally designed so that the nickel content of the bars was 10% by weight. A potential customer feels that the bars will be suitable for a particular application only if more than 75% of all bars produced contain at least 10% nickel (that is, that fewer than 25% of all bars contain less than 10% nickel). The customer will not agree to purchase the bars unless experimental evidence strongly indicates that this requirement is satisfied.

   If $f(x)$ denotes the density function of nickel content in a randomly selected bar, then the requirement is met only if the area under the graph of $f(x)$ to the right of 10 is more than .75. Equivalently, if θ denotes the twenty-fifth percentile of the nickel-content distribution, then we wish to test $H_0: \theta = 10$ versus $H_a: \theta > 10$ (so H_a states that more than 75% of the area under the density curve lies to the right of 10). Suppose that 20 bars are sampled and the observed values of nickel content are

   10.23, 10.95, 11.02, 10.14, 9.83, 10.30, 12.05, 11.26, 10.58, 10.18, 10.38, 11.49, 9.52, 10.86, 11.00, 10.19, 10.94, 11.13, 10.25, 11.02

   Develop a test procedure for H_0 versus H_a and carry out the test for this data. Your α should be as close to .10 as possible. *Hint:* Let $Y =$ the number of X_i's that exceed 10. If H_0 is true, what is the distribution of Y?

15.2 The Wilcoxon Signed-Rank Test

Consider again the problem of testing a hypothesis about the median $\tilde{\mu}$ of a continuous distribution (that is, population). If we wish to test $H_0: \tilde{\mu} = 0$ versus $H_a: \tilde{\mu} > 0$ using the sign test on the basis of ten observations from the population, then the rejection region $R = \{8, 9, 10\}$ has type I error probability $\alpha =$

.055. Suppose that our sample consists of observations $x_1 = -0.19$, $x_2 = 2.17$, $x_3 = 2.68$, $x_4 = -.57$, $x_5 = 2.02$, $x_6 = 2.46$, $x_7 = 3.02$, $x_8 = -.05$, $x_9 = .76$, and $x_{10} = 1.30$. Then the observed value of the test statistic Y, the number of positive observations, is $y = 7$, so at level $\alpha = .055$ we would not reject H_0 in favor of H_a.

A closer look at the data reveals that while there are indeed three negative observations, these three observations are quite small in magnitude compared with the seven positive observations. That is, x_1, x_4, and x_8 are barely negative, while the positive observations are relatively distant from zero. The nonrejection of H_0 for this data is due to the way in which the sign test extracts information from the sample; magnitudes of the observations are discarded and only the sign of each observation is retained. In a sense this is the price that must be paid for having a test that is valid for a very large class of populations (*all* continuous populations).

Symmetric Distributions and a New Test Statistic

Frequently an experimenter may be willing to assume more about the population of interest than just that it is continuous, and in particular may be willing to assume that the population distribution has a symmetrical shape. A population distribution is **symmetric** if the graph of the p.d.f. to the left of $\tilde{\mu}$ is a reflection or mirror image of the graph to the right of $\tilde{\mu}$. The graphs of several symmetric p.d.f.'s are pictured in Figure 15.3.

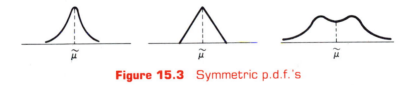

Figure 15.3   Symmetric p.d.f.'s

If we assume that the population yielding the above sample is symmetric, then the conclusion that $\tilde{\mu} = 0$ (actually ≤ 0) seems quite suspect, for the data suggests strongly that the point of symmetry lies to the right of zero ($\tilde{x} = 1.66$, which is much larger in magnitude than any of the negative observations). We now assume that *our sample comes from a continuous symmetric probability distribution,* and develop a test for making an inference about $\tilde{\mu}$ that is different from the sign test. When this new test is applied to the above sample, the data will argue strongly for the rejection of H_0 in favor of H_a. This is because the new test will take into account the magnitudes of the observations as well as the signs.

Figure 15.4 shows two different symmetric p.d.f.'s, one for which H_0 is true and one for which H_a is true. When H_0 is true then we expect that the negative observations in the sample should be comparable in magnitude to the magnitudes of the positive observations. If, however, H_0 is "grossly" untrue (Figure 15.4b), then observations of large absolute magnitude will tend to be positive rather than negative. For the ten observations in our sample, suppose we proceed as follows:

1. Disregarding the signs of the observations, rank the collection of 10 numbers in order of increasing magnitude.
2. Use as a test statistic $S_+ =$ sum of the ranks associated with the positive observations.

Absolute magnitude:	.05	.19	.57	.76	1.30	2.02	2.17	2.46	2.68	3.02
Rank:	1	2	3	4	5	6	7	8	9	10
Signed rank:	−1	−2	−3	4	5	6	7	8	9	10

The observed value of S_+ is $s_+ = 4 + 5 + 6 + 7 + 8 + 9 + 10 = 49$. H_0 is now rejected in favor of $H_a : \tilde{\mu} > 0$ when s_+ is too large because a large value of s_+ indicates that most of the observations with large absolute magnitude are positive, which in turn indicates a median greater than zero. If on the other hand, observations with negative signs are intermingled with positive observations when signs are disregarded, then the observed s_+ will not be very large.

(a)   (b)

Figure 15.4

The Distribution of S_+

To decide between H_0 and H_a, we must know whether 49 is a "surprisingly large" value of S_+ when the null hypothesis is true. This entails obtaining the probability distribution of S_+ when H_0 is true, and selecting as the rejection region a set of large s_+ values with small probability under H_0 (small α). If our observed s_+ is in the rejection region, we can then reject H_0 at the desired level α, and otherwise not reject H_0.

The probability distribution of S_+ is more difficult to obtain than was the distribution for the sign test. To illustrate, consider the case $n = 5$. The key observation in the derivation is this: When H_0 is true, *any* collection of signed ranks is *just as* likely as any other collection. Worded differently, the smallest observation in absolute magnitude is just as likely to be positive as negative, the second smallest observation in absolute magnitude is just as likely to be positive as negative, and similarly for the other observations. Thus for $n = 5$ the sequence $−1, +2, +3, −4, +5$ of signed ranks is just as likely to be observed (when H_0 is true) as is the sequence $−1, −2, −3, −4, +5$, or any other sequence. Table 15.1 lists the possible signed rank sequences for $n = 5$ and the s_+ value associated with each sequence. Since there are 32 such sequences (two possible signs for the smallest observation, two for the second smallest, and so on, or $2 \cdot 2 \cdot 2 \cdot 2 \cdot 2 = 32$ sequences), each sequence has probability $1/32$ of being observed when H_0 is true. From this the p.m.f. of S_+ under H_0 can immediately be obtained as in Table 15.2. Note that for $n = 10$ observations there are $2^{10} = 1024$ possible signed rank sequences, and to list these sequences would be very tedious. Each sequence, though, would receive probability $1/1024$ under H_0, from which the null distribution (distribution when H_0 is true) of S_+ can be easily obtained.

Table 15.1 Possible signed rank sequences for $n = 5$

Sequence	s_+	Sequence	s_+
-1 -2 -3 -4 -5	0	-1 -2 -3 $+4$ -5	4
$+1$ -2 -3 " "	1	$+1$ -2 -3 " "	5
-1 $+2$ -3 " "	2	-1 $+2$ -3 " "	6
-1 -2 $+3$ " "	3	-1 -2 $+3$ " "	7
$+1$ $+2$ -3 " "	3	$+1$ $+2$ -3 " "	7
$+1$ -2 $+3$ " "	4	$+1$ -2 $+3$ " "	8
-1 $+2$ $+3$ " "	5	-1 $+2$ $+3$ " "	9
$+1$ $+2$ $+3$ " "	6	$+1$ $+2$ $+3$ " "	10
-1 -2 -3 -4 $+5$	5	-1 -2 -3 $+4$ $+5$	9
$+1$ -2 -3 " "	6	$+1$ -2 -3 " "	10
-1 $+2$ -3 " "	7	-1 $+2$ -3 " "	11
-1 -2 $+3$ " "	8	-1 -2 $+3$ " "	12
$+1$ $+2$ -3 " "	8	$+1$ $+2$ -3 " "	12
$+1$ -2 $+3$ " "	9	$+1$ -2 $+3$ " "	13
-1 $+2$ $+3$ " "	10	-1 $+2$ $+3$ " "	14
$+1$ $+2$ $+3$ " "	11	$+1$ $+2$ $+3$ " "	15

Table 15.2 Null distribution of S_+ when $n = 5$

s_+	0	1	2	3	4	5	6	7
$p(s_+)$	1/32	1/32	1/32	2/32	2/32	3/32	3/32	3/32
s_+	8	9	10	11	12	13	14	15
$p(s_+)$	3/32	3/32	3/32	2/32	2/32	1/32	1/32	1/32

As an example of how the probabilities in Table 15.2 were obtained, consider the S_+ value 7; there are three signed rank sequences for which $S_+ = 7$, and each receives probability $1/32$ under H_0, so that $P(S_+ = 7$ when H_0 is true$) = 3/32$.

Table 15.2 can now be used to specify a rejection region for $H_0 : \tilde{\mu} = 0$ versus $H_a : \tilde{\mu} > 0$ for which α can be calculated. Consider the rejection region $R = \{13, 14, 15\}$. Then

$$\alpha = P(\text{reject } H_0 \text{ when } H_0 \text{ is true}) = P(S_+ = 13, 14, \text{ or } 15 \text{ when } H_0 \text{ is true})$$

$$= 1/32 + 1/32 + 1/32 = 3/32 = .094 \text{ from Table 15.2}$$

so that $R = \{13, 14, 15\}$ specifies a test with approximate level .1. For the rejection region $\{14, 15\}$, $\alpha = 2/32 = .063$. For the data set $x_1 = .58$, $x_2 = 2.50$, $x_3 = -.21$, $x_4 = 1.23$, $x_5 = .97$, the signed rank sequence is -1, $+2$, $+3$, $+4$, $+5$, so $s_+ = 14$ and at level .063 H_0 would be rejected.

A General Description of the Test

Because the underlying distribution is assumed symmetric, $\mu = \tilde{\mu}$, so we shall state the hypotheses of interest in terms of μ rather than $\tilde{\mu}$.*

Assumption

$X_1, X_2, \ldots, X_n$ is a random sample from a continuous and symmetric probability distribution with mean (and median) μ.

When the hypothesized value of μ is μ_0, the absolute differences $|x_1 - \mu_0|, \ldots,$ $|x_n - \mu_0|$ must be ranked from smallest to largest.

Null hypothesis: $H_0 : \mu = \mu_0$

Test statistic value: $s_+ =$ the sum of the ranks associated with positive $(x_i - \mu_0)$'s

Alternative hypothesis	Rejection region for level α test
$H_a : \mu > \mu_0$	$s_+ \geq c_1$
$H_a : \mu < \mu_0$	$s_+ \leq c_2$ [where $c_2 = n(n + 1)/2 - c_1$]
$H_a : \mu \neq \mu_0$	either $s_+ \geq c$ or $s_+ \leq n(n + 1)/2 - c$

where the critical values c_1 and c are obtained from Appendix Table A.9 and satisfy $P(S_+ \geq c_1) \approx \alpha$ and $P(S_+ \geq c) \approx \alpha/2$ when H_0 is true.

Example 15.5

A manufacturer of electric irons, wishing to test the accuracy of the thermostat control at the 500° F setting, instructs a test engineer to obtain actual temperatures at that setting for 15 irons using a thermocouple. The resulting measurements are

494.6, 510.8, 487.5, 493.2, 502.6, 485.0, 495.9, 498.2, 501.6, 497.3, 492.0, 504.3, 499.2, 493.5, 505.8

The engineer feels that it is reasonable to assume that a temperature deviation from 500° of any particular magnitude is just as likely to be positive as negative (the assumption of symmetry), but wants to protect against possible nonnormality of the actual temperature distribution, so decides to use the Wilcoxon signed-rank test to see whether the data strongly suggests incorrect calibration of the iron.

The hypotheses are $H_0 : \mu = 500$ versus $H_a : \mu \neq 500$, where $\mu =$ the true average actual temperature at the 500° F setting. Subtracting 500 from each x_i gives

---

*If the tails of the distribution are "too heavy," as was the case with the Cauchy distribution mentioned in Chapter 6, then μ will not exist. In such cases the Wilcoxon test will still be valid for tests concerning $\tilde{\mu}$.

-5.6, 10.8, -12.5, -6.8, 2.6, -15.0, -4.1, -1.8, 1.6, -2.7, -8.0, 4.3, $-.8$, -6.5, 5.8

The ranks are obtained by ordering these from smallest to largest without regard to sign:

Absolute magnitude:	.8	1.6	1.8	2.6	2.7	4.1	4.3	5.6	5.8	6.5	6.8	8.0	10.8	12.5	15.0
Rank:	1	2	3	4	5	6	7	8	9	10	11	12	13	14	15
Sign:	−	+	−	+	−	−	+	−	+	−	−	−	+	−	−

Thus $s_+ = 2 + 4 + 7 + 9 + 13 = 35$. From Appendix Table A.9, $P(S_+ \geq 95) = P(S_+ \leq 25) = .026$ when H_0 is true, so the approximately level .05 two-tailed test rejects when either s_+ is ≥ 95 or ≤ 25 (the exact α is $2(.026) = .052$). Since $s_+ = 35$ is not in the rejection region, it cannot be concluded at level .05 that μ is anything other than 500. Even at level .094 (approximately .1), H_0 is not rejected, since $P(S_+ \leq 30) = .047$ implies that s_+ values between 30 and 90 are not significant at that level. The *P*-value of the data is thus greater than .1. ■

Paired Observations

When the data consisted of pairs $(X_1, Y_1), \ldots, (X_n, Y_n)$, and the differences $D_1 = X_1 - Y_1, \ldots, D_n = X_n - Y_n$ were normally distributed, in Chapter 9 we used a paired t test to test hypotheses about the expected difference μ_D. If normality is not assumed, hypotheses about μ_D can be tested by using the Wilcoxon signed-rank test on the D_i's provided that the distribution of the differences is continuous and symmetric. *If X_i and Y_i both have continuous distributions that differ only with respect to their means* (so the Y distribution is the X distribution shifted by $\mu_1 - \mu_2 = \mu_D$), *then D_i will have a continuous symmetric distribution* (it is *not* necessary for the X and Y distributions to be symmetric individually). The null hypothesis is $H_0 : \mu_D = \Delta_0$, and the test statistic S_+ is the sum of the ranks associated with the positive $(D_i - \Delta_0)$'s.

Example 15.6   An article in the 1949 *Journal of Applied Physiology* ("Interval of Useful Consciousness at Various Altitudes") reported on the effect of a transfusion of a large amount of blood on the ability to perform a simple task at high altitude. The high-altitude condition (35,000 ft) was achieved by using a gas mask. Suppose it is believed that more blood would increase an individual's interval of useful consciousness following interruption of normal oxygen supply by an average of at most five seconds. Does the (slightly modified) data in Table 15.3 contradict this belief?

Since a five-second increase corresponds to $\mu_D = -5$, we wish to test $H_0 : \mu_D = -5$ versus $H_a : \mu_D < -5$.

Table 15.3

Subject	1	2	3	4	5
Seconds of useful consciousness before transfusion	38	65	68	59	67
Seconds of useful consciousness after transfusion	45	77	81	70	73
d_i	-7	-12	-13	-11	-6
$d_i - (-5)$	-2	-7	-8	-6	-1
Rank of $\lvert d_i - (-5) \rvert$	2	4	5	3	1
Sign	$-$	$-$	$-$	$-$	$-$

Since all signs are negative, the computed value of S_+ is $s_+ = 0$. From Appendix Table A.9 or Table 15.2, the test which rejects H_0 when $s_+ = 0$ has level .031, so the value 0 is significant at level .05. H_0 is thus rejected in favor of the conclusion that a transfusion increases time of useful consciousness by more than five seconds on average. ■

Ties in the Wilcoxon Test

Although a theoretical implication of the continuity of the underlying distribution is that ties will not occur, in practice they often do because of the discreteness of measuring instruments. If there are several data values with the same absolute magnitude, then they would be assigned the average of the ranks they would receive if they differed very slightly from one another. For example, if in Example 15.5 $x_8 = 498.2$ is changed to 498.4, then two different ($x_i - 500$)'s would have absolute magnitude 1.6. The ranks to be averaged would be 2 and 3, so each would be assigned rank 2.5.

A Large-Sample Approximation

Appendix Table A.9 provides critical values for level α tests only when $n \le 20$. For $n > 20$, it can be shown that S_+ has approximately a normal distribution with

$$\mu_{S_+} = \frac{n(n+1)}{4}, \quad \sigma^2_{S_+} = \frac{n(n+1)(2n+1)}{24} \quad \text{when } H_0 \text{ is true}$$

The mean and variance result from noting that when H_0 is true (the symmetric distribution is centered at μ_0), then the rank i is just as likely to receive a $+$ sign as it is to receive a $-$ sign. Thus

$$S_+ = W_1 + W_2 + W_3 + \cdots + W_n$$

where

$$W_1 = \begin{cases} +1 & \text{with probability } \dfrac{1}{2} \\ 0 & \text{with probability } \dfrac{1}{2} \end{cases}$$

$$W_n = \begin{cases} +n & \text{with probability } \dfrac{1}{2} \\ \\ 0 & \text{with probability } \dfrac{1}{2} \end{cases}$$

($W_i = 0$ is equivalent to rank i being associated with a $-$, so i does not contribute to S_+.)

S_+ is then a sum of random variables, and when H_0 is true, these W_i's can be shown to be independent. Application of the rules of expected value and variance gives the mean and variance of S_+. Because the W_i's are not identically distributed, our version of the Central Limit Theorem cannot be applied, but there is a more general version of the theorem that can be used to justify the normality conclusion.

The large-sample test statistic is now given by

$$Z = \frac{S_+ - n(n+1)/4}{\sqrt{n(n+1)(2n+1)/24}} \tag{15.1}$$

For the three standard alternatives, the critical values for level α tests are the usual standard normal values z_α, $-z_\alpha$, and $\pm z_{\alpha/2}$.

Example 15.7 A particular type of steel beam has been designed to have a compressive strength (pounds per square inch) of at least 50,000. For each beam in a sample of 25 beams, the compressive strength was determined and is given below. Assuming that actual compressive strength is distributed symmetrically about the true average value, use the Wilcoxon test to decide if the true average compressive strength is less than the specified value. That is, test $H_0 : \mu = 50,000$ versus $H_a : \mu < 50,000$ (favoring the claim that average compressive strength is at least 50,000).

Table 15.4

$x_i - 50,000$	Signed rank	$x_i - 50,000$	Signed rank	$x_i - 50,000$	Signed rank
-10	-1	-99	-10	165	$+18$
-27	-2	113	$+11$	-178	-19
36	$+3$	-127	-12	-183	-20
-55	-4	-129	-13	-192	-21
73	$+5$	136	$+14$	-199	-22
-77	-6	-150	-15	-212	-23
-81	-7	-155	-16	-217	-24
90	$+8$	-159	-17	-229	-25
-95	-9				

The sum of the positively signed ranks is $3 + 5 + 8 + 11 + 14 + 18 = 59$, $n(n + 1)/4 = 162.5$ and $n(n + 1)(2n + 1)/24 = 1381.25$, so

$$z = \frac{59 - 162.5}{\sqrt{1381.25}} = -2.78$$

The lower-tailed level .01 test rejects H_0 if $z \le -2.33$. Since -2.78 is ≤ -2.33, H_0 is rejected in favor of the conclusion that true average compressive strength is less than 50,000.   ■

When there are ties in the absolute magnitudes, so that average ranks must be used, it is still correct to standardize S_+ by subtracting $n(n + 1)/4$, but the following corrected formula for variance should be used:

$$\sigma_{S_+}^2 = \frac{1}{24} n(n + 1)(2n + 1) - \frac{1}{48} \Sigma(\tau_i - 1)(\tau_i)(\tau_i + 1) \tag{15.2}$$

where τ_i is the number of ties in the ith set of tied values and the sum is over all sets of tied values. If, for example, $n = 10$ and the signed ranks are 1, 2, -4, -4, 4, 6, 7, 8.5, 8.5, and 10, then there are two tied sets with $\tau_1 = 3$ and $\tau_2 = 2$, so the summation is $(2)(3)(4) + (1)(2)(3) = 30$ and (15.2) becomes $96.25 - 30/48 = 95.62$. The denominator in (15.1) should be replaced by the square root of (15.2), though as this example shows, the correction is usually insignificant.

Efficiency of the Wilcoxon Test

When the underlying distribution being sampled is normal, either the t test or the signed-rank test can be used to test a hypothesis about μ. The t test is the best test in such a situation because among all level α tests it is the one having minimum β. Since it is generally agreed that there are many experimental situations in which normality can be reasonably assumed, as well as some in which it should not be, there are two questions that must be addressed in an attempt to compare the two tests:

1. When the underlying distribution is normal (the "home ground" of the t test), how much is lost by using the signed-rank test?
2. When the underlying distribution is not normal, can a significant improvement be achieved by using the signed-rank test?

If the Wilcoxon test does not suffer much with respect to the t test on the "home ground" of the latter, and performs significantly better than the t test for a large number of other distributions, then there will be a strong case for using the Wilcoxon test.

Unfortunately there are no simple answers to the two questions. Upon reflection, it is not surprising that the t test can perform poorly when the underlying distribution has "heavy tails" (that is, when observed values lying far from

μ are relatively more likely than they are when the distribution is normal). This is because the t test depends in its behavior on the sample mean, which can be very unstable in the presence of heavy tails. The difficulty in producing answers to the two questions is that β for the Wilcoxon test is very difficult to obtain and study for *any* underlying distribution, and the same can be said for the t test when the distribution is not normal. Even if β were easily obtained, any measure of efficiency would clearly depend on which underlying distribution was postulated. A number of different efficiency measures have been proposed by statisticians; one which many statisticians regard as credible is called **asymptotic relative efficiency** (ARE). The ARE of one test with respect to another is essentially the limiting ratio of sample sizes necessary to obtain identical error probabilities for the two tests. Thus if the ARE of one test with respect to a second equals .5, then when sample sizes are large, twice as large a sample size will be required of the first test in order that it perform as well as the second test. Although the ARE does not characterize test performance for small sample sizes, the following results can be shown to hold:

1. When the underlying distribution is normal, the ARE of the Wilcoxon test with respect to the t test is approximately .95.
2. For any distribution, the ARE will be at least .86, and for many distributions will be much greater than 1.

We can summarize these results by saying that in large-sample problems, the Wilcoxon test is never very much less efficient than the t test and may be much more efficient if the underlying distribution is far from normal. Though the issue is far from resolved in the case of sample sizes obtained in most practical problems, studies have shown that the Wilcoxon test performs reasonably and is thus a viable alternative to the t test.

Exercises / Section 15.2 (8–14)

8. Reconsider the problem described in Example 15.2 and use the Wilcoxon test with $\alpha = .01$ to test the given hypotheses.

9. Use the Wilcoxon test to analyze the data given in Exercise 3.

10. The accompanying data is a subset of the data reported in the paper "Synovial Fluid pH, Lactate, Oxygen and Carbon Dioxide Partial Pressure in Various Joint Diseases" (*Arthritis and Rheumatism*, 1971, pp. 476–477). The observations are pH values of synovial fluid (which lubricates joints and tendons) taken from the knees of individuals suffering from arthritis. Assuming that true average pH for nonarthritic individuals is 7.39, test at level .05 to see whether the data indicates a difference between average pH values for arthritic and nonarthritic individuals.

7.02, 7.35, 7.34, 7.17, 7.28, 7.77, 7.09, 7.22, 7.45, 6.95, 7.40, 7.10, 7.32, 7.14

11. Both a gravimetric and a spectrophotometric method are under consideration for determining phosphate content of a particular material. Twelve samples of the material are obtained, each is split in half, and a determination is made on each half using one of the two methods, resulting in the following data.

Sample:	1	2	3	4
Gravimetric:	54.7	58.5	66.8	46.1
Spectrophotometric:	55.0	55.7	62.9	45.5
Sample:	5	6	7	8
Gravimetric:	52.3	74.3	92.5	40.2
Spectrophotometric:	51.1	75.4	89.6	38.4
Sample:	9	10	11	12
Gravimetric:	87.3	74.8	63.2	68.5
Spectrophotometric:	86.8	72.5	62.3	66.0

Use the Wilcoxon test to decide whether or not one technique gives on average a different value than the other technique for this type of material.

12. Analyze the data in Exercise 2 using the large-sample version of the Wilcoxon procedure, and compare the result with an analysis using a critical value from Appendix Table A.9.

13. The accompanying 26 observations on fracture toughness of base plate of 18% nickel miraging steel were reported in the paper "Fracture Testing of Weldments," *ASTM Special Publ. No. 381*, 1965, pp. 328–356. Suppose that a company will agree to purchase this steel for a particular application only if it can be strongly demonstrated from experimental evidence that true average toughness exceeds 75. Assuming that the fracture toughness distribution is symmetric, state and test the appropriate hypotheses at level .05, and also compute a *P*-value.

69.5, 71.9, 72.6, 73.1, 73.3, 73.5, 74.1, 74.2, 75.3, 75.5, 75.7, 75.8, 76.1, 76.2, 76.2, 76.9, 77.0, 77.9, 78.1, 79.6, 79.7, 80.1, 82.2, 83.7, 93.7

14. Suppose that observations $X_1, X_2, \ldots, X_n$ are made on a process at times $1, 2, \ldots, n$. On the basis of this data, we wish to test

H_0: the X_i's constitute an independent and identically distributed sequence

versus

H_a: X_{i+1} tends to be larger than X_i for $i = 1, \ldots, n$ (an increasing trend)

Suppose that the X_i's are ranked from 1 to n. Then when H_a is true, larger ranks tend to occur later in the sequence, while if H_0 is true, large and small ranks tend to be mixed together. Let R_i be the rank of X_i and consider the test statistic $D = \sum_{i=1}^{n} (R_i - i)^2$. Then small values of D give support to H_a (for example, the smallest value is 0 for $R_1 = 1, R_2 = 2, \ldots, R_n = n$), so H_0 should be rejected in favor of H_a if $d \leq c$. When H_0 is true, any sequence of ranks has probability $1/n!$. Use this to find c for which the test has level as close to .10 as possible in the case $n = 4$. *Hint:* List the 4! rank sequences and compute d for each one, and then obtain the null distribution of D. See the Lehmann book, p. 290, for more information.

15.3   The Wilcoxon Rank-Sum Test

When at least one of the sample sizes in a two-sample problem is small, the *t* test requires the assumption of normality (at least approximately). While the test is the best test in the sense of minimizing β for fixed α, there are situations in which an investigator would want to use a test that is valid even if the underlying distributions are quite nonnormal. We now describe such a test, called the Wilcoxon rank-sum test. An alternative name for the procedure is the Mann-Whitney test, though the Mann-Whitney test statistic is sometimes expressed in a slightly different form from that of the Wilcoxon test. The Wilcoxon test procedure is distribution-free because it will have the desired level of significance for a very large class of underlying distributions.

Assumptions

> $X_1, \ldots, X_m$ and $Y_1, \ldots, Y_n$ are two independent random samples from continuous distributions with means μ_1 and μ_2, respectively. The X and Y distributions have exactly the same shape and spread, the only possible difference between the two being in the values of μ_1 and μ_2.

When $H_0 : \mu_1 - \mu_2 = \Delta_0$ is true, the X distribution is shifted by the amount Δ_0 to the right of the Y distribution, while when H_0 is false, the shift is by an amount other than Δ_0.

Development of the Test When $m = 3$, $n = 4$

Consider first testing $H_0 : \mu_1 - \mu_2 = 0$. If μ_1 is actually much larger than μ_2, then most of the observed x's will fall to the right of the observed y's, while if H_0 is true then the observed values from the two samples should be intermingled. The test statistic will provide a quantification of how much intermingling there is in the two samples.

For concreteness, let $m = 3$ and $n = 4$. Then if all three observed x's were to the right of all four observed y's, this would provide strong evidence for rejecting H_0 in favor of $H_a : \mu_1 - \mu_2 \neq 0$, with a similar conclusion being appropriate if all three x's fall below all four of the y's. Suppose that we pool the X's and Y's into a combined sample of size $m + n = 7$ and rank these observations from smallest to largest, with the smallest receiving rank 1 and the largest, rank 7. If either most of the largest ranks or most of the smallest ranks were associated with X observations, we would begin to suspect H_0. This suggests the test statistic

$$W = \text{the sum of the ranks in the combined sample associated with } X \text{ observations} \qquad (15.3)$$

For the values of m and n under consideration, the smallest possible value of W is $w = 1 + 2 + 3 = 6$ (if all three x's are smaller than all four y's) and the largest possible value is $w = 5 + 6 + 7 = 18$ (if all three x's are larger than all four y's).

As an example of the calculation of the observed W, suppose that $x_1 = -3.10$, $x_2 = 1.67$, $x_3 = 2.01$, $y_1 = 5.27$, $y_2 = 1.89$, $y_3 = 3.86$, and $y_4 = 0.19$. Then the pooled ordered sample is $-3.10, 0.19, 1.67, 1.89, 2.01, 3.86,$ and 5.27. The X ranks for this sample are 1 (for -3.10), 3 (for 1.67), and 5 (for 2.01), so the computed value of W is $w = 1 + 3 + 5 = 9$.

The test procedure based on the statistic (15.3) is to reject H_0 if the computed value w is "too extreme"—that is, $\geq c$ for an upper-tailed test, $\leq c$ for a lower-tailed test, and either $\geq c_1$ or $\leq c_2$ for a two-tailed test. The critical constant(s) c (c_1, c_2) should be chosen so that the test has the desired level of significance α. To see how this should be done, recall that when H_0 is true, all seven observations come from the same population. This means that under H_0, any possible triple of ranks associated with the three x's—such as (1, 4, 5), (3, 5, 6), or (5, 6, 7)—has the same probability as any other possible rank triple. Since there are $\binom{7}{3} = 35$ possible rank triples, under H_0 each rank triple has probability $\frac{1}{35}$. From a list of all 35 rank triples and the w value associated with each, the probability distribution of W can immediately be determined. For example, there are four rank triples that have w value 11—(1, 3, 7), (1, 4, 6), (2, 3, 6), and (2, 4, 5)—so $P(W = 11) = \frac{4}{35}$. The summary of the listing and computations appears in Table 15.5.

Table 15.5 Probability distribution of W ($m = 3$, $n = 4$) when H_0 is true

w	6	7	8	9	10	11	12	13	14	15	16	17	18
$P(W = w)$	$\frac{1}{35}$	$\frac{1}{35}$	$\frac{2}{35}$	$\frac{3}{35}$	$\frac{4}{35}$	$\frac{4}{35}$	$\frac{5}{35}$	$\frac{4}{35}$	$\frac{4}{35}$	$\frac{3}{35}$	$\frac{2}{35}$	$\frac{1}{35}$	$\frac{1}{35}$

The distribution of Table 15.5 is symmetric about the value $w = (6 + 18)/2 = 12$, which is the middle value in the ordered list of possible W values. This is because the two rank triples (r, s, t) (with $r < s < t$) and $(8 - t, 8 - s, 8 - r)$ have values of w symmetric about 12, so for each triple with w value below 12, there is a triple with w value above 12 by the same amount.

If the alternative hypothesis is $H_a : \mu_1 - \mu_2 > 0$, then H_0 should be rejected in favor of H_a for large W values. Choosing as the rejection region the set of W values $\{17, 18\}$, $\alpha = P(\text{type I error}) = P(\text{reject } H_0 \text{ when } H_0 \text{ is true}) = P(W = 17 \text{ or } 18 \text{ when } H_0 \text{ is true}) = \frac{1}{35} + \frac{1}{35} = \frac{2}{35} = .057$; the region $\{17, 18\}$ therefore specifies a test with level of significance approximately .05. Similarly, the region $\{6, 7\}$, which is appropriate for $H_a : \mu_1 - \mu_2 < 0$ has $\alpha = .057 \approx .05$. The region $\{6, 7, 17, 18\}$, which is appropriate for the two-sided alternative, has $\alpha = \frac{4}{35} = .114$. The W value for the data given several paragraphs earlier was $w = 9$, which is rather close to the middle value 12, so H_0 would not be rejected at any reasonable level α for any one of the three H_a's.

General Description of the Wilcoxon Rank-Sum Test
The null hypothesis $H_0 : \mu_1 - \mu_2 = \Delta_0$ is handled by subtracting Δ_0 off each X_i and using the $(X_i - \Delta_0)$'s as the X_i's were previously used. Recalling that for any positive integer K, the sum of the first K integers is $K(K + 1)/2$, the smallest possible value of the statistic W is $m(m + 1)/2$, which occurs when the $(X_i - \Delta_0)$'s are all to the left of the Y sample. The largest possible value of W occurs when the $(X_i - \Delta_0)$'s lie entirely to the right of the Y's; in this case $W = (m + 1) + \cdots + (m + n) = (\text{sum of first } m + n \text{ integers}) - (\text{sum of first } m \text{ integers}) = m(m + 2n + 1)/2$. As with the special case $m = 3$, $n = 4$, the distribution of W is symmetric about the value that is halfway between the smallest and largest values; this middle value is $m(m + n + 1)/2$. Because of this symmetry, probabilities involving lower-tail critical values can be obtained from corresponding upper-tail values.

Null hypothesis: $H_0: \mu_1 - \mu_2 = \Delta_0$

Test statistic value: $w = \sum_{i=1}^{m} r_i$ where $r_i = $ rank of $(x_i - \Delta_0)$ in the combined sample of $m + n$ x's and y's.

Alternative hypothesis	*Rejection region*
$H_a: \mu_1 - \mu_2 > \Delta_0$	$w \geq c_1$
$H_a: \mu_1 - \mu_2 < \Delta_0$	$w \leq m(m + n + 1) - c_1$
$H_a: \mu_1 - \mu_2 \neq \Delta_0$	either $w \geq c$ or $w \leq m(m + n + 1) - c$

where $P(W \geq c_1$ when H_0 is true$) = \alpha$, $P(W \geq c$ when H_0 is true$) = \alpha/2$.

Because W has a discrete probability distribution, there will not always exist a critical value corresponding exactly to one of the usual levels of significance. Appendix Table A.10 gives upper-tail critical values for probabilities closest to .05, .025, .01, and .005, from which level .05 or .01 one- and two-tailed tests can be obtained. The table gives information only for $m = 3$, $4, \ldots, 8$ and $n = m, m + 1, \ldots, 8$ (that is, $3 \leq m \leq n \leq 8$). For values of m and n that exceed 8, a normal approximation can be used. To use the table for small m and n, though, *the X and Y samples should be labeled so that $m \leq n$.*

Example 15.8 The urinary fluoride concentration (parts per million) was measured both for a sample of livestock grazing in an area previously exposed to fluoride pollution and for a similar sample grazing in an unpolluted region.

Polluted:	21.3, 18.7, 23.0, 17.1, 16.8, 20.9, 19.7
Unpolluted:	14.2, 18.3, 17.2, 18.4, 20.0

Does the data indicate strongly that the true average fluoride concentration for livestock grazing in the polluted region is larger than for the unpolluted region? Use the Wilcoxon rank-sum test at level $\alpha = .01$.

The sample sizes here are 7 and 5. To obtain $m \leq n$, label the unpolluted observations as the x's ($x_1 = 14.2, \ldots, x_5 = 20.0$) and the polluted observations as the y's. Thus μ_1 is the true average fluoride concentration without pollution and μ_2 the true average concentration with pollution. The alternative hypothesis is $H_a: \mu_1 - \mu_2 < 0$ (pollution causes an increase in concentration), so a lower-tailed test is appropriate. Entering Appendix Table A.10 with $m = 5$ and $n = 7$, $P(W \geq 47$ when H_0 is true$) \approx .01$. The critical value for the lower-tailed test is $m(m + n + 1) - 47 = 5(13) - 47 = 18$; H_0 will now be rejected if $W \leq 18$. The pooled ordered sample appears below; the computed W is $w = r_1 + r_2 + \cdots + r_5$ (where r_i is the rank of x_i) $= 1 + 5 + 4 + 6 + 9 = 25$. Since 25 is not ≤ 18, H_0 is not rejected at (approximately) level .01.

x	y	y	x	x	x	y	y	x	y	y	y
14.2	16.8	17.1	17.2	18.3	18.4	18.7	19.7	20.0	20.9	21.3	23.0
1	2	3	4	5	6	7	8	9	10	11	12

■

A Normal Approximation for W

When both m and n exceed 8, the distribution of W can be approximated by an appropriate normal curve, and this approximation can be used in place of Appendix Table A.10. To obtain the approximation, we need μ_W and σ_W^2 when H_0 is true. In this case the rank R_i of $X_i - \Delta_0$ is equally likely to be any one of the possible values $1, 2, 3, \ldots, m + n$ (R_i has a discrete uniform distribution on the first $m + n$ positive integers), so $\mu_{R_i} = (m + n + 1)/2$. This gives, since $W = \Sigma R_i$,

$$\mu_W = \mu_{R_1} + \mu_{R_2} + \cdots + \mu_{R_m} = \frac{m(m + n + 1)}{2} \tag{15.4}$$

The variance of R_i is also easily computed to be $(m + n + 1)(m + n - 1)/12$. However, because the R_i's are not independent, $V(W) \neq mV(R_i)$. Using the fact that for any two distinct integers a and b between 1 and $m + n$ inclusive, $P(R_i = a, R_j = b) = 1/(m + n)(m + n - 1)$ (two integers are being sampled without replacement), $\text{Cov}(R_i, R_j) = -(m + n + 1)/12$, yielding

$$\sigma_W^2 = \sum_{i=1}^{m} V(R_i) + \sum_{i \neq j} \sum \text{Cov}(R_i, R_j) = \frac{mn(m + n + 1)}{12} \tag{15.5}$$

A Central Limit Theorem can then be used to conclude that when H_0 is true, the **test statistic**

$$Z = \frac{W - m(m + n + 1)/2}{\sqrt{mn(m + n + 1)/12}}$$

has approximately a standard normal distribution. This statistic is used in conjunction with the critical values z_α, $-z_\alpha$, and $\pm z_{\alpha/2}$ for upper-, lower-, and two-tailed tests, respectively.

Example 15.9

A paper that appeared in the *Journal of Applied Physiology* ("Histamine Content in Sputum from Allergic and Non-Allergic Individuals," 1969, pp. 535–539) reported the following data on sputum histamine level (micrograms per gram dry weight of sputum) for a sample of nine individuals classified as allergics and another sample of 13 individuals classified as nonallergics.

Allergics:	67.6, 39.6, 1651.0, 100.0, 65.9, 1112.0, 31.0, 102.4, 64.7
Nonallergics:	34.3, 27.3, 35.4, 48.1, 5.2, 29.1, 4.7, 41.7, 48.0, 6.6, 18.9, 32.4, 45.5

Does the data indicate that there is a difference in true average sputum histamine level between allergics and nonallergics?

Since both sample sizes exceed eight, we use the normal approximation. The null hypothesis is $H_0 : \mu_1 - \mu_2 = 0$, and observed ranks of the x_i's are $r_1 = 18$, $r_2 = 11$, $r_3 = 22$, $r_4 = 19$, $r_5 = 17$, $r_6 = 21$, $r_7 = 7$, $r_8 = 20$, and $r_9 = 16$, so $w = \Sigma r_i = 151$. The mean and variance of W are given by $\mu_W = 9(23)/2 = 103.5$ and $\sigma_W^2 = 9(13)(23)/12 = 224.25$. Thus

$$z = \frac{151 - 103.5}{\sqrt{224.25}} = 3.17$$

The alternative hypothesis is $H_a : \mu_1 - \mu_2 \neq 0$, so at level .01 H_0 is rejected if either $z \geq 2.58$ or $z \leq -2.58$. Because $3.17 \geq 2.58$, H_0 is rejected, and we conclude that there is a difference in true average sputum histamine levels (the paper also used the Wilcoxon test). ■

Ties in the Wilcoxon Rank-Sum Test

Theoretically the assumption of continuity of the two distributions ensures that all $m + n$ observed x's and y's will have different values. In practice, though, there will often be ties in the observed values. As with the Wilcoxon signed-rank test, the common practice in dealing with ties is to assign each of the tied observations in a particular set of ties the average of the ranks they would receive if they differed very slightly from one another.

If the sample sizes both exceed eight, the numerator of Z is still appropriate but the denominator should be replaced by the square root of the adjusted variance

$$\sigma_W^2 = \frac{mn(m + n + 1)}{12} - \frac{mn}{12(m + n)(m + n - 1)} \Sigma(\tau_i - 1)(\tau_i)(\tau_i + 1) \tag{15.6}$$

where τ_i is the number of tied observations in the ith set of ties and the sum is over all sets of ties. Unless there are a great many ties, there is little difference between (15.6) and (15.5).

Efficiency of the Rank-Sum Test

When the distributions being sampled are both normal with $\sigma_1^2 = \sigma_2^2$, both the two-sample t test and the Wilcoxon rank-sum test can be used, but the t test is known to be the best among all possible tests in the sense of minimizing β for a fixed α. Because the Wilcoxon test is valid for a much broader class of underlying distributions than just the normal, it might be used if the validity of the normality assumption is open to question. One aspect of efficiency relates to how much is lost by using W instead of T when normality is correct. The other aspect of efficiency is concerned with the performance of W relative to T in nonnormal situations.

In the previous section the notion of efficiency of test procedures was discussed in connection with the one-sample t test and Wilcoxon signed-rank test. The results for two-sample tests are the same as for the one-sample tests. When the normality assumption is correct, the rank-sum test is approximately 95% as efficient as the t test in large samples—meaning that the t test can give the same error probabilities as the Wilcoxon test using slightly smaller sample sizes. On the other hand, the Wilcoxon test will always be at least 86% as efficient as the t test and may be many times as efficient if the underlying distributions are very nonnormal, especially with heavy tails.

In summary, the Wilcoxon test never gives up very much to the t test and may be a distinct improvement on it, so if there is much doubt about the assumption of normality, the Wilcoxon test provides a good alternative means of analysis.

Lastly we note that β calculations for the Wilcoxon test are quite difficult, because the distribution of W when H_0 is false depends not only on $\mu_1 - \mu_2$ but also on the shapes of the two distributions. For most underlying distributions, the nonnull distribution of W is virtually intractable. This is why statisticians have developed large-sample (asymptotic relative) efficiency as a means of comparing tests.

Exercises / Section 15.3 (15–20)

15. In an experiment to compare the bond strength of two different adhesives, each adhesive was used in five bondings of two surfaces, and the force necessary to separate the surfaces was determined for each bonding. For adhesive 1 the resulting values were 229, 286, 245, 299, and 250, while the adhesive 2 observations were 213, 179, 163, 247, and 225. Let μ_i denote the true average bond strength of adhesive type i. Use the Wilcoxon rank-sum test at level .05 to test $H_0: \mu_1 = \mu_2$ versus $H_a: \mu_1 > \mu_2$.

16. The article "A Study of Wood Stove Particulate Emissions" (*J. Air Pollution Control Assn.*, 1979, pp. 724–728) reported the following data on burn time (hours) for samples of oak and pine. Test at level .05 to see whether or not there is any difference in true average burn time for the two types of wood.

Oak: 1.72, .67, 1.55, 1.56, 1.42, 1.23, 1.77, .48
Pine: .98, 1.40, 1.33, 1.52, .73, 1.20

17. A modification has been made to the process for producing a certain type of "time-zero" film (film that begins to develop as soon as a picture is taken). Because the modification involves extra cost, it will be incorporated only if sample data strongly indicates that the modification has decreased true average developing time by more than one second. Assuming that both developing time distributions differ only with respect to location if at all, use the Wilcoxon rank-sum test at level .05 on the accompanying data to test the appropriate hypotheses.

Original
  process: 8.6, 5.1, 4.5, 5.4, 6.3, 6.6, 5.7, 8.5
Modified
  process: 5.5, 4.0, 3.8, 6.0, 5.8, 4.9, 7.0, 5.7

18. The accompanying data resulted from an experiment to compare the effects of vitamin C in orange juice and in synthetic ascorbic acid on the length of odontoblasts in guinea pigs over a six-week period ("The Growth of the Odontoblasts of the Incisor Tooth as a Criterion of the Vitamin C Intake of the Guinea Pig," *J. Nutrition*, 1947, pp. 491–504). Use the Wilcoxon rank-sum test at level .01 to decide whether or not true average length differs for the two types of vitamin C intake. Compute also an approximate P-value.

Orange juice: 8.2, 9.4, 9.6, 9.7, 10.0, 14.5,
             15.2, 16.1, 17.6, 21.5
Ascorbic acid: 4.2, 5.2, 5.8, 6.4, 7.0, 7.3,
             10.1, 11.2, 11.3, 11.5

19. Test the hypotheses suggested in Exercise 18 using the following data:

Orange juice: 8.2, 9.5, 9.5, 9.7, 10.0, 14.5,
             15.2, 16.1, 17.6, 21.5
Ascorbic acid: 4.2, 5.2, 5.8, 6.4, 7.0, 7.3,
             9.5, 10.0, 11.5, 11.5

20. The paper "Measuring the Exposure of Infants to Tobacco Smoke," (*N. Engl. J. Med.*, 1984, pp. 1075–1078) reported on a study in which various measurements were taken both from a random sample of infants who had been exposed to household smoke and from a sample of unexposed infants. The accompanying data consists of observations on urinary concentration of cotanine, a major metabolite of nicotine (the values constitute a subset of the original data and were read from a plot that appeared in the paper). Does the data suggest that true average cotanine level is higher in exposed infants than in unexposed infants by more than 25? Carry out a test at significance level .05.

Unexposed: 8 11 12 14   20   43 111

Exposed: 35 56 83 92 128 150 176 208

15.4   Distribution-Free Confidence Intervals

The method we have used so far to construct a confidence interval can be described as follows: start with a random variable (Z, T, χ^2, F, or the like) that depends on the parameter of interest and a probability statement involving the variable, manipulate the inequalities of the statement to isolate the parameter between random endpoints, and finally substitute computed values for random variables. Another general method for obtaining confidence intervals takes advantage of a relationship between test procedures and confidence intervals; a $100(1 - \alpha)\%$ confidence inverval for a parameter θ can be obtained from a level α test for $H_0 : \theta = \theta_0$ versus $H_a : \theta \neq \theta_0$. This method will be used to derive intervals associated with the sign test, the Wilcoxon signed-rank test, and the Wilcoxon rank-sum test.

Before using the method to derive new intervals, reconsider the t test and the t interval. Suppose that a random sample of $n = 25$ observations from a normal population yields summary statistics $\bar{x} = 100$, $s = 20$. Then a 90% confidence interval for μ is

$$\left(\bar{x} - t_{.05, 24} \cdot \frac{s}{\sqrt{25}}, \quad \bar{x} + t_{.05, 24} \cdot \frac{s}{\sqrt{25}} \right) = (93.16, 106.84) \tag{15.7}$$

Suppose that instead of a confidence interval, we had wished to test a hypothesis about μ. For $H_0 : \mu = \mu_0$ versus $H_a : \mu \neq \mu_0$, the t test at level .10 specifies that H_0 should be rejected if t is either ≥ 1.711 or ≤ -1.711, where

$$t = \frac{\bar{x} - \mu_0}{s / \sqrt{25}} = \frac{100 - \mu_0}{20 / \sqrt{25}} = \frac{100 - \mu_0}{4} \tag{15.8}$$

Consider now the null value $\mu_0 = 95$. Then $t = 1.25$ so H_0 is not rejected. Similarly, if $\mu_0 = 104$, then $t = -1$, so again H_0 is not rejected. However, if $\mu_0 = 90$, then $t = 2.5$, so H_0 is rejected, and if $\mu_0 = 108$, then $t = -2$, so H_0 is again rejected. By considering other values of μ_0 and the decision resulting from each one, the following general fact emerges: *Every number inside the interval (15.7) specifies a value of μ_0 for which t of (15.8) leads to acceptance of H_0, while every number outside (15.7) corresponds to a t for which H_0 is rejected.* That is, for the fixed values of n, $\bar{x}$, and s, the set of all μ_0 values for which testing $H_0 : \mu = \mu_0$ versus $H_a : \mu \neq \mu_0$ results in acceptance of H_0 is precisely the interval (15.7).

Proposition

> Suppose that we have a level α test procedure for testing $H_0 : \theta = \theta_0$ versus $H_a : \theta \neq \theta_0$. For fixed sample values, let A denote the set of all values θ_0 for which H_0 is accepted. Then A is a $100(1 - \alpha)\%$ confidence interval for θ.

There are actually pathological examples in which the set A defined in the proposition is not an interval of θ values, but instead the complement of an interval or something even stranger. To be more precise, we should really replace the notion of a confidence interval with that of a confidence set. In the three cases of interest here, the set A does turn out to be an interval.

The Sign Interval for $\tilde{\mu}$

Let the sample be denoted by $X_1, \ldots, X_n$. Then for testing $H_0 : \tilde{\mu} = \tilde{\mu}_0$ versus $H_a : \tilde{\mu} \neq \tilde{\mu}_0$, the sign test statistic is $Y = $ number of X_i's which are $> \tilde{\mu}_0$ [the number of $(X_i - \tilde{\mu}_0)$'s which have positive signs]. H_0 is rejected if $y \geq c$ or $\leq n - c$; under H_0, Y has a binomial distribution with $p = .5$ and n, and c is determined by $P(Y \geq c) + P(Y \leq n - c) = \alpha$.

Example 15.10

The daily dietetic intake of nicotinic acid-equivalents (mg) was determined for a sample of 12 healthy individuals ("Assessment of Nicotinic Acid Status of Population Groups," *Amer. J. Clinical Nutrition,* 1964, pp. 169–174). The sample observations, listed in order of increasing magnitude, were

$$16.6, \ 19.4, \ 21.7, \ 23.4, \ 24.2, \ 26.9, \ 27.3, \ 27.4, \ 28.0, \ 29.5, \ 30.7, \ \text{and } 58.8$$

To obtain a $100(1 - \alpha)\%$ confidence interval for $\tilde{\mu} = $ the true median daily nicotinic acid-equivalent intake, we must use the sign test at level α to test $H_0 : \tilde{\mu} = \tilde{\mu}_0$ versus $H_a : \tilde{\mu} \neq \tilde{\mu}_0$ for various values of $\tilde{\mu}_0$. The confidence interval is then the set $A = \{\text{all } \tilde{\mu}_0 : H_0 \text{ is accepted}\}$.

For $n = 12$ and $p = .5$, $P(Y \leq 2) + P(Y \geq 10) = .0193 + .0193 = .0386 \approx .04$. If the rejection region for the sign test is $R = \{0, 1, 2, 10, 11, 12\}$, the test has (approximately) level .04, so the confidence level of the resulting interval will be 96%. The confidence interval is most easily identified from the plot of the data in Figure 15.5.

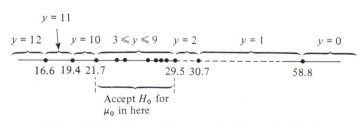

Figure 15.5   Plot of the data for Example 15.10

If $\tilde{\mu}_0 < 21.7$, then $y =$ number of x_i's $> \tilde{\mu}_0$ is at least 10 (10, 11, or 12), so H_0 would be rejected. Similarly, if $\tilde{\mu}_0 > 29.5$, then y is at most 2 (0, 1, or 2), so again H_0 is rejected. However, as Figure 15.5 indicates, if $\tilde{\mu}_0$ is between 21.7 and 29.5, then $3 \leq y \leq 9$ so H_0 is accepted. The set A is then $\{\tilde{\mu}_0 : 21.7 < \tilde{\mu}_0 < 29.5\}$ so our 96% confidence interval for $\tilde{\mu}$ is (21.7, 29.5). ∎

Example 15.10 shows that after the data values are ordered from smallest to largest, the interval for $\tilde{\mu}$ includes all numbers between two of these ordered values.

Proposition

> Denote the ordered sample values by $x_{(1)}, x_{(2)}, \ldots, x_{(n)}$, where $x_{(1)}$ is the smallest, $x_{(2)}$ the next smallest, and so on. Suppose that the level α sign test rejects $H_0 : \tilde{\mu} = \tilde{\mu}_0$ if either $y \geq c$ or $\leq n - c$. Then a $100(1 - \alpha)\%$ confidence interval for $\tilde{\mu}$ is $(x_{(n-c+1)}, x_{(c)})$.

Example 15.11
(Example 15.10 continued)

For $n = 12$ and $c = 10$, the level of significance is .04. Therefore the $100(1 - .04) = 96\%$ confidence interval is $(x_{(12-10+1)}, x_{(10)}) = (x_{(3)}, x_{(10)}) = (21.7, 29.5)$. The sample mean and sample standard deviation are easily computed as $\bar{x} = 27.83$ and $s = 10.61$, from which the 95% t interval for μ, assuming the distribution to be normal, is (21.09, 34.57). That this t interval is much longer than the sign interval is due primarily to the single outlying value 58.8. ∎

Notice that in contrast to the t interval, the sign interval is relatively insensitive to a few outlying values. In Example 15.10 the two extreme values on each end of the sample can be moved further from the middle without affecting the interval.

When the population is normal, $\tilde{\mu} = \mu$, so the sign interval is a competitor to the t interval. In this case the sign interval will usually be longer than the t interval, since outlying values in a normal sample are very unusual. The Wilcoxon interval, to be described next, is a better competitor of the t interval.

Finally, because of the discreteness of the binomial distribution, the usual confidence levels are not achievable. In Example 15.11 we settled for 96% rather than 95% confidence.

The Wilcoxon Signed-Rank Interval

To test $H_0 : \mu = \mu_0$ versus $H_a : \mu \neq \mu_0$ using the Wilcoxon signed-rank test, where μ is the mean of a continuous symmetric distribution, the absolute values $|x_1 - \mu_0|, \ldots, |x_n - \mu_0|$ are ordered from smallest to largest, with the smallest receiving rank 1 and the largest rank n. Each rank is then given the sign of its associated $x_i - \mu_0$, and the test statistic is the sum of the positively signed ranks. The two-tailed test rejects H_0 if s_+ is either $\geq c$ or $\leq n(n + 1)/2 - c$, where c is obtained from Appendix Table A.9 once the desired level of significance α is specified.

Just as was the case for the sign interval, for fixed $x_1, \ldots, x_n$ the $100(1 - \alpha)\%$ signed-rank interval will consist of all μ_0 for which $H_0 : \mu = \mu_0$ is accepted at level α. To identify this interval, it is convenient to express the test statistic S_+ in another form:

> $S_+ =$ the number of pairwise averages $(X_i + X_j)/2$ with $i \leq j$ which are $\geq \mu_0$ 　　　　　　　　　　　(15.9)

That is, if we average each x_j in the list with each x_i to its left, including $(x_j + x_j)/2$ (which is just x_j), and count the number of these averages that are $\geq \mu_0$, s_+ results. In moving from left to right in the list of sample values, we are simply averaging every pair of observations in the sample [again including $(x_j + x_j)/2$] exactly once, so the order in which the observations are listed before averaging is not important. The equivalence of the two methods for computing s_+ is not difficult to verify. The number of pairwise averages is $\binom{n}{2} + n$ (the first term due to averaging of different observations and the second due to averaging each x_i with itself), which equals $n(n + 1)/2$. If either too many or two few of these pairwise averages are $\geq \mu_0$, H_0 is rejected.

Example 15.12　The following observations are values of cerebral metabolic rate for rhesus monkeys: $x_1 = 4.51$, $x_2 = 4.59$, $x_3 = 4.90$, $x_4 = 4.93$, $x_5 = 6.80$, $x_6 = 5.08$, $x_7 = 5.67$. There are 28 pairwise averages which are, in increasing order

4.51, 4.55, 4.59, 4.705, 4.72, 4.745, 4.76, 4.795, 4.835, 4.90, 4.915, 4.93, 4.99, 5.005, 5.08, 5.09, 5.13, 5.285, 5.30, 5.375, 5.655, 5.67, 5.695, 5.85, 5.865, 5.94, 6.235, 6.80

The first few and the last few of these are pictured on a measurement axis in Figure 15.6.

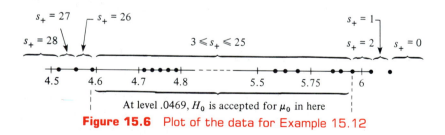

Figure 15.6　Plot of the data for Example 15.12

Because of the discreteness of the distribution of S_+, $\alpha = .05$ cannot be obtained exactly. The rejection region $\{0, 1, 2, 26, 27, 28\}$ has $\alpha = .0469$, which is as close as possible to .05, so the level is approximately .05. Thus if the number of pairwise averages $\geq \mu_0$ is between 3 and 25, inclusive, H_0 is accepted. From Figure 15.6 the (approximate) 95% confidence interval for μ is (4.59, 5.94). ■

In general, once the pairwise averages are ordered from smallest to largest, the endpoints of the Wilcoxon interval are two of the "extreme" averages. To express this precisely, let the smallest pairwise average be denoted by $\overline{x}_{(1)}$, the next smallest by $\overline{x}_{(2)}, \ldots$, and the largest by $\overline{x}_{(n(n+1)/2)}$.

Proposition

> If the level α Wilcoxon signed rank test for $H_0 : \mu = \mu_0$ versus $H_a : \mu \neq \mu_0$ is to reject H_0 if either $s_+ \geq c$ or $s_+ \leq n(n+1)/2 - c$, then a $100(1 - \alpha)\%$ confidence interval for μ is
>
> $$(\overline{x}_{(n(n+1)/2 - c + 1)}, \overline{x}_{(c)}) \qquad (15.10)$$

In words, the interval extends from the dth smallest pairwise average to the dth largest average, where $d = n(n+1)/2 - c + 1$. Appendix Table A.11 gives for $n = 5, 6, \ldots, 25$ the values of c that correspond to the usual confidence levels.

Example 15.13
(Example 15.12 continued)

For $n = 7$, an 89.1% interval (approximately 90%) is obtained by using $c = 24$ (since the rejection region $\{0, 1, 2, 3, 4, 24, 25, 26, 27, 28\}$ has $\alpha = .109$). The interval is $(\overline{x}_{(28 - 24 + 1)}, \overline{x}_{(24)}) = (\overline{x}_{(5)}, \overline{x}_{(24)}) = (4.72, 5.85)$, which extends from the fifth smallest to the fifth largest pairwise average. ∎

Other Uses and Properties of the Signed-Rank Interval

While the derivation of the interval depended on having a single sample from a continuous symmetric distribution with mean (median) μ, when the data is paired, the interval constructed from the differences $d_1, d_2, \ldots, d_n$ is a confidence interval for the mean (median) difference μ_D. In this case the symmetry of X and Y distributions need not be assumed; as long as the X and Y distributions have the same shape, the $X - Y$ distribution will be symmetric, so only continuity is required.

For $n > 20$, the large-sample approximation to the Wilcoxon test based on standardizing S_+ gives an approximation to c in (15.10). The result [for a $100(1 - \alpha)\%$ interval] is

$$c \approx \frac{n(n+1)}{4} + z_{\alpha/2} \sqrt{\frac{n(n+1)(2n+1)}{24}}$$

The efficiency of the Wilcoxon interval relative to the t interval is roughly the same as that for the Wilcoxon test relative to the t test. In particular, for large samples when the underlying population is normal the Wilcoxon interval will tend to be slightly longer than the t interval, but if the population is quite nonnormal (symmetric, but with heavy tails), then the Wilcoxon interval will tend to be much shorter than the t interval.

The Wilcoxon Rank-Sum Interval

The Wilcoxon rank-sum test for testing $H_0 : \mu_1 - \mu_2 = \Delta_0$ is carried out by first combining the $(X_i - \Delta_0)$'s and Y_j's into one sample of size $m + n$ and ranking them from smallest (rank 1) to largest (rank $m + n$). The test statistic W is then the sum of the ranks of the $(X_i - \Delta_0)$'s. For the two-sided alternative, H_0 is rejected if w is either too small or too large.

To obtain the associated confidence interval, for fixed x_i's and y_j's we must determine the set of all Δ_0 values for which H_0 is accepted. This is easiest to do if we first express the test statistic in a slightly different form. The smallest possible value of W is $m(m + 1)/2$, corresponding to every $(X_i - \Delta_0)$ less than every Y_j, and there are mn differences of the form $(X_i - \Delta_0) - Y_j$. A bit of manipulation gives

$$W = [\text{number of } (X_i - Y_j - \Delta_0)\text{'s} \geq 0] + \frac{m(m + 1)}{2}$$

$$= [\text{number of } (X_i - Y_j)\text{'s} \geq \Delta_0] + \frac{m(m + 1)}{2} \qquad (15.11)$$

Thus rejecting H_0 if the number of $(x_i - y_j)$'s $\geq \Delta_0$ is either too small or too large is equivalent to rejecting H_0 for small or large w.

Expression (15.11) suggests that we compute $x_i - y_j$ for each i and j, and order these mn differences from smallest to largest. Then if the null value Δ_0 is not either smaller than most of the differences or larger than most, $H_0 : \mu_1 - \mu_2 = \Delta_0$ is accepted. Varying Δ_0 now shows that a confidence interval for $\mu_1 - \mu_2$ will have as its lower endpoint one of the ordered $(x_i - y_j)$'s, and similarly for the upper endpoint.

Proposition

> Let $x_1, \ldots, x_m$ and $y_1, \ldots, y_n$ be the observed values in two independent samples from continuous distributions that differ only in location (and not in shape). With $d_{ij} = x_i - y_j$ and the ordered differences denoted by $d_{ij(1)}, d_{ij(2)}, \ldots, d_{ij(mn)}$, the general form of a $100(1 - \alpha)\%$ confidence interval for $\mu_1 - \mu_2$ is
>
> $$(d_{ij(mn - c + 1)}, d_{ij(c)}) \qquad (15.12)$$
>
> where c is the critical constant for the two-tailed level α Wilcoxon rank-sum test.

Notice that the form of the Wilcoxon rank-sum interval (15.12) is very similar to the Wilcoxon signed-rank interval (15.10); (15.10) uses pairwise averages from a single sample, while (15.12) uses pairwise differences from two samples. Appendix Table A.12 gives values of c for selected values of m and n.

Example 15.14   The paper "Some Mechanical Properties of Impregnated Bark Board" (*Forest Products J.,* 1977, pp. 31–38) reported the following data on maximum crushing strength (psi) for a sample of epoxy-impregnated barkboard and for a sample of barkboard impregnated with another polymer:

Epoxy (x's): 10,860, 11,120, 11,340, 12,130, 14,380, 13,070
Other (y's): 4,590, 4,850, 6,510, 5,640, 6,390

Obtain a 95% confidence interval for the true average difference in crushing strength between the epoxy-impregnated board and the other type of board.

From Appendix Table A.12, since the smaller sample size is five and the larger sample size is six, $c = 26$ for a confidence level of approximately 95%. The d_{ij}'s appear in Figure 15.7. The five smallest d_{ij}'s $[d_{ij(1)}, \ldots, d_{ij(5)}]$ are 4350, 4470, 4610, 4730, and 4830, while the five largest d_{ij}'s are (in descending order) 9790, 9530, 8740, 8480, and 8220. Thus the confidence interval is $(d_{ij(5)}, d_{ij(26)}) = (4830, 8220)$.

d_{ij}	4590	4850	y_j 5640	6390	6510
10,860	6270	6010	5220	4470	4350
11,120	6530	6270	5480	4730	4610
x_i 11,340	6750	6490	5700	4950	4830
12,130	7540	7280	6490	5740	5620
13,070	8480	8220	7430	6680	6560
14,380	9790	9530	8740	7990	7870

Figure 15.7   ∎

When m and n are both large, the Wilcoxon test statistic has approximately a normal distribution. This can be used to derive a large-sample approximation for the value c in (15.12). The result is

$$c \approx \frac{mn}{2} + z_{\alpha/2} \sqrt{\frac{mn(m + n + 1)}{12}} \tag{15.13}$$

As with the signed-rank interval, the rank-sum interval (15.12) is quite efficient with respect to the t interval; in large samples (15.12) will tend to be only a bit longer than the t interval when the underlying populations are normal, and may be considerably shorter than the t interval if the underlying populations have heavier tails than do normal populations.

Exercises / Section 15.4 (21–28)

21. Use the interval associated with the sign test to compute a 99% confidence interval for the true median travel time $\tilde{\mu}$ using the data in Exercise 2.

22. Compute a 98% confidence interval for the true median soil pH $\tilde{\mu}$ using the data in Exercise 3. For the given sample size, can a 95% confidence interval be obtained? Explain.

23. The paper "The Lead Content and Acidity of Christchurch Precipitation" (*New Zealand J. Science*, 1980, pp. 311–312) reported the accompanying data on lead concentration (μg/1) in samples gathered during eight different summer rainfalls. Assuming that the lead content distribution is symmetric, use the Wilcoxon signed-rank interval to obtain a 95% confidence interval for μ.

17.0, 21.4, 30.6, 5.0, 12.2, 11.8, 17.3, 18.8

24. Compute the 99% signed-rank interval for true average pH μ (assuming symmetry) using the data in Exercise 10. *Hint:* Try to compute only those pairwise averages having relatively small or large values (rather than all 105 averages).

25. Compute a 94% confidence interval for μ_D of Example 15.6 using the data given there.

26. The following observations are amounts of hydrocarbon emissions resulting from road wear of bias-belted tires under a 522 kg load inflated at 228 kPa and driven at 64 km/hr for six hours ("Characterization of Tire Emissions Using an Indoor Test Facility," *Rubber Chemistry and Technology,* 1978, pp. 7–25):

.045, .117, .062, .072

What confidence levels are achievable for this sample size using the signed-rank interval? Select an appropriate confidence level and compute the interval.

27. Compute the 90% rank-sum confidence interval for $\mu_1 - \mu_2$ using the data in Exercise 15.

28. Compute a 99% confidence interval for $\mu_1 - \mu_2$ using the data in Exercise 16.

Distribution-Free Analysis of Variance

The single-factor ANOVA model of Chapter 10 for comparing I population or treatment means assumed that for $i = 1, 2, \ldots, I$, a random sample of size J_i was drawn from a normal population with mean μ_i and variance σ^2. This can be written as

$$X_{ij} = \mu_i + \epsilon_{ij} \quad j = 1, \ldots, J_i; \quad i = 1, \ldots, I \tag{15.14}$$

where the ϵ_{ij}'s are independent and normally distributed with mean zero and variance σ^2. While the normality assumption was required for the validity of the F test described in Chapter 10, the validity of the Kruskal-Wallis test for testing equality of the μ_i's depends only on the ϵ_{ij}'s having the same continuous distribution.

The Kruskal-Wallis Test

Let $N = \Sigma J_i$, the total number of observations in the data set, and suppose that we rank all N observations from 1 (the smallest X_{ij}) to N (the largest X_{ij}). When $H_0 : \mu_1 = \mu_2 = \cdots = \mu_I$ is true, the N observations all come from the same distribution, in which case all possible assignments of the ranks $1, 2, \ldots, N$ to the I samples are equally likely and we expect ranks to be intermingled in these samples. If, however, H_0 is false, then some samples will consist mostly of observations having small ranks in the combined sample while others will consist mostly of observations having large ranks. More specifically, if R_{ij} denotes the rank of X_{ij} among the N observations and $R_{i.}$ and $\bar{R}_{i.}$ denote respectively the total and average of the ranks in the ith sample, then when H_0 is true

$$E(R_{ij}) = \frac{N + 1}{2}, \qquad E(\bar{R}_{i.}) = \frac{1}{J_i} \sum_j E(R_{ij}) = \frac{N + 1}{2}$$

The K-W test statistic is a measure of the extent to which the $\bar{R}_{i.}$'s deviate from their common expected value $(N + 1)/2$, and H_0 is rejected if the computed value of the statistic indicates too great a discrepancy between observed and expected rank averages.

Test statistic

$$K = \frac{12}{N(N+1)} \sum_{j=1}^{I} J_i \left(\bar{R}_{i\cdot} - \frac{N+1}{2} \right)^2$$

$$= \frac{12}{N(N+1)} \sum_{i=1}^{I} \frac{R_{i\cdot}^2}{J_i} - 3(N+1)$$

(15.15)

The second expression for K is the computational formula; it involves the rank totals ($R_{i\cdot}$'s) rather than the averages and requires only one subtraction.

If H_0 is rejected when $k \geq c$, then c should be chosen so that the test has level α. That is, c should be the upper-tail critical value of the distribution of K when H_0 is true. Under H_0, each possible assignment of the ranks to the I samples is equally likely, so in theory all such assignments can be enumerated, the value of K determined for each one, and the null distribution obtained by counting the number of times each value of K occurs. Clearly this computation is tedious, so while there are tables of the exact null distribution and critical values for small values of the J_i's, we shall use the following "large-sample" approximation.

Proposition

When H_0 is true and either

$$I = 3, \quad J_i \geq 6 \quad i = 1, 2, 3, \text{ or}$$
$$I > 3, \quad J_i \geq 5 \quad i = 1, \ldots, I$$

then K has approximately a chi-squared distribution with $I - 1$ d.f.

This implies that a test with approximate significance level α rejects H_0 if $k \geq \chi^2_{\alpha, I-1}$.

Example 15.15

The accompanying observations on axial stiffness index resulted from a study of metal plate connected trusses in which five different plate lengths—4 in., 6 in., 8 in., 10 in., and 12 in.—were used ("Modeling Joints Made with Light-Gauge Metal Connector Plates," *Forest Products J.*, 1979, pp. 39–44).

$i = 1$ (4"):	309.2	309.7	311.0	316.8	326.5	349.8	409.5
$i = 2$ (6"):	331.0	347.2	348.9	361.0	381.7	402.1	404.5
$i = 3$ (8"):	351.0	357.1	366.2	367.3	382.0	392.4	409.9
$i = 4$ (10"):	346.7	362.6	384.2	410.6	433.1	452.9	461.4
$i = 5$ (12"):	407.4	410.7	419.9	441.2	441.8	465.8	473.4

									$r_{i\cdot}$	$\bar{r}_{i\cdot}$
	$i = 1$:	1	2	3	4	5	10	24	49	7.00
	$i = 2$:	6	8	9	13	17	21	22	96	13.71
Ranks	$i = 3$:	11	12	15	16	18	20	25	117	16.71
	$i = 4$:	7	14	19	26	29	32	33	160	22.86
	$i = 5$:	23	27	28	30	31	34	35	208	29.71

Figure 15.8

The computed value of K is

$$k = \frac{12}{35(36)} \left[\frac{(49)^2}{7} + \frac{(96)^2}{7} + \frac{(117)^2}{7} + \frac{(160)^2}{7} + \frac{(208)^2}{7} \right] - 3(36) = 20.12$$

At level .01, $\chi^2_{.01,4} = 13.28$, and since 20.12 is ≥ 13.28, H_0 is rejected and we conclude that expected axial stiffness does depend on plate length. ∎

Friedman's Test for a Randomized Block Experiment

Suppose that $X_{ij} = \mu + \alpha_i + \beta_j + \epsilon_{ij}$ where α_i is the ith treatment effect, β_j is the jth block effect, and the ϵ_{ij}'s are drawn independently from the same continuous (but not necessarily normal) distribution. Then to test $H_0: \alpha_1 = \alpha_2 = \cdots = \alpha_I = 0$, the null hypothesis of no treatment effects, the observations are first ranked separately from 1 to I within each block, and then the rank average $\bar{r}_{i.}$ is computed for each of the I treatments. When H_0 is true, the $\bar{r}_{i.}$'s should be close to one another, since within each block all $I!$ assignments of ranks to treatments are equally likely. Friedman's test statistic measures the discrepancy between the expected value $(I + 1)/2$ of each rank average and the $\bar{r}_{i.}$'s.

Test statistic

$$F_r = \frac{12J}{I(I+1)} \sum_{i=1}^{I} \left(\bar{R}_{i.} - \frac{I+1}{2} \right)^2 = \frac{12}{IJ(I+1)} \sum R_{i.}^2 - 3J(I+1)$$

As with the K-W test, Friedman's test rejects H_0 when the computed value of the test statistic is too large. For the cases $I = 3$, $J = 2, \ldots, 15$ and $I = 4$, $J = 2, \ldots, 8$, Lehmann's book gives the upper-tail critical values for the test. Alternatively, for even moderate values of J, the test statistic F_r has approximately a chi-squared distribution with $I - 1$ d.f. when H_0 is true, so H_0 can be rejected if $f_r \geq \chi^2_{\alpha, I-1}$.

Example **15.16**   The paper "Physiological Effects During Hypnotically Requested Emotions" (*Psychosomatic Med.*, 1963, pp. 334–343) reported the following data on skin potential (millivolts) when the emotions of fear, happiness, depression, and calmness were requested from each of eight subjects.

Blocks (subjects)

x_{ij}	1	2	3	4	5	6	7	8
Fear	23.1	57.6	10.5	23.6	11.9	54.6	21.0	20.3
Happiness	22.7	53.2	9.7	19.6	13.8	47.1	13.6	23.6
Depression	22.5	53.7	10.8	21.1	13.7	39.2	13.7	16.3
Calmness	22.6	53.1	8.3	21.6	13.3	37.0	14.8	14.8

Ranks	1	2	3	4	5	6	7	8	$r_{i.}$	$r_{i.}^2$
Fear	4	4	3	4	1	4	4	3	27	729
Happiness	3	2	2	1	4	3	1	4	20	400
Depression	1	3	4	2	3	2	2	2	19	361
Calmness	2	1	1	3	2	1	3	1	14	196
										1686

Figure 15.9

Thus

$$f_r = \frac{12}{4(8)(5)}(1686) - 3(8)(5) = 6.45$$

At level .05, $\chi^2_{.05, 3} = 7.815$, and because 6.45 is not ≥ 7.815, H_0 is not rejected. There is no evidence that average skin potential depends on which emotion is requested. ∎

The book by Hollander and Wolfe discusses multiple comparison procedures associated with the Kruskal-Wallis and Friedman tests, as well as other aspects of distribution-free ANOVA.

Exercises / Section 15.5 (29–33)

29. The accompanying data refers to concentration of the radioactive isotope strontium-90 in milk samples obtained from five randomly selected dairies in each of four different regions.

	1:	6.4	5.8	6.5	7.7	6.1
Region	2:	7.1	9.9	11.2	10.5	8.8
	3:	5.7	5.9	8.2	6.6	5.1
	4:	9.5	12.1	10.3	12.4	11.7

Test at level .10 to see whether or not true average strontium-90 concentration differs for at least two of the regions.

30. The paper "Production of Gaseous Nitrogen in Human Steady-State Conditions" (*J. Applied Physiology*, 1972, pp. 155–159) reported the following observations on the amount of nitrogen expired (in liters) under four dietary regimens: fasting (1), 23% protein (2), 32% protein (3), and 67% protein (4). Use the K-W test at level .05 to test equality of the corresponding μ_i's.

1.	4.079	4.859	3.540	5.047	3.298
2.	4.368	5.668	3.752	5.848	3.802
3.	4.169	5.709	4.416	5.666	4.123
4.	4.928	5.608	4.940	5.291	4.674

1.	4.679	2.870	4.648	3.847
2.	4.844	3.578	5.393	4.374
3.	5.059	4.403	4.496	4.688
4.	5.038	4.905	5.208	4.806

31. The accompanying data on cortisol level was reported in the paper "Cortisol, Cortisone, and 11-Deoxycortisol Levels in Human Umbilical and Maternal Plasma in Relation to the Onset of Labor" (*J. Obstetric Gynaecology British Commonwealth*, 1974, pp. 737–745). Experimental subjects were pregnant women who were delivered between 38 and 42 weeks gestation. Group 1 individuals elected to deliver by Caesarean section before labor onset, group 2 delivered by emergency Caesarean during induced labor, and group 3 individuals experienced spontaneous labor. Use the K-W test at level .05 to test for equality of the three population means.

Group 1:	262	307	211	323	454	339
	304	154	287	356		
Group 2:	465	501	455	355	468	362
Group 3:	343	772	207	1048	838	687

32. In a test to determine if soil pretreated with small amounts of Basic-H makes the soil more permeable to water, soil samples were divided into blocks and each block received each of the four treatments under study. The treatments were (A) water with .001% Basic-H flooded on control soil, (B) water without Basic-H on control, (C) water with Basic-H flooded on soil pretreated with Basic-H, and (D) water without Basic-H on soil pretreated with Basic-H. Test at level .01 to see if there are any effects due to the different treatments.

	Blocks				
	1	2	3	4	5
A	37.1	31.8	28.0	25.9	25.5
B	33.2	25.3	20.2	20.3	18.3
C	58.9	54.2	49.2	47.9	38.2
D	56.7	49.6	46.4	40.9	39.4

	6	7	8	9	10
A	25.3	23.7	24.4	21.7	26.2
B	19.3	17.3	17.0	16.7	18.3
C	48.8	47.8	40.2	44.0	46.4
D	37.1	37.5	39.6	35.1	36.5

33. In an experiment to study the way in which different anesthetics affected plasma epinephrine concentration, 10 dogs were selected and concentration was measured while under the influence of the anesthetics isoflurane, halothane, and cyclopropane ("Sympathoadrenal and Hemodynamic Effects of Isoflurane, Halothane, and Cyclopropane in Dogs," *Anesthesiology,* 1974, pp. 465–470). Test at level .05 to see whether there is an anesthetic effect on concentration.

	\multicolumn{5}{c|}{*Dog*}				
	1	2	3	4	5
Isoflurane:	.28	.51	1.00	.39	.29
Halothane:	.30	.39	.63	.38	.21
Cyclopropane:	1.07	1.35	.69	.28	1.24
	6	7	8	9	10
Isoflurane:	.36	.32	.69	.17	.33
Halothane:	.88	.39	.51	.32	.42
Cyclopropane:	1.53	.49	.56	1.02	.30

Supplementary Exercises / Chapter 15 (34–40)

34. The paper "Effects of a Rice-rich versus Potato-rich Diet on Glucose, Lipoprotein, and Cholesterol Metabolism in Noninsulin-Dependent Diabetics" (*Amer. J. Clinical Nutr.,* 1984, pp. 598–606) gave the accompanying data on cholesterol synthesis rate for eight diabetic subjects. Subjects were fed a standardized diet with potato or rice as the major carbohydrate source. Participants received both diets for specified periods of time, with cholesterol synthesis rate (mmol/day) measured at the end of each dietary period. The analysis presented in this paper used a distribution-free test. Use such a test with significance level .05 to determine whether the true mean cholesterol synthesis rate differs significantly for the two sources of carbohydrates.

\multicolumn{9}{c}{*Cholesterol Synthesis Rate*}
Subject	1	2	3	4	5	6	7	8
Potato	1.88	2.60	1.38	4.41	1.87	2.89	3.96	2.31
Rice	1.70	3.84	1.13	4.97	.86	1.93	3.36	2.15

35. High-pressure sales tactics or door-to-door salespeople can be quite offensive. Many people succumb to such tactics, sign a purchase agreement, and later regret their actions. In the mid-1970s the Federal Trade Commission implemented regulations clarifying and extending rights of purchasers to cancel such agreements. The accompanying data is a subset of that given in the paper "Evaluating the FTC Cooling-Off Rule" (*J. Consumer Affairs,* 1977, pp. 101–106). Individual observations are cancellation rates for each of nine salespeople during each of 4 years. Use an appropriate test at level .05 to see if true average cancellation rate depends on the year.

\multicolumn{10}{c}{*Salesperson*}
	1	2	3	4	5	6	7	8	9
1973	2.8	5.9	3.3	4.4	1.7	3.8	6.6	3.1	0.0
1974	3.6	1.7	5.1	2.2	2.1	4.1	4.7	2.7	1.3
1975	1.4	.9	1.1	3.2	.8	1.5	2.8	1.4	.5
1976	2.0	2.2	.9	1.1	.5	1.2	1.4	3.5	1.2

36. The given data on phosphorus concentration in topsoil for four different soil treatments appeared in the article "Fertilisers for Lotus and Clover Establishment on a Sequence of Acid Soils on the East Otago Uplands" (*N. Zeal. J. Exper. Ag.,* 1984, pp. 119–129). Use a distribution-free procedure to test the null hypothesis of no difference in true mean phosphorus concentration (mg/g) for the four soil treatments.

Treatment					
I	8.1	5.9	7.0	8.0	9.0
II	11.5	10.9	12.1	10.3	11.9
III	15.3	17.4	16.4	15.8	16.0
IV	23.0	33.0	28.4	24.6	27.7

37. Refer back to the data of Exercise 36, and compute a 95% confidence interval for the difference between true average concentrations for treatments II and III.

38. The study reported in "Gait Patterns During Free Choice Ladder Ascents" (*Human Movement Sci.,* 1983, pp. 187–195) was motivated by publicity concerning the increased accident rate for individuals climbing ladders. A number of different gait patterns were used by subjects climbing a portable straight ladder according to specified instructions. The ascent times for seven subjects who used a lateral gait and six subjects who used a four-beat diagonal gait are given.

Lateral   .86 1.31 1.64 1.51 1.53 1.39 1.09
Diagonal 1.27 1.82 1.66 .85 1.45 1.24

a. Carry out a test using $\alpha = .05$ to see if the data suggests any difference in the true average ascent times for the two gaits.

b. Compute a 95% confidence interval for the difference between the true average gait times.

39. Suppose that we wish to test

H_0: the X and Y distributions are identical

versus

H_a: the X distribution is less spread out than the Y distribution

The accompanying figure pictures X and Y distributions for which H_a is true. The Wilcoxon rank-sum test is not appropriate in this situation because when H_a is true as pictured, the Y's will tend to be at the extreme ends of the combined sample (resulting in small and large Y ranks), so the sum of X ranks will result in a W value that is neither large nor small.

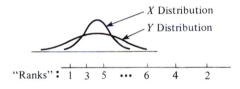

Consider modifying the procedure for assigning ranks as follows: After ordering the combined sample of $m + n$ observations, the smallest observation is given rank 1, the largest observation is given rank 2, the second smallest is given rank 3, the second largest is given rank 4, and so on. Then if H_a is true as pictured, the X values will tend to be in the middle of the sample and thus receive large ranks. Let W' denote the sum of the X ranks, and consider rejecting H_0 in favor of H_a when $w' \geq c$. When H_0 is true, any possible set of X ranks has the same probability, so W' *has the same distribution as does W when H_0 is true*. Thus c can be chosen from Appendix Table A.10 to yield a level α test. The data below refers to medial muscle thickness for arterioles from the lungs of both children who died from sudden infant death syndrome (x's) and a control group of children (y's). Carry out the test of H_0 versus H_a at level .05.

SIDS: 4.0,   4.4,   4.8,   4.9
Control: 3.7,   4.1,   4.3,   5.1,   5.6

See Lehmann for more information on this, called the Siegel-Tukey test.

40. The ranking procedure described in Exercise 39 is somewhat asymmetrical, since the smallest observation receives rank 1 while the largest receives rank 2, and so on. Suppose that both the smallest and the largest receive rank 1, the second smallest and second largest receive rank 2, and so on, and let W'' be the sum of the X ranks. The null distribution of W'' is no longer identical to the null distribution of W, so different tables are needed. Consider the case $m = 3$, $n = 4$. List all 35 possible orderings of the three X values among the seven observations (say, 1, 3, 7 or 4, 5, 6), assign ranks in the manner described, compute the value of W'' for each possibility, and then tabulate the null distribution of W''. For the test that rejects if $w'' \geq c$, what value of c prescribes approximately a level .10 test? This is the Ansari-Bradley test; see the book by Hollander and Wolfe.

Bibliography

Hollander, Myles, and Wolfe, Douglas, *Nonparametric Statistical Methods,* Wiley, New York, 1973. A very good reference on distribution-free methods with an excellent collection of tables.

Lehmann, Erich, *Nonparametrics: Statistical Methods Based on Ranks,* Holden-Day, San Francisco, 1975. An excellent discussion of the most important distribution-free methods, presented with a great deal of insightful commentary.

Marascuilo, Leonard, and McSweeney, Maryellen, *Nonparametric and Distribution-Free Methods for the Social Sciences,* Brooks/Cole, Monterey, Calif., 1977. A good survey, with particular attention given to multiple comparison methods.

Appendix Tables

Table A.1 Cumulative Binomial Probabilities

$$B(x; n, p) = \sum_{y=0}^{x} b(y; n, p)$$

a. $n = 5$

							p								
	0.01	0.05	0.10	0.20	0.25	0.30	0.40	0.50	0.60	0.70	0.75	0.80	0.90	0.95	0.99
0	.951	.774	.590	.328	.237	.168	.078	.031	.010	.002	.001	.000	.000	.000	.000
1	.999	.977	.919	.737	.633	.528	.337	.188	.087	.031	.016	.007	.000	.000	.000
x 2	1.000	.999	.991	.942	.896	.837	.683	.500	.317	.163	.104	.058	.009	.001	.000
3	1.000	1.000	1.000	.993	.984	.969	.913	.812	.663	.472	.367	.263	.081	.023	.001
4	1.000	1.000	1.000	1.000	.999	.998	.990	.969	.922	.832	.763	.672	.410	.226	.049

b. $n = 10$

							p								
	0.01	0.05	0.10	0.20	0.25	0.30	0.40	0.50	0.60	0.70	0.75	0.80	0.90	0.95	0.99
0	.904	.599	.349	.107	.056	.028	.006	.001	.000	.000	.000	.000	.000	.000	.000
1	.996	.914	.736	.376	.244	.149	.046	.011	.002	.000	.000	.000	.000	.000	.000
2	1.000	.988	.930	.678	.526	.383	.167	.055	.012	.002	.000	.000	.000	.000	.000
3	1.000	.999	.987	.879	.776	.650	.382	.172	.055	.011	.004	.001	.000	.000	.000
4	1.000	1.000	.998	.967	.922	.850	.633	.377	.166	.047	.020	.006	.000	.000	.000
x 5	1.000	1.000	1.000	.994	.980	.953	.834	.623	.367	.150	.078	.033	.002	.000	.000
6	1.000	1.000	1.000	.999	.996	.989	.945	.828	.618	.350	.224	.121	.013	.001	.000
7	1.000	1.000	1.000	1.000	1.000	.998	.988	.945	.833	.617	.474	.322	.070	.012	.000
8	1.000	1.000	1.000	1.000	1.000	1.000	.998	.989	.954	.851	.756	.624	.264	.086	.004
9	1.000	1.000	1.000	1.000	1.000	1.000	1.000	.999	.994	.972	.944	.893	.651	.401	.096

c. $n = 15$

							p								
	0.01	0.05	0.10	0.20	0.25	0.30	0.40	0.50	0.60	0.70	0.75	0.80	0.90	0.95	0.99
0	.860	.463	.206	.035	.013	.005	.000	.000	.000	.000	.000	.000	.000	.000	.000
1	.990	.829	.549	.167	.080	.035	.005	.000	.000	.000	.000	.000	.000	.000	.000
2	1.000	.964	.816	.398	.236	.127	.027	.004	.000	.000	.000	.000	.000	.000	.000
3	1.000	.995	.944	.648	.461	.297	.091	.018	.002	.000	.000	.000	.000	.000	.000
4	1.000	.999	.987	.836	.686	.515	.217	.059	.009	.001	.000	.000	.000	.000	.000
5	1.000	1.000	.998	.939	.852	.722	.403	.151	.034	.004	.001	.000	.000	.000	.000
6	1.000	1.000	1.000	.982	.943	.869	.610	.304	.095	.015	.004	.001	.000	.000	.000
x 7	1.000	1.000	1.000	.996	.983	.950	.787	.500	.213	.050	.017	.004	.000	.000	.000
8	1.000	1.000	1.000	.999	.996	.985	.905	.696	.390	.131	.057	.018	.000	.000	.000
9	1.000	1.000	1.000	1.000	.999	.996	.966	.849	.597	.278	.148	.061	.002	.000	.000
10	1.000	1.000	1.000	1.000	1.000	.999	.991	.941	.783	.485	.314	.164	.013	.001	.000
11	1.000	1.000	1.000	1.000	1.000	1.000	.998	.982	.909	.703	.539	.352	.056	.005	.000
12	1.000	1.000	1.000	1.000	1.000	1.000	1.000	.996	.973	.873	.764	.602	.184	.036	.000
13	1.000	1.000	1.000	1.000	1.000	1.000	1.000	1.000	.995	.965	.920	.833	.451	.171	.010
14	1.000	1.000	1.000	1.000	1.000	1.000	1.000	1.000	1.000	.995	.987	.965	.794	.537	.140

d. $n = 20$

							p								
	0.01	0.05	0.10	0.20	0.25	0.30	0.40	0.50	0.60	0.70	0.75	0.80	0.90	0.95	0.99
0	.818	.358	.122	.012	.003	.001	.000	.000	.000	.000	.000	.000	.000	.000	.000
1	.983	.736	.392	.069	.024	.008	.001	.000	.000	.000	.000	.000	.000	.000	.000
2	.999	.925	.677	.206	.091	.035	.004	.000	.000	.000	.000	.000	.000	.000	.000
3	1.000	.984	.867	.411	.225	.107	.016	.001	.000	.000	.000	.000	.000	.000	.000
4	1.000	.997	.957	.630	.415	.238	.051	.006	.000	.000	.000	.000	.000	.000	.000

(continued)

Table A.1 Cumulative Binomial Probabilities (*cont.*)

$$B(x; n, p) = \sum_{y=0}^{x} b(y; n, p)$$

d. $n = 20$ *(continued)*

x	0.01	0.05	0.10	0.20	0.25	0.30	0.40	0.50	0.60	0.70	0.75	0.80	0.90	0.95	0.99
5	1.000	1.000	.989	.804	.617	.416	.126	.021	.002	.000	.000	.000	.000	.000	.000
6	1.000	1.000	.998	.913	.786	.608	.250	.058	.006	.000	.000	.000	.000	.000	.000
7	1.000	1.000	1.000	.968	.898	.772	.416	.132	.021	.001	.000	.000	.000	.000	.000
8	1.000	1.000	1.000	.990	.959	.887	.596	.252	.057	.005	.001	.000	.000	.000	.000
9	1.000	1.000	1.000	.997	.986	.952	.755	.412	.128	.017	.004	.001	.000	.000	.000
10	1.000	1.000	1.000	.999	.996	.983	.872	.588	.245	.048	.014	.003	.000	.000	.000
11	1.000	1.000	1.000	1.000	.999	.995	.943	.748	.404	.113	.041	.010	.000	.000	.000
12	1.000	1.000	1.000	1.000	1.000	.999	.979	.868	.584	.228	.102	.032	.000	.000	.000
13	1.000	1.000	1.000	1.000	1.000	1.000	.994	.942	.750	.392	.214	.087	.002	.000	.000
14	1.000	1.000	1.000	1.000	1.000	1.000	.998	.979	.874	.584	.383	.196	.011	.000	.000
15	1.000	1.000	1.000	1.000	1.000	1.000	1.000	.994	.949	.762	.585	.370	.043	.003	.000
16	1.000	1.000	1.000	1.000	1.000	1.000	1.000	.999	.984	.893	.775	.589	.133	.016	.000
17	1.000	1.000	1.000	1.000	1.000	1.000	1.000	1.000	.996	.965	.909	.794	.323	.075	.001
18	1.000	1.000	1.000	1.000	1.000	1.000	1.000	1.000	.999	.992	.976	.931	.608	.264	.017
19	1.000	1.000	1.000	1.000	1.000	1.000	1.000	1.000	1.000	.999	.997	.988	.878	.642	.182

e. $n = 25$

x	0.01	0.05	0.10	0.20	0.25	0.30	0.40	0.50	0.60	0.70	0.75	0.80	0.90	0.95	0.99
0	.778	.277	.072	.004	.001	.000	.000	.000	.000	.000	.000	.000	.000	.000	.000
1	.974	.642	.271	.027	.007	.002	.000	.000	.000	.000	.000	.000	.000	.000	.000
2	.998	.873	.537	.098	.032	.009	.000	.000	.000	.000	.000	.000	.000	.000	.000
3	1.000	.966	.764	.234	.096	.033	.002	.000	.000	.000	.000	.000	.000	.000	.000
4	1.000	.993	.902	.421	.214	.090	.009	.000	.000	.000	.000	.000	.000	.000	.000
5	1.000	.999	.967	.617	.378	.193	.029	.002	.000	.000	.000	.000	.000	.000	.000
6	1.000	1.000	.991	.780	.561	.341	.074	.007	.000	.000	.000	.000	.000	.000	.000
7	1.000	1.000	.998	.891	.727	.512	.154	.022	.001	.000	.000	.000	.000	.000	.000
8	1.000	1.000	1.000	.953	.851	.677	.274	.054	.004	.000	.000	.000	.000	.000	.000
9	1.000	1.000	1.000	.983	.929	.811	.425	.115	.013	.000	.000	.000	.000	.000	.000
10	1.000	1.000	1.000	.994	.970	.902	.586	.212	.034	.002	.000	.000	.000	.000	.000
11	1.000	1.000	1.000	.998	.980	.956	.732	.345	.078	.006	.001	.000	.000	.000	.000
12	1.000	1.000	1.000	1.000	.997	.983	.846	.500	.154	.017	.003	.000	.000	.000	.000
13	1.000	1.000	1.000	1.000	.999	.994	.922	.655	.268	.044	.020	.002	.000	.000	.000
14	1.000	1.000	1.000	1.000	1.000	.998	.966	.788	.414	.098	.030	.006	.000	.000	.000
15	1.000	1.000	1.000	1.000	1.000	1.000	.987	.885	.575	.189	.071	.017	.000	.000	.000
16	1.000	1.000	1.000	1.000	1.000	1.000	.996	.946	.726	.323	.149	.047	.000	.000	.000
17	1.000	1.000	1.000	1.000	1.000	1.000	.999	.978	.846	.488	.273	.109	.002	.000	.000
18	1.000	1.000	1.000	1.000	1.000	1.000	1.000	.993	.926	.659	.439	.220	.009	.000	.000
19	1.000	1.000	1.000	1.000	1.000	1.000	1.000	.998	.971	.807	.622	.383	.033	.001	.000
20	1.000	1.000	1.000	1.000	1.000	1.000	1.000	1.000	.991	.910	.786	.579	.098	.007	.000
21	1.000	1.000	1.000	1.000	1.000	1.000	1.000	1.000	.998	.967	.904	.766	.236	.034	.000
22	1.000	1.000	1.000	1.000	1.000	1.000	1.000	1.000	1.000	.991	.968	.902	.463	.127	.002
23	1.000	1.000	1.000	1.000	1.000	1.000	1.000	1.000	1.000	.998	.993	.973	.729	.358	.026
24	1.000	1.000	1.000	1.000	1.000	1.000	1.000	1.000	1.000	1.000	.999	.996	.928	.723	.222

Source: Adapted from *Statistics for Management*, by Lincoln L. Chao. Copyright © 1980 by Wadsworth, Inc. Reprinted by permission of Brooks/Cole Publishing Company, Monterey.

Table A.2 Cumulative Poisson Probabilities

$$F(x; \lambda) = \sum_{y=0}^{x} \frac{e^{-\lambda}\lambda^{y}}{y!}$$

					λ					
	.1	.2	.3	.4	.5	.6	.7	.8	.9	1.0
0	.905	.819	.741	.670	.607	.549	.497	.449	.407	.368
1	.995	.982	.963	.938	.910	.878	.844	.809	.772	.736
2	1.000	.999	.996	.992	.986	.977	.966	.953	.937	.920
x 3		1.000	1.000	.999	.998	.997	.994	.991	.945	.981
4				1.000	1.000	1.000	.999	.999	.989	.996
5							1.000	1.000	.998	.999
6									1.000	1.000

	λ										
	2.0	3.0	4.0	5.0	6.0	7.0	8.0	9.0	10.0	15.0	20.0
0	.135	.050	.018	.007	.002	.001	.000	.000	.000	.000	.000
1	.406	.199	.092	.040	.017	.007	.003	.001	.000	.000	.000
2	.677	.423	.238	.125	.062	.030	.014	.006	.003	.000	.000
3	.857	.647	.433	.265	.151	.082	.042	.021	.010	.000	.000
4	.947	.815	.629	.440	.285	.173	.100	.055	.029	.001	.000
5	.983	.916	.785	.616	.446	.301	.191	.116	.067	.003	.000
6	.995	.966	.889	.762	.606	.456	.313	.207	.130	.008	.000
7	.999	.988	.949	.867	.744	.599	.453	.324	.220	.018	.001
8	1.000	.996	.979	.932	.847	.729	.593	.456	.333	.037	.002
9		.999	.992	.968	.916	.830	.717	.587	.458	.070	.005
10		1.000	.997	.986	.957	.901	.816	.706	.583	.118	.011
11			.999	.995	.980	.947	.888	.803	.697	.185	.021
12			1.000	.998	.991	.973	.936	.876	.792	.268	.039
13				.999	.996	.987	.966	.926	.864	.363	.066
14				1.000	.999	.994	.983	.959	.917	.466	.105
15					.999	.998	.992	.978	.951	.568	.157
16					1.000	.999	.996	.989	.973	.664	.221
17						1.000	.998	.995	.986	.749	.297
x 18							1.000	.999	.993	.819	.381
19								1.000	.997	.875	.470
20									.998	.917	.559
21									.999	.947	.644
22									1.000	.967	.721
23										.981	.787
24										.989	.843
25										.994	.888
26										.997	.922
27										.998	.948
28										.999	.966
29										1.000	.978
30											.987
31											.992
32											.995
33											.997
34											.999
35											.999
36											1.000

Source: Lincoln L. Chao, *Statistics: Methods and Analysis*, (2nd ed.), New York: McGraw-Hill Book Company, Copyright © 1974. Reprinted by permission.

Table A.3 Standard Normal Curve Areas

$\Phi(z) = P(Z \leq z)$

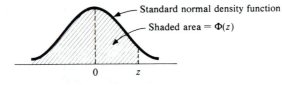

Standard normal density function

Shaded area $= \Phi(z)$

z	0.00	0.01	0.02	0.03	0.04	0.05	0.06	0.07	0.08	0.09
−3.4	0.0003	0.0003	0.0003	0.0003	0.0003	0.0003	0.0003	0.0003	0.0003	0.0002
−3.3	0.0005	0.0005	0.0005	0.0004	0.0004	0.0004	0.0004	0.0004	0.0004	0.0003
−3.2	0.0007	0.0007	0.0006	0.0006	0.0006	0.0006	0.0006	0.0005	0.0005	0.0005
−3.1	0.0010	0.0009	0.0009	0.0009	0.0008	0.0008	0.0008	0.0008	0.0007	0.0007
−3.0	0.0013	0.0013	0.0013	0.0012	0.0012	0.0011	0.0011	0.0011	0.0010	0.0010
−2.9	0.0019	0.0018	0.0017	0.0017	0.0016	0.0016	0.0015	0.0015	0.0014	0.0014
−2.8	0.0026	0.0025	0.0024	0.0023	0.0023	0.0022	0.0021	0.0021	0.0020	0.0019
−2.7	0.0035	0.0034	0.0033	0.0032	0.0031	0.0030	0.0029	0.0028	0.0027	0.0026
−2.6	0.0047	0.0045	0.0044	0.0043	0.0041	0.0040	0.0039	0.0038	0.0037	0.0036
−2.5	0.0062	0.0060	0.0059	0.0057	0.0055	0.0054	0.0052	0.0051	0.0049	0.0048
−2.4	0.0082	0.0080	0.0078	0.0075	0.0073	0.0071	0.0069	0.0068	0.0066	0.0064
−2.3	0.0107	0.0104	0.0102	0.0099	0.0096	0.0094	0.0091	0.0089	0.0087	0.0084
−2.2	0.0139	0.0136	0.0132	0.0129	0.0125	0.0122	0.0119	0.0116	0.0113	0.0110
−2.1	0.0179	0.0174	0.0170	0.0166	0.0162	0.0158	0.0154	0.0150	0.0146	0.0143
−2.0	0.0228	0.0222	0.0217	0.0212	0.0207	0.0202	0.0197	0.0192	0.0188	0.0183
−1.9	0.0287	0.0281	0.0274	0.0268	0.0262	0.0256	0.0250	0.0244	0.0239	0.0233
−1.8	0.0359	0.0352	0.0344	0.0336	0.0329	0.0322	0.0314	0.0307	0.0301	0.0294
−1.7	0.0446	0.0436	0.0427	0.0418	0.0409	0.0401	0.0392	0.0384	0.0375	0.0367
−1.6	0.0548	0.0537	0.0526	0.0516	0.0505	0.0495	0.0485	0.0475	0.0465	0.0455
−1.5	0.0668	0.0655	0.0643	0.0630	0.0618	0.0606	0.0594	0.0582	0.0571	0.0559
−1.4	0.0808	0.0793	0.0778	0.0764	0.0749	0.0735	0.0722	0.0708	0.0694	0.0681
−1.3	0.0968	0.0951	0.0934	0.0918	0.0901	0.0885	0.0869	0.0853	0.0838	0.0823
−1.2	0.1151	0.1131	0.1112	0.1093	0.1075	0.1056	0.1038	0.1020	0.1003	0.0985
−1.1	0.1357	0.1335	0.1314	0.1292	0.1271	0.1251	0.1230	0.1210	0.1190	0.1170
−1.0	0.1587	0.1562	0.1539	0.1515	0.1492	0.1469	0.1446	0.1423	0.1401	0.1379
−0.9	0.1841	0.1814	0.1788	0.1762	0.1736	0.1711	0.1685	0.1660	0.1635	0.1611
−0.8	0.2119	0.2090	0.2061	0.2033	0.2005	0.1977	0.1949	0.1922	0.1894	0.1867
−0.7	0.2420	0.2389	0.2358	0.2327	0.2296	0.2266	0.2236	0.2206	0.2177	0.2148
−0.6	0.2743	0.2709	0.2676	0.2643	0.2611	0.2578	0.2546	0.2514	0.2483	0.2451
−0.5	0.3085	0.3050	0.3015	0.2981	0.2946	0.2912	0.2877	0.2843	0.2810	0.2776
−0.4	0.3446	0.3409	0.3372	0.3336	0.3300	0.3264	0.3228	0.3192	0.3156	0.3121
−0.3	0.3821	0.3783	0.3745	0.3707	0.3669	0.3632	0.3594	0.3557	0.3520	0.3483
−0.2	0.4207	0.4168	0.4129	0.4090	0.4052	0.4013	0.3974	0.3936	0.3897	0.3859
−0.1	0.4602	0.4562	0.4522	0.4483	0.4443	0.4404	0.4364	0.4325	0.4286	0.4247
−0.0	0.5000	0.4960	0.4920	0.4880	0.4840	0.4801	0.4761	0.4721	0.4681	0.4641
0.0	0.5000	0.5040	0.5080	0.5120	0.5160	0.5199	0.5239	0.5279	0.5319	0.5359
0.1	0.5398	0.5438	0.5478	0.5517	0.5557	0.5596	0.5636	0.5675	0.5714	0.5753
0.2	0.5793	0.5832	0.5871	0.5910	0.5948	0.5987	0.6026	0.6064	0.6103	0.6141
0.3	0.6179	0.6217	0.6255	0.6293	0.6331	0.6368	0.6406	0.6443	0.6480	0.6517
0.4	0.6554	0.6591	0.6628	0.6664	0.6700	0.6736	0.6772	0.6808	0.6844	0.6879
0.5	0.6915	0.6950	0.6985	0.7019	0.7054	0.7088	0.7123	0.7157	0.7190	0.7224
0.6	0.7257	0.7291	0.7324	0.7357	0.7389	0.7422	0.7454	0.7486	0.7517	0.7549
0.7	0.7580	0.7611	0.7642	0.7673	0.7704	0.7734	0.7764	0.7794	0.7823	0.7852
0.8	0.7881	0.7910	0.7939	0.7967	0.7995	0.8023	0.8051	0.8078	0.8106	0.8133
0.9	0.8159	0.8186	0.8212	0.8238	0.8264	0.8289	0.8315	0.8340	0.8365	0.8389

(continued)

Table A.3 Standard Normal Curve Areas (cont.) $\Phi(z) = P(Z \le z)$

z	0.00	0.01	0.02	0.03	0.04	0.05	0.06	0.07	0.08	0.09
1.0	0.8413	0.8438	0.8461	0.8485	0.8508	0.8531	0.8554	0.8577	0.8599	0.8621
1.1	0.8643	0.8665	0.8686	0.8708	0.8729	0.8749	0.8770	0.8790	0.8810	0.8830
1.2	0.8849	0.8869	0.8888	0.8907	0.8925	0.8944	0.8962	0.8980	0.8997	0.9015
1.3	0.9032	0.9049	0.9066	0.9082	0.9099	0.9115	0.9131	0.9147	0.9162	0.9177
1.4	0.9192	0.9207	0.9222	0.9236	0.9251	0.9265	0.9278	0.9292	0.9306	0.9319
1.5	0.9332	0.9345	0.9357	0.9370	0.9382	0.9394	0.9406	0.9418	0.9429	0.9441
1.6	0.9452	0.9463	0.9474	0.9484	0.9495	0.9505	0.9515	0.9525	0.9535	0.9545
1.7	0.9554	0.9564	0.9573	0.9582	0.9591	0.9599	0.9608	0.9616	0.9625	0.9633
1.8	0.9641	0.9649	0.9656	0.9664	0.9671	0.9678	0.9686	0.9693	0.9699	0.9706
1.9	0.9713	0.9719	0.9726	0.9732	0.9738	0.9744	0.9750	0.9756	0.9761	0.9767
2.0	0.9772	0.9778	0.9783	0.9788	0.9793	0.9798	0.9803	0.9808	0.9812	0.9817
2.1	0.9821	0.9826	0.9830	0.9834	0.9838	0.9842	0.9846	0.9850	0.9854	0.9857
2.2	0.9861	0.9864	0.9868	0.9871	0.9875	0.9878	0.9881	0.9884	0.9887	0.9890
2.3	0.9893	0.9896	0.9898	0.9901	0.9904	0.9906	0.9909	0.9911	0.9913	0.9916
2.4	0.9918	0.9920	0.9922	0.9925	0.9927	0.9929	0.9931	0.9932	0.9934	0.9936
2.5	0.9938	0.9940	0.9941	0.9943	0.9945	0.9946	0.9948	0.9949	0.9951	0.9952
2.6	0.9953	0.9955	0.9956	0.9957	0.9959	0.9960	0.9961	0.9962	0.9963	0.9964
2.7	0.9965	0.9966	0.9967	0.9968	0.9969	0.9970	0.9971	0.9972	0.9973	0.9974
2.8	0.9974	0.9975	0.9976	0.9977	0.9977	0.9978	0.9979	0.9979	0.9980	0.9981
2.9	0.9981	0.9982	0.9982	0.9983	0.9984	0.9984	0.9985	0.9985	0.9986	0.9986
3.0	0.9987	0.9987	0.9987	0.9988	0.9988	0.9989	0.9989	0.9989	0.9990	0.9990
3.1	0.9990	0.9991	0.9991	0.9991	0.9992	0.9992	0.9992	0.9992	0.9993	0.9993
3.2	0.9993	0.9993	0.9994	0.9994	0.9994	0.9994	0.9994	0.9995	0.9995	0.9995
3.3	0.9995	0.9995	0.9995	0.9996	0.9996	0.9996	0.9996	0.9996	0.9996	0.9997
3.4	0.9997	0.9997	0.9997	0.9997	0.9997	0.9997	0.9997	0.9997	0.9997	0.9998

Table A-4 The Incomplete Gamma Function $F(x; \alpha) = \int_0^x \frac{1}{\Gamma(\alpha)} y^{\alpha-1} e^{-y} \, dy$

x＼α	1	2	3	4	5	6	7	8	9	10
1	.632	.264	.080	.019	.004	.001	.000	.000	.000	.000
2	.865	.594	.323	.143	.053	.017	.005	.001	.000	.000
3	.950	.801	.577	.353	.185	.084	.034	.012	.004	.001
4	.982	.908	.762	.567	.371	.215	.111	.051	.021	.008
5	.993	.960	.875	.735	.560	.384	.238	.133	.068	.032
6	.998	.983	.938	.849	.715	.554	.398	.256	.153	.084
7	.999	.993	.970	.918	.827	.699	.550	.401	.271	.170
8	1.000	.997	.986	.958	.900	.809	.687	.547	.407	.283
9		.999	.994	.979	.945	.884	.793	.676	.544	.413
10		1.000	.997	.990	.971	.933	.870	.780	.667	.542
11			.999	.995	.985	.962	.921	.857	.768	.659
12			1.000	.998	.992	.980	.954	.911	.845	.758
13				.999	.996	.989	.974	.946	.900	.834
14				1.000	.998	.994	.986	.968	.938	.891
15					.999	.997	.992	.982	.963	.930

Table A.5 Critical Values $t_{\alpha, \nu}$ for the t Distribution

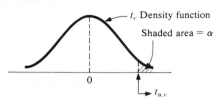

t_ν Density function

Shaded area = α

				α			
ν	.10	.05	.025	.01	.005	.001	.0005
1	3.078	6.314	12.706	31.821	63.657	318.31	636.62
2	1.886	2.920	4.303	6.965	9.925	22.326	31.598
3	1.638	2.353	3.182	4.541	5.841	10.213	12.924
4	1.533	2.132	2.776	3.747	4.604	7.173	8.610
5	1.476	2.015	2.571	3.365	4.032	5.893	6.869
6	1.440	1.943	2.447	3.143	3.707	5.208	5.959
7	1.415	1.895	2.365	2.998	3.499	4.785	5.408
8	1.397	1.860	2.306	2.896	3.355	4.501	5.041
9	1.383	1.833	2.262	2.821	3.250	4.297	4.781
10	1.372	1.812	2.228	2.764	3.169	4.144	4.587
11	1.363	1.796	2.201	2.718	3.106	4.025	4.437
12	1.356	1.782	2.179	2.681	3.055	3.930	4.318
13	1.350	1.771	2.160	2.650	3.012	3.852	4.221
14	1.345	1.761	2.145	2.624	2.977	3.787	4.140
15	1.341	1.753	2.131	2.602	2.947	3.733	4.073
16	1.337	1.746	2.120	2.583	2.921	3.686	4.015
17	1.333	1.740	2.110	2.567	2.898	3.646	3.965
18	1.330	1.734	2.101	2.552	2.878	3.610	3.922
19	1.328	1.729	2.093	2.539	2.861	3.579	3.883
20	1.325	1.725	2.086	2.528	2.845	3.552	3.850
21	1.323	1.721	2.080	2.518	2.831	3.527	3.819
22	1.321	1.717	2.074	2.508	2.819	3.505	3.792
23	1.319	1.714	2.069	2.500	2.807	3.485	3.767
24	1.318	1.711	2.064	2.492	2.797	3.467	3.745
25	1.316	1.708	2.060	2.485	2.787	3.450	3.725
26	1.315	1.706	2.056	2.479	2.779	3.435	3.707
27	1.314	1.703	2.052	2.473	2.771	3.421	3.690
28	1.313	1.701	2.048	2.467	2.763	3.408	3.674
29	1.311	1.699	2.045	2.462	2.756	3.396	3.659
30	1.310	1.697	2.042	2.457	2.750	3.385	3.646
40	1.303	1.684	2.021	2.423	2.704	3.307	3.551
60	1.296	1.671	2.000	2.390	2.660	3.232	3.460
120	1.289	1.658	1.980	2.358	2.617	3.160	3.373
∞	1.282	1.645	1.960	2.326	2.576	3.090	3.291

Source: This table is reproduced with the kind permission of the Trustees of Biometrica from E. S. Pearson and H. O. Hartley (eds.), *The Biometrica Tables for Statisticians,* vol. 1, 3rd ed., *Biometrica,* 1966.

Table A.6 Critical Values $X^2_{\alpha,\nu}$ for the Chi-Squared Distribution

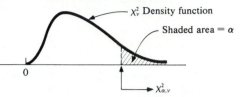

χ^2_ν Density function

Shaded area $= \alpha$

$\chi^2_{\alpha,\nu}$

ν	.995	.99	.975	.95	α .90	.10	.05	.025	.01	.005
1	0.000	0.000	0.001	0.004	0.016	2.706	3.843	5.025	6.637	7.882
2	0.010	0.020	0.051	0.103	0.211	4.605	5.992	7.378	9.210	10.597
3	0.072	0.115	0.216	0.352	0.584	6.251	7.815	9.348	11.344	12.837
4	0.207	0.297	0.484	0.711	1.064	7.779	9.488	11.143	13.277	14.860
5	0.412	0.554	0.831	1.145	1.610	9.236	11.070	12.832	15.085	16.748
6	0.676	0.872	1.237	1.635	2.204	10.645	12.592	14.440	16.812	18.548
7	0.989	1.239	1.690	2.167	2.833	12.017	14.067	16.012	18.474	20.276
8	1.344	1.646	2.180	2.733	3.490	13.362	15.507	17.534	20.090	21.954
9	1.735	2.088	2.700	3.325	4.168	14.684	16.919	19.022	21.665	23.587
10	2.156	2.558	3.247	3.940	4.865	15.987	18.307	20.483	23.209	25.188
11	2.603	3.053	3.816	4.575	5.578	17.275	19.675	21.920	24.724	26.755
12	3.074	3.571	4.404	5.226	6.304	18.549	21.026	23.337	26.217	28.300
13	3.565	4.107	5.009	5.892	7.041	19.812	22.362	24.735	27.687	29.817
14	4.075	4.660	5.629	6.571	7.790	21.064	23.685	26.119	29.141	31.319
15	4.600	5.229	6.262	7.261	8.547	22.307	24.996	27.488	30.577	32.799
16	5.142	5.812	6.908	7.962	9.312	23.542	26.296	28.845	32.000	34.267
17	5.697	6.407	7.564	8.682	10.085	24.769	27.587	30.190	33.408	35.716
18	6.265	7.015	8.231	9.390	10.865	25.989	28.869	31.526	34.805	37.156
19	6.843	7.632	8.906	10.117	11.651	27.203	30.143	32.852	36.190	38.580
20	7.434	8.260	9.591	10.851	12.443	28.412	31.410	34.170	37.566	39.997
21	8.033	8.897	10.283	11.591	13.240	29.615	32.670	35.478	38.930	41.399
22	8.643	9.542	10.982	12.338	14.042	30.813	33.924	36.781	40.289	42.796
23	9.260	10.195	11.688	13.090	14.848	32.007	35.172	38.075	41.637	44.179
24	9.886	10.856	12.401	13.848	15.659	33.196	36.415	39.364	42.980	45.558
25	10.519	11.523	13.120	14.611	16.473	34.381	37.652	40.646	44.313	46.925
26	11.160	12.198	13.844	15.379	17.292	35.563	38.885	41.923	45.642	48.290
27	11.807	12.878	14.573	16.151	18.114	36.741	40.113	43.194	46.962	49.642
28	12.461	13.565	15.308	16.928	18.939	37.916	41.337	44.461	48.278	50.993
29	13.120	14.256	16.147	17.708	19.768	39.087	42.557	45.772	49.586	52.333
30	13.787	14.954	16.791	18.493	20.599	40.256	43.773	46.979	50.892	53.672
31	14.457	15.655	17.538	19.280	21.433	41.422	44.985	48.231	52.190	55.000
32	15.134	16.362	18.291	20.072	22.271	42.585	46.194	49.480	53.486	56.328
33	15.814	17.073	19.046	20.866	23.110	43.745	47.400	50.724	54.774	57.646
34	16.501	17.789	19.806	21.664	23.952	44.903	48.602	51.966	56.061	58.964
35	17.191	18.508	20.569	22.465	24.796	46.059	49.802	53.203	57.340	60.272
36	17.887	19.233	21.336	23.269	25.643	47.212	50.998	54.437	58.619	61.581
37	18.584	19.960	22.105	24.075	26.492	48.363	52.192	55.667	59.891	62.880
38	19.289	20.691	22.878	24.884	27.343	49.513	53.384	56.896	61.162	64.181
39	19.994	21.425	23.654	25.695	28.196	50.660	54.572	58.119	62.426	65.473
40	20.706	22.164	24.433	26.509	29.050	51.805	55.758	59.342	63.691	66.766

For $\nu > 40$, $x^2_{\alpha,\nu} \doteq \nu \left(1 - \dfrac{2}{9\nu} + z_\alpha \sqrt{\dfrac{2}{9\nu}}\right)^3$

Source: This table is reproduced with the kind permission of the Trustees of Biometrica from E. S. Pearson and H. O. Hartley (eds.), *The Biometrica Tables for Statisticians,* vol. 1, 3rd ed., *Biometrica,* 1966.

Table A.7 Critical Values F_{α, ν_1, ν_2} for the F Distribution

$\alpha = .05$

$\nu_2 \backslash \nu_1$	1	2	3	4	5	6	7	8	9	10	12	15	20	24	30	40	60	120	∞
1	161.4	199.5	215.7	224.6	230.2	234.0	236.8	238.9	240.5	241.9	243.9	245.9	248.0	249.1	250.1	251.1	252.2	253.3	254.3
2	18.51	19.00	19.16	19.25	19.30	19.33	19.35	19.37	19.38	19.40	19.41	19.43	19.45	19.45	19.46	19.47	19.48	19.49	19.50
3	10.13	9.55	9.28	9.12	9.01	8.94	8.89	8.85	8.81	8.79	8.74	8.70	8.66	8.64	8.62	8.59	8.57	8.55	8.53
4	7.71	6.94	6.59	6.39	6.26	6.16	6.09	6.04	6.00	5.96	5.91	5.86	5.80	5.77	5.75	5.72	5.69	5.66	5.63
5	6.61	5.79	5.41	5.19	5.05	4.95	4.88	4.82	4.77	4.74	4.68	4.62	4.56	4.53	4.50	4.46	4.43	4.40	4.36
6	5.99	5.14	4.76	4.53	4.39	4.28	4.21	4.15	4.10	4.06	4.00	3.94	3.87	3.84	3.81	3.77	3.74	3.70	3.67
7	5.59	4.74	4.35	4.12	3.97	3.87	3.79	3.73	3.68	3.64	3.57	3.51	3.44	3.41	3.38	3.34	3.30	3.27	3.23
8	5.32	4.46	4.07	3.84	3.69	3.58	3.50	3.44	3.39	3.35	3.28	3.22	3.15	3.12	3.08	3.04	3.01	2.97	2.93
9	5.12	4.26	3.86	3.63	3.48	3.37	3.29	3.23	3.18	3.14	3.07	3.01	2.94	2.90	2.86	2.83	2.79	2.75	2.71
10	4.96	4.10	3.71	3.48	3.33	3.22	3.14	3.07	3.02	2.98	2.91	2.85	2.77	2.74	2.70	2.66	2.62	2.58	2.54
11	4.84	3.98	3.59	3.36	3.20	3.09	3.01	2.95	2.90	2.85	2.79	2.72	2.65	2.61	2.57	2.53	2.49	2.45	2.40
12	4.75	3.89	3.49	3.26	3.11	3.00	2.91	2.85	2.80	2.75	2.69	2.62	2.54	2.51	2.47	2.43	2.38	2.34	2.30
13	4.67	3.81	3.41	3.18	3.03	2.92	2.83	2.77	2.71	2.67	2.60	2.53	2.46	2.42	2.38	2.34	2.30	2.25	2.21
14	4.60	3.74	3.34	3.11	2.96	2.85	2.76	2.70	2.65	2.60	2.53	2.46	2.39	2.35	2.31	2.27	2.22	2.18	2.13
15	4.54	3.68	3.29	3.06	2.90	2.79	2.71	2.64	2.59	2.54	2.48	2.40	2.33	2.29	2.25	2.20	2.16	2.11	2.07
16	4.49	3.63	3.24	3.01	2.85	2.74	2.66	2.59	2.54	2.49	2.42	2.35	2.28	2.24	2.19	2.15	2.11	2.06	2.01
17	4.45	3.59	3.20	2.96	2.81	2.70	2.61	2.55	2.49	2.45	2.38	2.31	2.23	2.19	2.15	2.10	2.06	2.01	1.96
18	4.41	3.55	3.16	2.93	2.77	2.66	2.58	2.51	2.46	2.41	2.34	2.27	2.19	2.15	2.11	2.06	2.02	1.97	1.92
19	4.38	3.52	3.13	2.90	2.74	2.63	2.54	2.48	2.42	2.38	2.31	2.23	2.16	2.11	2.07	2.03	1.98	1.93	1.88
20	4.35	3.49	3.10	2.87	2.71	2.60	2.51	2.45	2.39	2.35	2.28	2.20	2.12	2.08	2.04	1.99	1.95	1.90	1.84
21	4.32	3.47	3.07	2.84	2.68	2.57	2.49	2.42	2.37	2.32	2.25	2.18	2.10	2.05	2.01	1.96	1.92	1.87	1.81
22	4.30	3.44	3.05	2.82	2.66	2.55	2.46	2.40	2.34	2.30	2.23	2.15	2.07	2.03	1.98	1.94	1.89	1.84	1.78
23	4.28	3.42	3.03	2.80	2.64	2.53	2.44	2.37	2.32	2.27	2.20	2.13	2.05	2.01	1.96	1.91	1.86	1.81	1.76
24	4.26	3.40	3.01	2.78	2.62	2.51	2.42	2.36	2.30	2.25	2.18	2.11	2.03	1.98	1.94	1.89	1.84	1.79	1.73
25	4.24	3.39	2.99	2.76	2.60	2.49	2.40	2.34	2.28	2.24	2.16	2.09	2.01	1.96	1.92	1.87	1.82	1.77	1.71
26	4.23	3.37	2.98	2.74	2.59	2.47	2.39	2.32	2.27	2.22	2.15	2.07	1.99	1.95	1.90	1.85	1.80	1.75	1.69
27	4.21	3.35	2.96	2.73	2.57	2.46	2.37	2.31	2.25	2.20	2.13	2.06	1.97	1.93	1.88	1.84	1.79	1.73	1.67
28	4.20	3.34	2.95	2.71	2.56	2.45	2.36	2.29	2.24	2.19	2.12	2.04	1.96	1.91	1.87	1.82	1.77	1.71	1.65
29	4.18	3.33	2.93	2.70	2.55	2.43	2.35	2.28	2.22	2.18	2.10	2.03	1.94	1.90	1.85	1.81	1.75	1.70	1.64
30	4.17	3.32	2.92	2.69	2.53	2.42	2.33	2.27	2.21	2.16	2.09	2.01	1.93	1.89	1.84	1.79	1.74	1.68	1.62
40	4.08	3.23	2.84	2.61	2.45	2.34	2.25	2.18	2.12	2.08	2.00	1.92	1.84	1.79	1.74	1.69	1.64	1.58	1.51
60	4.00	3.15	2.76	2.53	2.37	2.25	2.17	2.10	2.04	1.99	1.92	1.84	1.75	1.70	1.65	1.59	1.53	1.47	1.39
120	3.92	3.07	2.68	2.45	2.29	2.17	2.09	2.02	1.96	1.91	1.83	1.75	1.66	1.61	1.55	1.50	1.43	1.35	1.25
∞	3.84	3.00	2.60	2.37	2.21	2.10	2.01	1.94	1.88	1.83	1.75	1.67	1.57	1.52	1.46	1.39	1.32	1.22	1.00

(continued)

Table A.7 Critical Values F_{α, ν_1, ν_2} for the F Distribution (cont.)

$\alpha = .01$

$\nu_2 \backslash \nu_1$	1	2	3	4	5	6	7	8	9	10	12	15	20	24	30	40	60	120	∞
1	4052	4999.5	5403	5625	5764	5859	5928	5981	6022	6056	6106	6157	6209	6235	6261	6287	6313	6339	6366
2	98.50	99.00	99.17	99.25	99.30	99.33	99.36	99.37	99.39	99.40	99.42	99.43	99.45	99.46	99.47	99.47	99.48	99.49	99.50
3	34.12	30.82	29.46	28.71	28.24	27.91	27.67	27.49	27.35	27.23	27.05	26.87	26.69	26.60	26.50	26.41	26.32	26.22	26.13
4	21.20	18.00	16.69	15.98	15.52	15.21	14.98	14.80	14.66	14.55	14.37	14.20	14.02	13.93	13.84	13.75	13.65	13.56	13.46
5	16.26	13.27	12.06	11.39	10.97	10.67	10.46	10.29	10.16	10.05	9.89	9.72	9.55	9.47	9.38	9.29	9.20	9.11	9.02
6	13.75	10.92	9.78	9.15	8.75	8.47	8.26	8.10	7.98	7.87	7.72	7.56	7.40	7.31	7.23	7.14	7.06	6.97	6.88
7	12.25	9.55	8.45	7.85	7.46	7.19	6.99	6.84	6.72	6.62	6.47	6.31	6.16	6.07	5.99	5.91	5.82	5.74	5.65
8	11.26	8.65	7.59	7.01	6.63	6.37	6.18	6.03	5.91	5.81	5.67	5.52	5.36	5.28	5.20	5.12	5.03	4.95	4.86
9	10.56	8.02	6.99	6.42	6.06	5.80	5.61	5.47	5.35	5.26	5.11	4.96	4.81	4.73	4.65	4.57	4.48	4.40	4.31
10	10.04	7.56	6.55	5.99	5.64	5.39	5.20	5.06	4.94	4.85	4.71	4.56	4.41	4.33	4.25	4.17	4.08	4.00	3.91
11	9.65	7.21	6.22	5.67	5.32	5.07	4.89	4.74	4.63	4.54	4.40	4.25	4.10	4.02	3.94	3.86	3.78	3.69	3.60
12	9.33	6.93	5.95	5.41	5.06	4.82	4.64	4.50	4.39	4.30	4.16	4.01	3.86	3.78	3.70	3.62	3.54	3.45	3.36
13	9.07	6.70	5.74	5.21	4.86	4.62	4.44	4.30	4.19	4.10	3.96	3.82	3.66	3.59	3.51	3.43	3.34	3.25	3.17
14	8.86	6.51	5.56	5.04	4.69	4.46	4.28	4.14	4.03	3.94	3.80	3.66	3.51	3.43	3.35	3.27	3.18	3.09	3.00
15	8.68	6.36	5.42	4.89	4.56	4.32	4.14	4.00	3.89	3.80	3.67	3.52	3.37	3.29	3.21	3.13	3.05	2.96	2.87
16	8.53	6.23	5.29	4.77	4.44	4.20	4.03	3.89	3.78	3.69	3.55	3.41	3.26	3.18	3.10	3.02	2.93	2.84	2.75
17	8.40	6.11	5.18	4.67	4.34	4.10	3.93	3.79	3.68	3.59	3.46	3.31	3.16	3.08	3.00	2.92	2.83	2.75	2.65
18	8.29	6.01	5.09	4.58	4.25	4.01	3.84	3.71	3.60	3.51	3.37	3.23	3.08	3.00	2.92	2.84	2.75	2.66	2.57
19	8.18	5.93	5.01	4.50	4.17	3.94	3.77	3.63	3.52	3.43	3.30	3.15	3.00	2.92	2.84	2.76	2.67	2.58	2.49
20	8.10	5.85	4.94	4.43	4.10	3.87	3.70	3.56	3.46	3.37	3.23	3.09	2.94	2.86	2.78	2.69	2.61	2.52	2.42
21	8.02	5.78	4.87	4.37	4.04	3.81	3.64	3.51	3.40	3.31	3.17	3.03	2.88	2.80	2.72	2.64	2.55	2.46	2.36
22	7.95	5.72	4.82	4.31	3.99	3.76	3.59	3.45	3.35	3.26	3.12	2.98	2.83	2.75	2.67	2.58	2.50	2.40	2.31
23	7.88	5.66	4.76	4.26	3.94	3.71	3.54	3.41	3.30	3.21	3.07	2.93	2.78	2.70	2.62	2.54	2.45	2.35	2.26
24	7.82	5.61	4.72	4.22	3.90	3.67	3.50	3.36	3.26	3.17	3.03	2.89	2.74	2.66	2.58	2.49	2.40	2.31	2.21
25	7.77	5.57	4.68	4.18	3.85	3.63	3.46	3.32	3.22	3.13	2.99	2.85	2.70	2.62	2.54	2.45	2.36	2.27	2.17
26	7.72	5.53	4.64	4.14	3.82	3.59	3.42	3.29	3.18	3.09	2.96	2.81	2.66	2.58	2.50	2.42	2.33	2.23	2.13
27	7.68	5.49	4.60	4.11	3.78	3.56	3.39	3.26	3.15	3.06	2.93	2.78	2.63	2.55	2.47	2.38	2.29	2.20	2.10
28	7.64	5.45	4.57	4.07	3.75	3.53	3.36	3.23	3.12	3.03	2.90	2.75	2.60	2.52	2.44	2.35	2.26	2.17	2.06
29	7.60	5.42	4.54	4.04	3.73	3.50	3.33	3.20	3.09	3.00	2.87	2.73	2.57	2.49	2.41	2.33	2.23	2.14	2.03
30	7.56	5.39	4.51	4.02	3.70	3.47	3.30	3.17	3.07	2.98	2.84	2.70	2.55	2.47	2.39	2.30	2.21	2.11	2.01
40	7.31	5.18	4.31	3.83	3.51	3.29	3.12	2.99	2.89	2.80	2.66	2.52	2.37	2.29	2.20	2.11	2.02	1.92	1.80
60	7.08	4.98	4.13	3.65	3.34	3.12	2.95	2.82	2.72	2.63	2.50	2.35	2.20	2.12	2.03	1.94	1.84	1.73	1.60
120	6.85	4.79	3.95	3.48	3.17	2.96	2.79	2.66	2.56	2.47	2.34	2.19	2.03	1.95	1.86	1.76	1.66	1.53	1.38
∞	6.63	4.61	3.78	3.32	3.02	2.80	2.64	2.51	2.41	2.32	2.18	2.04	1.88	1.79	1.70	1.59	1.47	1.32	1.00

Source: This table is reproduced with the kind permission of the Trustees of Biometrica from E. S. Pearson and H. O. Hartley (eds.), *The Biometrica Tables for Statisticians*, vol. 1, 3rd ed., Biometrica, 1966.

Table A.8 Critical Values $Q_{\alpha,m,\nu}$ for the Studentized Range Distribution

ν	α	2	3	4	5	6	7	8	9	10	11
5	.05	3.64	4.60	5.22	5.67	6.03	6.33	6.58	6.80	6.99	7.17
	.01	5.70	6.98	7.80	8.42	8.91	9.32	9.67	9.97	10.24	10.48
6	.05	3.46	4.34	4.90	5.30	5.63	5.90	6.12	6.32	6.49	6.65
	.01	5.24	6.33	7.03	7.56	7.97	8.32	8.61	8.87	9.10	9.30
7	.05	3.34	4.16	4.68	5.06	5.36	5.61	5.82	6.00	6.16	6.30
	.01	4.95	5.92	6.54	7.01	7.37	7.68	7.94	8.17	8.37	8.55
8	.05	3.26	4.04	4.53	4.89	5.17	5.40	5.60	5.77	5.92	6.05
	.01	4.75	5.64	6.20	6.62	6.96	7.24	7.47	7.68	7.86	8.03
9	.05	3.20	3.95	4.41	4.76	5.02	5.24	5.43	5.59	5.74	5.87
	.01	4.60	5.43	5.96	6.35	6.66	6.91	7.13	7.33	7.49	7.65
10	.05	3.15	3.88	4.33	4.65	4.91	5.12	5.30	5.46	5.60	5.72
	.01	4.48	5.27	5.77	6.14	6.43	6.67	6.87	7.05	7.21	7.36
11	.05	3.11	3.82	4.26	4.57	4.82	5.03	5.20	5.35	5.49	5.61
	.01	4.39	5.15	5.62	5.97	6.25	6.48	6.67	6.84	6.99	7.13
12	.05	3.08	3.77	4.20	4.51	4.75	4.95	5.12	5.27	5.39	5.51
	.01	4.32	5.05	5.50	5.84	6.10	6.32	6.51	6.67	6.81	6.94
13	.05	3.06	3.73	4.15	4.45	4.69	4.88	5.05	5.19	5.32	5.43
	.01	4.26	4.96	5.40	5.73	5.98	6.19	6.37	6.53	6.67	6.79
14	.05	3.03	3.70	4.11	4.41	4.64	4.83	4.99	5.13	5.25	5.36
	.01	4.21	4.89	5.32	5.63	5.88	6.08	6.26	6.41	6.54	6.66
15	.05	3.01	3.67	4.08	4.37	4.59	4.78	4.94	5.08	5.20	5.31
	.01	4.17	4.84	5.25	5.56	5.80	5.99	6.16	6.31	6.44	6.55
16	.05	3.00	3.65	4.05	4.33	4.56	4.74	4.90	5.03	5.15	5.26
	.01	4.13	4.79	5.19	5.49	5.72	5.92	6.08	6.22	6.35	6.46
17	.05	2.98	3.63	4.02	4.30	4.52	4.70	4.86	4.99	5.11	5.21
	.01	4.10	4.74	5.14	5.43	5.66	5.85	6.01	6.15	6.27	6.38
18	.05	2.97	3.61	4.00	4.28	4.49	4.67	4.82	4.96	5.07	5.17
	.01	4.07	4.70	5.09	5.38	5.60	5.79	5.94	6.08	6.20	6.31
19	.05	2.96	3.59	3.98	4.25	4.47	4.65	4.79	4.92	5.04	5.14
	.01	4.05	4.67	5.05	5.33	5.55	5.73	5.89	6.02	6.14	6.25
20	.05	2.95	3.58	3.96	4.23	4.45	4.62	4.77	4.90	5.01	5.11
	.01	4.02	4.64	5.02	5.29	5.51	5.69	5.84	5.97	6.09	6.19
24	.05	2.92	3.53	3.90	4.17	4.37	4.54	4.68	4.81	4.92	5.01
	.01	3.96	4.55	4.91	5.17	5.37	5.54	5.69	5.81	5.92	6.02
30	.05	2.89	3.49	3.85	4.10	4.30	4.46	4.60	4.72	4.82	4.92
	.01	3.89	4.45	4.80	5.05	5.24	5.40	5.54	5.65	5.76	5.85
40	.05	2.86	3.44	3.79	4.04	4.23	4.39	4.52	4.63	4.73	4.82
	.01	3.82	4.37	4.70	4.93	5.11	5.26	5.39	5.50	5.60	5.69
60	.05	2.83	3.40	3.74	3.98	4.16	4.31	4.44	4.55	4.65	4.73
	.01	3.76	4.28	4.59	4.82	4.99	5.13	5.25	5.36	5.45	5.53
120	.05	2.80	3.36	3.68	3.92	4.10	4.24	4.36	4.47	4.56	4.64
	.01	3.70	4.20	4.50	4.71	4.87	5.01	5.12	5.21	5.30	5.37
∞	.05	2.77	3.31	3.63	3.86	4.03	4.17	4.29	4.39	4.47	4.55
	.01	3.64	4.12	4.40	4.60	4.76	4.88	4.99	5.08	5.16	5.23

(continued)

Table A.8 Critical Values $Q_{\alpha, m, \nu}$ for the Studentized Range Distribution (*cont.*)

				m						
12	13	14	15	16	17	18	19	20	α	ν
7.32	7.47	7.60	7.72	7.83	7.93	8.03	8.12	8.21	.05	5
10.70	10.89	11.08	11.24	11.40	11.55	11.68	11.81	11.93	.01	
6.79	6.92	7.03	7.14	7.24	7.34	7.43	7.51	7.59	.05	6
9.48	9.65	9.81	9.95	10.08	10.21	10.32	10.43	10.54	.01	
6.43	6.55	6.66	6.76	6.85	6.94	7.02	7.10	7.17	.05	7
8.71	8.86	9.00	9.12	9.24	9.35	9.46	9.55	9.65	.01	
6.18	6.29	6.39	6.48	6.57	6.65	6.73	6.80	6.87	.05	8
8.18	8.31	8.44	8.55	8.66	8.76	8.85	8.94	9.03	.01	
5.98	6.09	6.19	6.28	6.36	6.44	6.51	6.58	6.64	.05	9
7.78	7.91	8.03	8.13	8.23	8.33	8.41	8.49	8.57	.01	
5.83	5.93	6.03	6.11	6.19	6.27	6.34	6.40	6.47	.05	10
7.49	7.60	7.71	7.81	7.91	7.99	8.08	8.15	8.23	.01	
5.71	5.81	5.90	5.98	6.06	6.13	6.20	6.27	6.33	.05	11
7.25	7.36	7.46	7.56	7.65	7.73	7.81	7.88	7.95	.01	
5.61	5.71	5.80	5.88	5.95	6.02	6.09	6.15	6.21	.05	12
7.06	7.17	7.26	7.36	7.44	7.52	7.59	7.66	7.73	.01	
5.53	5.63	5.71	5.79	5.86	5.93	5.99	6.05	6.11	.05	13
6.90	7.01	7.10	7.19	7.27	7.35	7.42	7.48	7.55	.01	
5.46	5.55	5.64	5.71	5.79	5.85	5.91	5.97	6.03	.05	14
6.77	6.87	6.96	7.05	7.13	7.20	7.27	7.33	7.39	.01	
5.40	5.49	5.57	5.65	5.72	5.78	5.85	5.90	5.96	.05	15
6.66	6.76	6.84	6.93	7.00	7.07	7.14	7.20	7.26	.01	
5.35	5.44	5.52	5.59	5.66	5.73	5.79	5.84	5.90	.05	16
6.56	6.66	6.74	6.82	6.90	6.97	7.03	7.09	7.15	.01	
5.31	5.39	5.47	5.54	5.61	5.67	5.73	5.79	5.84	.05	17
6.48	6.57	6.66	6.73	6.81	6.87	6.94	7.00	7.05	.01	
5.27	5.35	5.43	5.50	5.57	5.63	5.69	5.74	5.79	.05	18
6.41	6.50	6.58	6.65	6.73	6.79	6.85	6.91	6.97	.01	
5.23	5.31	5.39	5.46	5.53	5.59	5.65	5.70	5.75	.05	19
6.34	6.43	6.51	6.58	6.65	6.72	6.78	6.84	6.89	.01	
5.20	5.28	5.36	5.43	5.49	5.55	5.61	5.66	5.71	.05	20
6.28	6.37	6.45	6.52	6.59	6.65	6.71	6.77	6.82	.01	
5.10	5.18	5.25	5.32	5.38	5.44	5.49	5.55	5.59	.05	24
6.11	6.19	6.26	6.33	6.39	6.45	6.51	6.56	6.61	.01	
5.00	5.08	5.15	5.21	5.27	5.33	5.38	5.43	5.47	.05	30
5.93	6.01	6.08	6.14	6.20	6.26	6.31	6.36	6.41	.01	
4.90	4.98	5.04	5.11	5.16	5.22	5.27	5.31	5.36	.05	40
5.76	5.83	5.90	5.96	6.02	6.07	6.12	6.16	6.21	.01	
4.81	4.88	4.94	5.00	5.06	5.11	5.15	5.20	5.24	.05	60
5.60	5.67	5.73	5.78	5.84	5.89	5.93	5.97	6.01	.01	
4.71	4.78	4.84	4.90	4.95	5.00	5.04	5.09	5.13	.05	120
5.44	5.50	5.56	5.61	5.66	5.71	5.75	5.79	5.83	.01	
4.62	4.68	4.74	4.80	4.85	4.89	4.93	4.97	5.01	.05	∞
5.29	5.35	5.40	5.45	5.49	5.54	5.57	5.61	5.65	.01	

Source: This table is abridged from Table 29 in *Biometrica Tables for Statisticians,* vol. 1, 3rd ed., by E. S. Pearson and H. O. Hartley (eds.). Reproduced with the kind permission of the Trustees of *Biometrica,* 1966.

Table A.9 Upper-Tail Critical Values and Probabilities for the Null Distribution of the Wilcoxon Signed-Rank Statistic S_+

$$P_0(S_+ \geq c_1) = P(S_+ \geq c_1 \text{ when } H_0 \text{ is true})$$

n	c_1	$P_0(S_+ \geq c_1)$	n	c_1	$P_0(S_+ \geq c_1)$
3	6	.125		78	.011
4	9	.125		79	.009
	10	.062		81	.005
5	13	.094	14	73	.108
	14	.062		74	.097
	15	.031		79	.052
6	17	.109		84	.025
	19	.047		89	.010
	20	.031		92	.005
	21	.016	15	83	.104
7	22	.109		84	.094
	24	.055		89	.053
	26	.023		90	.047
	28	.008		95	.024
8	28	.098		100	.011
	30	.055		101	.009
	32	.027		104	.005
	34	.012	16	93	.106
	35	.008		94	.096
	36	.004		100	.052
9	34	.102		106	.025
	37	.049		112	.011
	39	.027		113	.009
	42	.010		116	.005
	44	.004	17	104	.103
10	41	.097		105	.095
	44	.053		112	.049
	47	.024		118	.025
	50	.010		125	.010
	52	.005		129	.005
11	48	.103	18	116	.098
	52	.051		124	.049
	55	.027		131	.024
	59	.009		138	.010
	61	.005		143	.005
12	56	.102	19	128	.098
	60	.055		136	.052
	61	.046		137	.048
	64	.026		144	.025
	68	.010		152	.010
	71	.005		157	.005
13	64	.108	20	140	.101
	65	.095		150	.049
	69	.055		158	.024
	70	.047		167	.010
	74	.024		172	.005

Source: Adapted from W. J. Dixon and F. J. Massey, Jr., *Introduction to Statistical Analysis*, (3rd ed.), New York: McGraw-Hill Book Company, Copyright © 1969. Reprinted by permission.

Table A.10 Upper-Tail Critical Values and Probabilities for the Null Distribution of the Wilcoxon Rank-Sum Statistic *W*

$$P_0(W \geq c) = P(W \geq c \text{ when } H_0 \text{ is true})$$

m	*n*	*c*	$P_0(W \geq c)$	*m*	*n*	*c*	$P_0(W \geq c)$
3	3	15	.05			40	.004
	4	17	.057		6	40	.041
		18	.029			41	.026
	5	20	.036			43	.009
		21	.018			44	.004
	6	22	.048		7	43	.053
		23	.024			45	.024
		24	.012			47	.009
	7	24	.058			48	.005
		26	.017		8	47	.047
		27	.008			49	.023
	8	27	.042			51	.009
		28	.024			52	.005
		29	.012	6	6	50	.047
		30	.006			52	.021
4	4	24	.057			54	.008
		25	.029			55	.004
		26	.014		7	54	.051
	5	27	.056			56	.026
		28	.032			58	.011
		29	.016			60	.004
		30	.008		8	58	.054
	6	30	.057			61	.021
		32	.019			63	.01
		33	.010			65	.004
		34	.005	7	7	66	.049
	7	33	.055			68	.027
		35	.021			71	.009
		36	.012			72	.006
		37	.006		8	71	.047
	8	36	.055			73	.027
		38	.024			76	.01
		40	.008			78	.005
		41	.004	8	8	84	.052
5	5	36	.048			87	.025
		37	.028			90	.01
		39	.008			92	.005

Source: Adapted from W. J. Dixon and F. J. Massey, Jr., *Introduction to Statistical Analysis*, (3rd ed.), New York: McGraw-Hill Book Company, Copyright © 1969. Reprinted by permission.

Table A.11 Critical Constant c for the Wilcoxon Signed-Rank Interval

$$(\overline{x}_{(n(n+1)/2 - c+1)}, \overline{x}_{(c)})$$

n	Confidence level (%)	c	n	Confidence level (%)	c	n	Confidence level (%)	c
5	93.8	15	13	99.0	81	20	99.1	173
	87.5	14		95.2	74		95.2	158
6	96.9	21		90.6	70		90.3	150
	93.7	20	14	99.1	93	21	99.0	188
	90.6	19		95.1	84		95.0	172
7	98.4	28		89.6	79		89.7	163
	95.3	26	15	99.0	104	22	99.0	204
	89.1	24		95.2	95		95.0	187
8	99.2	36		90.5	90		90.2	178
	94.5	32	16	99.1	117	23	99.0	221
	89.1	30		94.9	106		95.2	203
9	99.2	44		89.5	100		90.2	193
	94.5	39	17	99.1	130	24	99.0	239
	90.2	37		94.9	118		95.1	219
10	99.0	52		90.2	112		89.9	208
	95.1	47	18	99.0	143	25	99.0	257
	89.5	44		95.2	131		95.2	236
11	99.0	61		90.1	124		89.9	224
	94.6	55	19	99.1	158			
	89.8	52		95.1	144			
12	99.1	71		90.4	137			
	94.8	64						
	90.8	61						

Source: Derived from W. J. Dixon and F. J. Massey, Jr., *Introduction to Statistical Analysis,* (3rd ed.), New York: McGraw-Hill Book Company, Copyright ©1969. Reprinted by permission.

Table A.12 Critical Constant c for the Wilcoxon Rank-Sum Interval

$$(d_{ij(mn-c+1)}, d_{ij(c)})$$

	Smaller Sample Size						
	5		6		7		8	
Larger Sample Size	Confidence Level (%)	c	Confidence Level (%)	c	Confidence Level (%)	c	Confidence Level (%)	c
5	99.2	25						
	94.4	22						
	90.5	21						
6	99.1	29	99.1	34				
	94.8	26	95.9	31				
	91.8	25	90.7	29				
7	99.0	33	99.2	39	98.9	44		
	95.2	30	94.9	35	94.7	40		
	89.4	28	89.9	33	90.3	38		
8	98.9	37	99.2	44	99.1	50	99.0	56
	95.5	34	95.7	40	94.6	45	95.0	51
	90.7	32	89.2	37	90.6	43	89.5	48
9	98.8	41	99.2	49	99.2	56	98.9	62
	95.8	38	95.0	44	94.5	50	95.4	57
	88.8	35	91.2	42	90.9	48	90.7	54
10	99.2	46	98.9	53	99.0	61	99.1	69
	94.5	41	94.4	48	94.5	55	94.5	62
	90.1	39	90.7	46	89.1	52	89.9	59
11	99.1	50	99.0	58	98.9	66	99.1	75
	94.8	45	95.2	53	95.6	61	94.9	68
	91.0	43	90.2	50	89.6	57	90.9	65
12	99.1	54	99.0	63	99.0	72	99.0	81
	95.2	49	94.7	57	95.5	66	95.3	74
	89.6	46	89.8	54	90.0	62	90.2	70

	Smaller Sample Size						
	9		10		11		12	
Larger Sample Size	Confidence Level (%)	c	Confidence Level (%)	c	Confidence Level (%)	c	Confidence Level (%)	c
9	98.9	69						
	95.0	63						
	90.6	60						
10	99.0	76	99.1	84				
	94.7	69	94.8	76				
	90.5	66	89.5	72				
11	99.0	83	99.0	91	98.9	99		
	95.4	76	94.9	83	95.3	91		
	90.5	72	90.1	79	89.9	86		
12	99.1	90	99.1	99	99.1	108	99.0	116
	95.1	82	95.0	90	94.9	98	94.8	106
	90.5	78	90.7	86	89.6	93	89.9	101

Source: Derived from W. J. Dixon and F. J. Massey, Jr., *Introduction to Statistical Analysis,* (3rd ed.), New York: McGraw-Hill Book Company, Copyright © 1969. Reprinted by permission.

Table A.13   Curves of $\beta = P$(Type II Error) for t Tests

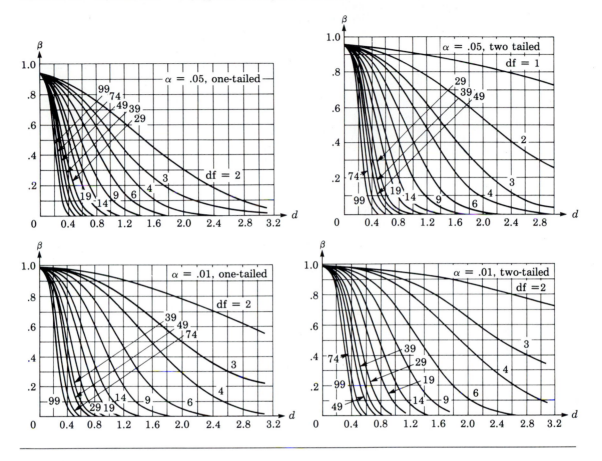

Table A.14 Critical Values c_α for the Test of Normality Based on r from a Normal Probability Plot

		a		
		.10	.05	.01
	5	.9033	.8804	.8320
	10	.9347	.9180	.8804
	15	.9506	.9383	.9110
	20	.9600	.9503	.9290
n	25	.9662	.9582	.9408
	30	.9707	.9639	.9490
	40	.9767	.9715	.9597
	50	.9807	.9764	.9664
	60	.9835	.9799	.9710
	75	.9865	.9835	.9757

Source: MINITAB Reference Manual

Answers to Odd-Numbered Exercises

Chapter 1

5. The eight frequencies are 2, 8, 12, 12, 3, 1, 0, and 2, respectively.

7. The ten relative frequencies are .01, .02, .09, .14, .17, .22, .20, .07, .07, and .01, respectively.

9. a. The eight relative frequencies are .117, .255, .138, .181, .117, .085, .096, and .011, respectively.
b. The height of each rectangle is the corresponding relative frequency divided by 5.
c. Yes.

11. No. The sum is 1.01.

13. Starting with the lowest stem, the numbers of leaves are 3, 6, 8, 0, 12, 7, 2, and 2. The gap.

15. a. 3.912   **b.** 3.90   **17. a.** 327.2   **b.** 341   **c.** 334.9

19. a. 125   **b.** 130   **21.** $\tilde{x} = 9.20$, $\bar{x}_{tr(25)} = 9.34$, $\bar{x}_{tr(10)} = 9.53$, $\bar{x} = 10.03$

23. .30, .60

25. a. 11.2   **b.** 24.12   **c.** 4.91   **27. a.–c.** $s^2 = .095$

29. .0485, .220, no.   **31. a.** .0272, .1649   **b.** .222   **c.** There are no outliers.

35. $s^2 = 485.44$. Divide by $(100)^2$.

37. a. 5   **b.** 84   **39.** 10.65   **41.** $\bar{y} = a\bar{x} + b$, $s_y^2 = a^2 s_x^2$

43. b. $\bar{x} = 9.65$, $s^2 = 24.12$

Chapter 2

1. a. $\{(A, A), (A, E), (A, J), (E, A), (E, E), (E, J), (J, A), (J, E), (J, J)\}$
b. $\{(A, E), (A, J), (E, A), (J, A)\}$ *Note:* A = American, J = Asian.

   c. $\{(A, E), (A, J), (E, A), (E, E), (E, J), (J, A), (J, E), (J, J)\}$;
      complement $= \{(A, A)\}$, a simple event.
3. a. $\mathcal{S} = \{(1, 1, 1), (1, 1, 2), (1, 1, 3), (1, 2, 1), (1, 2, 2), (1, 2, 3), (1, 3, 1),$
      $(1, 3, 2), (1, 3, 3), (2, 1, 1), (2, 1, 2), (2, 1, 3), (2, 2, 1), (2, 2, 2), (2, 2, 3),$
      $(2, 3, 1), (2, 3, 2), (2, 3, 3), (3, 1, 1), (3, 1, 2), (3, 1, 3), (3, 2, 1), (3, 2, 2),$
      $(3, 2, 3), (3, 3, 1), (3, 3, 2), (3, 3, 3)\}$
   b. $\{(1, 1, 1), (2, 2, 2), (3, 3, 3)\}$
   c. $\{(1, 2, 3), (1, 3, 2), (2, 1, 3), (2, 3, 1), (3, 1, 2), (3, 2, 1)\}$
   d. $\{(1, 1, 1), (1, 1, 3), (1, 3, 1), (1, 3, 3), (3, 1, 1), (3, 1, 3), (3, 3, 1), (3, 3, 3)\}$
5. a. $\{11, 22, 33\}$
   b. $\{1213, 1312, 1231, 1321, 2123, 2132, 2312, 2321, 3123, 3132, 3213, 3231\}$
7. a. There are 35 outcomes in $\mathcal{S}$.
   b. $\{AABABAB, AABAABB, AAABBAB, AAABABB, AAAABBB\}$

Section 2.2
(pages 37–45)

11. a. .8   **b.** .2   **c.** .4   **13. a.** .6   **b.** .3; $(A \cap B') \cup (A' \cap B)$
15. a. 4   **b.** .75   **c.** .25   **17.** .40   **19. a.** .2   **b.** .5   **c.** .3
21. a. $\frac{1}{4}, \frac{1}{2}, \frac{3}{13}, \frac{5}{13}$ **b.** $\frac{3}{4}, \frac{8}{13}$ **c.** $\frac{22}{52}, \frac{9}{13}$ **d.** $\frac{47}{52}$
23. a. .1   **b.** .7   **c.** .6

Section 2.3
(pages 46–54)

25. a. 20   **b.** 60   **c.** 10   **27. a.** 243   **b.** Approximately 10 years.
29. a. 2401   **b.** .050   **c.** .701   **d.** .408
31. a. 15,504   **b.** 56   **c.** .0578   **33. a.** 369,600   **b.** 5.01×10^{-8}
35. .0667, .333, .667

Section 2.4
(pages 54–64)

39. a. .333   **b.** .667   **c.** .143   **d.** .857   **e.** .5
41. a. .571   **b.** .80   **c.** .667   **d.** .50
43. a. .0111   **b.** .333   **c.** .200   **47. a.** .200   **b.** .509
49. a. .24   **b.** .34   **c.** .706, .206   **51.** .0588; .6735; .9996   **53.** .8182

Section 2.5
(pages 64–70)

57. a. No.   **b.** .30, .50   **59.** .248   **61. a.** .10, .20   **b.** 0
63. a. $p(2 - p)$ **b.** $1 - (1 - p)^n$ **c.** $(1 - p)^3$ **d.** $.9 + (1 - p)^3(.1)$
   e. $.1(1 - p)^3/[.9 + .1(1 - p)^3] = .0137$ for $p = .5$
65. .8588, .9949   **67.** .0625, .625

Supplementary
Exercises
(pages 70–73)

69. a. 1140   **b.** 969   **c.** 1020   **d.** .85
71. a. .25   **b.** .40   **c.** .60, .30   **d.** No.
73. .00833   **75.** .1074   **77.** .5267   **79.** .5
81. $1 - (1 - p_1)(1 - p_2) \cdots (1 - p_n)$ **83. a.** .0417   **b.** .375
85. $P(\text{hire } \#1) = \frac{6}{24}$ for $s = 0$, $\frac{11}{24}$ for $s = 1$, $\frac{10}{24}$ for $s = 2$, and $\frac{6}{24}$ for $s = 3$.

Chapter 3

Section 3.1
(pages 78–79)

1. $x = 0$ for FFF; $x = 1$ for SFF, FSF, and FFS; $x = 2$ for SSF, SFS, and FSS,
   and $x = 3$ for SSS.
3. Z = average of the two numbers, with possible values $2/2, 3/2, \ldots, 12/2$;
   W = absolute value of the difference, with possible values 0, 1, 2, 3, 4, 5.
5. No. In Example 3.4, let $Y = 1$ if at most three batteries are examined and let
   $Y = 0$ otherwise. Then Y has only two values.

7. a. $\{0, 1, \ldots, 12\}$, discrete.   **c.** $\{1, 2, 3, \ldots\}$, discrete.
  e. $\{0, c, 2c, \ldots, 10{,}000c\}$ where c is the royalty per book, discrete.
  g. $\{x : m \leq x \leq M\}$ where m (M) is the minimum (maximum) possible tension, continuous.
9. a. $\{2, 4, 6, 8, \ldots\}$, that is, $\{2(1), 2(2), 2(3), 2(4), \ldots\}$, an infinite sequence. Discrete.
  b. $\{2, 3, 4, 5, 6, \ldots\}$, that is, $\{1 + 1, 1 + 2, 1 + 3, 1 + 4, \ldots\}$, an infinite sequence. Discrete.

Section 3.2
(pages 86–88)

11. a. $p(4) = .25$, $p(6) = .40$, $p(8) = .35$, $p(x) = 0$ for $x \neq 4, 6,$ or 8.
  c. $F(x) = 0$ for $x < 4$, $= .25$ for $4 \leq x < 6$, $= .65$ for $6 \leq x < 8$, $= 1$ for $8 \leq x$.
13. a. .70   **b.** .45   **c.** .55   **d.** .71   **e.** .65   **f.** .45
15. a. $p(1) = \frac{1}{36}$, $p(2) = \frac{3}{36}$, $p(3) = \frac{5}{36}$, $p(4) = \frac{7}{36}$, $p(5) = \frac{9}{36}$, $p(6) = \frac{11}{36}$.
  b. $F(x) = 0$ for $x < 1$, $= \frac{1}{36}$ for $1 \leq x < 2$, $= \frac{4}{36}$ for $2 \leq x < 3$, $= \frac{9}{36}$ for $3 \leq x < 4$, $= \frac{16}{36}$ for $4 \leq x < 5$, $= \frac{25}{36}$ for $5 \leq x < 6$, $= 1$ for $6 \leq x$.
17. $p(0) = .16$, $p(1) = .33$, $p(2) = .32$, $p(3) = .19$, $p(y) = 0$ for $y \neq 0, 1, 2, 3$.
19. a. $p(114.9) = .30$, $p(115.9) = .24$, $p(117.6) = .18$, $p(119.9) = .28$.
  b. $F(v) = 0$ for $v < 114.9$, $= .30$ for $114.9 \leq v < 115.9$, $= .54$ for $115.9 \leq v < 117.6$, $= .72$ for $117.6 \leq v < 119.9$, $= 1$ for $119.9 \leq v$.
21. $p(y) = (1 - p)^y \cdot p$ for $y = 0, 1, 2, 3, \ldots$.
23. a. $p(x) = (\frac{2}{3})^{x-1} \cdot (\frac{1}{3})$ for $x = 1, 2, 3, \ldots$.
  b. $p(y) = (\frac{2}{3})^{y-2} \cdot (\frac{1}{3})$ for $y = 2, 3, 4, 5, \ldots$.
  c. $p(z) = (\frac{25}{54}) \cdot (\frac{2}{3})^{2z-2}$ for $z = 1, 2, 3, \ldots$.

Section 3.3
(pages 96–98)

27. a. 3.24   **b.** 1.785   **c.** .6224, .7889   **29. a.** p **c.** p
31. Expected net revenue $= \$1.98$ when 3 copies are ordered and $\$2.33$ when 4 copies are ordered.
33. $E(X) = (n + 1)/2$, $V(X) = (n^2 - 1)/12$
35. $E(X) = 2.3$, $V(X) = .81$, expected amount left $= 88.5$, variance $= 20.25$.

Section 3.4
(pages 105–107)

41. a. .124   **b.** .279   **c.** .635   **d.** .718
43. a. .585   **b.** .202   **c.** .909   **d.** .959   **e.** .762   **f.** .004
  g. Yes.   **h.** 15, 3.75
45. a. .944   **b.** .449   **c.** .816
47. a. 1.000, .988, .930, .678, .526   **c.** .996, .914, .736, .376, .244
  d. 1.000, .964, .816, .398, .236
49. The probability is .99 if A is chosen and .9963 when B is chosen. For $p = .5$, these become .75 and .6875, respectively.
53. a. 30, 21   **b.** 50, 25
55. For $p = .5$, the probabilities are .042 and .002, respectively, whereas the upper bounds are .25 and .111. For $p = .75$, the probabilities are .065 and .004, respectively.

Section 3.5
(pages 112–113)

57. a. .242   **b.** .030   **c.** .939   **59.** $h(x; 10, 12, 27)$, .282
61. a. $h(x; 6, 4, 11)$ **b.** 2.18
63. a. $h(x; 10, 15, 50)$ **b.** $b(x; 10, .30)$ **c.** $E(X) = 3$, $V(X) = 1.71$ for the hypergeometric model, and $E(X) = 3$, $V(X) = 2.1$ for the binomial model.

65. $p(3) = .250, p(4) = .375, p(5) = .375.$

67. $\binom{x-1}{9}\left(\frac{5}{6}\right)^{x-10}\left(\frac{1}{6}\right)^{10}$ for $x = 10, 11, 12, \ldots$; $E(X) = 60, V(X) = 300.$

Section 3.6
(pages 117–118)

69. a. .191   **b.** .526   **c.** .283   **d.** 8, 2.83
71. a. .163   **b.** .018   **c.** .671   **73. a.** 5, 2.236   **b.** .032   **c.** .007
75. a. .176   **b.** .875   **c.** 3.75   **77.** .033
79. a. .221   **b.** 6,800,000   **c.** $p(x; 1608.5)$

Supplementary
Exercises
(pages 118–121)

83. b. 3.12, .384, .620
85. a. $b(x; 15, .75)$ **b.** .764   **c.** .707   **d.** 11.25, 2.81   **e.** .310
87. a. .007   **b.** .098   **c.** .579
89. a. .986   **b.** .220   **c.** .044   **d.** .845, .109
     e. Reject the claim if either $x \le 5$ or $x \ge 20.$
91. $\text{Bin}(n, 1 - p_1 + p_1 p_2)$
93. a. 16   **b.** 4   **c.** .804   **d.** $h(x; 5, 4, 12)$; 1.67, .707
95. a. .135   **b.** .00144   **c.** $\sum\limits_{x=0}^{\infty} [p(x; 2)]^5$ **97.** 3.590
99. a. $(p_1)^{10} + (p_2)^{10} - (p_1 p_2)^{10}$
     b. $\left[\dfrac{p_1 p_2}{(1 - p_1)(1 - p_2)}\right]^{10} \cdot \sum\limits_{x=10}^{\infty} \binom{x-1}{9}^2 (1 - p_1)^x (1 - p_2)^x$
101. b. $.6p(x; \lambda) + .4p(x; \mu)$ **c.** $(\lambda + \mu)/2$ **d.** $(\lambda - \mu)^2/4 + (\lambda + \mu)/2$
103. $\sum\limits_{i=1}^{10} (p_{i+j+1} + p_{i-j-1})p_i$ where $p_k = 0$ if $k < 0$ or $k > 10.$

Chapter 4

Section 4.1
(pages 128–130)

1. a. .25   **b.** .50   **c.** .4375   **3. b.** .50   **c.** .6875   **d.** .6328
5. a. 3   **b.** .125   **c.** .109   **d.** .704
7. a. $f(x) = .1$ for $25 \le x \le 35$ and 0 otherwise.   **b.** .20   **c.** .40   **d.** .20
9. a. .562   **b.** .438, .438   **c.** .139

Section 4.2
(pages 137–138)

11. a. .25   **b.** .1875   **c.** .9375   **d.** 1.4142   **e.** $f(x) = x/2$ for $0 < x < 2.$
13. a. 1.33   **b.** .222, .471   **c.** 2
15. a. For $0 \le x \le 1$, $F(x) = 2[x - x^2/2].$ **b.** .75   **c.** .3125, .3125   **d.** .5
     e. .293   **f.** .333, .0556, .2357
17. a. For $2 \le x \le 4$, $F(x) = .25[3x - 7 - (x - 3)^3].$ **b.** 3   **c.** 3, .2
19. a. .597   **b.** .369   **c.** $f(x) = .0966 - .25 \ln(x)$ for $0 < x < 4$
21. 314.38   **23.** 248, 3.60

Section 4.3
(pages 148–150)

25. a. .4850   **b.** .3413   **c.** .4938   **d.** .9876   **e.** .9147   **f.** .9599
     g. .9104   **h.** .0791   **i.** .0668   **j.** .9876
27. a. 1.34   **b.** −1.34   **c.** .675   **d.** −.675   **e.** −1.55
29. a. .9772   **b.** .5   **c.** .9104   **d.** .8413   **e.** .2417   **f.** .6826
31. a. .1251   **b.** .9382   **c.** .0548   **33.** The second machine (.6826 vs. .9987).
35. a. .7745   **b.** .1587   **c.** .3085   **d.** 6.1645   **37.** 15.66

39. .3174 for $k = 1$, .0456 for $k = 2$, .0026 for $k = 3$, as compared to the bounds of 1, .25, and .111, respectively.

41. a. .7580 **b.** .6727 **43. a.** .9780 **b.** .8849 **c.** .8365

Section 4.4
(pages 156–158)

47. a. 120 **b.** 1.329 **c.** .371 **d.** .735 **e.** 0
49. a. .594 **b.** .092 **c.** .537
51. a. .424 **b.** .567, yes. **c.** 60 **d.** 66
53. a. .777 **b.** .368 **c.** .145 **d.** 13.863
55. a. n/λ, 20 **b.** .930 **c.** $1 - \sum\limits_{k=0}^{n-1} p(k; \lambda t)$ **57.** $-[\ln(1-p)]/\lambda$, $.693/\lambda$

Section 4.5
(pages 164–165)

61. a. .632, .632, .064 **b.** .470 **c.** 172.727 **65. a.** .8212 **b.** .1120
67. a. 149.157, 223.595 **b.** .9830 **c.** .0921 **d.** 148.41 **e.** 9.83
f. 125.90
69. a. .714, .0255 **b.** .0016 **c.** .0394 **d.** .286
71. a. $\alpha = \beta = 3$ **b.** .365 **c.** .635

Supplementary Exercises
(pages 175–178)

79. a. .4 **b.** .6 **c.** $F(x) = .04x$ for $0 \le x \le 25$
81. a. $f(x) = x^2$ for $0 \le x < 1$, $= 1.75 - .75x$ for $1 \le x < 7/3$, $= 0$ otherwise.
b. .917 **c.** 1.21
83. a. $F(x) = 1.5(1 - 1/x)$ for $1 \le x \le 3$. **b.** .90, .40 **c.** 1.648
d. .533 **e.** .267
85. a. 1.075, 1.075 **b.** .0614, .3331 **c.** 2.476
87. a. c/λ **b.** $(.5\lambda - a)/(\lambda - a)$
89. b. $F(x) = .5e^{-.2x}$ for $x \le 0$ and $= 1 - .5e^{-.2x}$ for $x > 0$.
c. .500, .665, .256, .670
91. a. $k = (\alpha - 1)5^{\alpha - 1}$; $\alpha > 1$. **b.** $[k/(\alpha - 1)] \cdot [1/5^{\alpha - 1} - 1/x^{\alpha - 1}]$ for $x \ge 5$.
c. $k/[(\alpha - 2)5^{\alpha - 2}]$

Chapter 5

Section 5.1
(pages 190–193)

1. a. .20 **b.** .42 **c.** At least one hose is in use at each pump; .70.
d. $p_X(x) = .16, .34, .50$ for $x = 0, 1, 2$, respectively; $p_Y(y) = .24, .38, .38$ for $y = 0, 1, 2$, respectively; .50. **e.** No; $p(0, 0) \ne p_X(0) \cdot p_Y(0)$.

3. a.

		y	
	0	1	2
0	1/4	1/4	1/16
x 1	1/4	1/8	0
2	1/16	0	0

b.

			y	
	1	2	3	4
1	1/16	1/16	1/16	1/16
2	1/16	1/16	1/16	1/16
x 3	1/16	1/16	1/16	1/16
4	1/16	1/16	1/16	1/16

5. a. .0509
b. $\binom{8}{x}\binom{10}{y}\binom{12}{6-x-y}\Big/\binom{30}{6}$ for $x = 0, 1, \dots$; $y = 0, 1, \dots$; $x + y \le 6$.
7. a. $f(x, y) = 1$ for $5 \le x \le 6$, $5 \le y \le 6$, and $= 0$ otherwise. **b.** .25 **c.** .306
9. a. .050
b. $f_X(x) = e^{-x}$ for $x \ge 0$; $f_Y(y) = 1/(1 + y)^2$ for $y \ge 0$; they are not independent because $f(x, y) \ne f_X(x) \cdot f_Y(y)$. **c.** .300

11. a. $(1 - e^{-\lambda t})^{10}$ **b.** $\binom{10}{k}(1 - e^{-\lambda t})^k (e^{-\lambda t})^{10-k}$

c. $\binom{9}{5}(1 - e^{-\lambda t})^5 (e^{-\lambda t})^4 (e^{-\mu t}) + \binom{9}{4}(1 - e^{-\lambda t})^4 (e^{-\lambda t})^5 (1 - e^{-\mu t})$

13. a. $f(x_1, x_3) = 72x_1(1 - x_3)(1 - x_1 - x_3)^2$ for $0 \le x_1, 0 \le x_3, x_1 + x_3 \le 1$.
b. .53125   **c.** $f(x_1) = 18x_1 - 48x_1^2 + 36x_1^3 - 6x_1^5$ for $0 \le x_1 \le 1$.

Section **5.2**
(pages 199–200)

15. a. 14.10   **b.** 9.60   **17.** L^2 **19.** 1/3   **21.** $-2/3$
23. a. -3.20 **b.** $-.21$

Section **5.3**
(pages 207–210)

29. a.

t	2	3	4	5	6	7	8
$p_{T_0}(t)$	.04	.12	.25	.28	.22	.08	.01

$\bar{x}$	1	1.5	2	2.5	3	3.5	4
$p_{\bar{x}}(\bar{x})$	.04	.12	.25	.28	.22	.08	.01

c. $E(T_0) = 4.8, E(\bar{X}) = 2.4$ **d.** $V(T_0) = 1.68, V(\bar{X}) = .42$

31. a.

t	320	360	400	440	480
$p_{T_0}(t)$	.16	.32	.32	.16	.04

$\bar{x}$	160	180	200	220	240
$p_{\bar{x}}(\bar{x})$	.16	.32	.32	.16	.04

(X_1, X_2) is a random sample.

b.

t	320	360	400	440	480
$p_{T_0}(t)$	.16	.24	.32	.20	.08

X_1 and X_2 do not have the same distribution.

c. $\mu_{T_0} = 392, \sigma_{T_0}^2 = 2176$.

d. X_1 and X_2 are not independent.

t	360	400	440
$p_{T_0}(t)$	16/30	7/30	7/30

e. $\mu_{T_0} = 388$.
33. a. 37.5   **b.** 52.083   **c.** $-2.5, 10.417$ **d.** $-12.5, 52.083$
35. a. 1300, 100, 10   **b.** 52, .16, .4   **37.** .80, .02667
39. a. $E(W) = n(n + 1)/4$ **b.** $V(W) = n(n + 1)(2n + 1)/24$

Section **5.4**
(pages 217–219)

41. .2912, .3463   **43. a.** .6026   **b.** .2981
45. a. .8342, .7888   **b.** .9957, .9938   **c.** .9914, .9876
47. a. .9803, .4803   **b.** 33   **49. a.** .9616   **b.** .0623   **51.** 10 : 52.76
53. .4826   **55.** .9616

Supplementary
Exercises
(pages 219–221)

57. a. 3/81,250
b. $f_X(x) = k(250x - 10x^2)$ for $0 \le x \le 20$ and $= k(450x - 30x^2 + \frac{1}{2}x^3)$ for $20 < x \le 30$; $f_Y(y)$ results from substituting y for x in $f_X(x)$. They are not independent.   **c.** .355   **d.** 25.969   **e.** 204.6154, $-.894$ **f.** 7.66
59. ≈ 1 **61.** 97   **63.** .8340   **65.** $E(X + Y) = 1.167$
67. b. 16   **c.** $\Sigma\Sigma a_i b_j \text{Cov}(X_i, Y_j)$ **69. b.** .855   **71.** 26.19, compared to 26

Chapter 6

Section **6.1**
(pages 235–238)

1. a. 65.72, 4.53   **b.** 65.10   **c.** 66.05   **d.** .70
3. a. 76.818; $\bar{X}$ **b.** 76.88   **5. a.** $\hat{\mu} = 5.102, \hat{\sigma} = .496$ **b.** 185.861
7. b. $1/n$ **11. a.** $\hat{\theta} = \Sigma X_i^2/2n$ **b.** 74.505   **13. b.** .444
15. a. $\hat{p} = 2\hat{\lambda} - .30 = .20$ **c.** $\hat{p} = (100\hat{\lambda} - 9)/70$

Section **6.2**
(pages 247–248)

17. b. $\hat{\alpha} = 5, \hat{\beta} = 28.0/\Gamma(1.2)$ **19.** $\hat{\lambda}_1 = \bar{x}, \hat{\lambda}_2 = \bar{y}, (\hat{\lambda_1} - \hat{\lambda_2}) = \bar{x} - \bar{y}$
21. a. 384.4, 18.86   **b.** 415.42
25. a. $\hat{\theta} = \min(X_i), \hat{\lambda} = n/\Sigma(X_i - \min(X_i))$ **b.** .64, .202

27. 27.38

31. With x_i = time between birth $i - 1$ and birth i, $\hat{\lambda} = 6/\sum_{i=1}^{6} ix_i = .0436$.

33. 9.55

Chapter 7

1. a. 99.5%   **b.** 85%   **c.** 2.96   **d.** 1.15
3. a. (4.52, 5.18)   **b.** (4.12, 5.00)   **c.** 55
5. a. By a factor of 4.   **b.** The length is decreased by a factor of 5.
7. a. $(\bar{x} - z_\alpha \cdot \sigma/\sqrt{n}, \infty); (4.57, \infty)$ **b.** $(-\infty, \bar{x} + z_\alpha \cdot \sigma/\sqrt{n}); (-\infty, 59.7)$
9. 950, .8714

11. (37.5, 39.9); none   **13.** (150.2, 154.4)   **15.** (.182, .316)
17. a. 385   **b.** 342   **19.** $\bar{x} \pm z_{\alpha/2} \cdot \sqrt{\bar{x}/n}$; a 95% C.I. is (3.50, 4.62).

21. a. 1.321   **b.** −1.321   **c.** −1.746   **d.** .99   **e.** .95   **f.** .002
23. (8.19, 10.85)   **25.** (33.53, 43.79)

27. a. 22.307   **b.** 34.381   **c.** 44.313   **d.** 46.925   **e.** 11.523   **f.** 10.519
29. (3.60, 28.98); (1.90, 5.38)

31. (.622, .678)   **33.** (9.28, 9.64)   **35.** (.198, .230)   **37.** (.185, .443)
39. (196.46, 223.04)   **41. b.** (1.44, 3.54)   **c.** $100(1 - n(.5)^{n-1})\%$

Chapter 8

1. a. Yes.   **b.** No.   **c.** No.   **d.** Yes.   **e.** No.   **f.** Yes.
5. $H_0: \sigma = .05$ versus $H_a: \sigma < .05$.
7. a. R_1 **c.** Bin(25, .5), .044   **d.** .488, .845, .845, .488   **e.** reject H_0
9. a. $H_0: \mu = 10$ versus $H_a: \mu \neq 10$.   **b.** .01   **c.** .5319, .0078
11. b. .0004, 0, < .01

13. a. .03   **b.** .003   **c.** .004
15. a. Reject H_0.   **b.** .8413   **c.** 143   **d.** .0052
17. a. $z = -2.27$, don't reject H_0.   **b.** .2266   **c.** 22
19. a. $z = -3.33$, so reject H_0 at level .01.   **b.** .1056   **c.** 217
21. Test statistic value $= -5.84$; reject H_0 and conclude that $\mu \neq 55$.
23. $z = 1.99$, $z_{.0005} = 3.27$, don't reject H_0.
25. a. $t = .50$; don't reject H_0.   **b.** .72
27. Test statistic value $= -1.24$, don't reject H_0.
29. a. Test statistic value $= -.63$, don't reject H_0.   **b.** $n = 18$

33. a. $z = 1.70$, reject H_0.   **b.** .6480   **c.** 600
35. a. $z = -1.01$, don't reject H_0; carry out the inventory.   **b.** .2981   **c.** ≈ 0
37. a. $z = 3.08$; using $\alpha = .01$, don't reject H_0.   **b.** .0322 (using $\alpha = .01$)   **c.** 637
39. a. $\{15, \ldots, 20\}$ **b.** .021, yes, yes   **c.** .874, .196   **d.** No.

Section 8.4
(pages 313–315)

41. a. Reject H_0.   **b.** Reject H_0.   **c.** Don't reject H_0.   **d.** Reject H_0.
e. Don't reject H_0.
43. a. .0808   **b.** .1841   **c.** .0287   **d.** .0082   **e.** .5398
45. a. $.025 < P\text{-value} < .05$ **b.** $P\text{-value} \approx .0005$ **c.** $.01 < P\text{-value} < .025$
d. $P\text{-value} > .10$ **e.** $P\text{-value} \approx .005$
47. $z = -3.71$, so $P\text{-value} < 2(.0002) = .0004$. Since $.0004 < .01$, reject H_0.
49. a. $P\text{-value} = .1660$; don't modify.   **b.** .9974
51. $t = .35$, $P\text{-value} > .20$, don't recalibrate.

Section 8.5
(page 318)

53. a. .8888, .1587, .0006   **b.** $P\text{-value} \ll .0002$; yes.   **c.** No.

Supplementary Exercises
(pages 318–321)

55. a. Test statistic value $= -3.11$; yes.   **b.** .0009   **57.** Yes.
59. a. Test statistic value $= 1.65$, so don't reject H_0; type II.   **b.** .099; yes
61. Test statistic value $= -3.32$; yes.
63. $H_a : \lambda > 4$, $z = 1.33$, so don't reject H_0. No.
65. Test statistic value $= -8.88$, so reject H_0. Yes.
67. $.01 < P\text{-value} < .025$, so H_0 is not rejected at level .01.
69. a. For $H_a : \mu > \mu_0$, reject H_0 if $2\Sigma X_i / \mu_0 \geq \chi^2_{\alpha, 2n}$.
b. Test statistic value $= 19.65 \nleq 8.260$, so don't reject H_0.
71. a. Yes. $B(5; 10, .9) = .002$ **c.** No.

Chapter 9

Section 9.1
(pages 331–333)

1. a. $-.4h$; it doesn't.   **b.** .0724, .2691   **c.** No.
3. Test statistic value $= 6.41$. Reject H_0.
5. a. $z = -2.90$, so reject H_0.   **b.** .0019   **c.** .8212   **d.** 66   **7.** $(-9.6, 0.0)$
9. a. $z = 2.89$, so don't use the high purity steel.   **b.** .2981
c. s_1 and s_2 replace σ_1 and σ_2; same conclusion.
11. $(-11.23, -6.31)$
13. $\hat{\theta} = -.97$, $\hat{\sigma}_{\hat{\theta}} = s_{\hat{\theta}} = 1.05$, $z = -.92$, so don't reject H_0. No.   **15.** They increase.

Section 9.2
(pages 341–343)

17. $t = -2.00$, so don't reject H_0.   **19.** $(-20.3, -6.9)$
21. Test statistic value $= 3.64$. Yes.
23. $t = -2.68$, so $.01 < P\text{-value} < .025$. Don't reject H_0. No.
25. $t = 4.25$, so $P\text{-value} < .001$. Yes, there does appear to be a difference.
27. $t = 3.04 > 2.131$, so conclude that $\mu_1 \neq \mu_2$.   **29.** 26

Section 9.3
(pages 350–352)

31. a. $t_{\text{paired}} = -3.05 < -2.947$, so reject H_0.
b. $t = -.57$, so the conclusion is opposite that of (a).
33. $t_{\text{paired}} = 1.87$, whereas $t_{.01, 8} = 2.896$, so don't reject H_0 at level .01 (at level .05, H_0 is barely rejected).
35. $t_{\text{paired}} = 3.59$, so $.005 < P\text{-value} < .01$. H_0 cannot be rejected at level .001.
37. Test statistic value $= 5.35$, so $P\text{-value} < .0005$. Yes, the data is highly significant.
39. Use the paired t test.

Section 9.4
(pages 359–360)

41. a. $z = -4.84$, so reject H_0.   **b.** .9988
43. a. $z = 1.43$. Conclude that there is no difference.   **b.** $n = 6582$
45. $(.021, .107)$
47. $(-.35, .07)$

Section **9.5**
(pages 363–364)

49. a. 3.69   **b.** 4.82   **c.** 2.07   **d.** 2.71   **e.** 4.30   **f.** .212
g. .95   **h.** .94
51. $f = .585$, so H_0 is not rejected. No.
53. $f = 1.22$, so it is quite plausible that $\sigma_1 = \sigma_2$.

Supplementary
Exercises
(pages 364–367)

55. Test statistic value $= .65$. Cessation does not appear to result in an increase.
57. Test statistic value $= -1.92 \not\le -2.896$. No.
59. Test statistic value $= 6.4$. Conclude that there is a difference.
61. a. $m = 141$, $n = 47$ **b.** $m = 240$, $n = 160$ **63.** $\hat{\sigma}^2 = .409$
65. Test statistic value $= .95$. No. No.
67. $z = -1.09$, which is certainly not ≥ 1.645. No.
69. $(-4.68, 6.23)$ **71.** $(-.055, .179)$
73. a. $t_{\alpha/2,\nu}$ (with ν estimated from the data) will replace $z_{\alpha/2}$ in the large-sample
z interval for $\mu_1 - \mu_2$.
b. $(.040, .530)$
75. $\hat{\lambda}_1 - \hat{\lambda}_2 \pm z_{\alpha/2} \sqrt{\dfrac{\hat{\lambda}_1}{m} + \dfrac{\hat{\lambda}_2}{n}}$ where $\hat{\lambda}_1 = \bar{x}$ and $\hat{\lambda}_2 = \bar{y}$; $(-1.19, .43)$.

Chapter 10

Section **10.1**
(pages 377–378)

1. $f = 1.85$, $F_{.05,4,15} = 3.06$, so don't reject H_0.
3. $f = 6.43 \ge 2.95 = F_{.05,3,28}$, so conclude that there are differences.
5.

Source	d.f.	SS	MS	f	$F_{.01}$
Treatments	3	509.13	169.71	10.85	4.4
Error	36	563.04	15.64		
Total	39	1072.17			

7. $f = 10.48$, $F_{.01,4,30} = 4.02$, so average axial thickness does appear to depend on
plate length.
9. a. μ **b.** $\sigma^2/J + \mu_i$ **c.** $\sigma^2/IJ + \mu^2$ **e.** σ^2; $>$

Section **10.2**
(pages 384–385)

11.

3	1	4	2	5
427.5	462.0	469.3	512.8	532.1

Brands 2 and 5 do not differ significantly from one another, but both differ
significantly from all three of the other brands. The only other significant difference
is between brand 3 and brand 4.
13. $w = 1.41$

	1	2	3	4
	4.39	4.52	5.49	6.36

The only significant differences are between 1 and 4 and between 2 and 4.
15. $w = 61.83$. The 2 in. thickness differs significantly from both the 8 and 10 in.
thicknesses. The 4 and 6 in. thicknesses both differ significantly from the
10 in. thickness.
17. $w = 6.35$, so 1 and 2 differ significantly from 4 and there are no other significant
differences.
19. a. $f = 3.49$, $F_{.05,4,15} = 3.06$, so reject H_0.
b. $w = 10.18$, and there are no significant differences.
21. No.

Section 10.3
(pages 394–395)

23. $MSTr = 152.17$, $MSE = 8.89$, $f = 17.12$. H_0 should be rejected.

25.

Source	d.f.	SS	MS	f
Groups	2	152.18	76.09	5.56
Error	71	970.96	13.68	
Total	73	1123.14		

27. a. $MSTr = 21.64$, $MSE = .273$, $f = 79.3$, $F_{.01, 5, 20} = 4.10$, so reject H_0.

b.

Pair	Interval	Pair	Interval	Pair	Interval
1, 2	1.30 ± 1.59	2, 3	-1.03 ± 1.59	3, 5	-3.31 ± 1.59
1, 3	$.27 \pm 1.67$	2, 4	$-.30 \pm 1.59$	3, 6	-4.27 ± 1.67
1, 4	1.00 ± 1.67	2, 5	-4.34 ± 1.50	4, 5	-4.04 ± 1.59
1, 5	-3.04 ± 1.59	2, 6	-5.30 ± 1.59	4, 6	-5.00 ± 1.67
1, 6	-4.00 ± 1.67	3, 4	$.73 \pm 1.67$	5, 6	$-.96 \pm 1.59$

c. $-4.16 \pm (2.336) \sqrt{.1719} = (-5.14, -3.18)$.

31. a. $\phi = 2$, $\beta \approx .10$ **b.** 9   **c.** $\phi = 1.26$, $\beta \approx .45$

33. Transforming by square roots and applying ANOVA yields $f = 3.23$, so conclude that brands do not differ significantly at level .01.

35. $\sigma^2 + J\sigma_A^2$

Supplementary
Exercises
(pages 396–397)

37. $f = .39$; no.

39. $MSE = .108$, $\Sigma c_i^2 = 1.25$, and a 95% C.I. is $(-.144, .474)$

41. $H_0 : \sigma_A^2 = 0$ versus $H_a : \sigma_A^2 > 0$, $SSTr = 31.5430$, $SSE = .7115$, $f = 118$, so reject H_0.

43. The two are identical.

Chapter 11

Section 11.1
(pages 410–412)

1. a. $f_A = 1.55$, so don't reject H_{0A}.   **b.** $f_B = 2.98$, so don't reject H_{0B}.

3. a. $f_A = 12.98$, $F_{.01, 3, 9}$, so conclude that there is a gas rate effect; $f_B = 105.31$, so conclude that there is a liquid rate effect.

b. $w = 95.44$; $\underline{231.75 \quad 325.25}$ 441.0   613.25, so only the lowest two rates do not differ significantly from one another.

c. $\underline{336.75 \quad 382.25 \quad 419.25}$ 473   so only the lowest and highest rates appear to differ significantly from one another.

5. $f_A = 2.65$, $F_{.01, 3, 12} = 5.95$, so there appears to be no effect due to angle of pull.

7. a.

Source	d.f.	SS	MS	f
Treatments	2	28.78	14.39	1.04
Blocks	17	2977.67	175.16	12.68
Error	34	469.55	13.81	
Total	53	3476.00		

True average adaptation score does not appear to depend on which treatment is given.

b. Yes; f_B is quite large, suggesting great variability between subjects.

9. $f_B = 8.87$, $F_{.01, 4, 8} = 7.01$, so conclude that $\sigma_B^2 > 0$.

Section 11.2
(pages 420–421)

13. a.

Source	d.f.	SS	MS	f
A	2	30,763.0	15,381.50	3.79
B	3	34,185.6	11,395.20	2.81
AB	6	43,581.2	7,263.53	1.79
Error	24	97,436.8	4,059.87	
Total	35	205,966.6		

b. $F_{.05, 6, 24} = 2.51, f_{AB} = 1.79$, so do not reject H_{0AB}.

c. $F_{.05, 2, 24} = 3.40, f_A = 3.79$, so reject H_{0A}.   **d.** Don't reject H_{0B}.

e. $w = 64.93$; only times 2 and 3 differ significantly from one another.

15. $f_{AB} = .32, f_A = 192.09, f_B = 8.96$, so interactions are not significant, but both main effects are significant.

17. $F_{.01, 8, 30} = 3.17, f_{AB} = 1.38$, so don't reject H_{0G}; $f_A = MSA/MSAB = 26.70$, $F_{.01, 2, 8} = 8.65$, so reject H_{0A} and conclude that at least one $\alpha_i \neq 0$; $f_B = 28.51$, so reject H_{0B} and conclude that $\sigma_B^2 > 0$.

21. a. $MSAB/MSE$.

b. $MSA/MSAB$ for testing H_{0A}; $MSB/MSAB$ for testing H_{0B}.

Section **11.3**
(pages 431–434)

23.

Source	d.f.	SS	MS	f
A	3	19,149.73	6,383.24	2.70
B	2	2,589,047.62	1,294,523.81	546.79*
C	1	157,437.52	157,437.52	66.50*
AB	6	53,238.21	8,873.04	3.75
AC	3	9,033.73	3,011.24	1.27
BC	2	91,880.04	45,940.02	19.40*
ABC	6	6,558.46	1,093.08	<1
Error	24	56,819.50	2,367.48	
Total	47	2,983,164.81		

An * appears beside each significant F ratio. Every F ratio involving A is insignificant, so factor A (drilling speed) appears to have no effect on thrust force.

25.

Source	d.f.	SS	MS	f
A	3	.22625	.075417	77.35
B	1	.000025	.000025	<1
C	1	.0036	.0036	3.69
AB	3	.004325	.001442	1.48
AC	3	.00065	.000217	<1
BC	1	.000625	.000625	<1
ABC = Error	3	.002925	.000975	
Total	15	.2384		

b. From the f column and F table, we conclude that only the A main effect is significant.

c. $Q_{.05, 3, 3} = 5.91$, so $w = 5.91(.000975/4)^{1/2} = .09227$.

$\bar{x}_{i..}$:   1.1200   1.3025   1.3875   1.4300

27. a. $MSABC/MSE$; MSC/MSE; $MSAB/MSABC$; $MSA/MSAC$

b.

Source	d.f.	SS	MS	f	$F_{.01}$
A	1	14,318.24	14,318.24	$MSA/MSAC = 19.85$	98.50
B	3	9,656.40	3,218.80	$MSB/MSBC = 6.24$	9.78
C	2	2,270.22	1,135.11	$MSC/MSE = 3.15$	5.61
AB	3	3,408.93	1,136.31	$MSAB/MSABC = 2.41$	9.78
AC	2	1,442.58	721.29	$MSAC/MSE = 2.00$	5.61
BC	6	3,096.21	516.04	$MSBC/MSE = 1.43$	3.67
ABC	6	2,832.72	472.12	$MSABC/MSE = 1.31$	3.67
Error	24	8,655.60	360.65		
Total	47				

At level .01, no H_0's can be rejected, so there appear to be no interaction or main effects present.

29.

Source	d.f.	SS	MS	f
A	5	6,475.80	1,295.16	
B	5	529.47	105.89	
C	5	508.47	101.69	1.59
Error	20	1,277.23	63.86	
Total	35	8,790.97		

Since $F_{.05, 5, 20} = 2.71$ and 1.59 is not ≥ 2.71, H_{0C} is not rejected; shelf space does not appear to affect sales.

Section 11.4
(pages 444–446)

31. a. $\hat{\beta}_1 = 54.38$, $\hat{\gamma}_{11}^{AC} = -2.21$, $\hat{\gamma}_{21}^{AC} = 2.21$.

b.

Source	Effect Contrast	MS	f
A	1,307	71,177.04	436.7
B	1,305	70,959.34	435.4
C	529	11,660.04	71.54
AB	199	1,650.04	10.12
AC	−53	117.04	<1
BC	57	135.38	<1
ABC	27	30.38	<1
Error		162.98	

33.

Source	SS	f
A	136,640.02	999.3
B	139,644.19	1,021.3
C	24,616.02	180.0
D	20,377.52	149.0
AB	2,173.52	15.90
AC	2.52	<1
AD	58.52	<1
BC	165.02	1.21
BD	9.19	<1
CD	17.52	<1
ABC	42.19	<1
ABD	117.19	<1
ACD	188.02	1.38
BCD	13.02	<1
ABCD	204.19	1.49
Error	4,375.33	
Total	328,607.98	

$F_{.05, 1, 32} \approx 4.15$, so only the four main effects and the AB interaction appear significant.

35.

Source	d.f.	SS	f
A	1	.436	1.12
B	1	.099	<1
C	1	.109	<1
D	1	414.12	1,067
AB	1	.497	1.28
AC	1	.078	<1
AD	1	.017	<1
BC	1	1.404	3.62
BD	1	.456	1.18
CD	1	2.190	5.64
Error	5	.388	

$F_{.05, 1, 5} = 6.61$, so only the factor D main effect is judged significant.

37. **a.** 1: (1), *ab, cd, abcd*; 2: *a, b, acd, bcd*; 3: *c, d, abc, abd*; 4: *ac, bc, ad, bd*.

b.

Source	d.f.	SS	f
A	1	14,028.13	53.89
B	1	92,235.13	345.33
C	1	3.13	<1
D	1	18.00	<1
AC	1	105.13	<1
AD	1	200.00	<1
BC	1	91.13	<1
BD	1	420.50	1.62
ABC	1	276.13	1.06
ABD	1	2.00	<1
ACD	1	450.00	1.73
BCD	1	2.00	<1
Blocks	7	898.88	<1
Error	12	3,123.72	
Total	31	111,853.88	

$F_{.01, 1, 12} = 9.33$, so only the *A* and *B* main effects are significant.

39. **a.** *ABFG*; (1), *ab, cd, ce, de, fg, acf, adf, adg, aef, acg, aeg, bcg, bcf, bdf, bdg, bef, beg, abcd, abce, abde, abfg, cdfg, cefg, defg, acdef, acdeg, bcdef, bcdeg, abcdfg, abcefg, abdefg*. {*A, BCDE, ACDEFG, BFG*}, {*B, ACDE, BCDEFG, AFG*}, {*C, ABDE, DEFG, ABCFG*}, {*D, ABCE, CEFG, ABDFG*}, {*E, ABCD, CDFG, ABEFG*}, {*F, ABCDEF, CDEG, ABG*}, {*G, ABCDEG, CDEF, ABF*}.

b. 1: (1), *aef, beg, abcd, abfg, cdfg, acdeg, bcdef*; 2: *ab, cd, fg, aeg, bef, acdef, bcdeg, abcdfg*; 3: *de, acg, adf, bcf, bdg, abce, cefg, abdefg*; 4: *ce, acf, adg, bcg, bdf, abde, defg, abcefg*.

41. $SSA = 2.250$, $SSB = 7.840$, $SSC = .360$, $SSD = 52.563$, $SSE = 10.240$, $SSAB = 1.563$, $SSAC = 7.563$, $SSAD = .090$, $SSAE = 4.203$, $SSBC = 2.103$, $SSBD = .010$, $SSBE = .123$, $SSCD = .010$, $SSCE = .063$, $SSDE = 4.840$. Error SS = sum of two-factor SS's = 20.568, Error $MS = 2.057$, $F_{.01, 1, 10} = 10.04$, so only the *D* main effect is significant.

Supplementary
Exercises
(pages 447–448)

43.

Source	d.f.	SS	MS	f
A main effects	1	322.667	322.667	980.38
B main effects	3	35.623	11.874	36.08
Interaction	3	8.557	2.852	8.67
Error	16	5.266	0.329	
Total	23	372.113		

$F_{.05, 3, 16} = 3.24$, so interactions appear to be present.

45. $MSA = 6.940$, $MSB = 1.870$, $MSC = 6.165$, $MSAB = 1.350$, $MSAC = 3.660$, $MSBC = 2.633$, $MSABC = 2.400$, $MSE = 4.37$, $f_{ABC} = 13.2$, so there appear to be three-way interactions.

$$SSA = \sum_i \sum_j (\overline{X}_{i\ldots} - \overline{X}_{\ldots})^2 = \frac{1}{N} \Sigma X^2_{i\ldots} - X^2_{\ldots}/N \text{ with similar expressions}$$

for *SSB, SSC,* and *SSD*, each having $N - 1$ d.f.

$$SST = \sum_i \sum_j (X_{ij(kl)} - \overline{X}_{\ldots})^2 = \sum_i \sum_j X^2_{ij(kl)} - X^2_{\ldots}/N, \text{ with } N^2 - 1 \text{ d.f.,}$$

leaving $N^2 - 1 - 4(N - 1)$ d.f. for error.

47.

Source	d.f.	SS	MS	f
A	4	285.76	71.44	.594
B	4	227.76	56.94	.473
C	4	2,867.76	716.94	5.958
D	4	5,536.56	1,384.14	11.502
Error	8	962.72	120.34	$F_{.05, 4, 8} = 3.84$
Total	24			

H_{0A} and H_{0B} cannot be rejected, while H_{0C} and H_{0D} are rejected.

Chapter 12

Section 12.1
(pages 455–456)

1. a. .5050   **b.** 1.3   **c.** 130   **d.** -130
3. a. .095   **b.** $-.475$ **c.** .830, 1.305   **d.** .4207, .3446   **e.** .0036
5. a. $-.01, -.10$ **b.** 3.00, 2.50   **c.** .3627   **d.** .4641

Section 12.2
(pages 467–470)

7. a. $10b_0 + 1269b_1 = 475$, $1269b_0 + 172,809b_1 = 62,631$
b. $\hat{\beta}_1 = .19990826 \approx .200$, $\hat{\beta}_0 = 22.13164131 \approx 22.13$
c. .200   **d.** 2.40   **e.** 46.13
9. a. $y = -45.5519 + 1.7114x$ **b.** 339.51   **c.** -85.57
d. the $\hat{y}_i$'s are 125.6, 168.4, 168.4, 211.1, 211.1, 296.7, 296.7, 382.3, 382.3, 467.9, 467.9, 553.4, 639.0, 639.0; a 45° line through (0, 0).
11. a. .0289926   **b.** .00414180, .06435682   **c.** .996
13. a. Using $\hat{y}_i$'s to one decimal place, $SSE = 16,213.64$. The computational formula gives $SSE = 16,205.45$.   **b.** 414,235.71; $r^2 = .961$
15. a. Yes.   **b.** $y = 137.88 + 9.3116x$ **c.** $r^2 = .990$
d. $y = 190.35 + 7.5515x$; yes.
19. a, b. $\hat{\beta}_1$ for "transformed" data $= \hat{\beta}_1$ for original data, $\hat{\beta}_0$ for transformed data $= \bar{y}$.
21. a. .1323   **b.** .9412   **c.** No.

Section 12.3
(pages 476–477)

23. $(-.0332, -.0170)$, yes.
25. a. .1101, .000262
b. $t = 3.06$, $t_{.05, 8} = 1.860$, so conclude that the data does contradict prior belief.
27. a. $t = 17.2$, P-value $< .001$, so judge the model useful.   **b.** (14.94, 19.28)

Section 12.4
(pages 483–484)

33. a. (48.97, 52.63)   **b.** $t = .78$; don't reject H_0.
35. $t = 3.42$, $t_{.005, 7} = 3.499$, so don't reject H_0.
37. (11.75, 14.75) when $x = 85$, and (17.30, 23.24) when $x = 95$; $\bar{x} = 82.67$, and 95 is further from $\bar{x}$ than is 85.
39. a. $t = 5.02 \geq 2.896 = t_{.01, 8}$, so reject H_0.
b. (3.236, 3.768) when $x = 200$, and (3.916, 4.448) when $x = 300$.
41. $t = .25$, P-value $> .2$, so don't reject H_0 at significance level .01.

Section 12.5
(pages 492–493)

43. a. .743   **b.** .542   **45. a.** .9963   **b.** .993
47. a. $t = -1.69$, and $t_{.025, 8} = 2.306$, so don't reject $H_0 : \rho = 0$. $.10 < P$-value $< .20$
b. $(-.877, .295)$
49. a. $(-.751, -.318)$ **b.** $z = -.49$, so don't reject H_0.   **c.** .328   **d.** .328
51. a. Reject H_0.
b. No. Statistical significance is not the same as practical significance.

Supplementary
Exercises
(pages 493–496)

53. a. $y = 81.1731 - .13326x$
 b. $t = -4.2$, $t_{.005, 6} = 3.707$, so conclude that the model is useful.
 c. $\Sigma(x_i - \bar{x})^2 = 1442.875$ for the given data, and 5000 and 3750 for the proposed experiments. Either proposal is preferable to the given experiment.
 d. (76.80, 78.88). Yes.
55. a. .836 **b.** $t = 8.44$, so the relationship does appear to be useful.
 c. $t = -3.11 \le -2.145$, so reject H_0 and conclude that $\mu_{Y \cdot 500} \ne 10$.
57. a. $y = 14.1904 - .14892x$ **b.** $t = -1.43$, so don't reject $H_0 : \beta_1 = -.10$.
 c. No; $\Sigma(x - \bar{x})^2 = 143$ here and 182 for the given data.
 d. A 95% C.I. for $\mu_{Y \cdot 28}$ is (9.599, 10.443).
59. $s_{\hat{\beta}_0} = .0995$, and the C.I. is (3.404, 3.838).
63. a. $y = .3813 + .1517x$ **b.** $y = .3894 + .1483x$, $\hat{y} = 1.84$

Chapter 13*

Section **13.1**
(pages 503–505)

1. a. 6.32, 8.37, 8.94, 8.37, and 6.32 **b.** 7.87, 8.49, 8.83, 8.94, and 2.83
 c. The deviation is likely to be much smaller for the x values of (b).
3. a. Yes. **b.** $-.75, .31, -.74, 1.13, .43, -.72, 1.43, .93, -1.51, -1.27, .90$
 c. No.
5. a. No.
 b. e_i's are $-16.60, 9.70, 19.00, -.70, -11.40$; e_i^*'s are $-1.55, .68, 1.25, -.05,$ -1.06; a quadratic function.
7. For set 1, simple linear regression is appropriate. A quadratic regression is reasonable for set 2. In set 3, (13, 12.74) appears very inconsistent with the remaining data. The estimated slope for set 4 depends largely on the single observation (19, 12.5), and evidence for a linear relationship is not compelling.
9. $V(\hat{Y}_i)$ increases and $V(Y_i - \hat{Y}_i)$ decreases. **11.** t with $n - 2$ d.f.; .02

Section **13.2**
(pages 512–514)

13. a. $\hat{\alpha} = 43,044.94$, $\hat{\beta} = -2.16$ **b.** Yes. **c.** Yes.
 d. $t = -1.53$, so don't reject H_0.
 f. $x = 40 \Rightarrow x' = 3.6889$; the C.I. for $\beta_0 + \beta_1 x'$ is (2.6203, 2.7833), so the desired interval is (13.74, 16.17).
15. a. $\Sigma x_i' = 15.501$, $\Sigma y_i' = 13.352$, $\Sigma(x_i')^2 = 20.228$, $\Sigma x_i' y_i' = 18.109$, $\Sigma(y_i')^2 = 16.572$, $\hat{\beta}_1 = 1.254$, $\hat{\beta}_0 = -.468$, $\hat{\alpha} = .626$, $\hat{\beta} = 1.254$
 c. $t = -1.07$, so don't reject H_0. **d.** $H_0 : \beta = 1$, $t = -4.30$, so reject H_0.
17. a. No. **b.** $Y' = \beta_0 + \beta_1 \cdot (1/t) + \epsilon'$, where $Y' = \ln(Y)$, so $Y = \alpha e^{\beta/t} \cdot \epsilon$.
 c. $\hat{\beta} = \hat{\beta}_1 = 3735.45$, $\hat{\beta}_0 = -10.2045$, $\hat{\alpha} = (3.70034) \cdot (10^{-5})$.
 $\hat{y}' = 6.7748$, $\hat{y} = 875.5$
 d. $SSE = 1.39587$, $SSPE = 1.36594$ (using transformed values),
 $f = .33 < 8.68 = F_{.01, 1, 15}$, so don't reject H_0.
19. a. $\hat{\mu}_{Y \cdot x} = 18.14 - 1485/x$ **b.** $\hat{y} = 15.17$
21. For the exponential model, $V(Y|x) = \alpha^2 e^{2\beta x} \sigma^2$, which does depend on x. A similar result holds for the power model.

---

*Some answers have been taken from computer output and may differ slightly from hand-calculated values.

Section 13.3   **23. a.** 39.41   **b.** 24.93   **c.** 217.82; 72.61; 8.52   **d.** .779
(pages 521–524)   **e.** $t = -7.88$, so reject H_0 and retain the quadratic predictor.
               25. a. $y = -251.719114 + 1000.202871x - 135.456884x^2$
                   b. 202,209.77; 18,382.71; .853   **c.** $(-227.84, -43.08)$
                   d. $t = -1.83 > -2.718 = -t_{.01, 11}$, so don't reject H_0.
               27. a. $.3463 - 1.2933(x - \bar{x}) + 2.3964(x - \bar{x})^2 - 2.3968(x - \bar{x})^3$
                   b. -2.3968; 33.6430   **c.** .1949
                   d. $t = -.97$, so the cubic term can be deleted.
               29. a. .873   **b.** .919   **c.** $1.200887 - .01048314x + .00002618x^2$
                   d. .00077118   **e.** $t = 1.40$, so delete the quadratic term; no.

Section 13.4   **31. a.** 277.50   **b.** \$20   **c.** .8407
(pages 536–539)   **33. a.** 1.96   **b.** 1.40   **c.** $-.0006$; $-.06$
                   d. The model does not include interaction predictors such as $x_1 x_4$.
               35. b. $\beta_2 + \beta_4(2x_2 + 1)$; 2.00
               37. b. 103.11; 96.85; -5.85 **c.** 5384.18; 24.46; .768
                   d. $f = 14.89 > 4.26 = F_{.05, 2, 9}$, so reject H_0.   **e.** (78.30, 115.40)
                   f. (38.50, 155.20)   **g.** $t = -1.36$, so x_1 can be eliminated.
               39. a. No.   **b.** $f = 5.04$, $F_{.05, 5, 8} = 3.69$, so reject H_0.   **c.** (16.67, 31.91)
                   d. $f = 3.44$, $F_{.05, 3, 8} = 4.07$, so the simpler model appears adequate.
               41. a. 85.390; $y_{16} - \hat{y}_{16} = .01$
                   b. With $x_1', \ldots, x_4'$ denoting the uncoded variables, substitute
                   $x_1 = 10x_1' - 3, x_2 = 10x_2' - 3, x_3 = x_3' - 2.5, x_4 = (x_4' - 160)/15$.
                   c. $H_0 : \beta_5 = \cdots = \beta_{14} = 0$, $f = 2.28$, $F_{.05, 10, 16} = 2.49$, so don't reject H_0;
                   all higher-order terms can be deleted.
                   d. $H_0 : \mu_{Y \cdot 0, 0, 0, 0} = 85.0$, reject H_0 if $t < -1.706$; since $t = 7.19$, don't reject H_0.

Section 13.5   **43. a.** .5386   **b.** $(-.0333, -.0139)$ **c.** $-.004334$; .000845   **d.** (.489, .575)
(pages 551–553)   **45.** First x_3 is eliminated, and x_4 is deleted at the second stage. The remaining t ratios
                   are large, so the procedure terminates.
               47. a. $h_{44} > 2(k + 1)/n = .6$, so the fourth observation is potentially influential.
                   b. No.   **c.** No.

Supplementary   **49.** With $y' = \log_{10}(y)$, relate y' to x via the simple linear regression model ($y' = \ln(y)$
Exercises   will give equivalent results). That is, relate y to x via an exponential model.
(pages 553–554) **51.** $f = 38.8$, so judge the model useful (compare to $F_{\alpha, 2, 6}$).
               53. a. $f = 106$, so the model is judged useful.   **b.** (.015, .067)
                   c. $t = 5.86$, so retain this variable.   **d.** $\hat{y} = 99.514$; very close.

Chapter 14

Section 14.1   **1.** $\chi^2 = 5.92 < 7.815 = \chi^2_{.05, 3}$, so don't reject H_0.
(pages 564–565)   **3.** $\chi^2 = 6.61 < 14.684 = \chi^2_{.10, 9}$, so don't reject H_0.
               5. a. [0, .2231), [.2231, .5108), [.5108, .9163), [.9163, 1.0694), and [1.0694, ∞)
                   b. $\chi^2 = 1.25 < \chi^2_{\alpha, 4}$ for any reasonable α, so the specified exponential distribution is
                   quite plausible.

7. a. $(-\infty, -.97), [-.97, -.43), [-.43, 0), [0, .43), [.43, .97),$ and $[.97, \infty)$

　　b. $(-\infty, .49806), [.49806, .49914), [.49914, .5), [.5, .50086), [.50086, .50194),$ and $[.50194, \infty)$

　　c. $\chi^2 = 5.53, \chi^2_{.10, 5} = 9.236,$ so P-value $> .10$ and the specified normal distribution is plausible.

Section 14.2
(pages 575–577)

9. $\hat{p} = .0843, \chi^2 = 280.3 > \chi^2_{\alpha, 1}$ for any tabulated α, so the model gives a poor fit.

11. The likelihood is proportional to $\theta^{233}(1 - \theta)^{367}$, from which $\hat{\theta} = .3883$. The estimated expected counts are 21.00, 53.33, 50.78, 21.50, and 3.41. Combining cells 4 and 5, $\chi^2 = 1.62$, so don't reject H_0.

13. $\hat{\lambda} = 3.167$, from which $\chi^2 = 103.98 \gg \chi^2_{\alpha, k-1} = \chi^2_{\alpha, 7}$ for any tabulated α, so the Poisson distribution provides a very poor fit.

15. $\hat{\theta}_1 = (2n_1 + n_3 + n_5)/2n = .4125, \hat{\theta}_2 = .2750, \chi^2 = 30.58, \chi^2_{.01, 3} = 11.344,$ so reject H_0.

17. $r = .967, c_{.10} = .9707$ for $n = 30$, so at this level the hypothesis of normality must be rejected (a very close call, though, and normality would not be rejected for $\alpha = .05$ or $.01$).

Section 14.3
(pages 588–590)

19. Estimated expected counts are 449.7, 7.3, 17.6, 8.8, 242.7 for long areas and 471.3, 7.7, 18.4, 9.2, 254.3 for short areas. $\chi^2 = 23.18, \chi^2_{.01, 4} = 13.277,$ so reject the hypothesis of homogeneity.

21. $\chi^2 = 44.98$, so conclude that sex and nature of foot asymmetry are related (proportions differ for the two sexes).

23. $\chi^2 = 6.49, \chi^2_{.10, 10} = 15.987,$ so don't reject H_0.

25. $\chi^2 = 14.15, \chi^2_{.05, 6} = 12.592,$ so at level .05 conclude that size of car and commuting distance are related.

29. $\hat{p}_{ij} = n_{ij.}/n, \hat{e}_{ijk} = (n_{ij.})(n_k)/n,$ d.f. $= 32 - 8 = 24$

Supplementary Exercises
(pages 590–591)

31. Expected counts are 7.75, 15.5, and 7.75, $\chi^2 = 3.65, \chi^2_{.05, 2} = 5.992,$ so the null hypothesis appears quite plausible.

33. $\chi^2 = 22.37, \chi^2_{.005, 4} = 14.860,$ so P-value $< .005$ and the independence hypothesis would be rejected at any reasonable significance level.

35. $r = .986, c_{.01} \approx .92,$ so normality appears quite plausible.

Chapter 15

Section 15.1
(pages 597–598)

1. a. $(a + b)/2$　　　**b.** $\ln(\lambda)/2$

3. At level .05, reject H_0 if either $y \geq 9$ or $y \leq 1$. Since $y = 2$, don't reject H_0.

5. a. $p = .50$　　　**b.** Bin(6, .50)

　　c. The rejection region $\{5, 6\}$ gives $\alpha = .1094 \approx .10$. Since $y = 5$, reject H_0.

7. When H_0 is true, Y has a Bin(20, .75) distribution, so $\{18, 19, 20\}$ gives $\alpha = .091 \approx .10$. Since $y = 18$, reject H_0.

Section 15.2
(pages 607–608)

9. Reject H_0 if either $s_+ \geq 47$ or ≤ 8. Since $s_+ = 14$, don't reject H_0.

11. Reject H_0 if either $s_+ \geq 64$ or ≤ 14. Only ranks 1 and 5 are negatively signed, so $s_+ = 72$ and H_0 is rejected.

13. Reject H_0 if $z \geq 1.645$. Because $z = (226.5 - 162.5)/37.16 = 1.72 \geq 1.645$, reject H_0. P-value $\approx 1 - \Phi(1.72) = .0427$.

Section **15.3**   **15.** $w = 38 \geq 36 = c_1$, so reject H_0.
(pages 614–615)   **17.** $w = 65$, whereas $c_1 = 84$, so H_0 is not rejected.
                    19. $z = 2.54 < 2.58$, so don't reject H_0.

Section **15.4**   **21.** $(x_{(5)}, x_{(16)}) = (44.1, 46.6)$ **23.** $(\bar{x}_{(5)}, \bar{x}_{(32)}) = (11.15, 23.80)$
(pages 621–622)   **25.** $(-13.0, -6.0)$ **27.** $(d_{ij(5)}, d_{ij(21)}) = (16, 87)$

Section **15.5**   **29.** $k = 14.06 \geq 6.251$, so reject H_0.   **31.** $k = 9.23 \geq 5.992$, so reject H_0.
(pages 625–626)   **33.** $f_r = 2.60 < 5.992$, so don't reject H_0.

Supplementary   **35.** $f_r = 9.62 > 7.815 = \chi^2_{.05, 3}$, so reject H_0.   **37.** $(-5.9, -3.8)$
Exercises   **39.** $w' = 26 < 27$, so don't reject H_0.
(pages 626–627)

Index